ESSENTIALS IN MODERN HPLC SEPARATIONS

ESSENTIALS IN MODERN HPLC SEPARATIONS

SERBAN C. MOLDOVEANU[1], VICTOR DAVID[2]

[1]R.J. Reynolds Tobacco Co., Winston-Salem, NC, USA

[2]University of Bucharest, Bucharest, Romania

ELSEVIER

AMSTERDAM • BOSTON • HEIDELBERG • LONDON • NEW YORK • OXFORD
PARIS • SAN DIEGO • SAN FRANCISCO • SYDNEY • TOKYO

Elsevier

225, Wyman Street, Waltham, MA 02451, USA

The Boulevard, Langford Lane, Kidlington, Oxford OX5 1GB, UK

Radarweg 29, PO Box 211, 1000 AE Amsterdam, The Netherlands

Notice

No responsibility is assumed by the publisher for any injury and/or damage to persons or property as a matter of products liability, negligence or otherwise, or from any use or operation of any methods, products, instructions or ideas contained in the material herein. Because of rapid advances in the medical sciences, in particular, independent verification of diagnoses and drug dosages should be made

Library of Congress Cataloging-in-Publication Data

Moldoveanu, Serban.

  Essentials in modern HPLC separations / Serban C. Moldoveanu, Victor David.

    p. cm.

  Includes bibliographical references and index.

  ISBN 978-0-12-385013-3

1. Separation (Technology) 2. High performance liquid chromatography. I. David, Victor, 1955 - II. Title.

  TP156.S45M65 2013

  660'.2842–dc23

                  2012016476

British Library Cataloguing in Publication Data

A catalogue record for this book is available from the British Library

The authors are personally responsible for the content, accuracy, and conclusions of this book.

ISBN: 978-0-12-385013-3

For information on all Elsevier publications
visit our web site at store.eslevier.com

Printed and bound by CPI Group (UK) Ltd, Croydon, CR0 4YY

Transferred to digital print 2013

Working together to grow
libraries in developing countries

www.elsevier.com | www.bookaid.org | www.sabre.org

ELSEVIER   BOOK AID International   Sabre Foundation

In memory of Professor Candin Liteanu, pioneer in implementing and developing high performance liquid chromatography in Romania

Contents

Preface

One may ask "why another book on HPLC?" The field is rapidly evolving, and new information is being accumulated from a large number of original studies published in scientific and technical journals but not reviewed yet in a book. From time to time, this information needs to be collected, classified, and presented systematically. This new book describes a number of such developments that more recently started to be utilized. One additional reason for a new book is that utilization of HPLC is widespread, and a large number of readers may have different needs and interests that are not completely addressed in other books. The main purpose of the present book is to provide practical guidance in the selection of columns, of mobile phases, and of separation conditions for different types of HPLC, together with justifying why a particular selection is recommended. Another purpose is to provide criteria for selecting specific HPLC methods. For example, the present volume shows how the octanol/water partition coefficient (log K_{ow}) of the analyte can be a very useful parameter for chromatographers. Discussions regarding the use of this parameter in HPLC have been previously published, but this book applies it consistently. Octanol/water partition coefficients for many molecules are readily available and are extensively used in the pharmaceutical field as well as for description of the environmental fate of compounds. A program available from the U.S. Environmental Protection Agency (EPA) containing both a database with experimental log K_{ow} values for many chemicals and a program for estimating log K_{ow} can be downloaded (free) from http://www.epa.gov/oppt/exposure/pubs/episuite.htm.

The main goal of the book is to provide material that describes useful information regarding HPLC. The challenge in making such a presentation is considerable, and the authors took advantage of the information from a number of other books available on the market. Among such books are *Introduction to Modern Liquid Chromatography* (L. R. Snyder, J. J. Kirkland, J. W. Dolan, Wiley, 2010), *HPLC for Pharmaceutical Scientists* (Y Kazakevich, R. LoBrutto, Wiley, 2007), *HPLC Columns, Theory, Technology, and Practice* (U. D. Neue, Wiley, 1997), and *Practical High-Performance Liquid Chromatography* (V. R. Meyer, Wiley, 2010). An enormous number of applications of HPLC have been published in peer-reviewed journals, in a number of books, and on the web. These sources of information are considered more useful for finding direct applications as compared to a new book with a limited number of pages. For this reason, except for examples, the present book does not contain recipes for particular analyses.

This book starts with an introduction that provides basic information about HPLC and HPLC instrumentation. The next chapter describes common parameters used for characterization of an HPLC separation; Chapter 3 is dedicated to equilibria in HPLC; Chapter 4 discusses interactions at the molecular level that take place during different types of HPLC separations; and Chapter 5 examines the separation mechanisms in different HPLC types. In the following chapters, the material shifts toward direct applications and covers columns and mobile phases in HPLC, as well as the characterization of analytes that determines the HPLC selection. The last chapter is dedicated to the practice of HPLC separations. While most books on HPLC focus the presentation on the types of liquid chromatography (reversed phase, normal phase, ion exchange, etc.), this book is organized based on the view that there are significant unifying points among all HPLC types. A more uniform presentation including all HPLC types has therefore been approached, which is believed to be easier to follow.

The authors wish to thank the editorial team from Elsevier, Linda Versteeg, Jill Cetel, Beth Campbell, and Mohanapriyan Rajendran, for their contribution to the publication of this book. Also, the authors express their thanks to Paul Braxton, Carol Moldoveanu, and Michael Davis for reviewing the manuscript and suggesting valuable corrections.

Basic Information about HPLC

Essentials in Modern HPLC Separations
http://dx.doi.org/10.1016/B978-0-12-385013-3.00001-X

1.1. INTRODUCTION TO HPLC

What is Chromatography?

The term *chromatography* designates several similar techniques that allow the separation of different molecular species from a mixture. Applications of chromatography are numerous and can be related to laboratory or industrial practices. The molecular species subjected to separation exist in a *sample* that is made of *analytes* and *matrix*. The analytes are the molecular species of interest, and the matrix is the rest of the components in the sample. For chromatographic separation, the sample is introduced in a flowing *mobile phase* that passes a *stationary phase*. The stationary phase retains stronger or weaker different passing molecular species and releases them separately in time, back into the mobile phase. When the mobile phase is a gas, the chromatography is indicated as gas chromatography (GC), and when it is a liquid, it is indicated as liquid chromatography (LC). Other types of chromatography include supercritical fluid, countercurrent, and electrochromatography. When the sample is present as a solution, its components are indicated as *solutes*. Sample dissolution and/or preliminary modifications are frequently necessary to have the analytes amenable for a chromatographic separation (see, e.g., [1]). In high performance (or pressure) liquid chromatography (HPLC), the stationary phase is typically in the form of a column packed with very small porous particles (1–5 μm in diameter), and the liquid mobile phase (or *eluent*) is moved through the column by a pump (at elevated pressure). Solutes are injected in the mobile phase as a small volume at the head of the *chromatographic column*. A schematic diagram of the separation process is shown in Figure 1.1.1.

As the mobile phase flows, the eluted molecules that are exiting the column can be detected by various procedures. The eluted molecules differ from the mobile phase components by certain physicochemical properties (UV-absorption, refractive index, fluorescence, molecular mass and fragmentation in a mass spectrometer, or others), which make them detectable. Finally, an electrical signal is typically associated with molecular detection, and the graphic output of this signal is known as a *chromatogram*. The separated components of a mixture eluting at different times (known as *retention times* t_R) are displayed as peaks in the chromatogram. Different peaks (or patterns) on the chromatogram belong to different components of the separated mixture. An example of a chromatogram with the retention times written above the peaks is shown in Figure 1.1.2. As shown in this figure, the separation of the peaks can be very good or only partial. Also, some compounds may not be separated at all. Separated peaks may indicate individual compounds only when each peak corresponds to a single molecular species.

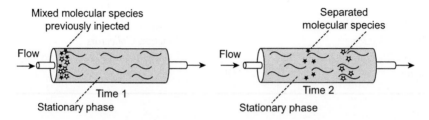

FIGURE 1.1.1 Simplified illustration of the separation process in chromatography (the black and white stars indicate two different molecular species).

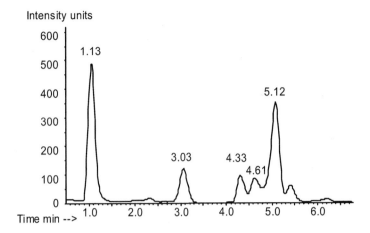

FIGURE 1.1.2 Picture of a chromatogram indicating the retention time for some of the peaks.

In HPLC the analytes are separated from each other and from the matrix as well as possible. The zones occupied by a specific analyte when it is eluted from the chromatographic column (peak width in a chromatogram) can be narrower or wider. The width of these zones affects the separation, and for two analytes with different retention times, the separation is better when the elution zones are narrower.

The peaks in the chromatogram may have different heights (and peak areas) depending on a number of factors such as the amount of compound in the mixture, amount of sample injected, and sensitivity of the detection procedure. Since peak areas are dependent on the amount of the compound, HPLC can be used for quantitation after a proper calibration. In this way, HPLC became an excellent technique for separation and quantitation of compounds even in very complex mixtures and is currently the most widely used analytical technique ever practiced.

As the previous short description of HPLC shows, the technique has two distinct parts: (1) separation of the analytes and of the matrix, (2) detection and measurement of the analytes. The discussion about the separation is the main subject of this book. Based on the nature of the analytes, the separation process is achieved depending on the choice of a chromatographic column and a specific mobile phase. The detection step is achieved using one or more detectors, and the sensitivity, selectivity, and stability of these detectors is essential for the success of the HPLC analysis. The present text does not include a detailed discussion of detection and measurement of the analytes, and focuses mainly on their separation.

Separation by HPLC can also be used for semipreparative or preparative purposes, some with industrial applications. In this case, the separated compounds of interest are collected for further utilization. However, the main focus of the present volume is analytical HPLC, and semipreparative and preparative HPLC are beyond the scope of this work.

Types of Equilibria in HPLC

The separation process in HPLC is based on an equilibrium established between the molecules present in the mobile phase and those retained in the stationary phase. The difference in the concentration of a molecular species in one phase and in another determines whether the species is retained or eluted with the mobile phase. When the concentration of the solute (analyte) is higher in the mobile phase than in the stationary phase, the solute is eluted faster from the chromatographic column. The opposite happens when the concentration of the solute is higher in the stationary phase. In this case, the solute is more strongly retained and the elution takes place after a longer period of time.

Common types of equilibria for a molecular species between two phases include, for example, the distribution of a compound between two immiscible solvents. Another common type takes place during the retention of a compound from a fluid on an adsorbing material such as charcoal. Chemical equilibrium in a solution, for example, between two ionic compounds, is also a known type. The main types of equilibria encountered in chromatography can be summarized as follows:

1) *Partition equilibrium.* This type of equilibrium takes place when the molecules of the solute are distributed between two liquid phases. In HPLC, one liquid phase is kept immobile on a solid material, and the other is mobile (the eluent). The immobilization of the liquid to become a stationary phase in partition chromatography is achieved, for example, when the liquid is highly polar and can establish hydrogen bonds with the solid support. One such example is water on a silica surface. In this case, the mobile phase should consist of a liquid less polar than water. However, the partition equilibrium can also be applied for a nonpolar stationary phase and a more polar mobile phase. The theory of separation in partition chromatography is based on liquid/liquid extraction principles. The different molecular species, being in continuous equilibrium between the mobile and stationary phase, will be separated based on their tendency to exist in higher concentration in the mobile liquid or in the stationary liquid, in accordance with their affinity for these

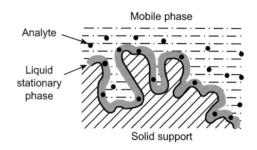

FIGURE 1.1.3 Schematic description of partition equilibrium.

phases. A schematic description of the partition chromatography process is shown in Figure 1.1.3. In partition chromatography, the concept of "immobile liquid" is commonly approached in a "loose" manner. For example, a layer of adsorbed water on the surface of a silica solid support, or a layer of bonded organic chains on a silica surface (such as in the common C18 chromatographic columns), or a layer of mechanically held polymer on an inert core are all considered liquid stationary phases for partition chromatography. The possibility of performing chromatography using two liquid phases without having one liquid phase immobilized is exploited in countercurrent chromatography. However, this subject is beyond the purpose of the present book (for details see, e.g., [2]).

2) *Adsorption equilibrium.* This type of equilibrium takes place when molecules are exchanged between a solid surface and a liquid mobile phase. Assuming that the stationary phase is very polar compared to the mobile phase, the polar molecules from the mobile phase are adsorbed on the solid stationary phase surface, while the less polar molecules are kept mainly in the mobile phase. Being in equilibrium between the solid and the liquid, the more polar molecules also elute from the chromatographic column, but later than the less polar compounds. A schematic description of the adsorption chromatography process is shown

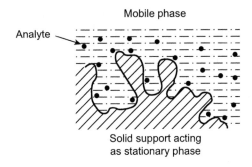

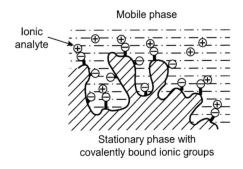

FIGURE 1.1.4 Schematic description of adsorption equilibrium.

FIGURE 1.1.5 Schematic description of ion exchange proces.

in Figure 1.1.4. The partition and the adsorption are utilized basically as models for describing the type of equilibrium, but a difference between the two processes is not commonly apparent from a thermodynamic point of view [3]. Also, in many instances the separation can be viewed either as a partition or as an adsorption, the differentiation being made only with the purpose of estimating differently the separation parameters, while the classification has no effect on the real process.

3) *Equilibria involving ions.* Equilibria between ions in solutions take place in numerous chemical reactions. For applications in HPLC, one ionic species must be immobilized, for example, by being connected through a covalent bond to a solid matrix. One example of this type of ion can be a sulfonic group connected to polystyrene. The ions in solution can be bound by ionic interactions to the immobilized counterion or may remain in solution. The equilibrium between solid phase and mobile phase, depending on the strength of the bond to the stationary phase, may provide a means for separation. A schematic description of the interactions in the ion-exchange chromatography process is shown in Figure 1.1.5.

4) *Equilibria based on size exclusion.* Size exclusion uses a stationary phase that consists of

a porous structure in which small molecules can penetrate and spend time passing through the long channels of the solid material, while large molecules cannot penetrate the pore system of the stationary phase and are not retained. Applied in HPLC, the large molecules elute earlier, while the small molecules are retained longer. An equilibrium can be envisioned between molecules in the mobile phase and those partly trapped in the solid matrix. A schematic description of the size exclusion process is shown in Figure 1.1.6.

5) *Affinity interactions.* This type of interaction is typical for protein binding and leads to equilibria that allow very specific separations. Examples of such interactions are protein-

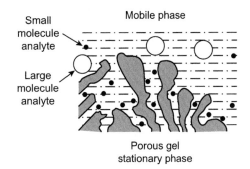

FIGURE 1.1.6 Schematic description of size exclusion process.

antibody and avidin-biotin. Affinity chromatography is widely used at low pressure for protein purification.

Criteria for the Classification of HPLC Procedures

HPLC comprises several similar techniques, all of which use a liquid mobile phase passing a stationary phase with the result of a "high-performance" separation of a mixture of compounds. Various versions of HPLC have specific differences and different applications. For this reason, the HPLC techniques are classified in various types; the classification is based on a number of criteria, such as the separation principle (in some separations more than one principle of separation may have a contribution) or the scale of the utilization. Additional differences have been used to distinguish more types of HPLC (see Section 1.2). This may include the nature of the stationary phase and mobile phase and/or the type of interactions (mechanism) that describe the energetics of the process. It should be noted that the classification based on the separation mechanism indirectly includes differences in the stationary and mobile phase. For example, ion-exchange chromatography is practiced specifically on an ion-exchange stationary phase and not on a reversed or a bioaffinity phase. However, the mechanism alone was not viewed as sufficient for differentiation of some types of HPLC. Based on the nature of the stationary phase and mobile phase, several types of HPLC can be differentiated and are discussed further in Section 1.2.

Other HPLC characteristics are also used for differentiating the HPLC types. One such characteristic is related to the composition of the mobile phase, which can be kept constant during the separation or can be modified. The HPLC performed at a constant composition of the mobile phase is known as *isocratic HPLC*, while that performed with a mobile phase that changes in composition during the separation is known as *gradient HPLC*. Gradient HPLC allows a change in the polarity and/or in the pH of the mobile phase during the separation and significantly increases the versatility of HPLC. When a sample contains solutes with very different properties and when a constant composition of the mobile phase (isocratic conditions) is used, the solutes may leave the chromatographic column at very different times. This may be seen as an advantage for the separation, but when the retention time of some of the solutes becomes unacceptably long, the change in the solvent composition (by using gradients) is necessary to speed up the separation.

The differences in the size of the particles used in the chromatographic column offer one more criterion for HPLC classification. The size of the particles that fill the chromatographic column affect the peak width and therefore the separation. In common HPLC the size of the particles in the chromatographic column is usually 3 to 10 μm. The HPLC techniques that use very small particles (e.g., with a diameter below 2.5 μm) in the chromatographic column (or cartridge) are usually referred to as *ultra performance liquid chromatography* or *UPLC*. The very small particles require additional modifications in the UPLC technique, such as significantly higher operation pressure for the mobile phase. The use of small particles in the columns can lead not only to better separations, but also to faster ones, and is one of the modern developments of HPLC. Another development in HPLC is the use of monolithic columns made of a single piece of a solid porous material.

Temperature can be another criterion to differentiate HPLC processes. Based on this parameter, the HPLC types are classified as: (1) *low temperature* (below freezing point of water, down to $-10\,^\circ$C), (2) *usual range temperature* (20–60 $^\circ$C), and (3) *high temperature* (up to

250 °C). Most separations are performed at usual temperatures. Low-temperature techniques are applied, especially for certain chiral separations. High temperature separations can be used for a number of applications, taking advantage of the modification of solvents properties as temperature increases. Water, for example, shows a decrease in polarity at temperatures between 100 and 250 °C and can be used as solvent in RP-HPLC. This particular mode of separation has been denoted as super-heated water, pressurized water, or subcritical water chromatography (as the temperatures used are lower than the critical temperature of water at 374 °C) [4]. The use of water as a mobile phase can provide an environmentally friendly "green" method of chromatographic analysis.

A different classification, important for practical purposes, is based on the scale of the HPLC equipment. Three general types of chromatography can be distinguished in this way: (1) *analytical HPLC*, (2) *semipreparative HPLC*, and (3) *preparative (large scale) HPLC*. Each of these types covers in fact a range of dimensions. For example, analytical HPLC can be further differentiated based on the dimensions of the HPLC column in the following subtypes: conventional, narrowbore, microbore, micro LC-capillary, and nano LC-capillary. In the present book, most discussions will refer to conventional, narrowbore, and microbore analytical HPLC. Analytical HPLC performed using microcapillary and nanoscale columns (still filled with a bed of particles) are less used in practice and require specialized equipment.

Role of Polarity in HPLC

One common concept related to HPLC separations is that of polarity. *Polarity* refers to an asymmetrical charge distribution in a molecule, which causes the molecule to act as an electric dipole. However, charge distribution is a very complex concept, and calculation of the values for charge density, used to characterize charge distribution in a molecule, is usually a difficult task. Also, polarity can refer to the analytes, to the mobile phase or to the stationary phase. The compounds are typically indicated as polar when opposite partial charges are known to be present in the molecule, when specific physical properties such as water solubility or solubility on solvents miscible with water are known, or when specific functional groups known to be "polar" such as $-COOH$, or $-NH_2$ are present in the molecule. In addition, during molecular interactions the charge distribution of the molecules suffers changes (expressed by *polarizability*). In any interaction, the polarizability affects the charge distribution of the molecules. For these reasons, comparing molecules or phases as "more polar" or "less polar" is not a quantitative assessment. The opposite to the polar character is the *hydrophobic character (lipophilic character)*. Nonpolar compounds that do not have polar groups and are not water soluble, or materials on which surface the water does not adhere are commonly indicated as hydrophobic. Both the polar and the hydrophobic character of a compound are reasonably described by the partition constant (also indicated as partition coefficient) between octanol and water K_{ow} (or P_{ow}). This parameter represents the ratio of concentrations of a (not ionized) compound between two phases, one being octanol and the other water, and is described by the formula (square brackets indicate molar concentrations):

$$K_{ow} = \frac{[solute]_{octanol}}{[solute]_{water}} \qquad (1.1.1)$$

Positive values for log K_{ow} indicate some hydrophobic character, and larger values show more hydrophobicity. Molecules with low or negative values for log K_{ow} are frequently indicated as polar, although there is not a direct relation between K_{ow} and the charge distribution in the molecule.

The experimental values for K_{ow} are known for many compounds, and several computer programs are available for their evaluation (e.g., MarvinSketch 5.4.0.1, ChemAxon Ltd. [5], EPI Suite [6]) as well as extensive tables [7, 8].

In the present text, some of these concepts will be further discussed and clarified. However, the concept of polar and nonpolar (hydrophobic) compounds or materials will be frequently used in an "imprecise" manner. For mobile phases, the extension of the concept of polarity is immediate, being based on the polarity of the molecules of the phase. For stationary phases, the polarity refers to the nature of the stationary phase surface or of phase active moiety.

As a conclusion, HPLC techniques can be differentiated based on a number of criteria. Each selected criterion has advantages and disadvantages. For example, the mechanism for a particular separation is not always well understood, and the "polar" or "nonpolar" nature of the stationary and mobile phase is sometimes difficult to quantify. The physical criteria such as particle dimension in the chromatographic column or the scale of the HPLC equipment are available in a range and the limits used for the classification are subjective. For this reason, the HPLC classifications should be viewed mainly as an attempt to have models, and sometimes a particular HPLC type can be classified in more than one way.

Qualitative Analysis and HPLC Main Use as a Quantitative Analytical Technique

HPLC analysis starts with a separation of components of interest from a sample. The separated analytes are represented by the peaks in the chromatogram. The analyte detection can be performed using a variety of instrumental devices (detectors), some of which provide qualitative information for the compound generating the peak. Some detectors are "universal," and either they provide no qualitative information (such as refractive index detectors) or they provide only partial information that is not in itself sufficient for positive identification of an analyte (e.g., UV absorption). In such cases, the chemical nature of the analyte must be known, and its peak in the chromatogram is identified based on its retention time established using standards that were previously analyzed. For this reason, the stability (reproducibility) of the retention time in a chromatogram (obtained in identical conditions) is very important.

Other detectors such as mass spectrometers (MS) or MS/MS offer more detailed insight into qualitative peak identification. However, the dependence on operational conditions of the mass spectra obtained using LC/MS or LC/MS/MS instrumentation makes the process of compound identification quite difficult even when using these techniques. Progress in MS identification of unknown compounds has been done, for example, by using very high accuracy in mass measurement for the parent ion of the analyte and for its fragments (e.g., using Orbitrap or cyclotron technologies). Also, specific computer programs (Mass Frontier, SmileMS) provide help in identifying unknown compounds. However, in most cases, the compound identification capability, even using MS or MS/MS detection, is not used for discovery of the composition of an unknown compound, but for positive identification of a known analyte, corroborated with the retention time of its standard, previously analyzed. The confirmation of the peak for a specific analyte in a chromatogram (typically using MS and three confirmation ions) is an important and common element in HPLC practice. The use of standards with labeled isotopes for the analytes (e.g., deuterated analyte) spiked in the sample is also a common practice for peak identification in LC/MS and LC/MS/MS. Although the retention time of the isotope labeled standards may vary slightly from that of the analyte itself, peak identification is significantly facilitated using this technique.

Quantitave analysis is the main use of HPLC. Once separated, the concentration of the analytes in the sample can be obtained from the chromatographic peak area (or height).

Peak areas (or peak heights) in the chromatogram are proportional to the concentration of the analytes, and quantitation is done using calibration curves with standards, or other procedures. Depending on the detection technique and the analyte properties, some HPLC analyses can provide results even for ultra-low traces of a compound (below ng/mL level). The versatility and high sensitivity of HPLC have contributed to its success and widespread use.

An exceptionally large number of methods using HPLC quantitation procedures has been published. These methods can be found in a variety of sources, including papers in scientific journals, books, web articles, and proceedings of conferences. The goal of this book is to describe the principles used for developing HPLC methods and to provide information for potential improvements regarding these techniques; detailed descriptions of analytical methods are beyond its purpose.

1.2. MAIN TYPES OF HPLC

A Classification of HPLC Types

A variety of HPLC types have been differentiated in the literature; some of these types are similar, and others exhibit significant differences. The differentiation was based not only on various criteria such as the nature of the stationary and mobile phases and the type of interactions assumed to lead to the separation, but also on the range of concentration of specific solvents in the mobile phase (e.g., of water) and so on. This section presents a common classification of the main types of HPLC. Because different HPLC types have different characteristics and applications, it is important to understand these differences and select the appropriate HPLC type for solving a specific separation/analysis problem.

1) *Reversed-phase HPLC* (or *RP-HPLC*) is the most common HPLC technique, and a very large number of compounds can be separated by RP-HPLC. This type of chromatography is performed on a nonpolar stationary phase with a polar mobile phase. A wide variety of nonpolar stationary phases is available, and RP-HPLC is very likely the most common type of chromatography used in practice. The stationary phase for RP-HPLC can be obtained, for example, by chemically bonding long hydrocarbon chains on a solid surface such as silica. The most common chain bound to silica is C18 (it contains 18 carbon atoms), which has a high hydrophobic character. The bonded phase hydrophobicity may vary depending on the nature of the substituent. For example, C18 bonded phase has a higher hydrophobicity than C8 bonded phases. Polymeric materials are also used as the RP-HPLC stationary phase. The mobile phase in RP-HPLC is typically a mixture of an organic solvent (CH_3CN, CH_3OH, isopropanol, etc.) and water, with a range of content in the organic solvent. Small amounts of buffers can also be added to the mobile phase in RP-HPLC. The interactions in RP-HPLC are considered to be the hydrophobic forces. These forces are caused by the energies resulting from the disturbance of the dipolar structure of the solvent. The so called solvophobic effect is caused by the force of "cavity-reduction" in water around the analyte and the nonpolar stationary phase when the two are interacting. The retention of the analyte on the

stationary phase is dependent on the contact surface area between the nonpolar moiety of the analyte molecule and the stationary phase, both immersed in the aqueous eluent. For this reason an analyte with a larger hydrophobic surface area (and usually with a large log K_{ow}; see rel. 1.1.1) is more retained on the stationary phase, resulting in longer retention time compared with an analyte with a smaller hydrophobic surface (and low K_{ow}). In RP-HPLC the separation is typically considered to be based on the partition of the analyte between the stationary phase (viewed as an immobilized liquid) and the mobile phase, although some experiments can be explained by adsorption equilibrium. The exceptional utility of RP-HPLC is based on the fact that most compounds have at least some hydrophobic moiety in their structure.

2) *Ion-pair chromatography* (IPC) is applied in particular to ionic or strongly polar compounds. This type of chromatography is very similar to RP-HPLC, with the difference of having a special mobile phase (ion-pair RP). In the mobile phase of ion-pair chromatography, a reagent is added, which interacts with the ions of the analytes and forms less polar compounds that can be separated based on hydrophobic interactions with the stationary phase. For example, acids that are ionized (or very polar) can be coupled with a reagent that produces "ion pairs" amenable to separation by RP-HPLC.

3) *Hydrophobic interaction chromatography* (HIC) is a type of RP-HPLC, sometimes indicated as a milder RP-HPLC, applied to the separations of proteins and other biopolymers. The technique is based on interactions between nonpolar moieties of a protein with solvent-accessible nonpolar groups (hydrophobic patches) on the surface of a hydrophilic stationary phase (e.g., hydrophobic ligands coupled on cross-linked agarose). The promotion of the hydrophobic effect by the addition of salts (such as ammonium sulfate) in the mobile phase drives the adsorption of hydrophobic areas from the protein to the hydrophobic areas on the stationary phase. The reduction of the salting out effect by decreasing the concentration of salts in solution leads to the desorption of the protein from the solid support.

4) *Nonaqueous reversed-phase chromatography (NARP)* is a RP-HPLC type utilized for the separation of very hydrophobic molecules such as triglycerides. In this type of chromatography, the stationary phase is nonpolar (similar to RP), while the mobile phase, though less nonpolar than the stationary phase, is nonaqueous (usually a mixture of less polar and more polar organic solvents) and capable of dissolving the hydrophobic molecules.

5) *Hydrophilic Interaction Liquid Chromatography (HILIC)* is a type of HPLC applied for polar, weakly acidic, or basic samples. In this type of HPLC the stationary phase is polar, and the mobile phase is less polar than the stationary phase. HILIC is the "reverse" of RP-HPLC. For HILIC, the polar stationary phase is typically made by chemically bonding on a solid support molecular fragments with a polar end group (diol, amino, special zwitterionic, etc.). The chromatography performed on bare silica support with free silanol ($\equiv$Si-OH) groups can also be considered as HILIC, depending on the mobile phase. The mobile phase in HILIC is typically a less polar but water-soluble solvent such as CH_3OH or CH_3CN, which also contains a certain proportion of water. The separation is based on the difference in polarity between the molecules. Ion-polar interactions may also play a role in separation. Viewed as having the separation equilibrium based on the interaction of a solid

surface with the molecules from a liquid, HILIC is a type of adsorption chromatography. However, a (polar) bonded phase may be seen as a stationary liquid phase, and in this case HILIC is a type of partition chromatography. When the separation is done on zwitterionic phases, HILIC chromatography is sometimes indicated as ZIC (from zwitterionic chromatography).

HILIC separations can also be performed on an ion-exchange stationary phase, with the mobile phase containing a high proportion of an organic solvent. This type of separation is sometimes indicated as eHILIC or ERLIC (from electrostatic repulsion hydrophilic interaction chromatography). This technique can be cationic eHILIC or anionic eHILIC, depending on the nature of the ion-exchange stationary phase. In this type of chromatography, the ionic stationary phase repels the similar ionic groups of the analyte and allows HILIC type interactions with the neutral polar molecules of the analyte.

6) *Normal-phase chromatography (NPC)* is a chromatographic type that uses a polar stationary phase and a nonpolar mobile phase for the separation of polar compounds. The nonpolar mobile phases used in this type of chromatography are solvents such as hexane, CH_2Cl_2, and tetrahydrofuran that are not water soluble. In normal-phase chromatography, the most nonpolar compounds elute first and the most polar compounds elute last. Normal-phase chromatography does not have a major difference from HILIC. Because NPC was identified as a separate type for a much longer time than HILIC, it is common in the literature to identify HILIC as a subtype of normal-phase chromatography and not the other way around. The difference consists in the use in HILIC of a mobile phase that contains some proportion of water. A polar organic normal phase is sometimes mentioned as a type of chromatography when the nonaqueous solvent contains polar additives such as trifluoroacetic acid.

7) *Aqueous normal-phase chromatography* (ANPC or ANP) is a technique performed on a special stationary phase (silica hydride), and the mobile phase covers the range including the types used in reversed-phase chromatography and those used in normal-phase chromatography. The mobile phases for ANP are based on an organic solvent (such as methanol or acetonitrile) with a certain amount of water such that the mobile phase can be both "aqueous" (water is present) and "normal" (less polar than the stationary phase). Polar solutes are most strongly retained in ANP, with retention decreasing as the amount of water in the mobile phase increases.

8) *Cation-exchange chromatography* is a type of HPLC used for the separation of cations (inorganic or organic). In this HPLC type the retention is based on the attraction between ions in a solution and the opposite charged sites bound to the stationary phase. In ion-exchange chromatography (IEC or IC) the ionic species are retained on the column based on coulombic interactions. In cation-exchange chromatography the ionic compound consisting of the cationic species M^+ in solution is retained by ionic groups covalently bonded to a stationary support of the type $R - X^-$. The ion-exchange material (e.g., an organic polymer with ionic groups) is not electrically charged, and therefore the initial form of the cation exchange already has an ionically retained cation in the form $R - X^- \ C^+$. The separation is achieved when different molecules in solution have different acidic or basic strength. For example, for a cation-exchange material, one species (e.g., C^+) that is bound to the

$R - X^-$ substrate is replaced by a stronger cationic species (e.g,. M^+) such that M^+ is retained from the solution, while C^+ passes into the mobile phase. Two different cations from solution, M_1^+ and M_2^+, can be separated based on their retention strength.

9) *Anion-exchange chromatography* is a type of HPLC used for the separation of anions (inorganic or organic). This HPLC is similar in principle to the cation-exchange type, but the anionic species B^- from solution are retained by covalently bonded ionic groups of the type $R - Y^+$. Similarly to cation exchange stationary phases, an anion exchange is initially in the form $R - Y^+ A^-$. For an anion-exchange material the anion A^- previously bound is replaced on the resin by the anion B^-, and two different anions B_1^- and B_2^- are separated based on their different retention strengths. The mobile phase in ion-exchange chromatography frequently consists of buffer solutions.

10) *Ion-exchange on amphoteric or zwitterionic phases* is a type of IEC that is very similar in principle to the cation-exchange or anion-exchange IEC. The stationary phase of this type of IEC contains groups that have an amphoteric character or, in the case of zwitterionic phases, both anionic and cationic groups. The mobile phase in these types of chromatography also consists of buffer solutions.

11) *Ion-exclusion chromatography* is an HPLC technique in which an ion-exchange resin is used for the separation of neutral species between them and from ionic species. In this technique, ionic compounds from the solution are rejected by the selected resin (through the so-called Donnan effect), and they are eluted as nonretained compounds. Nonionic or weakly ionic compounds penetrate the pores of the resin and are retained selectively as they partition between the liquid inside the resin and the mobile phase. (The Donnan effect or Gibbs-Donnan effect describes the distribution of ions in solution in two compartments separated by a semipermeable membrane).

12) *Ligand-exchange chromatography* is a type of chromatography in which the stationary phase is a cation-exchange resin loaded with a metal ion (e.g., of a transitional metal) that is able to form coordinative bonds with the molecules from the mobile phase. The elution is done with a mobile phase able to displace the analyte from the bond with the metal, and the separation is based on the differences in the strength of the interaction (of coordinative type) of these solutes with the bonded metal ion.

13) *Immobilized metal affinity chromatography* is closely related to ligand-exchange chromatography and uses a resin-containing chelating groups that can form complexes with metals such as Cu^{2+}, Ni^{2+}, and Zn^{2+}. The metal ions loaded on the resin still have coordinative capability for other electron donor molecules such as proteins. The retained analytes can be eluted by destabilizing the complex with the metal, for example, by pH changes or addition of a displacing agent such as ammonia in the mobile phase.

14) *Ion-moderated chromatography* is an HPLC technique similar to ligand-exchange chromatography, with the difference that the stationary phase loaded with the metal ion (e.g., Ca^{2+}, Na^+, K^+, Ag^+, or even H^+) does not form coordinative bonds with the analyte, the interactions being based mainly on polarity.

15) *Gel filtration chromatography* (GFC) is a type of size-exclusion chromatography (SEC) in which the molecules are separated based on their size (more correctly, their

hydrodynamic volume). In gel filtration an aqueous (mostly aqueous) solution is used to transport the sample through the column and is applied to molecules that are soluble in water and polar solvents. Size-exclusion chromatography uses porous particles with a variety of pore sizes to separate molecules. Molecules that are smaller than the pore size of the stationary phase enter the porous particles during the separation and flow through the intricate channels of the stationary phase. Small molecules have a long path through the column and therefore a long transit time. Some very large molecules cannot enter the pores at all and elute without retention (total exclusion). Molecules of medium size enter only some larger pores and not the small ones, and are only partly retained, eluting faster than small molecules and slower than the very large ones. The separation of small molecules between themselves is not typically achieved, and the technique is utilized mainly for the separation of macromolecules and of macromolecules from small molecules. GFC is sometimes indicated as aqueous SEC.

16) *Gel permeation chromatography* (GPC) is another type of size-exclusion chromatography (SEC), the only difference from gel filtration being the mobile phase, which in this case is an organic solvent. The technique is used mainly for the separation of hydrophobic macromolecules (such as solutions of certain synthetic polymers). GPC is sometimes indicated as nonaqueous SEC.

17) *Displacement chromatography* is a chromatographic technique where all the molecules of a sample are initially retained on a chromatographic column (loading phase). After the sample is loaded, a "displacement" reagent dissolved in the mobile phase is passed through the column and elutes the specific retained molecule. The method is more frequently applied as a preparative chromatographic technique than as a HPLC analytical method.

18) *Affinity chromatography* is a liquid chromatographic technique typically used for protein and other bio-molecule separation and commonly indicated as *bioaffinity chromatography*. It can be practiced on a variety of specifically made stationary phases that allow selective retention of the analytes based on affinity interactions.

19) *Chiral chromatography on chiral stationary phases* is a type of HPLC used to separate chiral compounds. Only specific applications require the separation of chiral compounds, and regular chromatography is much more common than chiral chromatography. Chiral chromatography still has numerous applications, particularly in the analysis of pharmaceutical compounds. The technique typically requires chiral stationary phases containing chiral selector groups.

20) *Chiral chromatography on achiral stationary phases* is also possible for some chiral solutes by using chiral modifiers in the mobile phase, although the stationary phase is not chiral.

21) *Multimode HPLC* is a type of chromatography in which the column contains by purpose more than one type of stationary phase, for example, some with bonded nonpolar groups (e.g. C18), and some with ionic groups (e.g., SO_3). This type of character can be encountered unintentionally on columns made using as a stationary phase a silica support covered with silanol groups, and also with hydrophobic groups (such as C18). In most cases, the presence of two types of interactions (e.g., polar and hydrophobic) is not desirable, but in some

TABLE 1.2.1   Separation Principle and Main Types of HPLC

Separation mechanism	Types of HPLC
Hydrophobic forces	1) Reversed phase (RP)
	2) Ion pair
	3) HIC
	4) Non-aqueous reversed phase (NARP)
Difference in polarity	5) HILIC
	6) Normal phase (NPC)
	7) Aqueous normal phase (ANP)
Ion interaction	8) Cation exchange
	9) Anion exchange
	10) Ion exchange on amphoteric and zwitterionic phases
	11) Ion exclusion
	12) Ligand exchange
	13) Immobilized metal affinity
	14) Ion moderated
Size exclusion	15) Gel filtration
	16) Gel permeation
Displacement	17) Displacement
Bioaffinity	18) Bioaffinity (not always HPLC)
Chiral	19) Chiral stationary phase
	20) Chiral mobile phase
Various principles together	21) Multi-mode

etc.) for each type of HPLC is not a straightforward subject. It is possible that more than one such mechanism takes place in a specific HPLC type, and in some cases it is difficult to decide, based on experimental data, which equilibrium mechanism is involved in the separation. However, an association between different types of HPLC and different equilibrium mechanisms can be observed. The main separation principles and the corresponding HPLC types are summarized in Table 1.2.1. The relation between the HPLC main groups and the equilibrium mechanism should not be viewed as rigid, and more than one type of equilibrium may take place in a specific type of HPLC.

1.3. PRACTICE OF HPLC

General Aspects

Viewed as a combination of information and operations, any chemical analysis including HPLC follows the typical scheme: input → process → output [1]. The input consists of initial information about the sample, such as origin, nature, and purpose of analysis. A special part of the input is related to the selection of the analytical procedure (information regarding the process). For the analysis of complex samples, chromatographic analysis is ideal because it has the advantage of combining a separation with the measurement. The output is formed by the results, when the purpose of the analysis is achieved. The process consists of various steps. In chromatographic methods of analysis, these steps usually follow the sequence: sample collection → sample preparation → analytical chromatography → data processing. The process is conducted based on a number of decisions regarding sample collection (procedure, quantity, number of replicates), sample preparation (choice of

instances dual properties of a stationary phase can be used to the advantage of the separation.

Relation between the Type of HPLC and Equilibrium Mechanism

The identification of equilibrium mechanism (e.g., partition, adsorption, ionic, size exclusion, cleanup, concentration, and/or derivatization), type of chromatographic analysis (HPLC, GC, etc.), as well as type of data processing (qualitative or quantitative measurements, statistical analysis, etc.). The analytical chromatography step can be considered the core of the process, and it includes the identification and measurement of the analytes. The choice of HPLC as the analytical step is done for numerous types of samples, such as small molecules with medium and low volatility, as well as larger molecules, including a wide range of synthetic and biopolymers. HPLC has the capability of separating complex mixtures and performs accurate quantitation with extreme sensitivity.

Selection of the Type of HPLC for a Particular Application

An important part of the information step in chromatographic analysis is the choice of the type of HPLC that should be used. This selection is made based on the nature of the sample, instrument availability, as well as other factors such as cost and time of analysis. Once the HPLC technique is selected as the core analytical procedure, further decisions should be made regarding the type of HPLC. The selection of an HPLC type for analysis of a particular set of samples is not always simple. However, some general rules may be used as guidance. This choice is determined primarily by the nature of the sample with its analytes and matrix. Reversed-phase chromatography, for example, is commonly used for a wide range of compounds, including various organic molecules that have some hydrophobic moiety. More polar molecules are typically analyzed using HILIC and ion-pair chromatography. In some instances even RP-HPLC can still be used for the separation. Ions (inorganic or small organic) are typically analyzed by IC. The separations of large molecules based on their molecular weight (in fact hydrodynamic volume) are

performed by size exclusion. Bioaffinity chromatography is widely utilized for the separation of biological macromolecules. (Further discussion of the dependence of the HPLC type on the chemical nature of the sample can be found in Chapter 9.)

The purpose of analysis is another determining factor. In the choice of a specific type of HPLC, it is important to know if the analysis is performed for the separation by molecular weight, for specific identification and quantitation of components, or for separation and quantitation of enantiomers. Other factors also influence the choice of HPLC, such as availability of equipment, requirements regarding analysis time, number of samples to be analyzed, availability of specific materials required for the analysis (columns, solvents, etc.), restrictions regarding safety (e.g., the nature and volume of solvents to be disposed), and level of training of the operator. This section provides only general guidance regarding the selection of the HPLC type, and this selection is based solely on the nature of the sample.

The selection of a particular type of chromatography for a specific analysis is a complex process, the previous discussion being only a schematic guide that is far from being comprehensive. Numerous sample/analyte details may determine the final choice of a specific chromatographic separation. The present book mainly discusses aspects of separation in conventional analytical HPLC.

Sample Collection and Sample Preparation for HPLC

Sample collection is a very important step for the success of any chemical analysis. This subject is discussed in various books and papers, but since it is outside the scope of the present book, the reader should refer to the dedicated literature (see e.g., [1, 9]). After sample collection, the analysis proceeds with the sample preparation, in accord with the selected analysis type. Again, numerous procedures are described in the literature for sample preparation [1]. Sample preparation may target the matrix of the sample, the analytes, or both. One common operation in sample preparation is the dissolution of the sample if the sample is solid. Then, the matrix is usually modified during cleanup, fractionation, and concentration of the sample. Proper processing of the sample may have considerable importance for the success of the HPLC analysis. A sample that contains a "dirty" matrix, having numerous other solutes that can impede the separation or destroy the chromatographic column must be avoided as much as possible. Also, the sample preparation may have a considerable part in increasing the analytes' concentration. The increase in the concentration of the analytes is very important especially when traces of specific compounds must be quantitated, as is necessary in many practical applications. This concentration can be done by a variety of procedures such as solid-phase extraction (SPE) and liquid-liquid extraction (LLE) [1]. The analytes can also be modified by chemical reactions (derivatization, etc.) in order to obtain better properties for the chromatographic analysis. The process of sample preparation is schematically shown in Figure 1.3.1. Sample preparation is usually described for each HPLC analytical procedure when applied for a practical analysis and covers a large part of the published literature on HPLC. Several books describing the general principles and various aspects of sample preparation for chromatography also have been published (see, e.g., [1, 10]).

Injection

Sample delivery for analysis in HPLC is achieved using injection. This is done with the purpose of introducing in the mobile phase the sample containing the analytes and matrix that are going to be separated and analyzed. The sample is dissolved in a solvent, and the choice of this solvent in connection with the volume of sample volume may have an effect on HPLC separation. The sample solvent must be soluble in the mobile phase. Larger volumes of the sample solvent may affect for a short period of time the composition of the mobile phase, influencing the separation, in particular the peak shape. For this reason, a sample volume is

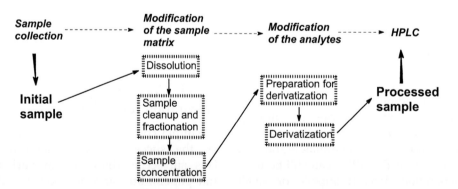

FIGURE 1.3.1 Diagram of a sample preparation involving dissolution, cleanup, fractionation, concentration, and derivatization.

usually limited to the range of 5 to 25 µL for common HPLC techniques. Smaller injection volumes can be used for mini or micro HPLC, and larger volumes than typical are sometimes used in order to obtain better sensitivity for the analysis. For semipreparative or preparative HPLC the injection volume is much larger. More details regarding sample injection are given in Section 9.2.

Column Selection in HPLC

The column is a major component for the HPLC separation, and its selection is critical for the success of the analysis. Numerous types of columns are commercially available. Columns may be different regarding: (1) the nature of the active stationary phase (RP, HILIC, IEC, SEC, bioaffinity, etc.), (2) the type of phase (porous particles, superficially porous particles, monoliths, etc.), (3) physical characteristics of particles (dimension, porosity, strength, etc.), (4) column dimensions (length, diameter), (5) mechanical construction (columns, cartridges, compressible columns), etc. The column is basically selected in agreement with the type of separation that was chosen, equipment availability, analysis requirements, and external information available. Column choice has a significant effect in achieving a desired separation, and a detailed discussion about chromatographic columns and their separating capabilities is given in Chapter 6.

Mobile Phase Selection

In all HPLC analyses, the choice of mobile phase is another critical step for effecting a successful separation. The solvents are selected depending on the type of HPLC, the nature of the analytes, the choice of the stationary phase, and also the type of detection used for the analyte measurement. The solvents in HPLC can be pure compounds such as water, methanol, ethanol, acetonitrile, tetrahydrofuran, hexane, or methylene chloride. However, more commonly solvent mixtures are used, and in some separations various additives are present in the mobile phase such as salts, acids, and bases (at low concentrations) that provide a specific pH and ionic strength of the mobile phase. The separations can be performed in isocratic conditions, but gradient separations are very common, in particular in RP-HPLC, HILIC, and NPC. Chiral and size-exclusion separations are typically not performed with gradient elution. Gradients (solvent composition modifications during the separation) are usually achieved by mixing two solutions with different composition. Gradients can also be achieved by mixing more than two solutions, but three or four solvent gradients are not common. The role of gradient is to increase certain components of the mobile phase, for example, an organic solvent in a partially aqueous solution. Solvent/solvent composition is selected with the purpose of (1) achieving separation, (2) achieving a fast separation, and (3) delivering the analytes to the detection without interfering with measurement. Mobile phase capability to elute an analyte at shorter retention times compared to other solvents is typically referred to as solvent "strength." In RP-HPLC, the increase in the concentration of organic solvent in the mobile phase leads to faster elution of the analytes. The nature of the mobile phase also affects other parameters of the HPLC process, such as the selection of the detector or the acceptable flow rate in the HPLC system. The chromatographic column generates backpressure during the mobile phase flow, and this backpressure is related to the mobile phase viscosity. The mobile phase nature and properties, as well as its delivery conditions as isocratic or gradient, are further discussed in Chapter 7.

Detection in HPLC and Quantitation Procedures

Detection in HPLC is typically based on a specific physicochemical property of the analyte. The detector is capable of transforming this property in an electrical signal or *detector response* represented by the chromatogram. Various detectors are available for HPLC, and their short description is given in Section 1.4. Also discussed in Section 1.4 are the qualities required for a detector, as well as the criteria for the detector selection. The intensity of the electrical signal generated by the detector in the form of peaks corresponding to each compound is commonly used for quantitation purposes. This signal (response) depends on the instantaneous concentration of the analyte that is introduced into the detector and produces the peak. Use of peak areas in the chromatogram is the most common way of quantitation. Ideally, for a given volume V_{inj} of sample injected into the chromatograph, the peak area is linearly dependent on the concentration of the sample. The quantitation procedure requires a calibration curve obtained with the compound to be analyzed. The concentration of interest c_i (or c if index i is neglected) can be determined from the areas using the relation:

$$c = b A \qquad (1.3.1)$$

where A is the peak area and b is the slope of the calibration curve for the analyzed compound and is sometimes indicated as sensitivity. The value b of the slope for the calibration curve may be different for different compounds. For this reason, the generation of calibration curves is usually necessary for each analyte that must be quantitated. For generating the calibration curve, it is possible to use solutions of different concentrations made using the pure compound to be analyzed as a *calibration standard*. The calibrations are done independently of the sample, and the calibration standard can in this case be considered an *external calibration standard*. (An external standard is analyzed in a different run from the sample, while an internal standard is added and analyzed together with the sample.) In many practical applications it is preferable to make the calibrations by adding different levels of the calibration compound to a blank sample that does not contain the analyte. This procedure makes the analysis of the samples containing the calibration standards as close as possible to the analysis of a real sample and allows the subtraction of the overall influence of the matrix in the analysis. For compounds that have similar structures, the calibration curve for only one of the compounds is sometimes utilized, and different compounds are quantitated based on the same calibration. However, this practice is not recommended and should be used only when the calibration standards of all compounds are not available.

Some linear calibrations do not pass through the origin, and the calculation of the peak area must be done using a relation of the form:

$$c = a + b A \qquad (1.3.2)$$

This type of dependence may indicate some problems with the particular analytical method, such as sample decomposition, loss of sample in the chromatographic process due to selective adsorption, and interfering signal from the blank sample. A negative value for the parameter a indicates in general a loss of analyte, while a positive value indicates background or interferences. In cases when the equation of the calibration curve is obtained from the equation of the trendline passing through the calibration points, it is highly recommended that the trendline be forced through zero.

Most HPLC detectors provide a linear dependence of the peak areas with the sample concentration. However, nonlinear dependences are sometimes encountered. Nonlinearity may be due to overloading of detectors that otherwise are expected to have a linear response. Also, very low levels of analyte may lead to a nonlinear response. For this reason, linearity must be verified for a whole range of concentrations, and particular attention must be paid to very low and very high concentrations. The linearity of fluorescence detection is valid only for low concentrations. A nonlinear calibration curve is sometimes necessary for fluorescence and chemiluminescence detectors. This still allows calculation of the concentration directly from the nonlinear calibration curve or from use of a corresponding relation. However, a linear calibration curve is preferable to other cases.

In many quantitative techniques, an *internal standard* is also introduced with every chromatographic run. Internal standards are compounds absent in the real samples, which are added in a constant amount at a chosen point during the analysis for verifying reproducibility, accounting for sample losses, and so on. The internal standard must be chosen in such a way as to behave in the analytical process as close as possible with the analytes, to not interfere with the analyte determination, to give a chromatographic peak convenient to integrate, and the like.

Besides the internal standards that are added in the sample, such that they go through the sample preparation process, it is sometimes useful to add a *chromatographic standard* in the processed sample. This standard is a type of internal standard used only for verifying that the chromatographic process works properly; it is introduced in the processed sample that is ready to be injected in the chromatograph.

A different quantitation technique besides the external calibration is that of standard addition. The standard addition method can be used to analyze an unknown sample of concentration c_x without the use of a calibration curve obtained in separate runs. It must be assumed, however, that the dependence of the peak area of the concentration follows rel. 1.3.1 and not 1.3.2. A set of known amounts of analyte

$\{q_i\}_j = 0,1,2....n$ with $q_0 = 0$ are added to the unknown sample, leading to the concentrations $c_i = (q_x + q_i) / (V_x + V_i)$ where V_x is the known volume of the sample to be analyzed, V_i is the volume of the added solution with the i standard, and $c_x = c_0 = q_x/V_x$. The relation between the concentration c_i and the signal (peak areas A_i) is in this case given by a relation of the type 1.3.1:

$$c_i = b\,A_i \ (i = 0, 1, 2...) \qquad (1.3.3)$$

The values for c_0 and b (as parameters) can be obtained from the added amounts and peak area measurements $\{q_j, A_j\}_j = 0,1..N$ using, for example, least-square fitting. The standard addition method can be used even with a single added amount to the unknown sample. If one single addition q_1 is made to the sample, and the unknown sample is considered as having $q_0 = 0$, two peak areas A_0 and A_1 are generated corresponding to c_0 and c_1. The two equations of the form 1.3.3 for c_0 and c_1 are $q_x/V_x = b\,A_0$ and $(q_x + q_1)/(V_x + V_1) = b\,A_1$, and they lead to the result:

$$\frac{(q_x + q_1)V_1}{(V_x + V_1)q_x} = \frac{A_1}{A_0} \qquad (1.3.4)$$

This relation can be easily rearranged to give:

$$c_x = \frac{q_1 A_0}{(V_x + V_1)A_1 - V_x A_0} \qquad (1.3.5)$$

When the addition of the standard does not dilute the sample ($V_1 = 0$), rel. 1.3.5 can be written in the form:

$$c_x = \frac{c_1 A_0}{A_1 - A_0} \qquad (1.3.6)$$

Other procedures can also be used for quantitation. One of these procedures is based solely on the peak area ratios for two compounds. For this procedure, a response factor F_x must be obtained initially. This response factor using an internal standard is calculated from the peak area A^*_{is} of the internal standard and the peak area A^*_x of the compound to be analyzed, both of which are added to a blank sample in equal amounts (concentration). The ratio of the two areas, usually obtained as an average of several measurements, gives the response factor:

$$F_x = A^*_{is}/A^*_x \qquad (1.3.7)$$

Ideally, the value for F_x remains constant for an interval of values for the pair of concentrations of the standard and the sample. The concentration of the unknown is then obtained by measuring in the same run the peak area of the compound to be analyzed (at unknown concentration) and the peak area of the standard using the formula:

$$c_x = F_x \, (A_x/A_{is}) \, c_{is} \qquad (1.3.8)$$

where A_x is the area of the compound x at unknown concentration, A_{is} is the area of the standard at the concentration c_{is}, and F_x is the response factor. In order to achieve a constant value for the response factor F_x in a range of concentrations, it is recommended that the two compounds, the internal standard and the analyte, be chemically similar or even identical except for use of a labeled compound for the standard.

1.4. OVERVIEW OF HPLC INSTRUMENTATION

General Comments

The HPLC instrument physically separates the components of a sample, typically in solution, and provides information about the concentration of each separated component. For this purpose, the instruments allow the injection of a measured small volume of sample in a mobile phase. This mobile phase flows through a chromatographic column where the separation takes place, and further through a detector (or detectors) capable of generating a signal proportional with the analyte concentration. Using calibrations, the concentration of the analytes can be determined. This process can be achieved using a large number of models of HPLC systems. The construction of these instruments depends significantly on the intended function and size of the HPLC separation. Modern HPLC instrumentation is sophisticated and is in continuous development [11]. For this reason, this section is intended only to give a basic and simplified view regarding the HPLC equipment.

Schematic Description of an HPLC Instrument

When used for analytical purposes, an example of an HPLC instrument consists of several components that are schematically shown in Figure 1.4.1. The system may include: (1) a solvent supply system (solvent container and degasser), (2) a high-pressure pumping system (shown as a dual piston mechanical pump), (3) an injector (shown with a syringe containing the sample that can be loaded in a loop, and a switching valve in two positions **A** load loop with sample, **B** connect loop in circuit flow to inject sample), (4) a chromatographic column (possibly with a guard column or precolumn), (5) one or more detectors (a spectrophotometric detector is schematized), and (6) a controller/data processing unit.

Some details on each component of an HPLC system are further discussed in this section. However, the description of HPLC instrumentation is not the main goal of this book, and more information on the subject can be obtained from various other sources such as instrument manuals (e.g. [12, 13]) or from other dedicated publications (e.g. [14]).

Solvent Supply Systems

The solvent supply provides the solvent(s) necessary as a mobile phase for the HPLC.

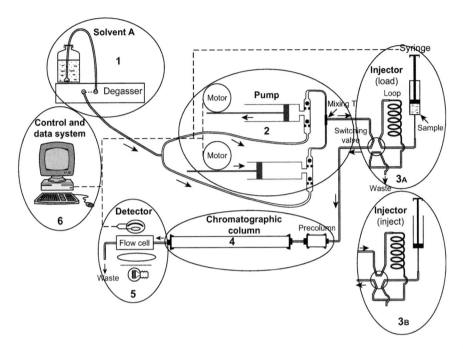

FIGURE 1.4.1 *Schematic description of a simple HPLC system.* 1) A solvent supply system (solvent container and degasser), 2) a high pressure pumping system (a dual piston mechanical pump is pictured), 3) an injector (a syringe with the sample and a switching valve in two positions **A** load loop, **B** inject), 4) a chromatographic column (possibly with a precolumn or guard column), 5) one or more detectors (a spectrophotometric detector is schematized), 6) a controller/data processing unit.

Some solvent supply systems may also have the capability to remove the gasses dissolved in solvents (the degassing capability). The solvent(s) are transferred through low pressure tubing to the pumping system. The tubes used for passing the mobile phase through the system need to fulfill mainly the requirement of being inert to the utilized solvents and to stand pressures up to about 50 psi (1 psi = 6.89476 kPa = 6.89476 10^{-2} bar = 6.80460 10^{-2} atm; 1 bar = 14.5037738 psi). Fluorocarbon polymers such as Teflon® are common materials used for this type of tubing, but polypropylene is also sometimes used. The solvent supply system of an HPLC has one or more reservoirs for the solvents used as the mobile phase. For HPLC performed in isocratic conditions and using a pure or a premixed solvent, only one reservoir is necessary. However, it is common in HPLC to

use gradient separations, or to use an isocratic separation but to generate the mixture of solvents using the pumps. In this case, two (or more) solvents that are mixed with the pumping system in variable proportions are required. For this reason, two or more solvent reservoirs are common in modern HPLC instruments. The reservoirs must be clean and inert to the solvents they contain. The solvents from the reservoirs must be free of particles, and they are either purchased as HPLC grade or/ and filtered through 0.45 μm filters before use. The filter selected for the filtration must be inert to the solvent. In the case of solvent mixtures containing a buffer, the general rule indicates that the buffer solution is made in water, then filtered, and only then mixed with the organic solvent (assuming correct concentrations and no precipitation after organic solvent addition).

The tubing transferring the liquid to the pump(s) typically has a frit at its mouth.

Degassing in HPLC systems is necessary because even very pure solvents may have small quantities of oxygen dissolved in them. This oxygen may be released in the form of very small bubbles in the HPLC system when a drop in pressure occurs (e.g., between the chromatographic column and the detector), or when a solvent with high solubility for oxygen (e.g., water) is mixed with another solvent with low solubility for this gas. When this mixing is done before the high-pressure pump, the gas bubbles may lead to variations in the pressure of the liquid delivered by the pump (pressure fluctuations). Dissolved gasses in the mobile phase may also influence the injection volume when small sample volumes (e.g., 2 μL) are injected. Also, the reading of the detectors can be perturbed by dissolved gasses. For example, oxygen may affect the reading of electrochemical detectors, the fluorescence intensity of certain compounds, and the UV absorption at very low wavelength range. The elimination of gasses dissolved in solvents is accomplished through two common procedures: helium sparging and use of a degasser apparatus. Helium sparging consists of passing a small flow of helium through the solvent. Although this procedure is useful for reducing the oxygen content, it poses a problem in the case of premixed solvents. Premixed solvents are frequently used as one of the mobile phase components in HPLC. If the premixed solvent contains a volatile component, the solvent composition may be changed in time by the preferential evaporation of the volatile component due to sparging. In particular, when ammonia is used in a mobile phase to adjust the pH of the solution, sparging is not recommended since drastic changes in the pH occur in time by the preferential elimination of ammonia from the solution.

A degasser apparatus is a device in which the solvent passes through a piece of a special polymeric tubing placed in a vacuum chamber. The tubing material (membrane) has selective permeability to gasses, and the vacuum created by a small pump reduces the content of the gasses from the solvent. The degassers, although popular in HPLC equipment, also may pose problems in specific applications. The polymeric tubing may absorb selectively specific components from the solvents and may be a source of contamination when changing from one solvent to another. It should also be noted that large bubbles coming from the solvent reservoir cannot be eliminated by the degasser apparatus; these bubbles make their way into the pumps, affecting their function.

Pumping Systems

The main pumping system consists of pump(s) able to deliver a constant flow of solvent through the injector, chromatographic column, and through the detector(s). The pumps must be able to generate a high pressure, which is needed mainly to overcome the resistance to flow of the chromatographic column. This flow is characterized by the *volumetric flow rate U*. In conventional HPLC systems, the pumps are usually capable of delivering U between 0.1 mL/min and 10 mL/min fluid and can generate up to 6000 psi (about 400 bar). New developments in using very fine particles in the chromatographic column require higher pressure and sometimes capability to produce flows at less than 0.1 mL/min. These instruments can generate up to 8500 psi (about 600 bar) or higher (e.g., 1200 bar for a flow up to 5 mL/min) and are indicated as UPLC or U-HPLC. For the HPLC systems used for other purposes than analytical, pumping parameters can vary significantly. The flow from the pumps (volumetric flow rate U) must be constant, without fluctuations or only with very small ones. This requirement is necessary mainly for the detectors, where the signal may fluctuate when the flow rate varies.

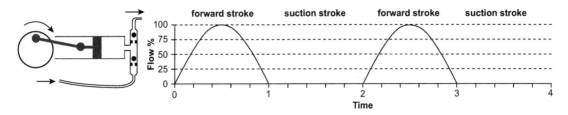

FIGURE 1.4.2 Flow from a single piston reciprocating pump when the piston is moved by a circular motion of a driving cam.

Most high-pressure pumps used in analytical HPLC are reciprocating pumps. A single-piston reciprocating pump consists of a cylinder with a reciprocating plunger in it, together with two valves mounted in the head of the cylinder. The liquid enters the cylinder through an inlet (suction) valve and is pushed through a discharge valve. During the suction the plunger retracts and the inlet valve opens, causing the admission of fluid into the cylinder, while the discharge valve is closed. In the forward stroke, the plunger pushes the liquid out through the discharge valve while the inlet valve is closed. However, the fluid flow from a single-piston reciprocating pump (and therefore the pressure in the system) has a pulsating profile. When the piston is moved by a circular motion of a driving cam, the flow rate has a half sinusoid shape, as shown in Figure 1.4.2. This type of flow is not suitable for HPLC. Dual-piston pumps consisting of two reciprocating pumps that alternate the forward stoke

are able to generate flow with only one zero flow point per cycle. However, with this setup the flow is still fluctuating. The use of specially shaped driving cams or of stepper-driven motors allows the generation of an almost continuous flow of liquid. The dual-piston pumps may be connected in parallel or in series. A dual-piston design working in parallel was already schematically shown in Figure 1.4.1. An accumulator-piston design with the pumps in series is shown schematically in Figure 1.4.3. This type of system also requires two pistons but only three valves in order to achieve the task of generating a continuous flow.

Modern systems are able to deliver flow with a precision of about 0.07% relative standard deviation (RSD%) and a flow accuracy of less than 1% from the nominal value. Because the pumps must deliver flow at high or very high pressure, their construction requires special materials such as inert steel body, sapphire or ceramic pistons, high-precision valves that do

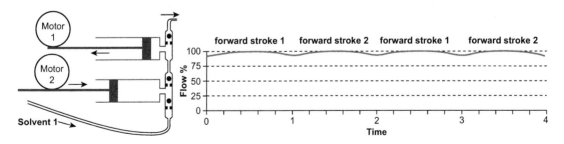

FIGURE 1.4.3 Flow from a dual accumulator-piston reciprocating pump when the pistons are moved by stepper-driven motors that allows low pulsation.

not have any leaks, special polymeric seals, and the like. For ion chromatography, the whole pumping system (except the piston) is typically made of strong polymeric materials such as polyetheretherketone (PEEK).

In addition to the specially designed pumps and valves, the flow without fluctuations from the pumping system can be achieved using a pressure pulsation damper. Pressure dampening is done, for example, by passing the fluid through a cell with a diaphragm wall that compensates the pressure variation. The pulsation with dampening can be reduced to less than 2% variation in pressure. Various models of dampers are available, and most of them have a volume of around 500 µL, in order to assure a small delay volume in the delivered fluid. Other sources of delay volumes, beside the damper, are present in an HPLC system.

A dual-piston pump can handle only one solvent and can be applied for isocratic separations that use a pure or a premixed solvent. However, since in HPLC it is frequently necessary to use gradient separation, instruments that handle more than one solvent have been developed. This type of instrument is also frequently used to generate a solvent mixture of a desired composition, even when this composition is not changed during the separation (isocratic conditions). There are three basic procedures to physically achieve the mixing of solvents: (1) low-pressure mixing, where the solvents necessary for the gradient are premixed with a low-pressure pump connected in front of the high-pressure pump, (2) high-pressure mixing that uses two (or more) high-pressure pumps, with each one dedicated to one solvent and with the mixing of the flows in a low-volume mixer, and (3) hybrid mixing that uses a high-pressure pump with two or three proportioning inlet valves.

In low-pressure mixing, two or more (usually four) solvents can be blended at the desired composition by using a low-pressure pump with several proportioning inlet valves controlled by a computer. The valves open repeatedly for a short period of time (typically less than one second); the duration of time the valve is opened is proportional to the desired mobile phase composition. The mixed solvents are delivered further to one (dual-piston) high-pressure pump. Low-pressure mixing has the advantage of using a single high-pressure pump (that is typically expensive), and has more flexibility in choosing a variety of solvents (in systems with four proportioning valves). However, the changes in the mobile phase composition when using a low-pressure mixing system are taking place more gradually than for the other systems (the change in composition is not instantaneous). Also, low-pressure mixing may be prone to the formation of small bubbles of gas in the mixed solvent, when the solubility of oxygen, for example, is higher in one solvent than in the mixture. These bubbles may enter the high-pressure pump and generate flow fluctuations.

In high-pressure mixing, two high-pressure (dual-piston) pumps are used, and the ratio of solvents in the mobile phase is controlled by the flow rate of the high-pressure pumps. The mixing of the two solvents A and B can be achieved either in a specially designed mixer (typically with volume of less than 500 µL) or in a mixing T (with virtually zero volume). High-pressure mixing is thought to provide a more precise control of the composition of the mobile phase (with a typical composition precision of less than 0.15% RSD% at 1 mL/min flow) and does not have the problem of formation of bubbles caused by the difference in solubility of oxygen in the mixed solvent compared to that in one of the components. However, the cost of high-pressure pumps is a disadvantage for this type of mixing. Most modern HPLC systems with high-pressure mixing have two pumps as well as the capability of switching between two solvent pairs (A1, B1, and A2, B2).

Since instruments with low-pressure and high-pressure mixing are common in laboratories, when transferring an established analytical technique from one instrument to another, attention should be paid to the type of instrument. In both types of chromatographic systems, there is a specific volume passing the system from the point at which the mobile phase solvents are mixed until they reach the head of the chromatographic column. This volume is known as *dwell volume* V_D. The dwell volume is typically different in low-pressure mixing systems (2–4 mL) and in high-pressure mixing systems (1–3 mL). Special instruments, such those used in microscale HPLC may have smaller dwell volumes (less than 300 μL). For certain applications using gradient separations on common HPLC systems, differences can be seen when working with one type of instrument or with the other, although the gradient program is the same. This is in particular caused by the differences in the dwell volume from one system to another.

Hybrid mixing uses a system of two reciprocating high-pressure pumps with proportioning inlet (suction) valves. Low-pressure mixing systems with four proportioning valves and one dual-piston high-pressure pump, as well as high-pressure mixing systems with two high-pressure pumps (each one dual piston), are much more common than hybrid systems.

To modify the composition of the mobile phase in gradient HPLC, the solvents that are blended in specified proportions should be perfectly miscible. Particular care must be paid to the solubility differences of certain additives present in one solvent when another solvent is added. For example, it is common in HPLC to use buffer solutions. These solutions can be easily made in water by adding mixtures of salts and acids or salts and bases. When a water solution containing these types of additives is mixed with an organic solvent (such as CH_3OH or CH_3CN), the solubility of the additives in the mixed organic/aqueous solution is drastically diminished. The formation of precipitates following the mixing must be carefully avoided and only buffers at low concentration of salts (typically less than 100 μmol) should be used when organic solvents are to be added to the aqueous buffer.

During the separation, the composition of the mobile phase can be kept constant for some intervals of time and modified for other periods of time. The modern instruments are commonly controlled using a computer with a dedicated program that assists in generating a specific gradient using a gradient time table. Based on the gradient timetable, the computer controls the pumping system that physically generates the desired mobile phase composition by mixing in the correct proportion of the solvents from the solvent supply system. The gradient starts when the sample is injected. After the gradient ends, the HPLC chromatograph is made ready for the next injection. The total runtime of the chromatogram, starting with the moment of injection until the end of the chromatographic run, is sometimes referred to as *total cycle time*. In a gradient separation, the dwell volume V_D of the system creates a dwell time t_D. Because of the dwell time, there is a delay between the change of composition at the point of solvent mixing (set in the time table) and the change in composition at the head of chromatographic column. Therefore, when attempting to modify the retention time of a peak by using a "stronger" solvent, this change should be done in the gradient timetable ahead of the peak retention time.

The modification in concentration between two changing points of the gradient can be linear. For a gradient starting at time t_1 with the concentration c_1 of component A and ending at time t_2 with the concentration c_2, at an intermediate point at time t the concentration of component A can be obtained using the formula:

$$c(t) = c_1 + \frac{(t - t_1)}{(t_2 - t_1)}(c_2 - c_1) \qquad (1.4.1)$$

In most HPLC separations the mobile phase is changed from one content in an organic modifier to another one. The change in the concentration of the organic modifier in a period of time is indicated as *gradient slope*. For a linear change in concentration, the gradient slope can be defined by the expression:

$$\Delta = \frac{(c_2 - c_1)}{(t_2 - t_1)} \quad (1.4.2)$$

Some HPLC pumping systems allow both a linear change in the gradient and a nonlinear modification of the concentration (see, e.g., [15]). This change can be achieved using a variation in concentration as a function of time given by the formula:

$$c(t) = c_1 + \left(\frac{t - t_1}{t_2 - t_1}\right)^n (c_2 - c_1) \quad (1.4.3)$$

where n is larger than 1 for curves of increase that "hold water" and is between 0 and 1 for the "does not hold water" type of curve. This type of gradient variation is illustrated in Figure 1.4.4 for $t_1 = 0$, $c_1 = 0$ and $t_2 = 2$, $c_2 = 100$ and different n values. A different type of nonlinear curve can be obtained using

a variation in concentration with a function given by the formula:

$$c(t) = c_1 + \left[1 - \left(1 - \frac{t - t_1}{t_2 - t_1}\right)^n\right](c_2 - c_1)$$

$$(1.4.4)$$

When n is larger than 1, the curve of increase "does not hold water," and when n is between 0 and 1, the curve of increase "holds water". This type of gradient variation is illustrated in Figure 1.4.5 for $t_1 = 0$, $c_1 = 0$ and $t_2 = 2$, $c_2 = 100$ and n values higher than 1.

Much less frequently than reciprocating pumps, syringe pumps are sometimes used in HPLC. Only the recent developments in UPLC made the syringe pumps more attractive. UPLC requires low flow rates (of the order of 0.1 mL/min) and less solvent compared to conventional HPLC. In syringe pumps, a cylinder is loaded with the mobile phase that is delivered at a specified flow rate by the movement of a piston. The flow from a syringe high pressure pump can be virtually fluctuation free as compared to that from a reciprocating pump. The price of a syringe pump can also be lower.

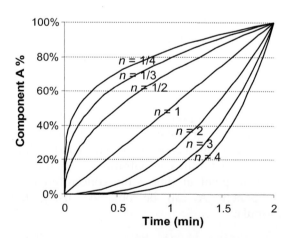

FIGURE 1.4.4 Non-linear gradient variation using formula 1.4.3.

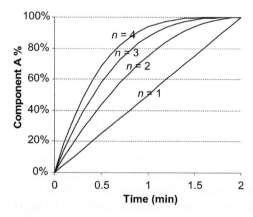

FIGURE 1.4.5 Non-linear gradient variation using formula 1.4.4.

Injectors

The role of the injector is to add in the mobile phase a small, precisely measured volume of a solution containing the sample. The injection must be done reproducibly and accurately. Reproducibility of injection is of particular importance, and modern injectors typically show less than 0.5% RSD% in the injected volume. The accuracy errors in injection volume are important mainly when comparing different instruments since for the same instrument the use of standards for quantitation may compensate small variations from a nominal volume. However, for a specific method it is recommended to keep the same injection volume when injecting different samples in order to avoid accuracy problems.

Figure 1.4.1 shows one of many possible injection systems in which a loop of a precise volume is filled first with the sample and then connected to the flow circuit using a switching valve. This system allows only the injection of a fixed volume of sample, equal with the loop volume. However, the volume of sample solution to be injected in different analyses may need to be varied. For example, for a very diluted sample, or for detection that is not very sensitive, a larger volume of sample may be necessary in order to assure a good measurement. Samples that generate a high detector signal need to be injected at lower volumes. Conventional HPLC systems have injectors capable to inject between 1 μL up to 100 μL sample solution (or even up to 1000 μL in special systems), typical volumes being between 2 μL and 20 μL. Injection of different sample volumes can be achieved using a larger loop (e.g. of 100 μL) that is only partially filled with sample (partial loop injection). The loop is typically filled initially with the mobile phase, and then the sample is introduced in the loop, occupying only a small portion of its volume. The placing of the loop in the mobile phase main circuit is then done in a similar manner as for fixed volume loops. The sample volume is typically controlled using a computer. For UPLC, the injection volumes can be between 20 and 500 nL, which is significantly smaller than for common analytical HPLC; for semipreparative or preparative HPLC, the volumes can be much larger. These systems may need specially designed injectors.

The sample is ideally introduced in the mobile phase flow as a zone (plug) with the concentration of the sample following a perfectly rectangular profile. The force pushing the fluid through the tube is given by the cross-sectional area of the tube multiplied by the pressure drop Δp, or $\pi r_t^2 \, \Delta p$ (where r_t is the tube radius). In a laminar flow, the layer adjacent to the wall is held virtually stationary by adhesion, and the inner fluid layers slide over the ones closer to the wall creating a shear force. This force is given by the formula:

$$F_\eta = \eta \, A \, du/dr \qquad (1.4.5)$$

where η is the liquid viscosity, u is the fluid linear velocity, dr is the infinitesimal distance between the layers, and A is the area of contact between the layers of the fluid [16]. Since the two forces are equal in a steady flow, at a distance r from the center of the tube, and taking $A = 2 \, \pi \, r \, L$ (L is the length of contact between layers), the following relation should be satisfied:

$$\pi r^2 \, \Delta p = -2 \, \pi \, r \, L \, \eta \, du/dr \ \text{ or } \ du = \frac{-\Delta p}{2L\eta} r dr$$
$$(1.4.6)$$

By integration, rel. 1.4.6 gives the fluid velocity at any point at distance r from the tube center:

$$u(r) = \frac{\Delta p}{4L\eta}(r_t^2 - r^2) \qquad (1.4.7)$$

Relation 1.4.7 shows that due to the friction with the tubing walls of a viscous fluid, downstream of injector (in a laminar flow), the shape of the sample plug is changing and generates a parabolic profile. This process, as well as the diffusion and other convection effects, contribute to deviations of the chromatographic peak from the ideal Gaussian shape, introducing tailing and asymmetry.

Two important parameters must be selected by the operator regarding injection: (1) the nature of the solvent for the sample, and (2) the injection volume. Besides the obvious requirement that the solvent for the sample should dissolve the sample completely, this solvent must also be soluble in the mobile phase. The injection volume is selected depending on a number of factors, including the type of instrumentation (HPLC, UPLC, detector type, etc.), the sensitivity of the detector, the loading capacity of the column (maximum amount of sample that still allows separation), and the effect of sample solvent on peak shape. Both too small volumes and too large volumes of injection pose a number of problems. Too small volumes may lead to problems with injection reproducibility, sensitivity of the detector, or sample loss in the chromatographic column, but they offer better peak shape, sometimes resulting in better separation. Too large volumes may affect the peak shape (broadened, with flat top, asymmetrical) that affects separation. A large volume of sample does not necessarily mean a large amount of analyte (e.g., when the sample is very diluted), but when the large volume is also associated with too much analyte, this can be associated with an overload of the column (problems with the separation) and of the detector (leading to a nonlinear response). (Further discussion on injection volume will be presented in connection with the nature of sample solvent in Section 9.2.)

Injector systems with automation capability are common. From a large number of samples in different vials (or well plates), these automatic systems (computer-controlled autosamplers) have the capability to select the desired sample vial from a tray, and to repeat the injection at a specified time or upon receiving an electrical signal from the computer. In autosamplers, since several samples are injected one after the other, it is possible to see carryover effects. Carryover effects represent the contamination of a sample with small amounts of components from the previous sample that remained in the autosampler after an injection. This problem is typically solved using a needle wash. Some autoinjectors have the capability of mixing the sample with specific reagents from different vials, in the event derivatization is necessary before the separation and detection of analytes. Also, cooling and heating capability is frequently present in modern autoinjectors.

The injection operation is usually unselective. However, special online sample preparation methods require more special injection techniques. In case of a multidimensional HPLC, for example, a fraction from eluted sample from a first column can be transferred by an interface to become an injected sample to a second column. Injection devices normally do not affect retention parameters unless operation problems occur (see, e.g., [17–19]). Also, in most HPLC applications, there is one injection for each chromatographic run. However, multiple injections in a single run (MISER) are possible for specific analyses, such that a larger number of samples can be analyzed within a specific interval of time and with a lower solvent usage [20].

Tubing and Connectors

The tubing used after the high-pressure pumps must be inert and amust also be able to withstand the high pressure generated by the pumps. Typical materials for the tubing are stainless steel (316 stainless steel) and polyetheretherketone (PEEK). Stainless steel is inert in most solvents, while PEEK may become very stiff after using solvents such as tetrahydrofuran or dimethylsulfoxide. Also, stainless steel

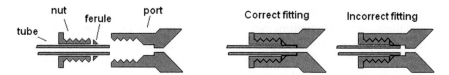

FIGURE 1.4.6 Tubing connection to a port, correct fitting, and incorrect fitting with a void space.

tubing can be used even at very high pressure, while some restrictions to pressure are applied to the PEEK tubing (as a function of internal diameter). However, for IC chromatography, PEEK is the material of choice for tubes and connectors. Tubes of several internal diameters (i.d.) are available, such as 0.12 mm i.d. (0.005 in.), 0.17 mm i.d. (0.007 in.), 0.25 mm i.d. (0.010 in.), 0.50 mm i.d (0.020 in.), and so on (for both stainless steel tubing and PEEK tubing a color code is available to designate the i.d.). The choice of the tubing after the injector starts to play a role in the shape of the sample plug. Tubing with 0.12 mm i.d. (0.005 in) is in general recommended to connect the injector with the chromatographic column. This tubing has a volume of about 0.13 μL/cm such that a sample of 5 μL will spread over about 38 cm, diminishing the effects of sample plug shape modifications. Another factor contributing to dilution and modifications in sample plug shape are the "void spaces" in fittings that connect the injector and the chromatographic column. Loss of resolution by peak broadening due to large void spaces and to turbulent flow either along the tubing or in fittings must be avoided. The fittings typically use a nut that connects to a port, and a ferule secures the end of the tubing in the fitting port. Common connector parts, correct fitting, and incorrect fitting of tubing with formation of a void volume are shown in Figure 1.4.6.

Chromatographic Columns

The chromatographic column is designed for performing the separation in HPLC. Its role and properties are related to various subjects discussed in this book; the present section gives only a simplified overview of the subject (see Chapter 6). The column typically consists of a tube made from metal (stainless steel) or plastic (e.g., polyetheretherketone, PEEK) that is filled with a stationary phase. At the two ends inside the column are special frits that keep the stationary phase from moving, and outside are fittings that allow the connection with high-pressure tubing. The physical dimensions of common analytical chromatographic columns vary, and values for length (internal) L can be between 30 mm and 250 mm (the common length being 50, 100, 150, or 250 mm), and internal diameters d can be between 1 mm and 10 mm (the common diameters being 2.1, 3.0, or 4.6 mm). Other dimensions are possible, particularly when the column is designed for special tasks. The newer columns tend to be shorter and narrower, as the solid particles that form the stationary phase are made smaller. Special cartridges (microfluidic chips) are also available as containers for the stationary phase. Based on the internal diameter of the analytical column, they are sometimes classified in the literature as (1) standard (3.0–4.6 mm i.d.), (2) minibore (2.0–3.0 mm i.d.), (3) microbore (0.5–2.0 mm i.d.), (4) capillary (0.2–0.5 mm i.d.), and (5) nanoscale (0.05–0.2 mm i.d.). Larger columns are used for semipreparative and preparative purposes. The empty volume of the column can be easily calculated as $V = (\pi/4) \, d^2 \, L$ (volume of a cylinder), and for analytical columns V ranges between 0.02 mL and 20 mL.

The nature of the stationary phase is selected based on the type of chromatography utilized for the separation (normal phase, reversed phase, ion exchange, size exclusion, etc.). A large assortment of types of stationary phases (column packings) are available. The stationary phase usually consists of small, solid particles with special properties. Besides small particles, porous polymeric materials and monolithic materials can be used as the stationary phase. When the stationary phase is made from small particles, the particles should have specific physical and chemical characteristics to serve as a stationary phase. The surface area of particles is one of the most important physical characteristics, being directly related to the retention in the column of the compounds to be separated. The effect on separation of the particle size is also important. Particle size influences in particular the *eddy diffusion*, which appears when local small streams of liquid follow different channels in the column. The effect is common within porous particles. This generates a broadening of the chromatographic bands (discussed in Section 2.2), which is not a desired feature. When the stationary phase is made from small porous particles, these particles are frequently obtained from an inert substrate material (such as silica) that is covered with the active phase. The particles can be of three main types: porous, superficially porous (core-shell), and pellicular. Porous particles are still the most common type of stationary phase used in HPLC. They are made from particles that are usually of 3–5 μm diameter, with a specific porous structure (e.g. of silica) and with the surface of this structure covered with an active constituent. This constituent can be physically or chemically bonded on the solid inert support, the bonded phases being the most common type. Since reverse-phase (RP-HPLC) is the most utilized technique, the largest variety of columns is of RP type. These columns have a hydrophobic active phase, for example, with octadecyl groups (C18), or with octyl groups (C8) bonded on silica. For other types of chromatography, the stationary phase may

be made in various forms. For example, ion-exchange HPLC can use particles from a substrate inert material that are covered with the active phase but also various types of ion-exchange polymers. Size-exclusion chromatography typically uses perfusion particles made from silica or special types of polymers. These particles contain very large pores (400–800 nm) connected with a network of smaller pores (30–100 nm). The structural rigidity of these particles is not as good as that of common porous particles made from silica, and restrictions to the maximum pressure that can be used with the columns made with these particles are typically indicated by manufacturer. At higher pressures than recommended, the stationary phase may "collapse" and the column can be damaged.

The chemical properties of the particles forming the stationary phase include (1) the chemical nature of the active surface, (2) chemical stability, (3) surface reactivity, and (4) density and distribution of surface reactive centers. The nature of stationary phases used in various analyses is fully discussed further in this book (see Chapter 6). In some systems, more than one chromatographic column is necessary for achieving the desired separation. In size exclusion chromatography, for example, two to four columns may be connected in series. In other types of separation, the use of more than one column is less common. The nature of the columns used in series may be the same or may be different. More than one column is also used in multidimensional HPLC, where a portion of the initial separation is further submitted for a second separation in a different column.

Some chromatographic columns require a specific temperature for performing a good separation, and for this purpose special column ovens are used. Common column ovens have the capability to control the column temperature in a range from about 10 °C below ambient to 80–100 °C. Higher temperatures can be achieved with special ovens used in high-temperature HPLC. In addition to heating the column, the ovens typically are able to heat the solvent entering the column, since peak shape distortions may be noticed when the column and the entering solvent have different temperatures.

In order to protect the stationary phase from the analytical HPLC column, it is common to use small pore frits (e.g., with 0.45 µm pores) as well as guard columns (cartridges) in the path of the mobile phase before the column. The frits have the capability of mechanical filtration of the mobile phase. For column protection, more important than frits are the guard columns. The guard (cartridge) columns are selected to match the stationary phase of the analytical column (the same active material), but their length is much shorter (e.g., a few mm), and in some cases their stationary phase has larger particle size. In the analyzed samples, there are sometimes components that are very strongly retained by a specific stationary phase. These components do not elute and tend to accumulate at the head of the column, deteriorating its performance in time. The use of a guard column allows selective retention of these components without affecting the efficiency of the column, retention time, backpressure, or the level of analytes. Guard columns are changed from time to time, while the analytical columns have longer service life (see Section 6.3).

Setups for Multidimensional Separations

Some separations are performed in "bidimensional" mode. For such separations, switching devices are used to divert, for a specific time interval, the eluent from the first column into the second column. The volume of diverted eluent from the first column containing the unseparated analytes of interest is indicated as a "heart-cut." The mobile phase from the first column can be the same in the second column, or they can be different and can be generated using the second pump. The schematic diagram of a heart-cut system working in three stages is shown in Figure 1.4.7. In stage "A" the flow from Pump 1 (isocratic or gradient) containing the sample goes into column 1 and into the detector. At a certain desired time the heart-cut is taken by switching to stage "B" where the flow from Pump 1 goes into column 1, then into column 2, and then into the detector. After the heart-cut is taken, the system is switched to stage "C" where the flow from Pump 2 (isocratic or gradient) goes into column 2 and then into the detector.

More complicated setups for achieving multidimensional separations are reported in the literature [21]. For example, the collection of the heart-cut can be done in a loop, after the solution passes the detector. After the collection of the heart-cut, this is submitted into the second column and into the detector, such that the detector will respond to both unseparated and separated compounds in the heart-cut. The choice of columns for multidimensional separation is typically done so that they have very different selectivity. The HPLC separations performed on different types of column and using different solvents that lead to a different separation are referred to as *orthogonal*.

Other Devices that are Part of the HPLC System

The multiple use of HPLC has promoted a variety of instrumental setups and developments. For example, the capability of column switching (sometimes part of the column oven) is a relatively common feature of HPLC instrumentation. The switching valves can be used only for selecting one of several columns by changing the flow to the desired one. Another common use of column switching is that of sample enrichment and cleanup. In this use, the flow with the sample is sent to a column

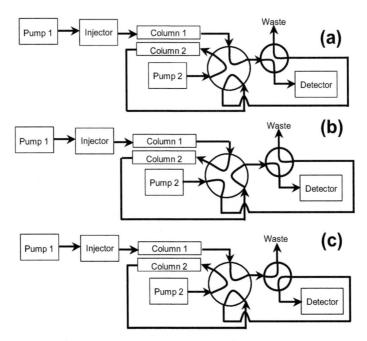

FIGURE 1.4.7 Schematic diagram for a bidimensional HPLC system.

where the analytes are strongly retained for the specific solvent composition. The flow to this column may carry a large volume of a diluted sample. While the analytes are accumulated (usually at the head of the column), the flow from this column is typically sent to waste. After enough analytes are accumulated, the flow in the column is reversed (with a switching valve), and the solvent is changed such that the analytes are eluted from the head of the first column and sent to a different column where the separation takes place. The flow from this second (analytical) column goes to the detectors. This setup is schematically shown in Figure 1.4.8.

Also, a specific sample derivatization may be selected for the increase in detection capability. When this derivatization must be done

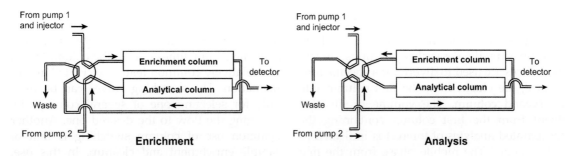

FIGURE 1.4.8 HPLC setup with two columns allowing enrichment of diluted samples.

after the separation of the analytes, "post-column derivatization" techniques are utilized. In this type of technique, reagents are added to the flow emerging from the analytical column. These reagents need thorough mixing with the analytes' flow, which can be achieved in a mixing T. In some instances, where the reaction with the derivatization reagent is not instantaneous or may need heating, "reactors" or mixing/reacting coils are added in the flow circuit after the analytical column. The reacting coils are sometimes heated or UV irradiated. The volume as well as the delay time in the coil must be adjusted depending on reaction requirements.

Another device sometimes used in HPLC is a flow splitter. This device is necessary when the flow from a conventional HPLC pump is too high to be used with either a capillary LC column or a specific detector (such as a mass spectrometer). A flow setting below a specific limit at a conventional HPLC pump may lead to undesirable fluctuations or to a difficult to control gradient program. The flow splitter allows the selection of a desired fraction from the total flow from the pump by using a waste outlet with controllable backpressure.

General Comments on Detectors

In HPLC, the mobile phase passes through a detector capable of performing measurements. The detection is based on the fact that the molecules of the sample have physicochemical properties different from those of the mobile phase. The measured properties are determined by the nature of the compound to be analyzed and that of the mobile phase. The choice of a specific property for detection depends on factors such as the extent of difference in a property from that of the mobile phase, sensitivity of the detector to the specific property, and availability of the detector. The selection of a specific detector is also correlated with the separation conditions

used for the analysis. Some detectors have specific requirements for the nature of mobile phase, selection of isocratic or gradient separation, selection of a specific temperature for the column and mobile phase, and other.

The detection of the molecular species eluted from the chromatographic column can be done using a variety of principles and techniques. Among these detection techniques are (1) UV-Vis absorption, (2) fluorescence, (3) refractive index (RI), (4) chemiluminescence, (5) various types of electrochemical detection, (6) mass spectrometry (MS), (7) evaporative light-scattering (ELS), and (8) other detection techniques. The qualities of the detectors should include (1) sensitivity, (2) reproducibility of the response, (3) linearity in a wide range of concentrations of sample, (4) capability for detection in a small volume of sample (5) capability of not contributing to peak broadening, and (6) stability to changes in flow and environmental parameters. Some detectors are designed to respond to all analytes (such as the RI detector) and are indicated as "universal" detectors. Other detectors can be compound type selective or even specific compound selective, or they can have settings that make them compound selective. Some detectors can generate only quantitative information, while others offer both qualitative and quantitative information, such as the MS detectors.

Modern HPLC systems commonly have more than one detector available. For example, UV-Vis and fluorescence detectors are frequently coupled in series, although they are not necessarily used simultaneously. Since the flow through a detector may pose some backpressure, when using detectors coupled in series, it should be verified that the flow-cell of the detector upstream can handle the backpressure generated by the second detector. For the detector downstream it should be verified that undesirable peak broadening does not occur because of the upstream detector. The coupling of detectors in parallel is also possible, but care must be

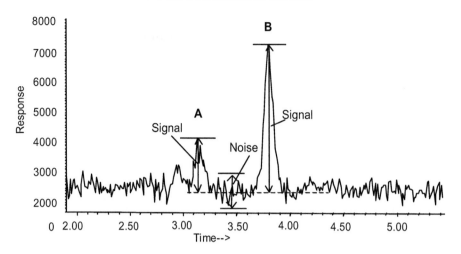

FIGURE 1.4.9   Detector response showing the measurement of noise and of two signals (A and B).

taken to ensure that appropriate flow goes through all the detectors. Since different detectors may pose different backpressures to the flow, the risk exists that most (or even all) of the flow goes to a single detector.

Detector sensitivity is a very important factor in HPLC analysis. This sensitivity depends on several factors such as analyte properties, sample matrix, mobile phase properties, detector settings, and detector manufacturer. Therefore, it is difficult to present a specific discussion on detector sensitivity. For these reasons, in analytical methods using HPLC, parameters such as limit of detection (LOD) and limit of quantitation (LOQ) are reported. They characterize globally the sensitivity of the method, and they consider a number of factors, including detector sensitivity [22].

The detector response (output) is typically dependent on the instantaneous concentration of the detected species, and for this reason quantitation is based on the area of the chromatographic peak. The output appears as an electrical signal recorded in the form of a chromatogram. In addition to the signal, all detectors are affected by the "noise," which is the random oscillation of the detector (electric) output.

Figure 1.4.9 shows the measurement of the noise, which is done on a flat area of the chromatogram (baseline) close to the place of a peak, and the measurement of the signal from the middle of baseline noise to the top of the peak.

A signal is typically considered usable for detection when the value of the signal S is at least three times larger than the value for the noise N (S/N > 3). In Figure 1.4.9, the signal for the peak B is usable for detection, while that for peak A is not. Other disturbances in the detector signal may include long-term noise and the drift. The long-term noise appears as a fluctuation of the signal with a wider frequency. The drift is a continuous variation (increase or decrease) of the signal for a period of time comparable with the length of the chromatogram. Drift may be caused not by the detector itself, but by changes, for example, in the composition of the mobile phase during gradient elution.

Spectrophotometric Detectors

Spectrophotometric HPLC detectors are basically UV-Vis spectrophotometers equipped with a flowthrough cell. In these instruments, a beam of monochromatic light (more correctly a beam of light with a narrow wavelength range) is sent through the eluent flowing through a cell of small volume (e.g., 1 μL for a micro cell, 5 μL for a semi-micro, and 14 μL for a standard cell, with path length of 5 mm to 10 mm depending on the cell, some cells having a conical shape). The monochromatic beam is split (using a beam splitter), one part going through the cell with the sample and to the detector, and the other to a reference detector. The baseline is obtained from the reading for the eluate when no solute is emerging from the column.

Two related quantities, transmittance T and absorbance A, are measurable for the light passing through the solution. Transmittance is defined as follows:

$$T = I_1/I_0 \qquad (1.4.8)$$

In rel. 1.4.8, I_0 is the intensity of the radiant energy incoming to the sample and I_1 is the intensity of the emerging light (T can also be expressed as percent). As expected, T is a function of the wavelength λ (or of frequency $\nu = c_{light} / \lambda$, where c_{light} is the speed of light) of the radiation that is absorbed. Absorbance is defined by the logarithm in base 10 of the inverse of transmittance as follows:

$$A = \log (1/T) = \log (I_0/I_1) \qquad (1.4.9)$$

Absorbance is related to the molar concentration [X] of the absorbing species X by the Lambert-Beer law:

$$A_\lambda = \varepsilon_\lambda [X] \, \mathcal{L} \qquad (1.4.10)$$

where ε_λ is the molar absorption (absorbance) coefficient at the specific wavelength λ and $\mathcal{L}$ is the path length of the light through the sample. For quantitation purposes, the absorbance is commonly used because it is proportional with the concentration. The quantitation can be done using calibration curves or standard addition method [1]. The absorbance of the liquid eluting from the separation system is typically measured at specific (small) time intervals (not continuously), generating points that form the chromatogram. The peaks in the chromatogram indicate an increase in the absorbance A_λ when an absorbing species elutes from the chromatographic column. For variable-wavelength detectors, the wavelength can be selected (by rotating a grating inside the spectrophotometer) and is usually set where a strong absorption of light by the analyte occurs (possibly at the maximum). This type of detector is one of the most commonly used in HPLC. Older detectors were made to measure the absorbance at a fixed wavelength such as 254 nm. This wavelength corresponds to the maximum emission (253.7 nm) of a mercury lamp that has been used as a UV source in simpler spectrophotometers. The area under the chromatographic peak of the analyte is proportional to the analyte amount injected into the column and it is used for quantitation with the help of a (linear) calibration curve, or a response factor between peak area and the analyte concentration. The peak height can also be used for quantitation, assuming equal peak widths for all concentrations. Some UV-Vis detectors have the capability to capture the absorbance for a whole spectrum range (UV-Vis diode array detectors or DAD). If the entire UV-Vis spectrum is measured in different points across a chromatographic peak, this can be used for evaluating peak purity and can also be a guide for qualitative identification of the analyte, although UV-Vis spectra are very seldom sufficient for compound identification. The UV region of spectrum starts at about 190 nm, and modern UV-Vis instruments have a typical working range between 190 nm and 600 nm. However, the common range of practical utility in UV spectrophotometric measurements starts at about 205 nm or higher. At lower values than this wavelength, a strong light absorption usually takes place because the solvents used as the mobile phase start absorbing. The wavelength cutoff of various solvents can be found in tables (see for example, [23]). Depending on the nature of the analyzed material, the detection limit of the UV-Vis detection in HPLC can be

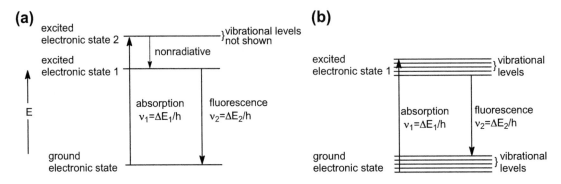

FIGURE 1.4.10   (a) Simplified scheme of electronic transitions during fluorescence. Vibrational energetic levels not shown. (b) Simplified scheme of electronic transitions during fluorescence. The difference in emission comes from differences in vibrational levels.

0.1–1.0 ng, with a linear range of five orders of magnitude. With an appropriate solvent that does not absorb in the range of UV-Vis measurement, the use of elution gradient can be applied for separation.

Fluorescence and Chemiluminescence Detectors

Fluorescence (FL) is the process of emission of light by a molecule after absorbing an initial radiation (excitation light). A molecule M goes from a lower energetic state (commonly ground state) to an excited state M^* by absorbing energy. The emission process may take place by the molecule bouncing back to the initial state without the change in the wavelength of the absorbed light. In this case, the process is difficult to use for analytical applications. However, it is possible that part of the energy of the excited molecule M^* is lost by nonradiative processes such as collisions with other molecules. In this case, the electron may go to another excited electronic state with lower energy and then, emitting a photon, reach the ground state. It is also possible that no intermediate electronic state is present, but the molecule acquires a lower vibrational energetic level and jumps to the ground state by emitting a photon of lower energy than the absorbed one. In both

cases, the fluorescence radiation has a lower frequency than the excitation radiation. These two processes are pictured in Figures 1.4.10a and 1.4.10b.

Fluorescence by emission of radiation at higher frequency than the absorbed one is also possible (anti-Stokes radiation), but it is uncommon. The average lifetime of an excited state of a molecule M^* undisturbed by collisions is about 10^{-8} s, and fluorescence can take place within this length of time. For some special compounds, the molecules can remain for a longer time in a metastable excited state. In this case fluorescence can be observed long after the initial radiation is interrupted. This type of fluorescence is commonly called phosphorescence. Fluorescence is less frequently observed than expected because it is very common that only nonradiative loss of energy takes place. The theory of fluorescence emission shows that the intensity of fluorescence F_{int} at the emission wavelength λ_2 can be expressed as a function of the intensity I_o of the excitation radiant energy with wavelength λ_1 incoming into the sample, by the expression:

$$F_{int\,\lambda2} = I_{o\lambda1}\left[1 - \exp\left(-\varepsilon_{\lambda2}\left[X\right]\mathcal{L}\right)\right]\Phi \qquad (1.4.11)$$

In rel. 1.4.11, Φ is the (quantum) fluorescence yield of the process, the other parameters being the same as defined for UV-Vis. For low concentrations, $1 - \exp(-\varepsilon_{\lambda 2}[X]\mathcal{L}) \approx \varepsilon_{\lambda 2}[X]\mathcal{L}$, and the intensity of fluorescence $F_{int\ \lambda 2}$ is related to the concentration $[X]$ by the approximation relation:

$$F_{int\ \lambda 2} = I_o\,_{\lambda l}\,\varepsilon_{\lambda 2}[X]\,\mathcal{L}\ \Phi \qquad (1.4.12)$$

In reality, only a part of emitted fluorescence is measured in the analytical instrument, and the intensity of this measured fluorescence $F'_{int\ \lambda 2}$ is given by the expression:

$$F'_{int\ \lambda 2} = \mathbf{a}\,I_0[X] \qquad (1.4.13)$$

where $\mathbf{a}$ is a constant coefficient that incorporates all the constants, including the losses due to partial fluorescence measurement. Measurement of fluorescence intensity (usually at the maximum of the emission band) is the base of quantitation of the fluorescent species. In practice, the fluorescence intensity is measured using sensitive fluorescence light detectors (FLD) that generate an electrical signal of intensity depending on a calibration constant for the instrument. The output is given in luminescence (or light) units (LU) that are arbitrary units proportional with the fluorescence intensity, but specific for the measuring instrument [24].

Similar to UV-Vis detection in HPLC, the fluorescence is measured in a flowthrough cell that is connected to the flow of the eluent incoming from the chromatographic column. Modern fluorescence detectors may have the capability of recording a tridimensional emission spectrum of the analyte by stopping the mobile phase flow at a chosen retention time (selected for the analyte) and performing a scan of the entire UV range used for excitation and for the entire emission band. In this way, the tridimensional fluorescence spectrum is displayed as a dependence of fluorescence intensity on emission wavelength and excitation wavelength. It is common in modern fluorescence HPLC detectors that the excitation beam is generated by a high power lamp that flashes at a specific number of times per second (e.g., 296 times in an Agilent 1200 Ser. detector), such that the signal is modulated in time. Also, the systems typically have a reference detector that measures the excitation light and corrects the flash lamp fluctuations. The detection in fluorescence methods encounters several difficulties because of nonlinearity of fluorescence due to self-absorption effects, difficulty in discriminating between overlapping broad spectra of interfering molecules, quenching produced by oxygen dissolved with the solvent, and so on. Because the intensity of fluorescence increases linearly with the intensity of the initial radiation, laser-induced fluorescence (LIF) detection is a successful technique applied in HPLC. For HPLC, lasers are a convenient excitation source because they have intense light focused onto a small volume, they are highly monochromatic, and the associated Raman light has a well-defined wavelength that can be avoided with the monochromator used for observing fluorescence. However, laser-induced fluorescence is still affected by background interference commonly arising from the Raman effect in the blank (molecular scattering) or from the low level of solid impurities in the solvent producing Rayleigh light scattering. Use of fluorescence detection (FLD) in HPLC is common. Detectors with constant excitation wavelength and variable absorption or with variable wavelength excitation and absorption are commercially available. Depending on the nature of the analyzed material, the detection limit using FLD can be as sensitive as $10^{-2} - 10^{-3}$ ng/mL, with a linear range of four orders of magnitude. When appropriately selected, the use of elution gradient can be applied for separation without interfering with the fluorescence. Different factors related to the mobile phase influence fluorescence such as pH, solvent nature, temperature (as much as 2%), presence of impurities, as well as the flow rate.

Chemiluminescence (CL), another technique used for HPLC detection, is the emission of light as a result of a chemical reaction. Certain compounds achieve excited energy states in specific chemical reactions and emit light following a transition to ground state. The wavelength of the light emitted by a molecule in chemiluminescence is the same as in its fluorescence, the energy levels of the molecules being the same. The difference comes from a different excitation process. If the energy of the chemical reaction is lower than that required for attaining the excited state, the chemiluminescence does not occur. Also, the deactivation of the excited molecule by nonradiative processes such as collisions with other molecules takes place for chemiluminescence similarly to fluorescence. The chemiluminescence intensity follows the same law as fluorescence, with the difference that quantum yield Φ from fluorescence must be replaced with a different quantum yield Φ_{CL}, which is defined as the proportion of analyte molecules that emit a photon during chemiluminescence. The Φ_{CL} increases with the efficiency of the chemical reaction producing the excitation (such as the oxidation process). Higher energies required by molecules to achieve the excited state diminish Φ_{CL}. In analytical uses of chemiluminescence, one more factor that must be taken into account is the time frame of the light emission. Certain chemiluminescent systems, though having very good Φ_{CL}, may emit the light for a period of 40–50 min. Much shorter times can be achieved using a catalyst. Because no excitation light is needed in chemiluminescence, the interfering light from Raman effect or light scattering by trace particles is nonexistent. In addition, the development of detectors virtually able to detect single photons makes the technique highly sensitive. Concentrations as low as a few hundred amol/mL of material were detected using chemiluminescence for certain analytes [25]. However, the luminescent molecules are not very common, and usually they are obtained by post-column derivatization with proper reagents.

Refractive Index Detectors

Refractive index detection (RI or RID) is another common technique in HPLC. Due to the modification of the refractive index of a solution as a function of the concentration of the solute, RI can be used for the quantitation of a variety of analytes. A schematic diagram of a RI detector is shown in Figure 1.4.11. This detector is based on the deviation of the direction of a light beam when passing under an angle from one medium to a medium with a different refractive index. This deviation depends on the difference in the refractive index between the two media. The change in the location of the beam on the (photoelectric) detector is made to generate a difference in the detector output. This output is electronically modified to provide a signal proportional to the concentration of the solute in the sample cell. The refractive index depends on the wavelength of the incident beam, and the most accurate RI measurements are done with monochromatic light (usually 589 nm, the sodium D line). With optical corrections, white light can still be used for the measurements.

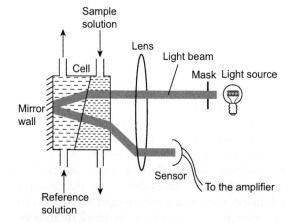

FIGURE 1.4.11 Schematic diagram of a refractive index detector.

This type of detection can be applied without the need for chromophore groups, fluorescence bearing groups, or other specific properties in the molecule of the analyte. In many cases the sensitivity of RI detection is not, however, as good as that of other types of detection. Also, it is not possible to use elution with gradient for the mobile phase, since this is associated with large variations in the refractive index of the mobile phase. The refractive index is sensitive to temperature changes, and a constant temperature must be maintained during measurements. The response of the detector is given in arbitrary RI units, depending on the detector settings, but proportional with the concentration of the analyte.

Electrochemical Detectors

Various types of electrochemical analytical techniques can be adapted for HPLC detection. Among these are amperometric, coulometric, potentiometric, and conductometric techniques. The techniques more commonly applied in HPLC are the conductometric, amperometric, and, to a lesser extent, coulometric procedures [26]. These techniques can have very high sensitivity, and the price of the detectors is relatively low. In ion chromatography, conductometric measurements are the most common. In amperometric techniques, the current intensity is measured in an electrochemical cell when a specific potential E is applied between two electrodes. Usually, only the reaction at one electrode is of interest, and a cell can be composed of a working electrode coupled with a nonpolarizable electrode (one that does not modify its potential upon passing of a current). This is known as the reference electrode, and examples are the saturated calomel electrode (SCE) and Ag/AgCl electrode. More frequently, a three-electrode cell arrangement is used. In this arrangement, the current is passed between a working electrode (made, for example, from glassy carbon) and an auxiliary electrode, while

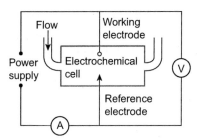

FIGURE 1.4.12a. Schematics of a two-electrode flow-cell.

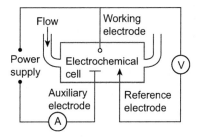

FIGURE 1.4.12b. Schematics of a three-electrode flow-cell.

the potential of the working electrode is measured relative to a separate reference electrode. The two types of cells are shown in Figures 1.4.12a and 1.4.12b.

Any overall cell reaction comprises two independent half-reactions, and the cell potential can be broken into two individual half-cell potentials. The half-reaction of interest that takes place at the working electrode surface can be either an oxidation or a reduction. A simple reduction reaction is written as follows:

$$Ox + n\,e^- \rightleftarrows Red$$

The electrode potential E for this half-reaction is reported to the potential of a reference standard hydrogen electrode (NHE), which is taken as zero. For a diffusion controlled process (static), a reversible reduction reaction with both Ox and Red species soluble and only with the oxidant initially present in the solution, the variation of the current intensity i as a function of the working electrode potential E follows an equation of the form:

$$E = E_{1/2} + \frac{RT}{nF}\ln\frac{i_{lim} - i}{i} \qquad (1.4.14)$$

In rel. 1.4.14, $E_{1/2}$ is a potential dependent on the nature of the oxidant and reduced molecules and is indicated as half-wave potential, R is the gas constant ($R = 8.31451$ J deg^{-1} mol^{-1} = 1.987 cal deg^{-1} mol^{-1}), T is temperature (in K deg.), n is the number of electrons involved in the electrochemical reaction, F is the Faraday constant (9.6485309×10^4 C mol^{-1}; C = coulomb), and i_{lim} is the limiting current intensity, which is a value determined by the largest rate of mass transfer to the electrode for the Ox species. The graph of expression 1.4.14 for a hypothetical reduction with $E_{1/2} = 0.5$ V is shown in Figure 1.4.13.

For the case of the electrochemically active species flowing over the surface of an electrode, which is the case of electrochemical detection in HPLC, the current-potential dependence is determined by a convective diffusion process (not only by diffusion). This makes the limiting current intensity for a Nernstian process dependent on the mobile phase flow rate and on channel and electrode geometry. For a rectangular channel

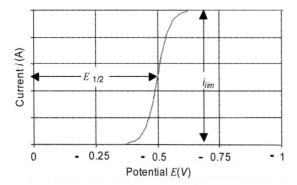

FIGURE 1.4.13 The current-potential curve of a Nernstian reaction involving two soluble species and only with the oxidant present initially. In this example, $E_{1/2} = -0.5$ V.

flow electrode in steady-state laminar flow with the working electrode at one wall, the limiting current intensity is given by the relation:

$$i_{lim} = 1.467\, n\, F\, c^* \, (A\, D\, b^{-1})^{2/3} U^{1/3} \qquad (1.4.15)$$

where c^* is the bulk concentration of the analyte, A is the electrode area, D is the diffusion coefficient of the analyte, b is the channel height, and U is the volumetric flow rate [27–29]. For different channel and electrode shapes, the expression for the current intensity is different [30, 31]. In amperometric detection, the current passing through the cell is measured at a fixed potential E, commonly chosen higher in absolute value than $E_{1/2}$ specific for the analyte. In these conditions, the desired electrochemical process takes place, but also all other species present in solution and having $E_{1/2}$ lower (in absolute value) than the chosen E value can become electrochemically active species. This may include even the solvent if the working potential E is very high. For eliminating this type of interference, compounds with low electrochemical potentials (in absolute value) are preferred for electrochemical detection. In HPLC, amperometric detection is frequently used for oxidation reactions. The quantitation can be done by calibration of the measured current i_{lim} versus different concentrations of analyte while maintaining strictly controlled flow conditions. Also, instead of a constant oxidation potential, a pulsed amperometric detection (PAD) can be used, alternating the oxidation analytical potential with a reducing pulse used for cleaning the electrode (depositions on the electrode may modify the nature of its surface and therefore the cell potential). The application of different working potentials is done at specific time intervals, and the measurement is made only when the active species are oxidized.

In coulometric detection, the amount of electricity (in coulombs) is measured during the electrochemical process. Potentiometric measurement can be applied, for example, in the case of ion concentration gradients across a semipermeable membrane. In this case the measured potential E is given by an expression of the form:

$$E = Const. - \frac{RT}{nF}\ln(c_X) \qquad (1.4.16)$$

The concentration c_X can be obtained by measuring the potential E (following expression 1.4.16) and using previously generated calibration curves.

The conductivity detectors are used for the measurement of concentrations of electrolytes in aqueous solutions. The molar concentration can be obtained based on the formula:

$$c = C_{cell}\frac{1}{\Lambda_m}\frac{1}{R} \qquad (1.4.17)$$

where C_{cell} is a constant depending of the measuring cell, R is the electrical resistance measured with the instrument from Ohm's law $R = I/V$, and Λ_m is the equivalent conductivity for the ionic species. Although Λ_m can be approximated for practical purposes as constant, it varies with the concentration following Kohlrausch's law $\Lambda_m = \Lambda_m^0 - A\sqrt{c}$, where A is a constant and Λ_m^0 is the limiting molar conductivity specific for each ion and temperature dependent. Values for limiting molar conductivities can be found tabulated.

In ion chromatography, the mobile phase frequently contains acids, bases, or salts (such as carbonates) that may interfere with the electrochemical detection. Bases such as KOH or acids such as methanesulfonic acid can be generated electrochemically (see Section 7.9). Eluent suppressors for ion chromatography are frequently necessary, in particular when the detection procedure is based on conductivity measurements. For this reason, before reaching the detector in IC, it is common to use a *suppressor*, which can reduce the conductivity caused by the eluent components by virtually eliminating the ions belonging to the mobile phase, and increasing the conductivity due to the analytes.

Two general types of suppressors are commonly used for IC, one being based on the addition in the flow of an ion-exchange cartridge (e.g., commercially available from Metrohm or Altech Co.) that eliminates the conductive ions of the mobile phase, and the other is based on semipermeable membranes. Special devices are commercially available that use a pair of ion-exchange cartridges used alternatively in the eluent flow. One cartridge is used for suppressing the eluent ions while the other is regenerated, such that no interruption in the operation is incurred. Various models of suppressors based on semipermeable membrane technology are available (e.g., self-regenerating suppressor SRS, micromembrane suppressor MMS, capillary electrolytic suppressor CES, or Atlas electrolytic suppressor AES, which are commercially available from Dionex/Thermo Scientific [32]).

In a separation where the eluent is, for example, a solution of NaOH or KOH, the suppressor (a resin or a semipermeable membrane) provides immobilized $R-SO_3H$ groups. In this case, the reaction of the NaOH from the eluent through the suppressor is described by the following scheme:

$$\underset{\substack{\text{mobile phase with}\\\text{high conductivity}}}{NaOH} + \underset{R}{SO_3H} \longrightarrow \underset{R}{SO_3Na} + \underset{\substack{\text{mobile phase with}\\\text{low conductivity}}}{H_2O}$$

$$(1.4.18a)$$

At the same time, an analyte in the form of a Na salt (e.g., a chloride) passing through the suppressor undergoes the following change:

$$\underset{\substack{\text{analyte with}\\\text{high conductivity}}}{NaX} + \underset{R}{SO_3H} \longrightarrow \underset{R}{SO_3Na} + \underset{\substack{\text{analyte with}\\\text{high conductivity}}}{HX}$$

$$(1.4.18b)$$

As a result of suppression, the conductivity caused by the analyte is not modified, while that caused by the mobile phase is eliminated, as shown for NaOH from the eluent, which is replaced by H_2O. Aqueous buffers containing

NaHCO$_3$ and Na$_2$CO$_3$ are also frequently used as eluents. The conductivity caused by such buffers is eliminated using a chemical suppressor by transforming the alkaline carbonates into H$_2$CO$_3$ that decomposes into H$_2$O and CO$_2$. Other anions (e.g., F$^-$, Cl$^-$, SO$_4^{2-}$) generate strong acids that are easily detected based on conductivity. Various other suppression techniques are used in practice [33]), including electrolytic suppressors used in anion LC. Similar to a resin, this type of suppressor eliminates cations such as K$^+$ or Na$^+$ from the eluent by replacing them with H$^+$ formed by electrolysis of water. The hydroxides from the mobile phase are transformed into water, while other salts (Cl$^-$, F$^-$, etc.) form strong acids.

In the case of a cation analysis, suppression can be achieved with a resin in OH$^-$ form. For example, metanesulfonic acid (MSA) used as an additive to the mobile phase can be retained by an anion-exchange resin with the formation of H$_2$O in the solution. At the same time, a salt (e.g., of an alkaline ion) passing through the resin can generate a strong base with high conductivity. Semipermeable membrane suppressors for cation exchange separations may contain tetra-butyl-ammonium hydroxide that reacts with mobile phase acids such as H$_2$SO$_4$ or CH$_3$SO$_3$H. By the interaction with the OH groups of the reagent in the semipermeable membrane, the SO$_4^{2-}$ ions, for example, are removed from the eluent and replaced by OH$^-$ (to form water with the H$^+$ of sulfuric acid), while other ions (e.g., Na$^+$, K$^+$, etc.) form hydroxides (e.g., NaOH, KOH) that have high conductivity.

Mass Spectrometric Detectors

Liquid chromatography-mass spectrometry (LC-MS) and liquid chromatography-mass spectrometry/mass spectrometry (LC-MS/MS) are two techniques that are becoming more and more common [33–37]. This was possible mainly due to the progress made in developing LC-MS interfaces able to convert efficiently the dissolved analyte from the mobile phase of the LC into gas-phase ions that are further analyzed with the mass analyzer. LC-MS and LC-MS/MS can provide exceptional sensitivity and selectivity compared to other detection techniques. The capability to easily measure ng/mL levels of compounds in the sample and to differentiate between molecules with different mass and fragmentation patterns, as well as the potential identification capability, makes LC-MS and LC-MS/MS invaluable techniques. The good resolving power required for the separation in other detection techniques can become less necessary with MS detection when coeluting compounds can be differentiated by their MS signal.

Two main techniques for ion formation in LC/MS are electrospray ionization (ESI) and atmospheric pressure chemical ionization (APCI). In these techniques, the goal is to generate charged molecules of the analyte, while the molecules of the solvent remain neutral. The charged molecules are further attracted to the ion mass analyzer entrance, while the solvent is eliminated as much as possible. The ionization process in LC-MS is mild compared to electron impact (EI) ionization typically used in GC-MS, and it is similar to CI ionization. The ionization in LC-MS can be conducted to form either positive ions or negative ions from the analyte neutral molecules. The choice between the positive and negative ionization mode depends on the nature of the compounds to be analyzed. Molecules with a basic character (e.g., amines, heterocycles containing nitrogen, etc.) have a tendency to form positive ions and are typically analyzed in positive mode. Other molecules such as acids and some oxygenated compounds (e.g., carbohydrates) are typically analyzed in negative mode in which better sensitivity for the analysis is obtained. A schematic drawing of an electrospray source is shown in Figure 1.4.14.

In the ESI source, the HPLC effluent is introduced through a capillary at 3–5 kV potential.

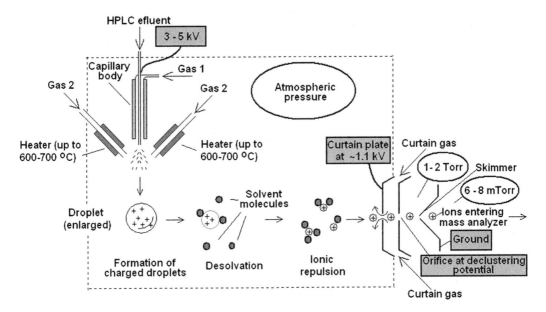

FIGURE 1.4.14 Schematic diagram of an electrospray ionization (ESI) inlet for a LC/MS instrument (positive ion mode). The decrease in electrical potential and in pressure are also indicated on the diagram.

The spray is changed into small droplets, and some of the solvent is vaporized using a current of heated gas. The heated droplets are charged and eliminate most of the solvent. Due to ionic repulsion the droplets generate individual ions, most of them still solvated. Since water is typically present in the HPLC effluent and H^+ ions are abundant, the formation of positive ions is the result of formation of molecular ions of the type $[M–H]^+$ where M are the molecules of the analyte. The ionization process with the formation of positive ions can be written in the form (not indicating the solvation molecules):

$$M + H^+ \rightarrow [M - H]^+ \qquad (1.4.19)$$

Multiple charged molecules can also be formed in this process, in particular for compounds such as peptides. Reaction 1.4.19 is related to the tendency of a compound to bind a proton. This property is also known as gas phase basicity (GPB). This is characterized by free enthalpy of formation ΔG^0_{GPB} given by the expression:

$$\Delta G^0_{GPB} = \Delta G^0([M - H]^+) - \Delta G^0(M)$$
$$- \Delta G^0(H^+) \qquad (1.4.20)$$

Similar to the formation of positive ions, molecules of the type MH can be ionized to generate negative ions in a reaction as follows:

$$MH \rightarrow [M]^- + H^+ \qquad (1.4.21)$$

Reaction 1.4.21 is related to the tendency to lose a proton, a property also known as gas phase acidity (GPA), described by the variation of free enthalpy, ΔG^0_{GPA}:

$$\Delta G^0_{GPA} = \Delta G^0([M]^-) - \Delta G^0(MH) + \Delta G^0(H^+) \qquad (1.4.22)$$

Usually, ΔG^0_{GPB} for most organic compounds lies between 500 and 1000 kJ mol^{-1}, while ΔG^0_{GPA} is situated between 1300 and 1650 kJ mol^{-1}.

These values can be useful in predicting ionization and fragmentation under CI-like conditions, and can explain why more compounds form positive ions more easily than negative ions in LC/MS [38].

The difference in the polarity between the solvent and the analyte molecules favors the formation of positive ions (or negative ions when working in negative mode), particularly from the analyte and much less from the solvent. A volatile acid such as HCOOH is typically added in the mobile phase to favor the process of positive ion formation. The positive ions are attracted to the curtain plate in the ion source, while the solvent molecules that are not charged are not attracted. Further desolvation and elimination of solvent molecules occur as the positive ions are directed toward the skimmer and further into the ion mass analyzer. For negative ion formation, salts such as $HCOONH_4$ or CH_3COONH_4 are added to favor the ionization of the analyte.

Adducts between analyte molecule and different ions reaching the MS interface are often observed in LC-MS. They are formed by ion-dipole, ion-induced dipole, hydrogen bonds, and even by van der Waals interactions. In the adduct formation, the molecules of the solvent S_{olv} are frequently involved. Due to their high concentration, it is common that solvent molecules are initially ionized to form in positive mode, for example, $[S_{olv}\text{-}H]^+$ ions. Further ionization of the analyte can be described as a proton transfer from species $[S_{olv}\text{-}H]^+$ to the molecule of the analyte, by a reaction shown as follows [38]:

$$M + [S_{olv}\text{-}H]^+ \Leftrightarrow [M\text{-}H\text{-}S_{olv}]^+ \Leftrightarrow [M\text{-}H]^+ + S_{olv}$$

$$(1.4.23)$$

The intermediate positive ion $[M\text{-}H\text{-}S_{olv}]^+$ is sometimes stable enough to be seen in the MS spectrum.

A similar type of reaction may take place in negative ionization mode, when the reaction can be written as follows:

$$MH + [S_{olv}]^- \Leftrightarrow [M\text{-}S_{olv}\text{-}H)]^- \Leftrightarrow [M]^- + S_{olv}H$$

$$(1.4.24)$$

Reactions 1.4.23 and 1.4.24 show that the solvent has an important role in the ionization process, and the adducts can be seen in the mass spectrum instead of the analyte molecular ion. These equilibria are influenced by ΔG^0_{GPB}, or ΔG^0_{GPA} values for the analytes and the solvents and by the concentration of $[S_{olv}\text{-}H]^+$, or $[S_{olv}]^-$, the electric fields applied into the source, and so on.

Besides adducts with solvent molecules, other adducts can be formed between the analyte M and other species in the eluted material. For example, in positive ionization mode, ions such as $[M\text{-}Na]^+$, $[M\text{-}K]^+$, $[M\text{-}NH_4]^+$, $[M\text{-}H_2O\text{-}Na]^+$, $[M\text{-}S_{olv}\text{-}Na]^+$, $[M\text{-}2S_{olv}\text{-}Na]^+$ can be seen. Negative ionization is less favorable to adduct formation, but is still possible with ions of the following types: $[2M]^-$, $[3M]^-$, $[2M\text{-}Na]^-$ [39].

In atmospheric pressure chemical ionization (APCI), the effluent from the HPLC is sent through a capillary that is heated and in a flow of gas, but not under an electrical potential. The jet of molecules of solvent and analyte in gas form, and those of an added gas (N_2, O_2) flow by (close to) a needle charged at a high voltage (3–5 kV) that generates a corona discharge. Some of the molecules are being loaded with positive charges (when working in positive mode). Due to the difference in polarity between the molecules of the analyte on one hand and those of solvent and gasses on the other, the charges tend to migrate to the analyte molecules. These charged molecules are attracted to the curtain plate of the ion source and further into the ion mass analyzer, as previously described for the ESI source.

Both ESI and APCI ionization techniques offer very reproducible generation of ions; they can be used with a wide range of solvents as HPLC eluent, they can work in a range of flow rates (e.g., 0.05 to 1.0 mL/min), and they do not involve problems with capillary plugging.

Other procedures used to form ions from the analyte molecule include the use of an intense beam of UV light for the analyte ionization instead of a corona discharge. This technique is known as atmospheric pressure photoionization (APPI). Older techniques that interface an LC with a mass spectrometer include: (1) the particle beam (PB), which consists of an aerosol generator from the LC flow (at 0.1–1 mL/min) followed by a desolvation chamber and a separator that directs the aerosols through a series of apertures separating the volatile compounds including the solvent from the solid aerosols, (2) continuous flow FAB, where the effluent is introduced directly into a vacuum region (with a flow rate of 5–10 μL/min) mixed with a matrix material such as glycerol, and the ionization is achieved using a beam of ions at 5–8 keV.

The ions generated in the source are further separated by mass and measured using a mass analyzer. Mass separation is usually achieved using either a quadrupole or an ion trap. Quadrupole type mass spectrometers separate the ions by passing them along the central axis of four parallel equidistant rods that have a fixed voltage (DC) and an alternating (RF) voltage applied to them. The field strengths (voltage) can be set such that only ions of one selected mass can pass through the quadrupole, while all other ions are deflected to strike the rods. By varying (with a precise rate) the strength and frequencies of the electric fields, different masses can be filtered through the quadrupole.

With the quadrupole instruments, a low-resolution type spectrum is obtained. For a m/z = 200, for example, the minimum mass difference that can be separated could be around 0.2 mass unit (resolution 1000). The mass range for the quadrupoles can go as high as 2000 Dalton, but common commercial quadrupole instruments have a mass range between 2 and 1100. The mass analyzers in LC can be utilized as a detector providing a very sensitive means for analyte measurement. The abundance of ions of a measured analyte is proportional to the analyte concentration, and the MS response can be calibrated for quantitative measurements. However, the formation of molecular ions in the ionization source and the absence of molecular fragments limis the identification capability of LC/MS.

Very useful in the analysis of eluates in LC are the MS/MS analyzers. Triple quadrupole systems are common MS/MS detectors. In a triple quadrupole, the first quadrupole (Q1) has the role of separating the (parent) ions generated in the ion source. These ions (usually molecular ions) are selected for further interactions in the collision cell (Q2) where they can be fragmented. For this purpose, a gas (such as N_2 or Ar) is introduced in the cell, and a specific voltage is applied to the Q2 quadrupole (or hexapole in some instruments). Depending on the collision gas pressure and on the voltage applied to the collision cell, the parent ions undergo different degrees of fragmentation (fragmentation by collisionally activated dissociation—CAD). The fragmentation strongly depends on the structural characteristics of the analyte. When adducts are formed during the ionization process, the ability of adducts to be fragmented also depends on their structure. The stability of cation adducts decreases in the following order: $[M-H]^+$ ~ $[M-NH_4]^+ > [M-Li]^+ > [M-Na]^+ > [M-K]^+ > [M-Cs]^+$ [40]. The third quadrupole Q3 is used for the separation of the resulting (daughter) ions following fragmentation in the collision cell. Several utilization techniques are common for MS/MS analyzers, such as (1) the product ion scan, when the whole range of ions generated by fragmentation of the precursor ions (parent ions) is analyzed, (2) the precursor ion scan, when only one ion is selected for the detector by Q3, while Q1 is scanning the whole range of ions produced by the source, (3) the neutral loss scan, when the instrument scans for a specific mass difference between the ions from Q1 and Q3, and (4) multiple reaction monitoring (or MRM) where a specific ion is selected by Q1 and a specific fragment is detected by Q3 (more than one pair of ions can be analyzed by MRM at the same time). Qualitative information from LC/MS/MS can be generated based on fragmentation of the parent ions. However, LC/MS/MS is more frequently used for highly selective quantitative analysis. Quantitative information using the MRM mode is also characterized by exceptional sensitivity (as low as fmol/mL concentration). Other instrument developments are continuously made in the field of LC-MS/MS. Examples are coupling an ion trap with a collision cell and a quadrupole, coupling a triple quadrupole with a fourth ion analyzer such as an Orbitrap (see, e.g., [41]) that achieves high sensitivity and also high resolution (e.g,. $\Delta M/M \approx 60,000$ for $m/z = 400$ or even up to 100.000), or using special techniques such as Fourier transform ion cyclotron resonance (FT-ICR-MS) capable of 1,000,000 resolution.

Other Types of Detectors

One other detector utilized in HPLC, in particular for compounds that do not have good light absorbance in UV, are not fluorescent, and may be difficult to ionize, is the evaporative light-scattering detector (ELSD) [42,43]. In this technique, the eluent is injected in the form of a spray from a nebulizer into a drift tube where a nebulizer gas is also introduced. The drift tube is heated and the solvent is evaporated, forming a fine mist from the nonvolatile molecules. This mist passes through a cell, where the scattered light from a beam that illuminates the cell is recorded. The gas generated from the solvent does not influence light scattering. The intensity of the scattered light is dependent on the analyte

concentration (within a certain range of concentrations since the linearity is not followed for a wide range). This detector has the advantage over the RI of being usable with gradient elution. However, the presence of any salts or nonvolatile materials in the mobile phase disturbs the measurements. Also, changes in the temperature of the eluent do not affect ELSD, while with RI detectors a careful temperature control must be applied since the refractive index varies with temperature. ELSD can be more sensitive than RI in specific applications. Modifications of ELSD were attempted to further improve its sensitivity, such as by adding a saturated stream of solvent to the mist of the analyte in order to grow the particles by condensation nucleation and detect them better (the technique is known as CNLSD) or to use lasers as light source (LLSD). Light scattering can also be used for detection directly on the liquid effluent when the analytes are polymers.

An alternative to light-scattering detection is a corona-charged aerosol detector (CAD or cCAD). This detector is also based on nebulization of column effluent (e.g., with N_2) and on drying of resulted droplets to remove the mobile phase components, producing analyte particles. A secondary stream of N_2 is made positively charged by passing it by a high-voltage platinum corona wire, which is sent to the opposing stream of analyte particles. The charged particles are detected, generating an electric signal with the intensity proportional to the amount of analyte eluted from the column. The CAD is more sensitive than ELSD and has a wider dynamic range [44].

For the target compounds that contain at least one nitrogen atom, the chemiluminescent nitrogen detector (CLND) can be employed. The principle of this detector relies on the combustion of the column effluent in a high-temperature furnace that converts the N-containing compounds into NO. The dried gas stream is passed into a chamber where it reacts with O_3, a reaction that is associated with chemiluminescence (measured by a photomultiplier). This detector has a high sensitivity but is not compatible with acetonitrile in the mobile phase [45].

Other known detection techniques include Fourier-transform infrared spectrometry (FTIR) [46, 47], nuclear magnetic resonance (NMR) [48], inductively coupled plasma-mass spectrometry (ICP-MS) [49], circular dicroism, optical rotary dispersion, polarimetry, and radioactivity.

Selection of a Detector for the HPLC Separation

Specific properties of detectors impose some restrictions on their selection for analyzing a given sample. Among the criteria used in selecting a detector for a particular application are the following: (1) availability of the detector, (2) purpose of analysis, (3) capabilities of the detector, (4) properties of the analyte, (5) type of elution (isocratic or gradient), (6) properties of the mobile phase used in separation, (7) stability/reliability of the detector, and (8) ease of maintenance/ operation. Besides the choice of a specific detector depending on the analysis, the selection of a specific method is sometimes decided based on the properties of the available or optimum detector. When changes other than that of the detection type are easier to make (e.g., mobile phase composition), it is common to modify the method instead of selecting a different detector. More than one detector can be present in an HPLC system. When several detectors are used, they are usually connected in series, although parallel connection is also possible.

1) Availability of the detector is a straightforward requirement. Some detectors are more expensive than others, and even if all are available, some consideration regarding their cost and cost of maintenance may play a role in their selection.

2) The purpose of analysis is essential in the detector selection. All detectors are designed to allow quantitative measurements, but some are not meant to provide any qualitative information, such as refractive index (RI) detectors, others give some hints on qualitative nature of the analytes, such as ultraviolet (UV) detectors, and others provide good information that can be used for compound identification such as mass spectrometric (MS) detectors. Therefore, when only a quantitative analysis is necessary and the nature of the analyte is known, the peak identification for the known compound can be done based on the retention time alone, the only concern being an efficient chromatographic separation with no interference. The selection of the detection type in such cases should consider criteria other than qualitative identification capability. For qualitative analysis of samples containing a mixture of compounds with a suspected structure that require only confirmation, techniques such as mass spectrometry MS or MS/MS must be used, working in specific modes (e.g., MRM). Also, specific fluorescence properties or UV absorption wavelength can be used to enhance the detector's selectivity.

For qualitative analysis of a sample with unknown composition, even with mass spectral detection, HPLC is not always very informative. The mass spectra in LC, when using a single mass spectrometer as a detector, do not offer much information about the fragments of an analyzed molecule, typically indicating only the molecular ion. Better information is obtained with the use of MS/MS instruments, but the fragmentation provided can vary considerably depending on the instrument and acquisition method. Although some library searches are available (e.g., SmileMS) as well as some mass spectral libraries for LC, the qualitative information on unknown compounds is not always easy to interpret. As an alternative to searching based on molecular fragmentation, the nature of an unknown compound can also be obtained from its precise molecular mass. Instruments generating high-resolution mass spectra (see Section 1.4) are often equipped with search programs (e.g., Mass-Frontier) that help in identifying the molecular composition and potential structures.

The selection of a specific detector for quantitation depends on the needed precision and accuracy, the nature of the analyte, the level of the analyte in the sample, and the methods used for sample preparation. Depending on the purpose of analysis, the selection of a detector or of a detector setting must be done in such a way to cover the analysis needs and be capable of achieving a required limit of quantitation (LOQ).

3) The detector capability is a key factor in selecting a specific one for an HPLC analysis. Some detectors respond to most analytes and are indicated as universal detectors [50]. Other detectors are specific for a class of compounds, and some have relatively limited applicability (e.g., radioactivity detectors and chiral detectors). The detectors used in LC can have very different sensitivities, which depend on the type of detector and on the nature of the analyte. For example, the sensitivity of RI detectors is typically lower than that of fluorescence (FL) detectors for a fluorescent compound, but in the absence of fluorescence, the RI detector can be utilized while the FLD is useless. Another characteristic of a detector is its capability to offer qualitative information in addition to the quantitative information (peak areas). Except for MS/MS detectors, the qualitative information provided by other detector types is quite limited. Detector settings, model, and manufacturer play an important role in detector sensitivity. Sensitivity of the detector is very important, but other characteristics also must be considered in its selection, such as stability of the signal, frequency of the measurement per unit time (that assures an accurate evaluation of the peak shape), resistance to acids and bases in the mobile phase, capability to be used in series with other detectors, and acceptable backpressure in case of connection in series with another detector.

4) The properties of the analytes either before or after sample preparation (e.g., purified, concentrated, derivatized) are essential in the selection of a detector. The physical and chemical characteristics of the analytes, as well as their differences from other sample components and from the mobile phase, must be carefully evaluated in order to select the ones to be used for detection. For example, the presence or absence of chromophore groups typically determines the use of UV/visible absorption detectors (fixed wavelength, variable wavelength, or diode array detector—DAD), which are among the most common detectors used in HPLC. The same is valid for FLD, which in some cases provides excellent sensitivity and is preferred to UV absorption. In some cases, such as in the analysis of carbohydrates, the absorption in UV is very low (except for very low wavelengths) and the compounds are not fluorescent. In such cases RI, electrochemical (amperometric), evaporative light scattering (ELSD), or corona-charged aerosol (CAD or cCAD) detectors have to be used. The MS detectors that are becoming more common are basically universal detectors, but their sensitivity is highly dependent on the nature of the analyte. The derivatization is frequently applied for modifying the initial analyte properties such that it can be amenable for a specific detection. The detection technique is always described for an analytical method when this is reported in the literature. Selection of the settings of the detectors obviously depends on the nature

of the analyte. This may include the choice of wavelength of absorption for UV, the choice of excitation and emission wavelength for FLD, or the choice of several parameter settings and masses to be monitored for MS.

5) Type of elution also plays a key role in selection of a detector, and in many cases the choice is made the other way around, the available detector determining the choice of isocratic or gradient elution. Several techniques are not applicable for gradient elution, or do not respond very well to the change of the composition of mobile phase. The refractive index detector, for example, must be used only with isocratic elution. Other detectors such as ELSD or CAD can be used with gradient elution, but the solvents must be volatile and the change in solvent composition may also generate some drift in the baseline. Even MS and MS/MS detectors may show differences in sensitivity at one solvent composition or the other, and the choice of gradient versus isocratic elution is sometimes influenced by this difference. Detectors such as UV can be used with gradient separation without any problem (as long as the mobile phase does not have absorption by itself).

6) The properties of the mobile phase also contribute to the decision regarding the choice of a specific detector. Mobile phase composition plays a crucial role in most separations (the subject of mobile phase selection will be presented in detail in Chapter 7). A specific mobile phase may determine the type of detector that can be used, and in some instances, the method is developed particularly to be used with a specific detector.

Physical and chemical properties of the mobile phase must be considered in relation to the requirements of a specific detector. For example, when detection is done in UV, the cutoff wavelength of the solvent must be considered, such that the solvent is "transparent" at the measuring wavelength. For the detection using ELSD or CAD, the mobile phase should be totally volatile, and in the case of the need of salt buffers in the mobile phase, these detectors are not usable. In LC/MS (and LC/MS/MS) the mobile phase composition influences the detection, and the presence of nonvolatile salts in the mobile phase is not recommended. Since LC/MS and LC/MS/MS detection may offer particular benefits to the analysis (very good sensitivity, qualitative information), the selection of the mobile phase needs to be done such that it is suitable for the MS detector, and not the other way around (choose the detector to accommodate the mobile phase). Also, the sensitivity of the detection may be drastically influenced by the mobile phase composition in MS and MS/MS. When a method has been developed for another type of detection (e.g., using nonvolatile buffers) and must be applied with MS detection, for example, for enhancing sensitivity or for obtaining some qualitative information on the analytes, the mobile phase is commonly modified to be suitable for the new detection technique.

7) Detector reliability is another factor to be considered in its selection. Typically, some detectors such as those based on UV absorption are extremely stable and reliable. Other detectors may be more prone to problems, which also may depend on instrument age, manufacturer, environmental conditions, and the like. For example, electrochemical detectors may show some drift or unreproducible results depending on the cleanliness status of the measuring electrode. In some instances, a choice between sensitivity and stability of the detector must be made.

8) Some detectors require special maintenance as well as more effort for establishing the proper operating conditions, such as MS or MS/MS detectors. Other detectors are very

simple to operate and require virtually no adjustments. The advantages of such detectors must be weighed versus their disadvantages, such as loss of sensitivity or lack of qualitative information.

Fraction Collectors

In some instances, in particular when the HPLC separation is performed for semipreparative or preparative purposes, some fractions of interest need to be collected as the effluent exits the detectors. This collection can be done using a fraction collector. Automated systems for fraction collection are available. They direct the flow emerging from the last detector to specified vials either at a given time or upon receiving a signal from the detector (indicating an eluting peak) [51].

Controlling and Data Processing Units

The controlling and data system in HPLC is a computer with data acquisition capability and with a program package that allows the user to perform (1) control of the hardware and (2) data acquisition and processing. In older systems, these functions could be performed manually/mechanically, without the use of a computer. However, new instruments have extensive capability of computer control for functions such as: flow rate and gradient composition generated by the pumps, maximum pressure that should be delivered by the pumps, vial from where the injection must be made (in case of autosamplers), volume of the sample injection, timing of the sample injection, reagent mixing with the sample (if desired), temperatures in the autoinjector storing the samples, temperature of the column, tracking of column usage, and parameters for the detector(s). The other function of the computer is the capture, signal processing, and storing of the data generated by the detector(s). The program packages also offer a user interface (sometimes very complex) allowing the user to further process the data and interpret the result. This part may include peak recognition to generate retention times, area measurement, data averaging, calibrations, peak shape, and other peak parameters characterization, and in the case of detectors that give qualitative information in the form of spectra, the computer may contain spectral libraries helping with compound identification.

References

[1] Moldoveanu SC, David V. Sample Preparation in Chromatography. Amsterdam: Elsevier; 2002.
[2] Ito Y, Bowman RL. Countercurrent chromatography: Liquid-liquid partition chromatography without solid support. Science 1970;167:281–3.
[3] Vailaya A, Horváth C. Retention in reversed-phase chromatography: Partition or adsorption? J. Chromatogr. A 1998;829:1–27.
[4] Smith RM. Superheated water chromatography—a green technology for the future. J. Chromatogr. A 2008;1184:441–5.
[5] http://www.chemaxon.com
[6] http://www.epa.gov/oppt/exposure/pubs/episuite.htm
[7] Hansch C, Leo A. Exploring QSAR, Fundamentals and Applications in Chemistry and Biology. Washington, DC: ACS; 1995.
[8] Hansch C, Leo A, Hoekman D. Exploring QSAR, Hydrophobic, Electronic and Steric Constants. Washington, DC: ACS; 1995.
[9] Baiulescu GE, Dumitrescu P, Zugravescu PG. Sampling. E Horwood, Chicester 1991.
[10] Lunn G, Hellwig LC. Handbook of Derivatization Reactions for HPLC. New York: John Wiley; 1998.
[11] http://www.hplcweb.com/HPLC_Equipment/HPLC_Manufacturers/
[12] Agilent 1200 Series. User Manuals. Agilent Technologies 2006.
[13] http://www.waters.com
[14] Snyder LR, Kirkland JJ, Dolan JW. Introduction to Modern Liquid Chromatography. 3rd ed. Hoboken, NJ: John Wiley; 2010.
[15] Haky JE. Gradient Elution. In: Cazes J, editor. Encyclopedia of Chromatography. Marcel Dekker; 2002.
[16] Giddings JC. Unified Separation Science. New York: John Wiley; 1991.
[17] Dolan JW. Autosampler carryover. LCGC 2001;19:164–8.
[18] Dolan JW. Autosamplers, Part I—Design features. LCGC 2001;19:386–91.

[19] Dolan JW. Autosamplers, Part II —Problems and solutions. LCGC 2001;19:478–82.

[20] Welch CJ, Gong X, Schafer W, Pratt EC, Birkovic T, Pirzada Z, et al. MISER chromatography (multiple injections in a single experimental run): The chromatogram is the graph,. Tetrahedron: Assim 2010;21:1674–81.

[21] Sheldon EM. Development of a LC-LC-MS complete heart-cut approach for the characterization of pharmaceutical compounds using standard instrumentation. J Pharm Biomed Anal 2003;31:1153–66.

[22] Mitchell CR, Bao Y, Benz NJ, Zhang S. Comparison of the sensitivity of evaporative universal detectors and LC/MS in the HILIC and the reversed-phase HPLC modes. J. Chromatogr. B 2009;877:4133–9.

[23] Sadek PG. The HPLC Solvent Guide. New York: John Wiley; 1996.

[24] Gaigalas AK, Li L, Henderson O, Vogt R, Barr J, Marti G, et al. The development of fluorescence intensity standards,. J. Res. Nat. Inst. Standards Technol 2001;106:381–9.

[25] Novak TJ, Grayeski ML. Acridinium-based chemiluminescence for high-performance liquid chromatography detection of chlorophenols. Microchemical J. 1994;50:151–60.

[26] Erickson BE. Electrochemical detectors for liquid chromatography. Anal. Chem. 2000;72:353A–7A.

[27] Matsuda H. Zur theorie der stationären strom-spannunos-kurven von redox-elektrodenreaktionen in hydrodynamischer voltammetrie: II. laminare rohr- und kanalstkömungen. J. Electroanal. Chem. Interf. Elect. 1967;15:325–36.

[28] Moldoveanu S, Anderson JL. Amperometric response of a rectangular channel electrode. J. Electroanal. Chem. Interf. Elect. 1984;175:67–77.

[29] Moldoveanu S, Handler GS, Anderson JL. On convective mass transfer in laminar flow between two parallel electrodes in a rectangular channel. J. Electroanal. Chem. Interf. Elect. 1984;179:119–30.

[30] Yamada J, Matsuda H. Limiting diffusion currents in hydrodynamic voltammetry: III. Wall jet electrodes. J. Electroanal. Chem. 1973;44:189–98.

[31] Moldoveanu S, Anderson JL. Numerical simulation of convective diffusion at a microarray channel electrode. J Electroanal Chem 1985;185:239–52.

[32] http://www.dionex.com/en-us/webdocs/4270-DS-Eluent_Suppressors-23Nov10-LPN1290-10.pdf

[33] Fritz JS, Gjerde DT. Ion Chromatography. Weinheim: John Wiley-VCH; 2009.

[34] Niessen WMA. State-of-the-art in liquid chromatography–mass spectrometry. J. Chromatogr. A 1999; 856:179–97.

[35] Niessen WMA. Liquid Chromatography-Mass Spectrometry. Weinheim: John Wiley-VCH; 2009. 206.

[36] Willooughby R, Sheehan E, Mitrovich S. A Global View of LC/M: How to Solve Your Most Challenging Analytical Problems. Pittsburgh: Global View Pub.; 2002.

[37] Ardrey RE. Liquid Chromatography—Mass Spectrometry: An Introduction. Chichester: John Wiley; 2003.

[38] van Baar BLM. Ionization methods in LC-MS and LC-MS-MS (TSP), APCI, ESP, and cf-FAB, in Application of LC-MS in Environmental Chemistry. In: Barcelo D, editor. Journal of Chromatography Library Series, vol. 59; 1996. p. 71–126. Amsterdam.

[39] Schug K, McNair HM. Adduct formation in electrospray ionization. Part 1: Common acidic pharmaceuticals. J. Sep. Sci. 2002;25:759–66.

[40] Medvedovici A, Albu F, David V. Handling drawbacks of mass spectrometric detection coupled to liquid chromatography in bioanalysis. J. Liq. Chromatogr. Rel. Technol. 2010;33:1255–86.

[41] LTQ XL. Orbitrap Operation Manual. Thermo-Fisher Scientific; 2009.

[42] Bünger H, Kaufner L, Pison U. Quantitative analysis of hydrophobic pulmonary surfactant proteins by high-performance liquid chromatography with light-scattering detection. J. Chromatogr. A 2000;870:363–9.

[43] Voress L, editor. Instrumentation in Analytical Chemistry 1988-1991. Washington, DC: ACS; 1992.

[44] Górecki T, Lynen F, Szucs R, Sandra P. Universal response in liquid chromatography using charged aerosol detection. Anal. Chem. 2006;78:3186–92.

[45] Fujinari EM, Courthaudon LO. Nitrogen-specific liquid chromatography detector based on chemiluminescence: Application to the analysis of ammonium nitrogen in waste water. J. Chromatogr. A 1992;592:209–14.

[46] Norton KL, Lange AJ, Griffiths PR. A unified approach to the chromatography-FTIR interface: GC-FTIR, SFC-FTIR, and HPLC-FTIR with subnanogram detection limits,. J. High. Res. Chromatogr. 1991;14:225–9.

[47] Somsen GW, Gooijer C, Th UA. Brinkman, Liquid chromatography-Fourier-transform infrared spectrometry. J. Chromatogr. A 1999;856:213–42.

[48] Albert K. Liquid chromatography-nuclear magnetic resonance spectroscopy. J. Chromatogr. A 1999; 856:199–211.

[49] Sutton LK, Caruso JA. Liquid chromatography-inductively coupled plasma mass spectrometry. J. Chromatogr. 1999;856:243–58.

[50] Zhang B, Li X, Yan B. Advances in HPLC detection—Towards universal detection. Anal Bioanal Chem 2008;390:299–301.

[51] http://www.labx.com

CHAPTER

2

Parameters that Characterize HPLC Analysis

OUTLINE

2.1. PARAMETERS RELATED TO HPLC SEPARATION

General Comments

Every analytical technique is characterized by its specificity/selectivity, reproducibility and repeatability, accuracy, range of linearity between the quantity of analyte and the response of the analytical instrument, limit of detection (LOD), limit of quantitation (LOQ), recovery yield of the sample processing, robustness, ruggedness, and stability [1]. In order to achieve optimum method

http://dx.doi.org/10.1016/B978-0-12-385013-3.00002-1

characteristics, the HPLC must be properly conducted and controlled. A number of parameters are used in HPLC for this purpose. This chapter focuses on parameters describing the separation. Some of these parameters are related to physical characteristics of the HPLC system and others to the separation itself. Separation parameters can be obtained by knowing the conditions in which the chromatogram was generated and by inspecting the generated chromatograms.

Flow Rate

One example of a physical characterization parameter is the flow rate of the mobile phase (already mentioned in Section 1.3). This parameter shows how fast the mobile phase moves through the column, and is also useful for calculation of the consumption of the mobile phase in a given time interval. The flow can be described by the linear flow rate u (velocity of a point in the fluid, expressed as length per time) and the volumetric flow rate U (volume of fluid that flows per unit time, expressed, e.g., in mL/min). The volumetric flow rate and the linear flow rate are related by an expression of the form $U = \mathcal{A} u$ where $\mathcal{A} = (1/4)\pi\varepsilon^* d^2$ is the area of the channel in which the flow takes place, and d is the internal diameter (i.d.) of the column. It should be noticed that the surface area for the empty column is not the same as $\mathcal{A}$ since the column is filled with the stationary phase. The volumetric flow rate U is controlled by the pump and can be easily set. The linear flow rate u (in the column) depends on the *column packing porosity* indicated as ε^*. (*Note:* In some texts, the notation for the volumetric flow rate U is F.) In common HPLC procedures, U is typically selected between 0.3 and 3 mL/min. However, when using microbore or nanobore columns (0.1−0.2 mm i.d.), the flow can be as low as a few μL/min, and for semipreparative purposes the flow can be significantly larger than 3 mL/min.

Retention Time

The *retention time* t_R can be defined as the time from the injection of the sample to the time of compound elution, taken at the maximum (apex) of the peak that belongs to the specific molecular species (known or not). The concept of retention time was already introduced in Section 1.1. It indicates how long it takes for a compound to elute from the column, and the retention time of the last peak in a chromatogram is used to estimate the necessary length of the chromatographic run. For a molecular species X, the retention time can be indicated as $t_R (X)$ and is usually measured in min. The (X) in the notation is sometimes omitted, but t_R is always related to a specific molecular species. Another notation specifying the analyte uses an index (e.g., $t_{R,i}$). Retention time depends not only on the structure of the specific molecule, but also on factors such as the nature of the mobile and stationary phases, the flow rate of the mobile phase, and dimensions of the chromatographic column. Retention time is usually characteristic for a specific compound in a given separation. For this reason, the retention time is critical in identifying analytes once their retention time is known (e.g., by using standards).

Of particular interest in a separation is the *dead time* t_0, which is the time a nonretained molecular species needs to elute from the chromatographic column. The dead time is also known as *void time* or *holdup time*. The dead time t_0 can be interpreted as part of the retention time $t_R(X)$ for the analyte X, which the analyte spends in the mobile phase moving through the column. This parameter is not related to the retention process and depends on the flow rate and physical characteristics of the column (length, diameter, porosity of stationary phase). The difference between the retention time and the dead time represents the time the analyte X is retained on the stationary phase. This difference is indicated as *reduced retention time* t'_R and is expressed by the formula:

$$t'_R = t_R - t_0 \qquad (2.1.1a)$$

or more precisely by the formula:

$$t'_R(X) = t_R(X) - t_0 \qquad (2.1.1b)$$

The value for t_0 is typically obtained as an approximation by using compounds that are very slightly retained, since it can be difficult to find a compound that is not retained at all on a chromatographic column. For example, the solvent used for injecting the sample (when different from the mobile phase) can be such a compound, and the retention time of this solvent peak can be taken as dead time. Other procedures for estimating t_0 are known. One procedure uses the minor disturbances in the background signal created by the sample injection. Other procedures use injections of nonretained compounds such as a deuterated solvent, which is the same as nondeuterated mobile phase. For RP-HPLC, the use of uracil or of inorganic salts that are assumed to be not retained on a hydrophobic column is also a common procedure for t_0 estimation. Another more elaborate procedure involves a "homologous series" in which a plot of retention times for a homologous series of compounds that are retained less and less as the number of carbon atoms is decreasing is extrapolated to zero [2]. From the length of the chromatographic column and t_0, it can be easy to evaluate u for a certain separation by taking $u = L/t_0$, where L is the length of the chromatographic column.

The time during a chromatographic separation is indicated as the *run-time*. The total time necessary for completing a chromatographic separation is slightly longer than the retention time of the last peak in the chromatogram. This time is sometimes referred to as *total run-time*, or length of the chromatogram. In practice,

when multiple samples are analyzed, the total run-time is an important parameter since its value is related to the number of samples analyzed within the same length of time.

Retention Volume

For a specific molecular species X, the *retention volume* $V_R(X)$ is defined as the volume of the mobile phase flowing from the injection time until the corresponding retention time $t_R(X)$ of a molecular species. The V_R and t_R are related by the simple formula:

$$V_R = U t_R \qquad (2.1.2a)$$

or more precisely by the formula:

$$V_R(X) = U t_R(X) \qquad (2.1.2b)$$

The retention volume corresponding to the dead time t_0 is known as *dead volume* V_0, or *void volume*. This volume corresponds to the volume of liquid in the column (and in the transfer lines from the injector to the column and from the column to the detector). The chromatographic column has a "volume not occupied by the stationary phase," which is the space between the stationary phase particles and inside their pores. A not-retained molecule has to travel through the tubing from the injector to the column (which is very small), through the volume not occupied by the stationary phase, and through the tubing from the column to the detector (also very small), which accounts for the t_0 and for the dead volume V_0. Corresponding to the reduced retention time t'_R, a *reduced retention volume* V'_R can be defined by the formula:

$$V'_R = V_R - V_0 \qquad (2.1.3)$$

The dead volume V_0 of a chromatographic column can be found not only by multiplying the dead time with the volumetric flow, but also by direct measurement. For this purpose, a column is sequentially filled with solvents of different densities and weighed. From the difference in the weight and the difference in the density of the solvents, it is possible to calculate the dead volume of the column.

The value of V_0 can be considered proportional to the volume of the empty column, the proportionality constant depending on the dimension (and shape) of the stationary phase particles and also on the way they are packed. For a column of length L and inner diameter d, the empty column volume is $(\pi/4)d^2 L$. However, the column is filled with the stationary phase, and only a fraction of this volume will give the dead volume. This can be expressed with the use of column packing porosity ε^*, a constant with the approximate value $\varepsilon^* \approx 0.7$ (for 5 µm particles column). The value for ε^* may vary depending on the stationary phase particle size and structure such that values that are somewhat different from $\varepsilon^* \approx 0.7$ are possible. Including the packing porosity, the column dead (void) volume is $V_0 = \varepsilon^*(\pi/4)d^2 L$. Since the column i.d. and the column length are typically expressed in mm while the volume is expressed in mL (cm^3), a factor of 10^{-3} must also be included in the calculation. Table 2.1.1 gives some typical values for the empty volume and dead volume for an HPLC column, depending

on its dimensions. At $U = 1$ mL/min the void time t_0 is numerically equal to the void volume V_0 as given in Table 2.1.1. Precise void volume of a column must be experimentally measured with an unretained compound.

Migration Rate

The migration rate of species X indicated as $u_R (X)$ can be defined as the velocity at which the species X moves through the column. The migration rate is inversely proportional to the retention times and therefore:

$$\frac{u_R(X)}{u} = \frac{t_0}{t_R(X)} \qquad (2.1.4)$$

If during the separation all the molecules of compound X would be all the time in the mobile phase, then $u_R(X)$ is equal to u. However, some of the molecules are retained and do not move, and only a fraction of molecules of compound X that are present in the mobile phase are moving. The value of $u_R(X)$ is determined by this fraction. Assuming that during the separation process the number of molecules of compound X that are in the mobile phase is $v_{mo}(X)$ and in the stationary phase is $v_{st}(X)$, then $u_R(X)$ will be given by the expression:

$$u_R(X) = \frac{v_{mo}(X)}{v_{mo}(X) + v_{st}(X)} u \qquad (2.1.5)$$

TABLE 2.1.1　Typical Values for the Void Volume of HPLC Columns.

Dimensions (i.d × length in mm)	Empty volume mL	Void volume V_0 mL	Dimensions (i.d × length in mm)	Empty volume mL	Void volume V_0 mL
2.1 × 100	0.35	0.24	4.6 × 250	4.15	2.90
2.1 × 150	0.52	0.37	4.6 × 300	4.99	3.49
2.1 × 250	0.87	0.61	10.0 × 100	7.85	5.50
2.1 × 300	1.04	0.73	10.0 × 150	11.78	8.25
4.6 × 100	1.66	1.16	10.0 × 250	19.63	13.75
4.6 × 150	2.49	1.75	10.0 × 300	23.56	16.49

Using the notation $v = \dfrac{v_{mo}(X)}{v_{mo}(X) + v_{st}(X)}$ for the fraction of molecules of compound X that is present in the mobile phase, from rel. 2.1.4. it can be seen that the relation between the retention time t_R and the dead time t_0, can be written as follows:

$$t_R(X) = (1/v)\, t_0 \qquad (2.1.6)$$

The relation 2.1.6 shows that for a solute present only in the mobile phase, $v = 1$ and $t_R = t_0$, and for a solute completely retained on the stationary phase $v = 0$ and $t_R = \infty$. The common situation is in between these two limits and, for example, if in a separation 25% of the molecules of an analyte are present in the mobile phase, $t_R(X) = 4\, t_0$. A similar relation with rel. 2.1.6 is valid between the retention volume V_R and the dead volume V_0.

Capacity Factor (Retention Factor)

The ratio of reduced retention time and the dead time is an important chromatographic descriptor named *retention factor* or *capacity factor k*. The formula for k is:

$$k = \frac{t_R - t_0}{t_0} = \frac{t'_R}{t_0} \qquad (2.1.7)$$

Since $V = U\,t$, rel 2.1.7 can also be written in the form:

$$k = \frac{V_R - V_0}{V_0} = \frac{V'_R}{V_0} \qquad (2.1.8)$$

(*Note:* The notation for retention factor k may vary in the literature regarding chromatographic parameters, the notation k' being sometimes used.)

The capacity (retention) factor $k(X)$ has the advantage of being dimensionless and independent of the flow rate of the mobile phase or the dimensions of the column; for this reason it is a very common and useful parameter for

peak characterization. From rel. 2.1.6 and 2.1.7, the retention (capacity) factor k can be related to other parameters regarding the separation, such as the fraction v of molecules of compound X that is present in the mobile phase. In this case the expression for k can be written as follows:

$$k = \frac{1 - v}{v} \qquad (2.1.9)$$

Expression 2.1.9 indicates that k is equal to the ratio between the fraction of molecules of compound X that are present in the stationary phase (equal to $1 - v$) and the fraction that is present in the mobile phase (equal to v). Since this ratio is kept constant only in isocratic conditions in a chromatographic separation, it becomes obvious that the retention factor is a parameter valid only for isocratic chromatographic separations (unchanged composition of the mobile phase). For a peak eluting in gradient conditions (with mobile phase composition changing during the run time), k is changing even across the peak, and an average retention factor k^* should be defined for gradient separations.

The retention factor k can be used to establish a relation between t_R and t'_R. Since from rel. 2.1.7 it results $t'_R = k\, t_0$, and since $t_R = t'_R + t_0 = (k + 1)\, t_0$, the following relation can be written:

$$\frac{t'_R}{t_R} = \frac{k}{k+1} \qquad (2.1.10)$$

In current applications, the retention factors of all analytes of interest are typically situated between 2 and 10. Retention factors $k(X)$ lower than 2 (indicating poor retention of component X) are sometimes acceptable, but care must be taken to ensure that the compound is separated from other analytes and from matrix components that may be not retained or poorly retained. If the separation is good, lower k values typically indicate faster elution of the peaks of interest. This is especially desirable in fast chromatography, where other factors such as ultra-high-pressure, small particles in the chromatographic column and use of elevated temperature are applied to speed up the separation. Use of these factors leads to shorter t_R but does not necessarily decrease the k values since they also shorten t_0. Generally, retention factors exceeding a value of 10 indicate strong retention. The corresponding peaks elute after a longer time and are typically wide. Retention factors up to 20 are sometimes necessary, mainly when very complex samples are studied. The retention factor k is frequently expressed in logarithmic form in base 10 or in base e (either $\log_{10} k = \log k$, or $\log_e k = \ln k$).

Equilibrium Constant and Phase Ratio

During the separation process, the molecules that are separated can be considered as being in a continuous equilibrium between the mobile phase and the stationary phase (this implies small amounts of samples such that the process is close to ideal). For a molecular species X, this equilibrium is the following:

$$X\text{(in mobile phase)} \rightleftarrows X\text{(in stationary phase)}$$

and can be considered governed by an equilibrium constant $K(X)$, defined as:

$$K(X) = \frac{(c_X)_{st}}{(c_X)_{mo}} \qquad (2.1.11)$$

where $(c_X)_{mo}$ is the molar concentration of species X in the mobile phase and represents the amount (in moles) of X in the volume V_0 of the mobile phase in the chromatographic column. This amount is proportional to the fraction v of molecules in the volume V_0 of the mobile phase. Similarly, the concentration in the stationary phase $(c_X)_{st}$ can be considered as representing the amount in moles of X from the stationary phase, proportional to $1 - v$ in a volume V_{st} of the immobilized liquid. As

a result, the equilibrium constant $K(X)$ can be written in the form:

$$K(X) = \frac{(1-v)/V_{st}}{v/V_0} = k(X)\frac{V_0}{V_{st}} \qquad (2.1.12)$$

Expression 2.1.12 indicates that the equilibrium of species X between the mobile and stationary phases is proportional to the retention factor k. From rel. 2.1.12 the formula for the retention factor k can be written as follows:

$$k = K\Psi \qquad (2.1.13)$$

where the following notation was used:

$$\frac{V_{st}}{V_0} = \Psi \qquad (2.1.14)$$

The parameter Ψ is known as the *phase ratio*. This parameter has a well-defined meaning in partition chromatography where species X is distributed between the liquid mobile phase (of volume V_0) and an immobilized liquid, this liquid having the volume V_{st}. In adsorption chromatography, the meaning of Ψ is more difficult to assess. The variation of K during a separation in gradient conditions is expected, as also k varies during a gradient separation. On the other hand, Ψ can be considered a stationary phase characteristic and should remain the same for a given column, regardless of isocratic or gradient separation. The precise value of Ψ is not easy to determine. However, an estimation of Ψ can be obtained from the averaged interstitial porosity of the column. As an example, for a C18 column with 5 μm particle size (average interstitial porosity), $V_{st} \approx 0.3$ $(\pi/4)d^2 L$. In this case, $V_0 \approx 0.7 (\pi/4)d^2 L$ and $\Psi \approx 1.5$ (dimensionless). The value of V_{st} may vary with the solute X and also with the mobile phase for the same column. For this reason, although expression 2.1.13 is useful from a theoretical point of view, in practice the separation of

k into the two components (K and Ψ) is difficult to obtain.

From rel. 2.1.7 and 2.1.13, it can be seen that the reduced retention time t'_R can be expressed by the formula:

$$t'_R = K(X) \, \Psi \, t_0 \qquad (2.1.15)$$

Formula 2.1.15 shows that the retention in a chromatographic column depends on the equilibrium constant $K(X)$ (of X from the mobile phase to the stationary phase), on the phase ratio Ψ, and on the dead time for the chromatographic column. A similar relation for the reduced retention volume V'_R can be written as follows:

$$V'_R = V_R - V_0 = K(X)\Psi V_0 = K V_{st} \quad (2.1.16)$$

This relation indicates that the retention volume depends on column parameters $K(X)$, Ψ, and V_0. Relation 2.1.16 also shows that the retention volume is proportional to the equilibrium constant $K(X)$ and the volume V_{st} (of immobilized liquid on the stationary phase).

General Equation of Solute Retention

Formula 2.1.15 describing the retention time in a chromatographic column can also be derived from the basic principle of mass conservation. For this purpose, an infinitesimal cross section with thickness dx and volume dV in the chromatographic column of length L will be considered. Taking an initial concentration $(c)_{mo}$ of the solute in the mobile phase, the change in the amount of solute after passing the infinitesimal volume dV of mobile phase is given by:

$$\frac{\partial (c)_{mo}}{\partial x} \, dx dV \qquad (2.1.17)$$

This total change is caused by the distribution of the solute between the mobile and stationary phases. The solute concentration c in the mobile phase will change by the following amount:

$$-\frac{V_0}{L} \, dx \frac{\partial (c)_{mo}}{\partial V} \, dV \qquad (2.1.18)$$

The change of the amount of solute in the stationary phase will take place by the amount:

$$-\frac{V_{st}}{L} \, dx \frac{\partial (c)_{st}}{\partial V} \, dV \qquad (2.1.19)$$

From the principle of mass conservation, the following relation must be satisfied (Vault equation):

$$\frac{\partial (c)_{mo}}{\partial x} \, dx dV = -\frac{V_0}{L} \frac{\partial (c)_{mo}}{\partial V} \, dx dV \\ -\frac{V_{st}}{L} \frac{\partial (c)_{st}}{\partial V} \, dx dV \qquad (2.1.20)$$

In rel. 2.1.20, the value for $(c)_{st}$ can be substituted using rel. 2.1.11, and the result can be written in the form:

$$\frac{\partial (c)_{mo}}{\partial x} + \frac{\partial (c)_{mo}}{\partial V} \left(\frac{V_0}{L} + K \frac{V_{st}}{L} \right) = 0 \qquad (2.1.21)$$

The differential equation 2.1.21 in $(c)_{mo}$ has the general solution in the form of an arbitrary function $\varphi(z)$[3]:

$$(c)_{mo} = \varphi(z) \quad \text{where} \quad z = V - \frac{x}{L} \, (V_0 + K V_{st})$$
$$(2.1.22)$$

The formula of function φ is determined by the initial conditions for the equation 2.1.21. When the sample is introduced in the column as a narrow plug, and the movement of this plug is studied without considering other effects, the expression for φ is given by the delta function $\delta(z)$:

$$\delta(z) = 0 \text{ if } z \neq 0, \text{ and } \int_{-\infty}^{+\infty} \delta(z)dz = 1 \quad (2.1.23)$$

Relation 2.1.23 indicates that $\varphi(z) = \delta(z)$ is nonzero only when $z = 0$ and therefore when:

$$V - \frac{x}{L}(V_0 + KV_{st}) = 0 \qquad (2.1.24)$$

The solute emerges at the end of the column at $x = L$ and at the retention volume V_R, when rel. 2.1.24 takes the form:

$$V_R = V_0 + K\,V_{st} \qquad (2.1.25)$$

Relation 2.1.25 is identical with rel. 2.1.16, which is also equivalent to rel. 2.1.15 obtained on an empirical basis.

Characteristics of Ideal Peak Shape and Definition of Efficiency

After the injection incorporates an extremely narrow band (plug) of an analyte in the flow stream of an HPLC system, this band is broadened as the time passes and leads to chromatographic peaks with a shape ideally described by a Gaussian bell curve of a given width. The peak broadening can be ideally considered as generated only by ordinary diffusion in time. The "transport" of the band of the analyte across the HPLC system supposedly does not directly affect the peak broadening (which is only an approximation). Being a diffusion process, peak broadening can be studied based on Fick's laws. In reality a number of other effects contribute to peak broadening, and they will be discussed later in Section 2.2.

Fick's second law (see, e.g., [4]) for diffusion in one direction (longitudinal diffusion in the direction of x) has the expression:

$$\frac{dc}{dt} = D\frac{\partial^2 c}{\partial^2 x} \qquad (2.1.26)$$

where t is time, c is concentration expressed in units of mass per units of length, and D is the diffusion coefficient (of the diffusing species in a specific solvent and at a specific temperature,

expressed typically in $cm^2 s^{-1}$). Fick's law can be directly applicable, for example, to the diffusion in a tube or a rectangular channel where the concentration c varies only along the channel and is the same across the channel. On the basis of the assumption that for $t = 0$ (initial condition) the concentration is described by a given function $c\,(x,0)$, the solution of equation 2.1.26 can be written as follows (see, e.g., [5]):

$$c(x,t) = \frac{1}{\sqrt{\pi Dt}} \int_{-\infty}^{+\infty} c(\eta, 0) \qquad (2.1.27)$$
$$\times \exp[-(\eta - x)^2/(4Dt)]d\eta$$

With the assumption that D is constant and the whole amount m of material was initially (at $t = 0$) contained in one point at $x = 0$, upon integration, expression 2.1.27 leads to the relation (see, e.g., [3]):

$$c(x,t) = \frac{m}{2\sqrt{\pi Dt}}\exp[-x^2/(4\,Dt)] \qquad (2.1.28)$$

By the introduction in rel. 2.1.28 of the notation:

$$\sigma^2 = 2\,D\,t \qquad (2.1.29)$$

the expression for $c(x,t)/m$ (where $t = $ const.) can be written as follows:

$$c(x,t)/m = \frac{1}{\sqrt{2\pi\sigma^2}}\exp(-x^2/2\sigma^2) \qquad (2.1.30)$$

Expression 2.1.30 characterizes a typical Gaussian bell curve (as a function of x), is called a normal probability density function, and is used to describe a random process. For example, the observational errors in an experiment are assumed to follow such a normal distribution. The parameter σ^2 is called the *variance*, and σ is called *standard deviation*. Graphs of c/m as a function of the distance x are shown in Figure 2.1.1 for $\sigma = 0.2$ (corresponding for example to $D = 10^{-2}$ and $t = 2$) and for

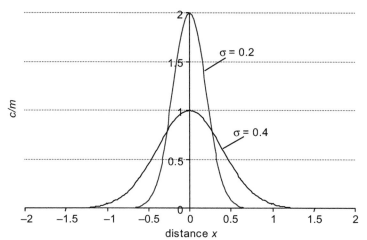

FIGURE 2.1.1 The variation of c/m as a function of distance x for two values of σ, namely $\sigma = 0.2$ (corresponding for example to $D = 10^{-2}$ and $t = 2$) and $\sigma = 0.4$. The graphs show the Gaussian bell shape of the concentration distribution. The whole amount m of material was initially contained at $x = 0$.

$\sigma = 0.4$ (corresponding for example to $D = 10^{-2}$ and $t = 8$). The parameter σ describes the width of the Gaussian curve, larger σ leading to wider bell shapes as seen in the figure.

The apex of the Gaussian curve is obtained for x = 0 and $c_{max}/m = (2\pi \sigma^2)^{1/2}$. The analytical expression of the Gaussian curve shows that for any chosen c with $0 < c < c_{max}$ the value for x is given by the expression:

$$x = \pm\sqrt{-2\sigma^2 \ln[(c/m)\sigma\sqrt{2\pi}]} \quad (2.1.31)$$

From rel. 2.1.31, the bell width $W = 2|x|$ is obtained as an increasing function of σ. Of particular importance in chromatography are the width at half height W_h and the width at the inflection point W_i of the Gaussian curve. Taking $c = 1/2\ c_{max} = 1/2\ m\ (2\pi\ \sigma^2)^{1/2}$ in rel. 2.1.31, the result is:

$$W_h = 2(2 \ln 2)^{1/2} \sigma \quad (2.1.32)$$

From the second derivative of the expression 2.1.30, the value for W_i, it is easily obtained as:

$$W_i = 2\sigma \quad (2.1.33)$$

The value for W_i for a given compound in a specific solvent can be obtained using the

value for D and for the diffusion time. As an example, for the diffusion of aniline in water at 25 °C, $D = 1.05\ 10^{-5}$ cm^2/s. For a diffusion time of 5 min, $W_i \approx 1.6$ mm.

In a chromatographic process that takes time, during the movement of the mobile phase (and assuming laminar flow), the analyte is diffusing, and if ξ is the distance from the origin to the center of the moving diffusion zone, the expression for c/m becomes:

$$c(x)/m = \frac{1}{\sqrt{2\pi\sigma^2}}\exp[-(x - \xi)^2/(2\sigma^2)] \quad (2.1.34)$$

This expression can be used for understanding peak broadening in a chromatographic process. For a separation where the mobile phase has a linear flow rate u, the distance from the origin to the center of the moving zone is $\xi = u\ t_R$. Therefore, from rel. 2.1.29 for a given ξ, the resulting σ is given by the expression:

$$\sigma^2 = 2 D \xi/u \quad (2.1.35)$$

With rel. 2.1.34 for c / m as a function of x, a chromatogram where c / m is measured shows peak broadening for an eluting analyte as illustrated in Figure 2.1.2.

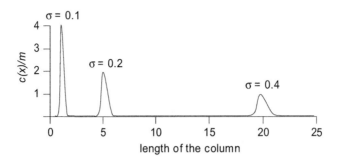

FIGURE 2.1.2   Peak broadening of an eluting analyte along a chromatographic column. In this ideal chromatogram, the broadening for the three peaks correspond to $\sigma = 0.1$, $\sigma = 0.2$, and $\sigma = 0.4$, which can be obtained, for example, for $D = 1$, $u = 250$ and $\xi = 1.25$, $\xi = 5$, and $\xi = 20$, respectively (arbitrary units).

For a chromatographic process, the concentration $c(x)$ in rel. 2.1.34 can be replaced with the peak height $h(x)$, as generated by the response to instant concentration of a chromatographic detector. The variation of h as a function of distance x from start is described in this case by the equation:

$$h(x) = A(2\pi\sigma^2)^{-1/2} \exp\left[-(x - \xi)^2/2\sigma^2\right]$$
$$(2.1.36)$$

where A is the total peak area (used for quantitation in HPLC), ξ is the distance to the middle of the moving zone (and the maximum of the Gaussian curve), and σ determines the extent of peak broadening. Also, because in most chromatographic processes the measured parameter is the retention time and not the length of the path in the column, σ (function of distance) should be replaced with σ_t, a function of time (time broadening) given by the formula:

$$\sigma_t = \frac{\sigma t_R}{t_0 u} \qquad (2.1.37)$$

With this replacement in rel. 2.1.36 the peak height h is expressed as a function of time t as follows:

$$h(t) = A(2\pi\sigma_t^2)^{-1/2} \exp\left[-(t - t_R)^2/2\sigma_t^2\right]$$
$$(2.1.38)$$

In typical chromatographic processes, rel. 2.1.38 remains valid, although in practice, the value for σ_t is determined by more factors besides longitudinal diffusion, which contributes very little to peak broadening. The maximum height of the peak is achieved when

$$t = t_R \text{ and } h_{max} = \frac{A}{\sigma_t\sqrt{2\pi}}.$$

The measurement of peak broadening on a chromatogram is not so frequently done at the inflection point of the Gaussian bell curve (as W_i), but at the half height of the curve (as W_h) or at the baseline (as W_b), where $W_b = 2\,W_i$ (all the values W_i, W_h, and W_b are now measured in units of time, but the same notation was maintained as for the values measured in length). The peak width W_b is measured between the points of intersection of baseline with the tangents to the curve at the inflexion points. Figure 2.1.3 shows the measurements of t_R, W_i, W_h, and W_b on an ideal chromatographic peak.

From rel. 2.1.32 or 2.1.33 it results that σ_t can be calculated from W_h or W_b using the expressions:

$$\sigma_t = (8 \ln 2)^{-1/2}\,W_h \approx 0.42466\,W_h \quad (2.1.39)$$

$$\sigma_t = 0.25\,W_b \qquad (2.1.40)$$

The value of σ (space broadening also indicated as σ_L) is related to another parameter

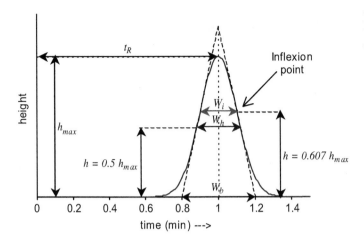

FIGURE 2.1.3 Measurements of retention time t_R and peak broadening W_i, W_h and W_b on a recorded chromatogram.

used to characterize zone spreading, namely, the *height equivalent to a theoretical plate H* (HETP), which is defined as:

$$H = \sigma^2/L \qquad (2.1.41)$$

This parameter is very useful in chromatography for the characterization of peak broadening per unit length of the column (since σ describes the width of the Gaussian curve). In addition to H, the peak broadening characterization in a column can be obtained using the *theoretical plate number N*. For a column of length L, N is defined as:

$$N = L/H \qquad (2.1.42)$$

Relation 2.1.42 indicates that N is proportional to the column length L and inversely proportional to H. The theoretical plate number N can be expressed as a function of length by a simple substitution of rel. 2.1.41 in 2.1.42 and:

$$N = L^2/\sigma^2 \qquad (2.1.43)$$

From rel. 2.1.37 and 2.1.40, and because $L = t_0 u$, the expression of N from rel. 2.1.43 can be written in the form:

$$N = t_R^2/\sigma_t^2 = 16\, t_R^2/W_b^2 \qquad (2.1.44)$$

The same expression and using rel. 2.1.39 and 2.1.43 can be written in the form

$$N = 5.5452\, t_R^2/W_h^2 \qquad (2.1.45)$$

In addition to the theoretical plate number N, an *effective plate number n* is defined by using t_R in rel. 2.1.44 instead of t'_R. The formula for N will be:

$$n = 16\, t_R'^2/W_b^2 \qquad (2.1.46)$$

It should be noted that $t'_R < t_R$ and therefore $n < N$ with a large range of differences (commonly 10 to 30%).

Relation 2.1.46 shows how n depends on chromatographic retention time t'_R, and since t'_R is compound related (index (X) omitted), it also shows that N (as well as n) are compound dependent (a correct notation for N should therefore be $N(X)$). Both rel. 2.1.44 and 2.1.46 can be used to measure the theoretical plate number or effective plate number based on experimental data obtained with a given column. This measurement is useful in practice to select columns (higher N gives lower peak broadening) and also to assess the loss in performance of a column after a certain period of usage when the N values start to decrease. Because N is related to the important characteristic of peak broadening, it is common to indicate it as a parameter to characterize the *efficiency* of a column. The values for N for HPLC columns can either be given for a specific column or reported as efficiency per meter. Also both N and n are used for the characterization of column efficiency, and n is sometimes named simply "efficiency". For modern HPLC analytical columns, the efficiency per meter N can be between 20,000 and 150,000, and for core-shell columns it can be as high as 300,000. The values for N (per unit length of the column) are influenced by physical properties of the stationary phase such as dimension of stationary phase particles, homogeneity of the particles dimensions, and structure of particles.

Since peak broadening is not caused solely by diffusion, if the peaks still maintain their Gaussian bell shape, it can be accepted that the peak width is determined by the sum of a number of independent random processes with normal distribution and:

$$W_i = 2\sigma = 2\sqrt{\sum_n \sigma_n^2} \qquad (2.1.47)$$

where σ_n are the standard deviations of these random independent processes.

Selectivity

Selectivity is another empirical parameter, typically indicated as α, which can be calculated from a given chromatogram. The value for α is calculated using the formula:

$$\alpha = \frac{t_R'(X)}{t_R'(Y)} \qquad (2.1.48)$$

where $t_R'(X) > t_R'(Y)$. Parameter α indicates the ratio of the distances in time between the apexes of two chromatographic peaks (for compounds X and Y). The selectivity factor is usually of interest for compounds that give adjacent peaks,

since peaks that are well distanced do not pose separation problems.

Using rel. 2.1.7, it can be easily noticed that α can also be expressed by the formula:

$$\alpha = \frac{k(X)}{k(Y)} \qquad (2.1.49)$$

where k is the retention factor for the two different compounds. In any chromatographic separation, larger α values are desirable for a better separation. However, the value of α alone cannot describe how good the separation of two compounds is. Even when α is large, peak broadening can be so large that the separation can be poor. The values for α are solute dependent, but also depend on the nature of the stationary phase and of the mobile phase. The chemical nature of the stationary phase is one of the two main factors influencing the separation for a given set of analytes, the second main factor being the choice of mobile phase. For this reason, the choice of a chromatographic column is frequently based on its selectivity α toward the analytes being separated. A value for $\alpha > 1.2$ is typically necessary for an acceptable separation. A special utilization of selectivity is related to the characterization of hydrophobic character of reversed phase chromatographic columns (see Section 6.4 for a discussion on methylene selectivity $\alpha(CH_2)$).

Resolution

Regardless of how far apart the apexes of two chromatographic peaks (as described by α) are, if the peaks are broad their separation can be compromised. A parameter that truly characterizes peak separation is the resolution R. This parameter is defined by the formula:

$$R = \frac{2[t_R(X) - t_R(Y)]}{W_b(X) + W_b(Y)} \qquad (2.1.50)$$

In formula 2.1.50 it can be easily seen that $t_R(X) - t_R(Y) = t_R'(X) - t_R'(Y)$, and this difference can be written as $\Delta t_R = t_R(X) - t_R(Y)$. The values for formula 2.1.50 can be obtained from the chromatogram as illustrated in Figure 2.1.3. A good peak separation is typically considered when $R > 1$ or even $R > 1.5$ when the two peaks are separated at the baseline ($\Delta t_R > \frac{1}{2}W_b(X) + \frac{1}{2}W_b(Y)$). The widths at the baseline of the two peaks can be different (as is also shown in Figure 2.1.4), but as an approximation it is possible to take $W_b(X) = W_b(Y) = W_b$. With these assumptions, the formula for R can be written in the form:

$$R = \Delta t_R / W_b \qquad (2.1.51)$$

The difference Δt_R can be written as a function of selectivity α in the following form:

$$\Delta t_R = (\alpha - 1)t'_R(Y) \qquad (2.1.52)$$

and omitting the index Y for t'_R, the expression for R becomes:

$$R = (\alpha - 1)t'_R / W_b \qquad (2.1.53)$$

In the ratio t'_R/W_b the value for t'_R can be replaced with t_R using rel. 2.1.10

$\left(t'_R = \left(\dfrac{k}{1+k}\right)t_R\right)$, and t_R/W_b can be expressed as a function of N as shown in rel. 2.1.44 ($t_R/W_b = 1/4N^{1/2}$). As a result, the following formula is obtained for R:

$$R = \frac{1}{4}(\alpha - 1)\left(\frac{k}{1+k}\right)N^{1/2} \qquad (2.1.54)$$

The value for Δt_R can also be written as a function of selectivity α in the following form:

$$\Delta t_R = (\alpha - 1)\alpha^{-1}t'_R(X) \qquad (2.1.55)$$

Omitting the index X for t'_R, the expression for R becomes:

$$R = (\alpha - 1)\alpha^{-1}t'_R / W_b \qquad (2.1.56)$$

And the formula for R is then written in the form:

$$R = \frac{1}{4}\left(\frac{\alpha - 1}{\alpha}\right)\left(\frac{k}{1+k}\right)N^{1/2} \qquad (2.1.57)$$

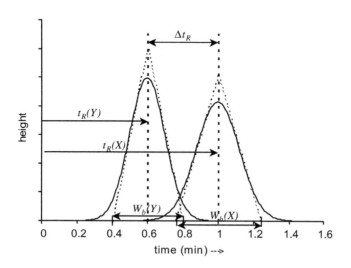

FIGURE 2.1.4 An idealized chromatogram showing the measurable parameters used for the calculation of resolution **R**.

It can be seen from rel. 2.1.57 that R depends on all three parameters: selectivity α, peak capacity k, and column efficiency N.

Both relations 2.1.54 and 2.1.57 are approximations, taking the peak width at the baseline as equal for the two peaks, and the theoretical plate number N as measured for one compound (Y in rel. 2.1.54) or for the other (X in rel. 2.1.57). An example of the variation of R as a function of α and k, assuming a chromatographic column with $N = 18,000$, is given in Figure 2.1.5.

The graph from Figure 2.1.5 shows that R is most sensitive to the parameter α, which is critical for obtaining a good separation. Larger k values are also useful, but as k increases, its importance for the increase in R is diminished.

Resolution depends as shown by formulas 2.1.54 (or 2.1.57) on α, k and N. By using rel. 2.1.13 for k, rel. 2.1.54 can also be written in the form:

$$R = \frac{1}{4}\left[\frac{K(X)}{K(Y)} - 1\right]\frac{K(Y)}{1/\Psi + K(Y)}N(Y)^{1/2}$$

$$(2.1.58)$$

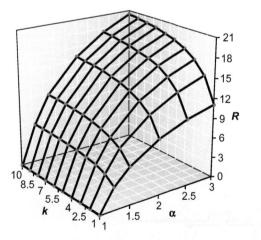

FIGURE 2.1.5 Graph showing the variation of R as a function of α and k assuming a chromatographic column with $N = 18,000$.

This expression shows that the resolution depends on the constants for the equilibrium between the mobile phase and the stationary phase for the two analytes to be separated, and on the phase ratio Ψ, and the number of theoretical plates N of the chromatographic column.

The requirement for the value of resolution R to be higher than 1 in order to have a good separation is translated for the theoretical plate number N in the requirement to satisfy the relation:

$$N > 16\frac{(1+k)^2}{k^2}\frac{\alpha^2}{(\alpha - 1)^2}$$

$$(2.1.59)$$

In many practical applications, the separation factor α between an analyte and other components of a specific matrix may be too close to unity. The increase in the number of theoretical plates of the column can be helpful for enhancing separation in these cases. A discussion on possibilities for increasing the values for N is given in Section 2.2.

Peak Capacity

The efficiency of an HPLC separation can be characterized by a parameter known as *peak capacity*. This parameter gives the number of peaks in a chromatogram that can be separated from one another with a resolution $R = 1$ for a predefined retention factor k. The requirement of a predefined k is equivalent to a given retention window, the restriction resulting from the fact that excessively large k values lead to very long retention times, which are not acceptable for practical purposes ($k < 20$ is a common limitation). The theory of a maximum peak capacity is based on the fact that an ideal peak (Gaussian shape) has the peak width $W_b = 4\sigma_t$ (see rel 2.1.40). As the peak width changes with the retention time, the peak capacity P can be defined by the formula:

$$P = 1 + \int_{t_0}^{t_{R\ max}} \frac{dt}{4\sigma_t}$$

$$(2.1.60)$$

where t_{Rmax} represents the retention time of the last peak in the chromatogram. Relation 2.1.60 can easily be interpreted as the length (in time) of a chromatogram divided by one peak width. The value of σ_t can be related (using rel. 2.1.37) to the theoretical plate number by the formula:

$$\sigma_t = \frac{t_0}{\sqrt{N}}(k_e + 1) \qquad (2.1.61)$$

where k_e is the retention factor k at the point of elution (and can be taken as $k_e = t/t_0 - 1$). For isocratic separations, introducing σ_t in rel. 2.1.60 and performing the integration, the resulting value for the peak capacity P is given by the formula:

$$P = 1 + \frac{\sqrt{N}}{4}\ln\left(\frac{t_{R\,max}}{t_0}\right) \qquad (2.1.62)$$

For gradient separations, a more complicated expression for P can be obtained [6,7].

Statistical Moments for the Description of Peak Characteristics

Some of the characteristics of chromatographic peaks can be obtained using statistical methods. The Gaussian shape of the peak can be seen as generated by a population of points with random normal (Gaussian) distribution. The estimation of population parameters such as mean and variance can be done in statistics by calculating the *population moments*. For a discrete distribution of measurements, the momentum results from a sum. For a continuous distribution described by a Gaussian function, the momentum will be given by an integral. Zero momentum for the chromatographic peak is given by the integral:

$$\Omega_0 = \int_0^\infty h(t)dt \qquad (2.1.63)$$

where $h(t)$ is given by expression 2.1.38. It can be seen that Ω_0 is the area of the chromatographic peak since:

$$\int_0^\infty (2\pi\sigma^2)^{1/2}\exp\left[-\frac{(t - t_R)^2}{2\sigma^2}\right]dt = 1 \qquad (2.1.64)$$

Peak area is proportional to the amount of analyte and is the main parameter used for quantitation in HPLC.

The first momentum is given by the expression:

$$\Omega_1 = \frac{1}{\Omega_0}\int_0^\infty t\,h(t)dt \qquad (2.1.65)$$

It can be shown that the first momentum is in fact the retention time t_R ($\Omega_1 = t_R$). The first momentum gives the expected value of a random variable, and assuming that no spreading of the chromatographic peak would occur, the expected value of time where the peak would be formed is t_R.

Higher moments can be defined by the formula:

$$\Omega_n = \frac{1}{\Omega_0}\int_0^\infty (t - \Omega_1)^n h(t)dt \qquad (2.1.66)$$

For example, the second momentum gives the variance and has the expression:

$$\Omega_2 = \frac{1}{\Omega_0}\int_0^\infty t^2 h(t)dt - t_R^2 \qquad (2.1.67)$$

Second momentum gives the value for σ_t^2 ($\Omega_2 = \sigma_t^2$). Also, it can be seen from rel. 2.1.43 that the theoretical plate number N can be obtained from the formula:

$$N = \frac{t_R^2}{\sigma_t^2} = \frac{\Omega_1^2}{\Omega_2} \qquad (2.1.68)$$

The third momentum is zero for a perfect Gaussian peak (as well as all higher momentum values). However, for a real peak that deviates from the Gaussian shape being asymmetrical, the third momentum Ω_3 describes this peak asymmetry also known as *skew*. A positive value for Ω_3 indicates tailing. The formula for the skew is the following:

$$\Omega_3 = \frac{1}{\Omega_0} \int_0^\infty t^3 h(t)dt - 3\Omega_0\Omega_2 + 2\Omega_0^3 \quad (2.1.69)$$

The fourth momentum (Ω_4) describes yet another property of peaks deviating from the perfect Gaussian shape indicating the extent of vertical flattening known as *excess* [8]. A positive value for Ω_4 indicates sharpening of the peak.

Description of Peak Characteristics for Gradient Separations

As previously mentioned, the formulas for the empirical peak characteristics were developed assuming that the HPLC separation takes place in isocratic conditions. In isocratic conditions, the retention factor k, the equilibrium constant K, and the diffusion coefficient D for a specific compound i are constant (for a given column and a given composition of the mobile phase). Gradient elution modes allow modifications of k_i's (therefore of retention times $t_{R,i}$), which depend on modification of the mobile phase's composition. For different types of chromatography, the change in mobile-phase composition may affect differently the values of k_i. For example, in RP-HPLC the k_i values for hydrophobic compounds are significantly higher for a mobile phase with a strong polar character (e.g., water) as compared to k_i values in a less polar mobile phase consisting, for example, of a high proportion of acetonitrile or methanol in water. In isocratic conditions, the value of k_i can be approximated as dependent on the *volume fraction* of organic component in the mobile phase ϕ by the expression:

$$\log k_i = \log k_{w,i} - S_i \phi \quad (2.1.70)$$

where $k_{w,i}$ is the (extrapolated) value of k_i for a specific analyte for $\phi = 0$, S_i is a specific constant for a solute, a solvent mixture, and a specific column, and does not depend on ϕ that is the ratio of volume of organic phase in the total volume of solvent (volume fraction of the organic solvent). The value $S = 3$ can be used for unknown analytes, but S can be obtained using graphs of measured $\log k_i$ for different ϕ values in isocratic conditions. The parameter is sometimes estimated using the approximation formula $S = 0.25 \, (M)^{1/2}$ (where M is the molecular weight of the analyte). Tables of this parameter are also available in the literature [9]. Several values for S for specific compounds are listed in Table 7.1.8.

For a linear gradient, the composition of the mobile phase is changed by rel. 1.4.1. For $t_1 = 0$ and $c_1 = \phi_0$ in rel. 1.4.1 (also assuming a dwell volume $V_D = 0$), for a specific time t the volume fraction of organic phase can be written in the form:

$$\phi = \phi_0 + \frac{\Delta\phi}{t_{grad}}t \quad (2.1.71)$$

With ϕ given by rel. 2.1.71, included in rel. 2.1.70, the expression for the capacity factor becomes:

$$\log k_i = \log k_{w,i} - S_i\phi_0 - S_i\frac{\Delta\phi t_0}{t_{grad}}(\frac{t}{t_0}) \quad (2.1.72)$$

Relation 2.1.72 shows that in linear gradient conditions, the capacity factor k_i depends across the chromatogram not only on the nature of the compound i, but also on the "instantaneous" mobile-phase composition. The variation of k_i during a gradient run-time for two compounds having different k_{wi} values and the same S is illustrated in Figure 2.1.6. The variation of $\log k_i$ during a (linear) gradient run-time is linearly dependent on the time t as seen from rel. 2.1.72. Relation 2.1.72 can also be written in the form:

$$\log k_i = \log k_{w,i} - S_i\phi_0 - b(\frac{t}{t_0}) \quad (2.1.73)$$

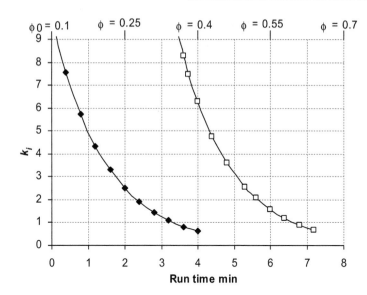

FIGURE 2.1.6 Variation of k_i during a linear gradient run time for two compounds with $k_{w1} = 25$ and $k_{w2} = 250$, $S = 4$, $\phi_0 = 0.1$, $\Delta\phi = 0.6$, $t_0 = 1$ min and $t_{grad} = 8$ min.

where b is known as *gradient steepness* and its value is given by the formula:

$$b = \frac{\Delta\phi S}{t_{grad}} t_0 \qquad (2.1.74)$$

Because a compound has different k_i values during the chromatographic run, it is not possible to use k_i as given by rel 2.1.73 for the calculation, for example, of t'_R (or t_R) following expressions similar to 2.1.7. Since k_i changes as the compound migrates across the chromatographic column, it was useful to define an "effective" value for k. This "effective" value is the *gradient retention (capacity) factor* k^* and is obtained using various approximations. The effective gradient capacity factor k^* is used for the estimation of several parameters in gradient elution such as $\alpha^* = k_i^*/k_j^*$ or the resolution R expressed by a formula similar to 2.1.57 as follows:

$$R^* = \frac{1}{4}\left(\frac{\alpha^* - 1}{\alpha^*}\right)\left(\frac{k^*}{1 + k^*}\right)N^{1/2} \qquad (2.1.75)$$

Further discussion of gradient elution, including the calculation of k^* and of other parameters for gradient elution, can be found in Section 7.6.

Quantitation in HPLC

The integration of $c(x)$ given by rel. 2.1.30 for x between $-\infty$ and $+\infty$ (since $\int_{-\infty}^{+\infty} e^{-x^2} dx = \sqrt{\pi}$) leads to the following result:

$$m = \int_{-\infty}^{+\infty} c(x)dx \qquad (2.1.76)$$

The same integral of $c(x)$ between $-\infty$ and $+\infty$ is equal to the total peak area A_{peak} (for compound i) under the curve representing the function $c(x)$. In chromatographic instruments, this peak area is obtained using instrumental detection/amplification procedures and the value for A_{peak} in a chromatogram will become only proportional to the amount of material injected in the HPLC system. Since the injection is performed using a given sample volume, and $m = c_i V_{inj}$ where c_i is the sample (initial) concentration, and V_{inj} is the injection volume, the following expression is used for quantitation in HPLC:

$$c_i V_{inj} = Const. A_{peak} \qquad (2.1.77)$$

The proportionality constant *Const.* depends on experimental conditions, and a common quantitation procedure is the use of calibration curves between peak area and the concentration of an analyte. These calibration curves are obtained using standards. In practice, deviations from rel. 2.1.77 may sometimes be encountered. For example, the relation between sample concentration and peak area can be of the form:

$$c_i V_{inj} = Const._1 + Const._2 A_{peak} \qquad (2.1.78)$$

In some cases, a quadratic equation fits better the relation between the concentration and peak area. These deviations from rel. 2.1.77 are caused by the background response of the detector or by its nonlinear response. Quantitation based on standard addition or peak area ratios with isotopic labeled compounds (used with MS detection) are also practiced and are based on proportionality given by rel. 2.1.77 (standard addition quantitation is not applicable if the dependence between the analyte concentration and peak area follow rel. 2.1.78 and *Const._1* is not known) (see, e.g., [1].).

Besides the peak area, peak height can also be used for quantitation. Relation 2.1.34 shows that $c(x)$ in a chromatogram is maximum when $x = \xi$, and then the following expression is valid:

$$c_{max} = \frac{V_{inj}c_i}{\sigma\sqrt{2\pi}} \qquad (2.1.79)$$

The maximum concentration of the analyte in the chromatogram corresponds to the apex of the chromatographic peak, and the peak height h_{max} is proportional to that maximum concentration c_{max}. Relation. 2.1.78 indicates that the initial concentration and the peak height in a chromatogram can also be related by a proportionality relation, which is expressed by the formula:

$$c_i V_{inj} = Const.' h_{max} \qquad (2.1.80)$$

The proportionality constant *Const.'* also depends on experimental conditions, and quantitation procedures involve the use of calibration curves.

Although measurement of the concentration of the injected sample seems to be equally possible using the peak area or the peak height, there are some differences between the two procedures. The formulas 2.1.78 and 2.1.79 were developed based on the assumption that the peak has an ideal Gaussian shape. This is not the case in most practical situations. Even if the peak shape deviates from Gaussian, rel. 2.1.77 remains valid, and the peak area in the chromatograms remains proportional to the amount of sample injected in the HPLC system. For peaks with a shape different from Gaussian, more variability regarding the proportionality between the peak height and the amount of the injected sample is typically seen.

The maximum concentration c_{max} for an HPLC separation is an important parameter since it determines the maximum signal in a selected detector and therefore is related to the detection limit of an analytical method. From rel. 2.1.79 a chromatographic dilution $\mathcal{D}$ can be defined using the formula:

$$\mathcal{D} = \frac{c_i}{c_{max}} = \frac{\sigma\sqrt{2\pi}}{V_{inj}} \qquad (2.1.81)$$

From rel. 2.1.41 it can be obtained that $\sigma = \sqrt{HL}$, and introducing this expression in rel. 2.1.81, a formula for the dilution $\mathcal{D}$ can be obtained (in microcolumn HPLC, other effects also contribute to the peak broadening, and more elaborate expressions for σ are utilized [10,11]). Relation 2.1.81 indicates that the maximum concentration of analyte c_{max} along a chromatographic peak is directly proportional to the initial concentration in the sample c_i, and inversely proportional to dilution $\mathcal{D}$. Smaller values for $\mathcal{D}$ are therefore preferable for obtaining larger detector signals. This can be achieved with larger injection volumes and columns that are shorter and with low values for the height equivalent to a theoretical plate H.

2.2. EXPERIMENTAL PEAK CHARACTERISTICS IN HPLC

Van Deemter Equation

The Van Deemter equation describes the factors influencing peak broadening in a chromatographic separation. The longitudinal diffusion of an analyte molecule in a solvent accounts for only a small proportion of the peak broadening. (Longitudinal diffusion was discussed in Section 2.1 to describe the Gaussian shape of an HPLC chromatographic peak.) Among the factors that contribute to peak broadening are the following: (1) longitudinal diffusion (already discussed), (2) eddy diffusion, (3) lateral movement of material due to convection, (4) mass transfer process in and out the stationary phase, and (5) contribution from the stagnant mobile phase in a porous material (mass transfer in and out mobile phase). For a chromatographic process involving all these broadening effects, the height equivalent to a theoretical plate H (HETP) can be written using rel. 2.1.41 and 2.1.47 in the following form:

$$H = \frac{\sum_n \sigma_n^2}{L} = H_L + H_E + H_C + H_T + H_S$$

$$(2.2.1)$$

The expression for the contribution to the theoretical plate of longitudinal diffusion is given by rel. 2.1.41, with σ given by rel. 2.1.29 (where σ is written in the form σ_L). However,

longitudinal diffusion in an HPLC column is hindered by the packing material, and an "obstruction factor" γ must be included in its formula. In this way, the formula for the contribution to plate height H_L becomes:

$$H_L = \frac{\sigma_L^2}{L} = \gamma \frac{2Dt}{L} = \gamma \frac{2D}{u}$$

$$(2.2.2)$$

The use of rel. 2.2.2 requires known values for parameters γ and diffusion coefficient D of the analyte in the mobile phase, and also for the value of linear flow rate u (dependent on the dimensions of the channel). The value for γ depends on the column packing and is typically around 0.625. The values for D (in $cm^2 s^{-1}$) are reported in the literature for various solutes and solvents and can also be estimated [12,13]. One additional observation regarding the value for D is that since this parameter depends on the nature of the solvent, during the gradient HPLC the value of D may change for a given solute. This change may contribute to a (small) variation of H_L across a peak that elutes in gradient conditions, contributing to deviations from a Gaussian peak shape. The value for u can be obtained from the volumetric flow U (set at the pump). For this purpose, it should be noticed that the value for the dead time t_0 of a column can be obtained either by dividing the length of the column L by u or by dividing the void volume of the column V_0 by U. Therefore, the following formulas can be written:

$$t_0 = \frac{L}{u} = \frac{\varepsilon^* \pi d^2 L}{4U} \quad \text{and} \quad u = \frac{4U}{\pi \varepsilon^* d^2} \quad (2.2.3)$$

where d is the inner diameter of the column, $\varepsilon^* \approx 0.7 \ 10^{-3}$ is the packing porosity (for a column filled with 5 μm diameter particles, u in mm/min, L and d in mm, and U in mL/min). Using all the necessary values, the H_L can be estimated as a function of the volumetric flow rate.

Eddy diffusion is caused by the fact that in a packed material the flow occurs through a tortuous channel system with various path lengths. The molecules of the same solute may randomly take different paths. This path will depend on the average diameter of a particle, and using the notation d_p for the particle diameter in the stationary phase, the contribution to the plate height of the eddy diffusion can be written in the form:

$$H_E = \Lambda \, d_p \qquad (2.2.4)$$

In rel. 2.2.4, Λ is a parameter depending on other characteristics of column packing and increases when the packing material is more irregular.

Lateral movement of material due to convection depends on column packing (through a parameter Γ), increasing when the particle diameter d_p increases and decreasing when the diffusion coefficient D increases. Also, this effect is stronger at higher linear flow rates. The contribution to the plate height of this type of diffusion is given by the expression:

$$H_c = \frac{\Gamma d_p^2 u}{D} \qquad (2.2.5)$$

The rate of transfer of solute into and out of the stationary phase is controlled by the rate of diffusion in the liquid stationary phase or by the adsorption-desorption kinetics in the case of adsorption processes. Using a random walk model (e.g. [14]), it can be shown that for a distribution process the contribution to the plate height due to this effect can be expressed by the formula:

$$H_T = \Theta \, \frac{k}{(1+k)^2} \frac{d_f^2}{D_s} u \qquad (2.2.6)$$

In rel. 2.2.6, Θ is a parameter depending on the particle shape, d_f is the thickness of the stationary phase, D_s is the diffusion coefficient

of the molecules of the solute in the stationary phase (different from D for the diffusion in the mobile phase), and k is the capacity factor for the solute.

The contribution to the plate height from the mass transfer in the stagnant mobile phase in the porous material is given by an expression similar to 2.2.6 where the parameter Θ is about the same, but d_f should be replaced by d_p (particle diameter related to the depth of the pores), and D_s by D the diffusion coefficient in the mobile phase. This effect, known as mobile phase mass transfer contribution, gives the following increase H_S to the theoretical plate height:

$$H_S = \Theta \, \frac{k}{(1+k)^2} \frac{d_p^2}{D} u \qquad (2.2.7a)$$

In size-exclusion chromatography, there are no sorption effects (ideally) to affect band broadening. However, a contribution to band broadening comes from the stagnant mobile phase in the porous material. For spherical porous particles it has been demonstrated that this contribution gives $\Theta \approx 1/30$, which increases the plate height with the quantity:

$$H_S = \frac{k}{30(1+k)^2} \frac{d_p^2}{D} u \qquad (2.2.7b)$$

When we combine all the factors contributing to the increase of the plate height, the result can be written in a general form known as the van Deemter equation [15]:

$$H = A + B\frac{1}{u} + Cu \qquad (2.2.8)$$

where A, B, and C are constants incorporating all the other parameters previously discussed. The Van Deemter equation provides information on the kinetic performance of a chromatographic column. The plot of the van Deemter equation for $A = 4$ μm, $B = 500$ μm^2/s, and

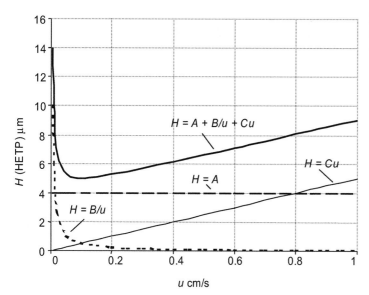

FIGURE 2.2.1 The plot of van Deemter equation and of its components for $A = 4$ μm, $B = 500$ μm^2/s and $C = 0.0005$ s.

$C = 0.0005$ s is given in Figure 2.2.1. An equivalent formula for H, is given by the expression:

$$H = A'd_p + B'\frac{D}{u} + C'\frac{d_p^2}{D}u \qquad (2.2.9)$$

The van Deemter equation given by formula 2.2.9 indicates the explicit contributions of: (1) the diffusion coefficient D of the solute in the mobile phase, and (2) the dimension of the particles of the stationary phase d_p. The contribution H_T, being very small, is neglected in 2.2.9.

The minimum of the van Deemter curve indicates the optimum flow rate of the mobile phase, for which the minimum plate height, and therefore the maximum number of theoretical plates N, can be obtained. This minimum is obtained from the condition of zero value for the differential $dH/du = 0$ that generates for the optimum u the formula:

$$u_{opt} = \frac{D}{d_p}\sqrt{\frac{B'}{C'}} \qquad (2.2.10)$$

The value for the optimum volumetric flow U can be easily obtained from rel. 2.2.3 and 2.2.10, as having the expression:

$$U_{opt} = \frac{\pi \varepsilon^* d^2 D}{4d_p}\sqrt{\frac{B'}{C'}} \qquad (2.2.11)$$

The value for minimum H can be obtained by including u_{opt} in expression 2.2.9, giving the following result:

$$H_{min} = d_p(A' + \sqrt{B'C'}) \qquad (2.2.12)$$

Relation 2.2.10 (and 2.2.11) indicates that the optimum flow rate depends on the nature of the analyte (through D) and on the column construction. In particular, the larger are the particles of the packing material (larger d_p), the lower is the optimum flow rate. On the other hand, the minimum plate height H_{min} is a function of column construction only, larger particles leading to larger HETP H, and therefore to a lower number of theoretical plates N (for the same column length).

Experimental attempts to verify equations 2.2.8 (and 2.2.9) showed that some deviations from the theoretical model occur. The main explanation for this effect is that eddy diffusion and mobile phase mass transfer are not totally independent effects. As described about eddy diffusion, intraparticle local streams that have different length and linear velocity lead to differences in the length of the path for molecules of the same species. However, the differences in the intraparticle velocity of local streams also affect the mobile phase mass transfer, and in return the eddy diffusion. The coupling of the two processes is captured in a different expression for H, given by the Knox equation [16, 17]. In the Knox equation it is common to replace the HETP H with a reduced value h (dimensionless) and the linear velocity u with a reduced linear velocity v (also dimensionless) given by the expressions:

$$ h = \frac{H}{d_p} \quad \text{and} \quad v = \frac{u d_p}{D} = \frac{4 U d_p}{\pi \varepsilon^* d^2 D} \quad (2.2.13) $$

With these notations, the Knox equation is written in the form:

$$ h = A'' v^{1/3} + B'' \frac{1}{v} + C'' v \quad (2.2.14) $$

Knox curves, as expected, have similar features with van Deemter curves, but at higher flow rates the increase of h (and therefore of H) is less pronounced. The graph of a Knox dependence with $A'' = 1$, $B'' = 2$ and $C'' = 0.05$ is shown in Figure 2.2.2. These values were chosen because they are close to many current analytical HPLC columns.

The values for the parameters A'', B'', and C'' in Figure 2.2.2. are common for columns with particle size 5 µm, and in optimum flow conditions it can be seen that $h \approx 2$. Based on rel 2.2.13, this indicates that the plate height H can be roughly approximated with the value $H \approx 2d_p$.

Even the Knox equation does not account for all the effects contributing to peak broadening. For example, it was shown that the mobile phase viscosity also influences some of the processes previously considered [18], and this aspect is not included (in explicit form) in rel. 2.2.8 or 2.2.14. The implicit dependence of H on the mobile phase viscosity results from the fact that the diffusion coefficient D is dependent on liquid phase

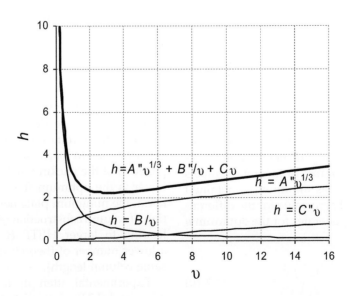

FIGURE 2.2.2 The plot of Knox equation for $A'' = 1$, $B'' = 2$ and $C'' = 0.05$. Individual contributions of different terms are also shown.

viscosity by the Stokes equation, which has the form:

$$D = \frac{RT}{6\pi\beta\,\eta\,r_{molecule}}\mathcal{N}^{-1} \qquad (2.2.15)$$

where η is the dynamic viscosity, β is a correction constant (close to 1), $r_{molecule}$ is the radius of the molecule (or particle), and $\mathcal{N}$ is Avogadro number ($\mathcal{N} = 6.02214179 \times 10^{23}$ mol^{-1}) (the units for dynamic viscosity are poise P, and 1 P = 0.1 Pa x s).

Other effects, such as extra column broadening, are not included in these equations [19]. Also, the processes taking place during the separation are considered ideal, and effects such as overloading of the stationary phase with sample that cannot be retained are not considered in the theory.

The applicability of the previous theory to monolith columns is a problem because all the parameters used for obtaining van Deemter and Knox equations are based on columns packed with porous (or core-shell) particles. The extension of these equations to describe monolith columns has been attempted using the concept of "domain" size [20, 21].

Kinetic Plots

Based on the van Deemter equation, a relation between the flow rate u used in a separation and the number of theoretical plates N can be immediately obtained ($N = L/H$ where L is the length of the column). In practice, a large number of theoretical plates N may be necessary for a separation, but at the same time a higher flow rate u is desirable in order to achieve a shorter elution time. Elution time depends on the nature of each separated compound, but information on how fast a chromatographic separation is can be obtained from the value of t_0 of an unretained compound. This is a consequence of rel. 2.1.7 where the retention of a given compound t_R can be obtained from t_0 by the expression $t_R = t_0\ (k + 1)$. The variation of theoretical plate number N as a function of elution time t_0 can be obtained by replacing in the van Deemter equation the values $u = L/t_0$ and $H = L/N$. With these substitutions, the van Deemter equation can be written in the form:

$$N = \frac{L^2 t_0}{ALt_0 + Bt_0^2 + CL^2} \qquad (2.2.16)$$

Such a dependence is shown in Figure 2.2.3 for three columns 50 mm, 100 mm, and 150 mm long with irregular particles of 5 μm. As shown in Figure 2.2.3, the optimum number of theoretical plates is obtained at different elution times and therefore at different flow

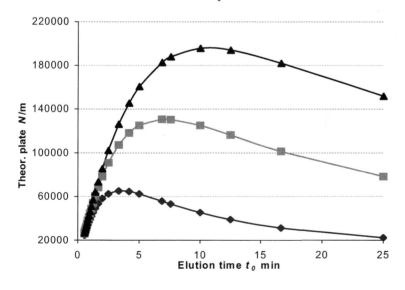

FIGURE 2.2.3 Dependence of the number of theoretical plates N/m for a column as a function of the elution time t_0 for three columns 50 mm, 100 mm and 150 mm long with irregular particles of 5 μm.

rates. These optimums can be unacceptable for a practical separation. For example, for the 150-mm column length the optimum value for N is reached at about 12.5 min, and a compound with capacity factor $k = 5$ will elute at about 50 min. For this reason, a higher flow rate is preferable even if the number of theoretical plates is diminished. However, the flow rate in a chromatographic column cannot be increased indefinitely, being limited by the increase in the pressure drop across the column Δp_{max}. The difference Δp between the pressure at the column inlet and that at the outlet of the column (filled with porous particles) when the linear flow is u is given by the following expression (the Darcy equation):

$$\Delta p = \frac{\eta u \phi_r L}{d_p^2} = \frac{\eta \phi_r L^2}{d_p^2 t_0} = \frac{\eta u \phi_r N H}{d_p^2} \quad (2.2.17)$$

where η is the mobile-phase viscosity, L is column length, d_p is the diameter of the particles in the bed, and ϕ_r is the column flow resistance factor. This type of variation of Δp as a function of the retention time t_0 (for $\eta = 2$ mPa/s and $\phi_r = 1$ mm^2) is shown in Figure 2.2.4 for the same three columns considered for Figure 2.2.3.

By setting a limiting backpressure for example, to about 300 bar (30,000 kPa), the limit of how low the elution time can be set can be calculated. For obtaining information on N at this limit of elution time (backpressure) for different columns (different length or different stationary phases), it is necessary to generate graphics displaying the variation of N for the points where each column reaches the limiting backpressure. These graphs are indicated as kinetic plots. One type of kinetic plot is the Poppe plot [22]. This is a plot of log (t_0/N) as a function of log N where t_0 is selected for a specific Δp value (maximum acceptable for the evaluated system). Other similar types of kinetic plots

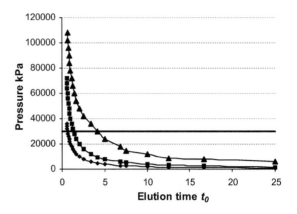

FIGURE 2.2.4   Variation of back pressure (pressure drop across the column) at different elution times t_0 for the three columns 50 mm, 100 mm and 150 mm long with irregular particles of 5 μm. A horizontal line is shown for 300 bar. (The variation of N as a function of elution time is shown in Figure 2.2.3).

were reported in the literature [23]. Since the maximum acceptable column backpressure Δp for a common silica based column can be set relatively high (e.g., 300 bar or for UPLC, 500–600 bar), the kinetic plots do not have a high utility. However, in size-exclusion chromatography, the maximum acceptable pressure is typically lower since this technique uses stationary phases that can collapse at higher pressures. For this reason, kinetic plots in SEC are more frequently useful.

Peak Asymmetry

Regardless of whether the variance σ^2 in peak broadening was caused by diffusion, eddy effects, or mass transfer effects, it was assumed that it characterizes a perfectly random process following a Gaussian distribution. However, it is not uncommon in practice to have HPLC peaks that do not have a perfect Gaussian shape. This deviation is particularly common for separations that take place in gradient conditions when various factors, including the value for D, are changing across the peak. Peak asymmetry was previously indicated as being measured by Ω_3 momentum known as *skew* (see rel. 2.1.69). The peaks may show either "tailing" (peak **a** in Figure 2.2.5) or fronting (peak **c** in Figure 2.2.5). A Gaussian peak (peak **b**) is also shown in Figure 2.2.5.

Besides the value for Ω_3 that can be calculated with the modern data processing capabilities of HPLC systems, two simpler parameters can be used for the characterization of peak asymmetry. One is the asymmetry parameter As (X), which is measured on the chromatographic peak of compound X by taking a perpendicular from the peak maximum and a parallel to the baseline at 10% peak height. The asymmetry is defined as the ratio of the rear (r) to front (f) segments cut on this parallel by the chromatographic peak and the perpendicular as shown in Figure 2.2.5 for peak **a**, and:

$$As(X) = \frac{r}{f} \qquad (2.2.18)$$

In addition to peak asymmetry parameter As (X), an additional parameter referred to as "tailing factor" TF (X) is sometimes used. The peak tailing is defined by the formula:

$$TF = \frac{f' + r'}{2f'} \qquad (2.2.19)$$

where f' and r' are measured in the same way as f and r, but at 5% of the peak height, and it is also compound dependent.

The asymmetry is generated in chromatography by various factors. One common cause is column overloading. In this case, due to the excessive amount of sample, the stationary phase is covered or saturated with the sample; the separation process is far from ideal, and a tail is formed $(r/f > 1)$. This type of tailing can be identified by the "triangular" peak shape (the tail does not have a slow "exponential-type" decrease). Also, when more than one type of interaction of an analyte with the stationary phase takes place, for example, due to "active spots" in the path of the analyte, peak "tailing" may take place $(r/f > 1)$. Peak "fronting" is less common, but it can be seen in separations with more than one type of interaction solute versus stationary phase.

The tailing or fronting of a chromatographic peak may adversely affect the separation. Corrections for the calculation of plate number N (see rel. 2.1.44) in case of peak tailing have been suggested [24]. Another problem regarding peak tailing or fronting is related to peak integration for the determination of peak areas. Modern HPLC systems depend on computer data processing capability for the detection and integration of chromatographic peaks for area measurement. Peaks that have tails or fronting are less accurately integrated by these programs, leading to errors in quantitation. The use of peak height for quantitation is also affected by tailing or fronting. When the peak shape in the chromatogram of standards used for a calibration is different (better) than the peak

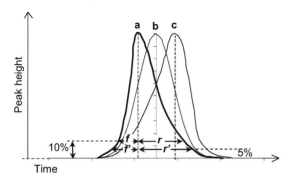

FIGURE 2.2.5 Peaks of three different shapes: **a** showing tailing, **b** with Gaussian shape, **c** showing fronting. (rear (r) and front (f) shown for peak **a**)

shape in the chromatogram of the real life samples, larger errors in accuracy of the analysis occur.

The most common cause of peak asymmetry is the existence on the stationary phase of more than one type of moiety that can lead to different types of interactions with the analytes. For example, on a silica base column that has C18 groups attached on its surface, it is possible that silanol groups ($\equiv$Si-OH) are also present. The interaction with the two types of groups may lead to peak asymmetry, and values of As around 1.5 are not uncommon for analytical HPLC columns. The manufacturer typically gives the asymmetry parameter for chromatographic column characterization. Peak asymmetry being measured for the peak of a specific compound is also a compound-dependent parameter.

Application of van Deemter Equation

The Van Deemter equation can be practically utilized for determining the optimum flow rate in a column in order to achieve maximum performance regarding the theoretical plate number N, as well as to verify the performance of a column when new or after a number of injections during its utilization. To verify the performance of a column, a "test mixture" of compounds can be used under specified conditions. An example for such a test mixture used for the performance evaluation of reverse-phase columns contains 6 μg/mL uracil, 10 μg/mL toluene, 25 μg/mL fluorene, and 40 μg/mL fluoranthene. The separation is recommended in isocratic conditions, 30% water and 70% acetonitrile. The separation should be repeated at several flow rates (e.g., between 0.4 mL/min and 2.0 mL/min, with a step increase of 0.2 mL/min), and the parameters t_0, t_R, and W_b are measured on the chromatograms (t_0 is typically taken from the retention time of uracil that is assumed not retained on the column). From the length of the column, the linear flow rate is calculated using $u = L/t_0$, the theoretical plate number N is calculated using rel. 2.1.44, and H (HETP) is calculated using rel. 2.1.42. The graph representing H as a function of u can be obtained. Having three different compounds in the test mixture, the values obtained for N and H will be slightly different, but they should not differ

FIGURE 2.2.6 Example of chromatograms obtained at different flow rates from a test mixture for the verification of a Zorbax Eclipse XDB-C8 new column performance in isocratic conditions, using van Deemter equation (see text for conditions).

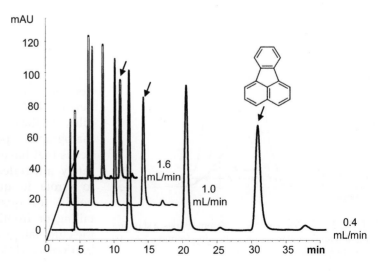

considerably. The results of such an experiment are shown as an example in Figure 2.2.6 for a Zorbax Eclipse XDB-C8 column, 3 µm particle size, 150 mm length, 4.6 mm i.d. The column was kept at 25 °C. The peak detection was done using UV absorption at 254 nm. The resulting chromatograms are shown in Figure 2.2.6 (only three traces shown from 7 measured). The experiment was performed for the new column and also after 800 injections of biological samples. Following the calculation of u from the retention time of uracil for different volumetric flow rates U and calculation of H from the broadening W_b of fluoranthene peak, the results were plotted and generated the graphs shown in Figure 2.2.7. The results shown in Figure 2.2.7 indicate that an optimum value for u is around 6.75 cm/min. From rel. 2.2.3 it can be written:

$$U = \frac{\varepsilon^* \pi d^2}{4} u \qquad (2.2.20)$$

Assuming for the column $\varepsilon^* \approx 0.7$, rel. 2.2.20 gives an optimum $U \approx 0.8$ mL/min (the i.d. of the column is 0.46 cm). Some degradation of column performance after 800 injections of

samples is also observed, although the H value for the old column is still relatively close to the initial value. Optimization of flow rate based on the van Deemter equation in order to obtain the best N values is not the only optimization of practical use. In practice, short run-times are also desirable. For this reason it is common that the flow rate in a column is selected higher than is indicated by the van Deemter equation, as long as the separation is still good and the column back pressure is acceptable. The flow rate is typically limited by the column backpressure (or HPLC maximum pressure) and not by the deviation from maximum column efficiency. However, column selection and flow rate optimization can be combined for obtaining optimum separation results [25].

Using $H = 15.5$ µm and $L = 150$ mm in rel. 2.1.42, the resulting value for the theoretical plate number N is 9677 and the effective plate number n can be calculated using the expression:

$$n = N \frac{t'^2_R}{t^2_R} \qquad (2.2.21)$$

For the evaluated column, n (for fluoranthene) is about 7354. The column efficiency n per meter for the column is about 50,000 (indicating a good column). A higher n would be generated with an earlier eluting peak such as toluene. The effective plate height n or the theoretical plate height N are typically indicated for commercially available chromatographic columns (as measured for a standard compound, which is usually toluene or ethylbenzene for C18 columns). As an example, the values for N for a number of C18 columns are listed in Table 2.2.1.

Peak Characterization Using Exponentially Modified Gaussian Shape

The Gaussian shape of a chromatographic peak is due mainly to the random spreading

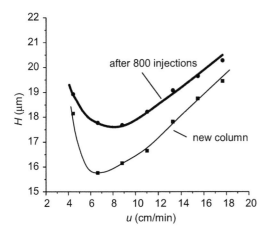

FIGURE 2.2.7 Plot of H as a function of u for a Zorbax Eclipse XDB-C8 column (see text for conditions).

TABLE 2.2.1   List of Theoretical Plate number N per m for Several C18 Columns that are Commercially Available (for Toluene in 10% H_2O 90% CH_3OH)

Column	N per m	Column	n per m
ACE C18	111,000	Nucleosil C18	101,000
ACE C18-300	103,500	Nucleosil C18AB	87000
ACE C18-HL	102,000	Partisil ODS	47,500
μBondapak C18	36,000	Partisil ODS2	41,000
Capcell Pak AG C18	51,000	Partisil ODS3	52,000
Develosil ODS-HG	85,500	Prodigy ODS2	48,000
Develosil ODS-MG	66,000	Prodigy ODS3	62,000
Develosil ODS-UG	92,000	Resolve C18	45,500
Exsil ODS	93,000	SunFire C18	91,500
Exsil ODS1	114,000	Symmetry C18	92,000
Gemini C18	75,500	Vydac 218TP	63,000
Hichrom RPB	97,500	Waters Spherisorb ODS1	100,500
Hypersil BDS C18	76,500	Waters Spherisorb ODS2	91,500
Hypersil GOLD	91,000	Waters Spherisorb ODSB	92,000
Hypersil HyPurity C18	73,000	YMC J'Sphere ODS H80	64,500
Hypersil ODS	94,500	YMC J'Sphere ODS M80	58,000
Inertsil ODS	73,500	YMC ODS A	99,500
Inertsil ODS3	60,500	YMC ODS AM	83,500
Inertsil ODS2	32,000	YMC Pro C18	105,000
Kromasil C18	99,000	Zorbax Extend C18	80,500
LiChrosorb RP-18	74,000	Zorbax ODS	85,500
LiChrospher RP-18	80,000	Zorbax Rx-C18	90,500
Luna 5 C18(2)	88,000	Zorbax SB-C18	103,000
Novopak C18	60,000	Zorbax XDB-C18	96,000

of the injection plug in the chromatographic column. However, the dead volumes along the chromatographic system and the viscous flow also contribute to the shape of the chromatographic peak [26]. It was suggested that a better equation indicated as exponentially modified Gaussian (EMG) describes more accurately the true chromatographic peak shape [27–31]. In this modified function, a Gaussian function with the variance σ^2 is combined with an exponential decay depending on a parameter τ. A Gaussian, an

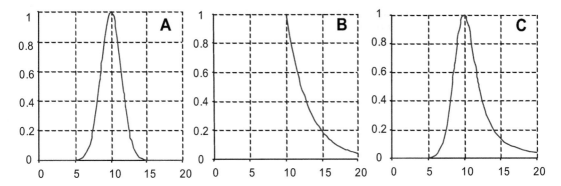

FIGURE 2.2.8 A Gaussian curve (A), an exponential decay (B) and the resulting combined curve (C).

exponential decay, and the combined curve are pictured in Figure 2.2.8. The EMG function describing the peak height (at origin) has the expression:

$$h(t) = \frac{A\sigma}{\tau\sqrt{2\pi}} \exp\left[\frac{1}{2}\left(\frac{\sigma}{\tau}\right)^2 - \frac{t - t_R}{\tau}\right] \int_{-\infty}^{z} \exp\left(-\frac{\xi^2}{2}\right) d\xi$$

(2.2.22)

where:

$$z = \frac{t - t_R}{\sigma} - \frac{\sigma}{\tau}$$

(2.2.23)

In this equation, A is the peak area, σ is the standard deviation of the Gaussian peak, and τ is the time constant of exponential axis. Unfortunately, the calculation of $h(t)$ for any t along the chromatogram using the EMG function given by rel. 2.2.22 is difficult. However, the expression from rel. 2.2.22 can be obtained in terms of error function $\mathrm{erf}(z) = \frac{2}{\sqrt{\pi}} \int_0^z \exp(-\xi^2) d\xi$ which allows an easier calculation, since erf (z) is tabulated and available in computer program packages [30].

From formula 2.2.22, different desired parameters for the chromatographic peak are obtained using the statistical moments (also

calculated using computer packages), as indicated in the following formulas:

$$\Omega_1 = t_R + \tau \quad \Omega_2 = \sigma^2 + \tau^2 \quad \Omega_3 = 2\tau^3$$
$$\Omega_4 = 3\sigma^4 + 6\sigma^2\tau^2 + 9\tau^4$$

(2.2.24)

From these expressions are calculated the following values:

$$skew = \frac{\Omega_3}{(\Omega_2)^{3/2}} \quad excess = \frac{\Omega_4}{(\Omega_2)^2} - 3$$

$$N = \frac{(\Omega_1)^2}{\Omega_2}$$

(2.2.25)

The use of EMG for the deconvolution (using appropriate computer programs) of the chromatographic peaks, in particular of those that are not well resolved, has been proven very efficient [24,32,33].

Summary of Chromatographic Peak Characteristics

A summary of peak characteristics can be obtained by processing the peaks in a chromatogram either manually or more commonly by using the capability of data processing programs from the computer that controls the HPLC. A summary of such characteristics is given in

TABLE 2.2.2   Typical Peak Characteristics Obtained using the Capability of Data Processing Program from the Computer that Controls the HPLC

No.	Characteristic	Notation	Common range
1	Retention time	t_R	1 - 30 min (longer times for complex chromatograms)
2	Void (dead) time	t_0	1 - 3 min
3	Capacity factor	k	2 - 10
4	Peak height	h	Depends on detector settings
5	Peak area	A	Depends on detector settings
6	Peak width at half height	W_h	0.05 to 1 min
7	Peak width at baseline	W_b	0.1 to 2 min
8	Peak start	t_{start}	Depends on t_R and W_b
9	Peak end	t_{end}	Depends on t_R and W_b
10	Skew	Ω_3	0 to 1
11	Excess	Ω_4	0 to 2
12	Symmetry (at 10% height)	As	1 to 1.3
13	Tailing	TF	1 to 1.3
14	Noise at peak baseline		Depends on detector settings
15	Signal to noise ratio	S/N	Depends on detector settings, compound nature and concentration, etc.
16	Integration type		Base to base, base to shoulder, etc.
17	Time increment		Depends on the rate of detector measurements of signal
18	Data points per peak		Depends on the rate of detector measurements of signal
19	Theoretical plate number (plates/ column)	N	4,000 − 40,000
20	Theoretical plate number (plates/meter)	N	30,000 − 300,000
21	Efficiency (plates/column)	n	4,000 − 40,000
22	Efficiency (plates/meter)	n	30,000 − 300,000
23	Foley Dorsey plates/column		N corrected for asymmetry (lower than N)
24	Foley Dorsey plates/meter		N corrected for asymmetry (lower than N)
25	Selectivity to previous peak	α	1.5 - 10 (depending on separation)
26	Selectivity to next peak	α	1.5 - 10 (depending on separation)
27	Resolution to previous peak:	R	1.5 - 15 (depending on separation)
28	Resolution to next peak	R	1.5 - 15 (depending on separation)

Table 2.2.2. The parameters described in this table offer a good description of various aspects of a separation relative to the column characteristics, specific separation quality, and detection.

References

[1] Moldoveanu SC, David V. Sample Preparation in Chromatography. Amsterdam: Elsevier; 2002.

[2] Rimmer CA, Simmons CR, Dorsey JG. The measurement and meaning of void volumes in reversed-phase LC. J. Chromatogr. A 2002;965:219−32.

[3] Karger BL, Snyder LR, Horvath C. An Introduction to Separation Science. New York: John Wiley; 1973.

[4] Berry RS, Rice SA, Ross J. Physical Chemistry. Oxford: Oxford University Press; 2000. pp. 526−527.

[5] Smirnov VI. Kurs Vischei Matematiki, vol. 2. Moscow: Gosudarstvennoe Izd.; 1952.

[6] Neue UD. Theory of peak capacity in gradient elution. J. Chromatogr. A 2005;1079:153−61.

[7] Davis JM, Giddings JC. Statistical theory of component overlap in multicomponent chromatograms. Anal. Chem. 1983;55:418−24.

[8] Poole CF. The Essence of Chromatography. Amsterdam: Elsevier; 2003.

[9] Dolan JW, Gant JR, Snyder LR. Gradient elution in high-performance liquid chromatography. II. Practical application to reversed-phase systems. J. Chromatogr. 1979;165:31−58.

[10] Vissers JPC. Recent developments in microcolumn liquid chromatography. J. Chromatogr. A 1999;856:117−43.

[11] Rieux L, Sneekes E-J, Swart R. Nano LC: Principles, evolution, and state-of-art of the technique. LCGC North America 2011;29:926−34.

[12] Hayduk W, Laudie H. Prediction of diffusion coefficients for non-electrolytes in dilute aqueous solutions. AIChE J 1974;20:611−5.

[13] Lyman WJ, Reehl WF, Rosenblatt DH. Handbook of Chemical Property Estimation Methods. Washington, DC: ACS; 1990.

[14] Reif F. Statistical Physics. New York: McGraw-Hill Book Co.; 1967.

[15] van Deemter JJ, Zuiderweg FJ, Kinkenberg A. Longitudinal diffusion and resistance to mass transfer as causes of nonideality in chromatography. Chem. Eng. Sci. 1956;5:271−89.

[16] Bristow PA, Knox JH. Standardization of test conditions for high performance liquid chromatography columns. Chromatographia 1977;10:279−89.

[17] Knox JH. In: Brown PR, Grushka E, editors. Advances in Chromatography, vol. 38. New York: Marcel Dekker; 1998.

[18] Giddings JC. Comparison of theoretical limit of separating speed in gas and liquid chromatography. Anal. Chem. 1965;37:60−3.

[19] Oláh E, Fekete S, Fekete J, Ganzler K. Comparative study of new shell-type, sub 2-µm fully porous and monolith stationary phases, focusing on mass-transfer resistance. J. Chromatogr. A 2010;1217:3642−53.

[20] Manakuchi H, Nagayama H, Soga N, Ishizuka N, Tanaka N. Effect of domain size on the performance of octadecylsilylated continuous porous silica columns in reversed phase liquid chromatography. J. Chromatogr. A 1998;797:121−31.

[21] Vervoort N, Gzill P, Baron GV, Desmet G. Model column structure for the analysis of the flow and band broadening characteristics of silica monoliths. J. Chromatogr. A 2004;1030:177−86.

[22] Poppe H. Some reflections on speed and efficiency of modern chromatographic methods. J. Chromatogr. A 1997;778:3−21.

[23] Desmet G, Clicq D, Gzil P. Geometry-independent plate height representation methods for the direct comparison of the kinetic performance of LC supports with different size morphology. Anal. Chem. 2005;77:4058−70.

[24] Foley JP, Dorsey JG. Equations for calculation of chromatographic figures of merit for ideal and skewed peaks. Anal. Chem. 1983;55:730−7.

[25] Meyer VR. The clever use of pressure in column liquid chromatography. Chromatographia 2010;72:603−9.

[26] Sternberg JC. In: Giddings JC, Keller RA, editors. Advances in Chromatography, vol. 2. New York: Marcel Dekker; 1966.

[27] Schmauch LJ. Response time and flow sensitivity of detectors for gas chromatography. Anal. Chem. 1959;31:225−30.

[28] Grushka E. Characterization of exponentially modified Gaussian peak in chromatography. Anal. Chem. 1972;44:1733−8.

[29] Dyson NA. Chromatographic Integration Methods. Cambridge, UK: Royal. Soc. Chem.; 1998.

[30] Naish PJ, Harwell S. Exponentially modified Gaussian function—a good model for chromatographic peaks in isocratic HPLC? Chromatographia 1988;26:285−96.

[31] Cai CP, Wu NS. Statistical moment analysis and deconvolution of overlapping chromatographic peaks. Chromatographia 1991;31:595−9.

[32] Foley JP. Equations for chromatographic peak modeling and calculation of peak area. Anal. Chem. 1987;59:1984−7.

[33] Foley JP, Dorsey JG. A review of the exponentially modified Gaussian (EMG) function: Evaluation and subsequent calculation of universal data. J. Chromatogr. Sci. 1984;22:40−6.

3.1. PARTITION EQUILIBRIUM

General Comments

In Section 1.1 it was indicated that HPLC separation is based on the equilibrium established between the molecules from the mobile phase and those present in the stationary phase, and that different types of equilibria are used for achieving separation in HPLC. This chapter presents a more detailed description of types of equilibria acting in HPLC separation. Each equilibrium is dependent on the analyte retention, which is commonly described in chromatography by the capacity factor k. The formal dependence of k on thermodynamic variables will be discussed in this chapter. Also, the influence on the equilibrium of parameters such as pH and temperature is described in further detail.

Liquid-Liquid Partition

Liquid-liquid partition represents the equilibrium type of various separations. These include solvent extractions and many HPLC separations. In liquid-liquid partition the component i is distributed between two nonmiscible liquid phases A and B in an equilibrium of the type:

$$i_B \rightleftarrows i_A \qquad (3.1.1)$$

In HPLC, the phase B represents the mobile phase, and A can be also considered a liquid that is immobilized on a solid support (the nature of A and B is different). A typical example of a liquid immobilized on a solid support is that of water on silica. In this case, the water molecules are "bound" to the solid silica frame by hydrogen bonds to the silanol groups from the surface (phase A). This type of stationary phase and a mobile phase exemplified as acetonitrile (phase B) are schematically shown in Figure 3.1.1.

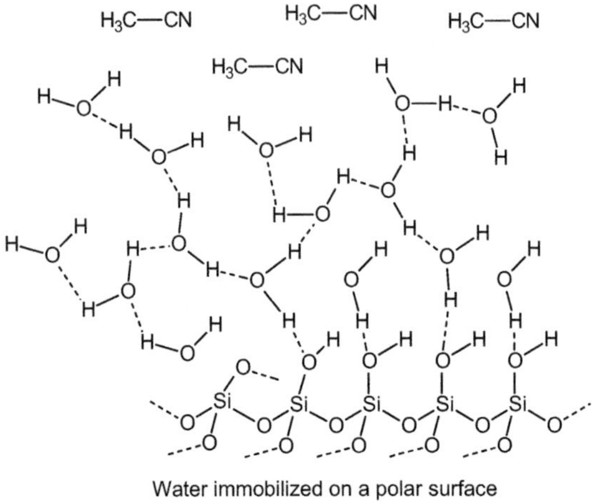

Water immobilized on a polar surface

FIGURE 3.1.1 Schematic description of water immobilized on a silica surface and acetonitrile as a mobile phase.

HPLC is more frequently practiced using a stationary phase made of a solid support covered with a bonded phase. If this bonded phase is equated with an immobilized liquid, the formal theory of liquid-liquid partition can be successfully applied for the description of the equilibrium taking place in different types of HPLC.

In the theory regarding the equilibrium, it is assumed that no "nonequilibrium" effects take place. For a sufficiently large amount of compound i, it is possible that the equilibrium cannot be attained due to the saturation of the stationary phase. Such cases are not discussed here, and a sufficient small amount of solute is assumed to participate to the equilibrium, such that the process is close to "ideal".

The detailed energetics of the process, describing the interactions at the molecular level in this distribution equilibrium, is not taken into consideration in the present treatment. When equilibrium is attained for "distributing" the compound i between phases A and B, the difference between the chemical potentials $\mu_{i,A}$ and $\mu_{i,B}$ of the component i in each of the two phases

A and B must be zero. This can be written in the following form:

$$\mu^0_{i,B} + RT \ln a_{i,B} = \mu^0_{i,A} + RT \ln a_{i,A} \quad (3.1.2)$$

In rel. 3.1.2, $\mu^0_{i,A}$ and $\mu^0_{i,B}$ are the standard chemical potentials of compound i, and $a_{i,A}$ and $a_{i,B}$ are the activities of analyte i in the two phases A and B, respectively, T is absolute temperature, and R is the gas constant ($R = 8.31451$ J deg^{-1} mol^{-1} = 1.987 cal deg^{-1} mol^{-1}). The rearrangement of rel. 3.1.2 leads to the expression:

$$\ln (a_{i,A}/a_{i,B}) = -(\mu^0_{i,A} - \mu^0_{i,B})/(RT)$$
$$= -\Delta\mu_i^0/(RT) \quad (3.1.3)$$

In rel. 3.1.3, $\Delta\mu_i^0$ represents the change in standard chemical potential when solute i is transferred from the mobile phase B into the stationary phase A. Relation 3.1.3 shows that for a constant temperature T the ratio of the activities of the analyte i in the two phases is always a constant expressed as follows:

$$K_i = a_{i,A}/a_{i,B} = \exp\left[-(\mu^0_{i,A} - \mu^0_{i,B})/(RT)\right] \quad (3.1.4)$$

The constant K_i is the *thermodynamic distribution constant* for the partition process of component i between liquid phases A and B (regardless of whether phase A, for example, is immobilized on a solid support).

The activities can be expressed by the product between activity coefficients γ and the molar concentrations c ($a = \gamma c$) such that rel. 3.1.4 can be written as follows:

$$K_i = (\gamma_{i,A} c_{i,A})/(\gamma_{i,B} c_{i,B}) = \exp\left[-\Delta\mu_i^0/(RT)\right] \quad (3.1.5)$$

Relation 3.1.5 can be written using base 10 logarithms in the form:

$$\log_{10} K_i = -\Delta\mu_i^0/(2.302\, RT) \quad (3.1.6)$$

Instead of using activities, it is more convenient to use molar concentrations and to introduce the formula:

$$K_i = c_{i,A}/c_{i,B} \quad (3.1.7)$$

The equilibrium constant K_i for the molecular species i between two phases A and B is typically called the *partition constant*. In HPLC, the partition constant is less utilized since the independent experimental measurement of K_i is not done frequently and the capacity factor $k_i = K_i \Psi$ is the common parameter characterizing the HPLC separations (see rel 2.1.13).

With formula 3.1.7, the relation between K_i and K_i is the following:

$$K_i = [(\gamma_{i,A})/\gamma_{i,B}] K_i \quad (3.1.8)$$

which shows that K_i is also a constant (depending on temperature and the nature of phases A and B and of compound i). The expression for K_i can be written as follows:

$$K_i = [(\gamma_{i,B})/(\gamma_{i,A})] K_i \quad (3.1.9)$$

and also:

$$K_i = [(\gamma_{i,B})/(\gamma_{i,A})] \exp\left[-\Delta\mu_i^0/(RT)\right] \quad (3.1.10)$$

The constant K_i is commonly used to characterize analytical partition processes. It gives the ratio of the concentrations of i in phases A and B at equilibrium, as compared to K_i that gives the ratio of activities. The value of K_i is dependent on temperature and also on the chemical nature of component i and of the two solvents A and B, although no chemical interaction is assumed in the system. Tables with K_i values for different systems and temperatures are available, and extra-thermodynamic techniques for practical estimations are also reported. For complex problems, an *effective equilibrium constant* K^{eff} is sometimes used instead of individual values for each equilibrium. The theory of liquid-liquid partition can be used to explain chromatographic processes where the stationary phase is assumed to be a liquid or a liquid-like material immobilized on an inert support. However, in many instances the classification of an HPLC separation as *partition chromatography* is more or less arbitrary. For this reason, the use of some of the parameters developed for liquid-liquid partition must be done only as an approximation.

For a constant pressure and temperature, $\Delta\mu^0 = \Delta G^0$, where ΔG^0 is the variation in the standard free enthalpy. The dependence of K_i on the variation of free enthalpy can be written in the form:

$$K_i = [(\gamma_{i,B})/(\gamma_{i,A})] \exp [-\Delta G_i^0/(RT)] \quad (3.1.11)$$

In diluted systems, the activity coefficients γ are very close to 1. With the common expression $\Delta G^0 = \Delta H^0 - T \Delta S^0$ where ΔH^0 and ΔS^0 are the standard enthalpy and entropy changes for the transfer of the analyte from the mobile to the stationary phase, the distribution constant can be written in the form:

$$K_i = \exp [-\Delta G_i^0/(RT)] = \exp [(-\Delta H_i^0 + T \Delta S_i^0)/(RT)] \quad (3.1.12)$$

Formula 3.1.12 can be used in deriving the dependence of capacity (retention) factor k on temperature and thermodynamic parameters. From rel. 2.1.13 showing that $k = K\Psi$, the following formula can be written:

$$\ln k = (-\Delta H^0 + T \Delta S^0)/(RT) + \ln \Psi \quad (3.1.13)$$

Expression 3.1.13 is known as the van't Hoff equation (sometimes rel. 3.1.13 is expressed as a logarithm in base 10 or log k, when the factor 1/2.303 multiplies the left side of the equality). This equation is useful in the estimation of thermodynamic parameters ΔH^0 and ΔS^0 for the transfer of the analyte from the mobile to the stationary phase, using experimental values of ln k at different column temperatures. By taking the derivative of ln k as a function of (1/T) (assuming ΔH^0, ΔS^0

and Ψ not dependent on temperature), the result is as follows:

$$\frac{d(\ln k)}{d(1/T)} = -\frac{1}{R} \Delta H^0 \quad \text{and} \quad \Delta H^0 = -R\frac{d(\ln k)}{d(1/T)} \quad (3.1.14)$$

Replacing ΔH^0 from rel. 3.1.14 in rel. 3.1.13, the formula for ΔS^0 can be immediately obtained when ln Ψ is known.

From rel. 2.1.16 and 3.1.12 an expression for the retention volume V_R can also be obtained as a function of standard enthalpy ΔH^0 and entropy ΔS^0 changes for the transfer of the analyte from the mobile to the stationary phase. This expression is the following:

$$V_R = V_0 + \exp [(-\Delta H^0 + T \Delta S^0)/(RT)] V_{st} \quad (3.1.15)$$

where V_{st} is the volume of the immobilized liquid on a solid support and V_0 is the volume of the mobile phase in the chromatographic column.

Since the value for ln Ψ in rel. 3.1.13 is not usually known, rel. 3.1.12 can be used more successfully for establishing a formula for the selectivity α in a separation process. From rel. 2.1.49 for the selectivity α, and from rel. 2.1.13 between the partition coefficient and the capacity factor, it can be seen that selectivity is also given by the formula:

$$\alpha = \frac{K_i}{K_j} \quad (3.1.16)$$

where the two species to be separated are indicated now by i and j ($K_i > K_j$). With rel. 3.1.12 for K, rel. 3.1.16 can be written in the form:

$$\ln \alpha = (-\Delta G_i^0 + \Delta G_j^0)/(RT)$$
$$= -\Delta(\Delta G^0)/(RT) \quad (3.1.17)$$

Formula 3.1.17 indicates that the selectivity in partition chromatography is given by the difference in the free enthalpies for the transfer of the solutes i and j from the mobile phase to the stationary phase.

Distribution Coefficient

The previous discussion regarding distribution of component i between the two nonmiscible liquid phases A and B is based on the assumption that the analyte i participates in the partition as a sole species and that it is not involved in any secondary chemical equilibrium. Although the analytes are not supposed to react during the HPLC separation, one common case of secondary chemical equilibrium is that in which a compound is present as several chemically modified species, but it is still identified as compound i. This type of equilibrium may include tautomerism, dimerization, ionization (electrolytic dissociation), ion-pairing, or complexation. For example, a simple acid can be present as molecular species RCOOH, or as $RCOO^-$ ion, and it is still identified as a unique compound. For such cases when the analyte i participates in other equilibria represented by species i_1, i_2, ..., i_n, all being still identified as compound i, the entire partition process is described better by a global parameter D_i given by the expression:

$$D_i = \frac{c_{i_1,A} + c_{i_2,A} + \ldots + c_{i_n,A}}{c_{i_1,B} + c_{i_2,B} + \ldots + c_{i_n,B}} \quad (3.1.18)$$

Parameter D_i, is named the *distribution coefficient*. (*Note:* The name *coefficient* and the name *constant* are sometimes confused, and the difference must result from the meaning of the parameter. Also, the *partition* and *distribution* names are sometimes used interchangeably.)

A system in which the compound i is present in equilibrium of two forms, indicated as i_1 and i_2, and is the subject of a chemical change in equilibrium in phase B only is shown in Figure 3.1.2. For the equilibrium shown in Figure 3.1.2, the distribution coefficient can be written in the form:

$$D_i = \frac{c_{i_1,A} + c_{i_2,A}}{c_{i_1,B} + c_{i_2,B}} = \frac{K_{i_1} + K_{i_2}K_{1,2}}{1 + K_{1,2}} \quad (3.1.19)$$

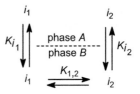

FIGURE 3.1.2 Equilibrium between phase A and phase B for a compound i present in two forms, i_1 and i_2, and subject of a chemical change in equilibrium in phase B.

The second side of expression 3.1.19 can be easily obtained by dividing each term of the middle side of the equation by $c_{i_1,B}$. The expression for the partition constants $K_{i,1}$ and $K_{i,2}$ are given by expressions of the form 3.1.7, and the equilibrium constant $K_{1,2} = c_{i_2}/c_{i_1}$ in phase B.

The individual constants in rel. 3.1.19 can be expressed as a function of standard enthalpies as follows: $K_{i_1} = \exp[-\Delta G^0_{i_1}/(RT)]$, $K_{i_2} = \exp[-\Delta G^0_{i_2}/(RT)]$, and $K_{1,2} = \exp[-\Delta G^0_{1 \leftrightarrow 2}/(RT)]$ where $\Delta G^0_{i_1}$ and $\Delta G^0_{i_2}$ are the free standard enthalpies for the retention of forms i_1 and i_2, respectively, and $\Delta G^0_{1 \leftrightarrow 2}$ is the free standard enthalpy for the equilibrium between i_1 and i_2.

However, a global equilibrium of total i in the mobile phase and in the stationary phase can be conceived as governed by the total free enthalpy ΔG^0_i. Also, since the capacity (retention) factor k for a particular type of analyte i (with more forms) depends on the fraction of molecules of compound i that are present in the mobile phase regardless of the form, rel. 2.1.13 must be replaced in case of a compound present in various forms with the expression:

$$k_i = D_i \Psi \quad (3.1.20)$$

where Ψ is the phase ratio. In logarithmic form (in base 10), rel. 3.1.20 will become:

$$\log k_i = \log D_i + \log \Psi \quad (3.1.21)$$

Peak Shape in Partition Chromatography

One application of the theory of partition chromatography is related to the shape of the chromatographic peak. As shown previously, the distribution constant K_i for a given system appears to depend only on temperature. The graph representing $c_{i,A}$ as a function of $c_{i,B}$ at a given temperature is called an isotherm and has the equation:

$$c_{i,A} = K_i c_{i,B} \qquad (3.1.22)$$

The isotherm is a straight line when K_i is a true constant. In this case, at a constant temperature, the equilibrium process maintains a constant ratio between $c_{i,A}$ and $c_{i,B}$ during the separation. However, in some cases K_i varies with $c_{i,A}$ and $c_{i,B}$, and the isotherm is no longer a straight line.

When $c_{i,A}$ represents the concentration of species i in the stationary phase, and $c_{i,B}$ the concentration of i in the mobile phase, rel. 3.1.22 is identical with rel. 2.1.11 (where species X is now indicated as i). As shown in Section 2.1, the retention time $t'_R(i)$ is given by expression 2.1.15, which can be written in the form:

$$t'_R = K_i \, \Psi \, t_0 \qquad (3.1.23)$$

Relation 3.1.23 indicates that for a chromatographic separation where K_i is a true constant, the concentrations of $c_{i,A}$ and $c_{i,B}$ do not affect K_i and the retention time $t'_R(i)$ is not affected by the concentration of the analyte. However, when K_i varies with the concentration of the analyte, the result will be that the retention time will also be modified. This is going to happen across the chromatographic peak where the analyte concentration starts at zero before the peak reaches a maximum at the peak apex, and then decreases again to zero. This change in the retention time is not large, but in some cases it can be large enough to modify the shape of the peak from the Gaussian shape to peaks showing fronting or tails. For example, it can be assumed that for a given analyte at the beginning of the peak $K = K^{\text{ideal}}$ (and does not depend on the concentration), but at the apex of a peak $K < K^{\text{ideal}}$. In this case the retention time t'_R for the apex will be slightly shorter than ideal and the peak will show deviation from Gaussian

shape and tailing. The reverse will happen for cases when at the apex of a peak $K > K^{\text{ideal}}$, and the peak will show deviation from Gaussian shape and fronting.

Evaluation of Capacity Factor k from Liquid-Liquid Distribution Constants

Different studies on liquid-liquid extraction (see, e.g., [1]) showed that the partition constants $K_{i,A}$ and $K_{i,B}$ for a compound i in two systems, (a) solvent A/water and (b) solvent B/water, are related by the expression:

$$\log K_{i,A} = a \log K_{i,B} + b \qquad (3.1.24)$$

where a and b are constants that are characteristic for the two solvents that are utilized and can be obtained using best regression fit for a large number of compounds. Relation (3.1.24) gives excellent results for systems that are somewhat similar such as octanol/water and hexanol/water. This type of correlation becomes weaker for systems that are very different. Further studies [2–4] showed that this type of correlation can be extended from liquid-liquid extraction data to values for the capacity factor k_i for chromatographic systems. The correlation between $K_{i,A}$ and k_i (for a specific stationary phase) is similar to rel. 3.1.24 and has the following expression:

$$\log k_i = a \log K_{i,A} + b \qquad (3.1.25)$$

The partition constant $K_{i,A}$ that is of particular interest in HPLC is the partition constant for octanol and water, typically indicated as K_{ow} (or P_{ow}). For this constant, extended information is available in the literature regarding its values, measurement, and applications [4–6]. Also, since K_{ow} describes the partition of a compound i between a polar solvent (water) and a solvent with a large hydrophobic moiety (octanol), its values are expected to correlate well with the capacity factor for reversed phase liquid chromatography that uses a C8 or C18 stationary phase. This assumption was proven correct for

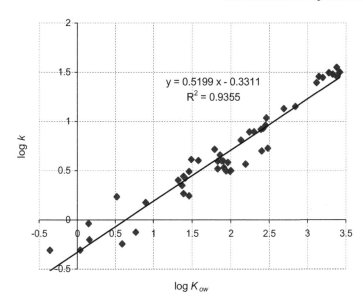

FIGURE 3.1.3 Dependence of log k_i vs. log K_{ow} for 72 mono and disubstituted aromatic compounds with k_i values obtained for a C18 stationary phase with water/methanol 50/50 (v/v) as a mobile phase [2].

numerous compounds, and the following relation is valid.

$$\log k_i = a \log K_{i,ow} + b \qquad (3.1.26)$$

Depending on the stationary phase and the mobile phase utilized, the coefficients a and b are obtained as best regression fit. As an example, the dependence of log k_i versus log K_{ow} for 72 mono and disubstituted aromatic compounds, with k_i values obtained for a C18 stationary phase using water/methanol 50/50 (v/v) as a mobile phase [2], is described by a linear expression of the form 3.1.26. Such a line, with $a = 0.5199$ and $b = 0.3311$, is shown in Figure 3.1.3. As shown in this figure, dependence has $R^2 = 0.9355$ indicating a very good correlation.

3.2. ADSORPTION EQUILIBRIUM

Liquid-Solid Equilibrium

In adsorption HPLC, the molecular species of the solutes from the mobile phase have the capability to be adsorbed on the surface of the solid stationary phase. An equilibrium is assumed to be established between the molecules of the solute (analyte) dissolved in the mobile phase and those adsorbed on the surface of stationary phase. Similar to the case of partition equilibrium, a small amount of solute is assumed to participate in this equilibrium. The stationary phase can be practically considered covered by a monolayer of molecules of either solute i or mobile phase A. The distribution of i between mobile phase and stationary phase can be expressed by the displacement equilibrium:

$$i_{mo} + n\, A_{st} \rightleftarrows i_{st} + n\, A_{mo} \qquad (3.2.1)$$

where the solute i can displace on the stationary phase n molecules of adsorbed phase A and where the index st is used for the stationary phase and mo for the mobile phase A. This type of process is exemplified by activated charcoal that adsorbs organic molecules. In HPLC the adsorption can be exemplified by a silica dry surface (covered with a monolayer of water), which adsorbs a monolayer of a solvent

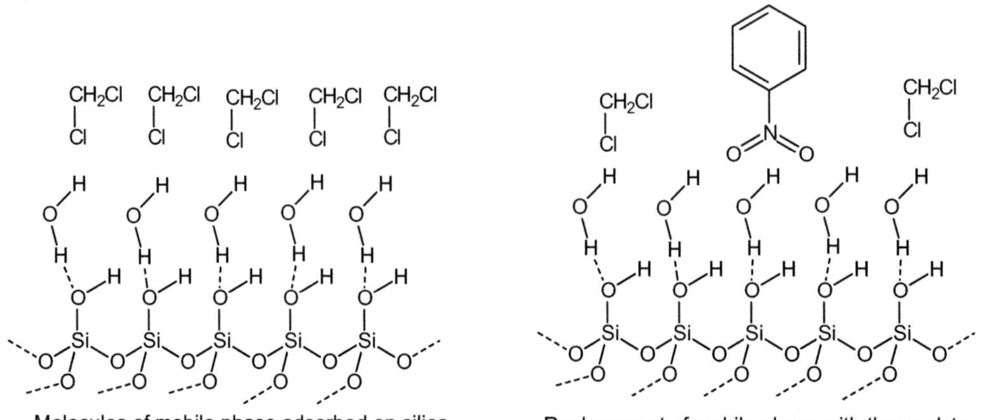

FIGURE 3.2.1 Schematic diagram of the replacement from a monolayer of three molecules of solvent with a molecule of analyte.

(e.g., CH_2Cl_2) followed by the replacement of n solvent molecules by a molecule of an analyte. Figure 3.2.1 shows a schematic picture of this process with nitrobenzene as the analyte molecule.

The process described by rel 3.2.1 can be characterized by a thermodynamic distribution constant K_i given by the ratio (a indicates activities):

$$K_i = (a_{i,st}/a_{i,mo})(a_{A,mo}/a_{A,st})^n \qquad (3.2.2)$$

Several assumptions are necessary to obtain a constant value for K_i (for a given temperature). These assumptions include the following: (1) the surface is uniform; that is, all adsorption sites are equivalent, (2) the adsorbed molecules do not interact such that K_i will be independent of the molecules concentration, (3) all adsorptions take place by the same mechanism, and (4) only a monolayer of molecules is formed even at maximum adsorption.

In practice, the mobile phase usually consists of a mixture of two or even more miscible solvents (the composition of this mixture is kept constant when the separation is performed in isocratic conditions). However, from a solvent mixture (e.g., for solvent A with the composition %A and for solvent B with the composition (100−%A), it can be assumed that only one of the solvents (e.g., A) is adsorbed on the surface of the stationary phase and is replaced by the analyte in an equilibrium of the type 3.2.1. The activity of solvent A in the mobile phase will be indicated by $a_{A,mo}$ and on the stationary phase by $a_{A,st}$. Not all these requirements are always fulfilled, and therefore the assumption that K_i is a constant should be considered only an approximation [7].

From the equality of chemical potentials for the components of the two sides of the equilibrium 3.2.1, the thermodynamic distribution constant can be written as follows (see rel. 3.1.4):

$$K_i = \exp[-(\mu^0_{i,mo} + n\,\mu^0_{A,st} - \mu^0_{i,st}$$
$$- n\,\mu^0_{A,mo})/(RT)] \qquad (3.2.3)$$

In rel. 3.2.3 the typical notation μ for standard chemical potentials was utilized. The two values for the standard chemical potentials in the mobile phase $\mu^0_{i,mo}$ and ($n\,\mu^0_{A,mo}$) are usually equal and cancel each other. This leads to the relation:

$$K_i = \exp[-(n\,\mu^0_{A,st} - \mu^0_{i,st})/(RT)] \qquad (3.2.4)$$

The concentration of solvent A in the mobile phase is not affected by the adsorption process; therefore the equilibrium can be considered only for the solute i between the stationary and mobile phases:

$$i_{mo} \rightleftarrows i_{st} \qquad (3.2.5)$$

This equilibrium is characterized as shown in Section 3.1 by an *equilibrium constant*:

$$K'_i = c_{i,st}/c_{i,mo} \qquad (3.2.6)$$

where $c_{i,st}$ is the number of moles of i per gram of adsorbent and $c_{i,mo}$ is the molar concentration of i in the mobile phase. If the quantity of adsorbed mobile phase per unit weight of adsorbent is indicated as C_A, the relation between K_i and K'_i will be the following:

$$K'_i = K_i C_A \qquad (3.2.7)$$

With rel. 3.2.7 and the assumption $\gamma_{i,st}\,/\,\gamma_{i,mo} = 1$ and $\gamma_{A,st} = 1$, the expression for K'_i becomes:

$$K'_i = \exp[-(n\,\mu^0_{A,st} - \mu^0_{i,st})/(RT)]\,C_A \qquad (3.2.8)$$

Similar to the case of partition type equilibrium, the detailed energetics of the adsorption process is not taken into consideration in this treatment.

Since the equilibrium constant in a separation process is typically difficult to measure, the common parameter describing the chromatographic behavior of a compound i in specific HPLC conditions is the capacity factor k. Using rel. 2.1.13, the expression of k for the compound i can be written in the following form:

$$k_i = K'_i\,\Psi \qquad (3.2.9)$$

Making the assumption that the ratio of activity coefficients $\gamma_{i,st}/\gamma_{i,mo} = 1$ and $\gamma_{A,mo}/\gamma_{A,st} = 1$,

the constant for the equilibrium 3.2.1 can be written in the form:

$$K_i = (c_{i,st}/c_{i,mo})(x_{A,mo}/x_{A,st})^n \qquad (3.2.10)$$

where $c_{i,st}$, for example, indicates the molar concentration of the solute i on the stationary phase, and $x_{A,mo}$ indicates the molar fraction of solvent A in the mobile phase. In rel. 3.2.10, it can be considered that $x_{A,st} = 1$ when the entire surface of the stationary phase is covered with A molecules. By replacing rel. 3.2.6 in rel. 3.2.10 followed by multiplication with Ψ, it can be seen that $\log k_i$ can be written in the form:

$$\log k_i = \log k_i(A) - n\,\log x_{A,mo} \qquad (3.2.11)$$

In rel. 3.2.11, $k_i(A)$ indicates the capacity factor of the solute i in pure solvent A as a mobile phase. Indeed, for the molar fraction $x_{A,mo} = 1$ its logarithm value is zero and $\log k_i = \log k_i(A)$. For lower concentrations (molar fraction < 1) of A in the mobile phase, the logarithm is negative and the capacity factor increases. Relation 3.2.11 is also known as the Soczewinski equation. It is common to express in practice this expression in the form:

$$\log k_i \approx \log k_i(A) - n\,\log \phi_{A,mo} \qquad (3.2.12)$$

where $\phi_{A,mo}$ is the volume fraction of solvent A in a mixture.

Relation 3.2.11 has been sometimes used for differentiating between the partition equilibrium and adsorption equilibrium in a chromatographic process. For stationary phases where a good linearity was noticed between $\log k_i$ and $\log \phi_{A,mo}$, it can be assumed that the chromatographic process is based on adsorption. In practice, however, the dependence curves $\log k_i$ versus $\log \phi_{A,mo}$ may have a variable degree of deviation from linearity, and the interpretation of the results is not always straightforward.

Regardless of the nature of equilibrium type, the expression for K'_i in rel. 3.2.9 is related to the thermodynamic variables of the system by an expression similar to rel. 3.1.12, such that:

$$\ln K' = (-\Delta H^0 + T \Delta S^0)/(RT) \quad (3.2.13)$$

where ΔH^0 and ΔS^0 are the standard enthalpy and entropy changes for the transfer of the analyte from the mobile to the stationary phase expressed by the equilibrium 3.2.5. This indicates that the van't Hoff equation is equally applicable regardless of partition or the adsorption mechanism.

Peak Shape in Adsorption Chromatography

Similar to the case of partition chromatography, the graph representing $c_{i,st}$ as a function of $c_{i,mo}$ at a given temperature is called an isotherm. The isotherm is a straight line when K'_i is a true constant at a constant temperature. However, in some cases K'_i varies with $c_{i,mo}$ and the isotherm is no longer a straight line. This can be seen, for example, at large loads of the column when, because of the limited surface capacity of the stationary phase, some part of

the sample is in contact with the stationary phase and another part is in contact with molecules already adsorbed. Also, when different molecules in a mixture are competing for the stationary phase surface, the adsorption of one component may depend on the adsorption of another component. Isotherms of different shapes deviating from linear have been reported in the literature (see, e.g., [8,9]).

Assuming for simplicity in rel. 3.2.10 that $n = 1$ and that the molar fraction of component i in solution is $x_{i,mo}$, the molar fraction of the solvent in solution will be $x_{A,mo} = 1 - x_{i,mo}$. Also, assuming that the molar fraction of the adsorbed compound i is $x_{i,st}$, the molar fraction of the adsorbed solvent is $x_{A,st} = 1 - x_{i,st}$. In this case, rel 3.2.10 can be written in the form:

$$K_i = \frac{x_{i,st}(1 - x_{i,mo})}{x_{i,mo}(1 - x_{i,st})} \quad (3.2.14)$$

Since typically the value for the molar fraction $x_{i,mo}$ is very low, $1 - x_{i,mo} \approx 1$ and rel. 3.2.14 can be rearranged to give:

$$x_{i,st} = \frac{K_i x_{i,mo}}{1 + K_i x_{i,mo}} \quad (3.2.15)$$

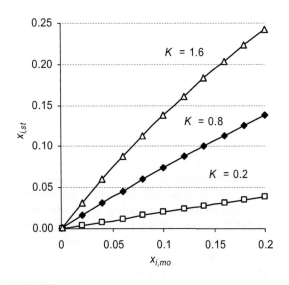
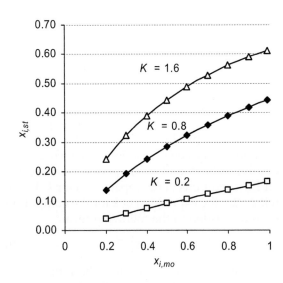

FIGURE 3.2.2 Hypothetic graphics for Langmuir isotherm $K_i = 0.2$, 0.8, and 1.6 for two concentration ranges.

Relation 3.2.14 is known as the Langmuir isotherm. The representation of this isotherm for hypothetic systems with $K_i = 0.2$, 0.8 and 1.6 on two concentration ranges for $x_{i,mo}$ (0 to 0.2, and 0.2 to 0.99) are shown in Figure 3.2.2. From Figure 3.2.2 it can be seen that at low concentrations for the analyte i and for lower K_i values, the dependence of the molar fraction of the adsorbed material practically depends linearly on the concentration in the mobile phase. This is not the case at higher concentrations or higher K_i values when the amount of adsorbed material is higher. Not all adsorption systems are well described by a simple Langmuir isotherm, and other formulas have been developed for a better description of the adsorption process [10]. For a system where dependence between $c_{i,st}$ and $c_{i,mo}$ is linear, the value of K'_i is a true constant and is not affected by the increase in concentration of the analyte.

Since the retention time in a chromatogram is related to the equilibrium, the retention time t'_R (i) for component i is given by expression 2.1.15, which can be written in the form:

$$t'_R = K'_i \Psi t_0 \qquad (3.2.16)$$

However, when K'_i varies with the concentration of the analyte, the peak shape is affected and deviates from a true Gaussian, as already explained in Section 3.1.

3.3. EQUILIBRIA INVOLVING IONS

General Aspects

Most equilibria involving ions are encountered in ion chromatography. Before discussing the separation in ion chromatography, it is important to understand the retention process of ions from solution onto the stationary phase. This retention process is in fact an exchange process between an ion species in the solution and another ion species in the stationary phase connected through ionic interactions with the stationary counterionic groups bonded to the stationary phase. Ion-exchange chromatography is subdivided into cation exchange chromatography and anion-exchange chromatography. In cation-exchange chromatography, the ions M^+ in solution are exchanged with the cations C^+ that were initially retained by stationary ionic groups of the type R-X$^-$. In anion exchange the anionic species B^- in solution are exchanged with the anions A^- that were initially retained by the stationary counterionic groups of the type R-Y$^+$. In this way, the electroneutrality of the system is always maintained (the ions in solution also have counterions). A special type of ion exchange may also take place on zwitterionic stationary phases (ZIC phases containing, for example, a sulfonic and an amino group in the bonded phase) when the mobile phase is a totally aqueous buffer. In this case, only one group is involved in the ion-exchange process, and the phase acts either as cation- or as anion-exchange material.

Retention Equilibrium in Ion Chromatography

The study of ion-exchange equilibria is usually based on a number of simplifying assumptions. One such assumption is that the maximum uptake of ions by an ion-exchange resin is constant and determined by the number of functional groups on the resin matrix [11]. Another assumption is that strict stoichiometric coupling takes place between the different components involved in the ion-exchange process. This implies that each ion from the resin phase is replaced by another one from the liquid phase, with an equivalent charge for maintaining electroneutrality. Also, it is assumed that cations do not penetrate an anion-exchange particle and anions do not penetrate a cation-exchange particle. These assumptions are useful for predicting the ion-exchange behavior of simple, dilute strong electrolyte solutions. However, when dealing with more complex systems, such as weak electrolyte solutions or solutions with high solute concentrations exceeding the resin capacity, the simplifying assumptions are not always valid. For example, the uptake of ions may exceed the total capacity of the resin in case of high concentrations.

For a given ion M^+ in solution and a resin in C^+ form, the exchange equilibrium is the following:

$$resX^- C^+ + M^+ \rightleftarrows resX^- M^+ + C^+ \qquad (3.3.1)$$

The constant for this equilibrium can be written using previous assumptions as follows:

$$K_{M,C} = \frac{[M^+]_{res}}{[M^+]_{mo}} \Big/ \frac{[C^+]_{res}}{[C^+]_{mo}} \qquad (3.3.2)$$

where the index res indicates resin and mo indicates mobile phase. The exchange constant $K_{M,C}$ indicates the degree to which an ion M^+ is preferred in the exchange process, compared to the ion C^+. Larger constants for $K_{M,C}$ indicate higher affinity for the resin of species M^+. The exchange constant can be expressed as a function of a distribution constant between the resin and the solution. The equilibrium described by rel. 3.3.2 can be viewed as equivalent with two independent equilibria:

$$C^+_{mo} \rightleftarrows resX^- C^+ \qquad (3.3.3)$$
$$M^+_{mo} \rightleftarrows resX^- M^+ \qquad (3.3.4)$$

which are described by the constants:

$$K_{res,C} = \frac{[C^+]_{res}}{[C^+]_{mo}} \qquad (3.3.5)$$

$$K_{res,M} = \frac{[M^+]_{res}}{[M^+]_{mo}} \qquad (3.3.6)$$

Each solute has a distribution constant (retention constant) for a specific resin. Using rel. 3.3.5

and 3.3.6, the equilibrium constants 3.3.2 can be expressed in the form:

$$K_{M,C} = K_{res,M}/K_{res,C} \qquad (3.3.7)$$

For a multicharge ion, the equilibrium taking place between the solution and the resin can be written in the form:

$$z \, resX^- C^+ + M^{z+} \rightleftarrows (resX)_z^{z-} M^{z+} + z \, C^+ \qquad (3.3.8)$$

and the equilibrium constant is:

$$K_{M,C} = \frac{[M^{z+}]_{res}}{[M^{z+}]_{mo}} \Big/ \left(\frac{[C^+]_{res}}{[C^+]_{mo}} \right)^z \qquad (3.3.9)$$

Based on rel. 3.3.2 or 3.3.9, it can be seen that the concentration of an ion in solution and the value of the equilibrium constant are important factors in the retention of an ion from solution. In practice, the ion-exchange column must be initially conditioned and made in the form C^+. The ion selected as C^+ is frequently H^+, and in this case rel. 3.3.9 will become:

$$K_{M,H} = \frac{[M^{z+}]_{res}}{[M^{z+}]_{mo}} \Big/ \left(\frac{[H^+]_{res}}{[H^+]_{mo}} \right)^z \qquad (3.3.10)$$

When a solution containing the ions M^+ having a relatively high $K_{res,M}$ is passed through the resin, the ions will attach to the resin and H^+ ions will be released in solution. The exchange of H^+ ions with M^+ ions can continue until the equilibrium described by rel. 3.3.10 is reached.

The change of the column into H^+ form before the separation can be accomplished by slowly flowing through it a solution of an inorganic acid. For example, a solution of 0.1 N HCl is passed through the stationary phase resin. This changes the resin into H^+ form. Relation 3.3.9 (or 3.3.10) is applicable only to describe an equilibrium, and this is not the case during the column conditioning when the concentration $[M^{z+}]_{mo} = 0$ and $[H^+]_{mo}$ is high. However, rel. 3.3.10 shows that $[H^+]_{res}$ will become as high as possible when $[H^+]_{mo}$ is high and since $[M^{z+}]_{mo}$ is practically zero,

$[M^{z+}]_{res}$ will also tend to zero After conditioning, the resin is thoroughly washed with water to eliminate the remaining HCl traces.

The same discussion for the equilibrium on a cation-exchange column can be easily applied for an anion-exchange column. In this case, a column in the form A^- exchanges a base B^- in the following equilibrium:

$$resY^+A^- + B^- \rightleftarrows resY^+B^- + A^- \quad (3.3.11)$$

This equilibrium is governed by the constant:

$$K_{B,A} = \frac{[B^-]_{res}}{[B^-]_{mo}} \Big/ \frac{[A^-]_{res}}{[A^-]_{mo}} \quad (3.3.12)$$

For any anion, a retention constant can be defined with the formula:

$$K_{res,B} = \frac{[B^-]_{res}}{[B^-]_{mo}} \quad (3.3.13)$$

Frequently, in anion-exchange chromatography A^- is OH^-, but other anions such as Cl^- are also common.

Equilibrium in the Presence of a Complexing Reagent

The use of various complexing agents in the solution interacting with an ion-exchange resin is another procedure used in ion-exchange HPLC for obtaining separations. The equilibrium between the complexing agent (ligand) in solution and the ions to be exchanged reduces the concentration of the free ions available for the exchange process. In this case two simultaneous equilibria take place in the mobile phase:

$$resX^-C^+ + M^+ \rightleftarrows resX^-M^+ + C^+ \text{ and}$$

$$L^- + M^+ \rightleftarrows LM$$

$$(3.3.14)$$

Assuming that the reaction with the ligand is present only in the mobile phase and is described by the equilibrium constant K_{LM} given by the expression:

$$K_{LM} = \frac{[LM]_{mo}}{[L^-]_{mo}[M^+]_{mo}} \quad (3.3.15)$$

the concentration of M^+ in the resin is given by the expression:

$$[M^+]_{res} = K_{res,M} \frac{[LM]_{mo}}{[L^-]_{mo} K_{LM}} \quad (3.3.16)$$

Expression 3.3.16 shows that a higher concentration of ligand or a high complexation constant diminishes the amount of species M^+ retained in the column. However, complexation can be used to favor retention on the resin. For example, specific ions form negatively charged complexes. Assuming that an ion M^{2+} forms with a ligand L^- four combinations ML^+, ML_2, ML_3^-, and ML_4^{2-}, the negatively charged complexes and the ligand can be retained on an anion-exchange resin, while the positive ions and the neutral molecules are not retained. In a mixture of ions in which only some have the complexing capability with the formation of negatively charged compounds, an anion exchanger can be used for separating the desired species. In this case, considering the constant K_{comp} describing the equilibrium:

$$4 L^- + M^{2+} \rightleftarrows ML_4{}^{2-} \quad (3.3.17)$$

and the equilibrium constant $K_{res,ML4}$, the distribution constant between the mobile phase and the resin for ML_4^{2-} species, the concentration of ML_4^{2-} in the resin can be estimated using the expression:

$$[ML_4{}^{2-}]_{res} = K_{res,ML4} K_{comp}[L^-]^4[M^{2+}] \quad (3.3.18)$$

Each complex ion has its specific distribution constants in the resin. The constant depends on factors such as bond strength, hydrophobic interactions, and steric hindrance. The adsorption of complex ions on resins is a more complicated process because in addition to the complexation in solution, the ligand adsorbed on the resin may still participate in complex formation. The donor electrons from the ligand retained as counterion are still available for complexation and may further retain M^{2+} ions from solution. The use of ion-exchange resins for the retention of metal ions as complexes in different pH conditions has been thoroughly studied and reported in the literature [12,13].

The application of rel. 2.1.25 for calculating the retention volume in a separation on an ion-exchange column (for cations) will have the expression:

$$V_R = V_0 + K_{res,M} V_{st} \qquad (3.3.19)$$

where $K_{res,M}$ is given by rel. 3.3.2. A similar formula will be obtained for the separation of anions. Based on expression 3.3.19, it can be seen that the adsorption and elution of two components depend on their distribution constants in the resin. Selecting properly the eluent, separation of various components of a sample can be achieved, for example, by gradual change of the eluent pH or composition. Ion separation using ion-exchange columns is commonly applied for analytical purposes [14].

Gibbs-Donnan Effect

When a semipermeable membrane separates two solution compartments and one side of the membrane contains water and the other side contains a solution of a dissociated salt consisting of small cations C^+ and anions A^- (e.g., Na^+ and Cl^-), the membrane is permeable to these ions, and the ions diffuse through the membrane. In time, the final solutions on both sides of the membrane contain equal concentrations of both positive and negative ions, and the system is electrically neutral. No charges are accumulated on the two sides of the membrane.

There are, however, ions that have large volumes and are not able to diffuse freely through the membranes. If a semipermeable membrane separates two compartments and

compartment No. 1 contains a solution of a dissociated salt consisting of small ions C^+ and A^-, and compartment No. 2 contains the same small cation C^+ and a large anion B^- to which the membrane is impermeable, the diffusion of ions is not uniform. Since compartment No. 2 does not contain anions A^-, they will diffuse to compartment No. 2, but the large B^- will not diffuse to compartment No. 1. This will result in a negative charge on compartment No. 2 since $[A^-]_2 + [B^-]_2 > [A^-]_1$. Because of the excess negative charge on compartment No. 2, the concentration $[A^-]_2$ will remain smaller than $[A^-]_1$. The negative electrical gradient will attract additional C^+ from compartment No. 1 to compartment No. 2 that already had C^+ and establish electrical neutrality. However, there will be an unequal concentration of diffusible ions with $[C^+]_2 > [C^+]_1$ and $[A^-]_1 > [A^-]_2$. This uneven distribution of ions on the two sides of the semipermeable membrane is known as the Gibbs-Donnan effect. The process is illustrated in Figure 3.3.1.

At equilibrium, the electrochemical potentials $\bar{\mu}_i$ of the ionic species i (A and C) on both two compartments No. 1 and No. 2 must be equal, which can be written as follows:

$$(\bar{\mu}_i)_1 = (\bar{\mu}_i)_2 \qquad (3.3.20)$$

From the expression of electrochemical potential $\bar{\mu}_i = \mu_i + z_i F \Phi$ (where z_i is the charge of ion i, F is the Faraday constant, and Φ is the local electrostatic potential), rel. 3.3.20 can be written as follows:

$$\Phi_1 - \Phi_2 = \frac{1}{z_i F}[(\mu_i)_2 - (\mu_i)_1] \qquad (3.3.21)$$

(where μ_i is the chemical potential of species i). The difference $\Phi_1 - \Phi_2$ is known as the Donnan potential and is indicated as E_{Donnan}. On the other hand, the chemical potential for the species i in a solution with pressure p is given by the formula:

$$\mu_i = \mu_i^0(p^0) + RT \ln(a_i) + (p - p^0)V_i \qquad (3.3.22)$$

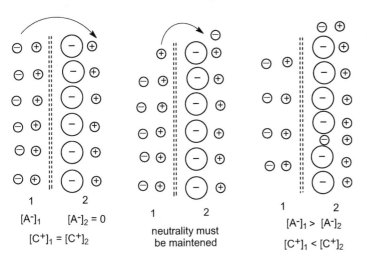

FIGURE 3.3.1 Illustration of Gibbs-Donnan effect. The tendency to have equal concentrations of diffusible ions on both sides of the membrane is compensated by the need for electrical neutrality.

where $\mu_i^0(p^0)$ is the chemical potential of species i in standard state, a_i is its activity, and V_i is its partial molar volume. From rel. 3.3.21 and 3.3.22, the expression for the Donnan potential for species i is given by the expression [11]:

$$E_{\text{Donnan}} = \{RT \ln [(a_i)_2/(a_i)_1] - \Pi V_i\}/(z_i F)$$

(3.3.23)

where Π is the osmotic pressure for the membrane (or the swelling pressure in the case of equilibrium between inside and outside of a resin).

For equilibrium the Donnan potentials of the two species A and C must be equal.

In ion exchange, the surface of the resin can be equated with the semipermeable membrane, the ions being able to migrate freely outside and inside the resin, while the bonded ionic groups can be equated with large ions (indicated above as B⁻) that cannot penetrate through the membrane. For this reason, the concept of equal electrochemical potentials $\bar{\mu}_i$ for the ionic species inside the resin and outside the resin can be immediately translated for the case of ion exchange equilibrium.

3.4. EQUILIBRIUM IN SIZE-EXCLUSION PROCESSES

General Aspects

Size-exclusion chromatography (SEC) is a chromatographic technique used for separating substances according to their molecular size, or more correctly, hydrodynamic volume. Size exclusion is mainly used for the separation of different polymers and for the separation of polymers from small molecules. The technique is also known as gel permeation chromatography when the mobile phase is an organic solvent, or gel filtration chromatography when the mobile phase is an aqueous (mostly aqueous) solution.

Polymeric molecules that are usually separated by SEC have various shapes and, typically, are not spherical. Also, the molecules during the separation may be solvated/hydrated such that the molecular shape is changed from that without solvent. For these reasons, the "true" size/shape of the molecule is difficult to define. The diffusional properties of the macromolecule can be selected to describe the molecular apparent size. The hydrodynamic volume of a molecule is the volume of a hypothetical hard sphere that diffuses with the same speed as the particle under examination. For SEC, the hydrodynamic volume is therefore the true parameter related to the separation.

Partition Equilibrium between Interstitial Mobile Phase and Pore Mobile Phase

The separation in SEC is done on a porous stationary phase. The small molecules from the sample moving with the mobile phase penetrate the pores of the stationary phase and have a long path through the column and therefore long retention times. Very large molecules cannot enter the pores at all and elute without retention (total exclusion). The macromolecules of medium size enter only some larger pores and are only partly retained, eluting faster than small molecules and slower than the very large ones, thus achieving separation.

Similar to any HPLC process, the retention process in SEC of a given molecular species can be formally characterized by the reduced retention volume V'_R given by rel. 2.1.16. From rel. 2.1.16 the following expression can be immediately obtained:

$$V_R = V_0 + K V_{st} \qquad (3.4.1)$$

In SEC the partition process takes place as a migration of the analytes between the interstitial volume of the column filled with the packing and the pores of the packing. In expression 3.4.1 the dead volume V_0 can be indicated for this process as the interstitial volume V_{inter}, and the volume V_{st} can be considered as equal to the volume of the pores of the packing V_{pores}. Therefore, the general equation 3.4.1 can be written for SEC separation in the form:

$$V_R = V_{inter} + K_{SEC} V_{pores} \qquad (3.4.2)$$

Using the same assumptions as for rel. 3.4.2, the phase ratio that was described in general for HPLC in Section 2.1 as having expression 2.1.14 will be given in SEC by the formula:

$$\Psi_{SEC} = \frac{V_{pores}}{V_{inter}} \qquad (3.4.3)$$

Using rel 3.4.3, the expressions for capacity factor k_{SEC} will be given by the formula:

$$k_{SEC} = K_{SEC} \frac{V_{pores}}{V_{inter}} \qquad (3.4.4)$$

The direct proportion between k_{SEC} and Ψ_{SEC} expressed by rel 3.4.4 has, however, a problem. The value of Ψ_{SEC} may increase by an increase in the number of pores and not only by the increase in the (average) pore volume. If the pores are numerous but keep a small average volume, these pores cannot contribute to the retention of larger macromolecular analytes. In this case, rel. 3.4.4 is no longer valid, and the proportionality remains valid only for a specific range of molecular sizes.

The partition constant K_{SEC} in rel. 3.4.4 represents the ratio of the concentrations of the macromolecule analytes in the pore (c_{pore}) and the concentration in the interstitial volume (c_{inter}):

$$K_{SEC} = c_{pore}/c_{inter} \qquad (3.4.5)$$

For understanding the variation of K_{SEC}, the elution process can be viewed as the movement of a sample zone of a solution containing the macromolecular analytes (a nonzero concentration of analyte) along the column packed with porous particles and filled with the mobile phase (with the analyte at zero concentration, $c = 0$). The initial concentration of macromolecules within the pores is also zero. The concentration gradient between the interstitial volume outside the pores during the sample zone ($c_{inter} > 0$) and within the pore (at $c_{pore} = 0$) "pulls" macromolecules into the pores because of the tendency to equalize the chemical potentials in the interstitial volume and in the

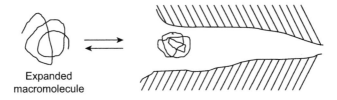

Expanded
macromolecule

FIGURE 3.4.1 Schematic representation of the partition of a coiled macromolecule between interstitial space and the pore of stationary phase (expansion of the molecule is associated with increase in entropy and shrinking is associated with decrease in entropy).

pore volume. Macromolecules outside of the pores are expanded. When pulled into the pore in order to equalize concentrations outside and inside the pore, the macromolecules are squeezed and their conformational entropy decreases. The macromolecules inside the pore are contracted and lose part of their conformational entropy. Some macromolecules cannot enter completely the pore volume because the loss in entropy would exceed the pulling force inside the pore. A schematic representation of the partition process in SEC separation is illustrated in Figure 3.4.1. This process indicates that the values of K_{SEC} are situated within the interval [0, 1]. If $K_{SEC} = 0$, the sample fraction will elute in the void volume (total exclusion), and when $K_{SEC} = 1$ the sample fraction elutes in the total column volume ($V_{inter.} + V_{pore}$). However, this process implies an ideal mechanism of SEC where the analytes do not exhibit any attractive or repulsive interaction with column packing except for the effects caused by the imperviousness of the pore walls. In some instances, values $K_{SEC} > 1$ are seen, indicating that other interactions take place between the analyte and the stationary phase. The expression for K_{SEC} using the general formula 2.1.12 is the following:

$$K_{SEC} = \exp\left[(-\Delta H^0 + T\,\Delta S^0)/(RT)\right] \quad (3.4.6)$$

where $-\Delta H^0$ is the standard enthalpy and ΔS^0 is the entropy change for the transfer of the analyte from the mobile to the stationary phase.

However, in the ideal case of no interaction with column packing, the value for ΔH^0 in rel. 3.4.6 is zero, and the separation mechanism is controlled exclusively by the entropy of the process. Therefore the expression for a "pure" size-xclusion process described by the constant K_{pure} is given by the formula:

$$K_{pure} = \exp(\Delta S^0/R) \quad (3.4.7)$$

Size exclusion of macromolecular analytes in the absence of any energetic interactions with the stationary phase is therefore an entropy-controlled process. Such a process is usually indicated as an entropic partition [15]. The loss of entropy when the molecules are trapped inside the stationary phase causes ΔS^0 to have negative values [16]. Expression 3.4.7 indicates that temperature should not significantly influence the exclusion processes. This was proven experimentally in several eluents that are good solvents for polymers [17]. The change in entropy is more significant for larger molecules (of polymers) and less important for smaller ones. This can be understood by starting with the following expression for entropy:

$$S = k_B \ln \Omega \quad (3.4.8)$$

where k_B is the Botzmann constant and Ω is the number of possible (equally probable) micromolecular states. The number of ways in which the individual molecules can occupy the space within the pore of a stationary phase is significantly larger for a small molecule than for a large one. This indicates that the number of micromolecular states for a small molecule in the pore is considerably larger than for a large molecule. Starting with similar number of states for large and small molecules in solution, the large molecules will have a smaller S value in the stationary phase, and therefore a considerable loss of entropy. At the same time, the small molecules will have only a minor loss of entropy. The result is that during the adsorption process, the large molecules will have a larger (in absolute value) negative ΔS^0. From rel. 3.4.7, it can be seen that a ΔS^0 larger in absolute value (and negative) leads to a smaller K_{pure} and consequently to a smaller V_R in rel. 3.4.2. The variation of the retention volume V_R as a function of entropy variation ΔS^0 for a hypothetical column where $V_{inter} = 5.9$ mL and $V_{pore} = 4.0$ mL is shown in Figure 3.4.2 (interaction entropy corresponding to polystyrene in the range of molecular weight (MW) from 2500 to 3 million [18]).

In some cases, the separation mechanism also has an enthalpic contribution besides the entropic change for the SEC mechanism. This happens in practice when some interactions between the packing and the macromolecular species are taking place. For this reason the SEC partition constant can be defined by two terms:

$$K_{SEC} = K_{pure}K_{interaction}$$
$$= \exp[(T\,\Delta S^0)/(RT)]\exp[(-\Delta H^0 + T\,\Delta S^{*0})/(RT)] \tag{3.4.9}$$

where $K_{interaction} = \exp[(-\Delta H^0 + T\,\Delta S^{*0})/(RT)]$, ΔH^0 is the enthalpic term, and ΔS^{*0} represents the entropy variation during the interaction process. For the SEC separations where $K_{SEC} > 1$ there is always an enthalpic contribution to the separation. For K_{SEC} in [0,1] interval, it is difficult to decide whether $\Delta H^0 = 0$ or it is only small. A more detailed discussion of thermodynamic factors affecting SEC separation can be found in the literature [18–20].

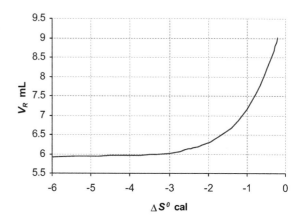

FIGURE 3.4.2 Variation of the retention volume V_R as a function of entropy variation ΔS^0 for a hypothetical system where $V_{inter} = 5.9$ mL and $V_{pore} = 4.0$ mL.

3.5. THE INFLUENCE OF pH ON RETENTION EQUILIBRIA

Preliminary Information

An important parameter in almost all analytical separations is the pH. The pH was initially defined for water as the solvating medium. The pH gives, in logarithmic form (base 10), the activity of hydrogen ions (a_{H^+}), or more precisely the activity ($a_{H_3O^+}$) of hydronium ions (H_3O^+). The pH expression is given by the formula:

$$pH = -\log a_{H_3O^+} = -\log c_{H_3O^+} - \log \gamma_{H_3O^+} \tag{3.5.1}$$

where ($\gamma_{H_3O^+}$) is the activity coefficient of hydronium ions and ($c_{H_3O^+}$) is their molar concentration. The reason for replacing hydrogen ions H^+ with hydronium (H_3O^+) or even with higher species (e.g. $H_9O_4^+$) is to account for the solvation of protons, but all these ions in fact describe the same entity (and notations H^+ or H_3O^+ are frequently used interchangebly).

In water, the pH scale is between 0 and 14 at 25 °C. The activity factor $\gamma_{H_3O^+}$ may be estimated with the Debye-Hückel equation, written in the form:

$$\log \gamma_{H_3O^+} = - \frac{A \cdot \sqrt{I}}{1 + a_s \cdot B \cdot \sqrt{I}} \quad (3.5.2)$$

In rel. 3.5.2, I represents the ionic strength of the solution, a_s is an ion-size parameter, and A and B are solvent- and temperature-dependent parameters, respectively. Their values are tabulated from empirical data. In an infinitely diluted water solution $\gamma_{H_3O^+} = 1$, and the second term from rel. 3.5.1 is zero. The ionic strength in a solution depends on the concentration of species i, c_i, and their net charges, z_i, by the formula:

$$I = \frac{1}{2} \sum_i c_i z_i^2 \quad (3.5.3)$$

Dependence of Compound Structure on pH

Compounds with acidic, basic, or amphoteric character may have structures that are strongly dependent on the pH of the solution (e.g., in the mobile phase). The structure of such compounds may change at different pH values. This change is caused by the interaction of the compound with the protons. Each ionization equilibrium between the protonated and deprotonated forms of the molecule can be described with its dissociation constant K_a or $pK_a = - \log K_a$. The pK_a value can be defined for both acids and bases. For simple acids HX, the dissociation takes place as follows:

$$HX \rightleftarrows H^+ + X^- \quad (3.5.4)$$

where the equilibrium constant K_a is defined by the well-known formula

$$K_a = \frac{[H^+][X^-]}{[HX]} \quad (3.5.5)$$

For simple bases BOH, the equilibrium can be considered to take place as follows

$$BOH_2^+ \rightleftarrows BOH + H^+ \quad (3.5.6)$$

and the dissociation constant K_a is defined by the formula:

$$K_a = \frac{[H^+][BOH]}{[BOH_2^+]} \quad (3.5.7)$$

A base dissociation constant can also be defined by:

$$K_b = \frac{[OH^-][B^+]}{[BOH]} \quad (3.5.8)$$

with the relation $K_a K_b = K_w$, where K_w is the water ion product and $K_w = 1.008 \ 10^{-14}$ at 25 °C. Acidic or basic character can be assigned to the molecule according to Brönsted's rule, which indicates that an acid is a proton donor and a base is a proton acceptor. As a result, a basic compound is protonated as a positive ion, while an acidic compound is protonated as a neutral molecule. An acidic compound will be deprotonated as a negative ion. Certain molecules can dissociate as both an acid or as a base and are known as amphoteric. Also, molecules can be multiprotic, having more than one ionizable group. The influence of pH on the structure of a molecule with amphoteric character is shown, as an example, for valine in Figure 3.5.1 where the percent of a specific form of the compound in function of solution pH is indicated. The presence of a compound in different forms as a function of pH is very common, and this significantly influences the separation process.

The Influence of pH on Partition

In a simple partition equilibrium between two liquid nonmiscible solvents, the distribution coefficient (D_i) and the distribution constant (K_i) are identical for apolar or polar solutes that have no acid or base functional groups in their

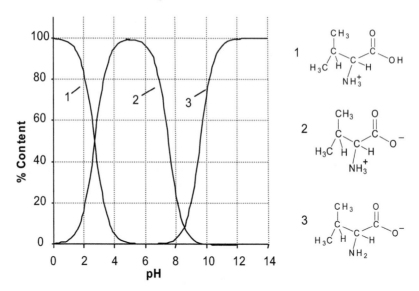

FIGURE 3.5.1 Variation in the proportion of different forms of valine in a solution at different pH values.

molecule that would lead to some dissociation in the aqueous or partially aqueous medium. When the solutes contain ionizable functional groups such as -OH, -SH, -COOH, -SO₃H, -NH₂, and others, it is possible for those solutes to have interactions into aqueous phase with proton or hydroxide ions leading to ionized species. In such cases, the two distribution parameters (D_i and K_i) are not identical, and the one that truly describes the partition process is the distribution coefficient D_i. In Figure 3.5.2 a schematic example is shown, which describes the partition process for a molecular weak acid, denoted by HX.

Similar to the case of solvents is the one where the organic phase role is taken by an organic stationary phase, and the aqueous phase is a mobile phase with a certain water content (indicated as w). In the case of the acids HX, the distribution constant K (see rel. 3.1.7) and distribution coefficient D (see rel. 3.1.18) are given by the following relations:

$$K_{HX} = \frac{c_{HX,o}}{c_{HX,w}} \qquad (3.5.9)$$

$$D_{HX} = \frac{\sum c_{HX,o}}{\sum c_{HX,w}} = \frac{c_{HX,o}}{c_{HX,w} + c_{X^-,w}} \qquad (3.5.10)$$

Expressions 3.5.9 and 3.5.10 show that higher K_{HX} or D_{HX} indicate a higher concentration of the analyte in the organic phase. This remains valid when the organic phase (o) is the stationary phase (e.g., in reversed phase chromatography). For these cases, a higher K_{HX} or D_{HX} is related to a higher capacity factor k based on rel. 2.1.13 and 3.1.20, respectively, and therefore to longer retention times in the chromatographic process.

Since in the case of ionizable groups the distribution process is controlled by the

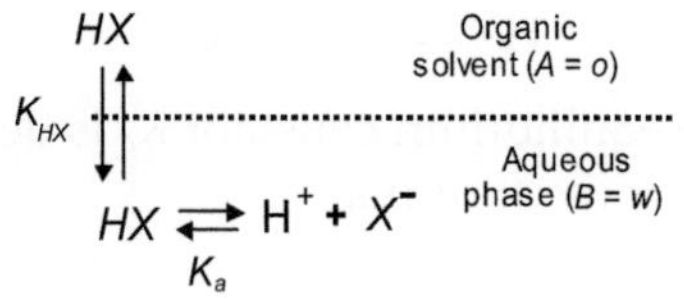

FIGURE 3.5.2 Representation of the partition process of a weak acid HX.

distribution coefficient D and not by the distribution constant K, it is important to analyze the variation of D for several particular cases. In the case of an acid HX, the concentration of the dissociated form in rel. 3.5.10 can be calculated from the dissociation constant (K_a) as:

$$c_{X^-,w} = K_a \frac{c_{HX,w}}{[H^+]_w} \qquad (3.5.11)$$

By substituting $c_{X^-,w}$ given by rel. 3.5.11 and K_{HX} given by rel. 3.5.16 in expression 3.5.10, the relation between K_{HX} and D_{HX} can be written as follows:

$$D_{HX} = \frac{c_{HX,o}}{c_{HX,w}\left(1 + K_a \dfrac{1}{[H^+]}\right)} = K_{HX} \frac{10^{-pH}}{K_a + 10^{-pH}}$$

$$(3.5.12)$$

The dependence given by rel. 3.5.12 is illustrated in Figure 3.5.3 for a compound with $K_a = 10^{-4.54}$, ($pK_a = 4.54$). At low pH values, $D_{HX} \approx K_{HX}$. At high pH values, $D_{HX} \approx 0$; and at pH $= pK_a$ which is the inflection point of the sigmoid curve of dependence, $D_{HX} \approx 0.5\, K_{HX}$.

A weak-base-type compound Y can be involved in an acid/base equilibrium as shown in the following:

$$Y + HOH \rightleftarrows YH^+ + HO^- \qquad (3.5.13)$$

This equilibrium is characterized by a basicity constant K_b ($K_b = K_w / K_a$). The same calculations as applied for the acids distribution lead to the following dependence of D_Y on pH:

$$D_Y = K_Y \frac{K_w}{K_w + K_b \cdot 10^{-pH}} \qquad (3.5.14)$$

(where K_w is the ionic product of water (10^{-14} at 25°C). The plot of this dependence is shown in Figure 3.5.4 for a base with $K_b = 10^{-4.64}$ ($pK_b = 4.64$).

The formulas 3.5.12 and 3.5.14 (as well as the graphs from Figures 3.5.3 and 3.5.4) indicate that for weak acidic or basic analytes, the pH plays a major role in the distribution of the analyte. Even if the distribution constant is highly favorable for the organic phase (large K values), since the parameter that truly describes the distribution process is D, its value can be very small if the pH of the aqueous phase is not selected properly. This effect can also be successfully used in a separation process. For example, the separation

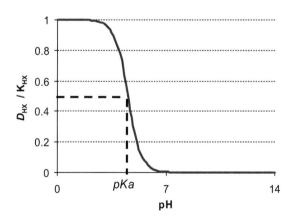

FIGURE 3.5.3 The dependence of D_{HX}/K_{HX} as a function of pH for an acid HX ($pK_a = 4.54$)

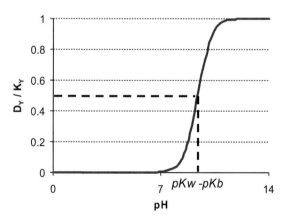

FIGURE 3.5.4 The dependence of D_Y/K_Y as a function of pH for a basic compound Y ($pK_b = 4.64$).

of a number of acids with different pK_a values can be easily separated using a gradient elution with the mobile phase changing its pH value.

Compounds with amphoteric character contain both acidic and basic functional groups (for example, amino acids). If such a compound is indicated as HXY and participates in a distribution process, but in the mobile phase (partly aqueous) is also part of two acid–base equilibria, these equilibria and the corresponding K_a and K_b constants can be written as follows:

$$HXY \rightleftarrows {}^{-}XY + H^{+} \qquad (3.5.15)$$

$$K_a = \frac{c_{-XY,w} \cdot c_{H^{+},w}}{c_{HXY,w}} \qquad (3.5.16)$$

$$HXY + HOH \rightleftarrows HXYH^{+} + OH^{-} \qquad (3.5.17)$$

$$K_b = \frac{c_{HXYH^{+},w} \cdot c_{HO^{-},w}}{c_{HXY,w}} \qquad (3.5.18)$$

The distribution of the neutral species between the two phases (o and w) is characterized by partition constant (K_{HXY}) and distribution coefficient (D_{HXY}):

$$K_{HXY} = \frac{c_{HXY,o}}{c_{HXY,w}} \qquad (3.5.19)$$

$$D_{HXY} = \frac{c_{HXY,o}}{c_{-XY,w} + c_{HXYH^{+},w} + c_{HXY,w}} \qquad (3.5.20)$$

Combining rel. 3.5.16, 3.5.18, and 3.5.19 with 3.5.20, the following expression for D can be obtained:

$$D_{HXY} = \frac{K_{HXY}}{1 + \dfrac{K_a}{[H^{+}]} + \dfrac{K_b \cdot [H^{+}]}{K_w}} \qquad (3.5.21)$$

Expression 3.5.21 can be written in the following form:

$$D_{HXY} = K_{HXY} \frac{K_w \cdot 10^{-pH}}{K_a \cdot K_w + K_w \cdot 10^{-pH} + K_b \cdot 10^{-2pH}} \qquad (3.5.22)$$

The graph showing the variation of D_{HXY} / K_{HXY} as a function of pH is given in Figure 3.5.5 for a compound with $pK_a = 2.47$ and $pK_b = 4.55$ (phenylalanine).

Taking into account that $K_a \cdot K_w \ll 10^{-14}$, $K_w = 10^{-14}$ and $K_b \gg K_w$, the limit value of D for the acidic pH domain is given by the expression:

$$D_{HXY}^{pH \to 0} = K_{HXY} \frac{K_w}{K_b} \qquad (3.5.23)$$

In the example shown in the graph from Figure 3.5.5, $K_w \ll K_b$ and the value for D is much lower than that for K (practically zero).

For pH $\to$ 14 and taking into account that $K_a \gg K_w$, the limit value of D_{HXY} for basic pH domain is given by the expression:

$$D_{HXY}^{pH \to 14} = K_{HXY} \frac{K_w}{K_a} \qquad (3.5.24)$$

For the example shown in the graph from Figure 3.5.5, $K_w \ll K_a$, and the value for D is again much lower than that for K (practically zero).

The maximum value for the ratio D/K can be obtained from the condition $\partial D_{HXY} / \partial [H^{+}] = 0$,

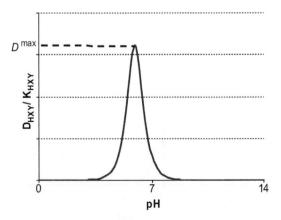

FIGURE 3.5.5 The dependence of D/K for an amphoteric compound with $pK_a = 2.47$ and $pK_b = 4.55$ as a function of the pH of the aqueous phase.

which leads to the following expression for D_{HXY}:

$$D_{HXY}^{max} = K_{HXY} \frac{1}{1 + 2\left(\dfrac{K_a K_b}{K_w}\right)^{1/2}} \qquad (3.5.25)$$

for the pH value:

$$pH_{HXY}^{max} = 7 + \frac{1}{2}(pK_a - pK_b) \qquad (3.5.26)$$

The result for D^{max} for the example taken for Figure 3.5.5 ($pK_a = 2.47$ and $pK_b = 4.55$) shows that $D^{max} \approx 1.55 \ 10^{-4} \ K$. Therefore much less compound will be present in the stationary phase than is predicted from its K value.

Relations 3.5.12, 3.5.14, or 3.5.22 allow the evaluation of the retention factor k_i for a compound i when the distribution constant K, the phase ratio Ψ, the nature of the analyte (acidic, basic, or amphoteric), and the pH of the mobile phase are known.

The previous discussion of the influence of pH on the distribution of a molecule that has acidic, basic, or amphoteric character was presented for simple systems, assuming that an equilibrium of different species of compound i takes place only in the mobile phase and that the ionized species are not at all partitioned in the stationary phase. However, an equilibrium as shown in Figure 3.5.6 is very possible.

For the case of the partition process of a weak acid HX when the ion X^- is also subject to the distribution, rel. 3.5.16 will remain valid for

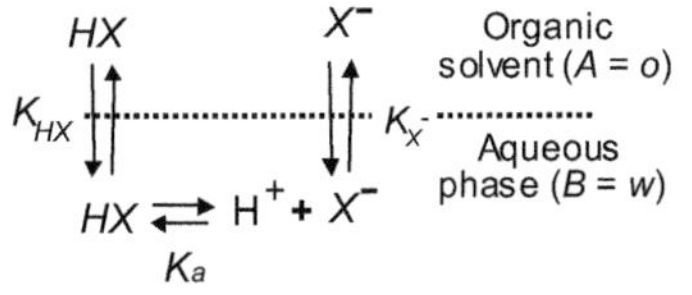

FIGURE 3.5.6 Representation of the partition process of a weak acid HX when the ion X^- is also subject to the distribution process.

K_{HX}, but the expression for D_{HX} should be replaced by the formula:

$$D_{HX} = \frac{\sum c_{HX,o}}{\sum c_{HX,w}} = \frac{c_{HX,o} + c_{X^-,o}}{c_{HX,w} + c_{X^-,w}} \qquad (3.5.27)$$

and an additional equilibrium constant will be defined as:

$$K_{X^-} = \frac{c_{X^-,o}}{c_{X^-,w}} \qquad (3.5.28)$$

By substituting in rel. 2.5.27 the following terms: $c_{X^-,w}$ obtained from rel. 3.5.11, $c_{X^-,o}$ obtained from rel. 3.5.28, and $c_{HX,o}$ from rel. 3.5.9, upon simplification with $c_{HX,w}$ the expression for D_{HX} can be written as follows:

$$D_{HX} = \frac{K_{HX} \ 10^{-pH} + K_{X^-} K_a}{K_a + 10^{-pH}} \qquad (3.5.29)$$

Similar expressions can be developed for a basic compound, amphoteric compound, dibasic acid, and the like [21].

Octanol/Water Partition Constant K_{ow} and Distribution Coefficient D_{ow}

The octanol/water partition constant K_{ow} is a parameter widely used in QSAR and related drug-designed techniques as a measure of molecular hydrophobicity. The values of this parameter for a large number of compounds are available in the literature [4, 5, 22], and various computer programs are available for their calculation (e.g., MarvinSketch 5.4.0.1, ChemAxon Ltd., [6], EPI Suite [23]). As shown in Section 3.1, log K_{ow} shows a good correlation with capacity factor in RP-HPLC for many organic compounds [2, 3]. For these reasons, the values for K_{ow} represent a convenient resource for characterizing the hydrophobicity of organic compounds with application to HPLC. However, compounds having ionizable groups exist in solution as a mixture of different forms (some ionic and some neutral). In this case, a distribution coefficient D_{ow} is defined by the formula (similar to rel 3.1.18):

$$D_{i,ow} = \frac{c_{i_1,o} + c_{i_2,o} + \dots + c_{i_n,o}}{c_{i_1,w} + c_{i_2,w} + \dots + c_{i_n,w}} \quad (3.5.30)$$

where $\{c_{i,o}\}$ $i = 1,2,\dots n$ represent all the forms of the compound i present in octanol and $\{c_{i,w}\}$ $i = 1,2,\dots n$ represent all the forms of compound i present in water. (*Note:* The terms *partition* and *distribution* are used interchangeably, and the difference must result from the meaning of the parameter). The general discussion on the dependence of distribution coefficient D on pH given in this section is equally applicable to the distribution coefficient D_{ow}, which depends on the water pH for ionizable compounds. For compounds having no ionizable groups, $K_{ow} = D_{ow}$. Variation of D_{ow} with the pH can be obtained using computing programs (e.g., MarvinSketch 5.4.0.1).

Since the structure of ionizable compounds depends on pH, the distribution coefficient D_{ow} is also pH dependent. For example, in an acidic pH, an acid may be very little or not ionized at all. In this case, at pH values where the compound is not ionized $D_{ow} = K_{ow}$, while for other pH values they are different and $D_{ow} < K_{ow}$. This is exemplified for butyric acid in Figure 3.5.7, where at low pH (pH < 3) the maximum value for log D_{ow} is 0.98 and $K_{ow} = D_{ow}$. For some amphoteric compounds, even in a pH where a neutral form would be expected, the compound is present in zitterionic form, and always $D_{ow} < K_{ow}$. Amino acids are this type of compound, and, for example, in the case of valine log $K_{ow} = 0.31$, and the dependence of log D_{ow} on pH shown in Figure 3.5.8 indicates that a maximum value is log $D_{ow} = -2.09$. The discrepancy is the result of assumption that valine can be a neutral molecule (log $K_{ow} = 0.31$) while in reality it always contains ionic groups.

The results pictured in Figures 3.5.7 and 3.5.8 indicate that for compounds that can be present in ionic form, in order to have information

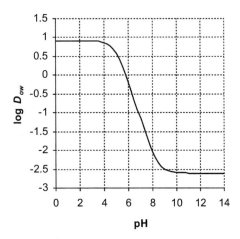

FIGURE 3.5.7 Variation of log D_{ow} with pH for butyric acid. log $K_{ow} = 0.98$ (max $D_{ow} = K_{ow}$)

regarding their octanol/water distribution, the values for D_{ow} should be considered at a specific pH. For amino acids, peptides, and proteins that are present in zwitterionic form, the value for log D_{ow} at the isoelectric point (pI) is important for their characterization (the isoelectric point (pI) is the pH value at which the molecule carries no electrical charge). For molecules with multiple zwitterionic points, log D_{ow} at

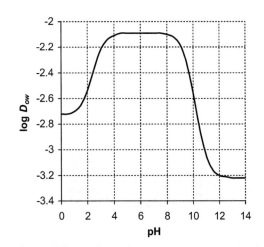

FIGURE 3.5.8 Variation of log D_{ow} with pH for valine. log $K_{ow} = 0.31$ (max $D_{ow} < K_{ow}$)

the isoelectric point may be in between the lowest and the highest value of log D_{ow} at different pH values.

3.6. THE INFLUENCE OF TEMPERATURE ON RETENTION EQUILIBRIA

General Aspects

Similar to all equilibria, the separation process is influenced by temperature. For a separation equilibrium, the capacity factor k is dependent on temperature following rel. 3.1.13, known as the van't Hoff equation:

$$\ln k = -\Delta H^0/(RT) + \Delta S^0/R + \ln \Psi \quad (3.6.1)$$

In rel. 3.6.1, ΔH^0 and ΔS^0 are the standard enthalpy and entropy changes, respectively, for transfer of the analyte from the mobile to the stationary phase (the formula 3.1.13 was developed based on distribution equilibrium but can be used in principle for any HPLC process). Since the retention is an exothermic process, ΔH^0 has negative values. Formula 3.6.1 indicates that for typical separations, the value of k decreases as the temperature increases (deviations from this rule are discussed further in this chapter). When the stationary phase, the analyte, and the mobile phase properties do not change with a temperature change, ΔH^0 and ΔS^0 can be considered temperature invariants. However, this assumption is only an approximation, and more correct estimations of the values for ΔH^0 and ΔS^0 at different temperatures can be made. Corrections as a function of temperature T of thermodynamic functions ΔH^0 and ΔS^0 are known (see, e.g., [1]). Nevertheless, for numerous separations, the linear dependence of $\ln k$ on $1/T$ can be experimentally verified [24]. The graphs representing $\ln k$ as a function of $1/T$ are known as van't Hoff plots.

Evaluation of Thermodynamic Parameters of a Separation from van't Hoff Plots

Linear van't Hoff plots are those in which the variation of $\ln k$ can be written in the simple form:

$$\ln k = a + b\frac{1}{T} \quad (3.6.2)$$

These plots can be successfully used for evaluating enthalpy change associated with the retention process, and also when the phase ratio Ψ of the column can be estimated, it is possible to evaluate the entropy changes [25]. In such cases, the following formula can be used:

$$\Delta H^0 = -R\,b \quad (3.6.3)$$
$$\Delta S^0 = R(a - \ln \Psi) \quad (3.6.4)$$

As an example, these parameters can be obtained from the retention study of two hydrophobic vitamins, retinol and calciferol (with structures given in Figure 3.6.1), performed on a C18 stationary phase and mobile phase containing water and methanol in different proportions. The van't Hoff plots for the experiments are shown in Figure 3.6.2 [26]. Using an estimation for $\Psi \approx 1.5$, from the plots shown in Figure 3.6.2, it can be possible to evaluate ΔH^0 and ΔS^0 for the separations. These values are given in Table 3.6.1. The results obtained for ΔH^0 and ΔS^0 given in the table are in fair agreement with thermodynamic data reported in the literature, for example, for the interaction between two hydrocarbon side chains of a protein (1.2 − 7.5 kJ/mol for ΔH^0 and 7 − 45 J/(K mol) for ΔS^0 at 298 K) [27]. The value of ΔH^0 reflects the degree of interaction between analyte and stationary phase, and a more negative ΔH^0

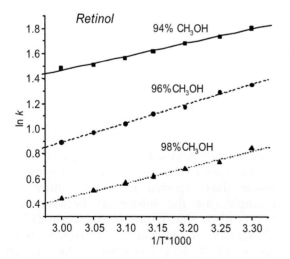

FIGURE 3.6.1 Chemical formulas for retinol (Vitamin A) and cholecalciferol (Vitamin D$_3$).

(larger absolute value) indicates a stronger interaction [28].

Nonlinear Dependence of the Capacity Factor on 1/T

When a compound i can be present during the separation in several forms i_1, i_2, i_3, and so on, that are in equilibrium and account for the total level of i, the capacity factor can be expressed by formula 3.1.21, where the distribution coefficient D_i replaces the simple equilibrium constant K_i. In such cases, the linear dependence of $\ln k$ on $1/T$ is not necessarily obeyed. The simple expression of the form 3.6.2 cannot be written (without including approximations). Replacing in rel. 3.1.21 the expression 3.1.19 for D_i for the simple case of

two forms of compound i, the result for $\ln k_i$ is the following:

$$\ln k_i = \ln \frac{K_{i_1} + K_{i_2} K_{1,2}}{1 + K_{1,2}} + \ln \Psi \qquad (3.6.5)$$

A replacement of the expressions for individual equilibrium constants in rel. 3.6.5 leads to a complicated formula, and for simplicity a power series can be used to express $\ln k_i$. In such cases, the following formula can be used:

$$\ln k_i = a + b\frac{1}{T} + c\frac{1}{T^2} + \dots + \ln \Psi \qquad (3.6.6)$$

The number of terms in expression 3.6.6 can be larger, but a polynomial form limited to $(1/T^2)$ is sometimes sufficient to describe the

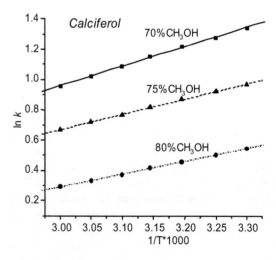

FIGURE 3.6.2 Van't Hoff plots for retinol and cholecalciferol separated on a C18 stationary phase and mobile phase containing water and methanol in different proportions [26].

TABLE 3.6.1   Evaluation of thermodynamic parameters ΔH^0 and ΔS^0 for the separation of retinol and calciferol on a C18 column [26].

Compound	%CH₃OH in the mobile phase	Regression parameters		Thermodynamic parameters	
		a	b	ΔH^0 (kJ·mol⁻¹)	ΔS^0 (J·mol⁻¹·K⁻¹)
Retinol	94	−1.871	1111.82	−9.24	−18.92
	96	−3.791	1559.81	−12.97	−34.88
	98	−3.408	1281.04	−10.65	−31.70
Calciferol	70	−2.903	1287.19	−10.70	−27.50
	75	−2.354	1007.32	−8.37	−22.94
	80	−2.249	846.88	−7.04	−22.06

variation of $\ln k_i$ with the temperature [29]. Several examples of such cases have been reported in the literature [30, 31] and are exemplified here for the drugs vincamine, epivincamine, and drotaverine. The formulas for these compounds are shown in Figure 3.6.3. The variation of $\ln k$ with temperature for vincamine and epivincamine separated on a Zorbax XDB-C18 column 150 mm L × 4.6 mm i.d. and 3.5 μm d.p., with a mobile phase consisting of a mixture of 0.2% triethylamine brought to pH = 6.0 with H₃PO₄ and acetonitrile, in the ratio 65/35 (v/v), and a flow rate 1 mL/min is shown in Figure 3.6.4A. The same type of plot for drotaverine on the same column and a mobile phase consisting of aqueous component containing 25 mM ammonium formate brought to pH = 4.5 with formic acid and ACN in the ratio 62.5/37.5 (v/v) at a flow rate of 1 mL/min is shown in Figure 3.6.4B.

In such cases where the nonlinear dependence of $\ln k$ on temperature occurs, from the values of parameters a, b, c and $\ln \Psi$ in formula 3.6.6, a total ΔG_i^0 can be calculated (and is temperature dependent). This is done by using the derivative of $\ln k$ as a function of $(1/T)$ (similar to the calculations shown for rel. 3.1.14) to obtain a total ΔH_i^0, followed by an estimation for ΔS_i^0 using a simple subtraction (the value for $\ln \Psi$ must be known). Similar nonlinear dependences as shown in Figures 3.6.4A and 3.6.4B have been reported in the

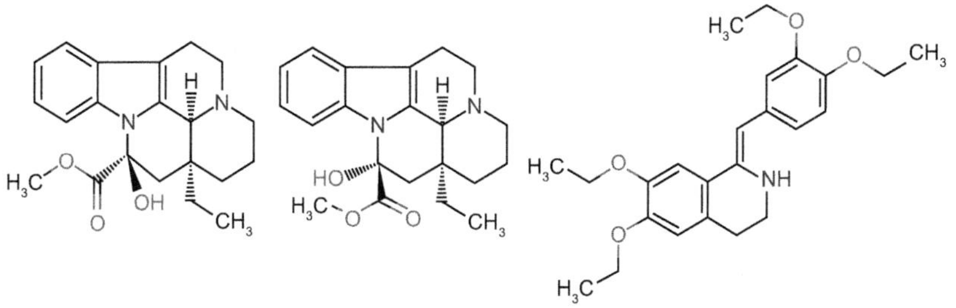

FIGURE 3.6.3   Chemical formulas for vincamine, epivincamine and drotaverine.

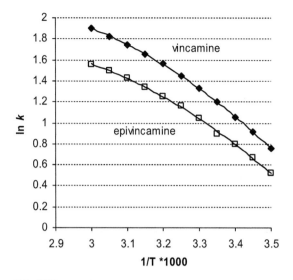

FIGURE 3.6.4A Variation of ln k with temperature for vincamine and epivincamine.

literature for other compounds, and their shape depends on parameters such as temperature interval, type of analyte and buffer, composition of the mobile phase [30].

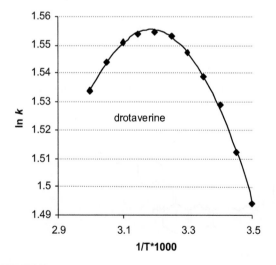

FIGURE 3.6.4B Variation of ln k with temperature for drotaverine.

Evaluation of Enthalpy-Entropy Compensation from van't Hoff Plots

In the retention process, a higher interaction of the retained molecule with the stationary phase is associated with a more negative enthalpy of the interaction. However, this stronger interaction may also be associated with a limitation of the movement of the molecule (in terms of the ability to rotate, vibrate, etc.) and hence with a decrease in the molecule's entropy. Since the change in enthalpy (ΔH) and the change in entropy (ΔS) have opposite signs, the value for ΔG will change very little if both enthalpy (in absolute value) and entropy will increase. This effect is known as enthalpy-entropy compensation and takes place in adsorption processes. The enthalpy-entropy compensation has been studied for different chromatographic separations, including reversed phase HPLC [32]. The effect can be assumed to occur when, for a series of homologous compounds (e.g., alkylbenzenes or alkylphenols), the calculated values of ΔH and ΔS using van't Hoff plots show linear dependence. However, it has been suggested that measurement errors may lead to a linear dependence without a true physical correlation between the two thermodynamic functions [33]. Specific tests were developed for proving a true physical compensation [34], and in HPLC retention this type of compensation was demonstrated to be real. The temperature where the compensation takes place was found to be different for different separation mechanisms. For example, in NP-HPLC the compensation process takes place (calculated temperatures) around 150 K, while in RP-HPLC the compensation takes place at higher temperatures and in a range between 500 K and 1000 K [32].

References

[1] Moldoveanu SC, David V. Sample Preparation in Chromatography. Amsterdam: Elsevier; 2002.

[2] El Tayar N, van de Waterbeemd H, Testa B. The prediction of substituent interactions in the lipophilicity of disubstituted benzenes using RP-HPLC,. Quant. Struct.-Act. Relat. 1985;4:69–77.

[3] Kaliszan R. Quantitative Structure-Chromatographic Retention Relationship. New York: John Wiley; 1987.

[4] Hansch C, Leo A. Exploring QSAR. Fundamentals and Applications in Chemistry and Biology. Washington, DC: ACS Prof. Ref. Books, ACS; 1995.

[5] Hansch C, Leo A, Hoekman D. Exploring QSAR: Hydrophobic, Electronic, and Steric Constants. Washington, DC: ACS Prof. Ref. Books, ACS; 1995.

[6] http://www.ChemAxon.com

[7] Wang M, Mallette J, Parcher JF. Strategies for the determination of the volume and composition of the stationary phase in reversed-phase liquid chromatography. J. Chromatogr. A 2008;1190:1–7.

[8] Fornstedt T. Characterization of adsorption process in analytical liquid-solid chromatography. J. Chromatogr. A 2010;1217:792–812.

[9] Sing KSW, Everett DH, Haul RAW, Moscou L, Pierotti RA, Rouquérol J, Siemieniewska T. Reporting physisorption data for gas/solid systems with special reference to the determination of surface area porosity. Pure & Appl. Chem. 1985;57: 603–19.

[10] Gritti F, Guiochon G. Surface heterogeneity of six commercial brands of end-capped C18-bonded silica. RPLC separations. Anal. Chem. 2003;75: 5726–38.

[11] Helferich F. Ion Exchange. New York: McGraw-Hill; 1962.

[12] Rothbart HL, Weymouth H,W,, Rieman III W. , Separation of oligophosphates. Talanta 1964;11:33–41.

[13] Nachod FC, Shubert I, editors. Ion Exchange Technology. New York: Academic Press; 1956.

[14] Haddad PR, Jackson PE. Ion Chromatography, Principles and Applications. Amsterdam: Elsevier; 1990.

[15] Berek D. Size exclusion chromatography—A blessing and a curse of science and technology of synthetic polymers. J. Sep. Sci. 2010;33:315–30.

[16] Myers AL. Characterization of nanopores by standard enthalpy and entropy of adsorption of probe molecules. Colloid and Surface A: Phys. Eng. Asp. 2004;241: 9–14.

[17] Gorbunov AA, Skvortsov AM. Statistical properties of confined macromolecules. Adv. Colloid Interface Sci. 1995;62:31–108.

[18] Striegel AM. Thermodynamic equilibrium of the solute distribution in size-exclusion chromatography. J. Chromatogr. A 2004;1033:241–5.

[19] Netopilík M. Relation between the kinetic and equilibrium quantities in size exclusion chromatography. J. Chromatogr. A 2004;1038:67–75.

[20] Wu C-S, editor. Handbook of Size Exclusion Chromatography. New York: Marcel Dekker; 1995.

[21] Horvath C, Melander W, Molnar I. Liquid chromatography of ionogenic substances with nonpolar stationary phases. Anal. Chem. 1977;49:142–54.

[22] Viswanadhan VN, Ghose AK, Revankar GR, Robins RK. Atomic physicochemical parameters for three dimensional structure direct quantitative structure-activity relationships. 4. Additional parameters for hydrophobic and dispersive interactions and their application for an automated superposition of certain naturally occurring nucleoside antibiotics. J. Chem. Inf. Comput. Sci. 1989;29:162–72.

[23] http://www.epa.gov/oppt/exposure/pubs/episuite.htm

[24] Galaon T, Mihailciuc C, Medvedovici A, David V. The influence of mobile phase flow-rate in RP-LC on thermodynamic parameters studied for polar compounds. J. Liq. Chromatogr. Rel. Technol. 2011;34: 521–36.

[25] Chester TL, Coym JW. Effect of phase ratio on van't Hoff analysis in reversed-phase liquid chromatography, and phase ratio independent estimation of transfer enthalpy. J. Chromatogr. A 2003;1003: 101–11.

[26] David V, Bala C, Rotariu L. Thermodynamic parameters of the reversed-phase liquid chromatography retention for some lipid-soluble vitamins. Chem. Anal. (Warsaw) 2004;49:191–9.

[27] Hobza P, Zahradnik R. Weak Intermolecular Interactions in Chemistry and Biology. Amsterdam: Elsevier; 1980. p. 220.

[28] Vervoort RJM, Ruyter E, Debets AJJ, Claessens HA, Cramers CA, de Jong GJ. Characterization of reversed-phase stationary phases for the liquid chromatographic analysis of basic pharmaceuticals by thermodynamic data. J. Chromatogr. A 2002;964: 67–76.

[29] Haidacher D, Vailaya A, Horvath C. Temperature effects in hydrophobic interaction chromatography. Proc. Natl. Acad. Sci. USA 1996;93:2290–5.

[30] Heinisch S, Puy G, Barrioulet MP, Rocca JL. Effect of temperature on the retention of ionizable compounds in reversed-phase liquid chromatography: Application to method development. J. Chromatogr. A 2006;1118:234–43.

[31] Galaon T, David V. Deviation from van't Hoff dependence in RP-LC induced by tautomeric interconversion observed for four compounds. J. Sep. Sci. 2011;34:1423–8.

[32] Miyabe K, Guiochon G. Thermodynamic interpretation of retention equilibrium in reversed-phase liquid chromatography based on enthalpy-entropy compensation. Anal. Chem. 2002;74:5982–92.

[33] Krug RR, Hunter WG, Grieger RA. Enthalpy-entropy compensation.1. Some fundamental statistical problems associated with the analysis of van't Hoff and Arrhenius data. J. Phys. Chem. B 1976;80:2335–41.

[34] Krug RR, Hunter WG, Grieger RA. Enthalpy-entropy compensation.2. Separation of the chemical from statistical effect. J. Phys. Chem. B 1976;80:2341–51.

CHAPTER

4

Intermolecular Interactions

OUTLINE

4.1. FORCES BETWEEN IONS AND MOLECULES

General Comments

The evaluation of molecular interactions is important for understanding and even predicting the behavior of the solutes in the separation process. Although this procedure is in many instances successful in predicting the directional behavior of sample components in a specific separation, the quantitative results regarding the separation are not always precise. This is mainly because of the complexity of the chromatographic process and the approximations frequently used to evaluate molecular interactions and properties. Typically, alternative paths are employed for obtaining quantitative information. These paths may combine the calculations of the interaction intensity with empirical data and/or with physicochemical or molecular

Essentials in Modern HPLC Separations
http://dx.doi.org/10.1016/B978-0-12-385013-3.00004-5

parameters of the participants in the separation process (solutes, mobile phase, stationary phase). Also, statistical techniques that do not provide a cause/effect relation between a specific property and a parameter (or set of parameters) are frequently utilized for generating the desired information. For example, a common procedure for predicting a given chromatographic property P_r as a function "f" of several physicochemical or molecular characteristics y_i is to express the property P_r as follows:

$$P_r = f(a_1 y_1 + a_2 y_2 + \ldots a_n y_n) \qquad (4.1.1)$$

where a_1, $a_2 \ldots a_n$ are coefficients calculated by multiple (linear) regression. Among the molecular characteristics y_i are structural additive parameters (carbon number, molecular mass, parachor, molar volume, molar refractivity, polarizability), physicochemical parameters (boiling point, dipole moment, ionization potential, formal charges in the molecule, n-octanol-water partition coefficient), and topological parameters related to the molecule shape (see, e.g., [1]).

Study of specific interactions that may take place between the species participating in the chromatographic process remains an important tool for understanding the separation process, and some aspects of molecular interactions are discussed in this section. Since molecular interactions are discussed at the level of particles (ions, molecules) and not at the molar level (gram quantities), the thermodynamic functions such as energy E, enthalpy H, free energy A, free enthalpy G, and chemical potential μ obtained for a molecule must be multiplied with Avogadro number $\mathcal{N} = 6.02214179 \times 10^{23}$ mol^{-1} for obtaining the corresponding molar functions. Also, for interactions at the molecular level, instead of gas constant $R = 8.31451$ J deg^{-1} mol$^{-1} = 1.987$ cal deg^{-1} mol^{-1}, the Boltzmann contant $k_B = R/\mathcal{N} = 1.3806504\ 10^{-23}$ J K^{-1} will be used. However, for simplicity, the notation

E, H, A, G, μ, will not be different for quantities at the molecular level compared to those used for moles of material (bulk level). The SI units are typically used for expressing various physical parameters, although occasionally other units are indicated.

Some of the different thermodynamic functions at the molecular level may have identical values. For example, the relation between the energy E and free energy A is given by the formula $E = A - T\left(\dfrac{\partial A}{\partial T}\right)_V$. Assuming A is not temperature dependent, the result is $E = A$, and therefore the two thermodynamic functions can be used interchangeably. Also, at constant pressure and volume, rel. $\Delta G = \Delta A + p\,\Delta V$ indicates $\Delta G = \Delta A$. The assumptions that lead to the equality of different thermodynamic functions must be considered carefully when extending the findings at the molecular to bulk level. Also, a problem regarding the application of the finding at the molecular level to the bulk level is that some properties of the continuum, such as the dielectric constant, should also be considered for the interaction at the molecular level. The two basic approaches to estimating interactions from continuum characteristics or from molecular characteristics should lead to consistent results, and this may impose corrections to each of the two approaches.

An additional problem that arises when considering intermolecular interactions is that of a reference state. The expressions obtained for the interactions between molecules (or atoms, or ions) typically have a reference state where the distance between participating species is infinite ($r = \infty$), and all the values of energies (or free energies) must be considered as values for the change in energy (ΔE) or free energy (ΔA). Nevertheless, in the interest of simplicity, the sign Δ for difference is omitted in most cases.

In the formulas further developed for evaluating the interactions, besides one-to-one electrostatic interactions that would be expected in a particle model, some concepts from the continuum model (bulk property) are also included. The explanation for this approach comes from the idea that the interaction energy indicates the energy value of bringing the two particles at the distance of interaction as they approach each other through the bulk medium and not through vacuum.

Charge to Charge Interaction

The free energy of the interaction between two charges q_1 and q_2 separated by the distance r is described by the Coulomb law:

$$A = \frac{q_1 q_2}{4\pi\varepsilon_0\varepsilon r} = \frac{z_1 z_2 e^2}{4\pi\varepsilon_0\varepsilon r} \qquad (4.1.2)$$

where z_1 and z_2 are the charges given in terms of elementary charge ($e = 1.602\ 10^{-19}$ C where C indicates coulomb), ε_0 is the vacuum permittivity ($\varepsilon_0 = 8.854\ 10^{-12}$ C V^{-1} m^{-1}), and ε is the dielectric constant of the medium relative to vacuum ($\varepsilon = 1$ for vacuum) (a few values for ε are given in Table 4.1.3.) (*Note:* The dielectric constants ε_0 and ε should not be confused with elutropic strength ε^0.) Relation 4.1.2 shows that the ionic interactions are very strong. Since all the interactions are assumed to take place at constant volume (in condensed phase), this free energy of interaction can be assumed to be equal with the free enthalpy of interaction ($\Delta G = \Delta A + p\,\Delta V$). Therefore, $A = G$ for the specified system. Also, since no temperature T appears in the expression 4.1.2, it can be concluded that $A = E$, the total energy of the system.

In rel. 4.1.2 the dielectric constant ε is that of the bulk of the medium (and not that of the region between the charges). This value for ε brings to the particle model a bulk property, since the particles are brought at the distance of interaction through the medium with dielectric constant ε.

Taking as an example the energy between two ions with the charges $+1$ and -1, at the distance $r = 0.276$ nm $= 2.76$ Å (that corresponds to the distance between the centers of Na$^+$ and Cl$^-$ in NaCl), the calculation gives the following value:

$$A = \frac{-(1.602\ 10^{-19})^2}{4\pi(8.854\ 10^{-12})(0.276\ 10^{-9})}$$

$$= -8.4\ 10^{-19}\ \text{J} \qquad (4.1.3)$$

The value given by rel. 4.1.3 is negative, indicating the attraction of ions. For 1 mole of material, the calculation using rel. 4.1.3 gives $A = -8.4\ 10^{-19} \cdot 6.02214179\ 10^{23} = 505.85$ kJ/mol. Since 1 cal $= 4.184$ J, the interaction energy is 120.9 kcal/mol. The true energy in an ionic crystal should take into consideration all the interactions for the ion pairs in the crystal lattice, but by just comparing the energy of two ions with that of an harmonic oscillator at temperature T ($A = k_B\,T$), it results that at room temperature the ion energy is (in absolute value) about 200 times higher than that of the harmonic oscillator made by the two particles ($k_B\,T = 1.3806504\ 10^{-23} \cdot 300 \approx 4.1\ 10^{-21}$ J). In a solvent where the dielectric constant ε is significantly higher than 1 (e.g., for water $\varepsilon = 77.46$), the interaction energy is diminished. The variation of the free energy of interaction for two ions ($z_1 = -1$ and $z_2 = 1$ or 2) at a distance varying from less than 10 Å up to 100 Å in water is shown in Figure 4.1.1. This variation assumes that no other interactions take place in solution and that no other ions are present. In reality, the decrease (in absolute value) of the interaction energy is much more rapid due to the screening caused by other particles present in solution (other ions and solvent molecules). Nevertheless, the interactions between ions in solution extend at distances much longer than molecular radii of small molecules. The value of the interaction force f between two charges can be easily obtained from rel. 4.1.2 using the expression:

$$f = -\frac{dE(r)}{dr} = -\frac{dA(r)}{dr} = \frac{z_1 z_2 e^2}{4\pi\varepsilon_0\varepsilon r^2} \qquad (4.1.4)$$

(attractive force is indicated by a negative sign, which is obtained if z_1 and z_2 are of opposite

FIGURE 4.1.1 Hypothetical variation of the free energy of interaction for two ions ($z_1 = -1$ and $z_2 = +1$ or $+2$) at a distance varying from less than 10 Å up to 100 Å in water.

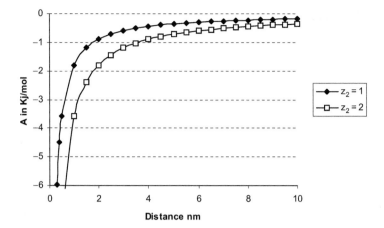

sign). Since the force is a vector, the interacting force between charges should have a sense and a direction. These are the same as the radius vector between the charges. In other words, expression 4.1.4 should be written in vector form as follows:

$$\vec{f} = \frac{z_1 z_2 e^2}{4\pi\varepsilon_0 \varepsilon r^2} \frac{\vec{r}}{r} \qquad (4.1.5)$$

The expression for the force also allows evaluation of the electrical field created by a charge q. The intensity of this field is given by the expression:

$$\vec{E} = \frac{q}{4\pi\varepsilon_0 \varepsilon r^2} \frac{\vec{r}}{r} \qquad (4.1.6)$$

and the field $\vec{E}_1$ generated by the charge $q_1 = z_1 e$ acting on the charge $q_2 = z_2 e$ generates the force given by the formula 4.1.5. Similar to the force, the electrical field is a vector, and its direction is the same as that of the radius vector between the charges.

In reality, as previously indicated for the interaction energy, in a solution containing ions, the electric field is strongly perturbed by other ions and polar particles in the surrounding solution, and the resulting energy decreases faster than $1/r^2$ as predicted by rel. 4.1.4 [2].

Energy of an Ion in a Continuous Medium

An ion, even if it is not interacting with other ions (or electrical charges), has a specific free energy, equal to the electrostatic work done for forming that ion. For an ion with the charge q and radius r, the increase of the charge with dq will require the energy given by the expression:

$$dA = \frac{q dq}{4\pi\varepsilon_0 \varepsilon r} \qquad (4.1.7)$$

The total free energy for charging one ion to the final charge ze will be:

$$A = \int_0^{ze} \frac{q dq}{4\pi\varepsilon_0 \varepsilon r} = \frac{(ze)^2}{8\pi\varepsilon_0 \varepsilon r} \qquad (4.1.8)$$

The energy given by rel. 4.1.8 is also known as Born energy. This energy is positive, indicating that the formation of an ion requires external work. In a medium with dielectric constant ε, the energy necessary to form a mole of ions will be given by the expression:

$$A = \mathcal{N} \frac{(ze)^2}{8\pi\varepsilon_0 \varepsilon r} \qquad (4.1.9)$$

The Born energy given by rel. 4.1.9 with $\varepsilon = 1$, but with a negative sign, can be considered as approximating the energy necessary to bring a mole of ions from vacuum into a solvent with a high dielectric constant. The expression of the free energy (at constant volume) for this transfer is given by the formula:

$$\Delta A = -\mathcal{N}\frac{(ze)^2}{8\pi\varepsilon_0 r}\left(1-\frac{1}{\varepsilon}\right) \approx -\mathcal{N}\frac{(ze)^2}{8\pi\varepsilon_0 r} \quad (4.1.10)$$

For this reason, Born energy is also indicated as solvation energy of ions (not considering the energy necessary to create the cavity in the solvent for accommodating the ions).

The change of the ionic interactions in different solvents due to the differences in the dielectric constant ε can be used to explain the difference in salt solubility in two solvents with different dielectric constants ε_1 and ε_2. In this case, using rel. 4.1.9 for the expression of the free energy, the change of the ionic interactions in different solvents can be written:

$$\Delta A = \mathcal{N}\frac{(ze)^2}{8\pi\varepsilon_0 r}\left(\frac{1}{\varepsilon_1}-\frac{1}{\varepsilon_2}\right) \quad (4.1.11)$$

The value of ΔA is negative when $\varepsilon_1 > \varepsilon_2$, indicating that the solubility is higher in a solvent with higher dielectric constant. Although this approach represents only a rough approximation of reality, an (approximate) linear dependence (with negative slope) was observed between the values for the solubility of monovalent ions and $1/\varepsilon$ of the solvent [3].

Ions in solutions or ionized groups connected to a stationary phase are typically encountered in ion chromatography and in ion-exchange processes. Also, in other types of chromatography, ion formation and interactions play a role in the separation process. For example, in RP HPLC performed on a silica-based stationary phase, the residual silanol groups on the stationary phase may ionize when the pH value of the mobile phase is high, and the silanol groups may bring additional types of interactions besides those involving hydrophobic forces. The comparison of interacting energies for ionic forces with the energies of other forces (polar-polar, between polarizable molecules, etc.) indicates that ionic forces are comparably strong. Also, comparing the interacting distances, the ionic forces act at longer distances. This is an additional reason to eliminate as much as possible ionic interactions in RP HPLC that may affect in an unpredictable way the results of a separation that is assumed to be based on a unique mechanism of separation.

Polar Molecules

Polarity in a molecule refers to the separation of the center of positive charges from that of negative charges, leading to an electric dipole of the molecule. The dipole of a polar molecule is characterized by the dipole moment m defined by the formula:

$$m = q\,d \quad (4.1.12)$$

where d is a vector with the length equal to the distance (in nm) between the two separated point charges $+q$ and $-q$ directed from the negative charge to the positive charge. The vector character of the dipole moment is frequently neglected. The unit of dipole moment is the debye (D), and $1\,D = 3.336\ 10^{-30}$ C m. Besides molecular dipole moments, the value for m can be calculated for chemical bonds and also for molecular fragments. Extensive tables with dipole moments are available in the literature [4]. Water, for example, has a dipole moment $m = 1.85\,D = 6.1716\ 10^{-30}$ C m. This value allows the calculation of the dipole moment of the OH bond. Considering that the H-O-H angle in water is $\theta = 104.5°$, the OH bond dipole moment will be $m_{OH} = m_{H2O}\ /\ [2\cos(0.5\,\theta)] = 1.51$ D.

The generation of the dipole moment can be explained by the displacement of the center of all electrons in the valence shell of the molecule, which is not located over the center of positive charges in the molecule.

For example, considering the water molecule with 8 valence electrons, (core electrons of oxygen are not included), the distance separating the charges (of $8e$) is given by $d = m/q = 6.2 \ 10^{-30}/(8 \cdot 1.602 \ 10^{-19}) = 4.84 \ 10^{-12}$ m ≈ 0.005 nm (ten times smaller distance compared to 0.05 nm the Bohr radius of hydrogen). This indicates for water a very small charge separation compared to the molecule diameter. Using this model of creating the dipole moment, a very small separation in space of the charge centers is typical for most polar molecules.

More commonly, the charges that generate the dipole moment can be considered as partial point charges centered at the extreme atoms in the molecule. (Various notations for the partial point charges can be encountered in the literature.) For a water molecule as an example, the partial charge $q = 0.41$ separated by a distance of 0.094 nm will generate the dipole moment $m = (0.41 \cdot 1.602 \ 10^{-19}) \cdot (0.094 \ 10^{-9}) = 6.1741 \ 10^{-30} \approx 1.85$ D.

The self-energy for a dipole can be calculated from the sum of the Born energies of the two partial charges $+q$ and $-q$ (on ions assumed to have radius $a = d/2$) minus the energy to bring the two charges together (separated by the distance $r = d$). Therefore, Born energy for a dipole is given by the expression:

$$A = \frac{1}{4\pi\varepsilon_0\varepsilon}\left(\frac{q^2}{2a} + \frac{q^2}{2a} - \frac{q^2}{r}\right) = \frac{q^2}{8\pi\varepsilon_0\varepsilon a}$$

$$= \frac{m^2}{4\pi\varepsilon_0\varepsilon d^3} \qquad (4.1.13)$$

This indicates that the self-energy for dipoles is about the same as that of an ion where the charge $z\,e$ is replaced by the partial charge q. This energy (with negative sign) can also be related to the solvation energy of a molecule with dipole moment m.

Relation 4.1.13 also allows an understanding of differences in the solubility of a polar compound in different solvents. Similar to the ionic compounds, for polar compounds an (approximate) linear dependence (with negative slope) was observed between the values for the solubility and $1/\varepsilon$ of the solvent. This indicates that the solubility of polar compounds parallels that of ionic compounds and that they have a higher solubility in solvents with higher dielectric constant.

While the definition of the dipole moment using the concept of two point charges is simple, in reality the charge distribution in a molecule is a more complicated concept. For this reason, it is more convenient to use for dipole moment definition the volume charge density $\rho(\vec{r})$, which is the amount of electric charge in an infinitesimal volume at a point $\vec{r}$. In this case, the dipole moment is defined by the formula:

$$m = \int_V \rho(\vec{r})\,\vec{r}\,dV \qquad (4.1.14)$$

The charge density (as well as the dipole moment) can be obtained, for example, using quantum chemical calculations [5] (V is here the volume of the molecule). For a molecule, the charge density can also be evaluated from an array of point charges q_i, using the expression:

$$\rho(\vec{r}) = \sum_{i=1}^{n} q_i \delta(\vec{r} - \vec{r_i}) \qquad (4.1.15)$$

where δ is the Dirac delta function ($\delta = 0$ for $\vec{r} \neq \vec{r_i}$ and ($\delta = 1$ for $\vec{r} = \vec{r_i}$) and $\vec{r_i}$ is the vector to the charge q_i from a reference point. With this observation, the dipole moment is given by the formula:

$$m = \sum_{i=1}^{n} q_i \int_V \delta(\vec{r} - \vec{r_i}) r dV = \sum_{i=1}^{n} q_i \vec{r_i} \quad (4.1.16)$$

Relation 4.1.16 indicates that the dipole moment is the vector sum of individual dipole moments created by the point charges.

Ion-to-Dipole Interaction

The interaction between an ion and a dipole can be reduced to the interaction between a charge ze and the two partial charges $+q$ and $-q$ of the dipole. Assuming the length of the dipole is d and the distance between the charge and the middle of the dipole is r, the Coulomb energy of the interaction is given by the expression:

$$A = \frac{ze\,q}{4\pi\varepsilon_0\varepsilon}\left(\frac{1}{AB} - \frac{1}{AC}\right) \tag{4.1.17}$$

where the distances AB and AC are shown in Figure 4.1.2. From simple geometric considerations it can be shown that the distances AB and AC can be calculated with the formulas:

$$AB = \left[(r - 0.5\,d\cos\theta)^2 + (0.5\,d\sin\theta)^2\right]^{1/2}$$

$$\approx r - 0.5\,d\cos\theta \tag{4.1.18a}$$

$$AC = \left[(r + 0.5\,d\cos\theta)^2 + (0.5\,d\sin\theta)^2\right]^{1/2}$$

$$\approx r + 0.5\,d\cos\theta \tag{4.1.18b}$$

(the approximation in rel. 4.1.18a and b being valid when $r \gg d$). Replacing the expressions

for AB and AC in rel. 4.1.17, the interaction energy becomes:

$$
\begin{aligned}
A &= -\frac{(ze)q}{4\pi\varepsilon_0\varepsilon}\left(\frac{1}{r - 0.5\,d\cos\theta} - \frac{1}{r + 0.5\,d\cos\theta}\right) \\
&= -\frac{(ze)q}{4\pi\varepsilon_0\varepsilon}\left(\frac{d\cos\theta}{r^2 - 0.25\,d^2\cos^2\theta}\right)
\end{aligned}
$$

$$\tag{4.1.19}$$

Since $r \gg d$, rel. 4.1.19 can be reduced to the following expression:

$$A = -\frac{(ze)m\cos\theta}{4\pi\varepsilon_0\varepsilon r^2} \tag{4.1.20}$$

Relation 4.1.20 allows the calculation of the energy of interaction between a dipole and an ion. This energy is attractive if the negative partial charge of the dipole points toward a positive ion ($0° < \theta < 90°$) showing a maximum for $\theta = 0°$, and it is repulsive if the positive charge points toward the positive partial charge of the dipole ($90° < \theta < 180°$), showing a minimum for $\theta = 180°$.

The ion-dipole interaction is weaker than the ion-ion interaction, but when the separation distance ion-dipole is shorter than 0.2–0.4 nm, it is still significantly higher than $k_B T$ ($k_B T \approx 4.1\ 10^{-21}$ J for $T = 300$ K). For example, using rel. 4.1.20 with $\theta = 0$ and considering a water molecule ($m = 1.85$ D gas phase) with the distance to an ion $r = 0.235$ nm (the estimated distance between a Na^+ ion and a water molecule in vacuum), the interaction energy in vacuum has the following value:

$$
\begin{aligned}
A &= -\frac{(1.602\ 10^{-19})(1.85 \cdot 3.336\ 10^{-30})}{4\pi(8.854\ 10^{-12})(0.235\ 10^{-9})^2} \\
&= -1.61\ 10^{-19}\text{J}
\end{aligned}
$$

This resulting value for A is about 40 times (in absolute value) higher than $k_B T$. However, at an ion-dipole distance of about 1.5 nm, $A \approx k_B T$ (in absolute value). For a molecule with $m = 1.5$ D

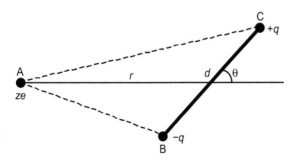

FIGURE 4.1.2 Schematic drawing of the interaction between a charge ze and a dipole generated by the partial charges q.

interacting at the distance of 0.4 nm with an ion with $z = 1$, the free energy $A \approx 4.50\ 10^{-20}$ J, which is equivalent to 27.1 kJ/mol. For the same interacting distance, this energy is higher than the sum of other interactions not involving ions (see rel. 4.1.46) and shows the importance of the presence of ions in systems involved in various separations. Multiplying with $\mathcal{N}$ the previous value for A, the resulting value is about 96 kJ mol^{-1} or about 23 kcal mol^{-1}. This high energy explains the orientation of dipoles of the solvent around the ions in solution. In the case of water, this process is known as hydration of the ions, and a hydration number can be associated with different ions. A solvation zone is also present around the ions in a solution.

The assumption made for obtaining rel. 4.1.20 for the free energy A that the dipole-ion interaction is present in a fixed position is valid when the ions and the dipoles are in close proximity, but it becomes a strong simplification when the distances are larger. When A given by rel. 4.1.20 (for $\theta = 0$) is of the order of $k_B T$, a Boltzmann angle-averaged expression for A should replace rel. 4.1.20. This expression is obtained from *potential distribution theorem* (see, e.g., [3]), and in this case the interaction energy between an ion and a dipole has the expression:

$$A \approx -\frac{(ze)^2\ m^2}{6k_B T (4\pi\varepsilon_0\varepsilon)^2\ r^4} \qquad (4.1.21)$$

The values for A generated using rel. 4.1.21 are smaller than those obtained using rel. 4.1.20. At a distance $r = 1.5$ nm between the ion and dipole, using rel. 4.1.21, the resulting value is about $A = -6.3\ 10^{-22}$ J (almost 7 times smaller in absolute value than $k_B T$ for T 300 K). This result indicates that outside of the first layer of solvating molecules that surround an ion, the molecules of a polar solvent are likely to rotate freely and not be further clustered around the ion.

One important observation regarding the expression of A given by rel. 4.1.21 is its dependence on temperature. This indicates that the total energy E for the dipole-ion interaction is no longer equal with A. Applying the formula:

$$E = A - T\left(\frac{\partial A}{\partial T}\right)_V \qquad (4.1.22)$$

to the expression of total energy E_{i-d}, and noticing that $\left(\dfrac{\partial A}{\partial T}\right) = \dfrac{(ze)^2\ m^2}{6k_B T^2 (4\pi\varepsilon_0\varepsilon)^2\ r^4}$, the result for the total energy is given by the expression:

$$E_{i-d} \approx -\frac{(ze)^2 m^2}{3k_B T (4\pi\varepsilon_0\varepsilon)^2 r^4} \qquad (4.1.23)$$

The part of energy that is not available as free energy of interaction is associated with the entropy in the system.

Dipole-to-Dipole Interaction

Two polar molecules interact through Coulomb-type forces, similarly to the interaction of ions and dipoles. Assuming the dipoles in a fixed position (not free rotating), the same procedure as for the interaction between an ion and a dipole can be followed. The resulting energy for two dipoles is given by the formula:

$$A = -\frac{m_1 m_2}{4\pi\varepsilon_0\varepsilon r^3}(2\cos\theta_1\cos\theta_2 - \sin\theta_1\sin\theta_2\cos\varphi)$$

$$(4.1.24)$$

where $m_1 = d_1 q_1$, $m_2 = d_2 q_2$, and the angles θ_1, θ_2 and φ are defined in Figure 4.1.3. The expression 4.1.24 of A for $\theta_1 = 0$, $\theta_2 = 0$, (φ irrelevant) becomes:

$$A = -\frac{m_1 m_2}{2\pi\varepsilon_0\varepsilon r^3} \qquad (4.1.25)$$

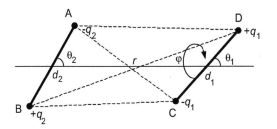

FIGURE 4.1.3 Schematic drawing of the interaction between two fixed dipoles m_1 and m_2.

The evaluation of this energy, for example, for two adjacent water molecules (at a distance $r = 0.28$ nm), gives $A \approx -3.1\ 10^{-20}$ J, indicating a quite strong interaction for the water molecules. The same calculation for $r = 0.55$ nm gives $A \approx -4.1\ 10^{-21}$ J. The comparison of these results with $k_B T \approx 4.1\ 10^{-21}$ J for $T = 300$ K shows that the dipole-dipole interaction becomes closer to the thermal energy of a harmonic oscillator as the distance between the interacting molecules increases. For this reason it is expected that the dipole orientation fails to be optimal when the energy A calculated with formula 4.1.25 becomes closer to $k_B T$, and an angle-averaged expression for A should replace rel. 4.1.25. The angle-averaged expression of energy A for the dipole-dipole interaction is the following:

$$A = -\frac{m_1^2 m_2^2}{3k_B T (4\pi\varepsilon_0\varepsilon)^2 r^6} \qquad (4.1.26)$$

For a distance $r = 0.28$ nm, $A \approx -2.0\ 10^{-20}$ J, only slightly lower than the one generated with rel. 4.1.25, but the energy calculated with rel. 4.1.26, has a much faster decrease as the distance r increases. For $r = 0.55$ nm, when rel 4.1.25 indicates an energy equal to that of thermal energy of a harmonic oscillator $k_B T$, the value of energy calculated with formula 4.1.26 is $A \approx -3.4\ 10^{-22}$ J. The angle-averaged energy between two dipoles is usually referred to as the *Keesom interaction*.

Similar to the case of ion-dipole interaction, the angle-averaged expression for the energy of two dipoles is temperature dependent. From expression 4.1.22 which relates E and A, the total energy of the interaction is double that given by rel. 4.1.26, which can be written as:

$$E_{d-d} = -\frac{2m_1^2 m_2^2}{3k_B T (4\pi\varepsilon_0\varepsilon)^2 r^6} \qquad (4.1.27)$$

The decrease (in absolute value) of the calculated E_{d-d} for two molecules as a function of temperature is shown, as an example, in Figure 4.1.4. The calculations showing the molar energy (where $E_{d-d} = 2\mathcal{N}A$) were based on the value $m = 1.5$ D and on an intermolecular distance of 0.3 nm. As seen from Figure 4.1.4, the increase in temperature decreases the interaction energy.

Polarizability of Molecules

Under the influence of an electrical field $\overrightarrow{E}$, molecules as well as atoms show a displacement of the electron cloud relative to the center of positive charges (nucleus for the atoms), generating an induced dipole moment m_{ind}.

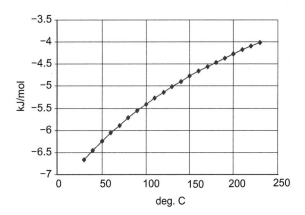

FIGURE 4.1.4 The temperature dependence of dipole-dipole interaction energy E_{d-d} for two molecules at 3 Å distance.

Polarizability (in electric field) α is defined by the ratio:

$$\alpha = \frac{\overrightarrow{m}_{ind}}{\overrightarrow{E}} \qquad (4.1.28)$$

The SI units of polarizabilty α are $C^2\,m^2\,J^{-1}$ ($C^2m^2J^{-1} = C\,m^2\,V^{-1}$), but more common values are expressed in units of $4\pi\varepsilon_0(\text{Å})^3 = 4\pi\varepsilon_0 10^{-30}$ $m^3 = 1.11\,10^{-40}\,C^2\,m^2\,J^{-1}$. Tables of polarizability, as well as procedures for its estimation using semiempirical formulas or quantum-mechanical calculations [5], are reported in the literature. As defined by rel. 4.1.28, polarizability is a scalar quantity indicating that the electric field produces polarization only in the direction of the field. However, polarizability can also be defined as a tensor when the electrical field generates moments of dipole in different directions from that of field $\overrightarrow{E}$.

Besides the displacement of the electron cloud (electronic polarizability), the molecules that have a permanent dipole also have an *orientational polarizability* resulting from orientation of the rotating dipoles.

A simple evaluation of electronic polarizability can be obtained considering that an electron is displaced by the length d in the electrical field E (vector sign not shown), as indicated schematically in Figure 4.1.5. In this case, the induced dipole moment will be $m_{ind} = \alpha_0\,E = d\,e$ (where d is the projection of the distance between charges $+e$ and $-e$ in the molecule). The force (vector sign not shown) between the

two charges in the presence of a field is given by expression 4.1.4, which can be written as follows:

$$f = -\frac{e^2}{4\pi\varepsilon_0 L^2}\sin\theta = \frac{e^2 d}{4\pi\varepsilon_0 L^3} = \frac{e m_{ind}}{4\pi\varepsilon_0 L^3}$$
$$(4.1.29)$$

Since the force is given by $f = e\,E$, from rel. 4.1.29, the result for the value for the induced dipole moment is $m_{ind} = 4\pi\varepsilon_0 L^3\,E$. From this relation the value for the electronic polarizability is:

$$\alpha_0 = 4\pi\varepsilon_0 L^3 \qquad (4.1.30)$$

This formula can be used for the electronic polarizability not only for atoms, but also for molecules where L is the radius of the molecule. It is common to express the electronic polarizability in units $(4\pi\varepsilon_0)\,10^{-30}\,m^3$ (L being given in Å).

The expression for orientational polarizabilty for a molecule with the permanent dipole moment m is obtained from the potential distribution theorem as the Boltzmann angle-averaged expression in the form:

$$\alpha_{orient} = m^2/3k_B T \qquad (4.1.31)$$

The total polarizability is therefore obtained from the formula:

$$\alpha = 4\pi\varepsilon_0 L^3 + m^2/3k_B T \qquad (4.1.32)$$

Relation 4.1.32 is known as the Debye-Langevin equation. Molecular polarizabilities can be obtained from molecular orbital calculations [5] or by semiempirical evaluations [6].

Ion-to-Molecule Interaction

The influence of an ion placed at the distance r from a neutral molecule with α polarizability has as a result the induction of a dipole moment $m_{ind} = \alpha\,E$ in the molecule. This induced dipole moment can be assumed as created by the

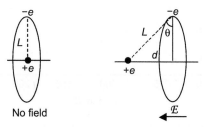

No field

FIGURE 4.1.5 Schematic diagram of the displacement of an electron in an electric field.

displacement of a charge q at a distance d in the molecule. The resulting force (vector sign not shown) on the molecule is therefore given by the formula:

$$f = q\frac{d\mathcal{E}}{dr} = \alpha\mathcal{E}\frac{d\mathcal{E}}{dr} \qquad (4.1.33)$$

The free energy for the interaction of the ion and the molecule is therefore given by the expression:

$$A = -\int f dr = -0.5\alpha\mathcal{E}^2 \qquad (4.1.34)$$

The intensity $\mathcal{E}$ of the field created by an ion with the charge ze is given using rel. 4.1.6 by the formula:

$$\mathcal{E} = \frac{ze}{4\pi\varepsilon_0\varepsilon r^2} \qquad (4.1.35)$$

Introducing expression 4.1.35 for $\mathcal{E}$ in rel. 4.1.34, the result is:

$$
\begin{aligned}
A &= -\frac{\alpha(ze)^2}{2(4\pi\varepsilon_0\varepsilon)^2 r^4}\\
&= -\frac{(ze)^2}{2(4\pi\varepsilon_0\varepsilon)^2 r^4}\left(\alpha_0 + \frac{m^2}{3k_BT}\right) \qquad (4.1.36)
\end{aligned}
$$

The second term in rel. 4.1.36 is identical to the value from rel. 4.1.21, indicating the interaction between an ion and a permanent dipole.

Dipole-to-Molecule Interaction

The interaction of a dipole with a polarizable molecule can be evaluated using the same procedure as previously used for an ion-to-molecule interaction, considering that the induced dipole is generated by a charge q of the dipole. For a fixed dipole m oriented at an angle θ to the line joining it to a polarizable molecule, the intensity of the electrical field generated is given by the expression:

$$\mathcal{E} = \frac{m(1 + 3\cos^2\theta)^{1/2}}{4\pi\varepsilon_0\varepsilon r^3} \qquad (4.1.37)$$

The free energy of interaction is given by rel. 4.1.34, which will take the form:

$$A = -\frac{m^2\alpha_0(1 + 3\cos^2\theta)}{2(4\pi\varepsilon_0\varepsilon)^2 r^6} \qquad (4.1.38)$$

For the angle averaged free energy, expression 4.1.38 becomes:

$$A = -\frac{m^2\alpha_0}{(4\pi\varepsilon_0\varepsilon)^2 r^6} \qquad (4.1.39)$$

It can be observed that replacing α_0 in rel 4.1.39 with α_{orient} given by rel. 4.1.31, the resulting energy is Keesom energy given by rel. 4.1.26.

For two molecules each possessing permanent dipole moments m_1 and m_2 and polarizabilities $\alpha_{0,1}$ and $\alpha_{0,2}$, the expression for the free energy of interaction becomes:

$$A = -\frac{m_1^2\alpha_{0,2} + m_2^2\alpha_{0,1}}{(4\pi\varepsilon_0\varepsilon)^2 r^6} \qquad (4.1.40)$$

This energy is known as *Debye interaction energy*. Since the free energy in rel. 4.1.40 does not depend on temperature, it can be concluded that the energy of the system is given by the expression:

$$E_{d-id} = -\frac{m_1^2\alpha_{0,2} + m_2^2\alpha_{0,1}}{(4\pi\varepsilon_0\varepsilon)^2 r^6} \qquad (4.1.41)$$

Nonpolar Molecule-to-Molecule Interaction

A considerably large number of molecules do not possess a permanent dipole moment (or have ionic character) or have a very small dipole moment. An example of a nonpolar molecule and the calculated point charges of its atoms is given in Figure 4.1.6 for octane. Point charges in a molecule can be calculated by different techniques and may lead to different results [5]. The charges indicated in Figure 4.1.6 were calculated using

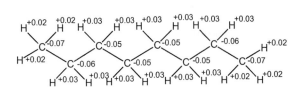

FIGURE 4.1.6 Calculated point charges in a molecule of octane (with MarvinSketch 5.4).

MarvinSketch 5.4 Plug-ins [7]. The same charges calculated, for example, with an empirical molecular orbital method (e.g., MOPAC-7 with AM1 parametrization [8]) can lead to charges as high as $+0.088$ for hydrogens and -0.2 for carbons. Since the electrons forming the C−H bonds are not perfectly distributed between the two atoms, small partial point charges are present for each atom. However, the dipole moment of such molecule is zero. Since water is the most common polar solvent and the nonpolar molecules or molecules with a very small dipole moment are not water soluble (and their surface does not get wet with water), such nonpolar compounds are frequently referred to as *hydrophobic*.

The interaction between two molecules that do not possess a dipole moment can be approached by considering that one molecule has the polarizability α_0, and the other possesses an instantaneous dipole moment m_{inst} created by the orbiting frequency of its electrons. In this case, rel. 4.1.40 can be applied for evaluating the free energy of their interaction. From quantum mechanical evaluations, it can be shown that $m_{inst} \approx {}^{3}/_{4}\,\alpha_0\,I$ for atoms and small molecules with the polarizability α_0 and I ionization potential. (The ionization potential is typically expressed in e.V, 1 e.V. $= 1.60218\ 10^{-19}$ J). This indicates that the free energy of interaction between two identical small molecules is given by the expression ($\varepsilon = 1$):

$$A = -\frac{3\alpha_0^2 I}{4(4\pi\varepsilon_0)^2\,r^6} \qquad (4.1.42)$$

For two dissimilar small atoms or molecules, the formula of free energy interaction becomes:

$$A = -\frac{3\alpha_{0,1}\alpha_{0,2}}{2(4\pi\varepsilon_0)^2\,r^6}\frac{I_1 I_2}{(I_1 + I_2)} \qquad (4.1.43)$$

The energy of interactions between nonpolar but polarizable molecules is known as *London dispersion energy*.

Expression 4.1.43 for the free energy of interaction of two molecules is independent of temperature, and therefore the dispersion energy $E_d = A$, where A is given by rel. 4.1.43.

The application of London formulas to the interaction of larger molecules or to asymmetric molecules leads to badly underestimated results mainly because the dispersion forces no longer act between the centers of the molecules but between various centers of electronic polarization within each molecule. More elaborate calculations, including those in evaluating the sum of each local interaction, can be done to obtain more accurate results. For this reason, the total dispersion energy E_d for a polyatomic molecule composed from j polarizable groups interacting with a surrounding phase composed from k polarizable groups (connected or not in a molecule) is obtained as a sum of all possible interactions:

$$E_d = \sum_i \sum_k (E_d)_{i,k} \qquad (4.1.44)$$

Hydrophobic character can be attributed not only to a whole molecule, but also to a moiety of a molecule such as in the case of a fatty acid that has a long hydrocarbon chain attached to a carboxyl group. The hydrocarbon chain can be referred to as a hydrophobic moiety and the carboxyl group as a polar one, although the whole molecule has a permanent dipole moment. An example of such a molecule is octanoic acid. The calculated point charges for octanoic acid (using MarvinSketch 5.4 Plug-ins [7]) are shown in Figure 4.1.7. As shown on the

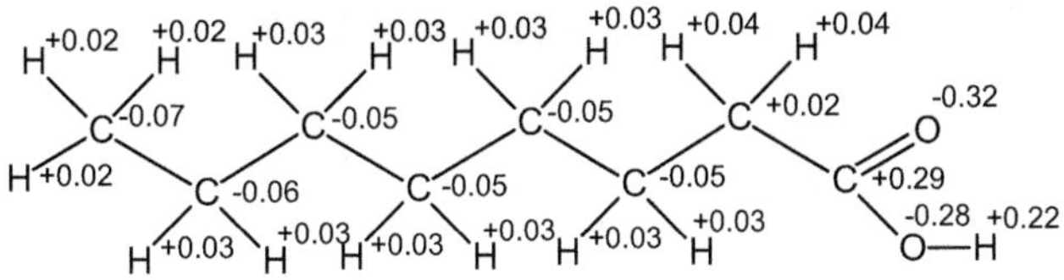

FIGURE 4.1.7 Calculated point charges in a molecule of octanoic acid (with MarvinSketch 5.4).

octanoic acid molecule, relative large point charges are present on the carboxylic group, but very rapidly along the hydrocarbon chain the point charges become equal to those in octane molecule. The dipole moment for octanoic acid is $m = 2.05$ D (calculated using MOPAC 7). However, though polar, the interactions attributed to the long hydrophobic chain cannot be neglected. The same separation of interactions from different parts of a polyatomic molecule as described for the polarizable interactions can be extended to other interactions. For example, when considering a molecule with polar groups and with a hydrophobic moiety (as in the example of octanoic acid), the polar part of interaction and the dispersion energy caused by the polarizability of specific molecular moieties must be added together.

Unified View of Interactions in the Absence of Ions

In the absence of ions, the total interaction of molecules in a medium with dielectric constant ε is given by the expression:

$$E_T = E_{d-d} + E_{d-id} + E_d \qquad (4.1.45)$$

$$E_T = -\frac{1}{(4\pi\varepsilon_0)^2}\left(\frac{2m_1^2 m_2^2}{3k_B T \varepsilon^2} + \frac{m_1^2 \alpha_{0,2} + m_2^2 \alpha_{0,1}}{\varepsilon^2}\right.$$

$$\left. + \frac{3\alpha_{0,1}\alpha_{0,2}}{2}\frac{I_1 I_2}{I_1 + I_2}\right)\frac{1}{r^6}$$

$$(4.1.46)$$

The evaluation of E_T at 300 K for an idealized pair of molecules with $m = 1.5$ D, $\alpha_0 = 4.2$

$(\cdot 1.11\ 10^{-40}\ C^2 m^2 J^{-1})$, $\varepsilon = 1$, and $I = 10.9$ e.V, for a distance varying between 3 Å and 7 Å generated the graph shown in Figure 4.1.8 (energy expressed per mol after multiplication of rel. 4.1.46 with $\mathcal{N}$). As shown in Figure 4.1.8, the dispersion energy E_d has the largest contribution to E_T (about 69%), followed by the dipole-dipole energy (E_{d-d} represents about 25% of E_T) and by dipole-induced dipole interaction E_{d-id}, which accounts for only about 6% of the total energy. As expected, the interaction energy decreases (in absolute value) as the distance between the molecules increases. For small molecules in the gas phase, the above calculations were proven relatively close to experimental data obtained based on the deviation from perfect gas law (van der Waals coefficients). For this reason, the forces leading to the energy E_T given by rel. 4.1.46 are frequently indicated as van der Waals forces. However, for the condensed phase, the theory must be further reevaluated.

Lennard-Jones Potential

Even assuming no free charges in a medium, based on rel. 4.1.46, the result seems to indicate that the only interaction between molecules would be the attraction with an energy proportional with $1/r^6$. However, in reality, the attraction between molecules is maximum to a specific distance r, and at very small molecular distances a strong repulsive force intervenes, not allowing the overlapping of the electronic clouds of the two molecules. One approach to explain why atoms cannot come close to each

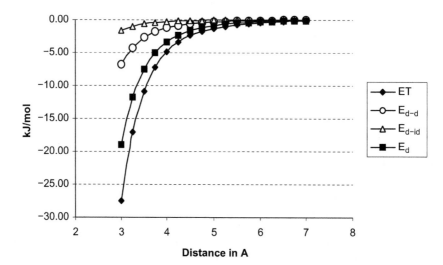

FIGURE 4.1.8 Theoretical calculation of E_{d-d}, E_{d-id}, E_d, and E_T for an idealized pair of molecules with $m = 1.5$ D, $\alpha_0 = 4.2$ ($\cdot$ 1.11 10^{-40} $C^2m^2J^{-1}$), $\varepsilon = 1$, and $I = 10.9$ e.V, at 300 K.

other beyond a specific distance is to consider that the atoms can be viewed as hard, incompressible spheres. The radius of this sphere is known as the *van der Waals radius* of the atom. A specific distance between atoms is also maintained between covalently bound atoms. The distance between two atoms in a molecule can be expressed as the sum of their covalent bond radii, equal with the *bond length*. Single-bond covalent radii of atoms are about 0.08 nm shorter than their van der Waals radii, which can be explained by a slight deformation of the atom sphere when they are connected by a covalent bond. Based on the atomic van der Waals radii and the bond length, the van der Waals radii, volumes, and surface area of molecules can be calculated [9]. The volume of the molecule is no longer a sphere, but in some instances molecular volumes are approximated with that of spheres. The molecular dimensions allow the calculation of a van der Waals *distance* between the centers of the interacting molecules (or atoms), and as an approximation, this distance can be obtained as the sum of van der Waals radii of the two "spherical" molecules.

However, the true shape of molecules can be very far from that of a sphere.

In order to account for the impossibility of two atoms or molecules coming closer to each other than their van der Waals radii, a repulsion potential must be included in the description of the total intermolecular interaction. One such potential is given by the semiempirical formula known as the Lennard-Jones potential, which describes the interaction between molecules as a function of the distance r by the expression:

$$E_{L-J}(r) = -\frac{A'}{r^6} + \frac{B'}{r^{12}} \qquad (4.1.47)$$

(where A' and B' are compound specific parameters). An alternative to rel. 4.1.47 is the formula:

$$E_{L-J}(r) = -4E_T(r_0)\left[\left(\frac{r_0}{2^{1/6}\,r}\right)^{12} - \left(\frac{r_0}{2^{1/6}\,r}\right)^6\right]$$

$$(4.1.48)$$

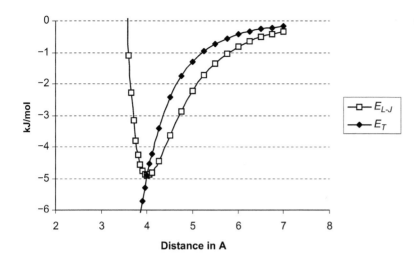

FIGURE 4.1.9 Variation of the interaction energy $E_{L\text{-}J}$ calculated with rel 4.1.48, and of E_T calculated with rel. 4.1.46 between two molecules with $m = 1.5$ D, $\alpha_0 = 4.2$ ($\cdot\ 1.11\ 10^{-40}\ C^2m^2J^{-1}$), $\varepsilon = 1$, and $I = 10.9$ e.V, at 300 K, when $r_0 = 4$ Å.

where E_T (having a negative value) is calculated using rel. 4.1.46 for r_0 equal with van der Waals distance between the interacting particles (multiplication with $\mathcal{N}$ is necessary to obtain values per mol). The dependence of $E_{L\text{-}J}$ energy on the distance r between molecules is shown in Figure 4.1.9, for $r_0 = 4$Å, together with the values for E_T for various distances obtained with rel. 4.1.46 (also shown in Figure 4.1.8). From rel. 4.1.48, it can be seen that for $r = r_0$, the coefficients are $(1/2^{1/6})^{12} = 0.25$ and $(1/2^{1/6})^6 = 0.5$, such that $E_{L\text{-}J} = E_T$. The energy of van der Waals interaction as given by rel. 4.1.48 depends on the distance r and also on the dipole moment m, polarizability α, and the ionization potential I, through E_T.

Hydrogen-Bond Interactions

Another important type of molecular interaction is caused by hydrogen bonds. The hydrogen bond appears to be partly electrostatic and partly covalent. In this bond, a hydrogen atom covalently bonded by atom A (A-H) is attracted by an atom containing a free electron pair :B. This leads to a strong polarization of the A-H bond and to

electrostatic interactions between $H^{\delta+}$ and $:B^{\delta-}$. This type of interaction occurs, for example, between a molecular group, such as O-H or S-H, which carries a marked electric dipole moment and O or N atoms from another molecule. This latter atom is characterized by the presence of at least one nonbonding orbital that can point toward the H-atom of the polar group of the molecule and is filled with a lone pair of electrons. The polar group that carries the H-atom is called the donor, while the group containing O or N with a nonbonding orbital is called the acceptor. This H-bond can be encountered in HPLC between different species from the mobile phase or between molecules interacting with groups in the stationary phase (residual silanol, amino, carboxyl, hydroxyl, etc.). Several hydrogen bond interactions are schematically depicted as follows:

Generally, according to their energy of formation, H-bonds are classified into three main categories: weak, intermediate (or medium-strength), and strong [10]. Weak H-bonds are characterized by low molar energy ($\Delta E < 20$ kJ/mol). They can be easily disrupted, and their formation is reversible. The average energy of a hydrogen bond for an OH group is 20–25 kJ/mol [1] and is 8–12 kJ/mol for NH_2 groups. Weak H-bonds occur when acceptors are not atoms with nonbonding orbitals, but a set of atoms with polarizable orbitals, such as π-orbitals extending, for instance, over aromatic rings. Such an H-bond, actually known as the π hydrogen bond, occurs between a water molecule, as donor, and an aromatic ring, as an acceptor.

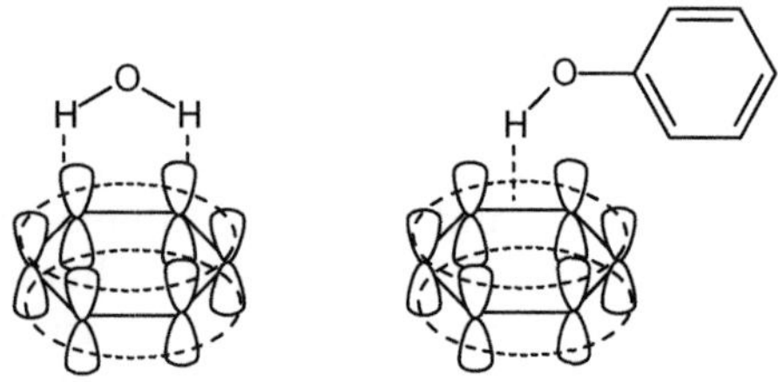

This type of H-bond is also involved when the aromatic acceptor is replaced by other types of molecule that also exhibit π-orbitals, such as those containing double or triple carbon–carbon bonds. This may explain why the introduction of double, triple, or aromatic bonds in chemical structures leads to a decrease in the hydrophobic character of the molecule.

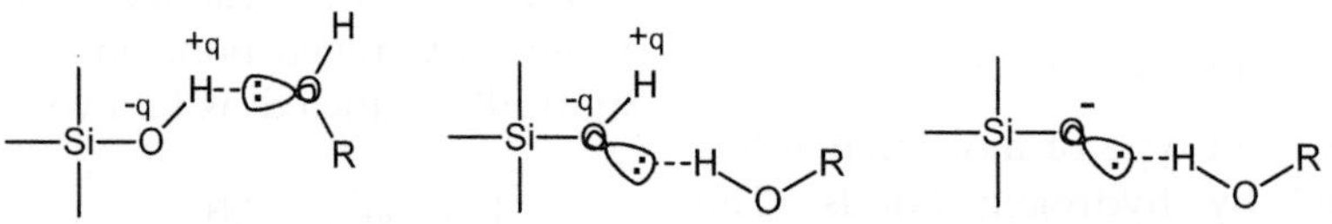

their vapor form. Two examples of such bonds are depicted as follows:

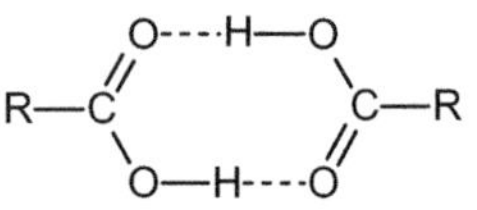

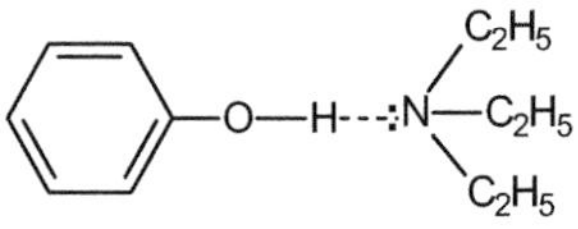

Strong H-bonds ($\Delta E > 40$ kJ/mol) are not very common, and they occur when the acceptor is found in ionic form, as given in the following example:

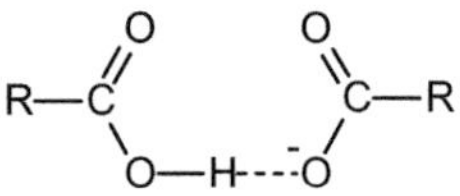

The H-bonds are always present when water is present. Mobile phases in RP-LC typically have an aqueous component; therefore, these types of forces are present together with polar-polar interactions, whether or not the target compounds are dissociable. H-Bonds can interfere in the RP-HPLC retention mechanism through the presence of free silanol groups present on the surface of silica base for the bonded stationary phases. In such a case, there are several possibilities for the formation of H-bonds between the OH groups from the stationary phase and the solute that participates in the retention mechanism, as schematically indicated in the following models:

Intermediate H-bonds (20 kJ/mol $< \Delta E < 40$ kJ/mol) are usually encountered when carboxyl or phenolic groups are involved. The first example given here refers to the centrosymmetric dimers of carboxylic acids, which are found in

This type of interaction may be blocked, for example, by the addition of triethylamine in the mobile phase, which may bind to the residual silanol groups and improve the peak symmetry in HPLC.

In certain systems, the energy of hydrogen bonds can be much higher (e.g., in HF), approaching 150 kJ/mol, which is comparable to that of a covalent bond (210–420 kJ/mol). A comparison of the energy of hydrogen bonds with other types of interactions can be seen in Table 4.1.1.

Charge Transfer or Donor-Acceptor Interactions

One more type of molecular interaction is the one that occurs between electron pair donor and electron pair acceptor compounds (EPD-EPA or charge transfer interaction). This type of interaction leads to complexes with a mesomeric structure between one formed by two noninteracting molecules (except for van der Waals forces) D and A and one with two components with ionic interactions D^+A^-. The electron donor molecules (Lewis bases) and the electron acceptor molecules (Lewis acids) can be n, σ, or π donors or, respectively, acceptors. Their interacting energy can be as low as about 10 kJ/mol but can reach up to 180 kJ/mol. An example of this type of interaction is that between tetracyanoethylene and an alkylbenzene, the formula of

the complex being written as a resonance between the structures:

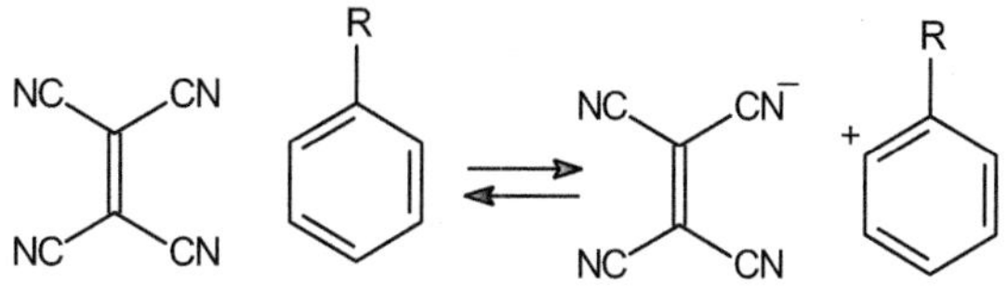

In most cases when donor-acceptor interactions take place, the ionic forms as shown in the previous reaction cannot be detected. Only very strong Lewis acids and strong Lewis bases have the tendency to form bonds mainly based on electrostatic interactions. For most organic molecules with a π electronic system, the molecular orbitals involved in the charge transfer are the π orbitals. These π orbitals are those with the highest energy in the molecule. For this reason the π donor π acceptor systems are frequently mentioned between compounds, although they are weak Lewis acids and bases [11]. Only partial charges can be assumed in these cases as existent in the molecule. For example, a NO_2 is an electron-withdrawing group, while OCH_3 or Cl are electron donor groups. The π system of an aromatic ring containing one or more electron-withdrawing groups will become a π-acceptor system and will be prone to donor-acceptor type interactions with a π-donor compound, although they are weak Lewis acids and bases, respectively.

Stacking and Inclusion in Supermolecular Systems

Intermolecular interactions may lead to specific arrangements for certain molecules such as in the stacking of consecutive aromatic base pairs in DNA. The multiple types of intermolecular forces that lead to stacking are difficult to differentiate, but since the effect is encountered, for example, in molecules with flat aromatic ring systems, the interaction is

TABLE 4.1.1 Typical values for different interaction energies.

Interaction type	Energy (kJ/mol)	Energy (kcal/mol)
Dispersion	8–30	2–7
Dipole – induced dipole	4–8	1–2
Dipole – dipole	4–13	1–3
Hydrogen bonding	20–40	5–10
Ion-dipol	60–130	15–30
Ionic	200–800	50–200
Covalent	200–400	50–100

attributed to π-π interactions. Ab-initio calculations indicate that the interaction of localized aromatic orbitals from different molecules make a minor contribution to stacking, and only common van der Waals forces of types previously described are responsible for this process. It is possible that since the molecules with extended π electron systems have flat surfaces, this improves the steric closeness for action of typical van der Waals forces and influences the molecular association. Nevertheless, π-π interactions even between molecules with small differences in the electron densities are indicated sometimes as contributing to the retention process in HPLC. When more significant differences are present between the electron densities of two molecules with π-electron systems, for example, when the stationary phase contains bonded dinitrophenyl groups and the solute is an aromatic compound with electron donor substituents, the π-π interactions are more important for explaining molecular stacking. The presence of π-π interactions for phases with phenyl or cyano groups may or may not play a role in the separation.

Inclusion is another type of supermolecular association, besides stacking. Specific molecules indicated as host, such as cyclodextrins (α, β, and γ) and cucurbituril, contain cavities with dimensions comparable to that of small molecules and are able to form inclusion complexes with these small molecules. An example of a complex between a β-cyclodextrin (7 glucose units) and guest molecule trans-1,4-bis[(4-pyridyl)ethenyl]benzene is schematically shown in Figure 4.1.10. Numerous other inclusion complexes are obtained from cyclodextrins.

The shape of cyclodextrins is that of a truncated cone. The two diameters of the internal cavity for α-cyclodextrin (with 6 glucose units) are 0.45 nm (narrow) to 0.57 nm (large), for β-cyclodextrin (with 7 glucose units) they are 0.62 − 0.78 nm, and for γ-cyclodextrin (with 8 glucose units) they are 0.79 − 0.95 nm (although the cyclodextrins are fairly rigid molecules, these values are averages since molecular shape is changing due to vibrations and rotations). The depth of the cavity is about 0.79 nm, and the wall thickness is about 0.38 nm. The shape of the inclusion molecule is typically very close to that of the cavity. The interactions in inclusion complexes comprise polar interactions, hydrogen bonding, and possibly charge transfer, with the steric restriction of molecular

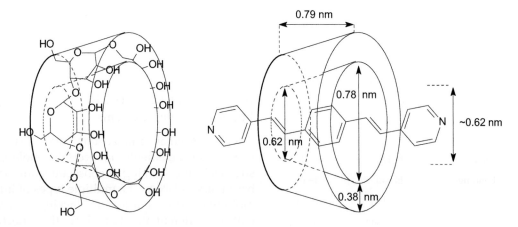

FIGURE 4.1.10 Schematic structure of the inclusion complex between a β-cyclodextrin and a guest molecule trans-1,4-bis [(4-pyridyl)ethenyl]benzene.

movements (also based on the same types of interactions) leading to stable complexes. The reason for including these types of interactions separately from the typical ones, although these interactions are basically reduced to polar, hydrogen bonding, or charge transfer, is that only the special shape of the host molecule allows the formation of a stable complex. The interactions by themselves are rather weak to account for the high stability of the complex. Since the formation of the complex is highly dependent on the geometry of the guest molecule, differences can be seen in the formation of inclusion complexes between enantiomers.

The Effect of a Solvent on Molecular Interaction

The effect of a solvent medium on molecular interactions was taken into consideration in the previous discussion solely by including the dielectric constant ε in the interaction formulas (ε is the dielectric constant of the medium relative to vacuum). However, in reality the solvent effect is more complicated because the intrinsic dipole moment and polarizability of an isolated gas molecule may be different from that in a liquid (liquid state or solution), and because all the effects are taking place in a surrounding with additional interactions besides the ones previously considered. Since dispersion forces (expressed as E_d energy) typically make the highest contribution to the molecular interactions, it is important to evaluate the true polarizability in a condensed medium. A solution to this problem is offered by replacing individual polarizability with that of a small sphere of radius d_i with a dielectric constant ε_i, in a field $\mathcal{E}$. The value of the induced dipole moment in this case is given by the formula [12]:

$$m_{ind} = 4\pi\varepsilon_0\varepsilon\left(\frac{\varepsilon_i - \varepsilon}{\varepsilon_i + 2\varepsilon}\right)d_i^3 \mathcal{E} \qquad (4.1.49)$$

For a larger molecule that can be correctly approximated with a sphere, rel. 4.1.49 leads to the following formula for polarizability:

$$\alpha = \varepsilon_0\left(\frac{\varepsilon - 1}{\varepsilon + 2}\right)\frac{3M}{\rho\mathcal{N}} \qquad (4.1.50)$$

where M is the molecular weight (MW), ρ is the density, and ε is the dielectric constant of the bulk of the molecule. The formula does not give very good agreement for small, highly polar molecules, but the agreement is good for larger polar molecules or weakly polar ones.

The polarizability can also be obtained using the Clausius-Mossoti expression:

$$\alpha = \varepsilon_0\left(\frac{n^2 - 1}{n^2 + 2}\right)\frac{3M}{\rho\mathcal{N}} + \frac{m^2}{3k_BT} = \varepsilon_0 R_m\frac{3}{\mathcal{N}} + \frac{m^2}{3k_BT} \qquad (4.1.51)$$

where n is the refractive index of the bulk material in the visible range of frequencies, R_m is the molar refraction, and the other parameters are the same as previously defined (see the Lorenz-Lorentz equation, e.g., in [12]). Numerous other studies attempting to account for the influence of a solvent on the molecular interactions have been published [13–15]. One simple procedure for accounting for the solvent contributions is the use of Born model (see, e.g., [5, 16]). In this model, a solvation energy E_{solv} is obtained based on the expression:

$$E_{solv} = -\left(\sum_A \frac{q_A^2}{2r_A} + \sum_A \sum_{B>A} \frac{q_A q_B}{r_{AB}}\right)\left(1 - \frac{1}{\varepsilon}\right) \qquad (4.1.52)$$

where q_A and q_B are the partial point charges on the molecule, r_{AB} are the distances between atoms, r_A is an effective radius of each charged atom, and ε is the dielectric constant of the solvent [17]. The effective atomic radii represent a measure of the distance from the nucleus to the boundary of the surrounding cloud of electrons and are reported in the literature [18, 19].

Successful results in the estimation of solvent effects were obtained considering interactions proportional with the solvent accessible area of the molecule. The van der Waals surface area of the molecule (VSA) can be obtained from geometric considerations. The atoms are considered spheres of van der Waals radius, and the van der Waals area of the molecule is the sum of atomic areas minus the areas of interlocked adjacent atoms. However, the accessible area of a molecule is different from VSA since it may contain small gaps, pockets, and clefts that are sometimes too small to be penetrated even by a solvent molecule. To account for true interactions, the concept of solvent accessible area (ASA or SASA) was introduced [20, 21]. This surface is obtained by hypothetically rolling a spherical probe of a diameter corresponding to the size of a solvent molecule on the original van der Walls surface (e.g., of a solvent molecule of radius 1.4 Å corresponding to water) [22]. Interaction energy can be evaluated using a solvation term proportional to the solvent-accessible area of the molecule [23, 24]. In this theory the effective energy of interactions for a solute j having n atoms is divided in two contributions, one due to the solutes interaction, E_j, and the other due to the interactions with the solvent, E_{sol}, in a relation of the form:

$$E(r) = E_j(r) + E_{solv}(r) \tag{4.1.53}$$

where $r = (r_1, r_2, r_3, \ldots, r_n)$ represents a vector indicating the position of each atom in the molecule j. The solvation component $E_{solv}(r)$ was found to be related to SASA by a relation of the form:

$$E_{solv}(r) = \sum_{i=1}^{n} \sigma_i A_i^{SASA} \tag{4.1.54}$$

TABLE 4.1.2 Several parameters utilized in the calculation of A^{SASA} for organic molecules [15].

Atom type	R_i	p_i	σ_i
Carbonyl carbon	1.72	1.554	0.012
Aliphatic carbon with 1 hydrogen	1.8	1.276	0.012
Aliphatic carbon with 2 hydrogens	1.9	1.045	0.012
Aliphatic carbon with 3 hydrogens	2	0.88	0.012
Aromatic carbon with 1 hydrogen	1.8	1.073	0.012
Amide nitrogen	1.55	1.028	20.06
Aromatic nitrogen with no hydrogens	1.55	1.028	20.06
Primary amine nitrogen	1.6	1.215	20.06
Nitrogen bound to 3 hydrogens	1.6	1.215	20.06
Guanidinium nitrogen	1.55	1.028	20.06
Proline nitrogen	1.55	1.028	20.06
Hydroxyl oxygen	1.52	1.08	20.06
Carbonyl oxygen	1.5	0.926	20.06
Carboxyl oxygen	1.7	0.922	20.06
Sulfur	1.8	1.121	0.012
Sulfur with 1 hydrogen	1.8	1.121	0.012
Polar hydrogen	1.1	1.128	0

where σ_i is an atomic solvation parameter and A_i^{SASA} is the solvent-accessible area of the atom i. Selected values for σ_i for atoms in a molecule are given in Table 4.1.2 [15]. The values for A_i^{SASA} can be obtained using the following analytical expression:

$$A_i^{SASA} = S_i \prod_{j \neq i}^{n} \left[1 - \frac{p_i p_{ij} b_{ij}(r_{i,j})}{S_i} \right] \tag{4.1.55}$$

In rel. 4.1.55 the parameters are as follows: S_i is the SASA of an isolated atom of radius R_i and is given by the formula:

$$S_i = 4\pi \left(R_i + R_{probe} \right)^2 \qquad (4.1.56)$$

R_{probe} is the radius of the solvent probe. Some values for R_i are given in Table 4.1.2. The parameter $b_{ij}(r_{i,j})$ represents the SASA removed from S_i due to overlapping between atoms i and j at the distance $r_{i,j}$. The calculation of $b_{ij}(r_{i,j})$ can be done using the expressions:

$$b_{ij}\left(r_{ij}\right) = 0 \quad \text{for} \quad r_{ij} > R_i + R_j + R_{probe}$$
$$(4.1.57a)$$

$$b_{ij}\left(r_{ij}\right) = \pi\left(R_i + R_{probe}\right)\left(R_i + R_j + 2R_{probe}\right)$$
$$- r_{ij})\left[1 + \left(R_j - R_i\right)r_{ij}^{-1}\right] \quad \text{otherwise}$$
$$(4.1.57b)$$

For a solvent with $R_{probe} = 1.4$ Å, the parameter p_i is given in Table 4.1.2, and $p_{ij} = 0.8875$ if the atoms i and j are covalently bonded, and $p_{ij} = 0.3516$ otherwise. The formulas for the calculation of SASA for organic molecules previously described are used in computer programs that provide values for this parameter for a variety of molecules (e.g., MarvinSketch 5.4.01 [7]). In addition to SASA, other similar parameters were introduced such as ASA_H, which is the solvent-accessible surface area of all hydrophobic atoms (with formal partial charges $|q_i| < 0.125$) and ASA_P, which is the solvent-accessible area of all polar atoms (with formal partial charges $|q_i| > 0.125$) [24]). The estimation of E_{solv} values was successfully used for simulation of the interactions of small proteins in solution and also can be used to understand the interaction of small polar molecules with a solvent.

One additional complication related to molecules in solution is the formation of molecular or ionic associations with the solvent molecules. For example, ions in aqueous solutions are not free but solvated [25]. Within the solvent, the interactions between solvated molecules can be different from the one between free molecules, and attempts to calculate the interaction energies frequently lead to incorrect results.

Solvophobic Interaction

The solvophobic interactions are commonly invoked for explaining the higher solubility of a nonpolar compound in a nonpolar solvent than in a polar one, although the interactions at the molecular level may indicate that the energy between a nonpolar molecule with another nonpolar molecule is smaller (in absolute value) than between a nonpolar molecule and a polar one. In order to dissolve a nonpolar molecule in a polar solvent, a nonpolar $\leftrightarrow$ polar interaction must replace a polar $\leftrightarrow$ polar one between the solvent molecules. Therefore, it is favored energetically to have polar $\leftrightarrow$ polar and nonpolar $\leftrightarrow$ nonpolar interactions instead of nonpolar $\leftrightarrow$ polar ones. This process is schematically suggested in Figure 4.1.11. In this

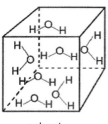

solvent

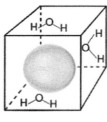

create cavity

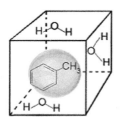

place solute

FIGURE 4.1.11 Schematic illustration of placing a hydrophobic molecule in a polar solvent

figure, a pure solvent (H_2O) is initially shown, typical polar-polar interactions taking place between the solvent molecules. In stage 2 of the process, a cavity is created in the solvent, and this would require energy. After the cavity is formed, a solute molecule is placed in it, leading to new, different interactions.

The change in standard free energy $A^{sol}_{j,S}$ (symbol Δ and the index "0" for standard expressions are omitted) necessary for placing a molecular species j into a solution formed by molecules S can be expressed by the free energy required to create the cavity in the solvent to accommodate the species j, indicated as $A^{cav}_{j,S}$ plus the free energy of van der Waals interactions between the molecule j and the surrounding molecules S, indicated as $A^{vdw}_{j,S}$. A third term that accounts for the change in the free volume of the system (excluded volume expansion) should also be added. In conclusion, the change in the free energy should be given by the expression:

$$A^{sol}_{j,S} = A^{cav}_{j,S} + A^{vdw}_{J,S} + RT \ln(RT/p_0 V_S)$$

$$(4.1.58)$$

where p_0 is the atmospheric pressure and V_S is the molar volume of the solvent equal to molecular mass (weight) divided by the density ρ. The expressions for both terms $A^{cav}_{j,S}$ and $A^{vdw}_{j,S}$ have been reported in the literature [26–28]. The free energy (change) for the cavity formation in the solvent by the molecular species j can be expressed by the formula [29]:

$$A^{cav}_{j,S} = \kappa^e_{j,S} \, \mathcal{A}_j \gamma'_S (1 - W_S) \, \mathcal{N} \qquad (4.1.59)$$

In rel. 4.1.59 $\kappa^e_{j,S}$ is a special parameter (dependent on S) further defined, $\mathcal{A}_j$ is the molecular surface area of the molecule j, γ'_S is the surface tension of the solvent S, and W_S is a correction factor given by the expression:

$$W_S = \left(1 - \frac{\kappa^s_{j,S}}{\kappa^e_{j,S}}\right)\left(\frac{d \ln\gamma'_S}{d \ln T} + \frac{2}{3}C_{exp,j}T\right) \quad (4.1.60)$$

where $C_{exp,j}$ is the coefficient of thermal expansion for the species j. The coefficient $\kappa^e_{j,S}$ expresses the ratio between the energy required for the formation of a suitably shaped cavity with a surface area $\mathcal{A}_j$ in the solvent S and the energy required to expand the planar surface of the solvent by the same area, which is approximately given by $\mathcal{A}_j \gamma'_S$. A similar coefficient must be developed for entropies. The coefficient $\kappa^s_{j,S}$ expresses the ratio between the entropy required for the formation of a suitably shaped cavity with a surface area $\mathcal{A}_j$ in the solvent and the entropy required to expand the planar surface of the solvent by the same area. The expression for $\kappa^e_{j,S}$ is the following:

$$\kappa^e_{j,S} = 1 + (\kappa^e_S - 1)\left(V_S/V_j\right)^{2/3} \qquad (4.1.61)$$

where V_j is the molar volume for the species j, and where κ^e_S corresponds to the pure solvent and is given by the expression:

$$\kappa^e_S = \frac{\mathcal{N}^{1/3}\Delta H_{vap,S}}{V_S^{2/3}\gamma'_S\left(1 - \dfrac{d \ln\gamma'_S}{d \ln T} - \dfrac{2}{3}C_{exp,S}T\right)}$$

$$(4.1.62)$$

The values for $\Delta H_{vap,S}$ (the heat or enthalpy of vaporization of the solvent) are available in the literature [30] or can be estimated [31]. Regarding entropies, a similar expression with 4.1.61 is valid for $\kappa^s_{j,S}$, and a similar one with 4.1.62 is valid for κ^s_S, when $\Delta H_{vap,S}$ is replaced by the vaporization entropy variation $\Delta S_{vap,S}$ of the pure solvent. Tables for (dimensionless) values of κ^e_S and for κ^s_S have been reported in the literature [27] and are reproduced in Table 4.1.3. The resulting values for W_S following these calculations are relatively small.

The expression for the interacting free energy $A^{vdw}_{j,S}$ of the molecule j with the surrounding solvent S when j is placed in the

TABLE 4.1.3 Parameters used for the calculation of solvophobic forces published in the literature for different solvents [27].

Solvent	$\kappa^e{}_S$	$\kappa^s{}_S$	D_S	γ'_S dyn/cm	ε_S (at 25°C)	$\dfrac{d \ln\gamma'_S}{d \ln T}$	ω_S	$C_{exp,S} \, 10^3$
Heptane	0.687	0.542	0.234	19.80	1.92	0.100	0.348	1.250
Isooctane	0.672	0.527	0.238	18.32	1.91	0.089	0.313	1.207
CCl$_4$	0.629	0.475	0.272	26.15	2.23	0.126	0.255	1.220
Cyclohexane	0.621	0.466	0.254	24.38	2.01	0.120	0.262	1.200
Benzene	0.629	0.469	0.293	28.20	2.27	0.139	0.215	1.220
Toluene	0.679	0.529	0.291	27.92	2.37	0.121	0.256	1.066
Aniline	0.972	1.127	0.335	42.79	6.85	0.105	0.256	0.850
n-Octanol	0.827	0.831	0.238	29.40	3.40	0.103	0.288	0.890
n-BuOH	1.089	0.931	0.241	24.20	17.10	0.108	0.252	0.940
EtOH	1.543	1.508	0.220	21.85	24.44	0.084	0.147	1.080
Nitromethane	0.808	0.580	0.231	36.47	38.20	0.168	0.192	1.192
MeOH	1.776	1.609	0.202	22.20	32.63	0.086	0.100	1.190
Water	1.277	1.235	0.205	72.00	77.46	0.157	0.023	0.257

cavity is a relatively complicated problem, since this interaction must account for the interaction of the molecule j with the whole surrounding solvent. This energy consists of two different terms, one caused by electrostatic forces $A^{es}{}_{j,S}$, and the other by dispersion forces $A^{disp}{}_{j,S}$ such that:

$$A^{wdv}{}_{j,S} = A^{es}{}_{j,S} + A^{disp}{}_{j,S} \quad (4.1.63)$$

The expression for $A^{es}{}_{j,S}$ is obtained from the Onsager reaction field [32] and is given by the formula:

$$A^{es}{}_{j,S} = -\frac{\mathcal{N} m_j^2}{2\, v_j} D_S P_{j,S} \quad (4.1.64)$$

where:

$$D_S = \frac{2(\varepsilon_S - 1)}{2\varepsilon_S + 1} \qquad P_{j,S} = \frac{v_j}{4\pi\varepsilon_0 \left(v_j - D_S \alpha_j \right)} \quad (4.1.65)$$

In rel. 4.1.65 the common notations m_j for the dipole moment, α_j for polarizability of species j, and ε_S for the dielectric constant of the solvent were used (index s added to previous notation). The parameter v_j is the volume of a molecule, and $v_j = V_j/\mathcal{N}$ for the species j. (*Note:* The volume of a molecule $v_j = V_j/\mathcal{N}$ is not the same as van der Waals volume V_{vdW} since the volume of voids and the changes in the volume due to interactions must be included in the value of v, which can be obtained using the expression $v = V_{vdW} + V_{void} + \Delta V_{int}$).

The expression for $A^{disp}{}_{j,S}$ is obtained by using an effective pair potential that contains a correction from the gas phase potential of interaction of Lennard-Jones type [27, 33]. Following a relatively elaborate calculation, the formula obtained is:

$$A^{disp}{}_{j,S} = -\frac{27(1-x)}{8\pi}\left(Q'_{j,S} + Q''_{j,S} \right) \Upsilon_{j,S} D_j D_S \quad (4.1.66)$$

In rel. 4.1.66, x is a proportionality constant that relates the entropy of interaction to the energy of interaction and has a typical value of $x \approx 0.436$. The other parameters in rel. 4.1.66 are as follows:

$$Y_{j,S} = 1.35 \frac{I_j I_S}{I_j + I_S} \qquad (4.1.67)$$

where I_j is the ionization potential of species j and I_S is the ionization potential of the molecule of solvent. The expression for D_j is given by the formula:

$$D_j = \frac{n_j^2 - 1}{n_j^2 + 2} = \frac{4\pi}{3v_j} \mathcal{N}\alpha_j \qquad (4.1.68)$$

where D_j is the so-called Clausius-Mosotti function for the molecular species j and where n_j is the refractive index for the compound j (see rel. 4.1.51) and α_j the (average) polarizability. A formula identical to 4.1.68 gives the value for D_S, where n_j and α_j are replaced, respectively, by n_S and α_S of the solvent.

The expression for $Q'_{j,S}$ (dimensionless) is the following:

$$Q'_{j,S} = v_j \left[\frac{\bar{\sigma}^6}{(\bar{d} - \bar{l})^9} \left(\frac{t^2}{11} + \frac{t}{5} + \frac{1}{9} \right) \right.$$
$$\left. - \frac{1}{(\bar{d} - \bar{l})^3} \left(\frac{t^2}{5} + \frac{t}{2} + \frac{1}{3} \right) \right] \qquad (4.1.69)$$

where:

$$t = \frac{\bar{l}}{\bar{d} - \bar{l}} \qquad \bar{d} = 1/2(d_j + d_S) \qquad \bar{l} = 1/2(l_j + l_S)$$
$$\bar{\sigma} = 1/2(\sigma_j + \sigma_S) \qquad (4.1.70)$$

and d_j is the diameter of molecule j, d_S is the diameter of solvent molecule, l is Kihara parameter (see, e.g., [34]), and σ is London parameter (for solute j or solvent S). All these parameters can be estimated using the following formulas (shown for species j only,

the formulas for the solvent molecules having the same expressions):

$$d_j = 1.74 \left(\frac{3v_j}{4\pi} \right)^{1/3} \qquad l_j = d_j \frac{0.24 + 7\omega_j}{3.24 + 7\omega_j}$$

$$\sigma_j = d_j \frac{2.66}{3.24 + 7\omega_j} \qquad (4.1.71)$$

where v_j can be determined from the molar volume, and ω_j is the acentric factor that can be calculated for nonpolar compounds based on the expression:

$$\omega_j = -\log p_{red} - 1 \qquad (4.1.72)$$

where p_{red} is a reduced vapor pressure with $p_{red} = p/p_c$ (p_c is the critical pressure) at a reduced temperature $T/T_c = 0.7$. For polar compounds the acentric factor is taken from a nonpolar model compound with similar geometry.

The value for $Q''_{j,S}$ is typically taken as $Q''_{j,S} \approx 0.1\, Q'_{j,S}$.

Including all the necessary expressions in rel. 4.1.58, the resulting free energy of hydrophobic interactions leads to the following formula:

$$A^{sol}_{j,S} = A^{cav}_{j,S} + A^{disp}_{j,S} + A^{es}_{j,S}$$
$$+ RT \ln \left(RT/p_0 V_S \right) \qquad (4.1.73)$$

where:

$$A^{cav}_{j,S} = \kappa^e_{j,S} A_j \gamma'_S (1 - W_S) \mathcal{N} \qquad (4.1.74)$$

$$A^{es}_{j,S} = -\frac{\mathcal{N} m_j^2}{2v_j} D_S P_{j,S} \qquad (4.1.75)$$

$$A^{disp}_{j,S} = -\frac{15.23}{8\pi} \left(Q'_{j,S} + Q''_{j,S} \right) Y_{j,S} D_j D_S \qquad (4.1.76)$$

The verification of the above formula was performed and reported in the literature [27–29]. Some values of the parameters used for calculating solvophobic free energy were published in the literature for different solvents [27] and are given in Table 4.1.3.

The evaluation of hydrophobic interactions by direct measurements is not very common. One of the few examples of direct measurements is that of two methane molecules. The van der Waals forces between two methane molecules in vacuum is about -2.5×10^{-21} J. The same interaction in water is about -14.0×10^{-21} J. This effect is caused by the reduction of the surface exposed to water when the two hydrophobic molecules are together. Similar results were reported for other molecules, such as the interaction of two molecules of benzene in water with a free energy of -8.4 kJ mol^{-1} and that of cyclohexane with -11.3 kJ mol^{-1} when they form dimers [3].

The problem of solvophobic interactions is more complicated than is summarized in this section, and there is no simple theory to offer a complete picture of the subject. Polyatomic molecules, for example, can have one or more polar moieties and a hydrophobic part. These two different group characters in the same molecule are difficult to treat in a unified mathematical model. When attempting to apply the theory of interactions to separation problems as encountered in HPLC, the type of calculations presented in this section can be used only as directional information and no truly quantitative data can be generated.

When the solvent is capable of forming hydrogen bonds, such as in the case of water, the hydrogen bonds are disrupted in the solvent when a solute is dissolved. This component was not addressed completely in the previous theory, except through the value of γ'_S, the surface tension. The accommodation in solution of a molecule that cannot form hydrogen bonds (a hydrophobic molecule) does not necessarily break the hydrogen bonds, leading only to their reorientation. Not breaking the hydrogen bonds leads to a smaller enthalpic change than in the case of breaking the bonds. However, the

molecule reorientation causes the entropic factor to be strongly affected during the dissolution of a nonpolar molecule in a polar solvent, at the same time as the enthalpic factor. For the dissolution of molecules in a polar solvent such as water, the free energy of transfer is roughly estimated as proportional to the surface area of the dissolved molecule. The number of reoriented molecules of the solvent is also influenced by the surface area of the hydrophobic molecule that is dissolved. The entropy change during the dissolution process is therefore expected to depend on the surface area of the dissolved molecule as well.

Solvophobic interactions are key to understanding the retention process in RP-HPLC. They also explain numerous solubility effects. For example, the theory can explain why perfluorinated alkanes have much lower water solubility than the corresponding hydrocarbons, in spite of the fact that the polarity of the C-F bond is much higher than that of the C-H bond (and a stronger interaction of the solute with the water molecules is expected). The dispersion forces between water and fluorocarbons are not very different from those between water and hydrocarbons. This is caused by the fact that although the fluorine atoms are larger than the hydrogen atoms (van der Waals radius for H is 1.2 Å and for F is 1.47 Å), and are assumed to have higher polarizability, the electronegativity of fluorine atoms reduces the polarizability of the electron system in the fluorinated compounds. The point of difference between hydrocarbons and fluorocarbons is that the C-F bond has a much larger dipole moment than does the C-H bond, leading to a stronger interaction with the dipolar water. Therefore, it would be expected that a fluorocarbon surface is more hydrophilic than that of the corresponding hydrocarbon. This effect is indeed noticed, for example, when bonded fluorinated hydrocarbons are used as stationary phases in RP-HPLC chromatography. These phases are still hydrophobic, but their hydrophobic character is lower than that of the corresponding hydrocarbon. The explanation for the poor fluorocarbon water solubility is that fluorocarbons have a larger molar volume (and molecular surface area) than the corresponding hydrocarbons. For example, a fluorocarbon with a surface area of 299.9 $\mathring{A}^2$ (perfluorohexane) corresponds to a hydrocarbon (hexane) with surface area of 215.6 $\mathring{A}^2$. The work done to form a cavity large enough to accommodate a fluorocarbon (A^{cav}) offsets the anticipated free-energy benefit from enhanced energetic interactions with water. An entropic effect due to movement restrictions imposed to the solvent molecules by the polar fluorocarbon may also contribute to the low fluorocarbon solubility in water [35].

Chaotropic Interactions

Besides solvent molecules that affect molecular interactions in solutions, these interactions can also be affected by the presence of other solutes. For example, hydrogen bonds are stronger in a nonpolar medium than in one with some polarity. The addition of certain solutes that increase the polarity of a solution may affect significantly other solute interactions that form, for example, hydrogen bonds or aggregation through hydrophobic effects. These solutes are known as *chaotropes*. Some inorganic ions can act as chaotropes, their disruptive character growing in the order: $H_2PO_4^- < HCOO^- < CH_3SO_4^-, < CF_3COO^-, < BF_4^-, < ClO_4^-, < PF_6^-$. In solutions of mixed solvents containing water, the chaotropes may also disrupt the solvation shell of other solutes, modifying their interactions.

4.2. FORCES BETWEEN MOLECULES AND A SURFACE

General Comments

The evaluation of molecule and surface interactions may provide some insight into the adsorption process. Similarly to the case of intermolecular interactions, the evaluation of free energy in different processes takes into consideration both intramolecular forces and concepts from the continuum model. A simple theory for estimating the forces between the molecules and the surface is, however, difficult to develop except for a few models. For this reason, only estimations of such interactions are typically available in the literature (for a more detailed discussion on the subject see, e.g., [11]).

Charge to Surface-Charged Interaction

The charge on a surface is characterized by its charge density σ_s. This charge density generates an electric field perpendicular to the surface. The intensity of this field can be evaluated considering a circular strip of radius r and width dr on the charged surface (see Figure 4.2.1) positioned around A, the perpendicular projection of a charge ze on the surface acting on a charge ze at the distance x from the surface. The field generated by this narrow strip on the point ze is determined by the distance $(r^2 + x^2)^{1/2}$ and also by the angle θ (since the field is perpendicular to the surface). The total intensity of the field can be obtained by integrating over

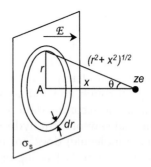

FIGURE 4.2.1 Schematic diagram of the interaction between a charged surface (charge density σ_s) and a charge ze at distance x.

r (from 0 to ∞ for an infinite surface). The expression for $\mathcal{E}$ will be:

$$\mathcal{E} = \sigma_s \int_0^\infty \frac{2\pi r \cos\theta \, dr}{4\pi\varepsilon_0\varepsilon(x^2 + r^2)}$$

$$= \frac{x\sigma_s}{2\varepsilon_0\varepsilon} \int_0^\infty \frac{r dr}{(x^2 + r^2)^{3/2}} = \frac{\sigma_s}{2\varepsilon_0\varepsilon} \qquad (4.2.1)$$

This relation indicates that the force between the charge ze and the surface is given by the expression:

$$f = \frac{\sigma_s ze}{2\varepsilon_0\varepsilon} \qquad (4.2.2)$$

Expression 4.2.2. indicates that the interaction between a charged surface and a particle with the charge ze is independent of the distance x between the particle and the charge. The free energy to move a charge ze for the distance r in the field created by the surface will be given by the formula $A = f \, r$, and this energy is given by the formula:

$$A = \frac{\sigma_s zer}{2\varepsilon_0\varepsilon} \qquad (4.2.3)$$

In rel. 4.2.3 the sign of energy depends on the type of charges on the surface, a negative value indicating attraction and a positive one repulsion. The calculation of A for a charge e at the distance of 0.6 nm from a surface with the charge 0.03 C/m^2 indicates an energy of 98 kJ/mol, showing that the interaction between a surface and a charge is of the same order of magnitude as the intermolecular interactions. In practice, since the medium always contains other charges with interfering fields, the true force between the charge and the surface decreases with the distance and it is much smaller.

Neutral Molecule-to-Surface Interaction

The interaction between a solid surface of body made from identical molecules and a molecule j can be evaluated in a similar manner as was done for the interaction between a charged surface and an ion with the charge ze. For this purpose, it is useful to consider first an interacting potential of the form (where C' is a constant):

$$E(r) = \frac{C'}{r^n} \qquad (4.2.4)$$

An infinitesimal interaction should be further considered between a circular ring of cross section $dx \, dy$ in the solid surface and the molecule j, as shown in Figure 4.2.2. This ring volume is $2\pi y dxdy$. Using the notation ρ for the number density of molecules in the solid, the number of molecules in the ring will be $2\pi\rho y dxdy$. The interaction between the molecule j at distance r and the surface will be given by the expression [36]:

$$A(r) = -2\pi\rho C' \int_r^\infty dx \int_0^\infty \frac{ydy}{(x^2 + y^2)^{n/2}}$$

$$= \frac{\pi\rho C'}{(n-2)} \int_r^\infty \frac{dx}{x^{n-2}} = -\frac{\pi\rho C'}{(n-2)(n-3)r^{n-3}}$$

$$(4.2.5)$$

Relation 4.2.5 indicates that considering the Lennard-Jones potential given by rel. 4.1.47, the interaction between the molecules from

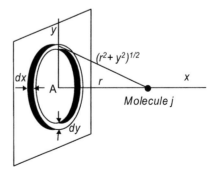

FIGURE 4.2.2   A circular ring of cross section $dx \, dy$ taken in the solid surface interacting with molecule j.

a solid surface should be expressed by the formula:

$$E(r) = -\frac{\pi \rho A'}{6 r^3} + \frac{\pi \rho B'}{90 r^9} \qquad (4.2.6)$$

Regarding rel. 4.2.5, it should be observed that it derives from a Lennard-Jones type potential that does not imply the existence of charges, but only of all other types of interactions. Using an expression of the type 4.1.48 for estimating parameters A' and B' in the Lennard-Jones potential for molecules having optimum molecule-to-molecule interaction at $r_0 = 6$Å for an idealized pair of molecules with $m = 1.5$ D, $\alpha_0 = 4.2$ (*1.11 10^{-40} $C^2m^2J^{-1}$), $\varepsilon = 1$, and $I = 10.9$ e.V, at 300 K the variation of Lennard -Jones molecule-to-molecule energy E_{L-J} and that between surface-to-molecule energy E_{surf} (for $\rho = 1$) for a distance varying between 3 Å and 7 Å is shown in Figure 4.2.3. The results shown in Figure 4.2.3 indicate that the interaction of a surface compared to that of a single molecule is basically of the same order of magnitude, with the surface acting more strongly than a single molecule, as expected. However, in reality, since the medium always contains other molecules with interfering interactions, the true force between the molecule and the surface

decreases with the distance faster than is indicated by rel. 4.2.6. On the other hand, the interactions of ions in solution take place at larger distances. This was shown in Figure 4.1.1 (even if the figure exaggerates to a certain extent the interaction range).

The interaction of the charges on a surface with individual molecules in a medium may explain some exclusion process seen in reversed-phase HPLC. The typical phases in RP-HPLC consist of a silica surface covered with the hydrophobic bonded phase, but also with a large number of silanol groups. The silanol groups have a slight acidic character, and the ionization process leads to the accumulation on the silica surface of small negative charges. These charges may act through Coulombic forces toward ionic species such as naphthalene sulfonic acids (that are almost completely ionized in solution). These types of molecules are excluded from the pores of the stationary phase. For this reason, for example, the retention times for naphthalene sulfonic acids are shorter than holdup time (dead time) t_0 for the column, as measured with small compounds that are not retained but also are not excluded from the stationary phase pores (such as uracil or thiourea) [37].

FIGURE 4.2.3 Variation of the interaction energy E_{L-J} calculated with rel 4.1.48, and of E_{surf} calculated with rel. 4.2.6 with $m = 1.5$ D, $\alpha_0 = 4.2$ (⋅1.11 10^{40} $C^2m^2J^{-1}$), $\varepsilon = 1$, and $I = 10.9$ e.V, at 300 K, when $r_0 = 6$ Å and $\rho = 1$.

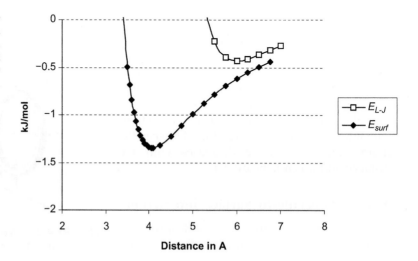

References

[1] Kaliszan R. Quantitative Structure-Chromatographic Retention Relationship. New York: John Wiley; 1987.

[2] van Oss CJ. The Properties of Water and Their Role in Colloidal and Biological Systems. Amsterdam: Academic Press; 2008.

[3] Israelachvili J. Intermolecular & Surface Forces. Amsterdam: Academic Press; 1991.

[4] McClellan AL. Tables of Experimental Dipole Moments. San Francisco: W. H. Freeman and Co. 1963.

[5] Moldoveanu S, Savin A. Aplicatii in Chimie ale Metodelor Semiempirice de Orbitali Moleculari, Edit. Academiei RSR, Bucuresti 1980.

[6] Miller KJ, Savchik J. A new empirical method to calculate average molecular polarizabilities. J. Am. Chem. Soc. 1979;101:7206−13.

[7] http://www.ChemAxon.com

[8] Stewart JJP. MOPAC-7, QCPE 113. Bloomington: Indiana University; 1994.

[9] Bondi A. Van der Waals volumes and radii. J. Phys. Chem. 1964;68:441−51.

[10] Marechal Y. The Hydrogen Bond and the Water Molecule. Amsterdam: Elsevier; 2007. p. 6.

[11] van Oss CJ, Chaudhury MK, Good RJ. Interfacial Lifshitz-van der Waals and plolar interactions in macroscopic systems. Chem. Rev. 1988;88:927−41.

[12] Murgulescu IG, Sahini VM. Introducere in Chimia Fizica. vol. I, 2, Edit Acad. R.S.R., Bucuresti; 1978. pp. 127−166.

[13] Davis ME, Madura JD, Luty BA, McCammon JA. Electrostatics and diffusion of molecules in solutions: Simulation with the University of Huston Brownian dynamics program. Comp. Phys. Comm. 1991;62:187−97.

[14] Davis ME, McCammon JA. Calculating electrostatic forces from grid-calculated potentials. J. Comput. Chem. 1990;11:401−9.

[15] Ferrara P, Apostolakis J, Caflisch A. Evaluation of a fast implicit solvent model for molecular dynamics simulations. Prot. Struct. Funct. Genet. 2002;46:24−33.

[16] Still WC, Tempczyk A, Hawley RC, Hendrickson T. Semianalytical treatment of solvation for molecular mechanics and dynamics. J. Am. Chem. Soc. 1990;112:6127−9.

[17] Hefter G. Ion solvation in aqueous-organic mixtures. Pure Appl. Chem. 2005;77:605−17.

[18] Clementi E, Raimondi DL, Reinhardt WP. Atomic screening constants from SCF functions. II. Atoms with 37 to 86 electrons. J. Chem. Phys. 1967;47:1300−7.

[19] Slater JC. Atomic radii in crystals. J. Chem. Phys. 1964;41:3199−204.

[20] Richards FM, Richmond T. Solvents, interfaces and protein structure. Ciba Found Symp. 1977;60:23−45.

[21] Connolly ML. Solvent-accessible surfaces of proteins and nucleic-acids. Science 1983;221:709−13.

[22] Shrake A, Rupley JA. Environment and exposure to solvent of protein atoms. Lysozyme and insulin. J. Mol. Biol. 1973;79:351−71.

[23] Eisenberg D, McLachlan AD. Solvation energy in protein folding and biding. Nature 1986;319:199−203.

[24] Lazaridis T, Karplus M. Effective energy function for proteins in solution. Prot. Struct. Funct. Genet. 1999;35:133−52.

[25] Hefter G. Ion solvation in aqueous-organic mixtures. Pure Appl. Chem. 2005;77:605−17.

[26] Sinanoğlu O. The C-potential surface for predicting conformations of molecules in solution. Theor. Chim. Acta 1974;33:279−84.

[27] Halicioğlu T, Sinanoğlu O. Solvent effects on *cis-trans* azobenzene isomerization: a detailed application of a theory of solvent effects on molecular association. Ann. N.Y. Acad. Sci. 1974;158:308−17.

[28] Horvath C, Melander W, Molnar I. Solvophobic interactions in liquid chromatography with nonpolar stationary phases. J. Chromatogr. 1976;125:129−56.

[29] Sinanoğlu O. In: Pullman B, editor. Molecular Associations in Biology. New York: Academic Press; 1968. p. 427−45.

[30] Svoboda V, Kehiaian HV. Enthalpies of Vaporization of Organic Compounds: A Critical Review and Data Compilation. In: IUPAC Chem. Data. Ser. 32, Blackwell Sci; 1985.

[31] Lyman WJ, Reehl WF, Rosenblatt DH. Handbook of Chemical Property Estimation Methods. Washington, DC: ACS; 1990.

[32] Onsager L. Electric moments of molecules in liquids. J. Amer. Chem. Soc. 1936;58:1486−93.

[33] Sinanoglu O. An intermolecular potential for use in liquids. Chem. Phys. Lett. 1967;1:340−2.

[34] Liu T-C. Application of Kihara parameters in conventional force fields. J. Math. Chem. 2010;48:363−9.

[35] Dalvi VH, Rossky PJ. Molecular origin of fluorocarbon hydrophobicity. Proc. Natl. Acad. USA−PNAS 2010;107:13603−7.

[36] Voicu VA, Mircioiu C. Mecanisme Farmacologice la Interfete Membranare. Edit. Academiei, Bucuresti; 1994; 149−150.

[37] Jandera P, Bunčeková S, Halama M, Novotná K, Nepraš M. Naphthalene sulphonic acids − new test compounds for characterization of the columns for reversed-phase chromatography. J. Chromatogr. A 2004;1059:61−72.

Retention Mechanisms in Different HPLC Types

Essentials in Modern HPLC Separations
http://dx.doi.org/10.1016/B978-0-12-385013-3.00005-7

145

5.1. RETENTION IN REVERSED-PHASE CHROMATOGRAPHY

General Comments

Reversed-phase chromatography is the most common HPLC separation technique and is used for separating compounds that have hydrophobic moieties and do not have a dominant polar character (although polarity of a compound does not exclude the use of RP-HPLC). In RP-HPLC, the stationary phase has a hydrophobic character, and the mobile phase has a polar or partially polar character. The RP-HPLC is frequently practiced using a bonded phase on silica, with the bonded moiety (ligand) being a hydrocarbon chain (e.g., C8, C18). The mobile phase in RP-HPLC is typically a partially aqueous, partially organic solvent and is more polar than the stationary phase. The solute molecules are in equilibrium between the hydrophobic stationary phase and partially polar mobile phase, the direction of the equilibrium determining a stronger or a weaker retention of the analytes. This equilibrium can be studied using a model in which the solute j interacts with a ligand L that represents the

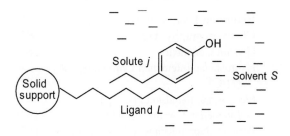

FIGURE 5.1.1 Schematic illustration of the interaction between a solute j with a ligand L in a solvent S.

stationary phase, the interaction taking place in a solvent S that plays an active role in the direction of the equilibrium. The process is schematically shown in Figure 5.1.1. The retention process can therefore be described in a solvent by the simple chemical equilibrium:

$$j + L \rightleftarrows jL \qquad (5.1.1)$$

with the understanding that the solvent has a major role in displacing the equilibrium in one direction or the other. Depending on the nature of solute j, ligand L, and the solvent S, it is possible to obtain an evaluation of the equilibrium constant K_j for equilibrium 5.1.1. Using a simplification of rel. 3.1.11, the equilibrium constant for equilibrium 5.1.1 can be expressed by the formula:

$$\ln K = \frac{-\Delta A^0}{RT} \qquad (5.1.2)$$

where ΔA^0 is the free energy for the equilibrium, the activity coefficients γ for the species in the equilibrium are considered to be 1, and the free enthalpy ΔG^0 is taken as equal to the free energy of the process ΔA^0 (assuming no volume change during the process).

The characterization of retention and separation processes in RP-HPLC can be done, as for all HPLC techniques, using parameters such as capacity factor k and selectivity α. The dependence of capacity factor k on thermodynamic variables was formally discussed in Sections 3.1 and 3.2, and an expression for k (or $\ln k$) can be obtained from rel. 5.1.2 as follows:

$$\ln k = \frac{-\Delta A^0}{RT} + \ln \Psi \qquad (5.1.3)$$

Relation 5.1.3 is similar to van't Hoff equation 3.1.13, where $\Delta G^0 = \Delta A^0$. This model can be in general used for understanding the retention process in several types of HPLC, and it is applied successfully in reversed-phase chromatography.

Evaluation of Equilibrium Constant Based on Solvophobic Theory

For the calculation of the equilibrium constant K_j for equilibrium 5.1.1, the detailed energetics of the process allowing the calculation of ΔA^0 must be taken into consideration. The types of possible interactions were described in detail in Section 4.1. ΔA^0 can be caculated starting with evaluation of the change in free energy in the equilibrium 5.1.1 assumed to take place in an ideal gas phase system with no intervention of the solvent, indicated as A^{gas} (symbol Δ and the index "0" for standard expressions are omitted for simplification of notation). To this free energy A^{gas} must be added the solvent interaction energy A^{sol} for each component of the equilibrium (j, L and jL). In other words, the total variation in free energy for the equilibrium 5.1.1 is given by the formula:

$$A \equiv \Delta A^0 = A^{gas} + \sum_{i=j,L,jL} A_i^{sol} \qquad (5.1.4)$$

Based on the description from Section 4.1, the value for A^{gas} can be obtained assuming the absence of ions in any of the participants using a formula of the type 4.1.46, and it can be written in the form:

$$A^{gas} = E_T \qquad (5.1.5)$$

where the interaction takes place between j and L (E_T can be written as $E_T(j,L)$). The same energy A^{gas} can also be evaluated based on quantum molecular calculations. The expression for the solvent interaction energy A_i^{sol} for each component can be obtained from the theory of solvophobic interactions using formula 4.1.58. For the component j, this formula is written again as (see Section 4.1):

$$A^{sol}_{j,S} = A^{cav}_{j,S} + A^{es}_{j,S} + A^{disp}_{j,S}$$
$$+ RT \ln(RT/p_0 V_S) \qquad (5.1.6)$$

where the index "S" indicates the surrounding molecules of solvent S. Similar formulas must be written for L and for jL. In this manner, the expression 5.1.4 for the total variation in free energy in equilibrium 5.1.1 becomes:

$$A = E_T(j,L) + \left(A^{cav}_{jL,S} + A^{es}_{jL,S} + A^{disp}_{jL,S}\right)$$
$$- \left(A^{cav}_{j,S} + A^{es}_{j,S} + A^{disp}_{j,S}\right) - \left(A^{cav}_{L,S}\right.$$
$$+ A^{es}_{L,S} + A^{disp}_{L,S}\right) - RT \ln(RT/p_0 V_S)$$
$$(5.1.7)$$

Relation 5.1.7 can be rearranged in the following form:

$$A = E_T(j,L) + A^{cav} + A^{es} + A^{disp}$$
$$- RT \ln(RT/p_0 V_S) \qquad (5.1.8)$$

where:

$$A^{cav} = A^{cav}_{jL,S} - A^{cav}_{j,S} - A^{cav}_{L,S} \qquad (5.1.9)$$
$$A^{es} = A^{es}_{jL,S} - A^{es}_{j,S} - A^{es}_{L,S} \qquad (5.1.10)$$
$$A^{disp} = A^{disp}_{jL,S} - A^{disp}_{j,S} - A^{disp}_{L,S} \qquad (5.1.11)$$

The expressions of A^{cav} for jL, j and L can be obtained from rel. 4.1.74, and similarly the expressions for A^{es} and A^{disp} can be obtained from rel. 4.1.75 and 4.1.76, respectively.

For the A^{cav} component of the total free energy, the following expression can be written:

$$A^{cav} = \mathcal{N}\gamma'_S\left[\kappa^e_{jL,S}\mathcal{A}_{jL}\left(1 - W_{jL,S}\right) - \kappa^e_{L,S}\mathcal{A}_L\left(1\right.\right.$$
$$\left.\left. - W_{L,S}\right) - \kappa^e_{j,S}\mathcal{A}_j\left(1 - W_{j,S}\right)\right]$$
$$(5.1.12)$$

This relation can be further simplified using the following approximations:

$$W_{jL,S} = W_{j,S} = W_{L,S} = 0 \qquad (5.1.13)$$

$$\kappa^e_{jL,S} = \kappa^e_{L,S} = 1 \tag{5.1.14}$$

$$A_{j,L} = A_j + A_L - \Delta A \tag{5.1.15}$$

where ΔA is the contact surface area between the solute j and the ligand L. With these approximations, the expression for A^{cav} can be written in the form:

$$A^{cav} = -\mathcal{N}\gamma'_S\left[\left(1 - \kappa^e_{j,S}\right)A_j + \Delta A\right] \tag{5.1.16}$$

From rel. 4.1.61 for $\kappa^e_{j,S}$ the following expression is obtained for A^{cav}:

$$A^{cav} = -\mathcal{N}\gamma'_S\left[\left(\kappa^e_S - 1\right)(V_S/V_j)^{2/3}A_j + \Delta A\right] \tag{5.1.17}$$

For the A^{es} component of the total free energy, the following expression can be written:

$$A^{es} = -\frac{\mathcal{N}\,\mathcal{D}_S}{2}\left(\frac{m^2_{jL}}{v_{jL}}\mathcal{P}_{jL,S} - \frac{m^2_L}{v_L}\mathcal{P}_{L,S} - \frac{m^2_j}{v_j}\mathcal{P}_{j,S}\right) \tag{5.1.18}$$

This relation can be further simplified using the following approximations:

$$m_{jL} = m_j \quad \text{and} \quad m_L = 0 \tag{5.1.19}$$

$$v_{jL} = \lambda\,v_j \tag{5.1.20}$$

$$\mathcal{P}_{jL,S} = \mathcal{P}_{j,S} \tag{5.1.21}$$

where λ is a proportionality factor between the molecular volume of the complex jL and the solute j. With these approximations, the expression for A^{es} can be written in the form:

$$A^{es} = \frac{\mathcal{N}(\lambda - 1)}{2\lambda}\frac{m^2_j}{v_j}\mathcal{D}_S\mathcal{P}_{j,S} \tag{5.1.22}$$

One further approximation is the following:

$$A^{disp}_{jL,S} = A^{disp}_{L,S} \tag{5.1.23}$$

With this approximation, it can be written:

$$A^{disp} = -A^{disp}_{j,S} \tag{5.1.24}$$

With all these estimations and simplifications, the expression for $\ln K$ is given by the formula:

$$\begin{aligned}
\ln K = {}& -\frac{E_T(j,L)}{RT} + \frac{A^{disp}_{j,S}}{RT} - \frac{\mathcal{N}(\lambda - 1)}{2\lambda RT}\frac{m^2_j}{v_j}\mathcal{D}_S\mathcal{P}_{j,S} \\
& + \frac{\mathcal{N}\gamma'_S}{RT}(\kappa^e_S - 1)\left(V_S/V_j\right)^{2/3}A_j + \frac{\mathcal{N}\gamma'_S}{RT}\Delta A \\
& + \ln\frac{RT}{p_0 V_S}
\end{aligned} \tag{5.1.25}$$

By adding the term $\ln\Psi$ to rel. 5.1.25, a formula for $\ln k$ is obtained. This formula can be written in the following form:

$$\ln k = \ln\Psi + \mathbf{a} + \mathbf{b} + \mathbf{c} + \mathbf{d}\,\gamma'_S\frac{A_j}{V^{2/3}_j} + \mathbf{e}\,\gamma'_S\,\Delta A_j$$
$$+ \ln\frac{RT}{p_0 V_S} \tag{5.1.26}$$

In rel. 5.1.26 the coefficients $\mathbf{a}$, $\mathbf{b}$, $\mathbf{c}$, $\mathbf{d}$, and $\mathbf{e}$ have the following expressions:

$$\mathbf{a} = -\frac{E_T(j,L)}{RT} \tag{5.1.27}$$

$$\mathbf{b} = \frac{A^{disp}_{j,S}}{RT} = -\frac{15.23}{8\pi RT}\left(Q'_{j,S} + Q''_{j,S}\right)Y_{j,S}D_jD_S \tag{5.1.28}$$

with $Q'_{j,S}$, $Q''_{j,S}$, Y_j, D_j, and D_S given by expressions discussed in Section 4.1:

$$\begin{aligned}
\mathbf{c} = {}& -\frac{\mathcal{N}(\lambda - 1)}{2\lambda RT}\frac{m^2_j}{v_j}\mathcal{D}_S\mathcal{P}_{j,S} \\
\approx {}& -\frac{\mathcal{N}}{4\pi\varepsilon_0 RT}\frac{\lambda - 1}{2\lambda}\frac{1}{1 - \left(\alpha_j/v_j\right)}\frac{2(\varepsilon_S - 1)}{2\varepsilon_S + 1}\frac{m^2_j}{v_j}
\end{aligned} \tag{5.1.29}$$

$$\mathbf{d} = \frac{\mathcal{N}}{RT}(\kappa^e_S - 1)V_S^{2/3} \tag{5.1.30}$$

$$\mathbf{e} = \frac{\mathcal{N}}{RT} \tag{5.1.31}$$

Partial calculations based on rel. 5.1.26 can be performed, and several results have been reported in the literature for different solvents and analytes [1]. However, some of the terms in rel. 5.1.26 are not easy to evaluate. Various approximations for expression 5.1.26 have been reported in the literature [1–3]. More difficulties arise because the treatment of entropy contribution in the estimation of ΔA^0 in rel. 5.1.26 is very likely incomplete. A decrease in the ability of adsorbed molecules on a stationary phase to rotate and vibrate may be associated with a large modification in the entropy during the adsorbtion process, and an enthalpy-entropy compensation was reported for RP-HPLC [4]. This effect is not captured in rel. 5.1.26. The importance of expression 5.1.26 lies in the identification of the main physical characteristics of the solute, stationary phase, and mobile phase that contribute to the value of k, rather than in its usefulness for direct calculations.

The value of k given by rel. 5.1.26 depends on several terms (**a**, **b**, **c**, **d**, **e**). A short discussion regarding each term contribution is as follows:

1) The first term (**a**) is given by rel 5.1.27 and represents the interactions in an ideal gas system consisting of solute molecule j and the ligand L of the stationary phase. Relation 5.1.26 shows that the value of k is higher when **a** is higher (since E_T has negative values, **a** is higher when the absolute value of E_T is higher). The expression for energy E_T was discussed in Section 4.1, and an approximation of the $E_T(j,L)$ value can be obtained using rel. 4.1.46 for the distance r equal with the van der Waals distance between the interacting molecules (j and L). In rel. 4.1.46, it was shown that the dispersion energy E_d (given, for example, by rel. 4.1.44) is an important part of the interaction. The term **a** makes some contribution in explaining why the retention of the same hydrophobic solute on a C18 column is stronger, for example, than on a C8 column. The longer

ligand on a C18 hydrophobic stationary phase leads to stronger wan der Waals forces with the solute.

2) The term **b** (see rel. 5.1.28) represents the dispersion forces between the molecule j and the molecules of the surrounding solvent S. As seen from rel. 4.1.76, $A^{disp}{}_{j,S}$ has negative values, and the larger those energies are, the lower is the value for k. This is expected since the stronger are the interactions of the solute with the solvent molecules, the lower is the expected value for k.

3) The term **c** contains a number of parameters that can vary. For most molecules $\lambda > 1$ and $\alpha_j / v_j < 1$. Therefore a larger **c** will have as a result a lower value for log k. The absolute value of **c** increases when the dipole moment of the solute j is higher. It is expected that molecules with higher dipole moment m_j will show lower retention compared to similar molecules regarding other properties, but with lower dipole moment.

4) The coefficient **d** is multiplied in rel. 5.1.26 with γ'_S and $A_j / V_j^{2/3}$. The value of **d** depends on $(\kappa^e{}_S - 1) V_S^{2/3}$. For nonpolar solvents such as heptane, CCl$_4$, and toluene, $\kappa^e{}_S < 1$, and therefore the whole term will be negative (see Table 4.1.3). This is in good agreement with lower log k values in less polar solvents. For water, methanol, and other polar solvents the factor $(\kappa^e{}_S - 1)$ is positive, leading to a larger log k, in agreement with experimental data. The term **d** γ'_S A_j / $V_j^{2/3}$ has a larger absolute value when γ'_S is larger and A_j is larger, and the term is smaller when V_j is larger.

5) The coefficient **e** is constant, and the term $\gamma'_S \Delta A_j$ has a positive value. This indicates that solvents with a higher value for γ'_S, the surface tension of the mobile phase, lead to a larger log k value, showing that the surface tension of the mobile phase is an important parameter in HPLC. Liquids with higher surface tension lead to higher

capacity factors. A larger ΔA is related to molecules with a larger hydrophobic moiety. This indicates that when all the other parameters are equal, the molecules that have a larger hydrophobic moiety will be retained more strongly. However, since steric effects also affect the value of ΔA_j, the steric fit will also play a role in the retention process. Molecules that can "present" a larger hydrophobic area to the stationary phase should be retained more strongly than those with a sterically hindered hydrophobic area.

The dependence given by rel. 5.1.26 of capacity factor k as a function of the van der Waals area of the molecule A_j indicates in a simplified form that ln K_j should have a linear dependence on A_j. This linear dependence is exemplified for several hydroxybenzenes in Figure 5.1.2, where log k for a separation on a LiChrosphaer 100 RP-18e column is plotted as a function of molecule van der Waals area A_j.

The mobile phase for the separation was methanol/water at different concentrations. The hydroxybenzenes used for the test were p-cresol, 2,6-dimethylphenol, phenol o-cresol, 3,5-dimethylphenol, β-naphtol, p-ethylphenol, and α-naphtol. The van der Waals areas were obtained using MarvinSketch Plug-ins [5]. As shown in Figure 5.1.2, very good correlations can be seen between the differences in the molecular areas and the differences in the logarithm of the capacity factors for molecules with similar structure as predicted by rel. 5.1.16. These correlations prove experimentally the dependence of retention on RP columns on the molecular area of the analyte. The values of k depend on the molecular surface area through the free energy required for creating the cavity in the solvent to accommodate the species j, indicating the major role of the solvent in RP separation. A schematic diagram picturing the "repulsion" of a hydrophobic molecule from a polar solvent and its "acceptance" by the hydrophobic phase at the

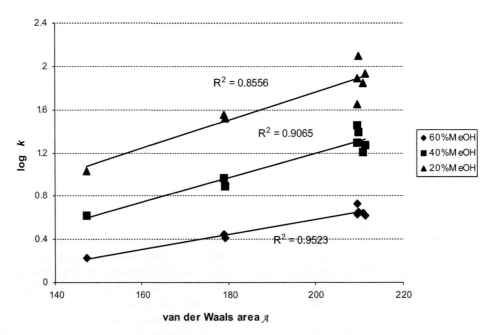

FIGURE 5.1.2 Linear dependence of log k for several hydroxybenzenes separated on a C18 column on van der Waals area A_j.

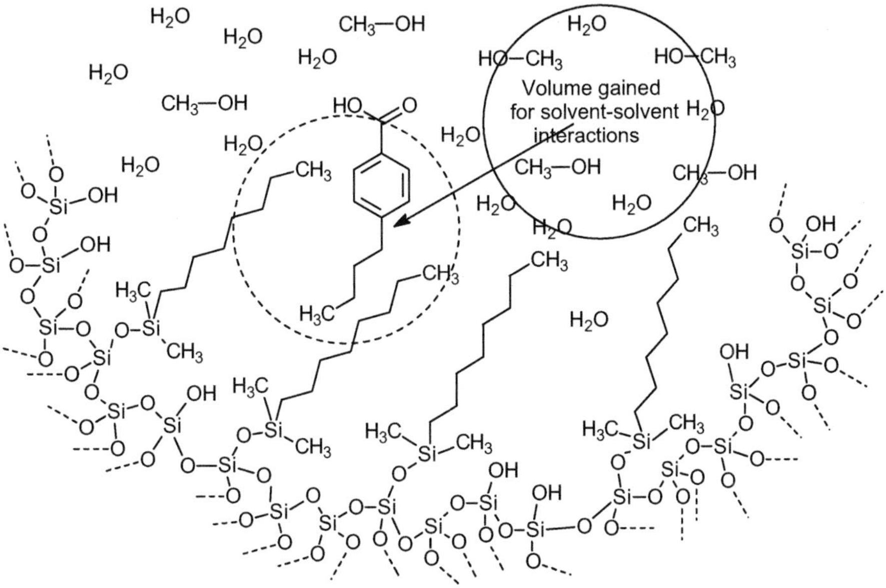

FIGURE 5.1.3 Schematic diagram picturing the "repulsion" of a hydrophobic molecule from a polar solvent and its "acceptance" by the hydrophobic phase at the interface between a mobile phase H_2O/CH_3OH and a C8 bonded phase.

interface between a mobile phase H_2O/CH_3OH and a C8 bonded phase is shown in Figure 5.1.3.

The hydrophobic character of the analyte, which is one of the determining factors regarding the hydrophobic interactions, is difficult to evaluate directly from molecular properties involved in rel. 5.1.26. However, the hydrophobic characterization of solutes can be obtained using the experimental parameter octanol/water partition constant log K_{ow} (see rel. 1.1.1). As previously shown in Section 3.1 (see Figure 3.1.3), a good correlation can be observed between log k and log K_{ow} for a number of aromatic compounds. Figure 5.1.4A displays a chromatogram of four analytes separated on a C18 column using water/acetonitrile 50/50 as a mobile phase. From their retention times, the values for log k can be calculated. Figure 5.1.4B shows the correlation between log k and log K_{ow} for the four analytes. The values of K_{ow} are strongly correlated with the nonpolar van der Waals area of the molecules (see Section 8.2), and indirectly, it can be considered that log K_{ow} is a measure of the nonpolar van der Waals molecular area. For this reason, further characterization of hydrophobicity will be conveniently discussed in terms of log K_{ow} values.

The variation of ln (RT/p_0V_S) with the solvent composition was evaluated for several solvent mixtures [1]. For the water/methanol and for water/acetonitrile mixture, for example, the value for ln (RT/p_0V_S) decreases with the increase in the concentration of the organic solvent. This is in agreement with the decrease in the value of log k with the increase in the organic solvent component in the mobile phase.

Qualitative Remarks Regarding Retention in RP-HPLC

The attempt to describe the retention process in RP-HPLC at the molecular level for a specific solute (analyte), stationary phase, and solvent can encounter difficulties due to the complex

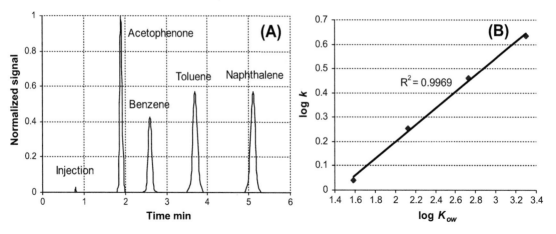

FIGURE 5.1.4 Chromatogram of four analytes on a C18 column and the correlation between their $\log k$ and $\log K_{ow}$ values.

calculations necessary and to the need for a number of parameters that are not always known or available for the separation participants. However, an excellent agreement exists between the theory and experimental data regarding the directional information. The solvophobic theory [1] shows that the factors leading to an increase in the capacity factor in RP are (1) the higher energy of interaction in the ideal "gas phase" between the solute and the stationary phase, (2) the larger surface area of contact between the hydrophobic moiety of the solute and the stationary phase, (3) the higher polarity of the mobile phase, and (4) the higher surface tension of the mobile phase. On the other hand, a lower $\log k$ value is caused by (1) the stronger interactions between the solute and the solvent molecules, (2) the higher dipole moment of the solute, and (3) a lower polarity (interactions among solvent molecules) of the mobile phase.

The result of the theory indicates that compounds with a larger hydrophobic moiety will have a higher retention (larger $\log k$ values). As expected, more hydrophobic stationary phases will also lead to higher $\log k$ for hydrophobic compounds. Characterization of the

influence of the mobile phase on the retention in RP-HPLC is relatively straightforward. On the other hand, the interaction of the solutes with the stationary phase is more difficult to characterize. This is caused mainly by the solid phase nature of the stationary phase, as well as by the presence of other active groups besides the hydrophobic ones. For example, phases containing long-chain hydrocarbons (e.g., C18) as hydrophobic groups may also have free silanol groups when the hydrocarbon chains are bound to a silica support (as is the case for numerous RP-HPLC columns). For this reason, even when the nature of the hydrophobic groups on the stationary phase is the same (e.g., C18), on different columns various retentions are achieved for the same analyte and assuming the same Ψ for the columns.

One suggested approximation for the hydrophobicity of the stationary phase would be use of the hydrophobicity parameter $\log K_{ow}$ for a "model" small molecule that would somehow simulate the stationary phase [6]. For example, a C18 phase bonded on silica could be simulated by the structure shown in Figure 5.1.5. For the structure from Figure 5.1.5, it is simple to calculate the corresponding $\log K_{ow}$. This may allow

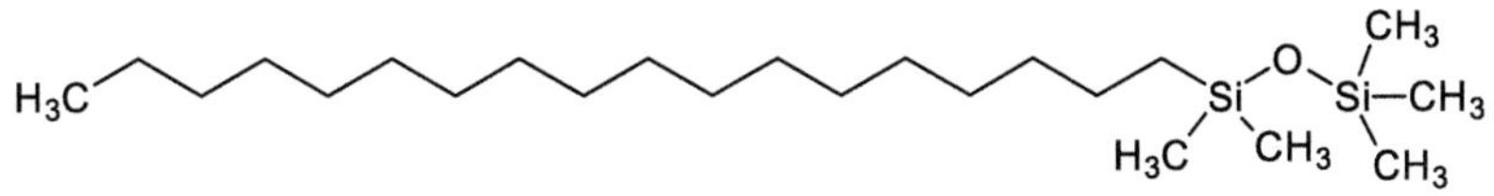

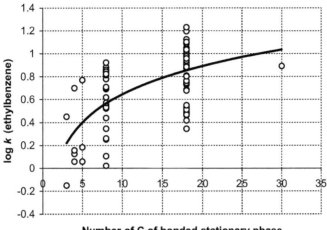

FIGURE 5.1.5 Structure of a small molecule that may simulate the hydrophobic character of a C18 bonded phase on silica.

FIGURE 5.1.6 Variation in hydrophobic character as described by log k_{EB} for columns with different lengths (carbon number) of hydrocarbon chain acting as stationary phase.

a comparison of the hydrophobicity for different stationary phases. However, this approximation is too simplistic to provide quantitative information. A more appropriate description of hydrophobic character of different columns is obtained using, for example, the capacity factor for a hydrophobic compound such as ethylbenzene (EB) with a specific mobile phase (acetonitrile/aqueous 60 mM phosphate buffer 50/50 v/v) (see Section 6.4) [7]. Figure 5.1.6 shows the variation of log k_{EB} with the length of the carbon chain in the bonded phase, for a number of 90 different columns used in RP-HPLC. The stationary phase of these columns contains chains of different lengths (C3, C4, C5, C8, C18, and C30). It can be seen that the hydrophobic character as described by log k_{EB} is generally increasing with the number of carbons, but a large variation in this character

is seen from column to column, even when the stationary phase has the same number of carbons. The results from Figure 5.1.6 indicate that other characteristics of the stationary phase besides the chemical structure of the ligand L influence the retention process.

Elution Process in RP-HPLC

The solvents used in RP-HPLC have a complex interaction with the solutes, and a more elaborate description of solvent properties is given in Chapter 7. In short, solvents with larger superficial tension γ'_S lead to larger log k values. Also, the solvent nature influences indirectly the retention by the very important component represented by the energy required to make a cavity in the solvent to accommodate the solute. The interaction energy among solvent

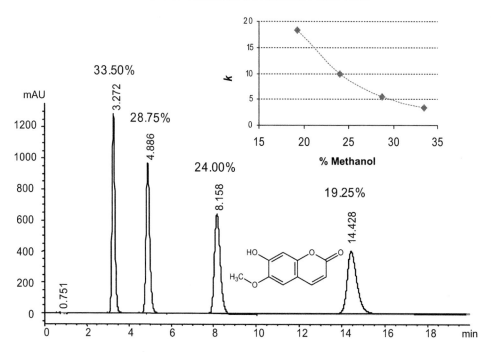

FIGURE 5.1.7 Separation of scopoletin on a C18 column with mobile phase an aqueous acetate buffer at pH = 4.4 and methanol. Different contents of methanol % in the mobile phase led to different retention times. The detection was done at λ = 340 nm. The variation of k with the methanol content is also shown.

molecules is higher in more polar solvents than in less polar solvents; this makes it more difficult for the analyte to go into the mobile phase, leading to a higher log k value for the analyte. An example of the influence on retention of the content of organic solvent is given in Figure 5.1.7 where a sample of scopoletin (7-hydroxy-6-methoxychromen-2-one) was subject to a chromatographic separation on a C18 column (Gemini 5u C18, 150 mm × 2.0 mm) in isocratic conditions at various contents of organic solvent in the mobile phase. The mobile phase consisted of an aqueous acetate buffer at pH = 4.4, and methanol as the organic phase. The detection was done by UV absorption at λ = 340 nm. As shown in Figure 5.1.7, the decrease in methanol content leads to an increase in the retention time and in the capacity factor k. The variation of k as a function of organic phase content shown in Figure 5.1.7 is typical in RP-HPLC. This type of

variation is also shown in Figure 5.1.8 for three compounds, acetophenone, propiophenone, and benzene, in a mixture of methanol/water on a Spherisorb ODS-2 column (100 × 5 mm, with 5 mm particles) [8].

The graphs from Figure 5.1.8 show that the retention (interaction with the stationary phase) is stronger when the solvent contains less organic component and the values for A^{cav} are higher. As the concentration in the organic component increases, the solvophobic interactions diminish and the retention is weaker. This effect is the basis of RP-HPLC separations in gradient conditions when the concentration of the organic component in the mobile phase is varied. At the beginning of the separation, a low content of organic solvent in the mobile phase is used such that all hydrophobic components are retained on the stationary phase. As the organic solvent content increases during

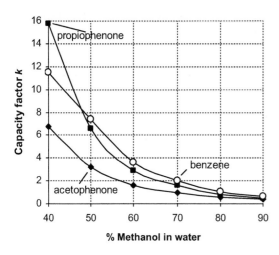
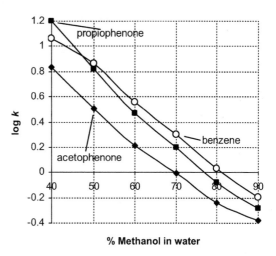

FIGURE 5.1.8 Typical variation of k and of $\log k$ for three compounds on a RP type chromatographic column, as the concentration of the organic component increases in the mobile phase.

the gradient, the hydrophobic compounds are eluted in the order of their hydrophobic character (the less hydrophobic compounds are eluted earlier). Hydrophobicity differences between the sample components combined with the variation in the content of organic constituent in the mobile phase (in gradient conditions) offer a very wide range of interaction intensities between the solute and the stationary phase. This leads to the extreme versatility of RP-type chromatography in separating compounds.

Separation in RP-HPLC Based on Solvophobic Theory

The separation is formally characterized by the selectivity α, which is dependent not only on the nature of the compounds to be separated but also on the stationary phase and on the mobile phase in the chromatographic process. The expression of α for two compounds X and Y to be separated can be obtained from rel. 2.1.49 and rel. 5.1.2, and can be written in the form:

$$\alpha = \exp\left[\frac{-\Delta A^0(X) + \Delta A^0(Y)}{RT}\right] \quad (5.1.32)$$

Using formula 5.1.7 for ΔA^0 (where j is X or Y), the difference in the free energies $-\Delta A^0(X) + \Delta A^0(Y)$ that determines the value of α in rel. 5.1.32 can be written in the form:

$$
\begin{aligned}
-\Delta A^0(X) + \Delta A^0(Y) = & -E_T(X, L) + E_T(Y, L) \\
& - (A^{cav}_{XL,S} - A^{cav}_{YL,S}) \\
& - \left(A^{vdw}_{XL,S} - A^{vdw}_{YL,S}\right) \\
& + (A^{cav}_{X,S} - A^{cav}_{Y,S}) \\
& + \left(A^{vdw}_{X,S} - A^{vdw}_{Y,S}\right)
\end{aligned}
$$

$$(5.1.33)$$

Assuming in rel. 5.1.33 that $(A^{cav}_{XL,S} - A^{cav}_{YL,S}) \approx 0$ and $(A^{vdw}_{XL,S} - A^{vdw}_{YL,S}) \approx 0$, rel. 5.1.33 leads to the following expression for $\ln \alpha$:

$$
\begin{aligned}
\ln \alpha = & \left[- E_T(X, L) + E_T(Y, L) + A^{cav}_{X,S} \right. \\
& \left. - A^{cav}_{Y,S} + A^{vdw}_{X,S} - A^{vdw}_{Y,S}\right]/RT
\end{aligned}
$$

$$(5.1.34)$$

Although rel. 5.1.34 does not provide an explicit expression for α, it indicates a set of points where differences between the two molecules and their relation to the stationary phase and mobile phase are important for achieving a larger α. The points of differences between the two analytes that are key factors in determining a good separation include the following: (1) the interaction "in vacuum" between the stationary phase and the analytes $- E_T(X,L) + E_T(Y,L)$, (2) the free energies required for creating the cavity in the solvent $A^{cav}_{X,S} - A^{cav}_{Y,S}$, and (3) the van der Waals interactions with the solvent molecules $A^{vdw}_{X,S} - A^{vdw}_{Y,S}$.

The interactions "in vacuum" between the stationary phase and the analyte and the van der Waals interactions with the solvent molecules are not specific for RP-HPLC. They are common for other types of interactions involved in the retention encountered in HPLC. In RP-HPLC these interactions have a certain role, but they are not the major component of the value of the free energy for the equilibriums. On the other hand, the energies required for creating the cavity in the solvent are specifically important for this type of separation. This emphasizes the role of the mobile phase in RP-HPLC.

Nonaqueous Reversed-Phase Chromatography (NARP)

Nonaqueous reversed-phase chromatography (NARP) does not differ as a mechanism from regular RP. In NARP, the main particularity is that the mobile phase is nonaqueous, but it is still more polar than the stationary phase. The absence of water in the mobile phase is usually required by the insolubility of the analytes in mixtures of solvents containing water. The technique is applied to compounds such as triglycerides, carotenoids, and hydrophobic vitamins. In principle, the types of interactions in NARP are not different from the hydrophobic interactions described for RP-HPLC. The absence of water from the mobile phase may only cause a lower importance of the cavity free-energy component in the separation.

Other Theories Applied for Explaining RP Separation

Solvophobic theory offers a reliable base for explaining the separation process in RP-HPLC [1,9]. However, a number of other theories have been applied for understanding intermolecular forces in RP-HPLC [9–11]. Among these are the theory explaining the retention as a combination of adsorption and partition models [12] and theories based on statistical thermodynamics [13–15]. These theories offer less practical utility compared to the one based on solvophobic interactions, and their presentation is beyond the scope of the present book.

5.2. RETENTION AND SEPARATION PROCESS IN ION-PAIR CHROMATOGRAPY

General Aspects

Ion-pair chromatography (IP) is a frequently utilized HPLC technique, applied for the separation of analytes (organic or inorganic) that contain ionizable or strongly polar groups, which lead these compounds to have a poor retention on hydrophobic columns (e.g. C18 or C8). Among these analytes are organic acids, amino acids, and amines. IP makes these compounds amenable to RP-HPLC type separation, using hydrophobic stationary phases (e.g., C18, C8). This can be achieved with the help of an ion-pairing agent (IPA or hetaeron), which is added to the mobile phase usually at concentrations between 10 and 100 mM/L. The ion-pairing agent (IPA) is an ion species selected so that it has an opposite charge to the analyte and is able to form a molecular association with it (or with a targeted solute). Compounds used as IPA can be, for example, quaternary amines (with various chain lengths of the substituent) in the case of analyzing acids, or strong organic acids (e.g., sulfonic acid with various chain substituents) in the case of analyzing amines. Two examples of IPA, one for acidic compounds analysis and the other for basic compounds analysis, and the formation of ion pairs are as follows:

$$R\text{-}COO^- + C_6H_{13}(CH_3)_3N^+ \rightarrow$$
$$R\text{-}COO^{-+}N(CH_3)_3C_6H_{13}$$

$$R\text{-}NH_3^+ + CH_3(CH_2)_7\text{-}SO_3^- \rightarrow$$
$$R\text{-}NH_3^{+-}O_3S\text{-}(CH_2)_7CH_3$$

The use of IPA with amphoteric character was also reported in the literature [16].

IPA molecules must contain a hydrophobic moiety, which allows its bonding to the stationary phase through hydrophobic interactions. After adding IPA, the separation on hydrophobic stationary phase becomes similar to that in RP-HPLC, and the increase in the hydrophobic character of the complex leads to stronger retention while the increase in the hydrophobic character of the mobile phase leads to the faster elution of the analytes. Although ion pairs are typically separated on

RP-type chromatographic columns, separations on silica with nonpolar solvents as a mobile phase were also reported [17]. This type of separation is a normal phase (NP) type.

The effect of adding IPA in the mobile phase is shown in Figure 5.2.1 for the retention of a basic polar compound, acyclovir (log $K_{ow} = -1.56$), on a common RP stationary phase without IPA and with the addition of IPA (sodium 1-heptansulfonate). Without the addition of IPA into the mobile phase, acyclovir elutes close to the void time of the column. The separation was done on a C18 150 mm × 4.6 mm i.d. × 5 μm column. The mobile phase was made of solutions A: 15% CH_3OH + 85% aq. solution with 0.1% H_3PO_4, and B: 15% CH_3OH + 85% aq. solution with 0.1% H_3PO_4 and 10 mM/L $C_7H_{15}SO_3Na$ as hetaeron (IPA). The flow rate was 1 mL/min, $T = 25°C$, and fluorescence detection was used, excitation 262 nm, emission 380 nm [18]. IPA enhances the retention of the analytes with the opposite charge to IPA, decreasing the retention of analytes with the same charge with IPA, and has a negligible effect on the retention of uncharged analytes. Since the interactions between the IPA and the analyte depend on the ionization state of the two participants, and the

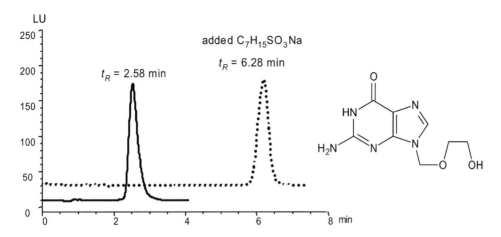

FIGURE 5.2.1 Chromatographic peaks belonging to acyclovir on a C18 column with and without $C_7H_{15}SO_3Na$ as hetaeron (IPA).

ionization state of both IPA and analyte depends on the pH, IP is a technique in which the mobile phase pH plays an important role. By increasing the retention (capacity factor k) for numerous polar and ionic compounds, the addition of IPA to the mobile phase found numerous applications in HPLC separations.

Ion-pairing Mechanisms

Two limiting model theories explain the mechanism of retention in IP: (1) the partition model in which the mechanism assumes the formation of ion pairs in solution, with the ion pair formed between the ionic or partially charged molecules of the analyte and the IPA followed by the retention of the preformed ion pair on the stationary phase; (2) the electrostatic model, which assumes that IPA is first bound to the hydrophobic stationary phase by its hydrophobic moiety, and once adsorbed it interacts with the components of the mobile phase by ionic/polar forces [19]. Both mechanisms indicate similar dependencies on pH, IPA, and analyte concentration, polarity of IPA, and analyte [20, 21]. Other more complex treatments of the retention process in IP were developed that attempt to include in the equilibria the formation of ion pairs in solution as well as after the IPA is adsorbed during the stationary phase [22−25].

Partition Model

The partition model for IP separation considers that the formation of ion pairs takes place first in solution. Once the ion pair is formed, the separation can be considered as being based on the interaction of the ion pair with the hydrophobic stationary phase in a polar solvent. For a basic compound and a strong acid IPA, the formation of the complex in the solution is favored by higher stability constant K of the complex, higher basicity of the analyte, higher concentration of IPA, and lower pH of the mobile phase. For an acidic analyte, the

same parameters will influence the formation of the ion pair, but the IPA should be an ionizable base, and the equilibrium is favored by a stronger acidity constant and a higher pH of the mobile phase.

The association mechanism of the analyte and the IPA is rather complicated. According to the partition model, two steps characterize the partition process of ion pairs between an aqueous phase (w) and an immiscible organic (bonded) phase (o) that is immobilized as a stationary phase:

$$A^{\pm} + IPA^{\mp} \rightarrow A^{\pm}IPA^{\mp} \qquad (5.2.1)$$

$$\left(A^{\pm}IPA^{\mp}\right)_w \rightleftarrows \left(A^{\pm}IPA^{\mp}\right)_o \qquad (5.2.2)$$

Examples of A^- can be compounds containing dissociated functional groups such as -O$^-$ (phenolic), -COO$^-$, -SO$_3^-$, -O-SO$_3^-$, and -S$^-$, and examples of A^+ can be compounds containing mainly amino groups (from primary -NH$_2$ to quaternary-NR$_3^+$). Compounds with zwitterionic character such as amino acids can also be included in the types of analytes A that can be separated by ion-pair chromatography.

The partition of ion pairs between two immiscible phases is further exemplified for an amine (indicated as A-NH_2), but can be equally discussed for any other compound capable of ion-pair association. The formation from an amine of ammonium cations in an aqueous phase is dependent on the concentration of H$^+$ in this phase; therefore pH plays a major role in the formation of ion pairs. The counterion in this case must be a compound that is able to form anions IPA^-. Examples of IPA^- are compounds containing sulfonic groups, or strong carboxylic acid groups (e.g., perfluorinated organic acids). In order to provide hydrophobicity to the ion pair, IPA^- must contain a hydrocarbon chain, denoted here by R, and IPA^- will have the form R-X$^-$ (e.g., a totally dissociated sulfonic acid R-SO$_3^-$). The schematic representation of liquid-liquid partition of these species is given in Figure 5.2.2.

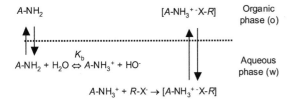

FIGURE 5.2.2 Representation of the partition process of a basic compound A-NH$_2$ (ionized as A-NH$_3^+$) as ion pairs formed with R-X$^-$ as IPA$^-$ [26].

As shown in Figure 5.2.2, two species can be present in the organic (stationary) phase, A-NH$_2$ and [A-NH$_3^{+-}$X-R]. Although the affinity for the organic phase of A-NH$_2$ is significantly lower than that of the ion pair [A-NH$_3^{+-}$X-R], the distribution process of A-NH$_2$ cannot be neglected. Therefore, two partition constants can be assigned to the distribution of two species between aqueous and organic phases:

$$K_{A-NH_2} = \frac{c_{A-NH_2,o}}{c_{A-NH_2,w}} \quad (5.2.3)$$

$$K_{[A-NH_3^+-X-R]} = \frac{c_{[A-NH_3^+-X-R],o}}{c_{[A-NH_3^+-X-R],w}} \quad (5.2.4)$$

The distribution coefficient for the extraction based on ion pairing $D_{ion\ pair}$ can be written as follows:

$$D_{ion\ pair} = \frac{c_{A-NH_2,o} + c_{[A-NH_3^+-X-R],o}}{c_{A-NH_2,w} + c_{[A-NH_3^+-X-R],w}} \quad (5.2.5)$$

Taking into account that the expression of K_b for the analyte A-NH$_2$ is given by the relation:

$$K_b = \frac{c_{A-NH_3^+}\,c_{HO^-}}{c_{A-NH_2}} = \frac{c_{A-NH_3^+}\,K_w}{c_{A-NH_2}\,c_{H^+}} \quad (5.2.6)$$

(see rel. 3.5.8), the dependence of the distribution coefficient on the proton concentration in aqueous phase becomes:

$$D_{ion\ pair} = \frac{1}{K_w + K_b \cdot 10^{-pH}} \left(K_w K_{A-NH_2} \right. \quad (5.2.7)$$
$$\left. + K_b K_{[A-NH_3^+-X-R]} 10^{-pH} \right)$$

With the replacement of the first product in 5.2.7 with the formula for D_Y given by rel. 3.5.14 in which Y is substituted with A-NH$_2$, the resulting expression for the distribution coefficient is given by:

$$D_{ion\ pair} = D_{A-NH_2}$$
$$+ K_{[A-NH_3^+-X-R]} \frac{K_b \cdot 10^{-pH}}{K_w + K_b \cdot 10^{-pH}} \quad (5.2.8)$$

Relation 5.2.8 indicates that for a very basic pH of the aqueous phase (i.e., [H$^+$] close to 10^{-14}), the second term of the above expression becomes 0, indicating that at this pH region the ion pair between analyte and IPA$^-$ does not play a role, and the partition is based only on the free analyte A-NH$_2$. This partition is negligible (according to the initial assumption), and therefore very little retention in the organic phase of the species A-NH$_2$ will take place. On the contrary, for a very acidic aqueous phase (pH close to 0), $K_b >> K_w$ (for an analyte with a strong basic character), and the value of distribution coefficient of the process becomes almost identical to the partition constant of the ion pair formed between the analyte and IPA$^-$. A comparison of two graphs showing the variation of $D_{ion\ pair}$ as a function of pH in the presence and the absence of IPA in aqueous phase is shown in Figure 5.2.3. For the organic phase being immobilized as a stationary phase, the higher value for $D_{ion\ pair}$ indicates larger capacity factors and longer retention times.

Relation 5.2.8 indicates a formal dependence of distribution $D_{ion\ pair}$ on the pH value. However, the pH also influences the concentration of A-NH$_2$ and [A-NH$_3^{+-}$X-R] in a solution, without the influence of the extraction equilibrium. For understanding this influence, two equilibria in solution should be considered:

$$\text{A-NH}_2 + \text{H}_2\text{O} \rightleftarrows \text{A-NH}_3^+ + \text{OH}^- \quad (5.2.9)$$
$$\text{A-NH}_3^+ + \text{R-X}^- \rightleftarrows \text{A-NH}_3^{+-}X - R \quad (5.2.10)$$

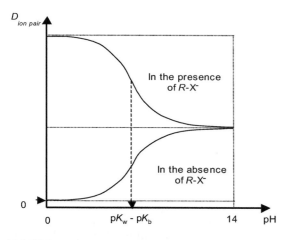

FIGURE 5.2.3 The dependence of the distribution coefficient of a basic analyte (A-NH$_2$) on pH for two situations (in the presence and in the absence of the ion pairing agent, R-X$^-$).

The first equilibrium is governed by the basicity constant K_b given by rel. 5.2.6. The second equilibrium depends on the stability of the [A-NH$_3^+$ $^-$X-R] complex and is governed by the stability constant K given by the expression:

$$K = \frac{c_{[A-NH_3^+{}^-X-R]}}{c_{A-NH_3^+}\, c_{R-X^-}} \quad (5.2.11)$$

From the initial concentration of the analyte $c_{init} = c_{A-NH_2} + c_{[ANH_3^{+-}X-R]}$, the concentration of the complex $c_{[ANH_3^+X-R]}$ can be obtained. For this purpose, the expression for the concentration c_{A-NH_2} is replaced using rel. 5.2.11 and 5.2.6. After a few simple calculations, the following expression is obtained:

$$c_{[ANH_3^{+-}X-R]} = \frac{KK_b c_{init} c_{R-X^-}\, 10^{-pH}}{K_w + KK_b c_{R-X^-}\, 10^{-pH}} \quad (5.2.12)$$

Relation 5.2.12 indicates that when the term $KK_b c_{R-X^-}$ [H$^+$] is large, K_w can be neglected, and the (molar) concentration of the complex [A-NH$_3$ $^{+-}$X-R] is basically equal with the initial concentration of the compound A-NH$_2$. The term is high when: 1) the constant K of the complex formation is high, 2) the compound

A-NH$_2$ has a high basicity constant K_b, 3) the concentration of IPA c_{R-X^-} is high, and 4) the pH is low. The equilibria 5.2.9 and 5.2.10 are further influenced by the distribution, and rel. 5.2.11 and 5.2.12 describe the process in the absence of the organic phase. However, the conclusions regarding the parameters increasing the formation of ion pairs remain valid even in the presence of the organic phase (hydrophobic solvent or stationary phase).

A similar discussion with that for A-NH$_2$ can be made for strong acidic analytes HX such as those containing carboxyl or sulfonyl groups. These analytes are found in aqueous phase more in the dissociated X$^-$ form than not in the dissociated. The ion-pairing agent in such cases should be a cation or IPA$^+$. Common IPA$^+$ compounds are quaternary amines with a hydrophobic moiety R, and common examples of such compounds are quaternary ammonium salts R-N(CH$_3$)$_3^+$. The ion pair formed between these ionic species and IPA can be written as [R-N(CH$_3$)$_3^+ X^-$]. These associates lead to the retention of the ionic X$^-$ on a hydrophobic phase. For these cases, the equivalent result with formula 5.2.8 will indicate a stronger retention of the ion pair at higher pH.

The distribution process for the ion pair expressed by rel. 5.2.8 indicates that two effects must be taken into consideration following the ion-pair formation. One is the distribution equilibrium for the free analyte, and the other is the distribution of the ion pair. Both analyte and ion pair can be considered hydrophobic components, and the theory developed for RP chromatography will apply for understanding their retention. Larger hydrophobic areas of the analyte and the complex analyte-IPA, more polar solvents and with higher superficial tension γ', and stationary phases with higher hydrophobic character will contribute to larger k values for the analytes. Since the ion pair contains the hydrophobic moiety of the IPA, larger capacity factors will be expected for the ion pair with IPA having larger hydrophobic moiety, for example, a longer hydrocarbon chain for an alkyl sulfonic acid in case of basic analytes, or for a quaternary amine in case of acidic analytes (amino acids are typically separated in IP using strong acids as IPA).

Electrostatic Model

The electrostatic model for IP separation also considers two stages, but the first process is the retention of IPA by hydrophobic interactions on the stationary phase, and the second process is the retention of the analyte by the ionic/polar forces between the ionic/polar groups of the IPA and those of the analyte. In support for this mechanism is the observation that the column used in IP requires for good reproducibility an equilibration time when the mobile phase containing IPA is flushed through the column (conditioning of the stationary phase). This step would suggest that the adsorption of IPA on the stationary phase is initially necessary.

The electrostatic model assumes the formation of two parallel layers of charges (double electrostatic layer) when electrically charged molecules of IPA are adsorbed on the stationary phase surface. The adsorption of IPA is assumed to take place by hydrophobic interactions, and since the concentration of IPA in the mobile phase is high (compared to any analyte), the surface of the stationary phase can be assumed covered with IPA. The ionic ends of IPA are creating the first adsorbed layer of charges [27]. In this theory, only the first layer of charges is compact, while the second layer is diffuse, and the electrical potential created by the first layer decreases exponentially away from the adsorbing surface (based on the Gouy-Chapman theory of electrical double layer; see, e.g., [28]). The secondary diffuse layer is due to the IPA counterions (the initial ones or those provided by the solutes). Both layers are assumed to be under dynamic equilibrium. The counterions of opposite signs to those of the adsorbed IPA are electrostatically attracted toward the charged surface, and those with the same electrical charges are repelled from the surface [29]. This process is illustrated in Figure 5.2.4.

Considering that the ion-pairing agent is a hydrophobic anion, IPA^-, the analyte i is represented by the cation $A-X^+$, and the stationary

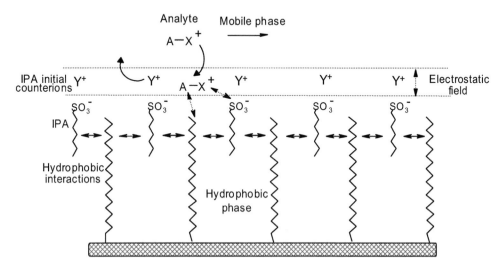

FIGURE 5.2.4 Illustration of the formation of electrical double layer during ion-pairing mechanism (the analyte i is represented by $A-X^+$, IPA is an alkyl sulfonate).

phase is considered a ligand L, the main equilibria within this mechanism are the following:

$$L + IPA^- \rightleftarrows L\,IPA^- \qquad (5.2.13)$$

$$L\,IPA^- + A\text{-}X^+ \rightleftarrows L\,IPA^{-\,+}X\text{-}A \qquad (5.2.14)$$

The electrostatic model is only an approximation, however. Experimentally, it can be proven that the analyte has its own interaction with the stationary phase, and some retention of the analyte takes place without IPA in the mobile phase. The effect of IPA is only that of enhancing the retention. The interactions between the analyte and the stationary phase can be considered as "mediated" by the addition of IPA.

Although the direct interaction of the solute i with the stationary phase independent of IPA can be small, this process must be taken into consideration when discussing the retention process. The change in free energy (Δ ommited in the notation) to bring a charged solute to a surface of the stationary phase that involves IPA can therefore be divided into two contributions:

$$A^0_{\ i} = A^0_{\ hydrophobic} + A^0_{\ el} \qquad (5.2.15)$$

In rel. 5.2.15, $A^0_{hydrophobic}$ represents total variation in free energy of the adsorption of analyte i in the absence of the electric field created by the adsorption of IPA on the stationary phase surface (as given by rel. 5.1.15). The contribution to the free energy caused by the electrostatic interaction between charged analyte i and the electric field of the stationary phase surface is given by A^0_{el}. Taking into account the relation between capacity factor k and equilibrium constant K for the partition of the species i between mobile and stationary phase, that is, $k_i = \Psi\,K$ (where Ψ is the phase ratio), the expression for k can be written as follows:

$$
\begin{aligned}
k_i &= \Psi\exp\left[-\left(A^0_{\ hydrophobic} + A^0_{\ el}\right)/RT\right]\\
&= \Psi\exp\left(-A^0_{\ hydrophobic}/RT\right)\exp\left(-A^0_{\ el}/RT\right)\\
&= k_i(0)\exp\left(-A^0_{\ el}/RT\right)
\end{aligned}
$$
$$(5.2.16)$$

In rel. 5.2.16 $k_i(0)$ represents the capacity factor of the species i in the absence of IPA in the mobile phase. The contribution to the retention factor due to the electrostatic interaction between charged analyte i and the electric field created by IPA adsorbed onto the stationary phase is represented in this equation by the term $\exp(-A^0_{el}/RT)$ [30].

The electrostatic contribution that represents the work involved in the transfer of an ionic analyte with charge z_i to the charged surface of the stationary phase (A^0_{el}) is given by the formula:

$$A^0_{\ el} = z_i\,F\,\Delta E \qquad (5.2.17)$$

where F represents the Faraday constant and ΔE is the difference in electrostatic potential between the bulk of the mobile phase and the stationary phase surface. With this formula rel. 5.2.16 can be written in the following form:

$$k_i = k_i(0)\exp\left(-\frac{z_i F \Delta E}{RT}\right) \qquad (5.2.18)$$

Relation 5.2.18 indicates an important contribution of $k_i(0)$ to the separation process, which corresponds to the experimental finding that the structure of the analyte strongly influences the separation. The electrostatic potential ΔE created by the adsorption of IPA on the stationary phase surface and its value can be obtained based on Gouy-Chapman's theory. The full expression for ΔE is reported in various publications (see, e.g., [31]). For IP that involves small charges and small electrical fields an approximation of ΔE formula is the following:

$$\Delta E = \frac{z_{IPA} F\, n_{IPA}^{s.p.}}{\kappa \varepsilon_0 \varepsilon_{mo}} \qquad (5.2.19)$$

where $n_{IPA}^{s.p.}$ represents the surface molar fraction of the charged species (IPA), ε_0 is the electrical permittivity of vacuum, ε_{mo} is the dielectric constant of the surrounding medium (mobile phase), and κ is a variable known as Debye length and is given by the formula:

$$\kappa = F\left[\frac{\sum\limits_{j} z_j^2 c_j}{\varepsilon_0 \varepsilon_{mo} RT}\right]^{1/2} \tag{5.2.20}$$

the sum being made over all the ionic species in the solution. The electrostatic contribution to the change in free energy depends basically on the electric charge of the analyte and IPA involved into the partition process. Thus, the sign ($+$ or $-$) for the A^0_{el} can be positive when the analyte and IPA have the same sign charges and negative when they are of opposite charges.

The electric field created on the surface of stationary phase due to the adsorption of IPA could influence further adsorption of IPA because the adsorbed molecules of this species may repel other molecules coming from the mobile phase, resulting in an equilibrium between adsorbed and free molecules of IPA. Therefore, as the surface concentration of IPA increases, the area accessible for additional molecules on the surface decreases, and these molecules will find it more and more difficult to find an adsorption site on the stationary phase surface. On the other hand, there is an equilibrium process between adsorption and desorption of IPA on the surface, determined by its concentration in mobile phase and its affinity toward hydrocarbon chains of the hydrophobic stationary phase. This affinity can be estimated based on the Langmuir model for adsorption, that gives the molar fraction $n_{IPA}^{s.p.}$ of the species adsorbed on the surface by rel. 2.2.15. Based on rel. 2.2.15 and with the appropriate expression for the equilibrium constant K_i, it can be established that $n_{IPA}^{s.p.}$ is given by the formula:

$$n_{IPA}^{s.p.} = n_{IPA,max}^{s.p.} K_{IPA} c_{IPA} e^{-\frac{z_{IPA} F \Delta E}{RT}} \Big/ \left(1 + K_{IPA} c_{IPA} e^{-\frac{z_{IPA} F \Delta E}{RT}}\right) \tag{5.2.21}$$

where K_{IPA} is the adsorption constant for IPA, c_{IPA} is the concentration of IPA in the mobile

phase (equivalent to c_{R-X^-} in rel. 5.2.11), and $n_{IPA,max}^{s.p.}$ is the maximum value for the IPA molar fraction on the surface. Expression 5.2.21 can be simplified since the denominator of the ratio is practically 1 and can be written in the form:

$$n_{IPA}^{s.p.} = n_{IPA,max}^{s.p.} K_{IPA} c_{IPA} e^{-\frac{z_{IPA} F \Delta E}{RT}} \tag{5.2.22}$$

This equation shows that the surface molar fraction of IPA is determined by the IPA concentration in mobile phase c_{IPA} and by the adsorption constant K_{IPA} (which at its turn is determined by IPA hydrophobicity).

Further use for the value of $n_{IPA}^{s.p.}$ as given by rel. 5.2.22 is in obtaining an expression for k_i in ion-pair chromatography. For this purpose, an expression relating $n_{IPA}^{s.p.}$ and k_i must be obtained. The first step for this is to express the potential ΔE as a function of capacity factors k_i and $k_i(0)$. By applying the natural logarithm to rel. 5.2.18, the following expression is obtained:

$$\Delta E = -\frac{RT}{z_i F} \ln \frac{k_i}{k_i(0)} \tag{5.2.23}$$

From rel. 5.2.23 introduced in the formula 5.2.22 of $n_{IPA}^{s.p.}$, the result is the following:

$$n_{IPA}^{s.p.} = n_{IPA,max}^{s.p.} K_{IPA} c_{IPA} \left(\frac{k_i}{k_i(0)}\right)^{\frac{z_{IPA}}{z_i}} \tag{5.2.24}$$

On the other hand, ΔE is given by rel. 5.2.19, which can be written using rel. 5.2.24 in the form:

$$\Delta E = \frac{z_{IPA} F}{\kappa \varepsilon_0 \varepsilon_{mo}} n_{IPA,max}^{s.p.} K_{IPA} c_{IPA} \left(\frac{k_i}{k_i(0)}\right)^{\frac{z_{IPA}}{z_i}} \tag{5.2.25}$$

Relations 5.2.23 and 5.2.25, both giving an expression for ΔE, imply that the following formula is valid:

$$-\frac{RT}{z_i F} \ln \frac{k_i}{k_i(0)} = \frac{z_{IPA} F}{\kappa \varepsilon_0 \varepsilon_{mo}} n_{IPA,max}^{s.p.} K_{IPA} c_{IPA} \left(\frac{k_i}{k_i(0)}\right)^{\frac{z_{IPA}}{z_i}} \tag{5.2.26}$$

Relation 5.2.26 can be rearranged in the form:

$$\left(\frac{k_i}{k_i(0)}\right)^{-\frac{z_{IPA}}{z_i}} \cdot \ln\left(\frac{k_i}{k_i(0)}\right)^{-\frac{1}{z_i z_{IPA}}}$$

$$= \frac{F^2}{RT\varepsilon_0\varepsilon_{mo}} \cdot \frac{n_{IPA,max}^{S.P.} K_{IPA} c_{IPA}}{\kappa} \qquad (5.2.27)$$

Some transformation of the first term in rel. 5.2.27 is necessary. Using the approximation $\ln(x+1) = x$ valid for small values of x, the following formula can be obtained:

$$\ln\left\{\left[\ln\left(\frac{k_i}{k_i(0)}\right)^{-\frac{1}{z_i z_{IPA}}} - 1\right]\right.$$

$$\left. +1\right\} \cong \ln\left(\frac{k_i}{k_i(0)}\right)^{-\frac{1}{z_i z_{IPA}}} - 1 \qquad (5.2.28)$$

After taking the logarithm of rel. 5.2.27 and introducing the approximation 5.2.28, the following expression can be written:

$$\ln\left(\frac{k_i}{k_i(0)}\right)^{-\frac{z_{IPA}}{z_i}} + \ln\left(\frac{k_i}{k_i(0)}\right)^{-\frac{1}{z_i z_{IPA}}} - 1$$

$$= \ln\frac{F^2}{\varepsilon_0\varepsilon_{mo}RT} + \ln\frac{n_{IPA,max}^{s.p.} K_{IPA} c_{IPA}}{\kappa} \qquad (5.2.29)$$

Relation 5.2.29 can be further modified into the following equation that describes the dependence in IP electrostatic model of the capacity factor k_i on the main experimental parameters [30]:

$$\ln k_i = \ln k_i(0) - \frac{z_i z_{IPA}}{z_{IPA}^2 + 1}\left[\ln c_{IPA}\right.$$

$$\left. + \ln\frac{n_{IPA,max}^{s.p.} K_{IPA}}{\kappa} + \ln\frac{F^2}{\varepsilon_0\varepsilon_{mo}RT} + 1\right]$$

$$(5.2.30)$$

According to this equation, the capacity factor k_i is dependent in IP on several parameters that are further discussed in detail (natural logarithms can be replaced in rel. 5.8.21 with logarithms in base 10 since $\ln x = 2.303 \log x$).

1) In rel. 5.12.18, the capacity factor $k_i(0)$ of the ionic species i in the absence of IPA can be for some compounds sufficiently high to play an important role. The value of $k_i(0)$ depends on the hydrophobicity of i and the phase ratio of the chromatographic column Ψ. The contribution of $k_i(0)$ to the total value of k_i can be a key factor in the separation. On columns that have additional interactions besides pure hydrophobic ones, polar compounds can be retained stronger or weaker, depending on their structure even in the absence of the ion-pair agent.

2) There is a linear dependence between $\ln k_i$ (and $\log k_i$) and $\ln c_{IPA}$. When changing $\ln c_{IPA}$, the charges $z_i z_{IPA}/(z_{IPA}^2 + 1)$ will give the slope of the change in $\ln k$. For example, if the ionic analyte and IPA have charges $+1$ and -1, respectively, the slope of the dependence of $\ln k_i$ on $\ln c_{IPA}$ is 0.5. This effect is exemplified in Figure 5.2.5 for three biguanidines where the linear dependence of $\ln k_i$ on $\ln c_{IPA}$ is shown and the slopes of the dependence lines are very close to 0.5 value. (Experimental conditions: C18 column; 50% MeOH in mobile phase; 50% aqueous solution containing 5–50 mM $C_6H_{13}SO_3Na$ as IPA at pH $= 2$) [32]).

3) The influence of the hydrophobic nature of IPA in the value of capacity factor is expressed in rel. 5.2.30 by the factor K_{IPA}. The hydrophobic character of IPA is determined by its structure and the moiety with hydrophobic character. For example, for a series of alkylsulphonates ($C_nH_{2n+1}SO_3^-$) used as IPAs, the value of K_{IPA} increases with the increase in the number of C atoms (n) in its molecule. As an example, the variation of $\log k$ with the length of the alkyl substituent

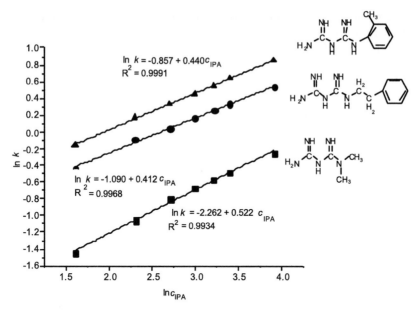

FIGURE 5.2.5 Dependence of the capacity factor for three biguanidines (tolylformin, phenformin, and metformin) on the concentration of sodium hexanesulphonate in mobile phase.

for several alkylsulfonates used as IPA in the analysis of four polar compounds belonging to biguanidine class is shown in Figure 5.2.6. The separation was done on a C18 column with the mobile phase composition: 40% MeOH; 60% aqueous component containing individual IPA as 10 mM/L, pH $= 2.0$, adjusted with H_3PO_4 [32].

In the selected example, the dependence of log k on the length of alkyl chain is linear for three of the compounds and quadratic for one, but this type of dependence is not predictable or universal. The only conclusion is that log k increases with the increase in the hydrophobicity of the organic moiety of IPA. The increase in this hydrophobicity is also associated with the decrease in the solubility of IPA in water, which is a limiting factor in selecting the length of the alkyl chain. As shown in

rel. 5.2.30, a linear dependence exists between log k_i and log c_{IPA}, and therefore a higher concentration of IPA is desired for increased retention.

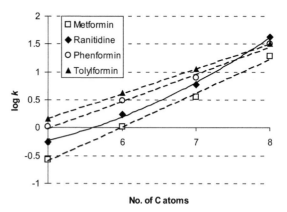

FIGURE 5.2.6 Dependence of log k on the number of C atoms in alkylsulfonates used as IPA for separation of four polar compounds.

The assumption that K_{IPA} does not depend on the analyte may be only an approximation, since formation of some association solute/IPA in the mobile phase is still possible, and the adsorption of the ion pairs from the mobile phase can be different depending on the analyte. Also, in the separation process, different interactions besides the hydrophobic ones may play a role. In the case of partially polar stationary phases such as phases containing silanol groups, polar groups for end-capping, or polar-embedded groups, the polar interactions may contribute to the separation.

4) The effect of the nature and concentration of the organic modifier in mobile phase (e.g., methanol) is found implicitly in the value of $k_i(0)$, and K_{IPA}. The influence of the organic modifier typically follows the same trend as described for RP-HPLC and exemplified in Figure 5.1.7, the higher organic content in the mobile phase leading to lower k_i values.

5) The influence of the electrolyte concentration in the mobile phase is included in the expression of inverse Debye length κ, and the value of the capacity factor decreases with the increase of the electrolyte concentration in the mobile phase. However, other effects (see chaotropic and salting out effects) may compensate for this decrease.

6) The influence of pH on the value of capacity factor is found in the term $\ln k_i(0)$, and when the analyte i becomes more dissociated by modifying the pH, its retention increases. Ionic compounds often have different ionization forms depending on pH. As an example, the % in a solution of the four possible forms of metformin (N,N-dimethylimidodicarbonimidic diamide) at different pH values is shown in Figure 5.2.7.

As expected, the formation of ion pairs of each ionized form of the analyte will be different leading to a different k value. The variation of k for the case of metformin separated on a C18 column with a mobile phase made of 35% MeOH

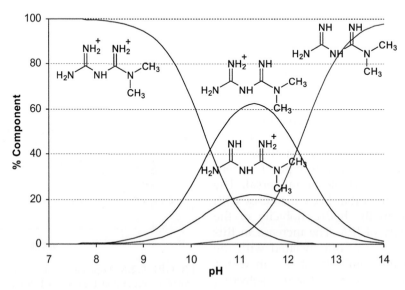

FIGURE 5.2.7   Different forms of metformin in a solution, depending on pH.

+ 65% aqueous solution of 10 mM sodium heptanesulphonate + 0.1% H_3PO_4 and the pH adjusted with 10% KOH solution is shown in Figure 5.2.8 [32].

For strong acidic/basic or for ionic analytes, the influence of pH on the separation is less important, unless the analyte is involved in a structure modification that is influenced by pH. For example, the retention of pralidoxime in an IP separation, with sodium octanesulfonate as IPA, is strongly influenced by pH. This ionic compound has a structure that participates in the following equilibrium:

The retention of pralidoxime at neutral or basic pH of the mobile phase on a C18 column (Eclipse XDB C18) takes place according to a simple RP mechanism, even in the presence of sodium octanesulfonate (IPA reagent). In acidic conditions, in the presence of IPA, the retention takes place following the ion-pair formation. This can be verified by studying at different pH values the variation of log k for this compound as a function of the concentration of the organic modifier (MeOH%) in the mobile phase. For IP-type retention (low pH), a linear dependence on the proportion of MeOH is noticed, while for RP retention (high pH), the dependence is not linear, proving the difference in the retention mechanism.

7) Expression 5.2.30 also accounts for the fact that when i and IPA have the same charge ($z_i = z_{IPA}$), the value of the capacity factor decreases with the increase of IPA concentration in the mobile phase since a positive quantity is subtracted from ln $k(0)$.

$$ \text{(5.2.31)} $$

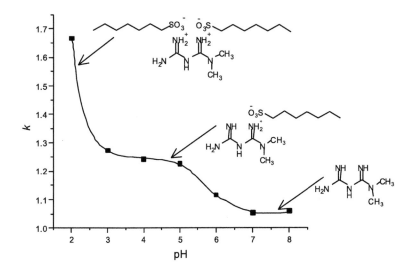

FIGURE 5.2.8 Dependence of the capacity factor k of metformin on the pH of the aqueous mobile phase in ion-pairing mechanism with sodium heptanesulphonate as IPA.

As seen from the discussion based on electrostatic model and that based on formation of ion pairs in solution, both models describe parameters that influence the IP retention process, but some dependences are more obvious from the partition model (e.g., pH dependence from the ion-pair formation in solution) and other dependences more obvious from the electrostatic model.

Chaotropes in Ion Pairing

In place of usual ion-pairing agents, a different type of additive has been used in IP, namely, chaotropic agents such as BF_4^-, ClO_4^-, or PF_6^-. Chaotropes have the capacity to disrupt the polar intermolecular forces such as those with water. These reagents are likely to form strong associations with polar components, such that the salts analyte/chaotrope have lower interaction with the mobile phase and can be retained on hydrophobic stationary phases. It is also possible that the solvation shell that typically surrounds the ionic analytes is disrupted by the addition of chaotropes that interact with the water. This solvation shell can suppress the ability of the analyte for hydrophobic interaction with the stationary phase, leading to a decrease in the retention [33–35]. If the mobile phase contains counteranions (chaotropes) with high polarizability and a low degree of hydration, they can disrupt the solvation shell of the positively charged basic analytes, and thus they enhance the ability of the ion-associated complex to interact with the stationary phase.

The chaotropic effect of different anions (BF_4^-, ClO_4^-, or PF_6^-) on the retention of positively charged basic analytes is different from the salting-out effect. The addition of salts changes the characteristics of the polar solvents (e.g., water) by increasing the polar interactions inside the solvent, decreasing the number of solvent (water) molecules available to interact with the charged part of the solute, and diminishing the interaction of the solvent with the solute. For proteins that were initially soluble in water, salting out produces an increase in the hydrophobic interactions between the proteins and leads to their precipitation. Certain ions have a stronger salting-out effect than others regarding protein solubility, and a specific order of their potency was established (Hoffmeister series). The Hoffmeister effect is stronger when the charge delocalization and polarizability are higher. Anions appear to have a larger effect than cations and are usually ordered as follows: $F^- \approx SO_4^{2-} < H_2PO_4^- < HCOO^- < CH_3SO_3^- < Cl^- < NO_3^- < CF_3COO^- < BF_4^- < ClO_4^- < SCN^- < PF_6^-$. For the cations, the series is: guanidinium $> Ca^{2+} > Mg^{2+} > Li^+ > Na^+ > K^+ > NH_4^+$. The chaotropicity of an inorganic additives is related to its position in the Hoffmeister series [36, 37].

5.3. RETENTION AND SEPARATION ON POLAR STATIONARY PHASES

General Comments

The compounds with a high polar character (very low or negative K_{ow}) are poorly retained on hydrophobic stationary phases. The compounds that are polar, acidic, or basic can be separated using ion pair chromatography (see Section 5.2), but another alternative leading to excellent separations is the use of phases with polar character. These polar phases, such as bare silica or bonded phases on silica with the ligand having polar moieties are used in several types of chromatography including normal phase chromatography (NPC), hydrophilic interaction chromatography (HILIC), and aqueous normal phase chromatography (ANPC or ANP). The main difference between NPC and HILIC is the utilization of different types of mobile phase, organic with no water in NPC, and organic polar with some water in HILIC. Both techniques have the mobile phase less polar than the stationary phase. However, the total absence of water in NPC and the presence of water in HILIC seems to lead to a difference in the retention mechanism. In HILIC the hydrophobic interactions have some role in the separation. This role is absent in NPC. ANP is different, being practiced on hydride columns with a wide range of solvent polarities. High polarity phases are also those used in ion exchange, but the separation process in this case is different (and is discussed in Section 5.4). The separation on ANP phases is not very well understood.

Retention Process in NPC and HILIC

The retention principles for polar stationary phases used in NPC and in HILIC have a number of similarities. In NPC, the main types of interactions are polar interactions, hydrogen bond formation, charge transfer, and also ionic interactions [30]. In HILIC, added to the types of interactions from NPC, are the hydrophobic interactions, which also may play a role in the retention process [38]. Based on thermodynamic concepts, the expression for log k in NPC and HILIC is given by a formula identical to rel. 5.1.3 (van't Hoff equation), which can be written in the form:

$$\log k = \log \Psi - \frac{\Delta A^0}{2.302\,RT} \qquad (5.3.1)$$

Formally, the retention process in both NPC and HILIC can be described by an equilibrium of the type 5.1.1 where the same notations were used, j for the solute (analyte), L for the bonded (or immobilized) ligand in the stationary phase, and jL for their union:

$$j + L \rightleftarrows jL$$

The evaluation of the change in free energy ΔA^0 for the process can be obtained by a formula similar to rel. 5.1.4, which was used for the

description of reversed phase retention process. The expression for ΔA^0 is given by formula:

$$\Delta A^0 = A^{gas} + \sum_{i=j,L,jL} A^{sol}{}_i \qquad (5.3.2)$$

where $A^{sol}{}_i$ indicates the interaction energies with the solvent. By following again the same reasoning as for RP, which was developed in Section 5.1, the expression for ΔA^0 can be written in a form similar to rel. 5.1.7:

$$\Delta A^0 = E_T(j, L) + (A_{jL}^{cav} + A_{jL,S}^{vdw}) - (A_j^{cav} + A_{j,S}^{vdw})$$
$$- (A_L^{cav} + A_{L,S}^{vdw}) - RT \ln (RT/p_0 V_S) \qquad (5.3.3)$$

The difference between rel. 5.3.3 and rel. 5.1.7 (describing the energies involved in reversed phase separations) is in the magnitude of the terms involved in this relation. One first observation for NPC/HILIC is that the values for A^{cav} for all the components involved in the separation, including the solute j, have a smaller contribution to the value of ΔA^0. Since the interaction among less polar solvent molecules is not high, the energy required to make a cavity in the solvent to accommodate any solute does not play a major role in the total energy of interaction, as it was the case for RP chromatography. With a series of simplifying approximations similar with those used in Section 5.1, and not attempting to develop a formula for the interaction with the solvent, rel. 5.3.3 can be written in the form:

$$\Delta A^0 = E_T(j, L) - A_{j,S}^{vdw} - A_j^{cav}$$
$$- RT \ln(RT/p_0 V_S) \qquad (5.3.4)$$

With rel. 5.3.4 the expression for log k can be written as follows:

$$\log k = \log \Psi - \frac{E_T(j, L)}{2.303\,RT} + \frac{A_{j,S}^{vdw}}{2.303\,RT} + \frac{A_j^{cav}}{2.303\,RT}$$
$$+ \log \frac{RT}{p_0 V_S} \qquad (5.3.5)$$

Most of the terms in rel 5.3.5 have directionally an opposite trend as compared to those developed for RP-HPLC. The values for $E_T(j,L)$ as well as $A_{j,S}^{vdw}$ term in rel. 5.3.4 will be more important in NPC/HILIC than in RP-HPLC (all energies in rel. 5.3.4 have negative values). This shows a stronger retention for stationary phases with active groups having higher polarity which will lead to larger $E_T(j,L)$ (in absolute value). In HILIC separations, the log k values for numerous polar compounds are for different stationary phases in the order: silica > amino > diol > cyano, which corresponds to the stationary phase polarity. Since more polar solvents lead to larger $A_{j,S}^{vdw}$ (in absolute value), this type of solvent will have the effect of a decrease in log k (or retention time of the analyte). In the theory developed for RP separation, the evaluation of the contribution of van der Waals interactions between the solute j and the solvent S was expressed by the terms **b** and **c** in rel. 5.1.26 and was based on solvophobic theory. For the case of interactions with polar molecules, this term is better evaluated using either the expression 4.1.52 or 4.1.54. From rel. 4.1.52 it can be seen that the values for E_{solv} are higher when the partial charges of the interacting molecules (solute and solvent) are higher and when the dielectric constant of the solvent is higher. This is in agreement with the fact that more polar solvents will decrease the retention time of the solute. The same result is obtained from rel. 4.1.54, which indicates that the analytes with a larger A_i^{SASA} have a stronger interaction with the solvent.

The analysis of the values for A_j^{cav} indicates that the solvents that are more polar will lead to higher absolute values for this term and therefore smaller log k values (faster elution). The compounds (solutes) that are separated by NPC/HILIC will be retained more strongly on the stationary phase when they have larger E_T values for the interaction with the stationary phase. As shown in Section 4.1,

the values for E_T (see rel. 4.1.46) are higher when the dipole moments m and the polarizabilties α are higher for the interacting species (ligand L and solute j). This explains why for a given stationary phase and a selected mobile phase, in both NPC and HILIC the more polar compounds are retained more strongly and have larger log k values. For HILIC, the solvent molecules having polar character, including water as a mobile phase component, will have a considerably higher A_j^{cav} in comparison with NPC, showing in this respect similarity with RP-HPLC. For this reason, hydrophobic interactions may be considered as having contribution to HILIC separation. This effect is more pronounced when the interaction between the solvent molecules is stronger than the one between the solvent and solute molecules.

In the evaluation of E_T for polar interactions, rel. 4.1.46 takes into consideration only dipole-dipole, dipole-induced dipole and dispersion energies. However, the polar molecules are frequently capable of establishing hydrogen bonds, electron donation or withdrawal interactions, and ion-exchange type interactions. These interactions may play a major role in the values of E_T and should be considered when evaluating the potential of an analyte to elute ahead of another. For this reason, the separation process in NPC and in HILIC can be considered as taking place based on polar interactions, hydrogen bond formation and ion exchange type interactions. Also, the geometry of the molecule may be important for determining the strength of the interaction of a molecule with the polar phase, steric hindrance affecting how close a polar group may come to the interacting ligand from the stationary phase.

Since HILIC separations imply the presence of water in the mobile phase, various studies showed that the adsorbed water on the stationary phase plays an important role in HILIC separations [39,40].

The change in the mobile phase composition from a high content of water to a low (or very low) content of water, when using a polar stationary phase, may lead to a shift from HILIC type separation to a NPC type separation. For example, the curve describing the dependence of t_0 on a polar column, when the content of acetonitrile (ACN) in the mobile phase is modified, has a U-shape with a minimum point situated at about 70% ACN. This point could be an indication of changing the retention mechanism (up to 70% ACN v/v) from a high component in hydrophobic interactions (HILIC) to NP between 70 and 100% ACN in mobile phase.

Besides other types of stationary phases used in HILIC are zwitterionic ones. Separations on these columns are sometimes indicated as zwitterion chromatography or ZIC. In ZIC, since the stationary phase contains ions, it is likely that ion-dipol (or ion-partial charge) interactions are also contributing to the values of ΔA^0. Since such interactions are in general stronger than van der Waals forces not involving ions, it is possible that they dominate the retention process [41,42].

The previous analysis of the interaction energies determining the value for log k in NPC show that the expected elution in this type of separation should be opposite to that in RP-HPLC. Indeed, relatively good negative correlation has been reported regarding log k values for a series of compounds separated by both NPC and RP-HPLC [43]. However, since the factors that contribute to the value of ΔA^0 in NPC are not directly related with the factors that were determining the expression of ΔA^0 in RP (see rel. 5.1.17), the R^2 value for such correlations are in general modest.

Comparing NPC/HILIC on the one hand with RP on the other, the theory as well as the practical results indicate that more polar phases in NPC/HILIC retain more strongly the more polar analytes, while the more hydrophobic phases in RP retain more strongly the more hydrophobic compounds. Regarding the

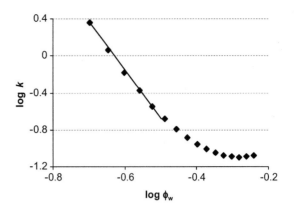

FIGURE 5.3.1 Typical variation of peak capacity log k as a function of volume fraction of water log ϕ_w in the mobile phase for a HILIC column and a polar compound.

mobile phase, the more polar mobile phases lead to a faster elution in NPC/HILIC, while in RP the less polar solvents lead to a faster elution. The typical capacity factor on a polar column (e.g. a HILIC column) decreases when the water content in the mobile phase increases. For example, using the notation ϕ_w for the volume fraction of water in the mobile phase, a typical variation of log k as a function of log ϕ_w is shown in Figure 5.3.1. The layer of water adsorbed on the polar surface of a polar stationary phase plays an important role in the separation. The linear interval of dependence of log k on log ϕ_w is typically interpreted as a region of water concentration in the mobile phase where the retention equilibrium is based on adsorption [44]. As shown in Section 3.2 (see rel. 3.2.12), linearity between log k and log ϕ_w is expected for adsorption equilibrium following Soczewinski equation.

Separation in NPC and HILIC

As previously indicated in Section 5.1, the value for selectivity α is given by expression 5.1.32. Replacing in this formula the values for ΔA obtained for two compounds X and Y, where

ΔA is given by rel. 5.3.4 (and $j = X$ or Y), an expression that is formally identical with rel. 5.1.34 will be obtained. However, the importance of different terms in that expression is not the same as in case of RP-HPLC. The importance of the difference $A^{cav}{}_{X,S} - A^{cav}{}_{Y,S}$ in the NPC and to a certain extent in HILIC is significantly less important than the same term in RP-HPLC. In NPC and HILIC interactions "in vacuum" between the stationary phase and the analytes, $- E_T(X,L) + E_T(Y,L)$, plays a much more important role for the retention process and therefore for the separation. Due to the fact that polarity of the analyte molecule is affected by the position of the group in the molecule, by the donor-withdraw electronic effects, as well as by the steric hindrance, separation of isomers is in general more successful when the separation is based on polar interactions.

5.4. RETENTION PROCESS IN ION-EXCHANGE CHROMATOGRAPHY

General Comments

Ion-exchange chromatography is dedicated to the separation of ionic species of the sample present in the mobile phase. These ionic species are reversibly retained by ionic groups covalently bonded to a stationary phase. The equilibrium between the mobile and the stationary phase is the key to the ion separation. Two types of mechanisms can be used for the separation. One type is based on the difference in the affinity for the column of the species that are separated. The elution in this mechanism uses a mobile phase that contains competing ions (driving ions) that replace the analyte ions and "push" them from the column. Because the affinity for the stationary phase of the diving ions is typically lower than that of the analytes, a higher concentration of the driving ions is necessary in the mobile phase for achieving elution. Another type of mechanism for retention and elution is

based on the addition of a complexing reagent in the mobile phase. This reagent changes the form in which the sample species are initially present (as simple ions) and modifies them to be retained differently or to be eluted by "pulling" the analyte down the column.

The Mechanism of Ion-Exchange Retention and Elution

Section 3.3 showed that the equilibrium between an ion in solution and an ion exchanger can be illustrated by reaction 3.3.8, and the equilibrium constant is given by expression 3.3.9. When the two ions involved in the equilibrium have the charges $+z_C$ and $+z_M$, respectively, a similar equation with 3.3.8 can be written as follows:

$$z_M\ res^{z_c-}\text{-}C^{z_c+} + z_C\ M^{z_M+} \rightleftarrows z_C\ res^{z_M-}\text{-}M^{z_M+}$$
$$+ z_M\ C^{z_c+}$$

$$(5.4.1)$$

In rel. 5.4.1 the counterions present in the solution are not specifically indicated, but the neutrality of the whole system is assumed as maintained. An equilibrium as described by rel. 5.4.1 takes place, for example, between a resin in Na^+ form $Res\text{-}SO_3^-\ Na^+$ and an ion in solution, e.g. Ca^{2+}, as $Ca^{2+}Cl^-{}_2$, when $z_C = +1$ (for Na^+ ion), and $z_M = +2$ (for Ca^{2+} ion). The constant for the equilibrium 5.4.1 can be written as follows:

$$K_{M,C} = \left(\frac{[M^{z_M+}]_{res}}{[M^{z_M+}]_{mo}}\right)^{z_C} \left(\frac{[C^{z_c+}]_{mo}}{[C^{z_c+}]_{res}}\right)^{z_M} \quad (5.4.2)$$

where the index res indicates resin and mo indicates mobile phase. The concentration in the resin can be obtained by considering the whole volume of the resin. The equilibrium 5.4.1 can also be described using a *thermodynamic equilibrium constant* $K_{M,C}$ defined by the expression:

$$K_{M,C} = \left(\frac{a_{res}(M)}{a(M)}\right)^{z_C} \left(\frac{a(C)}{a_{res}(C)}\right)^{z_M} \quad (5.4.3)$$

where a_{res} indicates the activity of the ions in the resin and a the activity in solution (the charges for the ions are not written to simplify notation). Similarly with the equilibrium for cations (charges z_C and z_M positive), an equivalent equilibrium with 5.4.1 and an equivalent equilibrium constants with 5.4.2 or 5.4.3 can be written for anions.

The quantitative treatment of the ion-exchange equilibria can be theoretically approached with the assumption that the electrostatic interactions between fixed charges (functional groups) in the resin and mobile charges (ions) are long-range interactions covering distances much longer than molecular radii, as shown in Section 4.1 (see Figure 4.1.1). Since in the resin phase there are numerous charges, the phase can be considered as a homogeneous phase instead of comprising numerous point charges. The ions that are exchanged are assumed to be distributed freely over the two phases, while the functional groups that have the retaining charge on the resin are covalently linked to the matrix and cannot leave the resin phase. A stationary boundary can be visualized as a semipermeable membrane, permeable to all species except the functional groups connected to the skeleton of the resin. With these assumptions, the theory for ion-exchange equilibria can be based on the Donnan membrane equilibrium theory. The presence of small pores in the resin may cause steric hindrance and create energetic inhomogeneities, but this is not expected to happen for the small-sized solutes. Physical inhomogeneities may also be caused by very large pores in the stationary phase. In these, the liquid is not part of the gel and should be regarded as being part of the mobile phase. In these conditions from rel. 3.3.23 it results that the Donnan membrane potential for the electrolyte with the activity a_{res} in the resin and activity a in the solution is given by the expression:

$$E_{Donnan} = [RT \ln(a_{res}/a) - \Pi V]/(z\,F) \quad (5.4.4)$$

where Π is the swelling (osmotic) pressure for the resin, z is the charge of exchanging ions, and V is their partial molar volume (F is Faraday's constant).

For the equilibrium 5.4.1, the Donnan potentials of the two species C^{z_C+} and M^{z_M+} that are exchanged must be equal, and as a result the following expression must be fulfilled (charges not written anymore for the ions to simplify notation):

$$RT \ln \left[\left(\frac{a_{res}(M)}{a(M)} \right)^{z_C} \left(\frac{a(C)}{a_{res}(C)} \right)^{z_M} \right]$$
$$= \Pi \left(z_M V_C - z_C V_M \right) \quad (5.4.5)$$

By substituting expression 5.4.3 for $K_{M,C}$ in rel. 5.4.5, the following expression is obtained:

$$RT \ln K_{M,C} = \Pi(z_M V_C - z_C V_M) \quad (5.4.6)$$

Higher thermodynamic equilibrium constants show higher affinity for the resin of species M^{z_M+}, and rel. (5.4.6) indicates that species with higher (absolute value) charges are favored by the resin over those with lower charges. The thermodynamic equilibrium constant can be transformed into an equilibrium constant, depending on concentrations instead of activities, by transforming the activities in concentrations with the help of activity coefficients γ of each species in the resin and in solution. The expression for the equilibrium constant can be written in this case in the form:

$$RT \ln K_{M,C} = \ln \left[\left(\frac{\gamma_{res}(C)}{\gamma(C)} \right)^{z_M} \left(\frac{\gamma(M)}{\gamma_{res}(M)} \right)^{z_C} \right]$$
$$+ \Pi \left(z_M V_C - z_C V_M \right)$$
$$(5.4.7)$$

Relation (5.4.7) shows that the exchange equilibria are influenced by the osmotic term, the charges of the ions, and the activity coefficients of each species. Similar expressions with 5.4.7 can be established for anions (the charges z in this case must be taken in absolute value). Experimental studies indicated that even in very diluted solutions, employing a common approximation of $\gamma = 1$ is not useful in rel. 5.4.7, since the expression contains the ratios of different activity coefficients, which can be very different from 1 for ionic species in solution and in the resin. Relation 5.4.7 is applicable for the cations C and M in equilibrium. The initial ions on the stationary phase are typically the same as those used in the mobile phase as driving ions since the column is conditioned with the mobile phase before starting the separation [45].

A similar formula with rel. 5.4.7 can be developed for the case of the anion-exchange process, where X is the initial form of the resin and A is an anionic analyte:

$$z_A\, res^{z_X+}X^{z_X-} + z_X A^{z_A-} \rightleftarrows z_X\, res^{z_A+}A^{z_A-}$$
$$+ z_A X^{z_X-}$$

$$(5.4.8)$$

An example of an equilibrium of the type 5.4.8 is shown below for an anionic exchange where the resin is in acid carbonate form:

$$Res\text{-}NR_3{}^+HCO_3{}^- + A^- \rightleftarrows Res\text{-}NR_3{}^+A^-$$
$$+ HCO_3{}^-$$

$$(5.4.9)$$

Assuming that the participating ions in the equilibrium have the charges $-z_A$ and $-z_X$, respectively, formula 5.4.7 can be written in the form:

$$\log K_{A,X} = \frac{1}{RT}\log\left[\left(\frac{\gamma_{res}(X)}{\gamma(X)}\right)^{z_A}\left(\frac{\gamma(A)}{\gamma_{res}(A)}\right)^{z_X}\right]$$
$$- \frac{1}{2.303RT}\Pi\left(z_A V_X - z_X V_A\right)$$

$$(5.4.10)$$

From the expression of equilibrium constant 5.4.10 and taking into consideration other effects influencing the retention and elution in ion-exchange chromatography, the expression for the capacity factor k for an anionic analyte under isocratic conditions becomes more complicated and is given by the following equation [46]:

$$\log k_A = \frac{1}{z_X}\log K_{A,X} + \frac{z_A}{z_X}\log\frac{\Omega}{z_X} + \log\frac{w_{sp}}{V_{mp}}$$
$$- \frac{z_A}{z_X}\log C_X$$

$$(5.4.11)$$

where z_A is the charge on the anionic analyte A, z_X is the charge of the eluting ion, C_X is the concentration of the anion X in the eluent, $K_{A,X}$ is the ion-exchange selectivity coefficient between the anionic analyte A and the mobile phase competing anion X, Ω is the ion-exchange capacity of the stationary phase, w_{sp} is the weight of the stationary phase, and V_{mp} is the volume of mobile phase in the column. For a given anionic analyte and competing anion X from mobile phase, this equation can be formally reduced to the following logarithmic dependence:

$$\log k_A = \alpha - \frac{z_A}{z_X}\log C_X = \alpha - \beta \log C_X$$

$$(5.4.12)$$

where the parameters α and β can be determined from experimental measurements of the capacity factor and concentration of the competing ion X. Relation 5.4.12 indicates that a decrease in the capacity factor is achieved when the concentration of the competing ions C_X is increased. This shows that elution can be achieved by increasing the concentration of the competing ions in the mobile phase, which is done practicing gradient separations (see Section 7.9).

The use of complexing reagents in the mobile phase complicates the retention and elution process in ion exchange. As shown in Section 3.3, two equilibria are established in the presence of a ligand, as expressed by rel. 3.3.14. Relation 3.3.16 shows that the concentration of the retained ions in the column ($[M^+]_{res}$) diminishes as the concentration of the ligand in the mobile phase and the complexation constant are higher. The separation can be modulated based on the retention of newly formed species, the ligand concentration, and the value of constant K_{LM} for the complexation equilibrium [47]. Negatively charged ligands L^- are sometimes used as complexing agents, and in the case of negatively charged complexes formed with metal ions, the separation of these complexes can be performed using anion-exchange columns.

Besides the equilibrium aspects necessary to understand the ion-exchange behavior, the kinetic factors are also very important in ion chromatography, since in the dynamic applications, in addition to the convective process of longitudinal flow of the fluid through the column and the movement of fluid in the void space between the resin particles, other factors are also important. Among these factors, the diffusion of the compound of interest through the solvent immobilized on the resin particles, the diffusion within the gel microchannels, and the kinetics of the exchange process in itself are factors determining the rate of exchange. Due to the complexity of the process, only the estimation of certain kinetic aspects is usually possible. An expression that estimates the time $t_{1/2}$ for half of the complete conversion of a resin from form C into form M, when the limiting factor is the diffusion in the particle, is given by the formula:

$$t_{1/2} = 0.0075 \, d_p^2 / D_{res} \qquad (5.4.13)$$

where d_p is the particle diameter and D_{res} is the diffusion coefficient in the resin. However, other expressions were developed to describe the kinetics of the ion-exchange process [48].

Separation in Ion Chromatography

Since the constant $K_{M,C}$ is dependent on the nature of the ion M^+, the separation of different ions can be achieved using the ion-exchange stationary phase. The selectivity α in this case for two ions M_1^{z+} and M_2^{z+} can be defined by the formula:

$$\alpha = K_{M_1,C}/K_{M_2,C} = \frac{[M_1^{z+}]_{res}}{[M_1^{z+}]_{mo}} \Big/ \frac{[M_2^{z+}]_{res}}{[M_2^{z+}]_{mo}}$$

$$(5.4.14)$$

Depending on the affinity for the resin of species M_1^{z+} and M_2^{z+}, they can be separated, provided that α is large enough to assure separation. This implies that at equal concentrations in the mobile phase, the concentration in the resin of an ion with higher affinity for the resin will be higher and its retention will be stronger, leading to longer retention times in the chromatographic process. The affinity for the resins of various inorganic ions is in the following order $M^+ < M^{2+} < M^{3+} < M^{4+}$. This order is also verified experimentally. For the same valence cations, it was found that the affinity varies in general as follows: $Li^+ < H^+ < Na^+ < NH_4^+ < K^+ < Rb^+ < Ag^+ < Tl^+$, and for divalent ions $Be^{2+} < Mn^{2+} < Mg^{2+} < Zn^{2+} < Co^{2+} < Cu^{2+} < Cd^{2+} < Ni^{2+} < Ca^{2+} < Sr^{2+} < Pb^{2+} < Ba^{2+}$. For trivalent ions the order is $Al^{3+} < Sc^{3+} < Y^{3+} < Eu^{3+} < Pr^{3+} < Ce^{3+} < La^{3+}$.

For the anions B^- the order of affinity for the resin is in the order $B^- < B^{2-} < B^{3-}$. For the ions with the same charge, the following order of affinity was established: $OH^- < F^- < CH_3COO^- < HCOO^- < H_2PO_4^- < HCO_3^- < Cl^- < NO_2^- < HSO_3^- < CN^- < Br^- < NO_3^- < HSO_4^- < I^-$. However, inversions in this order are possible, for example, due to the nature of the resin, or formation of complexes.

Retention of Neutral Molecules on Ion-Exchange Phases

In addition to the exchange of ions, ion-exchange stationary phases are able to retain specific neutral molecules. This process can take place in two different ways. The first is related to the retention based on the formation of complexes. Specific ions such as transition metals (Cu^{2+}, Zn^{2+}, Co^{2+}, Ni^{2+}, etc.) may be retained on a cation-exchange resin and still have the capability to accept lone-pair electrons from donor ligands such as amines. Using this mechanism, neutral ligand molecules can be retained on resins already treated with the transitional metal ions.

The second process of retention of neutral molecules is based on adsorption on an ion-exchange matrix without involving an ion-exchange process. Organic molecules can be adsorbed on the resin. For example, amines can be retained on a strong cation exchanger in K^+ form. The counterion is important in the adsorbing capability of the resin, and the elution is possible by using solutions of salts at different concentrations. A "salting out" effect is used to modify the adsorption, the variation of the distribution constant in the presence of the salt being described by a formula:

$$\ln K_{res,A}(c) = \ln K_{res,A}(c = 0) + \kappa c \quad (5.4.15)$$

where $K_{res,A}(c)$ is the distribution coefficient for the analyte A in the presence of a salt, $K_{res,A}(c = 0)$ is the distribution constant in water, κ is a constant specific for the system, and c is the molar concentration of the salt. As shown in rel. 5.4.15, the increase in salt concentration increases the adsorption, and in a chromatographic process using this effect, the elution is done by diluting the initial solution of the eluent [49].

5.5. SEPARATION PROCESS IN CHIRAL CHROMATOGRAPHY

General Comments

Molecular species having the same molecular formula but differences in their structure are called isomers. The isomers can be *structural* with the atoms bonded in different ways, and *steric* with the same bond structure, but having differences in the arrangement in space of their atoms. Structural isomers can have differences in the atom chain, in the position of functional groups, or in the functional groups. The stereoisomers have different interatomic distances between certain atoms that are not bound directly. One type of stereoisomers is geometric or conformation isomers. Examples of conformation isomers are *cis-trans* (Z/E) isomers of compounds containing C=C bonds and *syn-anti* isomers of compounds containing C=N bonds. Stereoisomers that are mirror images of each other and are not superimposable, although the atomic distances are the same in the molecules, are called *enantiomers*. These compounds have the property called *chirality*. Chirality, which is needed for the existence of enantiomers, is commonly caused by the existence in the molecule of at least one tetrahedral carbon atom substituted with groups that are different. *Diastereoisomers* are compounds with more than one chiral center that are not superimposable and are not mirror images of one another. One more steric isomer type is that of conformers. These are stereoisomers able to convert into each other by rotation of one fragment of the molecule around a single bond. Figure 5.5.1 indicates different types of isomers.

The chirality in an enantiomer is specified using the symbols R and S. For the assignment of a symbol R or S to a chiral carbon, its substituents are at first ranked in a sequence a, b, c, d. Keeping d in the back, when a, b, c are seen counterclockwise, the

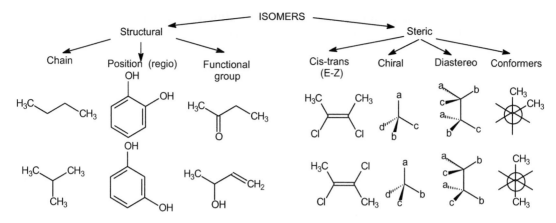

FIGURE 5.5.1 Different types of isomers with examples.

carbon is labeled S; when they are seen clockwise, the carbon is labeled an R, as indicated below:

asymmetry is due to a hindered rotation, as in the example of 2,2',6,6'-substituted biphenyl, with the substitution groups having

The ranking of the substituents is based on specific rules. For example, for the four atoms directly attached to the asymmetric carbon, a higher atomic number outranks the lower one (e.g., $O > N > C > H$). For the same atoms directly attached to the asymmetric carbon, the priorities are assigned at the first point of difference. More detailed rules are reported in the literature (see, e.g., [50]).

Besides having a chiral carbon, chiral molecules may be generated with other elements such as phosphorus or a sulfur chiral atom. Also, it is not only a chiral center (such as an asymmetric carbon) that generates enantiomers. A chiral axis or a chiral plane also can lead to enantiomers (helicoidal chirality is also known). In the case of a chiral axis, the

large volumes (NO_2, COOH, Br, etc.) as follows:

For carbohydrates, an older configurational notation is still in use. For glyceraldehyde, the S enantiomer is also called L, and the R enantiomer is called D. Monosaccharides are classified by convention based on the asymmetry type of the carbon most distant from the carbonyl group in the L series and D series when this carbon has the same stereochemistry as in L-glyceraldehyde or in D-glyceraldehyde, respectively.

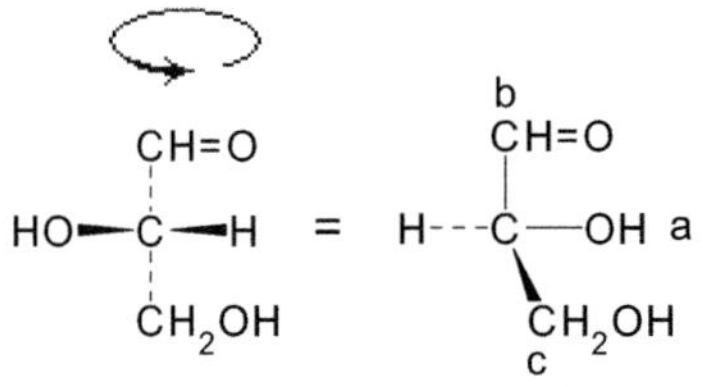

have different behavior toward a chiral stationary phase. For this reason the separation of enantiomers must be done on chiral

L-Glyceraldehyde (S)

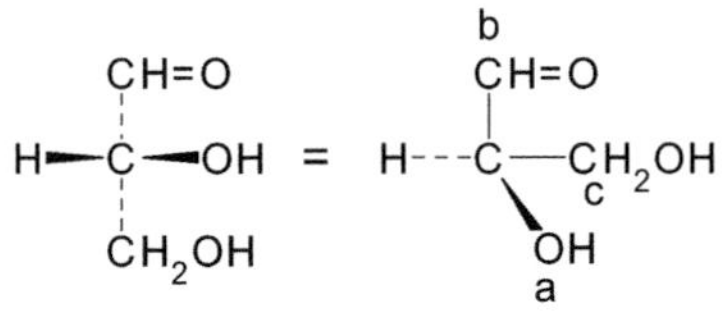

D-Glyceraldehyde (R)

More than one asymmetric carbon can be present in a molecule, as in the case of carbohydrates. The stereoisomers generated by more than one asymmetric carbon can be a mirror image one to the other (enantiomers) or may have different steric arrangements being diastereoisomers. Two diastereoisomers of this type are as follows:

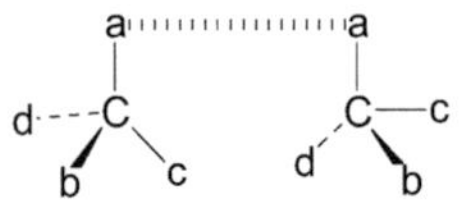

(S,S)-derivative

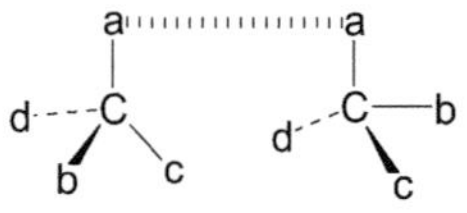

(S,R)-derivative

Stereoisomers can convert one to the other when the free enthalpy of interconversion is lower than about 23 kcal/mol.

The separation of enantiomers is typically not very simple, while the separation of other isomers including diastereoisomers can be done more easily using common chromatographic techniques. For this reason, one general procedure used before separation of enantiomers is their transformation into diastereoisomers by derivatization with a chiral reagent. Various such reagents are described in the literature [50]. The physicochemical behavior of enantiomer molecules can be different only in a chiral environment, and two enantiomers

stationary phases or using a chiral modifier in the mobile phase.

Types of Mechanisms in Chiral Separations

Several interaction mechanisms were suggested as being responsible for the separation of chiral molecules, including the ones listed below [51, 52]:

1) *Separation based on differences in the polar intermolecular interactions with the stationary phase of the enantiomers.* The intermolecular interactions are also affected when a chiral modifier is added in the mobile phase, but this type of separation is typically discussed separately. Since the stationary phases developed for chiral separations must provide a chiral environment, they contain chiral moieties. Chiral phases are developed to be used in a mobile phase less polar than the stationary one, similar to NPC and HILIC. The interactions causing the separation are of a polar nature, including polar-polar, hydrogen bonding, π-donor-π-acceptor, and π-π stacking. Since the enantiomer molecules have very similar physical properties such as polarizability and dipole moments, the reason for differences in the interactions with the stationary phase are of

a geometric (steric) nature, such as the spatial access to the chiral polar phase, access that is different for the two analyte enantiomers. A single point or two points of interaction (e.g., two hydrogen bonds formation) between the chiral stationary phase and the chiral analyte are not sufficient for providing a difference in the enantiomer separation. Three points of interaction that are different in nature (e.g., through one donor-type hydrogen bond, an acceptor-type hydrogen bond, and π-donor-π-acceptor interactions) between the stationary phase and the analytes are necessary for obtaining differentiation. This is explained in Figure 5.5.2 for a Pirkle-type stationary phase (Leucine) and the molecule of hexobarbital. As shown in this model, one

of the enantiomers (R) is able to establish (hypothetically) three different interactions with the bonded stationary phase, two different hydrogen bonds (one with the stationary phase donor and one with it acceptor), and a π-donor-π-acceptor interaction with the acceptor dinitrophenyl. The other enantiomer (S) can establish only two points of interaction (as also suggested schematically under each interacting system). Similar phases to the one shown in Figure 5.5.2 have been developed, with the dinitrophenyl moiety replaced by an electron donor π system (e.g., a naphthyl-amino acid group). Such phases are used for the separation of compounds with π-acceptor-type moieties.

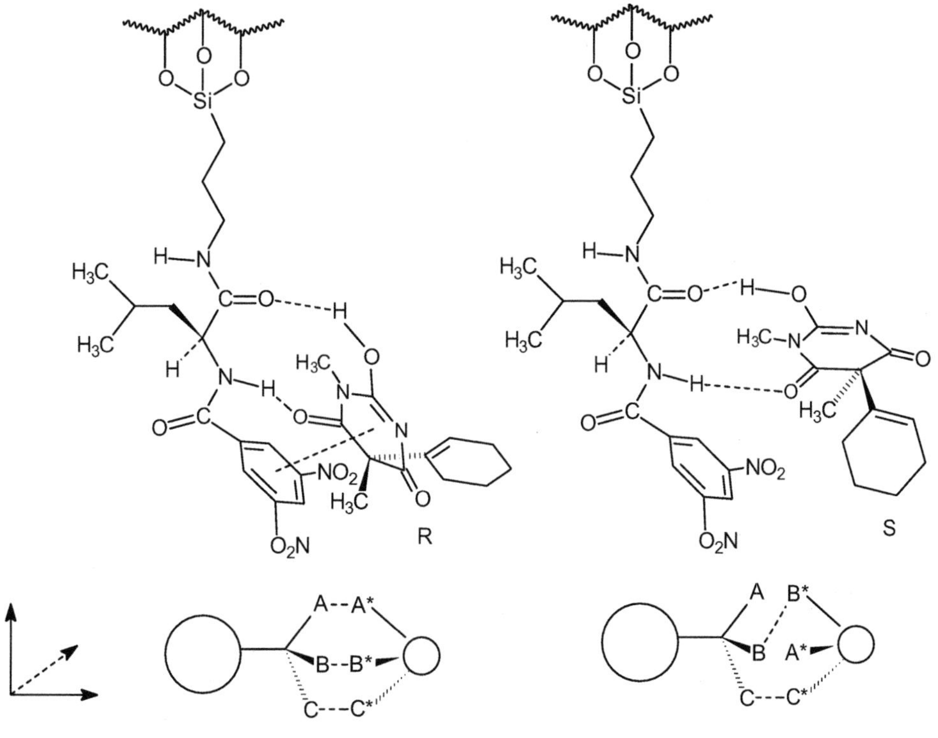

FIGURE 5.5.2 Hypothetical model of the interaction of a Pirkle type stationary phase (Leucine) and the molecule of hexobarbital in S form and R form (dotted line --- shows interactions).

2) *Separation based on inclusion properties and molecular interactions with the enantiomers.* This type of phase adds a geometric fit of the analyte molecule into the stationary phase, as one more capability to establish specific molecular interactions. The common examples of interactions of this type are the phases based on cellulose and cellulose derivatives. Cellulose-based polymers have a linear structure, but the polymeric chains form strong hydrogen bonds between them, which leads to an interstitial structure that allows molecular inclusions. The main types of interactions remain hydrogen bonding and polar interactions with the molecules of the solute. The specific orientation of OH groups in cellulose allows multipoint interactions that are different between the enantiomers.

3) *Separation based mainly on inclusion properties of the stationary phase.* The phases displaying this type of interaction are cyclodextrins, crown ethers, and amylose-type phases. These phases have the ability to produce inclusion complexes with numerous molecules, depending on the guest molecule geometry. Cyclodextrins, for example, have a truncated conical cavity (see Section 4.1) that forms an inclusion similar to a pseudorotaxane with one part of the chiral molecule. The wide side of the cyclodextrin cavity is encircled by secondary hydroxyl groups (two per glucopyranose unit), and the narrow side is encircled with primary hydroxyl groups (one per glucopyranose unit). These groups give the cyclodextrin molecule a hydrophilic exterior. In contrast, the inside cavity is relatively hydrophobic, giving the molecule the ability to complex a wide variety of molecular guests and limit the rotation of the included molecule [53]. The hydroxyl groups offer a large number of chiral centers that have different interactions with the remaining substituents of the chiral molecule. These interactions are of the well-known types, polar-polar and hydrogen bonding. Derivatized cyclodextrins (e.g., with acetyl groups) provide further differentiation in

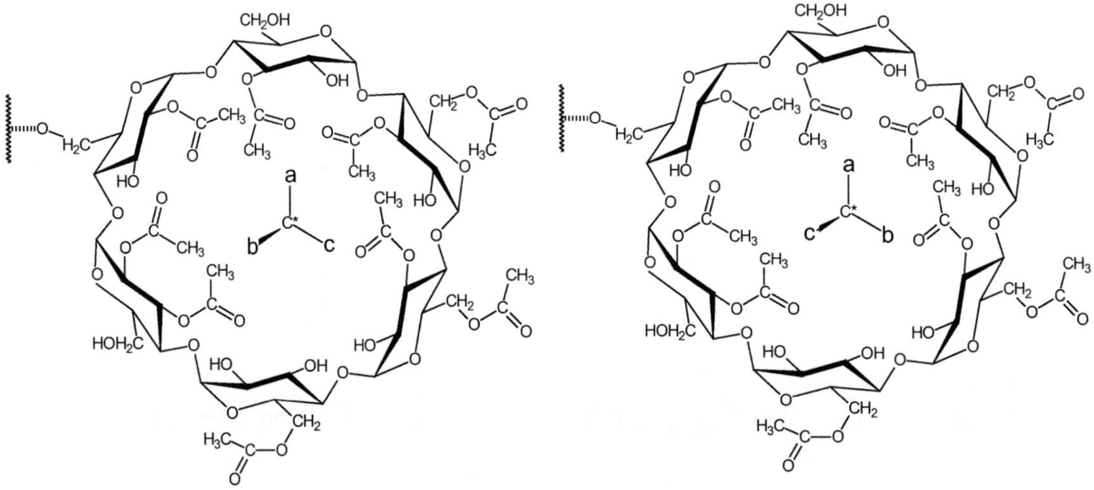

FIGURE 5.5.3 α-Cyclodextrin partially acetylated in interaction with the two enantiomers of a guest molecule (a fourth substituent, not shown, is assumed as included in the cyclodextrin cone).

the types of interactions with the chiral molecule. As an example, Figure 5.5.3 shows a α-cyclodextrin molecule (anchored, e.g., to a silica surface) and partially derivatized with acetyl groups, in interaction with two enantiomers. Due to the chrality of the guest, different groups from the two enantiomers and from the cyclodextrin come in close proximity, such that the overall molecular interactions of the enantiomers with the host molecule are different. This difference in the interaction intensity allows the HPLC separation. Amylose has a helical structure and weaker hydrogen bonding between the helices compared to cellulose. This structure leads to molecular inclusions with the solute molecules, and also to polar-polar and hydrogen bonding formation similar to that offered by cyclodextrins.

4) *Separation based on ligand exchange chromatography using a chiral resin loaded with a transitional metal (e.g., Cu^{2+}) capable of forming at the same time complexes with the solutes (enantiomeric analytes).* The separation of the solutes is based on the differences in the strength of the interactions (of coordinative and ionic type) of these solutes with the bonded metals ions in the asymmetric resin. The model of these interactions is shown in Figure 5.5.4.

5) *Separation based on both hydrophobic and polar interactions.* As previously indicated for the phases where the separation is based on

chelate

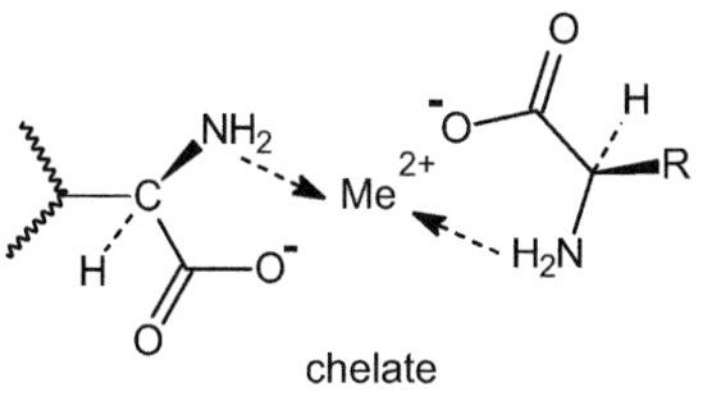

FIGURE 5.5.4   Model of a chelate formed with the chiral selector from the stationary phase and with an analyte molecule (e.g. an amino acid).

polar interactions, a single point or two points of interaction are not sufficient for providing good conditions for a chiral separation. The multiple points of interaction can be offered not only by polar ones, but by a combination of polar and hydrophobic interactions, where the solvent also plays a key role in the separation [54]. Specific polymers, such as proteins or macrocyclic antibiotics, are able to provide both polar and hydrophobic multipoint interactions that allow the separation of enantiomers.

5.6. RETENTION PROCESS IN SIZE-EXCLUSION CHROMATOGRAPHY

General Comments

Size-exclusion chromatography is used to separate molecules based on their size (more correctly, their hydrodynamic volume), and it is applied mainly for the separation of macromolecules and of macromolecules from small molecules. The equilibrium governing size exclusion (for both gel filtration or GFC and gel permeation or GPC) was previously discussed in Section 3.4. There it was shown that the retention volume for this separation is given by rel. 3.4.2 ($V_R = V_{inter} + K_{SEC} V_{pores}$), where V_{inter} is the interstitial volume of the column and V_{pores} is the volume of the pores of the packing. Several studies were focused on the relation between K_{SEC} and characteristics of the separation process. For instance, for slit-like pores of the stationary phase with $\bar{d}$ as an effective mean pore radius and for the (Gaussian-type) coiled macromolecules having the effective mean radius $\bar{r}$, the expression of K_{SEC} can be approximated by the following relations [55]:

$$K_{SEC} = 1 - \frac{2 \cdot \bar{r}}{\sqrt{\pi \cdot \bar{d}}} \quad \text{for} \quad \bar{r} << \bar{d} \quad (5.6.1)$$

and

$$K_{SEC} = \frac{8}{\pi^2} \exp\left[-\left(\frac{\pi \cdot \bar{r}}{2 \cdot \bar{d}}\right)^2 \right] \quad \text{for} \quad \bar{r} >> \bar{d}$$

(5.6.2)

The separation process in SEC depends, as expected, on the nature of the stationary phase and also on the solvent used as mobile phase. Depending on the nature of the polymer and its solubility, a specific choice of the solvent must be made such that the polymer has a good solubility in the mobile phase. Compared with HPLC of small molecules, the concentration of polymer solutions injected into the analytical SEC columns is much higher, at the level of several mg/mL. Besides the requirement to dissolve a larger amount of polymer, the solvent influences the extent of variation of the entropic factor ΔS^0 in the separation process, which is larger for solvents that well dissolve the polymer. Depending on the polymer solubility, choices for the stationary phase are made as hydrophilic packing for GFC, or hydrophobic for GPC. The packing material for gel filtration can be a hydrophilic polymer (such as dextrans treated with epichlorhydrin) or silica gels bonded with hydrophilic functional groups. Gel filtration uses mobile phases based on water, buffers, or partial aqueous solutions (up to 20% methanol or acetonitrile). In nonaqueous SEC, the stationary phase is typically a hydrophobic polymer and the mobile phase involves solvents such as tetrahydrofuran, toluene, or chloroform [56].

Dependence of Retention Volume on the Molecular Weight in SEC

Size exclusion (either GPC or GFC) is used for the separation of various types of macromolecules, as well as for the separation of macromolecules from small molecules. However, a very useful application of this technique is the evaluation of the molecular weight of polymers. For this purpose, it is important that the enthalpic interaction between the polymer and the stationary phase is very small or, ideally, nonexistent. Assuming no attractive or repulsive interaction with column packing except for the effects caused by the imperviousness of the pore walls, the expression of K_{SEC} is given by rel. 3.4.7, which, included in rel. 3.4.2 for the retention volume, gives the formula:

$$V_R = V_{inter} + \exp(\Delta S^0/R) \, V_{pores}$$

(5.6.3)

where ΔS^0 is the entropy change when the analyte passes from the mobile phase into the stationary phase. In this process ΔS^0 has negative values since the interaction between the solute and stationary phase is typically exothermic. The change in entropy is more significant for larger molecules (of polymers) and less important for smaller ones (see rel. 3.4.7). From rel. 5.6.3, it can be seen that a ΔS^0 larger in absolute value (and negative) leads to a smaller retention volume V_R. Therefore V_R has lower values for larger molecules than for smaller molecules that are retained. Since the molecular volume is related to molecular weight, it is expected that V_R will depend in some way on the molecular weight (MW) of the polymeric material. This dependence is verified in practice, and the following formula has been proven valid for a certain range of MW for various polymers:

$$V_R = A - B \log(MW)$$

(5.6.4)

In rel. 5.6.4, A and B are constants for polymers with different molecular weights but of the same type. An ideal variation of V_R with the molecular weight is shown in Figure 5.6.1.

In practice, expression 5.6.4 is valid only for a certain range of MW values. Also, different molecular structures lead to different slopes for the dependence of V_R on $\log(MW)$. Figure 5.6.2 shows four types of curves common for GPC. Curve B in Figure 5.6.2 approaches the ideal behavior. The deviation from ideal may

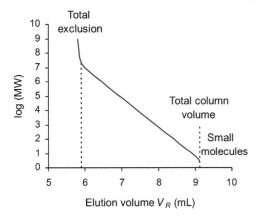

FIGURE 5.6.1 Ideal variation of elution volume V_R with log (MW) in SEC.

appear at the end of the calibration interval (curve A), at the beginning of the calibration interval (curve D), or at both ends (curve C). The linearity between log MW and the elution volume can be utilized for the characterization of molecular weight distribution in both synthetic and natural polymers. Due to possible deviations from linearity, as shown in Figure 5.6.2, the dependence between MW and the elution volume must be calibrated in the molecular range of interest using polymer

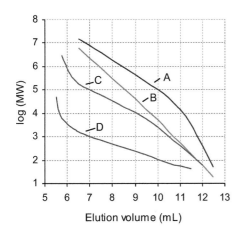

FIGURE 5.6.2 Experimental types of variation of elution volume V_R with log (MW) in GPC.

standards with known molecular weight and having similar structures to those that are analyzed. In some cases, the separation mechanism also has an enthalpic contribution besides the entropic changes for the SEC mechanism. These enthalpic contributions either bring deviations from the true value of MW or may lead to $K_{SEC} > 1$ and the elution of polymers at volumes larger than the total column volume. These effects are caused by various types of interactions of the analyte with the stationary phase. The presence of enthalpic contributions to the retention process can be detected based on chromatographic peak shape (sharp leading edge followed by tailing), unexpected retardation of the peak of polymers with high MW, poor reproducibility of retention time, and large changes in the retention time upon solvent changes.

5.7. RETENTION PROCESS IN OTHER CHROMATOGRAPHY TYPES

General Comments

The separation mechanisms for several common HPLC techniques have been discussed in this chapter. These techniques include reversed phase (RP, NARP, and ion-pair), polar (NP and HILIC), ion exchange (cation and anion exchange), size exclusion (GFC and GPC), and chiral separations. Another common technique is bioaffinity chromatography. The interactions in bioaffinity chromatography are complex and difficult to capture in one type of mechanism. The technique is used mainly for peptides and protein separation, and it is based on reversible interactions between a protein and a ligand covalently attached to a solid support. Compounds used as ligands are, for example, heparin, lectins, and nucleotides. These compounds can be bound to a support such as agarose, cellulose, or organic polymers and used as selective stationary phases. Specific immunoproteins also can be bound, for example, on agarose activated with cyanogen bromide or with other activation reagents and serve as a stationary phase. These types of materials have a very high specificity for a certain antigen that generated (e.g., in an animal) the bonded immunoprotein. Affinity resins with immobilized biotin or avidin can also be used as stationary phases. The operating pressure typically used in bioaffinity chromatography is low (1–2 bar), and this technique is sometimes not included as part of HPLC methods [57].

Besides the common HPLC types previously mentioned, there are several other chromatographic techniques with less widespread utilization. Among these can be listed hydrophobic interaction chromatography (HIC), electrostatic repulsion hydrophilic interaction chromatography (eHILIC or ERLIC), aqueous-normal-phase chromatography (ANPC or ANP), ion exclusion, ligand exchange, immobilized metal affinity chromatography (IMAC), ion moderated, displacement chromatography, and multimode HPLC. These techniques can be very successful in specific separations. A short description of the mechanism in each of these techniques is given in this section.

Hydrophobic Interaction Chromatography (HIC)

As indicated in Section 1.2, besides RP-HPLC, several types of chromatographic techniques involve hydrophobic interactions in the separation. One such technique is hydrophobic interaction chromatography (HIC). In HIC, involved in the interaction process are the hydrophobic moieties of a molecule with solvent-accessible nonpolar groups (hydrophobic patches), the surface of the hydrophilic stationary phase with some hydrophobic ligans attached, and the polar solvent. The molecules typically separated by this technique are certain biopolymers. The stationary phase in HIC can be either silica or an organic polymer with a hydrophilic coating on which are attached

hydrophobic ligands such as short-chain alkyl groups or phenyl groups. The modification of mobile-phase polarity in HIC by addition of inorganic salts enhances the adsorption of hydrophobic areas of the analyte at the hydrophobic areas on the stationary phase. The interactions involved in this type of separation are typically more complex, but the hydrophobic interactions play an important role in the separation. In protein separation, HIC is a useful technique since it typically does not modify the protein conformation.

Electrostatic Repulsion Hydrophilic Interaction Chromatography

HILIC-type separations (polar stationary phase, less polar mobile phase but containing water) can be performed on ion-exchange (cation-exchange or anion-exchange) stationary phases, when the technique is sometimes indicated as eHILIC or ERLIC. In this technique, in addition to water, the mobile phase always contains organic solvents. The retention/separation is based on both ion-interaction and HILIC-type interactions. Since electrostatic forces are about ten times stronger than nonionic polar interactions (see Section 3.1), ions from the solution that have the same polarity with those from the stationary phase are strongly repelled from it. On the other hand, neutral polar molecules can easily interact with the stationary phase by a HILIC-type mechanism. A cation-exchange column (surface negatively charged) will show no retention for anionic species, but will show typical polar interaction with neutral molecules of the analyte. Similarly, an anion-exchange column (with positively charged stationary phase) will show no retention for cations but retain neutral polar molecules. This type of effect can be controlled by adjusting the pH of the mobile phase that influences the ionization of the analytes or by the addition of IPA agents when ion-pair-type interactions may play a role in the separation.

TABLE 5.7.1   Types of retention on silica hydride stationary phase.

Stationary phase	Mobile phase	Water content	Term	Similar to
Weak polar	Non-polar	None	ONP	NPC
Weak polar	Some polarity but less than stationary phase	Some	ANP	HILIC
Weak polar	Polar (more than the stationary phase)	Added water	RP	RP

Aqueous-Normal-Phase Chromatography

This is a technique performed on a special stationary phase, silica hydride, which has Si-H groups bonded on a typical silica surface. The preparation of such phases is described in Section 6.1. The most important feature of a hydride silica phase is that water does not adsorb strongly to the phase. Regardless of the mechanism based on partition or adsorption in NPC and HILIC, the water from the surface of the polar stationary phase plays an important role in the types of interaction with the analyte and in the separation in general. For silica hydride surfaces, this water is either absent, or it is not strongly bound to the stationary phase. This allows a silica hydride phase to work as:

1) a normal phase (NPC) with organic nonaqueous mobile phase (a procedure sometimes indicated as ONP),
2) a HILIC phase, with a normal phase with a partially aqueous mobile phase, and with the stationary phase more polar than the mobile phase (sometimes indicated as ANP), and also
3) an RP phase, with the mobile phase containing water and also being more polar than the stationary phase. These retention modes are summarized in Table 5.7.1.

The different types of retention displayed by hydride-type stationary phases can also be manifested differently in function of the type of compounds separated on a silica hydride phase. Some hydrophobic compounds behave on the

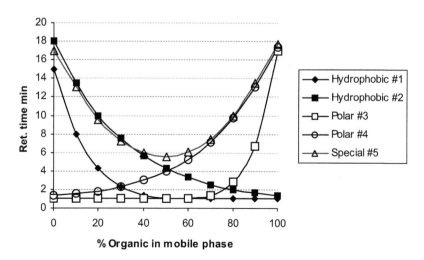

FIGURE 5.7.1 Retention of model compounds on a silica hydride column, exemplifying different types of behavior.

silica hydride in the same way as on a RP column, with significantly decreased retention when the organic component increases in the mobile phase. This is exemplified by hypothetical compound #1 in Figure 5.7.1. Other hydrophobic compounds show more retention than expected when the mobile phase is high in the organic constituent as exemplified by compound #2. Regarding polar compounds, some will behave similarly to a polar compound on a HILIC column, with low retention at low concentration or the organic constituent in the mobile phase (as exemplified by compound #3). However, some compounds show unexpected high retention when the content of the polar component in the mobile phase is still high. This is exemplified by compound #4 in Figure 5.7.1. There are also compounds with both polar groups and hydrophobic moieties in their structure, which have retention in both a high-organic and low-organic mobile phase, with a minimum of retention in partly organic partly polar mobile phases. This type is exemplified by compound #5. Since silica hydride can work in such various conditions, it is difficult to assign a specific mechanism for the separation [58–61].

Ion Exclusion

Ion exclusion is applied mainly for the separation of neutral molecules or weakly polar ones from ionic molecules. This type of separation can be practiced on ion-exchange stationary phases having, for example, strong acid groups bonded to a solid support. Strong acids are not retained at all due to repulsive electrostatic forces (Donnan exclusion). Such compounds will elute in the void volume of the chromatographic column. Weak or very weak acids are not subject to Donnan exclusion, and they penetrate the pores of the stationary phase and are separated based on differences in pK_a, size and hydrophobicity. The mechanism in size exclusion is more complex than in simple ion-exchange chromatography.

Ligand-Exchange and Immobilized Metal Affinity Chromatography

In ligand-exchange chromatography, the stationary phase is a cation exchanger that carries ions of a transition metal such as Cu^{2+} or Ni^{2+}. The transition metal can form coordinative bonds with electron donor molecules, and the mobile phase must contain donor molecules such as a low concentration of ammonia or even just water. These molecules are attached to the transition metal such that the stationary phase is loaded with the (weak) complexes of the transition metal. The analytes are also electron donor molecules such as amines, amino acids, or hydroxy compounds. The retention process consists of the replacement of the weak ligand (e.g., ammonia) with the analyte molecules that are stronger ligands. However, by increasing the concentration of ammonia in the mobile phase, or simply in isocratic mode by continuing to flush the column with the initial ammonia solution, the analyte ligand is eluted from the column (with the weaker ligand analytes eluting first [62]).

Ligand-exchange chromatography has been further developed into a similar technique known as immobilized metal affinity chromatography (IMAC). The difference from ligand exchange is that the stationary phase has covalently connected chelating groups able to form very strong coordinative complexes with transitional metals. These groups are typically derived from iminodiacetic acid or tricarboxymethylethylenediamine. The phase is pretreated with the metal ion solution, and during this process, weaker coordinative ligands such as water become part of the complex. The analytes are typically electron donor molecules such as amino acids, peptides, or proteins. These analyte molecules form relatively stable complexes with the inorganic metal. When the metal is connected to the resin, an immobilized complex involving the analyte, the metal and the resin functionality is formed. The retention process (the analyte being a peptide with a histidine residue) is schematically shown in the following reaction:

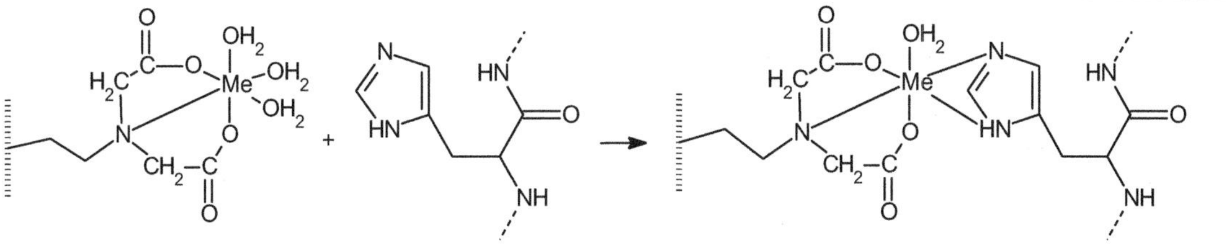

the loading step, a "displacement" reagent dissolved in the mobile phase is passed

The elution can be done with buffers that have a pH where the analyte to metal bond is weakened, or with an additive such as ammonia in the mobile phase, which competes with the analyte bonded on the column. The metal itself can also be eluted from the column using a strong complexing agent such as EDTA. The metal (the same or different) can be later replaced on the column.

Ion-Moderated Chromatography

Ion-exchange stationary phases containing, for example, sulfonic groups, either free or in the form of a salt (Na^+, K^+, Ag^+, Ca^{2+}, Pb^{2+}), are also used for separations indicated as ion-moderated chromatography. In ion-moderated chromatography, the separation is based mainly on polar interactions between the bonded metal and the analytes, and possibly size-exclusion effects. Also, ligand exchange-type interactions with the formation of very weak coordinative bonds between the analyte and the ions retained in the stationary phase (which is an ion exchanger) may explain the retention.

Displacement Chromatography

Displacement chromatography is applied mainly as a preparative technique. For separation, the sample is initially retained on a chromatographic column (loading phase). After

through the column and elutes the specific retained molecule. The displacement reagent is selected such that it has the capability to bind to the column more strongly than the analytes and can displace it. When the molecules of the analytes are bound strongly or weakly to the stationary phase, they are displaced with more difficulty or more easily, respectively, and therefore are eluted at different retention times. After all molecules of the sample are displaced, the column should be regenerated.

Affinity Chromatography

Affinity chromatography commonly refers to *bioaffinity* or *immunoaffinity chromatography*. The separation in affinity chromatography is based on the interaction of a specific (biological) compound present in the mobile phase with its natural biological complement immobilized on the surface of a stationary phase [63]. Examples of such interactions are between an antigen and its antibody, between lectins and glycoproteins, between an immobilized metal ion and proteins containing amino acid residues that have affinity for the metal ion (e.g., histidine), between biotin and avidin. The nature of the interaction is complex, and a variety of types of forces occur between the two participants (one participant being immobilized). In immunoaffinity, a sample that has passed through the column is separated into two bands. The first band elutes with a retention factor $k = 0$ and contains all the compounds of the sample that do not bind to the affinity ligand. At this time, the analyte of interest should be strongly adsorbed to the ligand and should not elute ($k = \infty$). A change in the mobile phase, such as a modification in pH, in the ionic strength, or in other parameters determines the analyte molecules to decouple from the complex with the immobilized complement and thus to elute and produce the second band [64].

Multimode HPLC

In many HPLC types, the separation mechanism is not unique. For example, in HILIC, besides the polar interactions, hydrophobic interactions may play a role in separation. Also, in ion-exchange chromatography, besides the ionic interactions other interactions may occur, in particular when the mobile phase contains an organic component. However, these multiple action modes of different stationary phases that have a uniform structure are not indicated as "multimode." In multimode HPLC, the stationary phase is purposely not uniform and contains two (or more) different materials with a specific interaction mode, such as ion exchange and reversed phase (e.g., Primesep D, Promix SP, or Promix AP columns [65]).

References

[1] Horvath C, Melander W, Molnar I. Solvophobic interactions in liquid chromatography with nonpolar stationary phases. J. Chromatogr. 1976;125: 129–56.

[2] Horvath C, Melander W. Liquid chromatography with hydrocarbonaceous bonded phases; theory and practice of reversed phase chromatography. J. Chromatogr. Sci. 1977;15:393–404.

[3] Galushko SV. Calculation of retention and selectivity in reversed-phase liquid chromatography. J. Chromatogr. 1991;552:91–102.

[4] Miyabe K, Guiochon G. Thermodynamic interpretation of retention equilibrium in reversed-phase liquid chromatography based on enthalpy-entropy compensation. Anal. Chem. 2002;74:5982–92.

[5] http://www.ChemAxon.com

[6] David V, Albu F, Medvedovici A. Structure-retention correlation of some oxicam drugs in reversed-phase liquid chromatography. J. Liq. Chromatogr. Rel. Technol. 2005;27:965–84.

[7] Carr PW, Dolan JW, Neue UD, Snyder LR. Contribution to reverse phase column selectivity I: Steric interaction. J. Chromatogr. A 2011;1218:1724–42.

[8] Smith RM, Burr CM. Retention prediction of analytes in reversed-phase high performance liquid chromatography based on molecular structure. I: Monosubstituted aromatic compounds. J. Chromatogr. 1989;475:57–74.

[9] Vailaya A, Horváth C. Retention in reversed-phase chromatography: partition or adsorption? J. Chromatogr. A 1998;829:1–27.

[10] Dill KA. The mechanism of solute retention in reversed-phase chromatography. J. Phys. Chem. B 1987;91:1980–8.

[11] Vailaya A. Fundamentals of reversed phase chromatography: Thermodynamic and exothermodynamic treatment. J. Liq. Chromatogr. Rel. Technol. 2005;28: 965–1054.

[12] Jeroniec M. Partition and displacement models in reversed-phase chromatography. J. Chromatogr. A 1993;656:37–50.

[13] Martire DE, Boehm RE. Unified theory of retention and selectivity in liquid chromatography. 2. Reversed-phase liquid chromatography with chemically bonded phases. J. Phys. Chem. 1983;87:1045–62.

[14] Scheutjens JMHM, Fleer GJ. Statistical theory of the adsorption of interacting chain molecules. 1. Partition function, segment density distribution, and adsorption isotherms. J. Phys. Chem. 1979;83:1619–53.

[15] Scheutjens JMHM, Fleer GJ. Statistical theory of the adsorption of interacting chain molecules. 2. Train, loop, and tail size distribution. J. Phys. Chem. 1980;84:178–90.

[16] Umemura T, Tsunoda K, Koide A, Oshima T, Watanabe N, Chiba K, Haraguchi H. Amphoteric surfactant-modified stationary phase for the reversed-phase high-performance liquid chromatographic separation of nucleosides and their bases by elution with water. Anal. Chim. Acta 2000;419:87–92.

[17] Pettersson C. Chromatographic separation of enentiomers of acids with quinine as chiral counter ion. J. Chromatogr. A 1984;316:553–67.

[18] Farca A, David V, Medvedovici A, Ionescu M. Application of ion-pair mechanism for the determination of acyclovir in plasma samples by HPLC with

fluorescence detection. Rev. Roum. Chim. 2003;48: 781—7.

[19] Horvath C, Lipsky SR. Use of liquid ion exchange chromatography for the separation of organic compounds. Nature 1966;211:748—9.

[20] Knox JH, Hartwick RA. Mechanism in ion-pair liquid chromatography of amines, neutrals, zwitterions, and acidis using anionic hetaerons. J. Chromatogr. A 1981;204:2—21.

[21] Horvath C, Melander W, Molnar I, Molnar P. Enhancement of retention by ion-pair formation in liquid chromatography with nonpolar stationary phases. Anal. Chem. 1977;49:2295—305.

[22] Cecchi T, Pucciarelli F, Passamonti P. Extended thermodynamic approach to ion interaction chromatography. Anal. Chem. 2001;73:2632—9.

[23] Cecchi T. Extended thermodynamic approach to ion interaction chromatography. Influence of the chain length of the solute ion: A chromatographic method for the determination of ion-pairing constants. J. Sep. Sci. 2005;29:549—54.

[24] Cecchi T, Pucciarelli F, Passamonti P. Extended thermodynamic approach to ion interaction chromatography: A mono-and bivariate strategy to model the influence of ionic strength. J. Sep. Sci. 2004;27:1323—32.

[25] Checchi T. Use of lipophilic ion adsorbtion isotherms to determine the surface area and the monolayer capacity of a chromatographic packing as well as the thermodynamic equilibrium constant for its adsorption. J. Chromatogr. A 2005;1072:201—6.

[26] David V, Medvedovici A. Partition model applied to the retention process of basic analytes in reversed-phase and ion-pairing liquid chromatography. Rev Roum Chim 2005;50:837—43.

[27] Ståhlberg J. The Gouy-Chapman theory in combination with Langmuir isotherm as a theoretical model for ion-pair chromatography. J. Chromatogr. 1986;256: 231—45.

[28] Dukhin SS, Derjaguin BV. Electrokinetic Phenomena. New York: John Wiley; 1974.

[29] Cecchi T. Ion pairing chromatography. Crit. Rev. Anal. Chem. 2008;38:161—213.

[30] Bartha A, Ståhlberg J. Electrostatic retention model of the reversed-phase ion-pair chromatography. J. Chromatogr. A 1994;668:255—84.

[31] Cecchi T, Passamonti P. Retention mechanism for ion-pair chromatography with chaotropic reagents. J. Chromatogr. A 2009;1216:1789—97.

[32] Radulescu M, David V. Partition versus electrostatic model applied to the ion-pairing retention process of some guanidine based compounds. J. Liq. Chromatogr. Rel. Technol. 2012;35 [in press].

[33] Dai J, Carr PW. Role of ion pairing in anionic additive effects on the separation of cationic drugs in reversed-phase liquid chromatography. J. Chromatogr. A 2005;1072:169—84.

[34] Hefter G. Ion solvation in aqueous-organic mixtures. Pure. Appl. Chem. 2005;77:605—17.

[35] Phechkrajang CM. Chaotropic effect in reversed-phase HPLC: A review. Mahidol. Univ. J. Pharm. Sci. 2010;37:1—7.

[36] Flieger J. The effect of chaotropic mobile phase additives on the separation of selected alkaloids in reversed-phase high performance liquid chromatography. J. Chromatogr. A 2006;1113:37—44.

[37] Kazakevich IL, Snow NH. Adsorption behavior of hexafluorophosphate on selected bonded phases. J. Chromatogr. A 2006;1119:43—50.

[38] Hemström P, Irgum K. Hidrophylic interaction chromatography. J. Sep. Sci. 2006;29:1784—821.

[39] Karatapanis AE, Fiamegos YC, Stalikas CD. A revisit to the retention mechanism of hydrophilic interaction liquid chromatography using model organic compounds. J. Chromatogr. A 2011;1218: 2871—9.

[40] McCalley DV, Neue UD. Estimation of the extent of the water-rich layer associated with the silica surface in hydrophilic interaction chromatography. J. Chromatogr. A 2008;1192:225—9.

[41] Cecchi T, Pucciarelli F, Passamonti P. Ion-interaction chromatography of zwitterions: The fractional charge approach to model the influence of the mobile phase concentration of the ion-interaction reagent. Analyst. 2004;129:1037—46.

[42] Cecchi T, De Marco C, Pucciarelli F, Passamonti P. The fractional charge approach in ion-interaction chromatography of zwitterions: Influence of the stationary phase concentration of the ion interaction reagent and pH. J. Liq. Chromatogr. Rel. Technol. 2005;28:2655—67.

[43] Snyder LR, Kirkland JJ, Dolan JW. Introduction to Modern Liquid Chromatography. 3rd ed. Hoboken, NJ: John Wiley; 2010. pp. 364.

[44] McCalley DV, Neue UD. Estimation of the extent of the water-rich layer associated with the silicasurface in hydrophilic interaction chromatography. J. Chromatogr. A 2008;1192:225—9.

[45] Foti G, Revesz G, Hajos P, Pellaton G, Kovats E. Classical retention mechanism in ion exchange chromatography: theory and experiment. Anal. Chem. 1996;68:2580—9.

[46] Shellie RA, Ng BK, Dicinoski GW, Poynter SDH, O'Reilly JW, Pohl CA, Haddad PR. Prediction of analyte retention for ion chromatography separations performed using elution profiles comprising multiple

isocratic and gradient steps. Anal. Chem. 2008;80: 2474−82.

[47] Fritz JS, Gjerde DT. Ion Chromatography. Weinheim: Wiley-VCH; 2009.

[48] Sahni SK, Reedijk J. Coordination chemistry of chelating resins and ion exchangers. Coord. Chem. Rev. 1984;59:1−139.

[49] Sargent R, Rieman III W. Salting-out chromatography: Amines. Anal. Chim. Acta 1957;17:408−14.

[50] Moldoveanu SC, David V. Sample Preparation in Chromatography. Elsevier; 2002.

[51] Wainer IW. Proposal for the classification of high performance liquid chromatographic chiral phases: How to choose the right column. Trends. in Anal. Chem. 1987;6:125−34.

[52] Lämmerhofer M. Chiral recognition by enantioselective liquid chromatography: Mechanism and modern chiral stationary phases. J. Chromatogr. A 2010;1217:814−56.

[53] Schimitzer AR. From the Past to the Future of Rotaxanes. In: Pignataro B, editor. Tomorrow's Chemistry Today. Weinheim: Wiley-VCH; 2008. p. 133.

[54] Beesley TE, W Scott RP. Chiral Chromatography. New York: John Wiley; 1998.

[55] Berek D. Size exclusion chromatography: A blessing and a curse of science and technology of synthetic polymers. J. Sep. Sci. 2010;33:315−35.

[56] Mori S, Barth HG. Size Exclusion Chromatography. Berlin: Springer Verlag; 1999.

[57] Rhemrev-Boom MM, Yates M, Rudolph M, Raedts M. Immunoaffinity chromatography: a versatile tool for fast and selective purification, concentration, isolation and analysis. J. Pharm. Biomed. Anal. 2001; 24:825−33.

[58] Pasek JJ, Matyska MT. Aqueous Normal-Phase Chromatography: The Bridge Between Reversed Phase and HILIC. In: Wang PG, He W, editors. Hydrophilic Interaction Chromatography (HILIC) and Advanced Applications. Boca Raton, FL: CRC Press; 2011. p. 1−26.

[59] Pesek JJ, Matyska MT, Larrabee S. HPLC retention behavior on hydride-based stationary phases. J. Sep. Sci. 2007;30:637−47.

[60] Pesek JJ, Matyska MT, Hearn MTW, Boysen RI. Temperature effects on solute retention for hydride-based stationary phases. J. Sep. Sci. 2007;30:1150−7.

[61] Pesek JJ, Matyska MT. A comparison of two separation modes: HILIC and aqueous normal phase chromatography. LCGC 2007;25:480−90.

[62] Walton HF. Ligand-exchange chromatography: A brief review. Ind. Eng. Chem. Res. 1995;34:2553−4.

[63] Urh M, Simpson D, Zhao K. Affinity Chromatography: General Methods, in Methods in Enzymology, vol. 463. Elsevier; 2009. p. 417−38.

[64] Azarkan M, Huet J, Baeyens-Volant D, Looze Y, Vandenbussche G. Affinity chromatography: A useful tool in proteomics studies. J. Chromatogr. B 2007; 849:81−90.

[65] http://www.columnex.com/sielc-technologies.php

CHAPTER

6

Stationary Phases and Their Performance

OUTLINE

Essentials in Modern HPLC Separations
http://dx.doi.org/10.1016/B978-0-12-385013-3.00006-9

6.1. SOLID SUPPORTS FOR STATIONARY PHASES

General Comments

The type of stationary phase selected for an HPLC analysis depends on the specific type of HPLC intended to be utilized (see, e.g., [1]). As discussed in Chapter 1, different types of HPLC may use very different stationary phases. Some of the most popular HPLC techniques use as a stationary phase small inorganic particles, usually made of porous silica, covered with an active bonded phase. Typically, the phase is bonded to the silica surface through covalent bonds obtained using a derivatization process, but solid supports only coated with the active phase are also known. Among the HPLC techniques that use this type of particle as stationary phase are reversed phase (RP), HILIC, normal phase (NP), ion pair, and nonaqueous reversed phase (NRP). Other materials such as porous polymers are used as stationary phases for these techniques, but less frequently. For HPLC types such as size exclusion or ion exchange, other types of stationary phases are more common than those based on silica particles. For ion-exchange stationary phases, porous organic polymeric materials are frequently utilized, although small silica particles with a bonded active phase having cation- or anion-exchange capability are also useful. For size-exclusion HPLC (SEC, either gel permeation or gel filtration), the separation mechanism being different from that of other types of chromatography, the main focus in making the stationary phase is its porous structure rather than the chemical structure of its surface. For obtaining specific porosities of the stationary phase to be used in SEC, either inorganic or organic porous polymers are utilized. Special procedures for making the stationary phases are also employed for those used in affinity chromatography.

The stationary phases made from small inorganic particles with an active phase bonded on its surface can be of three main types: porous, superficially porous (core-shell), and pellicular, the porous particles being the most common. The porous particles consist of a skeleton of solid material with the whole surface of the skeleton covered with an active stationary phase. The core-shell particles have a solid nonporous core surrounded by a porous outer shell covered with an active stationary phase, and the pellicular particles are solid spheres covered with a thin layer of stationary phase. The materials used as an inert support for the stationary phases are further discussed in this section.

Another type of stationary phase is the monolithic one. Monolithic stationary phases are made of a single piece of a solid porous material. Two main types of pores are characteristic for the monoliths: the mesopores and the "through pores". The monoliths can have either a silica or a polymeric structure, and the backbone of the monolith is usually covered with a bonded active phase (similar to that on particles).

Silica Support for Stationary Phases

Silica is the most common solid support for making HPLC stationary phases. This is due to the capability of silica materials to be made porous, with a large surface area covered with reactive groups (silanol $\equiv$Si-OH) and at the same time to have a high rigidity and resilience to crashing. Silica, or silicon dioxide (SiO_2), may exist in several crystalline polymorphic forms (quartz, tridymite, and cristobalite), or in amorphous forms. Amorphous silica lacks a crystal structure and does not generate a sharp X-ray diffraction pattern. There are several amorphous forms such as colloidal silica, precipitated silica, silica gels, and fumed silica (aerogel, silica nano-springs, etc., are also forms of colloidal silica). The surface of such amorphous silica contains numerous silanol groups, can contain adsorbed water molecules, or can be anhydrous (except for a monomolecular layer of water). Chemical reactions are possible with the silanol groups present on silica surface, allowing desired organic group attachments through covalent bonds. These attachments (ligands) form a layer that further acts as a stationary phase necessary for the HPLC separation. The high chemical stability and the rapid mass transfer effects obtained with these bonded phases played a major role in the development of HPLC techniques.

The formation process of the solid support of amorphous silica starts with a chemical reaction that generates silicic acids. In extremely diluted solutions, several simple silicic acids were identified, such as metasilicic acid (H_2SiO_3), ortosilicic acid (H_4SiO_4), disilicic acid ($H_2Si_2O_5$), and pyrosilicic acid ($H_6Si_2O_7$). All these acids are very weak acids (e.g., for ortosilicic acid, $pK_1 = 9.84$, $pK_2 = 13.2$). In more concentrated solutions, polysilicic acids with the general formula $[SiO_x(OH)_{4-2x}]_n$ are rapidly formed from the condensation reaction between the silanol groups that give rise to tridimensional structures (precipitates in gel form) containing siloxane (Si-O-Si) groups. The reaction of formation of silicic acid may start with acidic hydrolysis of salts such as Na_2SiO_3, Na_4SiO_4, and K_4SiO_4, or with the hydrolysis in the presence of an acid or a base (as catalysts) of several silicon compounds such as SiH_4, $SiCl_4$, or of alkoxysilanes ($Si(OR)_4$ where $R = CH_3$, C_2H_5, C_3H_7, etc.). Tetraethyl orthosilicate $Si(OC_2H_5)_4$, for example, is a common alkoxysilane used for the preparation of silica gels. The choice of the reactions and of the conditions used to obtain silica gels varies considerably, depending on the intended use of a specific material [2]. The reactions for silicic acid generation can be written as follows:

$$Na_2SiO_3 + HX + H_2O \rightarrow 2NaX$$
$$+ H_4SiO_4 \text{ (or } Si(OH)_4 \text{ or } H_2SiO_3 + H_2O)$$

$$(6.1.1)$$

$$Si(OR)_4 + 4\,H_2O + H^+ \rightarrow Si(OH)_4$$
$$+ 4\,ROH + H^+$$

$$(6.1.2)$$

$$Si(OR)_4 + 4\,H_2O + B: \rightarrow Si(OH)_4 + 4\,ROH + B:$$

$$(6.1.3)$$

The formation of polysilicic acids from the initial silicic acid can be described by the (idealized) reactions 6.1.4:

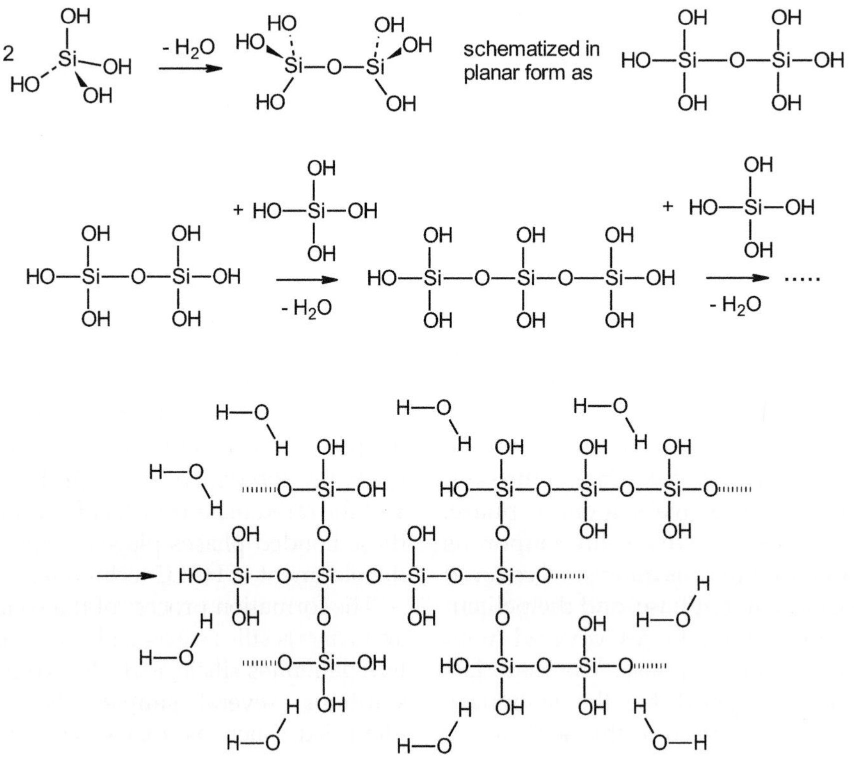

$$(6.1.4)$$

When the polysilicic acid is obtained by hydrolysis of alkoxysilanes, part of the alkoxy groups may remain attached to the silicon atom and further participate in the elimination reaction to form siloxane bonds.

After their formation, the initial smaller polysilicic acids are still present in solution form (sol form), and as suggested in reaction 6.1.4, a large number of water molecules are bound through hydrogen bonds or mechanically retained in the structure. As the molecules increase in size, the sol particles are growing and further changing in a system containing both a liquid phase and a solid phase with a morphology ranging from discrete particles to a continuous polymeric network filled with a large number of pores of different dimensions. This material generates a gel (hydrogel or alcogel depending on the groups still left unreacted) where most of the structures are crosslinked. The distinction between a material in sol form and that of a gel is not precise, and as the amount of water in the structure decreases and the polymeric network increases, the gel form becomes more obvious. The control of the size of sol particles and the elimination of water with the increase in the crosslinking of the polymeric structure can be achieved by selecting the concentration of reagents that form the polysilicic acid and also by controlling other parameters such as temperature, pH, gelling time, addition of electrolytes for flocculation, and addition of detergents. The process is very complex, and as a function of selected parameters the growth of sol particles or of the reticulation can be favored, such that the properties of the final material may be different. This process is schematically suggested in Figure 6.1.1 (see also [3]). When the gel continues to be kept in the initial liquid, its structure changes in a process known as ageing. Ageing affects the properties of the final material by several mechanisms known as polycondensation (further reactions of silanol groups to form siloxane bonds),

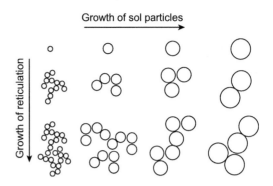

FIGURE 6.1.1 Growth of sol particles and of reticulation from the initial small units of polysilicic acid.

syneresis or pore narrowing (the shrinkage of the gel network with the elimination of liquid from the pores), and coursing (the dissolution of small particles and the growth of larger ones).

Following the gelling and ageing process, the hydrogel (or alcogel) is typically washed, and it is dried (converted into a xerogel) for obtaining a solid material. A silica gel, as first precipitated from water, may contain up to 300 moles of H_2O to 1 mole of SiO_2. The drying process consists of three stages. In the first stage the gel loses water, and its volume decreases with the volume of water that was evaporated. In this stage, the pore volume decreases, and the stiffness of the dried gel increases. In stage two, the liquid from the pores of the gel is eliminated by migration to the surface of the material where it is evaporated. In stage three, the liquid escapes from the pores directly by evaporation. When dried below 150 °C, surfaces containing a large number of silanol groups ($\equiv$Si-OH) are maintained. When heated at 300 to 1000 °C, the silanol-covered surfaces dehydroxylate to form siloxane (Si-O-Si) surfaces. The final properties of the dried material, such as density (porosity), hardness, active surface area, pore volume, and pore-size distribution, depend on a considerable number of parameters such as reagents choice and concentration, electrolyte addition, temperature of reaction, pH, gelling time, ageing, and rate and temperature of drying. The number of silanol groups on the material active surface is determined by the same parameters. The xerogel pores may still contain some water that is retained by strong adsorption forces. The silanol groups have weak acidic properties, which are slightly different from group to group depending on the fact that the groups are isolated, vicinal, or geminal. Schematic formulas for different types of silanols are shown in Figure 6.1.2.

Silanol groups present at the surface of the silica particle play a major role in the use of silica gel as a solid support for the stationary phase in chromatography. Bare amorphous silica can be used in direct-phase chromatography where a layer of water molecules adsorbed on the solid surface acts as a stationary phase. The silanol groups are active in the separation process, influencing this layer of adsorbed water. An adsorption mechanism on the silanol groups is also postulated as a separation potential mechanism. The main utilization of silica xerogels is related to the role of silanol groups as the place where other structural groups are introduced by derivatization in order to obtain modified silica for reversed-phase, chiral, hydrophilic, ionic-exchange, and other type of stationary phases. The main property used for these derivatizations is the acidic character of O-H groups, which allows them to react with different derivatization reagents. Different silica gels have various surface characteristics regarding the density of silanol groups, and these can be derivatized to get attached functionalities to the surface. In general, a higher concentration of reactive hydroxyl groups is present in smaller pore silicas, typically with a pore diameter less than 50 Å, whereas less reactive hydroxyls predominate on larger pore silicas having pore diameters greater than 150 Å.

The derivatization (see Section 6.2) is aimed mainly at binding the fragments (ligands) that act as stationary phase. However, other derivatizations are also practiced, such as end-capping which is further derivatization to attach small groups such as -Si(CH$_3$)$_3$ to the silanols left after the desired bonded phase was attached (see Section 6.4). The acidic character of silanol groups left after derivatization makes them reactive in particular toward strong bases. When made for further utilization as HPLC solid support for the stationary phase, this characteristic imposes strict limitations on the pH of mobile phase used in HPLC, which should not exceed pH $= 8-9$. Significant effort and progress was made in obtaining silica solid supports that are more resistant to higher pH solutions since typical xerogels can be easily destroyed by basic media following a reaction with the silanol acidic groups.

The pores considered for differentiating silicas are sometimes indicated as *mesopores*. Larger pores of $10-20$ μm forming channels (through pores) with larger dimensions can also be present in silicas, but they are in a small proportion. A common classification of silicas (silica gels) based on their average pore diameter is in types A, B, and C. The main characteristics of these silicas are given in Table 6.1.1. Silicas with characteristic values outside those indicated in this table are also utilized in practice, and they can be still classified in the same types, depending on the closeness of their characteristics to those indicated in the table. Type B (pore size) silicas are the most commonly used in HPLC, but Type A and Type C silicas are also used for specific purposes. The silicas for chromatographic use are available under various trade-names (Davisil, Astrosil, Lichrospher, Kromasil, etc.). Because the terminology Type

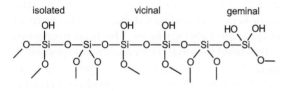

FIGURE 6.1.2 Schematic formulas for isolated, vicinal, and geminal silanols.

TABLE 6.1.1 Classification of silica gels based on pore size and some typical characteristics*.

Property	Type A (fine pore)	Type B (common in HPLC)	Type C (large pore)
Aspect	Transparent	Semi-transparent	Milky
Average pore diameter nm	2.0–3.0	4.0–10.0	80–125
Pore volume mL/g	0.35–0.45	0.6–0.9	0.7–1.0
Surface area m²/g	650–800	450–700	300–400

* Note: Some confusion in the nomenclature of silicas may be encountered, when the terms Type A and Type B are related to silica purity and not to pore size. Related to purity, Type A refers to less pure material while Type B refers to highly purified homogenous spherical silica.

A and Type B is also used for the caharacterization of purity, it is preferable to use other terms for the pore-size characterization. For example, Davisil silica is characterized by "Grade," which describes both pore size and particle size as indicated in Table 6.1.2 (all these silicas are of high purity).

The silica materials may be contaminated with traces of metallic ions such as Al^{3+}, Fe^{3+} and Ni^{2+}, depending on the synthesis of the silica gel or the manufacturing process. These metallic ions may be present either in the form of isolated oxides and hydrated oxides in silica or connected through oxygen bonds attached to a Si atom. These metallic ion impurities may affect the retention process, usually causing peak tailing in HPLC, owing to complexation with the functional groups from analyte molecules. The acidity of the surface silanols may also be increased in the presence of these metal impurities. For this reason, very high purity silica is desired. More recently, procedures for purification of silica have been developed, and a silica that has a less acidic surface, lower metal content, and a more homogeneous distribution of silanol groups was developed. It leads by derivatization to a denser and more reproducible bonding of the attached groups and is frequently utilized for the manufacturing of stationary phases. This type of silica is also termed Type B (not to be confused with Type B related to pore size) to differentiate it from the lower purity silica that has higher metal content and more acidity that is termed Type A silica (a term not related to pore size).

Besides the chemical properties of the surface, its physical properties are also extremely important in HPLC. These physical characteristics include: (1) particle surface area, (2) pore size, (3) pore volume, (4) pore size distribution, (5) particle shape, (6) particle size, (7) particle size distribution, and (8) structural rigidity. These properties play a major role in the retention process of the stationary phase, even after surface derivatization. For example, one parameter used for the characterization of column retention is the phase ratio Ψ that characterizes

TABLE 6.1.2 Classification of several high purity silicas using "Grade" description.

Grade (Davisil®)	Pore size Å	Particle size mesh
633	60	200–425
635	60	60–100
636	60	35–60
643	150	200–425
646	150	35–60
12	22	28–200
62	150	60–200
923	30	100–200

the volume of immobilized liquid on the solid support surface (the definition of Ψ is given by rel. 2.1.14). Phase ratio Ψ depends on the total pore volume and is indirectly related to particle surface area, smaller pores leading to larger surface area. Although a large surface area obtained when the pores are small is desirable, smaller pores are useful up to a point since too small pores may block some larger molecules to enter them and therefore may not allow the derivatization process inside these pores or do not allow larger molecules to interact with the surface. Although the lowest pore size may increase the phase ratio Ψ, a small pore size lower than about four times the hydrodynamic diameter of the molecule (diameter of a sphere that diffuses at the same rate as the molecule) can also hinder the diffusion and leads to lower N than expected for the calculated particle surface. For this reason, for the separation of analytes with molecules having the molecular weight up to about 10,000 Dalton, the pore size typically used is between 4 and 20 nm, leading to surface areas between about 300 m^2/g up to about 600 m^2/g (the weight refers to the final stationary phase). For the separation of proteins and of other polymers with the molecular weight above 10,000 Dalton, the typical pore size is about 30 nm or larger.

Also of considerable interest for HPLC use is the control of the xerogel particle size distribution and shape. Several technologies have been developed for controlling these parameters, mainly with the purpose of production of spherical particles "in situ". This can be achieved by: 1) the choice of the reagents used to be hydrolyzed, 2) choice of the medium of hydrolysis (use of ammonia as a morphological catalyst), 3) control of the way the gel particles are growing [4], 4) the use of seeding (procedure also used for the production of core-shell [5]). Another procedure for determining the size distribution and shape of particles is the use of a special spray-drying procedure that leads to the formation of spherical particles rather than irregular-shaped

particles and also of particles with similar sizes [6]. By these procedures the desired particle dimensions can be obtained. One example is the production of microamorphous silica gels made of particles with diameters less than 1 μm. These xerogels have very high surface areas, usually greater than 300 m^2/g.

The physical parameters of a stationary phase are related mainly to the extent of eddy diffusion in the chromatographic process. Eddy diffusion is determined by the difference in the path length taken by individual molecules of the same species when they migrate across the porous solid-phase material. The difference in the path length and therefore the eddy diffusion can be significantly higher when a wider range of long and short paths is possible. This is the case when the particles are of different sizes or when the particle shape is irregular. Figure 6.1.3 suggests this difference in length of the path for an irregular particle shape, a spherical particle, and a spherical smaller particle. For reducing eddy diffusion, considerable technological effort was involved in generating particles with a controlled size and shape [7, 8].

A reduction in the differences in potential path length of molecules of the same species when they go across a stationary-phase particle can be achieved by using uniformly shaped smaller particles, as suggested in Figure 6.1.3. The first step in achieving the goal of reduced eddy diffusion was the use of spherical particles instead of particles of irregular shape, with all particles having (about) the same diameter. The other step was the use of particles of smaller diameter. Spherical particles with 3 to 5 μm diameter are common for many analytical columns. On the other hand, the required pressure to pass the mobile phase increases for columns filled with small particles. In the past, when the pressure generated by the pumps was more limited, particles with a diameter around 5 μm were common, and the backpressure due to the column was not higher than about 240 bar (3500 psi), this depending on the

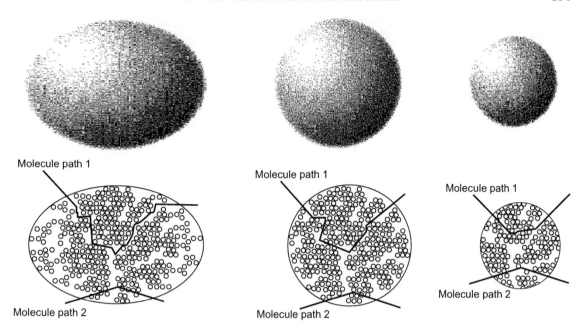

FIGURE 6.1.3 Three particles, one with an irregular shape, one spherical, and one spherical but smaller. The potential differences in the molecular path for each shape are also suggested.

solvent, column length, and column diameter. In modern HPLC, the tendency is to reduce the particle size (3 μm, 2.1 μm, or even lower); these columns require higher working pressure up to 600 bar (9000 psi) or even higher. The decrease in particle size is also related to other properties of the column such as column lifetime. Particle size influences the *interstitial volume*, which is the space between particles. Interstitial volume also contributes to the peak broadening in HPLC. This volume is typically around 70% of the total internal volume of the column.

Columns with particles around 2 μm and lower may generate pressures up to 15,000 psi, and typically they are used with lower flow rates and narrow columns [9]. Several SEM pictures at different magnifications of the particles used in a C18 column with 5 μm diameter particles are shown in Figure 6.1.4.

The same effect of paths with less variable length as achieved with small porous particles can be achieved using superficially porous particles or even pellicular particles. Superficially porous particles (core-shell) have a solid nonporous core surrounded by a porous outer shell. For particles of about 3 μm diameter, the porous shell may have 0.3–0.4 μm in depth. Pellicular particles are solid spheres covered with a thin layer of stationary phase. The shape and structure of fully porous particles, core-shell ones, and pellicular particles are schematically shown in Figure 6.1.5. The hypothetical paths of two molecules in a fully porous particle, core-shell particle, and a non-porous particle are also shown in the same figure. The difference in path is shorter in core-shell particles, and the path lengths are very close to each other for pellicular particles [10].

Since the interaction between the molecules in the mobile phase with the stationary phase takes place at the active surface of the particles, it is expected that porous particles will have the largest available surface (large Ψ values for the same other characteristics), while the nonporous

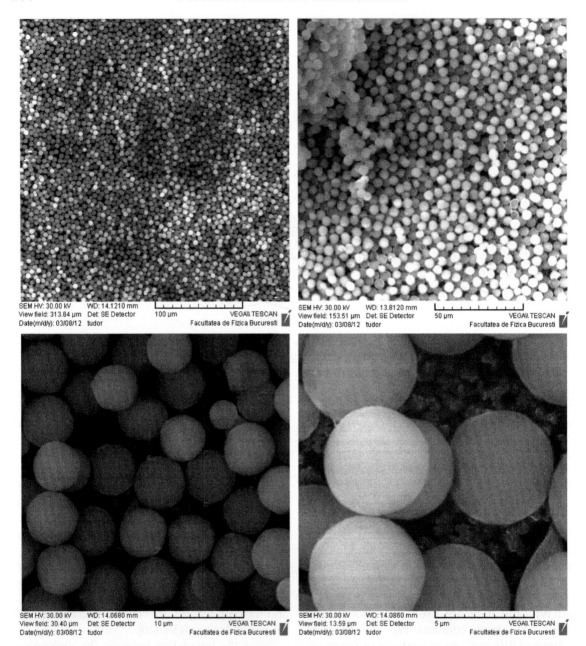

FIGURE 6.1.4 SEM pictures at different magnifications of 5 μm diameter particles used in a C18 column (the scale 100 μm, 50 μm, 10 μm and 5 μm is indicated on pictures).

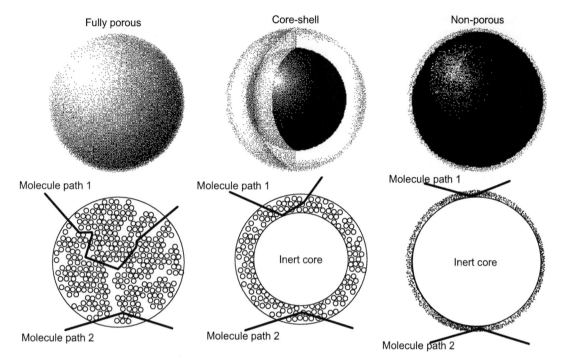

FIGURE 6.1.5 Schematic aspect of fully porous, core-shell, and pellicular (non porous) particles. Hypothetical paths of two molecules in each particle type showing length differences are also illustrated.

ones will have a very low available surface (small Ψ and resulting lower capacity factors k). On the other hand, the fully porous particles will have the largest eddy diffusion. Pellicular particles will show very little eddy diffusion, but the low Ψ may not be sufficient for some separations. The core-shell particles may be the optimum choice between fully porous and pellicular ones. Typical core-shell particles have similar standard diameters as porous particles (e.g., 5 μm, 3 μm, 2.7 μm, and around 2 μm). The pore size for the porous shell of core-shell particles is chosen similar to that for porous particles, and the resulting surface areas are typically around 150 m²/g.

Structural rigidity is another important parameter of the stationary phase. Silica-based particles typically offer a good structural rigidity (particle strength), standing values between 8,000 and 15,000 psi (depending also on the particle dimension). As smaller particles are used for making stationary phases, higher pressures of the mobile phase must be used for performing the chromatography. When the particle strength is exceeded, these may "mechanically collapse" and this affects the column performance. Organic polymeric phases have a lower structural rigidity as compared with silica; they also swell and shrink depending on the mobile-phase solvents, these being considerable disadvantages for the polymeric phases.

Organic/Inorganic Silica Supports

The hydrolytic instability of pure silica-based supports, even after a derivatization that covers the surface with desired groups (ligands) such as long hydrocarbon chains, represents a considerable problem when using mobile phases with a low pH or with a high pH. One path applied to alleviate this problem is the use of hybrid inorganic/organic materials. These materials contain some organic fragments able to protect the silica from the attack of excess of OH⁻ or H⁺ ions. The result is a silica base material that remains amenable to the derivatization necessary to make a bonded phase having attached organic fragments. One such material is obtained using the type of reaction described by 6.1.5:

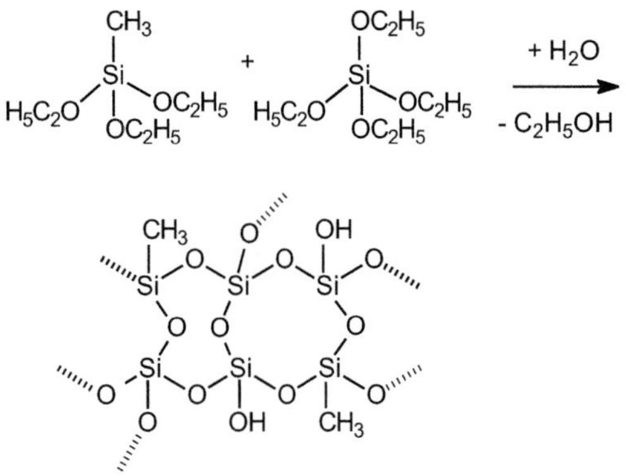

(6.1.5)

Materials with even better stability than those having CH_3 groups in the silica were obtained using *ethylene-bridged* structures (indicated as BEH technology by Waters) [11] or propylene-bridged materials. This type of silica-based support has the empirical formula $(SiO_2)_8$ $(O_3(SiCH_2CH_2Si)_2O_3)$ and is prepared in two steps. In the first step, a polyethoxy-siloxane oil phase is prepared from hydrolytic condensation in acidic conditions of bis(triethoxysilyl)ethane and tetraethoxysilane using a small amount of water. Highly spherical porous hybrid particles are then prepared by further condensation under alkaline conditions in an oil-in-water emulsion. The first step of these reactions can be schematically written as follows:

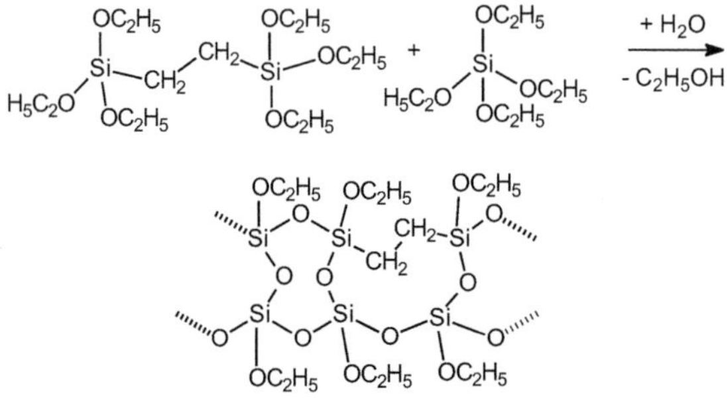

The resulting polyethoxysilane, after further hydrolysis, generates the desired material that can be derivatized to form the bonded phase. Once obtained, the silica-based organic/inorganic material is subjected to surface-ripening steps similar to those used for silica. Significant improvements at both low pH and high pH (up to pH = 12.5) were reported for stationary phases obtained with organic/inorganic phases. Other modifications of the organic/inorganic silica supports were reported, including a controlled surface charge procedure (CSH technology) [12]. Usually, the silica surface is slightly negatively charged due to the dissociation of silanols. This charge can be neutralized by adding specific reagents such that the surface reactivity is decreased.

Combination of inorganic support with organic bound groups capable of further polymerization can also be achieved by reacting the silica surface with reagents such as vinyltriethoxysilane or methacryloxypropyltriethoxysilane. In this case, a reactive group capable of further polymerization is attached to the silica surface, allowing a better shielding of the possible free silanol groups from the silica surface. Other procedures for covering the silica base material with an organic/inorganic layer are utilized for providing support materials with a wide range of pH stability (e.g., in Kromasil Eternity columns that have an extended pH range up to 12).

Coated or Immobilized Polymeric Stationary Phases on Silica

Besides the derivatization techniques frequently utilized for binding a phase with desired chemical properties on the silica surface, other techniques are also used for the same purpose. The goal is to take advantage of silica, which has excellent properties for a stationary phase, including a large surface area, mechanical strength, and pores of appropriate dimensions. These properties are difficult to achieve with other supports such as inorganic or organic polymers. One simple procedure to cover silica with the desired active stationary phase, and also to cover some of the silanol groups not wanted in the separation, is a simple coating. For example, derivatized cellulosic materials can be coated on silica by simple adsorption using a proper solvent for the organic material (e.g., acetone, dichloromethane, or tetrahydrofuran) [13]. The coating must not alter the porous structure of silica, but at the same time it may not be sufficient to assure stability of the coated silica material since some spots may remain unprotected (uncovered). For this reason, some chemical reactions that assure good bonding of the coat can be performed, using, for example, the immobilization of a presynthesized polymer. A presynthesized polymer (polybutadiene, polysiloxane), still having some reactivity by interrupting the polymerization process before it is complete, can be immobilized on silica using gamma radiation, microwaves, or thermal treatment, possibly in the presence of a radical initiator [14]. The main disadvantage of this procedure for obtaining immobilized polymers on silica particles is the inhomogeneity of the polymeric covering such that some parts of the silica still remain uncovered, and the overall stability of the stationary phase is not significantly improved compared to standard bonded phases. An alternative procedure that appears to be successful in obtaining stable phases with polymers immobilized on silica particles is to generate a layer of zirconia or titania on the silica before bonding the polymer [15, 16]. The method of attaching zirconium or titanium oxide on silica starts with dry silica gel, which is treated with $ZrCl_4$ or $TiCl_4$ followed by exposure to ammonia. An alternative procedure consists of the treatment of silica gel with zirconium or titanium tetrabutoxide followed by a hydrolysis step of the adsorbed layer of tetrabutoxide on silica surface. This type of support is known as metalized silica. The polymers bonded on metalized silica include poly(methyloctylsiloxane), poly(methyltetradecylsiloxane), poly(methyldecyl-(2-5%)-diphenylsiloxane), and poly(dimethylsiloxane). The bonding procedure involves the deposition on the metalized silica of the polymer dissolved in a solvent, followed by particle drying and exposure to gamma radiation, microwave radiation, or thermal treatment. Such phases show better resistance to mobile phases with high pH and good peak symmetry [17].

Other procedures for obtaining immobilized polymers on silica are known. One such procedure uses, for example, vapor deposition of a silicone polymer of the base silica gel particles. This procedure leads to a quite homogeneous monolayer of polymer. The polymer can be further derivatized to attach the desired bonded phase.

Hydride-based Silica

The presence of silanol groups on the surface of silica-based stationary phases even after derivatization and end-capping prompted continuous effort to generate new stationary phases that do not have this problem. One suggested solution to the problem is the use of hydride silica as support for further derivatization [18]. Silica hydride support material is sometimes indicated as Type C silica. The basic chemical reaction for this process is the following:

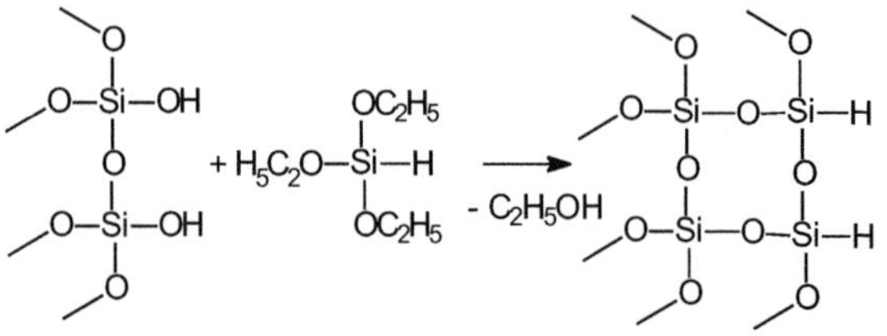

$$ (6.1.6) $$

Hydride-based silica can also be obtained by different reactions such as reduction of a material containing chlorine attached to a silica surface, as shown in reaction 6.1.7:

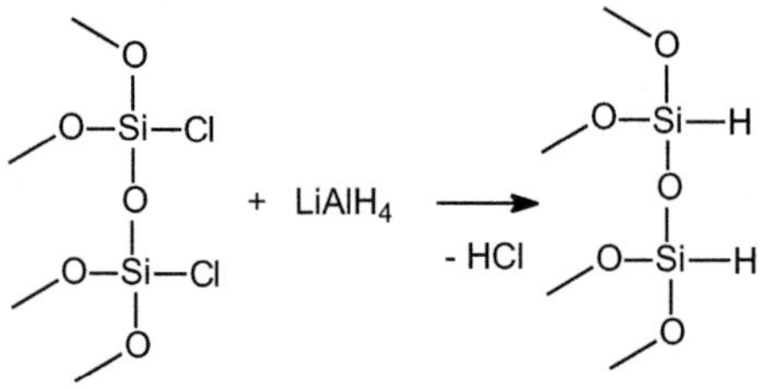

$$ (6.1.7) $$

It has been shown that the Si-H group on the silica surface is stable in aqueous solutions in the range of pH from about 2 up to about 9, and it is not hydrolyzable as would be a small molecule of silane [19]. Unbonded silica hydride can be used as a normal phase (similar to a bare silica material), but further derivatization of the silica surface can be done using hydrosilation reactions for the attachment of desired bonded phases (see Section 6.2). The silica hydride surface can be populated with hydride material up to 95%. A number of studies have been published in the literature regarding the preparation and use of silica hydride stationary phases [20–22].

Other Inorganic Support Materials

Another inorganic material used as solid support for stationary phases in HPLC is zirconia (hydrated). Zirconium dioxide (ZrO_2), also known as zirconia, may exist in several crystallographic and amorphous forms and possesses both acidic and basic properties (amphoter properties). Some physical properties (mechanical stability, high surface area, control of average pore diameter and of particle diameter, swelling capability) as well as chemical properties (chemical stability over a large pH range 1–14) recommend this compound as substrate for stationary phases in HPLC. ZrO_2 can be prepared by precipitation from zirconium salts, zirconyl salts ($ZrOX_2$), or zirconium alkoxides. The method of precipitation, pH, temperature, and the other parameters discussed at the formation of silica xerogels also influence the formation of zirconia xerogels that may consist of small (2–25 μm) spherical and rigid particles. Sol -gel derived zirconia xerogels prepared via the hydrolysis of zirconium n-propoxide in methanol, ethanol, or 2-propanol is one of the most common approaches to preparing this support [23]. Other procedures to prepare zirconia supports can use zirconium oxide that is further treated with a base for hydration and reprecipitated in the presence of a surfactant and in specific physical conditions (temperature, stirring) that allow control of the particle properties. It is also possible to embed hydrated zirconium oxide in colloidal form in a mass of an organic polymer that is further destroyed by combustion. Similar to hydrated zirconium dioxide, hydrated titanium dioxide ($TiO_2 \times H_2O$) or titania can be used as solid support for stationary phases in HPLC [24].

The surface of zirconia xerogels is formed from Zr atoms bound to O atoms and also to OH groups, similar to the case of amorphous silica surfaces. The average concentration of Zr atoms on zirconia surface is about 12 μmole/m^2. The surface is covered with hydroxyl groups in a concentration of about 20–25 μmole/m^2 and is the site for further reactions when the derivatization is desired. The specific surface area of zirconia supports is dependent on the thermal treatment in the drying process, and this area decreases sharply when the drying is performed at temperatures higher than 300 °C. Above this temperature two processes are responsible for the decrease of area surface: microcrystallite growth and intercrystallite sintering. Temperature treatment influences the porous structure of zirconia. When subjected to heat treatment of 150–300 °C, a transitionally microporous material is obtained; at temperatures of 300–600 °C a transitionally macroporous material is obtained; and above 700 °C macroporous adsorbents with pores of 20–500 Å are obtained. These materials obtained at higher temperatures do not have enough OH groups necessary for derivatization and surface modification. Generally, the pore volume of various zirconia-based materials is much lower than that of silica materials, which makes silica continue to be the preferred solid support used in manufacturing stationary phases for HPLC.

For the use of HPLC support, zirconia stationary phases pose a number of challenges that are different from those of silica. The OH groups bound to zirconium atom are much less acidic as compared to the silanol groups, although they have the capability to act as a Brönsted acid. The lower acidity of zirconia makes this phase resilient to a wider pH range, between pH = 1 to about pH = 10. Zirconia surface may also act as a Brönsted base, and also as Lewis acid due to the d electrons of zirconium atoms that can interact with various ligands. The simplified surface of zirconia surface (showing empty electronic levels, OH groups, and water) is as follows:

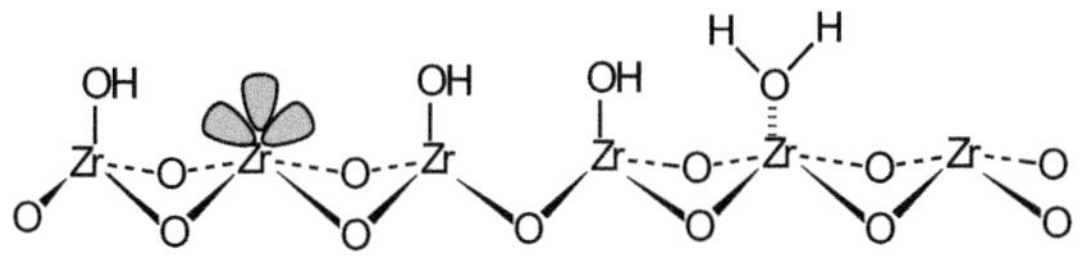

Besides the behavior determined by the substituents attached to the zirconia surface, for example, that of a reverse phase when the surface is covered with hydrophobic substituents (e.g. C18), zirconia also exhibits polar and ionic interactions. For this reason, deactivation of the zirconia surface, in particular the elimination of Lewis acid sites using a chelating agent, is applied to the newly developed zirconia stationary phases. The pore structure of zirconia is also somewhat different from that of silica and leads in general to lower chromatographic performance compared to silica. Phases coated on zirconia are also known.

Attempts have also been made to use other materials as solid support, which would have higher stability to high pH and to heating. Among these materials are Al_2O_3 (alumina), as well as ThO_2 and CeO_2. Although ThO_2 and CeO_2 have been studied, columns made with these materials are not commercially available. The surface of alumina and its corresponding chemistry is more complex than silica. The basic spinel structure of Al_2O_3 often possesses defects that result in various arrangements of aluminum ions. Therefore, the hydration of its surface as well as the number of hydroxyl groups per unit area is determined by the specific three-dimensional structure of the oxide [25]. The advantage of Al_2O_3-based stationary phases is their stability up to pH = 12. Al_2O_3 is mostly used as support for phases containing Al_2O_3 covered on the surface with a polymeric layer (e.g., polystyrene-divinylbenzene copolymer). These columns are more stable in acidic medium, but they pose problems during the analysis of acidic compounds due to the mixed types of interactions they offer, one with the polymer and the other with the alumina surface. Alumina support has also been evaluated for reaction with triethoxysilane (catalyzed by HCl) to obtain a silica hydride-covered alumina.

Among other inorganic compounds that are used in chromatography are specific materials that have ion-exchange properties. These materials include several groups of natural silicates such as zeolites. Zeolites are hydrated aluminosilicates with the general formula $M_xO \cdot Al_2O_3 \cdot nSiO_2 \cdot mH_2O$, where M is Na, K, Ca, or other similar metals. These compounds act as selective cation exchangers, the exchange process being associated with a molecular sieving mechanism. Several hydrated oxides also act as cation exchangers. Among these oxides are hydrated oxides of Si, Al, Zr, Fe, and Sn. These hydrated oxides have both adsorbing and ion-exchange properties. Other inorganic compounds, such as polymeric hydrated zirconium phosphate and hydroxyapatite, also have ion-exchange properties.

Inorganic Monolithic Supports (Silica-based Monoliths)

Monolithic columns have been more recently introduced in practice as compared to those having the stationary phase made from small particles. Modified silica rods are probably the most common type of monolithic columns, but organic polymers are also used as monoliths. The columns based on a silica rod are made of a single piece of porous silica. These rods are prepared by a polymerization or polycondensation process, either in situ in a column tube, such as in glass tubes or fused silica capillaries, or in a column mold. In situ preparation of monolithic columns has the advantage that no further encapsulation of the porous monolith in a tube resistant to the pressure of the solvent is needed. However, this approach is not compatible with monolithic silica columns having larger diameters (4.6 mm or more) due to the shrinkage that occurs during the sol-gel preparation process. When the monolithic silica rod is made in a mold, it has to be further clad with a suitable material such as PEEK (polyether-ether-ketone), to which the column end fittings can be attached for use in the HPLC process [26].

Monoliths have a porous structure characterized by mesopores (pores between 2 and 50 nm in diameter) and macropores (about 4,000 to 20,000 nm in diameter), with silica skeleton of approximately 1–2 μm thick and a void volume of almost 80% of the entire column volume. These pores provide monoliths with high permeability, a large number of channels (the macropores are typically interconnected, the reason for which they are also indicated as through pores), and a high surface area (generated mainly by mesopores) [26]. The backbone of a monolithic column can easily be chemically altered for specific applications. Figure 6.1.6 shows the SEM picture of the typical porous structure of a monolithic silica column. The monoliths' unique structure gives them several physicomechanical properties that enable them to perform competitively against traditionally packed columns. Monolithic columns show an efficiency equivalent to 3–5 μm i.d. silica particles, but with a 30–40% lower pressure drop. In general, the monolithic materials are prepared in macro- and microscopic formats. The first category includes disks, rods (continuous columns), and tubes, whereas the second category comprises capillary columns in fused-silica or silica steel tubing and in microfluidic chips.

In the case of particle-packed columns, the plate height H is roughly proportional to the particle diameter, whereas the backpressure is inversely proportional to the particle diameter. Usually, a linear increase in column efficiency due to the use of smaller particles is accompanied by a quadratic increase in column backpressure. In this respect, the monoliths have a major advantage over the particle-packed columns. The flow through monolithic channels is closer to laminar, and thus they allow less eddy diffusion. The transport of an analyte through the monolithic bed is based mainly on perfusion, and in the monolithic media with a low proportion of mesopores, the analyte diffusion into and out of the pores does not significantly contribute to the band broadening. Due to the reduced diffusion in solute transfer through the monolithic bed, the peak efficiency does not decrease as sharply as in particle-packed columns when increasing the linear flow rate of the mobile phase. This allows the application of higher flow rates and shorter analysis time, without a significant loss of plate numbers [27–29].

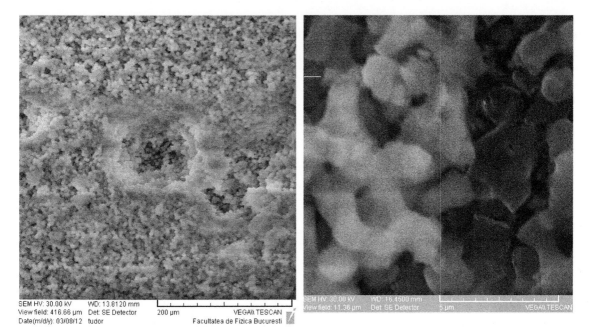

FIGURE 6.1.6 SEM picture of the typical porous structure of monolithic silica columns.

For the production of silica-based monoliths, the basic sol-gel process is similar to that used in the preparation of solid supports for porous silica materials. The process is conducted as a sequential hydrolysis followed by a polycondensation of silane derivatives. Used for this purpose are compounds such as tetraethoxysilane (TEOS), or tetramethoxysilane (TMOS) in aqueous acid or basic medium with an appropriate porogenic solvent (e.g., polyethyleneglicol (PEG)). The hydrolysis conditions and the choice of the porogenic solvent are essential for obtaining mesopores and macropores [30, 31]. Similarly to the preparation of porous silica particles, either hydrogels or alcogels can be obtained. The first step in the process is the formation of a sol (a colloidal suspension of solid species in a liquid). The sol is converted into a gel through polycondensation of the sol, leading to a wet structure. The process of gelation starts with the aggregation of particles into fractal clusters, and then the clusters interpenetrate to some extent and finally link together to form an infinite network. Xerogels are dried by evaporation of the liquid, and aerogels are usually obtained by removal of solvents under hypercritical conditions. The free silanols resulting from polycondensation are further derivatized with reagents generally used for derivatization of silica-based particles. However, monoliths having the surface covered with the desired functionalities can be directly obtained by hydrolysis of the appropriate compounds such as trimethoxyoctylsilane or trimethoxyoctadecylsilane. Nonsilica aerogels are notably weak and fragile in monolithic form; however, the synthesis of high-porosity monolithic alumina aerogels has been recently described.

Organic Polymers Used as Support for Stationary Phases

Porous polymers are frequently used as stationary phases for size-exclusion chromatography, and some are used for ion-exchange chromatography. Some polymers are also used for reversed-phase chromatography. For size-exclusion chromatography, the polymers do not need specific functionalities, and the main characteristic of those phases is their tridimensional structure. Perfusion particles usually contain very large pores (e.g., 600 to 1000 nm diameter) connected to a network of smaller pores (30 to 100 nm diameter). The flow of the mobile phase through a column filled with perfusion particles takes place through the particles and participates in a diffusion process. Polymers with controlled pore dimensions are frequently used as the stationary phase for size-exclusion chromatography. Also, the size of the polymer particles is important to control (similar to the case of silica particles). This can be achieved using monosized polymer seed particles [32].

One of the most common polymeric supports used as a stationary phase is polystyrene crosslinked with divinylbenzene (PS-DVB). This material is obtained by suspension polymerization using a two-phase organic/aqueous system. The crosslinking polymerization is performed in the presence of inert diluents that are miscible with the starting monomers but must not dissolve in the aqueous phase. Submicrometer particles (microbeads) form as the styrene-divinylbenzene polymerizes and precipitates out of solution. The formed microbeads fuse together to form macroporous particles. Initially, a network of microporosity may be present in the microbeads, and polymerization conditions must be controlled to minimize this type of porosity because it results in a soft polymer that has poor mechanical strength and high propensity of swelling. After the crosslinked PS-DVB porous particles are formed, any residual reactants, diluents, and surfactants must be removed by thorough washing.

Other polymers such as methacrylates and polyvinyl alcohol were also used. Organic polymers were utilized for preparation of both particles and monolithic columns. Also, the stationary phases that are organic polymers

and contain specific functionalities can be obtained either by derivatization or directly by polymerization of a monomer having already attached the desired functionality. For RP-HPLC columns having C18, NH_2 or CN groups are commercially available, and those having COOH, SO_3H, NH_2, and NR_3^+ are commercially available for ion-exchange chromatography.

A particular direction where efforts have been made to develop polymeric stationary phases is that of polymeric monoliths [33–36]. Monoliths were obtained from a variety of monomers, using or not using a crosslinker. Initial monoliths were simply obtained from styrene and divinylbenzene that were copolymerized in a porogenic solvent and in the presence of a radical initiator. A new convenient procedure starts with glycidyl methacrylate, which is polymerized by UV, thermal, or γ-radiation initiation, in the presence of a crosslinker such as ethylenedimethacrylate, and using a porogenic solvent and an initiator (e.g., 2,2′-azobis-isobutyronitrile) [37]. Solvents typically used as porogens are mixtures of cyclohexanol and dodecanol, a higher content of dodecanol leading to monoliths with larger pores [38]. The reactions taking place during polymerization are of the type described by 6.1.8:

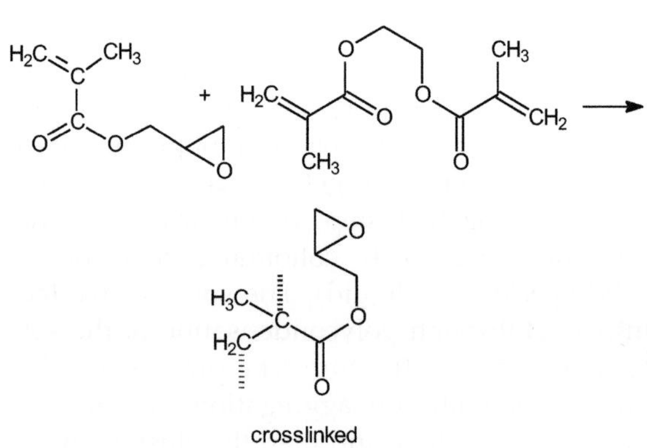

crosslinked

$$(6.1.8)$$

TABLE 6.1.3 Methacrylate polymers evaluated for producing monolith columns for HPLC [39].

Type of polymer	Crosslinking agent
Glycidyl methacrylate	Ethylene dimethacrylate
Glycidyl methacrylate with styrene	Ethylene dimethacrylate
Glycidyl methacrylate with butyl methacrylate	Ethylene dimethacrylate
Glycidyl methacrylate with octyl methacrylate	Ethylene dimethacrylate
Glycidyl methacrylate with dodecyl methacrylate	Ethylene dimethacrylate
Glycidyl methacrylate with N-vinyl-pyrrolidone	Ethylene dimethacrylate
Glycidyl methacrylate with 2-hydroxyethylmethacrylate	Ethylene dimethacrylate
Glycidyl methacrylate	Trimethylolpropane trimethacrylate with triethylene glycol dimethacrylate
Glycidyl methacrylate	Trimethylolpropane trimethacrylate
Glycidyl methacrylate with glycidyl methacrylate-peptide conjugate	Ethylene dimethacrylate
Glycidyl methacrylate with [2-(methacryloyloxy)ethyl]-trimethylammonium	Ethylene dimethacrylate
Methacrylic acid	Ethylene dimethacrylate or trimethylolpropanetrimethacrylate
2-Hydroxyethyl methacrylate	Ethylene dimethacrylate
Hydroxypropyl methacrylate	Ethylene dimethacrylate
Ethyl methacrylate	Ethylene dimethacrylate
Ethyl methacrylate	Trimethylolpropane triacrylate
Ethyl methacrylate with N,N-dipyrid-2-ylmethacrylate	Trimethylolpropane triacrylate
Butyl methacrylate with 2-acrylamido-2-methyl-1-propanesulfonic acid	Ethylene dimethacrylate
Butyl methacrylate	Glycerol dimethacrylate or 1,4-butanedioldimethacrylate
Butyl methacrylate with 2-hydroxyethyl methacrylate	1,4-Butanediol dimethacrylate
Butyl methacrylate	Ethylene dimethacrylate
Butyl methacrylate	Diethylene glycol dimethacrylate or triethylene glycol dimethacrylate
Hexyl methacrylate	Ethylene dimethacrylate
Octyl methacrylate	Ethylene dimethacrylate
n-Lauryl methacrylate	Ethylene dimethacrylate
Stearyl methacrylate	Ethylene dimethacrylate

(*Continued*)

TABLE 6.1.3   Methacrylate polymers evaluated for producing monolith columns for HPLC [39]. (*cont'd*)

Type of polymer	Crosslinking agent
n-Octadecyl methacrylate	Ethylene dimethacrylate
Polyethylene glycol methyl ether acrylate	Ethylene dimethacrylate or polyethyleneglycol dimethacrylate
Glycerol dimethacrylate	Glycerol dimethacrylate
Trimethylolpropane trimethacrylate	Trimethylolpropane trimethacrylate
N,N-Dimethyl-N-methacryloxyethyl-N-(3-sulphopropyl) ammonium betaine	Ethylene dimethacrylate
2-Diethylaminoethyl methacrylate	Ethylene dimethacrylate
Sulfoethyl methacrylate	Poly(ethylene glycol) diacrylate

A number of methacrylate polymers evaluated for generating monoliths are listed in Table 6.1.3 [39]. Some of the polymers obtained from the monomers listed in this table contain glycidyl groups. These groups are used for attaching ligands (such as C18 chains) and generate columns similar regarding the ligand with those having a silica base. The control of porosity of these polymers is achieved using specific porogenic solvents in specific proportions to the monomers. These porogens can be either good or bad solvents for the polymer. The initial reaction takes place in homogeneous solution. If the porogen is a good solvent for the polymer, the phase separation takes place late and the pores are smaller. Otherwise the polymer precipitates early, generating larger pores.

Several other polymers and copolymers based on methacrylate polymers or copolymers are reported in the literature as being used as monolith stationary phases in HPLC [39, 40]. Among these are copolymers of glycidyl methacrylate with styrene, other alkyl methacrylates, and N-vinyl-pyrrolidone. Besides glycidyl methacrylate, other methacrylate polymers containing hydroxy instead of epoxy groups were used as starting material for monoliths synthesis.

Crosslinking and porogenic solvents in which the polymerization takes place have an essential role in creating an appropriate porous structure of the polymer. The changes in the ratio of monomeric precursors and of the porogenic diluents (low-density/high-density materials), the rate of initiation (initiator concentration, temperature, light intensity), as well as copolymerization dynamics of the monomer and crosslinker, lead to fundamentally different materials, which pose problems with batch reproducibility.

A problem regarding organic polymer monoliths is related to the mass transfer properties of the material and not to the derivatization or the increase in the surface area of the material. The nanoporosity of solvated polymeric material is dependent on the solute/solvent–polymer interaction regardless of the crosslinking density, and the polymers show swelling propensity in the hydro-organic solvents used in HPLC. Such porosity can be described as gel porosity. The nature of the crosslinking polymerization reactions, and the formation of porous scaffolds in general, inherently lead to a significant amount of gel porosity, resulting in mass transfer resistance of small molecules. This depends on the retention and size of analytes, as well as the linear chromatographic flow velocity. The fluid transport properties in the heterogeneous macropore space (flow dispersion) and the transport of small molecules in the swollen and porous polymer matrix are factors that are different from those that are operative in silica-based monolith stationary phases that have a large proportion of meso pores (though pore size is about 10–20 μm and meso pore size is around 20 nm) [40].

6.2. REACTIONS USED FOR OBTAINING ACTIVE GROUPS OF STATIONARY PHASES

General Comments

The active part of a stationary phase can be obtained through a variety of procedures. A basic difference between these procedures is the fact that some active phases are obtained by the derivatization of a solid surface, usually that of silica, while other phases are synthesized directly. The widespread use of silica for obtaining stationary phases is based on its reactivity, large surface area, mechanical strength, and the possibility of generating porous silica particles of uniform desired dimensions. The most common type of silica used for preparing stationary phases is of Type B (related to purity). Regarding the pore size, silicas with medium pores (Type B related to pore size) are the most commonly used. Silicas with narrow pores (Type A related to pore size) and with large pores (Type C) are also sometimes used as

support for the modern stationary phases, for special purposes.

The derivatization of silica is based on the reactions of its silanol groups with specific reagents that contain the desired molecular fragment to be attached to the surface. This fragment can be the final functionality or an intermediate one toward the desired bonded phase on the surface. Some schematic examples of fragment attachments to a silica surface are shown in Figure 6.2.1.

The bonded phase obtained by derivatization can have a hydrophobic character, a polar character, and even an ionic or a mixed-type character (e.g., polar and nonpolar, or even nonpolar and ionic.) Different types of reagents and reactions can be used to derivatize the silanol groups from the silica surface; some of these are further described in this section. Examples of procedures used for the direct preparation of stationary phases without the need of starting with previously made silica (as solid support followed by derivatization) are also presented below (direct synthesis of the phase). The derivatization process can be performed on silica already shaped in particles of the desired dimensions, and it can also be performed on monoliths.

Derivatizations to Generate Bonded Silica-based Stationary Phases

Silica-based stationary phases can be obtained by attaching through covalent bonds the fragment (group) that will act as stationary

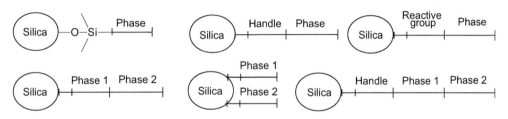

FIGURE 6.2.1 Schematic examples of fragment attachments to the silica surface.

phase on the silica xerogel surfaces. More recently, the same procedure has been successfully applied on organic/inorganic silica structures such as those having ethylene bridges (a bridged ethane-silicon hybrid or BEH-type stationary phase) or methyl groups already present in the silica structure.

For a successful reaction with the silanol groups from the silica surface, it is important that these groups are reactive. Silica surface can be activated either by physical treatment (thermal or hydrothermal) that changes the ratio of silanol and siloxane concentration of the silica surface or by chemical treatment that leads to changes in chemical characteristics of silica surface. One such procedure is the boiling of silica with acidic solutions that can generate on silica surface an adequate and dense layer of silanols and consequently of the bonded groups. A thermal treatment of the silica consists of heating at 250–400 °C before the derivatization [41].

The most common type of HPLC is reversed-phase HPLC (RP-HPLC) where the stationary phase is nonpolar. Nonpolar phases are frequently based on a derivatized silica support, although columns with a hydrophobic stationary phase can be made using other procedures. Long alkyl chains are most commonly used to form a hydrophobic surface on the porous silica skeleton. The first attempt to attach alkyl chains to a silica matrix was made using the reaction between the silanol groups on the silica surface and 1-octanol, leading to a Si-O-C bond. This reaction is indicated in 6.2.1 [42]:

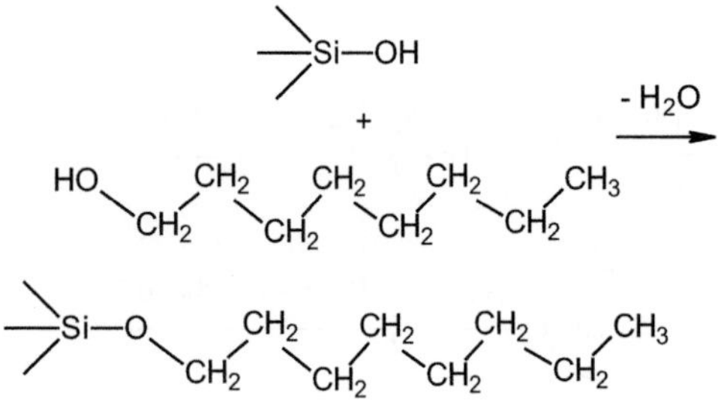

$$(6.2.1)$$

The problem with this procedure is that the Si-O-C bond obtained in a reaction of the type shown in 6.2.1 is not very stable, and procedures to obtain more stable derivatized surfaces were considered. One important desirable characteristic of HPLC chromatographic columns is the absence of the bleed from the bonded phase. This characteristic affects the column stability in time and matters during measurements, in particular when the detection is performed using a MS-type detector. The increased hydrolytic stability is one of the main features that allow minimum bleed from the column. The stability of the bonds in the stationary phase is a factor to be considered when using a specific column in a separation method. One procedure to obtain stable bonds is the use of reagents that can generate Si-O-Si bonds in which one silicon is part of the silica surface and the other is further connected to a hydrophobic moiety.

A stable Si-O-Si bond can be obtained, for example, using a monofunctional silanization reagent. One common type of reaction applied for the generation of covalently bonded hydrophobic groups to the silica surface through a Si-O-Si bond is indicated in 6.2.2 [1]:

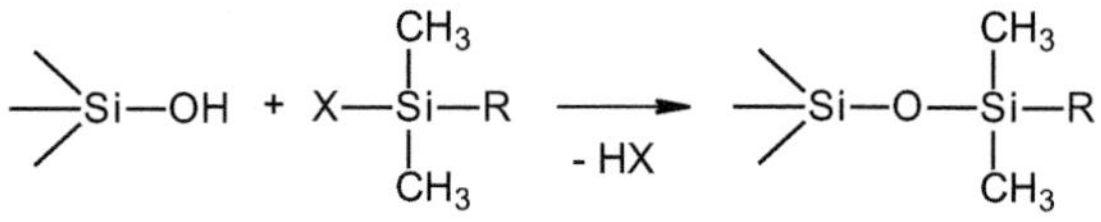

$$(6.2.2)$$

where X is most commonly −Cl, but other groups such as −OCH$_3$, −OCH$_2$CH$_3$, or even −NH$_2$ can be used, and R is typically a hydrophobic n-hydrocarbon chain such as octyl (C8) or octadecyl (C18). The derivatization of a silica surface with a monofunctional silanization reagent is typically referred to as *monomeric functionalization*. Monomeric functionalization attaches a monomolecular layer of active phase on the silica surface [43–45].

A variety of other R groups can be attached to the silica surface using this procedure, including a variety of hydrophobic groups (linear alkyl, cycloalkyl, phenyl, etc.) when reversed stationary phases are desired. However, the procedure is also applied to obtain phases containing ligands such as $-(CH_2)_3-NH_2$, $-(CH_2)_3-CN$, $-(CH_2)_3-NO_2$, and $-CH_2-CH(OH)-CH_2(OH)$, which are polar. The common reaction based on chlorosilanes as derivatization reagents needs the presence of organic bases (pyridine or triethylamine) in the reaction medium to remove HCl formed as a by-product. When alkoxydimethyl-R-silane is used as a silanization reagent, toluene is a proper solvent for the reaction. In this case, the alcohol formed as a reaction by-product should be removed from the surface because it can subsequently react with silanol groups leading to alkoxy-derivatized silica surface, which is unstable toward aqueous mobile phases. In both cases, the modified silica surface is not a homogeneous one. It may contain a significant number of underivatized silanol groups, which will further participate in the retention mechanism. The presence of free silanols after derivatization represents one of the major challenges in obtaining fully derivatized silica surfaces. Further derivatization is necessary to reduce the silanols number. A schematic description of a portion of a silica surface partly derivatized using a monofunctional reagent is shown in Figure 6.2.2. The structure shows the Si-O-Si bonds, the dimethylsilyloctyl groups, the OH groups (isolated, vicinal, geminal), as well as the crosslinking bonds.

The calculated point charges on a hypothetic molecule where the octyldimethylsilyloxy group is attached to a $Si(OH)_3$ group is shown in Figure 6.2.3. This molecule simulates a silica structure partly derivatized with dimethyloctylsilyl groups. Figure 6.2.3 shows that the

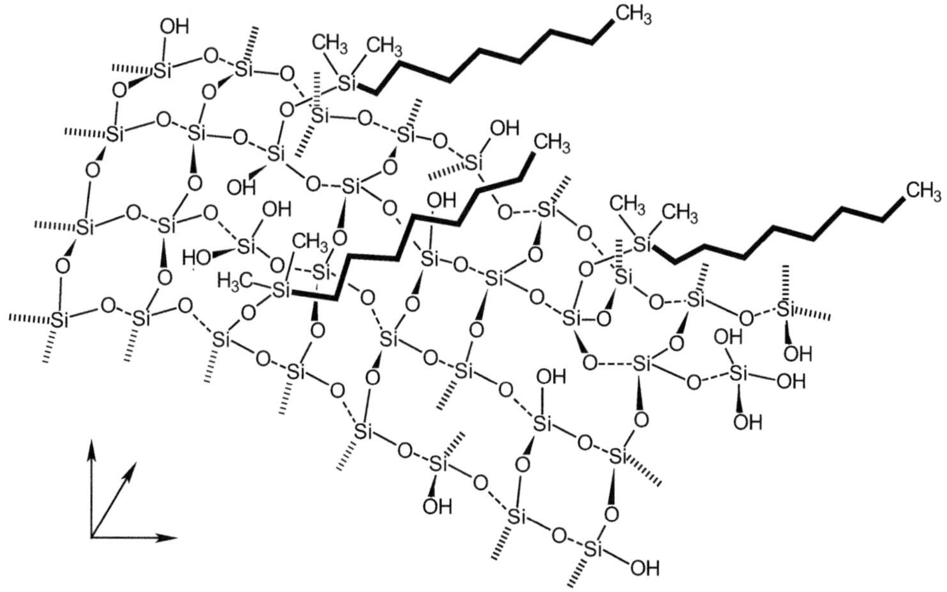

FIGURE 6.2.2 Schematic description of a portion of a silica structure partly derivatized with a monofunctional reagent, showing the Si-O-Si bonds, the dimethyloctylsilyl groups, the OH groups (isolated, vicinal, geminal), as well as the crosslinking bonds ().

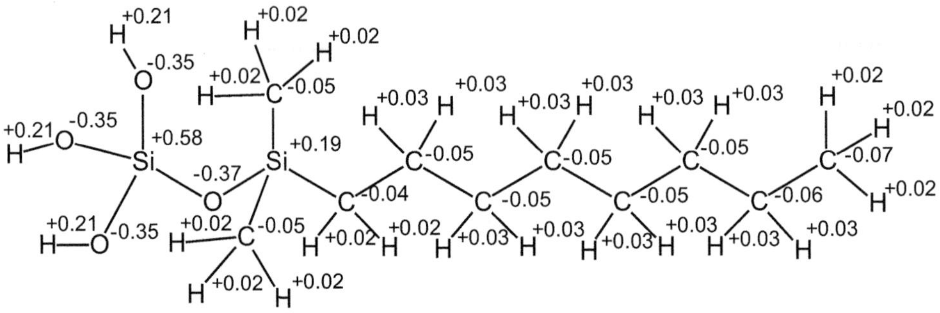

FIGURE 6.2.3 Point charges on a hypothetic molecule with octyldimethylsilyloxy group attached to a $Si(OH)_3$ unit.

long alkyl chain (C8) in the model molecule has point charges very similar to those in a long-chain hydrocarbon (see Figure 4.1.6), and very likely the charge distribution on a real bonded phase is similar to that in a long-chain hydrocarbon. On the other hand, the hydrogens from the silanols of the $Si(OH)_3$ group have a point charge of +0.21, indicating a quite strong acidic character typical for the silanols on the silica surface. The residual silanols remaining in the silica structure offer a secondary place of interactions besides the hydrophobic group. They produce peak tailing for highly polar compounds due to the interaction of the type H-bonding, dipole-dipole, or ion-dipole with different functional groups in analyte molecule. Above pH = 5, a pronounced tailing for basic compounds is observable owing to the dissociation of silanol groups and their involvment in ion-exchange-type interactions with polar or dissociated groups from analyte molecule. Even when the silanol group presence on the bonded phases can be useful for particular separations, the problem that remains is the reproducibility of their polar interactions, since the silanol groups retain water from the mobile phase and their polarity may vary depending on the composition of the mobile phase recently utilized.

The effort to reduce the number of residual silanols on the silica matrix is a common procedure in preparing chemically bonded stationary phases. This is performed by a separate procedure of blocking residual silanols by derivatization with small groups, such as trimethylsilyl, using trimethylchlorosilane or hexamethyldisilazane as reagents. This procedure is known as end-capping. The process is applied after the derivatization with the desired bonded phase is performed. Besides trimethylchlorosilane or hexamethyldisilazane, difunctional derivatization reagents are also used for end-capping. The reaction is typically performed at elevated temperatures to achieve a derivatization that is as complete as possible [46–48]. By end-capping, the carbon content of packing material does not significantly change, and thus its hydrophobic character for the reversed phase phases is kept almost constant (the carbon load % C is determined by elemental analysis).

The lack of reactivity of some silanol groups with a specific derivatization reagent is thought to be caused mainly by steric hindrance. When attempting to attach, for example, two long aliphatic chains on two vicinal or germinal silanol groups, the reaction may not occur because the reagent cannot penetrate to the reactive site. For this reason, end-capping is done with reagents having a small molecule. Another possibility of derivatizing more silanols from silica surface is through the use of reactions with di- or trifunctional reagents. These reagents have two

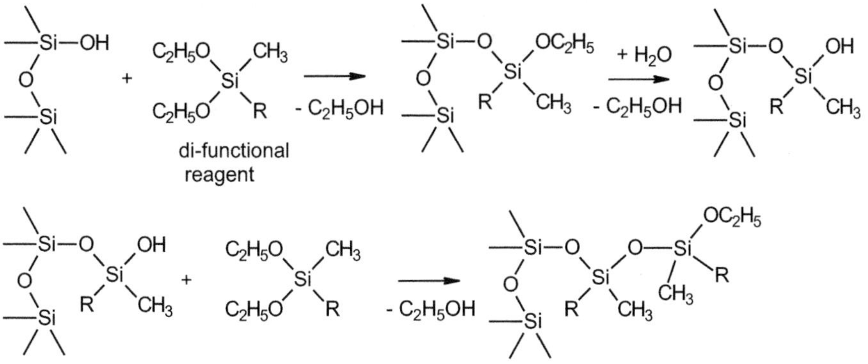

(6.2.3)

or three reactive functional groups such as ethoxy or chloro. A schematic description of the reactions of a di-functional reagent with two ethoxy groups is shown in reactions 6.2.3 for an isolated silanol. The reaction takes place by means of three-step sequences. In the first step the di-functional reagent reacts with silica to link methylalkylethoxysilyl or methyl-alkylchlorosilyl groups, which in the second step are hydrolyzed in order to substitute the chlorine atom or the ethoxy group with Si-OH, and then subjected to another derivatization reaction with di-functional silane. This sequence is repeated several times (8 to 10 times) and is schematically shown in reaction 6.2.3.

A reaction between vicinal silanols and a di-functional reagent is shown in 6.2.4.

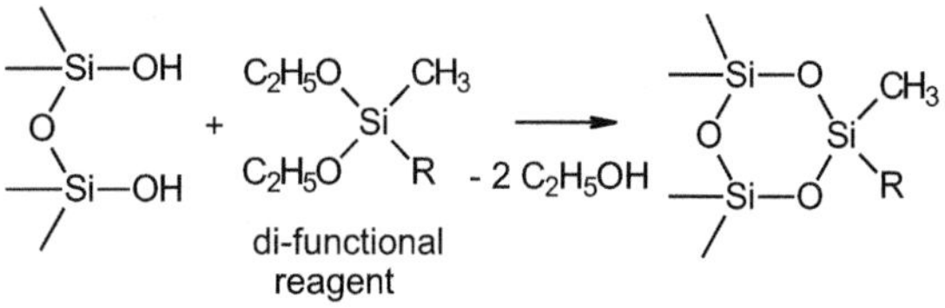

(6.2.4)

As seen from reactions 6.2.3 and 6.2.4, a di-functional reagent has better capability to react with vicinal silanols and also to give a good density of substituents for isolated silanols.

A model chain of reactions with a tri-functional reagent (triethoxysilane) is given in 6.2.5:

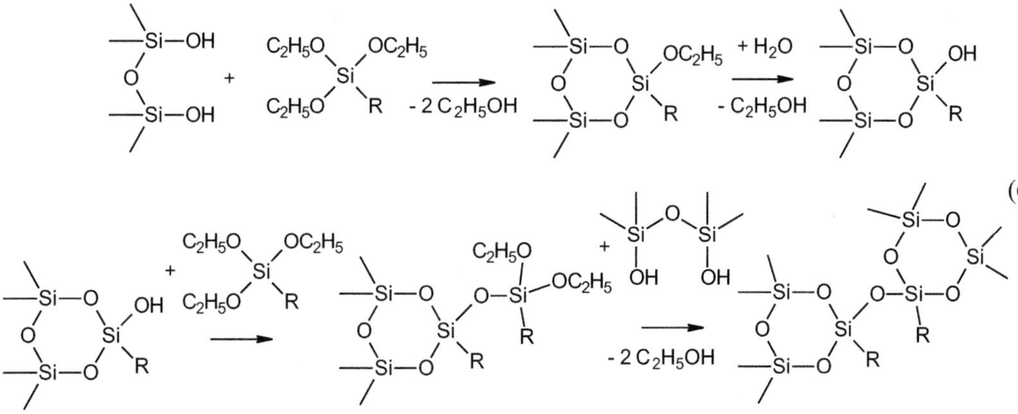

(6.2.5)

The possibility that a tri-functional reagent connects with the silica surface with all three reactive substituents is also possible, as shown in the following schematic reaction:

$$(6.2.6)$$

Derivatization with difunctional or trifunctional silanes typically leading to multiple bonds and tridimensional bonding on the silica surface is known as *polymeric functionalization*. Since reactions of the type 6.2.5 and 6.2.6 take place in an aqueous medium, they lead to a polycondensation process such that besides the direct bonding of groups O-Si-R to the silica surface, there are also aggregates of pre-reacted reagent that are attached, forming a polymeric thin layer of stationary phase bonded to the silica and containing a tridimensional structure. The polymeric functionalization attaches a multimolecular layer of bonded phase on the silica surface, differently from monomeric functionalization that attaches a monomolecular layer of active phase. Polymeric derivatization leading to a multimolecular layer of bonded phase is called vertical bonding (or *vertical polymerization*) [49]. The process is rather complex, particularly when two vicinal silanols are involved. Although this approach leads to a better derivatization yield, a significant number of silanols remain on the silica surface. Further derivatization by repeating the process or using reagents containing less bulky moieties can be performed, using, for example, trimethyl-chlorosilane or hexamethyldisilazane. Even so, not all the silanols are removed due to problems related to steric hindrance. These residual hydroxyl groups are made more inert by repeated derivatization with the reactive reagent with small-volume substituents and use of more intense reaction conditions (e.g., higher temperatures).

In contrast to vertical polymerization, *horizontal polymerization* can also be practiced with di- or tri-functional silanes. In this procedure, the tri-functional reagent is reacted with silica in anhydrous conditions. This procedure requires the use of a mixture of reagents with long R substituents (e.g., octadecyl) and short R (e.g., propyl) in order to obtain a more complete cover of the silica surface. Horizontal derivatization seems to be associated with a self-assembling process of the hydrocarbon chains that leads to a more uniform coverage of the silica surface [50]. Since in this case the layer of molecules covering the silica surface is closer to monomolecular, the carbon load achieved by horizontal polymerization is typically lower than that for vertical polymerization. Horizontal polymerization is less frequently utilized to obtain stationary phases. The resulting structures obtained by the two derivatization techniques are schematically shown in Figure 6.2.4.

In summary, according to the type of reagent used in the derivatization procedure (mono-, di-, and tri-functional silanes) and the conditions applied for derivatization (addition of water with the derivatization reagent or use of anhydrous conditions), the hydrophobic bonded phases can be classified in three categories: brush phases (monomeric), oligomeric phases, and bulk phases (polymeric). The brush phases are obtained using monochlorsilanes, such as octyldimethylchlorsilane (to attach dimethyloctylsilyl chains), or octadecyldimethylchlorsilane (to attach dimethyloctadecylsilyl chains). A brush phase is also obtained when using horizontal polymerization with reagents such as alkyltrichlorsilanes or alkyltrialcoxysilanes in anhydrous conditions. The alkyl chains in these cases are considered to stand out from the silica surface like bristles of a brush, a structure that is the result of a self-assembling

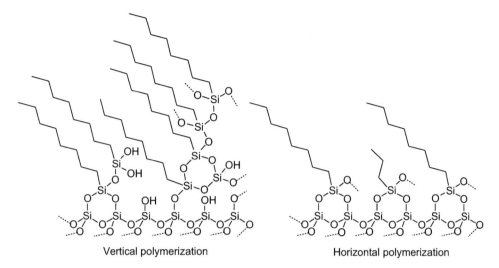

Vertical polymerization   Horizontal polymerization

FIGURE 6.2.4 Schematic illustration of vertical (polymeric) derivatization that generates a bonded polymer layer on silica surface and horizontal derivatization of silica.

process. The di-functional silanes (with limited amounts of water in the reaction) produce oligomeric bonded materials. The bulk phases are obtained from alkyltrichlorsilanes or alkyltrialcoxysilanes in vertical polymerization, which takes place in the presence of some water and leads to a complex crosslinking process that finally generates a polymeric layer covering the silica surface. Crosslinking is achieved not only involving the silanol groups from the silica surface, but also from reactions between the reagent molecules. The carbon load on the silica is in this case the highest, and bulk polymeric phases are very common.

A number of common reagents used for the derivatization of silica surfaces are shown in Table 6.2.1. These reagents may be used for attaching on the silica surface either hydrophilic or hydrophobic functionalities. Also, the same reagents used for the derivatization of silica gel particles can be used for the derivatization of silica monoliths. A much wider range of compounds than is indicated in Table 6.2.1 can be attached to the silica surface by direct reaction with the silanol groups. Some of the attached groups may act as the active bonded

phase on the silica surface. Among these groups are the alkyl, cyanopropyl, and phenyl. Other groups can be used as the active bonded phase, and also as intermediates for further derivatizations. Among such groups are glycidyl, amino, and azido.

In addition to the simple reagents used to derivatize the silica surface, more complicated compounds can be made reactive and used for reaction with the silanol groups. For example, cholesterol can react with allyl bromide to form a 3-allyl ether. In the presence of a catalyst (H_2PtCl_6), the double bond of the allyl ether reacts with the silanol groups attaching the cholesterol fragment to the silica surface (with a C3 spacer) [51].

Another method that has been reported for the modification of silica supports is based on a chlorination/organometalation two-step reaction sequence. In the first step, the silanols on the silica surface are converted to chlorides with thionyl chloride as shown in reaction 6.2.7:

$$\equiv Si\!-\!OH + SOCl_2 \longrightarrow \ \equiv Si\!-\!Cl + SO_2 + HCl$$

(6.2.7)

TABLE 6.2.1   Examples of reagents used for the derivatization of silica surface.

Reagent	Formula	Uses
Trimethylchlorosilane	$ClSi(CH_3)_3$	end-capping
Propyltrichlorosilane	$Cl_3SiCH_2CH_2CH_3$	bonded phase
Octyltrichlorosilane	$Cl_3SiCH_2(CH_2)_6CH_3$	bonded phase
Octadecyltrichlorosilane	$Cl_3SiCH_2(CH_2)_{16}CH_3$	bonded phase
Phenyltriethoxysilane	$(CH_3CH_2O)_3SiC_6H_5$	bonded phase
γ-Glicidoxypropyltrimethoxysilane	$(CH_3O)_3SiCH_2CH_2CH_2{-}O{-}CH_2{-}CH{-}CH_2$ $\diagdown\!O\!\diagup$	bonded phase
γ-Aminopropyltriethoxysilane	$(CH_3CH_2O)_3SiCH_2CH_2CH_2NH_2$	bonded phase
γ-Aminopropylmethyldiethoxysilane	$(CH_3CH_2O)_2CH_3SiCH_2CH_2CH_2NH_2$	bonded phase
γ-Aminopropyldimethylethoxysilane	$(CH_3CH_2O)(CH_3)_2SiCH_2CH_2CH_2NH_2$	bonded phase
Cyanopropyltriethoxysilane	$(CH_3CH_2O)_3SiCH_2CH_2CH_2CN$	bonded phase
Nitropropyltriethoxysilane	$(CH_3CH_2O)_3SiCH_2CH_2CH_2NO_2$	bonded phase
Dihydroxypropyltriethoxysilane	$(CH_3CH_2O)_3SiCH_2CH(OH)CH_2(OH)$	bonded phase
3-Mercaptopropyltriethoxysilane	$(CH_3CH_2O)_3SiCH_2CH_2CH_2SH$	bonded phase
3-Azidopropyltrimethoxysilane	$(CH_3O)_3SiCH_2CH_2CH_2N_3$	bonded phase
Hexamethyldisilazane	$(CH_3)_3SiNHSi(CH_3)_3$	end-capping

This step must be performed under extremely dry conditions (usually in a closed vessel purged with a dry gas such as nitrogen) because the presence of water traces leads to the reversal of the reaction, with hydroxyl groups replacing the chloride bonded to the silica and regeneration of silanols. Then the substitution chain R can be attached to the surface using a Grignard's reagent or an organolithium compound:

$$\overset{|}{\underset{/}{\equiv}}Si{-}Cl + RMgCl \longrightarrow \overset{|}{\underset{/}{\equiv}}Si{-}R + MgCl_2 \qquad (6.2.8)$$

$$\overset{|}{\underset{/}{\equiv}}Si{-}Cl + RLi \longrightarrow \overset{|}{\underset{/}{\equiv}}Si{-}R + LiCl \qquad (6.2.9)$$

Those structures that have R directly bonded to the siloxane matrix are more stable than structures based on substitution of H from silanol, which could be an advantage for the phases bonded by this procedure. However, due to the specific reaction condition from the first step and the formation of salts as by-products, the materials obtained using silica chlorination are less known as commercial packing materials for HPLC applications.

A common procedure for preparing more complex desired stationary phases is to further react a molecular fragment already attached to the silica surface with another reagent. The initially attached functionality is either itself transformed into the desired functionality or it can be used as a reactive "anchor" to attach a desired fragment that will form the stationary phase.

One example of changing one attached group into another group is shown in 6.2.10 for the case of obtaining a strong cation exchange stationary phase:

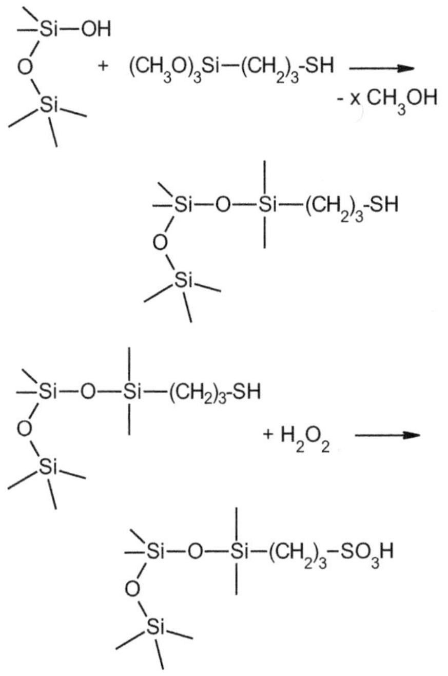

(6.2.10)

The use of fragments that are already attached to the phase and have a reactive group able to continue a coupling for obtaining more complicated stationary phases is much more widespread than the functionality change. The procedure can be used, for example, to obtain hydrophobic phases with embedded polar groups, phases having macrocycles attached to silica, special chiral phases (this procedure will be discussed and exemplified later in this chapter for particular types of stationary phases). The two-step approach for the synthesis of polar-embedded phases is common [52–54]. This approach is based on derivatizing the silica surface with a group such as aminopropyl, followed by further reactions of the amino silica. A typical reaction for preparation of amino silica is shown in reaction 6.2.11:

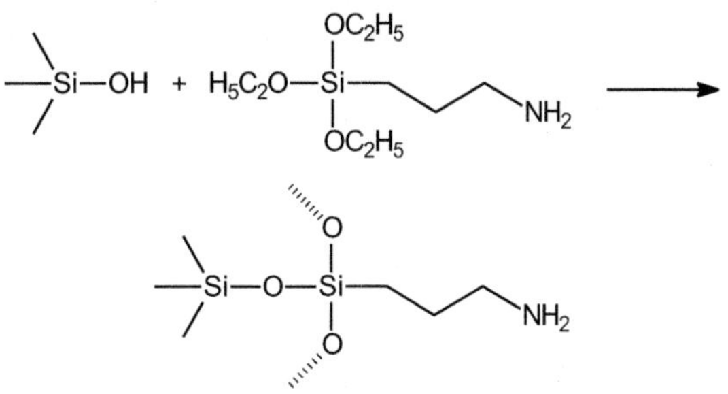

(6.2.11)

The amino silica is further reacted with an acid chloride or an isocyanate of a long carbon chain carboxylic acid (e.g., with R = C8, C18, etc.), as exemplified in the reaction 6.2.12 for an acid chloride:

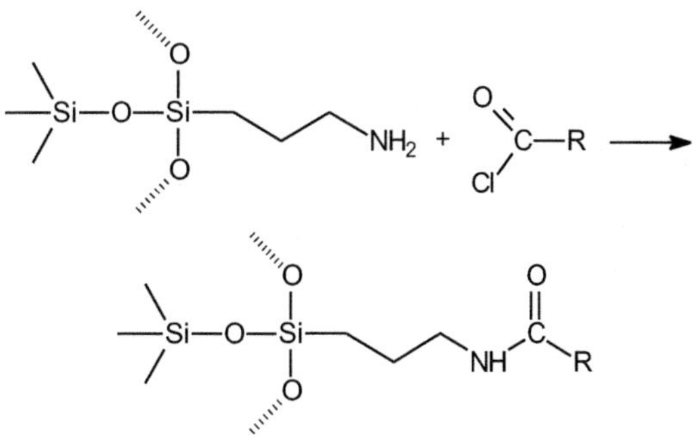

(6.2.12)

The reaction with an isocyanate is given by 6.3.13:

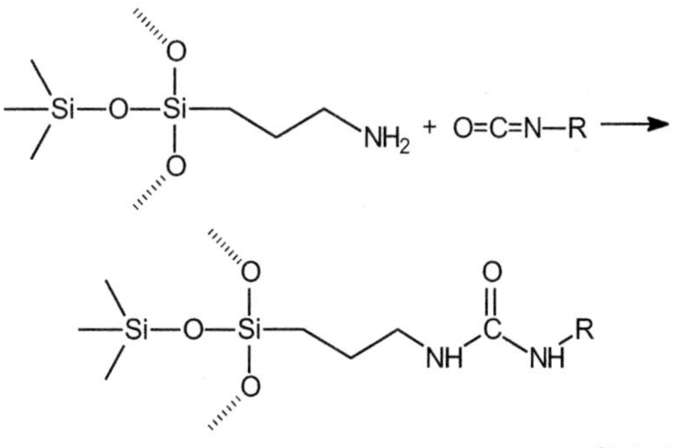

(6.2.13)

The preparation of octadecyl phases with polar-embedded groups was obtained, for example by a reaction similar to 6.2.12 where R is octadecyl [55].

The two-step derivatization can also start with a glycidyl group on silica surface. An example of such a reaction is given by 6.2.14:

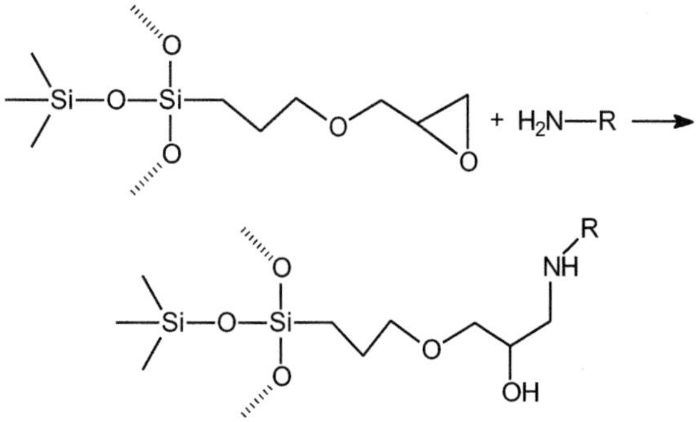

(6.2.14)

Silica derivatized with azido groups can participate in Huisgen cycloadditions (Click chemistry) as shown in reaction 6.2.15:

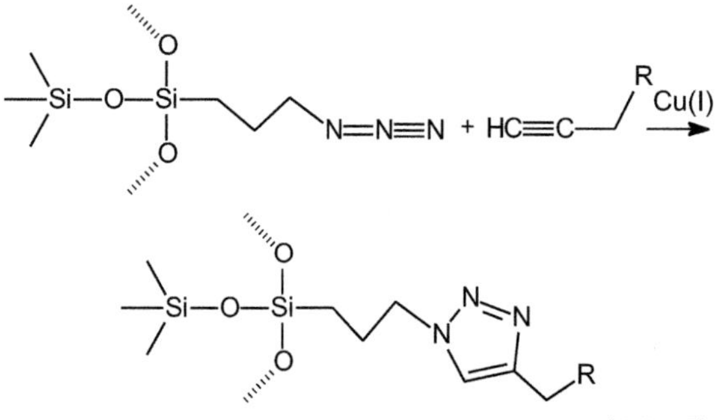

(6.2.15)

Numerous other phases obtained from silica surface derivatization are reported in the literature [56].

Direct Synthesis of Silica Materials with an Active Bonded Phase Surface

Derivatized silica surfaces also can be obtained directly through hydrolysis of the appropriate reagents, without having bare active silica prepared as a first step, followed by further derivatization. For example, the hydrolysis of tetramethyl orthosilicate (tetramethoxysilane or TMOS) in the presence of octyltrimethoxysilane in an acidic medium generates a silica gel that contains octyl groups attached to the silica structure. The same procedure can be applied to obtain silica having C18 chains attached on its surface. Another example of direct reaction for obtaining a functionalized silica surface is the preparation of propylsulfonic acid functionalized mesoporous silica. This material can be prepared in a unique step starting with tetraethyl orthosilicate, 3-mercaptopropyltrimethoxysilane, and H_2O_2, in strong acidic conditions [57].

Direct hydrolysis of a mixture of silane derivatives (co-hydrolysis) may generate the desired functional groups on the silica surface, but it can also make the process more difficult to control particularly in relation to the size and shape of stationary phase particles [58]. For this reason, co-hydrolysis has been used more frequently for preparation of monolith-type materials where the control of particle size is not a critical step. The procedure leads to so-called organic-inorganic monoliths. One example of co-hydrolysis reaction is given in 6.2.16:

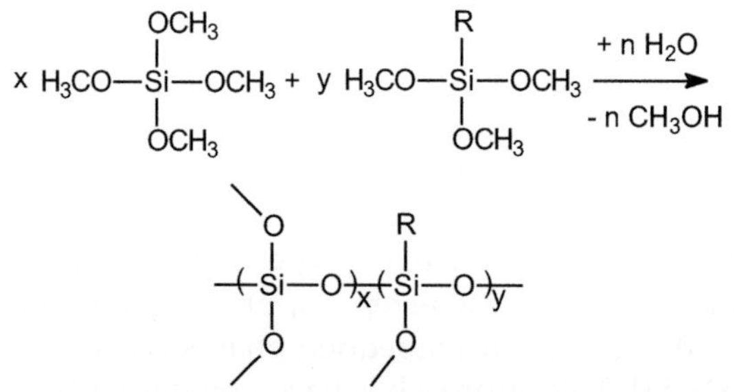

(6.2.16)

In reaction 6.2.16, the radical R can be octyl or octadecyl, but also other radicals. The procedure

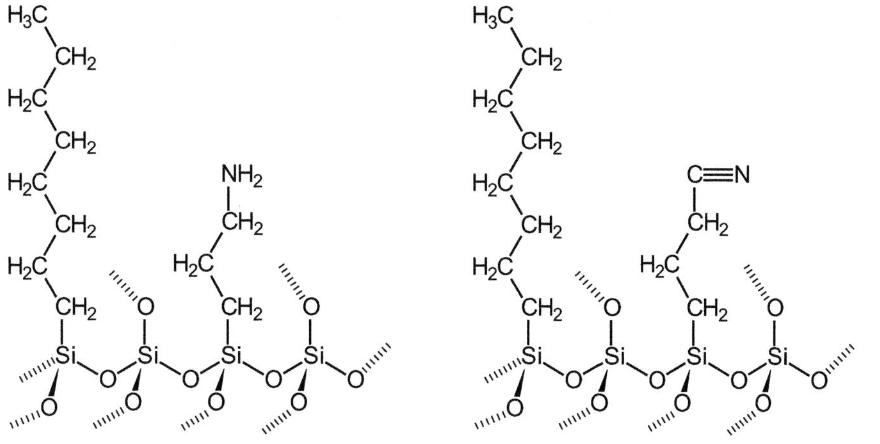

FIGURE 6.2.5 Schematic structure of a mixed mode phase with octyl and amino groups and a phase with octyl and cyano groups.

is also utilized for preparing monoliths with mixed-mode functionalities. For example, a mixture of tetramethoxysilane (TMOS), aminopropyltrimethoxysilane, and octyltrimethoxysilane can be hydrolyzed in acidic medium to generate a monolith with the active surface covered with both octyl and aminopropyl groups. In a similar manner, a mixed-mode phase having octyl and cyano groups can be prepared, starting with tetramethoxysilane (TMOS), cyanopropyl-trimethoxysilane, and octyltrimethoxysilane [59]. Both structures are schematically shown in Figure 6.2.5.

Derivatization of Silica Hydride Supports

Silica hydride as a support material for HPLC has the advantage of reduced presence of silanol groups. Besides bare silica hydride, functionalized silica hydride is also used as a stationary phase. The attachment by covalent bonds of different functionalities of the silica hydride support is typically performed using a hydrosilation reaction. This reaction can be schematically written as shown in 6.2.17:

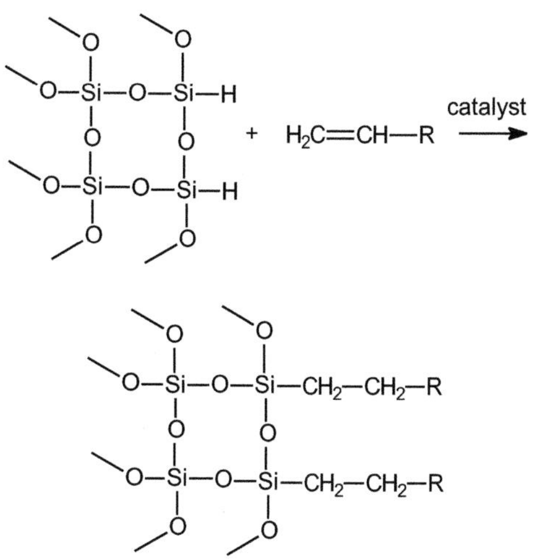

$$(6.2.17)$$

In reaction 6.2.17, the radical R can be selected to form a hydrophilic, hydrophobic, an ion exchange, or a chiral phase. The reaction takes place in the presence of a catalyst that can be $H_2[PtCl_6]$ or a free radical initiator. Double attachment of the bonded groups can be obtained when using alkynes in the hydrosilation as shown in the reaction 6.2.18 [60]:

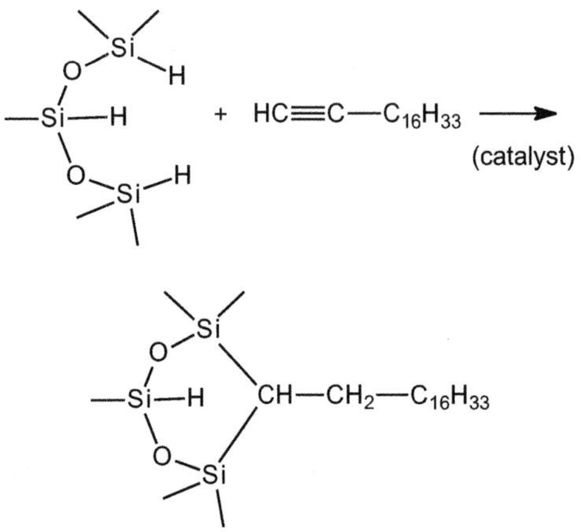

$$(6.2.18)$$

The same procedure can be used to attach other functional groups, such as cyano, amino, and fluorinated hydrocarbons [61−64]. Silica hydride phases can also be the subject of end-capping.

The functionalities bound to silica hydride are not limited to simple groups. For example, a very stable structure based on Si-C bonds can be obtained from the reaction of hydride silica with vinylbenzo-18-crown-6, in presence of $H_2[PtCl_6]$, leading to the following structure of the stationary phase:

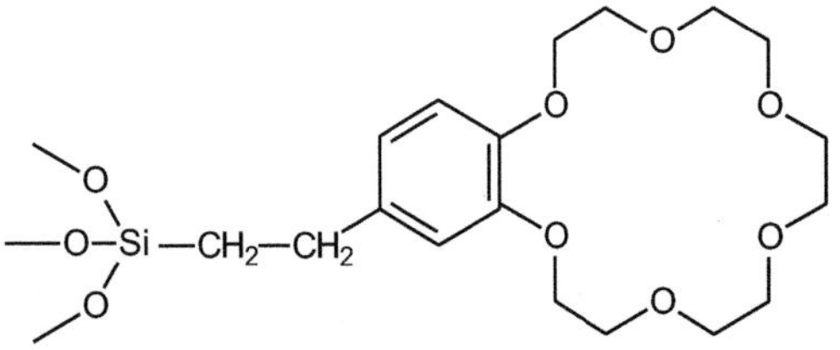

This type of phase can be used in ion-exclusion separations.

Derivatization of Presynthesized Organic Polymers

Various procedures were applied for the derivatization of a preformed polymer in order to obtain the desired functionalities on the surface of material forming the stationary phase. Specific derivatizations were performed on vinyl-type polymers that have attached on the polymeric backbone specific reactive groups, such as glycidy or alkyne. Such reactions were applied for the synthesis of particle-type phases as well as monoliths. One example of synthesis starts with the polymerization of ethylene dimethacrylate with propargyl acrylate as shown in 6.2.19:

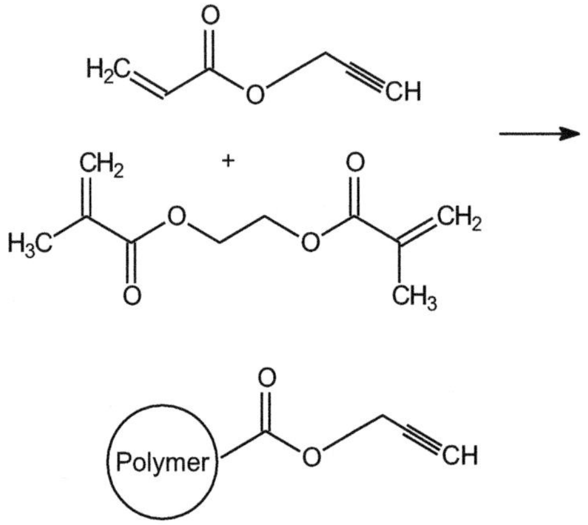

$$(6.2.19)$$

The polymer with alkyne groups is further derivatized using a copper(I)-catalyzed (3 + 2) azide alkyne cycloaddition that is shown schematically in reaction 6.2.20:

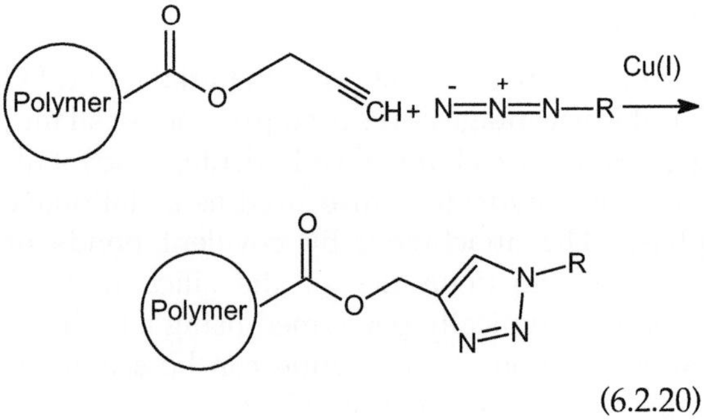

$$(6.2.20)$$

The radical R in reaction 6.2.20 can be, for example, a long hydrocarbon chain such as C18 [65].

Other examples include the synthesis of cation-exchange resins, or anion-exchange resins can be obtained by such derivatizations. The polymeric backbone of organic ion exchangers is usually obtained either by polycondensation or by polymerization. A common polycondensation reaction used to obtain polymeric resins is that between phenol

leads to a tridimensional thermorigid polymer. The inert resin can be modified by direct sulfonation into a strong cation exchanger.

The polycondensation in the presence of acid catalysts may leave a significant number of free $-CH_2OH$ groups. These can be further derivatized with $SOCl_2$, and the $-CH_2OH$ groups are changed into $-CH_2Cl$ groups. Upon treatment with $N(CH_3)_3$, the resin can be changed into a strong anion-exchange material as shown in the following reaction:

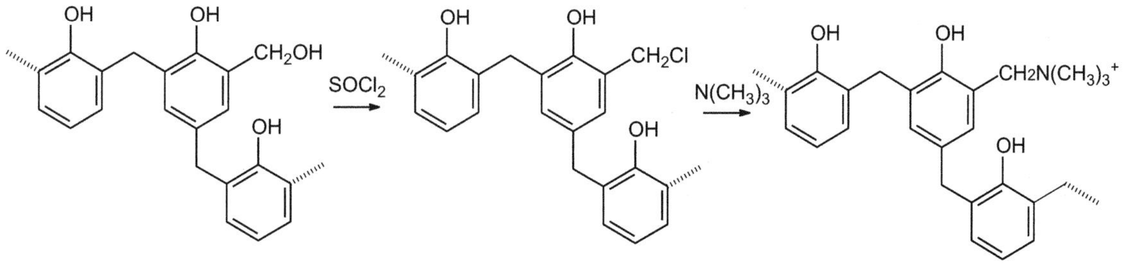

and formaldehyde. The reaction is written in 6.2.21:

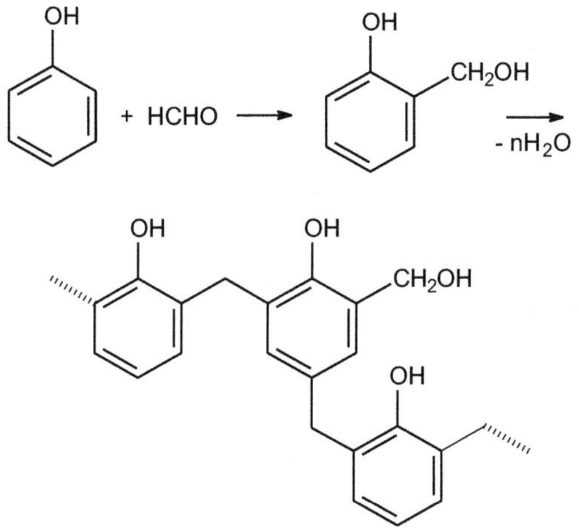

$$(6.2.21)$$

Both acid and basic catalysts can be used in this reaction. In the presence of basic conditions, the reaction takes place more completely and usually

Various other organic polymer-based stationary phases were synthesized. However, silica-based stationary phases have several advantages and are much more commonly used in HPLC [66].

Synthesis of Organic Polymers with Active Groups

Direct synthesis of polymers with active groups present in their molecular structure is a convenient procedure for obtaining polymer-based stationary phases. For this purpose, the polymerization is done starting with specific monomers that have the desired group. Sulfonated styrene can be used, for example, in a copolymerization reaction with divinylbenzene (usually in the presence of benzoyl peroxide) to form a polymer with sulfonic groups. Copolymerization between acrylic acid or methacrylic acid and divinylbenzene generates a resin with carboxyl

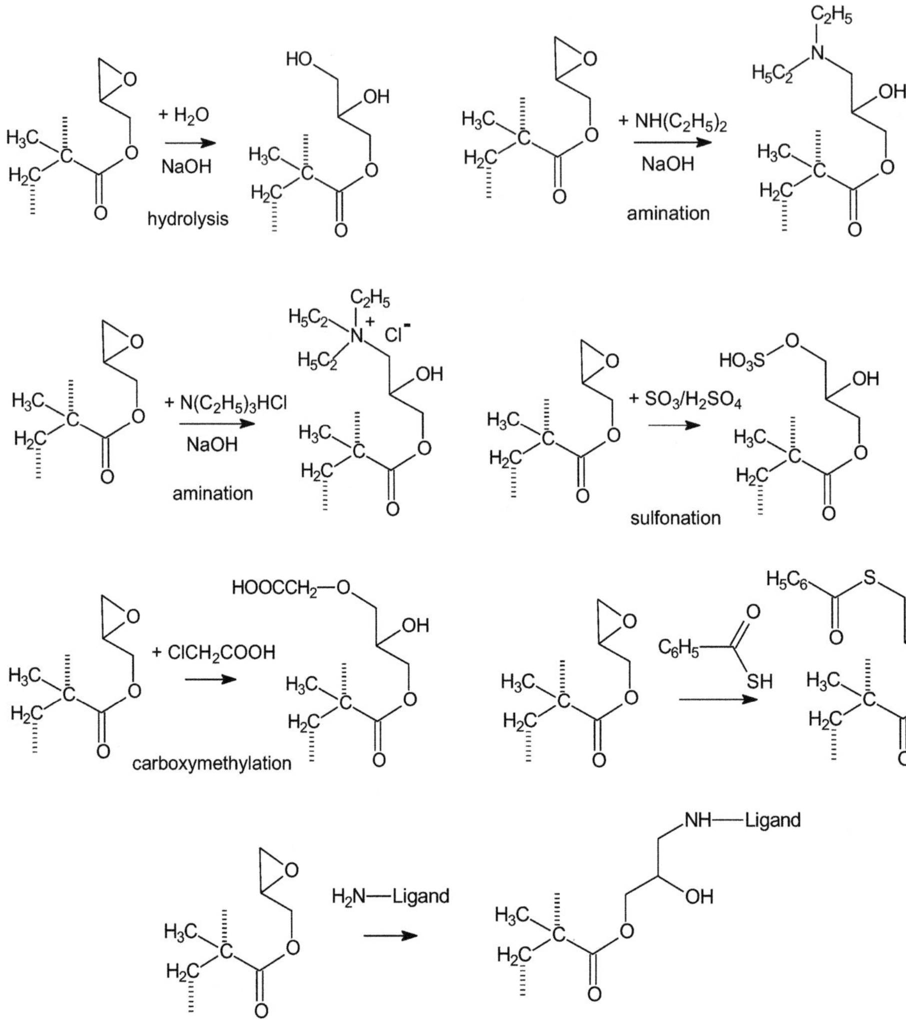

FIGURE 6.2.6 Various coupling reactions for polymers having glycidyl groups.

groups, as shown in reaction 6.2.22 (for meth-acrylic acid):

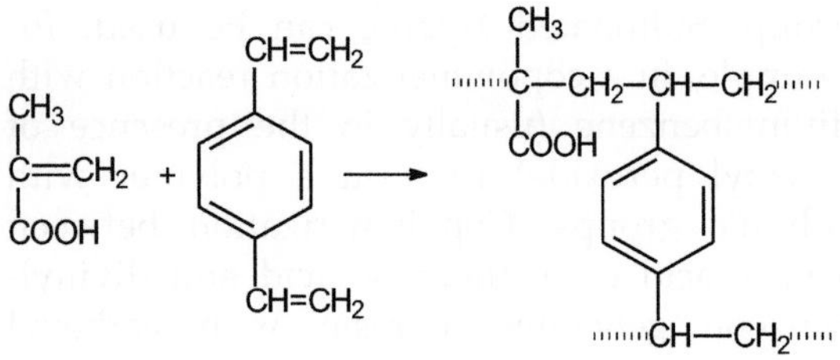

This resin can be used as a weak cation-exchange resin.

Synthesis of Organic Polymeric Monoliths with Active Functionalities

A number of polymeric monoliths that can be further derivatized with desired functional groups were synthesized and utilized as stationary phases in HPLC [37]. A convenient procedure to perform such derivatizations starts with a polymer having glycidyl groups. Such polymers can be further derivatized using

(6.2.22)

various reactions including hydrolysis, amination, alkylation, and sulfonation. Some of these
reactions are shown in Figure 6.2.6. Other
reagents besides those indicated in this figure
were used to react with the glycidyl groups
bonded to a polymeric backbone. Bio-
functionalization of the epoxy or hydroxyl-
bearing monoliths was also reported in the literature [67]. Reactions involving intermediate
derivatization using a diamine and glutaraldehyde, carbonyldiimidazole, disuccinimidyl
carbonate, and the like, were used to bind
ligands on the polymeric monolith.

Silica Covered with a Bonded Active Polymer

The stationary phases made with a polymer covering silica particles and at the

with the mobile phase. This type of phase
can be obtained, for example, when the
polymer covering the silica particles is
poly(methyloctyl-siloxane) or poly(methyltetradecylsiloxane). Covalent bonding on the
silica surface can be achieved by using
a derivatization procedure with a reactive
polymer. In this case, the silica is first derivatized, for example, with γ-aminopropyltrie-
thoxysilane. This reaction is followed by
derivatization with a polymer containing
specific reactive groups. As an example, silica
can be coated with poly(N-isopropylacryla-
mide), following the chain of reactions indicated in 6.2.23:

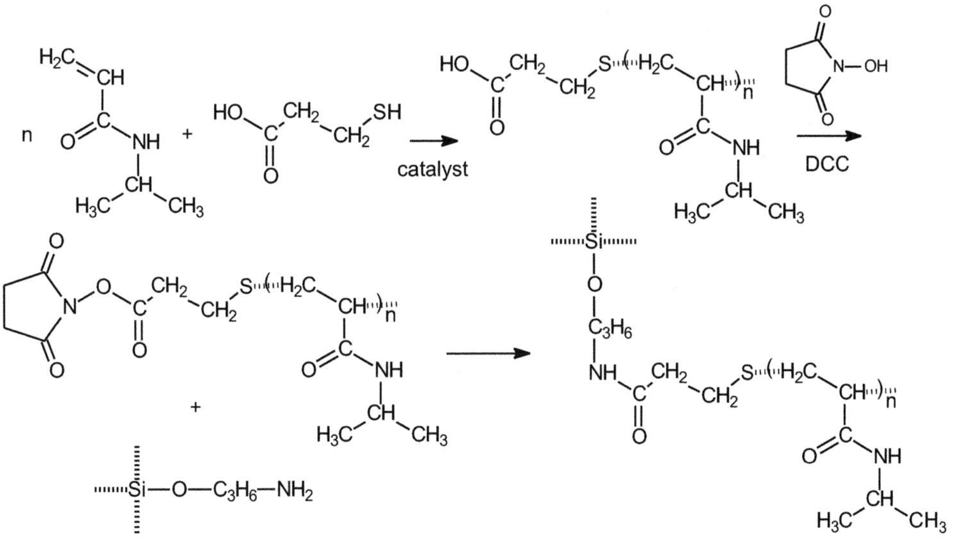

$$(6.2.23)$$

same time connected to the silica surface by
covalent bonds were previously described in
Section 6.1. Some of these phases can be
synthesized independently as a polymer
with the desired ligands in its structure and
then bonded to a silica backbone. The silica
will provide the mechanical resistance and
possibly a larger surface area of contact

The example of poly(N-isopropylacryla-
mide) was chosen because this polymer also
shows a change in hydrophilic-hydrophobic
character with temperature. The polymer
undergoes a coil-globule transition at its
lower critical solution temperature in water.
On the silica surface, at lower temperature
(15 °C) the polymer is hydrated and the

grafted stationary phase shows a weaker hydrophobic character. At higher temperatures (40 °C), the polymer loses water and covers the silica surface, the stationary phase showing a more accentuated hydrophobic character [68].

6.3. PROPERTIES OF STATIONARY PHASES AND COLUMNS

General Comments

The variety of HPLC techniques and the rapid evolution in technology that is taking place in this field make it difficult to describe specific properties desirable for stationary phases and columns. The properties of a column that can be considered very good for a particular application can be seen as inadequate for more demanding ones. While a considerable proportion of users of HPLC techniques are still working with relatively old technologies and columns, new developments such as the introduction of UPLC, high-temperature HPLC, micro-HPLC systems, columns with small particles (e.g., 1.7 μm diameter), columns with superficially porous particles (core-shell), or monolithic columns are increasingly more common [69]. Part of the delay in adopting the new improvements in HPLC is caused by some disadvantages (or perceived disadvantages) of the new technologies, such as lower reliability of methods that maximize specific parameters (e.g., working backpressure of the equipment), lower capability of handling "dirty" samples, necessity to use more elaborate sample cleanup steps, requirements for strict maintenance, and higher cost of the equipment. For this reason, the choice of a specific separation or column is highly dependent on individual conditions of the application, and only general aspects of the problem are discussed in this section.

Packing of the Stationary Phase in the Column and Flow Direction

The stationary phase is used to fill the chromatographic column (see Section 1.4 for some details regarding the empty column). Compact, uniform packing of the chromatographic column with the stationary phase is important since it affects the column plate number. When using high pressures with a specific column, the stationary phase should not collapse mechanically (and should not change its volume). If the volume of the stationary phase shrinks, a void volume can be formed at the head of the column, significantly affecting the column plate number. A denser bed of the stationary phase may also affect the backpressure under which the column must be operated, influencing chromatographic conditions. Also, for certain types of stationary phases, the separation itself can be drastically degraded if the stationary phase collapses.

Different procedures of packing the columns with the stationary phase are used, depending on the stationary phase. Most commonly, the stationary phase is introduced in the column as a slurry in a specially selected liquid that allows the formation of a homogeneous suspension of the stationary phase and hinders the particle aggregation. For this reason, the solvent used for preparing the slurry has to be good at wetting and dispersing the stationary phase. The particle concentration in the slurry is typically between 7 and 15%. For the case of rigid particles, either axial compression or radial compression can be applied to generate a dense bed of particles. Slurry packing involves the use of high pressure (usually 50% higher than the maximum pressure at which the column will be used) to push a dilute slurry of stationary phase through the column. The maximum pressure at which a column can be used depends on the particle strength (resistance to crushing) and may range between 9000 and 15,000 psi (600 to 1000 bar) for silica-based columns. Rigid polymeric particles typically stand lower pressure than silica-based ones (up to about 5000 psi). Soft gel columns must be used at even lower pressures, and they may be packed using gravity-based slurry sedimentation. More information on HPLC column packing can be found in the literature [70]. The filling of the chromatographic column with a slurry, deposits the particles of stationary phase in a specific arrangement. This leads to the recommendation of a unique direction of flow in the column during its use (the same in which the stationary phase was deposited). However, this restriction is less important for columns with very small particles that withstand high pressures. The packing procedure of the chromatographic column may not be of much interest when the columns are obtained from manufacturers, but the *maximum pressure* of column utilization indicated by the manufacturer is a parameter that should be noted and should never be exceeded.

Physical Properties of the Stationary Phase

In addition to other stationary phase properties such as the nature of the bonded phase, the physical properties of the stationary phase are also very important for their utilization. One of the important physical characteristic of the stationary phase is the dimension of phase particles, with smaller particles leading to smaller theoretical plate height H (and larger N). Smaller particles also lead to higher column backpressure, and this inconvenience is circumvented by the use of UPLC. Some of these physical properties are included in Table 6.3.1. Procedures for achieving some of specific properties indicated in Table 6.3.1 are discussed in Section 6.1. Significant effort also has been made to characterize the phase once obtained.

Physical Dimensions of the Chromatographic Column

Depending on the type of application, a variety of column dimensions are used in HPLC. The choice of column dimensions is based on certain requirements of the analysis, in particular related to the sample load. In analytical HPLC, the sample load is preferred to be low, but this load depends on the detector sensitivity, which limits the minimum amount

TABLE 6.3.1 Some physical properties of stationary phases.

Property	Range
Nature of phase support	Silica, organic/inorganic silica (ethylene bridged), polymer-coated silica, hydride silica, graphitic carbon, zirconia, organic polymers
Phase structure	Particles, monolithic
Particle shape	Irregular or spherical
Particle size	1 μm to 20 μm (Particles below 2.5 μm are typically used in UPLC)
Particle type	Porous, core-shell, pellicular
Particle size distribution	Narrow (1−2% size variation) to larger distributions used e.g in SEC
Porosity type	One type or multiple type of channels
Porosity (for silica)	50 Å to 4000 Å see Table 6.1.1. (Type A, Type B, Type C porosity)
Purity (for silica)	High purity (Type B purity), medium (Type A purity)
Surface area (for silica)	50 m^2/g to 500 m^2/g

TABLE 6.3.2   Classification of HPLC columns based on dimensions and purpose [71].

Type	Inner diameter (mm)	Length (mm)	Typical flow rate (mL/min)	Void volume (mL)	Sample loading
Preparative	>25	300, larger	>20	>50	>25 mg
Semi-preparative	10	250, larger	5–10	>5	10–20 mg
Conventional	3, 4.6	50, 100, 150, 250	0.5–2	>1	50–200 µg
Narrowbore	2, 2.1	50, 100, 150, 250	0.2–0.5	0.2	20–100 µg
Microbore	1, 1.7	50, 100	0.05–0.1	<0.1	<5 µg
Micro LC-capillary	<0.5	50,100	1–10 µL/min	10–20 µL	1 µg
Nano LC-capillary	<0.1	50	<1 µL/min	0.1–1 µL	<0.1 µg

of analyte that can be injected. The upper limit of the amount of sample to be loaded on the column is limited by the amount of stationary phase and its capacity to retain the solute. Approximate formulas were developed for estimating maximum column load (see rel. 9.2.8). Other factors that determine the dimension of the column include available equipment and required column resilience. A classification of columns based on their dimension is indicated in Table 6.3.2 [71]. The table also notes the approximate sample loading for the column. Except for preparative and semipreparative columns (used e.g. in flash chromatography), the other types of columns listed in Table 6.3.2 are used (mainly) for analytical purposes. Industrial preparative columns can be very large (compared to the dimension suggested in this table), and sample loading can be significantly higher than mg.

Types of Stationary Phases and USP Classification

Depending on the type of chromatography, a considerable number of stationary phases types were differentiated. Along with the classification of HPLC types given in Table 1.2.1, a similar classification can be indicated for the stationary phases. This classification is given in Table 6.3.3 and follows the same principles as the classification given in Table 1.2.1 (the principle of separation except that the classification of the HPLC types also included the type of mobile phase). A few examples of columns are also given in Table 6.3.3.

Although the stationary phases can be classified based on their main characteristics, the concept of "type" is broad, and a range of properties can be present for phases of the same type. As an example, the columns used in reversed-phase chromatography (RP-HPLC) are characterized by their hydrophobic character. Columns with the "hydrophobicity" within a wide range are available, starting with cyanopropyl stationary phases that have an intermediate character between polar and nonpolar (hydrophobic), and ending with C30 phases (having attached to the silica a hydrocarbon chain with 30 carbon atoms) that are highly hydrophobic. This range of a specific character (such as hydrophobicity) is essential in selecting a column for obtaining an optimum separation. The choice of a specific type of column (from a range of columns available), combined with the selection of a solvent with certain properties (e.g., polar character of the solvent used in RP-HPLC), allow the fine tuning of a separation.

Other, more detailed classifications of stationary phases are reported in the literature. One example is the classification proposed by US Pharmacopeia (USP) [72]. This classification

TABLE 6.3.3 List of several types of stationary phases.

Types of stationary phases	Type of column	Active groups
1) Hydrophobic (non polar, widely used in RP-HPLC)	Reversed phase bonded on silica non-polar	C4, C8, C18, endcapped C18, other n-alkyl (e.g. C30), mixed alkyl, phenyl, diphenyl, mixed alkyl-aryl, cyclohexyl, fullerene, carbon (graphytized), etc.
1) Hydrophobic	Reversed phase polymeric	Divinylbenzene (DVB), polystyrene-divinylbenzene (PS-DVB), PS-DVB-methacrylate, DVB/methacrylate, polymethacrylate, etc.
1) Hydrophobic	Reversed phase polymer-coated	Polybutadiene on silica, polybutadiene on alumina.
1) Hydrophobic	Reversed phase with polar embedded on silica	C18-carbamate, C8-carbamate, alkylamide, C8-amide, C14-amide, C18-amide, C16-sulfonamide, phenyl ether
1) Hydrophobic	Fluorinated phase	Pentafluorophenyl, perfluoroalkyl, etc.
2) Weakly polar	Reversed phase bonded with some polar character	Propyl-CN, fluoroalkyl, pentafluorophenyl, etc.
3) Normal phase (polar) 4) HILIC (polar)	Bare silica	Silanol
	Bonded phase	Amino, diol, zwitterionic, polyamine, etc.
5) Cation exchange (ionic) 6) Anion exchange (ionic) 7) Ligand exchange 8) Ion-mediated	Bonded on silica	-Sulfonic, -carboxyl, -ammonium quaternary, -NR$_3$X (R = CH$_3$, C$_2$H$_5$, X = Cl, OH), mixed mode anoin/cation, metal ion loaded.
5) Cation exchange (ionic) 6) Anion exchange (ionic) 7) Ligand exchange 8) Ion-mediated	Polymeric	-Sulfonic, -carboxyl, -ammonium quaternary, -NR$_3$X (R = CH$_3$, C$_2$H$_5$, X = Cl, OH), mixed mode anoin/cation, metal ion loaded.
9) Gel filtration 10) Gel permeation	Silica gel based	Silica gel
9) Gel filtration 10) Gel permeation	Cross-linked polymers	Polystyrene/divinylbenzene, vinyl alcohol copolymers, polyamide, poly(hydroxymethacrylate), etc.
11) Displacement	Special	
12) Bioaffinity, affinity	Numerous	
13) Chiral	Bonded on silca	Chiral substituents such as (S)-valine, other bonded chiral selectors on silica.
13) Chiral	Polymeric	Dextrans, cellulose, proteins, etc.
14) Other	Hydride, etc.	Hydride, etc.

TABLE 6.3.4   USP classification of HPLC chromatographic columns [73]*.

Code	Description	Examples
L1	Octadecyl silane (C18) chemically bonded to porous silica or ceramic microparticles, 1.5 to 10 μm in diameter, or a monolithic rod.	Over 250 columns such as: Luna C18(2), Luna C18(2)-HST, Gemini C18, Synergi Hydro-RP, Onyx C18 (M), Aquity UPLC BEH C18, Aquity UPLC Shield RP 18, Atlantis T3, μBondapack C18, Nova-Pak C18, Symmetry C18, ABridge C18, XTerra MS C18
L2	Octadecyl silane (C18) chemically bonded to silica gel of a controlled surface porosity that has been bonded to a solid spherical core, 30 to 50 μm in diameter.	Bondapak Prep C18, etc.
L3	Porous silica particles, 1.5 to 10 μm in diameter, or a monolithic silica rod.	Luna Silica(2), Aquity UPLC BEH HILIC, Atlantis HILIC Silica, Onyx Si (M), SunFire Silica, XBridge HILIC, Zorbax SIL, etc.
L4	Silica gel of controlled surface porosity bonded to a solid spherical core, 30 to 50 μm in diameter	Porasil Prep Silica
L5	Alumina of controlled surface porosity bonded to a solid spherical core, 30 to 50 μm in diameter.	
L6	Strong cation-exchange packing: sulfonated fluorocarbon polymer-coated on a solid spherical core, 30 to 50 μm in diameter.	
L7	Octyl silane (C8) chemically bonded to totally porous silica particles, 1.5 to 10 μm in diameter, or a monolithic silica rod.	Luna C8(2), Aquity UPLC BEH C8, Nova-Pak C8, Resolve C8, SunFire C8, Symmetry C8, XBridge C8, XTerra MS C8, XTerra RP8, Onyx C8 (M), Nucleosil C8, Zorbax C8, etc.
L8	An essentially monomolecular layer of aminopropylsilane (NH2) chemically bonded to totally porous silica gel support, 3 to 10 μm in diameter.	Luna 10 μm NH2, μBondapak NH2, Waters Spherisorb NH2, Zorbax NH2, etc.
L9	Irregular or spherical, totally porous silica gel having a chemically bonded, strongly acidic cation-exchange coating, 3 to 10 μm in diameter.	Partisil 10 μm SCX (I), Spherisorb SCX, Luna 10 μm SCX, Zorbax SCX
L10	Nitrile groups (CN) chemically bonded to porous silica particles, 3 to 10 μm in diameter.	Luna CN 100 Å, Capcell CN UG, μBondapak CN, Nova-Pak CN, Resolve CN, Waters Spherisorb CN
L11	Phenyl groups (C6H5) chemically bonded to porous silica particles, 1.5 to 10 μm in diameter.	Synergi Polar-RP, Luna Phenyl-Hexyl, Gemini C6-Phenyl, Prodigy PH-3 , Aquity UPLC BEH Phenyl, μBondapak Phenyl, XBridge Phenyl, XTerra Phenyl, etc.
L12	Strong anion-exchange packing made by chemically bonding a quaternary amine to a solid silica spherical core, 30 to 50 μm in diameter	AccelPlus QMA
L13	Trimethylsilane (C1) chemically bonded to porous silica particles, 3 to 10 μm in diameter.	Develosil TMS-UG (C1) 130 Å, TSKgel TMS-250, Waters Spherisorb C1

(Continued)

TABLE 6.3.4 USP classification of HPLC chromatographic columns [73]*. (*cont'd*)

Code	Description	Examples
L14	Silica gel having a chemically bonded, strongly basic quaternary ammonium anion-exchange coating, 5 to 10 μm in diameter.	Partisil 10 μm SAX (I), PartiSphere 5 μm SAX, Waters Spherisorb SAX
L15	Hexylsilane (C6) chemically bonded to totally porous silica particles, 3 to 10 μm in diameter.	PhenoSphere C6, Waters Spherisorb C6
L16	Dimethylsilane (C2)chemically bonded to totally porous silica particles, 5 to 10 μm in diameter.	Maxsil RP2 60 Å (I), Lichrosorb RP2, etc.
L17	Strong cation-exchange resin consisting of sulfonated cross-linked styrene-divinylbenzene copolymer in the hydrogen form, 7 to 11 μm in diameter.	Rezex RHM Monosaccharide, IC-Pak Ion Exclusion, IC-Pak Cation, Shodex RSpak DC-613, Rezex ROA
L18	Amino and cyano groups chemically bonded to porous silica particles, 3 to 10 μm in diameter.	Partisil PAC (I)
L19	Strong cation-exchange resin consisting of sulfonated cross-linked styrene-divinylbenzene copolymer in the calcium form, 9 μm in diameter.	Rezex RCM, Rezex RCU, Sugar-Pak 1, Shodex SC-1011
L20	Dihydroxypropyl groups chemically bonded to porous silica particles, 3 to 10 μm in diameter.	Luna HILIC Shodex PROTEIN KW-800 series, TSKgel QC-PAK 200 and 300, BioSuite 125, Insulin HMWP (I), Protein-Pak (I)
L21	A rigid, spherical styrene-divinylbenzene copolymer, 3 to 10 μm in diameter.	Polymerx RP-1, Phenogel 100 Å, IC-Pak Ion Exclusion, Shodex SP-0810, etc.
L22	A cation exchange resin made of porous polystyrene gel with sulfonic acid groups, about 10 μm in size.	Rezex ROA
L23	An anion exchange resin made of porous polymethacrylate or polyacrylate gel with quaternary ammonium groups, about 10 μm in size.	Shodex IEC QA-825, TSKgel BioAssist Q, TSKgel SuperQ-5PW, BioSuite Q AXC, BioSuite DEAE, Protein-Pak Q 8HR
L24	A semi-rigid hydrophilic gel consisting of vinyl polymers with numerous hydroxyl groups on the matrix surface, 32 to 63 μm in diameter.	YMC-Pac PVA-Sil, Toyopearl HW-type
L25	Packing having the capacity to separate compounds with a MW range from 100 to 5000 Daltons (as determined by polyethylene oxide), applied to neutral, anionic, and cationic water-soluble polymers. A polymethacrylate resin base, crosslinked with poly-hydroxylated ether (surface contained some residual carboxyl functional groups) was found suitable.	PolySep-GFC-P2000, Shodex OHpak SB-802.5HQ, Ultrahydrogel DP +120, TSK-gel G1000PW
L26	Butyl silane (C4) chemically bonded to totally porous silica particles, 1.5 to 10 μm in diameter.	Jupiter 300 C4, Aquity UPLC BEH300 C4, Delta-Pak C4, Symmetry300 C4, XBridge BEH300 C4
L27	Porous silica particles, 30 to 50 μm in diameter.	Sepra (I), Porasil (I), Nucleodur, YMS-Pack Silica
L28	A multifunctional support, which consists of a high purity, 100 Å, spherical silica substrate that has been bonded with anionic (amine) functionality in addition to a conventional reversed phase C8 functionality.	Altech mixed mode C8/anion, Generik C8/Amino, ProTec C8

(Continued)

TABLE 6.3.4   USP classification of HPLC chromatographic columns [73]*. (cont'd)

Code	Description	Examples
L29	Gamma alumina, reversed phase, low carbon percentage by weight, alumina-based polybutadiene spherical particles, 5 μm diameter with a pore diameter of 80 Å.	Gamabond ARP-1, Gamabond Alumina Potency
L30	Ethyl silane (C2) chemically bonded to a totally porous silica particle, 3 to 10 μm in diameter.	Maxsil RP2 60 Å (I), Nucleosuil C2, APEX Prepsil C2
L31	A strong anion-exchange resin-quaternary amine bonded on latex particles attached to a core of 8.5 μm macroporous particles having a pore size of 2000 Å and consisting of ethylvinylbenzene cross-linked with 55 % divinyl benzene.	Ion Pac AS 10, Ion Pack AS 16
L32	A chiral ligand-exchange packing- L-proline copper complex covalently bonded to irregularly shaped silica particles, 5 to 10 μm in diameter.	CHIRACEL WH, Astec CLD-D (or −L), Nucleosil Chiral-1
L33	Packing having the capacity to separate proteins of 4,000 to 400,000 daltons. It is spherical, silica-based and processed to provide pH stability.	BioSep-SEC-S2000, BioSep-SEC-S3000 BioBasic SEC 120, Nucleosil 125-5 GFC, Shodex KW-404
L34	Strong cation-exchange resin consisting of sulfonated cross-linked styrene-divinylbenzene copolymer in the lead form, about 9 μm in diameter.	Aminex Fast Carbohydrate, Rezex RPM Monosaccharide, Shodex Sugar SP0810, Nucleogel Sugar Pb
L35	A zirconium-stabilized spherical silica packing with a hydrophilic (diol-type) molecular monolayer bonded phase having a pore size of 150 Å.	Bio-Sep-SEC-S2000, Zorbax GF-250, Zorbax GF-450
L36	3,5-dinitrobenzoyl derivative of L-phenylglycine covalently bonded to 5 μm aminopropyl silica.	Nucleosil Chiral-3
L37	Polymethacrylate gel packing having the capacity to separate proteins by molecular size over a range of 2,000 to 40,000D.	PolySep-GFC-P3000, Shodex OHpak SB-803HQ, Ultrahydrogel 250
L38	Methacrylate-based size-exclusion packing for water-soluble samples.	PolySep-GFC-P1000, Shodex OHpak SB-802HQ, Ultrahydrogel
L39	Hydrophilic polyhydroxymethacrylate gel of totally porous spherical resin.	PolySep-GFC-P Series, Shodex OHpak SB-800HQ series, Shodex RSpak DM-614
L40	Cellulose tris-3,5-dimethylphenylcarbamate coated porous silica particles, 5 μm to 20 μm in diameter.	CHIRACEL OD, Lux cellulose 1, Nucleocel Delta
L41	Immobilized α-acid glycoprotein on spherical silica particles, 5 μm in diameter.	Chiral-AGP
L42	Octylsilane and octadecylsilane groups chemically bonded to porous silica particles, 5 μm in diameter.	Chromegabond PSC, HiChrom RPB-250A
L43	Pentafluorophenyl groups chemically bonded to silica particles, 5 to 10 μm in diameter.	Curosil-PFP, Ultra-PFP, Pinnacle DB PFP (Restek) Allure PFP Propyl

(Continued)

TABLE 6.3.4 USP classification of HPLC chromatographic columns [73]*. (*cont'd*)

Code	Description	Examples
L44	A multifunctional support, which consists of a high purity, 60 Å, spherical silica substrate that has been bonded with a cationic exchanger, sulfonic acid functionality in addition to a conventional reversed phase C8 functionality.	Chromegabond RP-SCX, Generik C8/SCX
L45	Beta cyclodextrin bonded to porous silica particles, 5 to 10 μm in diameter	Astec Cyclobond I, II or II ser., Chiral CD-Ph, Nucleodex Beta-PM, ChiralDex,
L46	Polystyrene/divinylbenzene substrate agglomerated with quaternary amine functionalized latex beads, 10 μm in diameter.	CarboPac PA1, Transgenomic AN1
L47	High capacity anion-exchange microporous substrate, fully functionalized with a trimethylamine group, 8 μm in diameter.	CarboPac MA1, Hamilton PRP-X100, X110, Hamilton RCX-10
L48	Sulfonated, cross-linked polystyrene with an outer layer of submicron, porous, anion-exchange microbeads, 15 μm in diameter.	IonPac AS5, IonPac AS7
L49	A reversed-phase packing made by coating a thin layer of polybutadiene on to spherical porous zirconia particles, 3 to 10 μm in diameter.	Zirchrom PBD, Discovery ZR-PBD
L50	Multifunction resin with reversed-phase retention and strong anion-exchange functionalities. The resin consists of ethylvinylbenzene, 55 % cross-linked with divinylbenzene copolymer, 3 to 15 μm in diameter, and a surface area of not less than 350 m^2/g, substrate is coated with quaternary ammonium functionalized latex particles consisting of styrene cross-linked with divinylbenzene.	OmniPac PAX-500, Proteomix SAX-POR
L51	Amylose tris-3,5-dimethylphenylcarbamate-coated, porous, spherical, silica particles, 5 to 10 μm in diameter.	Chiralpak ADS, Nucleocel Alpha
L52	A strong cation exchange resin made of porous silica with sulfopropyl groups, 5 to 10 μm in diameter.	TSKgel SP-2SW, BioBasic SCX, Supecosil LC-SCX
L53	Weak cation-exchange resin consisting of ethylvinylbenzene, 55 % cross-linked with divinylbenzene copolymer, 3 to 15 μm diameter. Substrate is surface grafted with carboxylic acid and/or phosphoric acid functionalized monomers. Capacity not less than 500 μEq per column.	IonPac CS14
L54	A size exclusion medium made of covalent bonding of dextran to highly cross-linked porous agarose beads, about 13 μm in diameter.	Superdex peptide HR 10/30
L55	A strong cation exchange resin made of porous silica coated with polybutadiene-maleic acid copolymer, about 5 μm in diameter.	IC-Pak C M/D, Waters Spherisorb SCX, Universal Cation

(*Continued*)

TABLE 6.3.4   USP classification of HPLC chromatographic columns [73]*. (*cont'd*)

Code	Description	Examples
L56	Isopropyl silane (C3) chemically bonded to totally porous silica particles, 3 to 10 μm in diameter.	Zorbax SB C3
L57	A chiral-recognition protein, ovomucoid, chemically bonded to silica particles, about 5 μm in diameter, with a pore size of 120 Å.	Ultron ES-OVM
L58	Strong cation-exchange resin consisting of sulfonated cross-linked styrene- divinylbenzene copolymer in the sodium form, about 7 to 11 μm diameter.	Rezex RNM-Carbohydrate, Aminex HPX-87N, Shodex SUGAR KS-801, -802, etc.
L59	Packing with the capacity to separate proteins by molecular weight over the range of 10 to 500 kDa. Spherical 10 μm, silica-based, and processed to provide hydrophilic characteristics and pH stability	BioSep-SEC-S3000, Biosuite 125, Nanofilm SEC-150, TSK-GEL G2000SW, G3000SW, etc
L60	Spherical, porous silica gel, 3 to 10 μm in diameter, surface has been covalently modified with palmitamidopropyl groups and endcapped.	Acclaim polar Advantage, Ascentis RP-Amide, HALO RP-Amide, Prism RP, Supecosil LC-ABZ, etc.
L61	Hydroxide-selective, strong anion-exchange resin consisting of a highly cross-linked core of 13 μm microporous particles, pore size less than 10 Å, and consisting of ethylvinylbenzene cross-linked with 55 % divinylbenzene with a latex coating composed of 85 nm diameter microbeads bonded with alkanol quarternary ammonium ions (6 %).	Ion Pac AS-11, Ion Pac AG-11
L62	C30 silane bonded phase on a fully porous spherical silica, 3 to 15 μm in diameter.	Develosil Combi-RP, Develosil RP-Aqueous, Develosil RP-Aqueous-AR, ProntoSil c30, YMC-Pack Carotenoid, Zodiac 120 C30
L63	Glicopeptide teicoplanin linked to spherical silica (chiral phase)	CHIROBIOTIC V, T, TAG, R.
L64	Strongly basic anion exchange resin consisting of 8% crosslinked styrene-divinylbenzene copolymer with a quaternary ammonium group in the chloride form, 45 to 180 μm in diameter	AG 1-X8
L65	Strongly acidic cation exchange resin, consisting of 8% sulfonated crosslinked styrene-divinylbenzene copolymer with a sulfonic group in hydrogen form, 63–250 μm diameter	AG 50W-X2
L66	Crown ether coated on 5 μm silica gel substrate	CrownPak CR
L67	Porous vinyl alcohol copolymer with C18 alkyl group attached to the hydroxyl group of the polymer, 2 to 10 μm diameter	Supelcogel ODP-50, apHera C18, Asahipak ODP-40
L68	Spherical porous silica containing a polar group within or intrinsic to the hydrocarbon bonded phase (e.g. alkylamide)	Ultra II IDB,

(Continued)

TABLE 6.3.4 USP classification of HPLC chromatographic columns [73]*. (*cont'd*)

Code	Description	Examples
L69	Polymeric (ethylvinilbenzene/divinylbenzene) with strong anion exchange (quaternary amine) on latex beads, about 6.5 μm diameter	CarboPac PA20
L70	Cellulose tris(phenylcarbamate) coated on 5 μm silica	Chiracel OC-H
L71	A rigid, spherical polymethacrylate, 4 to 6 μm diameter	RSpak DE-613
L72	(R)-Phenylglycine and 3,5-dinitroaniline urea linkage covalently to silica	Chirex 3012, Sumichiral OA-3300
L73	A rigid, spherical polydivinylbenzene particle 5 to 10 μm diameter	Jordi-gel DVB
L##	(Dalteparin sodium, anion exchange Dowex 1X8) —strongly basic (type I) anion exchange resin in chloride form	Dowex 1X8
L##	(Dalteparin sodium, cation exchange Dowex 50WX2) —strongly acidic cation exchange resin in H^+ form	Dowex 50WX2
L##	(Glucosamine, Shodex NH2P-50) Polyamine chemically bonded to cross-linked polyvinyl alcohol polymer, 5 μm diameter	apHera NH2 Amino, Shodex NH2P50
L##	(Ethylhexyl triazone, FluoFix)-Fluorocarbon chains chemically bonded to 5 μm spherical silica particles	Wakopak FluoFix-II 120E, 120E, 120N

* Note: All particles are spherical unless differently indicated (M) for monolith, (I) for irregular shape.

as code L1, L2, and so on, is shown in Table 6.3.4, including several examples of columns for each category [73]. In spite of its complexity, the USP code is not always appropriate for a column classification. For example, new core-shell type columns such as Kinetex XB-C18, Poroshell 120 SB-C18, or certain zirconia-based columns such as ZirChrom-EZ, are not easily classifiable using the criteria shown in Table 6.3.4.

Characterization of the Polarity of a Column Using K_{ow} for Models of Stationary Phase

The polarity of the stationary phase is an important parameter for predicting its behavior in a separation. As previously discussed in Section 5.1, this polarity is difficult to assess from column to column because it refers to a solid phase that is not even uniform, containing groups belonging to both the support and the bonded moieties. Characterization of columns using a set of parameters obtained from their behavior toward a set of "test" compounds has been developed for a considerable number of RP columns (see Section 6.4) as well as for HILIC columns (see Section 6.5). However, a simplified characterization can be obtained using K_{ow} (log K_{ow}) values for simplified models representing different stationary phases. This type of characterization is given in Table 6.3.5. The models from this table were selected only to illustrate the variation in the polarity of different columns. However, other models can be used for the same purpose. In the models from Table 6.3.5, the terminal -$Si(CH_3)_3$ group was selected to simulate an end-capped silica structure with hydrophobic character. This

TABLE 6.3.5   Simplified models for the stationary phase and the corresponding calculated log K_{ow}

Column	Model structure	log K_{ow}*
C18 end-capped	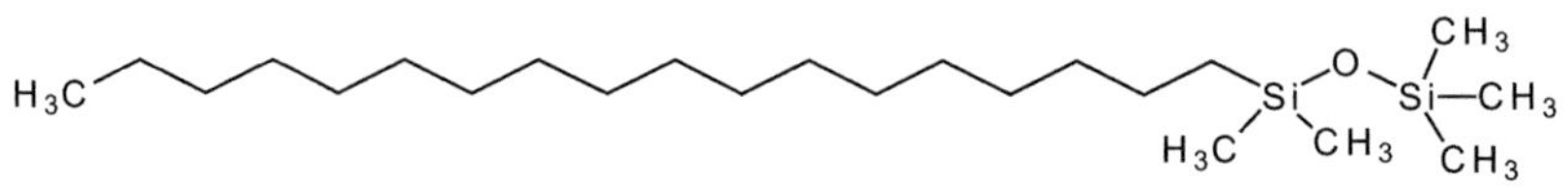	9.75
C18 not end-capped	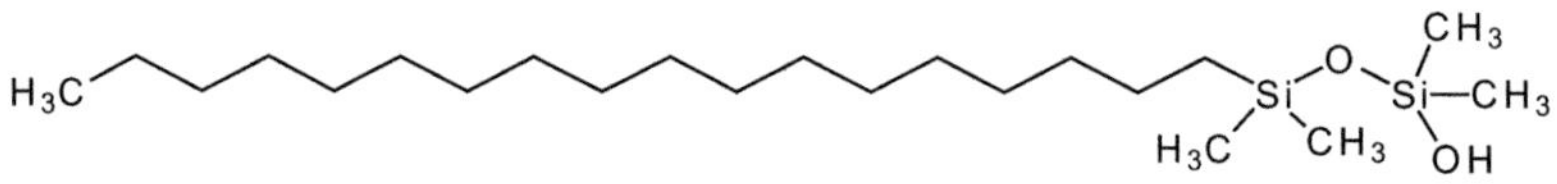	7.75
C8 end-capped	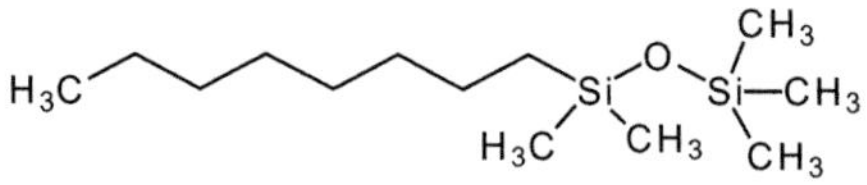	5.97
Fluorinated	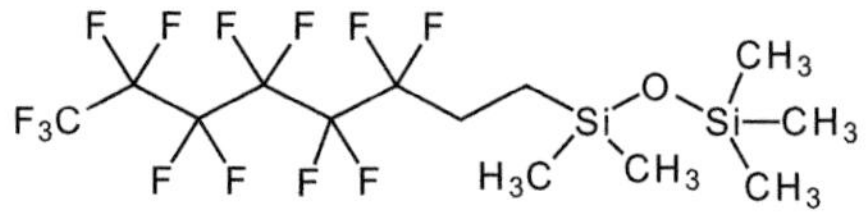	8.38
Amide embedded	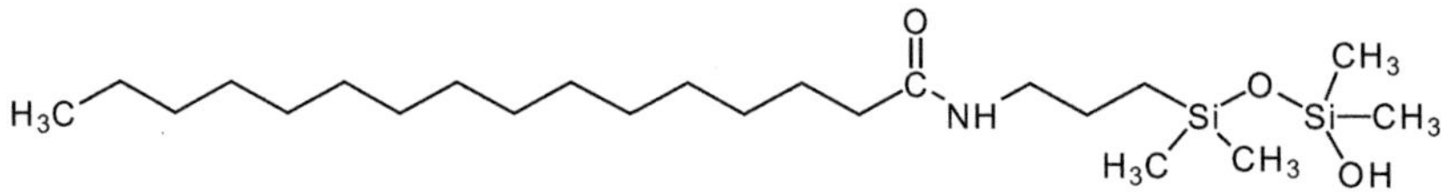	7.86
Butyl-phenyl	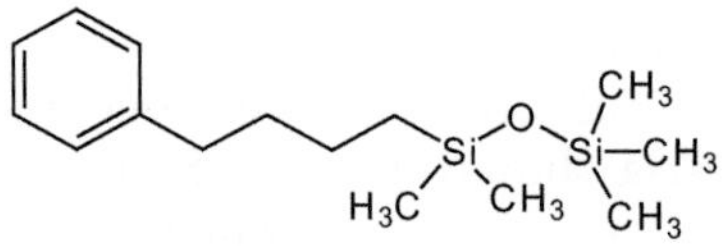	5.81
Cyanopropyl	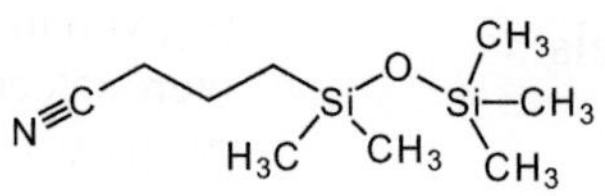	2.80

(Continued)

TABLE 6.3.5 Simplified models for the stationary phase and the corresponding calculated log K_{ow} (cont'd)

Column	Model structure	log K_{ow}*
HILIC amide	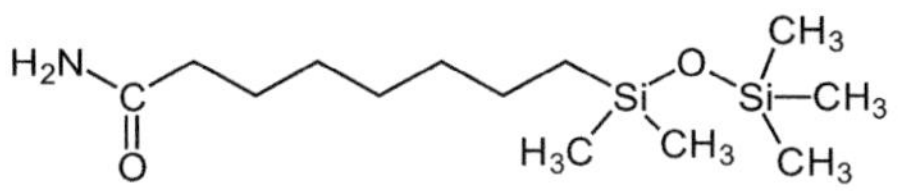 $H_2N-C(=O)-\cdots-Si(CH_3)(H_3C)-O-Si(CH_3)_2-CH_3$	3.69
HILIC amino	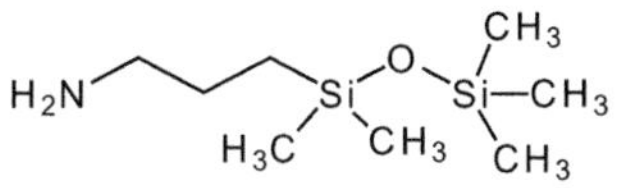 $H_2N-\cdots-Si(CH_3)(H_3C)-O-Si(CH_3)_2-CH_3$	2.25
HILIC diol	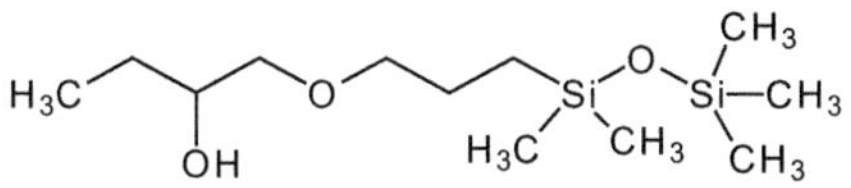 $H_3C-CH(OH)-CH_2-O-\cdots-Si(CH_3)(H_3C)-O-Si(CH_3)_2-CH_3$	1.46
Zwitterionic	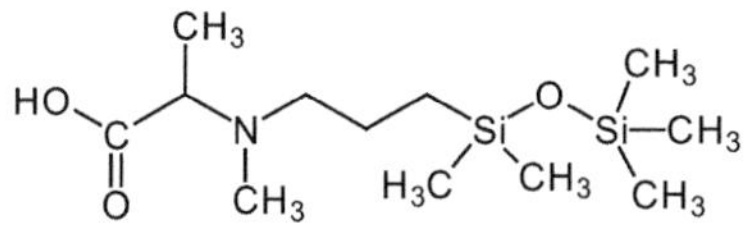 $HO-C(=O)-C(CH_3)-N(CH_3)-\cdots-Si(CH_3)(H_3C)-O-Si(CH_3)_2-CH_3$	0.43

Note: The K_{ow} values for the model compounds were calculated using a computer program from MarvinSketch 5.4.0.1, ChemAxon Ltd., [74].

group was replaced with -Si(CH$_3$)$_2$OH for surfaces that were not end-capped.

Performance of Stationary Phases and Columns

The criteria for comparing the performance of a stationary phase include a number of characteristics, some related to the chemical and physical properties of the phase and others related to external factors [75–77]. Among these criteria are those listed in Table 6.3.6.

The separation capability is typically characterized by the selectivity α, which is dependent, besides the column characteristics, on the nature of the compounds to be separated and on the mobile phase. A variety of tests were developed for assessing the selectivity of the column. A "hydrophobic selectivity" parameter is particularly useful for characterization of hydrophobic stationary phases (see Section 6.4). Selectivity α is also the parameter to which the resolution R is very sensitive; therefore the choice of a column with large α for the specific separation is very important. The resolution is also a function of k and N (as given by rel. 2.1.57, repeated in 6.3.1).

$$R = \frac{1}{4}\left(\frac{\alpha - 1}{\alpha}\right)\left(\frac{k}{1+k}\right)N^{1/2} \qquad (6.3.1)$$

The choice of a column for which the capacity factor k value is higher can also be advantageous for achieving a specific separation. Higher k values help the separation but also indicate

TABLE 6.3.6   Criteria used for selecting a good HPLC column.

No.	Criterion	*
1	Efficiency and ability to obtain separation of sample analytes, between them and from the sample matrix	V
2	Low skew/low asymmetry/low tailing	C
3	Large values for theoretical plate number N	I
4	Achieve separation within relatively low t_R (low t_R being related to low k values)	C
5	No bleed of the stationary phase	I
6	Stability to extreme operating conditions such as low pH, high pH, range of solvents (e.g. high water content for hydrophobic columns), various components in the sample matrix	D
7	Capability to maintain identical separation characteristics for a large number of injections	I
8	Stability under varying operating conditions	D
9	Rapid regeneration capability	D
10	Acceptable values for the backpressure and the capability to maintain this back-pressure constant in time when identical conditions of a separation are used	C
11	Ability to tolerate high pressure and a rapid change in pressure	D
12	Reproducibility of the separation from column to column of the same type but from different batches	D
13	Capability to achieve analytes separation within a wide range of experimental conditions (have versatility)	D
14	Availability in different formats that can be chosen to fit a wide range of available HPLC instruments, and ease of transferability from format to format of separation conditions	D
15	Low price and extensive information regarding the column characteristics	D

Note: Vital = V, Critical = C, Important = I, Desirable = D

longer retention times, which is not always desirable. Because the k values are both compound dependent and mobile phase dependent, the comparison of columns regarding the k value should be performed under similar conditions [78, 79].

The increase in resolution R is also dependent on the increase in the number of theoretical plates N (and n). This indicates that columns with high N values are preferable, as they produce narrower peaks. Higher N (and n) values can be achieved through specific properties of the stationary phase such as small particles, uniform shape (spherical), and narrow particle size distribution. The use of core-shell particles significantly increases the number of theoretical plates as compared to porous particles of the same dimensions. For example, in a comparison to a traditional 3 µm particle C18 column (150 mm length and 4.6 mm i.d.) that had $N = 166{,}502$ theoretical plate per m, a core-shell column with the same dimensions (Kinetex™ from Phenomenex) had $N = 295{,}343$ per m. A traditional 1.7 µm particle C18 column (50 mm length and 2.1 mm i.d.) had $N = 272{,}080$ per m, while a core-shell column with the same dimensions (Kinetex™ from Phenomenex) had $N = 318{,}680$ per m. On the other hand, core-shell columns may have lower k. The use of monolith columns may also reduce peak broadening.

The peak shape is another parameter of considerable importance in HPLC. Besides the decrease in the column resolution due to tailing or other effects seen in a column with high skew, the precision of the peak area integration is also affected by the peak shape. Correct measurement of peak area is essential in obtaining quantitative information in HPLC. For this reason, columns with lower skew/asymmetry are preferable to those generating peaks far from a Gaussian shape. The skew/asymmetry values are compound dependent, and a specific column may be perfectly suitable for the separation and measurement of some compounds and may not be good for others. Peak tailing is frequently caused by multiple types of interactions of the stationary phase with the analytes, such as hydrophobic interactions and polar interactions, which may take place together, for example, in a column with a hydrophobic phase on silica that also has numerous free silanol groups [80]. A variety of phases were designed to reduce this problem, for example, those using end-capping. Also, the very high purity of the silica base (low in trace metals) contributes to the reduction of tailing.

The pH range acceptable for the mobile phase used with a specific column is another parameter that characterizes the performance of a specific stationary phase. The separation of some solutes does not require mobile phases with the pH lower than 2 or higher than 8. Most HPLC columns with the stationary phase based on a silica support can be used without any problems in this range of pH, and the organic polymeric stationary phases are stable in a wider pH range. However, specific applications may require lower or higher pH for the mobile phase than the range 2 to 8. For this reason, columns that are stable in a wider pH range are preferred. The problem with the need for a wide pH range is encountered in particular for basic compounds that can be partially ionized in the pH range 4 to 8. The ionized components of the analyte interact with the free or ionized silanol groups of the stationary phase, producing tailing (undesirable peak shape). Lower pH of the mobile phase significantly decreases the ionization of the silanol groups still existent in the stationary phase, diminishing interactions and leading to better peak shape. The stability of different columns at the pH of the mobile phase is typically reported by the column manufacturer and/or is reported in the literature. Table 6.3.7 shows some of these reported results [75].

The range of accepted solvents is another important factor regarding the quality of a stationary phase / chromatographic column. For the stationary phases used in size exclusion, ligand exchange, displacement chromatography, bioaffinity chromatography, and chiral separations, the columns are much more sensitive, and special care must be taken regarding the nature of the mobile phase. The phases consisting of bonded organic moieties on a silica surface or on organic polymers, such as those used in RP-HPLC, are in general resistant to a large number of solvents. However, in some cases, the solvent may sufficiently modify the surface of the stationary phase to affect its properties regarding the separation. One such solvent is pure water (with no organic addition), which can affect conventional reversed stationary phases by a process known as dewetting and also phase collapse (not mechanically). Most separations on hydrophobic stationary phases take place using a water-organic solvent with a solvent ratio between 20 and 80%. Phases that can be used at higher water content (even in 100% water) have been developed and are preferable in numerous applications, in particular when the analytes are poorly retained by the hydrophobic stationary phase.

The stability in time of the stationary phase is another important characteristic. This characteristic can be estimated by the number of injections n that can be made on a column, with unaltered results for the chromatography. This number of injections depends, however, on numerous other factors besides the column quality. The type of sample, the nature of the matrix, the degree of cleanup performed on the sample before

TABLE 6.3.7   Stability of several columns to mobile phase pH [75]*.

Column	Manufacturer	pH 2	pH 3	pH 6	pH 7	pH 7	pH 8	pH 9	pH 10	pH 11
Medum**		TFA	phosph.	phosph.	acetat.	phosph.	bicarb.	ammon./ bicarb	ammon.	tri- phosph.
XBridge C18	Waters	1	?	?	?	1	1	1	1	2
XTerra MS C18	Waters	1	?	3	1	3	1	1	?	?
XBridge Phenyl	Waters	1	?	?	1	2	3	3	3	?
Luna C18	Phenomenex	1	?	1	2	2	2	2	?	?
Zorbax Bonus RP C18	Agilent	1	1	2	2	3	3	3	?	?
Sunfire C18	Waters	1	1	3	3	3	?	?	?	?
Zorbax SB C8	Agilent	1	?	?	?	1	2	?	?	?

Note: 1 = Stable to more than 500 injections, 2 = Stable to more than 200 but less than 500 injections, 3 = Stable only to less than 200 injections, ? no data available.

**Note: TFA = trifluoroacetic acid, phosph. = phosphate buffer, acetat. = acetate buffer, bicarb. = bicarbonate buffer, ammon. = ammonia, tri-phosph. = sodium triphosphate.*

injection, the injection volume, the sample concentration, and the like, are factors influencing n. Also, in order to protect the column from undesirable materials that may reach the stationary phase, filters and pre-columns are commonly utilized in the sample path. The efficiency of the pre-column and the frequency with which this pre-column is changed also influence the number of injections n, as does the nature of the mobile phase . When the column is utilized at extreme pH values (very close to the limits of the acceptable range), the lifetime of the column is diminished. The typical problems associated with columns that become "old" are loss of retention, peak broadening, increase in backpressure, and bleeding (observable especially in MS detection). Columns with predicted high n values are those resistant to a wider pH range, have a lower backpressure characterized as new, and/or have a high purity of the silica material. In practice, it can also be noted that columns with a larger amount of stationary phase appear to be more resilient in time and lead to larger n. Larger i.d.

for the chromatographic column is, however, associated with a slightly lower value for the number of theoretical plates N, and sometimes a compromise should be chosen between a more robust column and a column offering a better separation. One additional factor that must be considered when changing from a column with one diameter d_1 to a column with a different diameter d_2 is the change in linear flow rate in the column. Considering that the volumetric linear flow rate U is related to the linear flow rate by the formula $U = A\,u$, where A is the area of the channel in which the flow takes place, in order to maintain the same linear flow rate u in two columns with different diameters, the following condition must be fulfilled:

$$U_1 = \frac{A_1}{A_2} U_2 = \frac{d_1^2}{d_2^2} U_2 \qquad (6.3.2)$$

Relation 6.3.2 indicates that for narrower columns, a lower volumetric flow rate U is necessary to achieve the same linear flow rate

u in the two columns. Since the height of theoretical plate H is dependent on u (see van Deemter equation 2.2.8), the efficiency of the separation can be affected when changing columns with different diameters, even if all the other parameters are kept identical but the flow rate is not adjusted properly.

The acceptable value for the backpressure of a column when working in typical conditions is a very important parameter. This parameter is frequently limited not by the column but by the maximum possible pressure generated by the HPLC pumps. In UPLC, where smaller particles and column diameters are selected for achieving higher N values, the maximum pressure of the pumps is higher than in standard HPLC. The value of a column backpressure is determined by the particle size, the flow rate, the nature of the mobile phase, temperature, the column dimension, and the like. A formula that governs backpressure in a packed bad column (difference between the pressure at the column inlet and that at the outlet) is the following:

$$\Delta p = \frac{\eta U L}{\pi K^0 r^2 d_p^2} \qquad (6.3.3)$$

where Δp is the pressure necessary for a liquid to flow through the column with the volumetric flow rate U. In rel. 6.3.3 η is the mobile phase viscosity (typically given in mPa s), K^0 is specific permeability for the phase (adimensional), r is the column radius, L is its length, and d_p is the diameter of the particles in the bed (in µm) (see Darcy equation, rel. 2.2.17).

As expected, the backpressure for a chromatographic column is higher at higher viscosity of the mobile phase, higher flow rates, and longer columns. However, the column diameter and the diameter of the particles of the stationary phase are even more sensitive parameters affecting Δp since there are values at power 2. Since d_p is also related to the number of theoretical plates of the column, a compromise must be found

between the choice of smaller particles with higher backpressure or better separation [81, 82]. Since the viscosity of liquids typically decreases as the temperature increases, a higher temperature for the column typically leads to a lower backpressure. However, limitations on the temperature are imposed by other factors such as the effect of temperature on the separation and stability of the analytes, solvents, and stationary phase. Columns with core-shell particles and monolithic columns characterized by the same number of theoretical plates N compared to traditional columns have lower backpressure for similar flow conditions. For example, for a traditional C18 column, 1.7 µm particles, 50 mm long, 2.1 mm i.d, a typical working backpressure in CH_3CN/H_2O 50/50 is higher than 400 bar at 0.6 mL/min flow rate ($N \approx 270,000$ per m). A core-shell C18 column with 2.6 µm particles that has a similar N, and identical length and i.d., develops pressures below 300 bar.

From rel. 6.3.3, it can be seen that the specific permeability K^0 for a column can be obtained from the expression:

$$K^0 = \frac{\eta U L}{\pi \Delta p \, r^2 d_p^2} \qquad (6.3.4)$$

For practical purposes, instead of the specific permeability K^0, a column permeability K^* that can be directly measured and is defined by the formula:

$$K^* = \frac{u L}{\Delta p} \qquad (6.3.5)$$

where u is the linear flow rate. Since $u = L / t_0$ the expression for K^* can be written in the form:

$$K^* = \frac{L^2}{\Delta p t_0} \qquad (6.3.6)$$

The value for K^* can be measured for a given column based on Δp, t_0 and L and provides information on the column characteristics. From rel. 2.2.3 that connects linear flow rate u

with volumetric flow rate U, the relation between K^0 and K^* is the following:

$$K^0 = K^* \frac{\eta \varepsilon^*}{d_p^2} \qquad (6.3.7)$$

(where ε^* is the column packing porosity). This formula allows the evaluation of K^0 for a column, when the solvent viscosity η, particle dimensions d_p and ε^* are known.

Holding the backpressure of a column in time, after repeated injections, is dependent on the operating conditions, as well as on the quality of the stationary phase. Plugging the frits of the column with sample components, use of extreme pH values for the mobile phase that degrade the stationary phase, use of unfiltered solvents that contain small particles, and plugging the column with noneluted components are among the practices that must be avoided but are not directly related to column quality. The extreme case of column "plugging" such as the use of a mobile phase that generates a precipitate in the column (e.g., use of an inorganic buffer that precipitates during a gradient run) also must be avoided. However, the strength of the silica particles (for the column with a packed particle bed) as well as the absence of very small "dust" particles in the stationary phase are characteristics of the stationary phase necessary to avoid plugging in time of the frit that retains the stationary phase in the column, which can cause a significant increase in the backpressure.

One other aspect of column backpressure (when the pumps are able to deliver a specified one) is the ability of the stationary phase to tolerate that high pressure and a rapid change in pressure (e.g., during a gradient run when the mobile phase viscosity changes). This factor is highly dependent on the nature of the stationary phase. For size-exclusion chromatography, for example, where the stationary phase is typically made of highly porous organic polymers, the recommended working pressure of the column is very different from that for silica-based columns. For the silica-based columns, the rigidity of the silica support and the pressure at which the column was packed are important factors regarding the pressure limit for a specific column. This working pressure must be lower than the *specified maximum pressure,* a wider gap between the two values being desirable.

The reproducibility of the separation from column to column of the same type, but from different batches, is another important parameter. By using high-purity silica, novel derivatization procedures, standardized conditions for column packing, and rigorous column testing before being put on the market, the requirement to be reproducible is met by most modern HPLC columns.

The availability of the columns in different formats for the same stationary phase and for ease of transferring a separation method from one column to another is also desirable. Longer columns for the same phase lead to higher values for N (and n) as shown by rel. 2.1.42. Also, the decrease in the particle size (for phases of the same nature) has a significant impact on the N value. Since R is proportional with $N^{1/2}$, longer columns and columns with lower particle size are sometimes necessary for better resolution. Increasing the column length and decreasing the particle size have the effect of increasing the backpressure. For this reason, the availability of equivalent phases in different formats that can accommodate different requirements such as separation characteristics, column backpressure, and sample size is very useful.

Selection of a Column for an HPLC Separation

The main criterion in the choice of a chromatographic column is the adequacy for performing the proper separation intended. This separation must produce well-defined peaks (narrow, with no tailing) such that the peak processing can be done accurately. Peak recognition, peak area measurement, data averaging, and other data processing operations done by the computer programs controlling the HPLC instruments are always performed more accurately and reproducibly when the peak shape is good.

In general, the selection of a column is determined by (1) the properties of the compounds that have to be separated, (2) the specific column properties that are compound independent, (3) the requirements of the separation, independent of analytes and mobile phase, and 4) the mobile phase requirements.

1) The column choice depends mainly on the properties of the compounds to be separated. This should be associated with the selection of the mobile phase/elution program such that a successful separation is achieved. Further details regarding the selection of a chromatographic column in function of the compound/class of compounds that are subject to analysis are given in this book. The main goal in a separation is to obtain good R values for the compounds of interest. This is achieved by having large N, α, and k values for the separation. Further discussions will be found in this book regarding ways to obtain columns with large N values, good separation characteristics, and high k and α values for the compounds of interest.

Sometimes other special requirements are imposed on the analysis; for example, low t_R (that typically imply low k values) in a separation are necessary when rapid analyses are needed. Also, as the separation time increases, the broadening of the chromatographic peaks increases (see rel. 2.1.29 and Figure 2.1.2). Since an improvement in the resolution R is related to the increase in the k value, other procedures are involved that attempt to reduce both t_0 and t_R, such that the value of $k = (t_R - t_0)/t_0$ is not affected, although t_R is smaller. This

can be achieved using higher flow rates, and narrower and shorter chromatographic columns. This type of columns is used, for example, in UPLC-type techniques.

2) The main property of the column that is basically independent of a specific compound can be considered the number of theoretical plates N (although N varies to a certain extent from compound to compound, this variation is not significant and the choice of a specific molecular species for measuring N eliminates this variability). As previously discussed, columns with smaller and more uniform particles lead to higher N values. Also the columns with core-shell particles are highly recommended for their high N values although they have lower Ψ and therefore lower k values. Geometric properties of the column also contribute to the success of a separation. Longer columns lead to a higher N, but at the same time the analysis time is typically increased. Also, the column backpressure is higher for longer columns. The linear variation of the retention time with the column length is shown in Figure 6.3.1 for seven test compounds on a Zorbax SB-C18; 3.5 μm particle diameter column.

The change in column diameter also may affect the separation. Relation 6.3.2 shows the requirement in volumetric flow rate change for achieving identical linear flow rates in columns of different diameter. Columns with a larger diameter offer better resilience regarding the number of samples that can be injected without affecting results but in general have a slightly lower N (for the same length) and require larger volumes of mobile phase.

Besides the properties of a column related to the separation in itself, other characteristics related to the column utilization are important. For example, the column should be chemically stable and show no bleed. The bleed indicates column

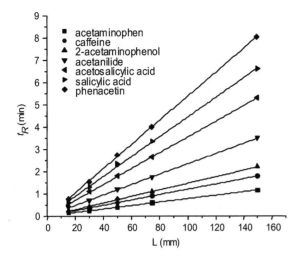

FIGURE 6.3.1 The variation of retention time for seven test compounds with the variation of the column length (Zorbax SB-C18; 3.5 μm particle diameter) (as shown in [71]).

degradation but can also interfere with detection, in particular for specific detectors (such as MS). The resilience to a wide pH range is another desirable property. The backpressure produced by the column is another practical parameter to consider. High backpressures are not desirable for a number of reasons, including frequent instrument failure associated with it.

The HPLC separations are typically performed repeatedly, and analyzing numerous samples by the same method is common in many laboratories. The analysis of numerous samples and the associated data analysis is typically performed using computer controlled instruments. Maintaining the same separation characteristics when numerous samples are analyzed is another important property for the chromatographic column. This will allow peak identification without human intervention and facilitates accurate data reporting.

3) The analysis requirements such as speed, number of samples to be analyzed, desire to collect the separated analytes or not, and restrictions regarding the nature and the volume of solvents to be used in the mobile phase also play a role in the choice of the column. The type of available HPLC instrumentation regarding solvent supply system, maximum pressure delivered by the pumps, types of detectors, etc. are also factors influencing or determining the choice of the chromatographic column.

4) Mobile phase requirements such as pH, water content in the mobile phase, mobile phase viscosity may also influence the column selection.

A number of other column characteristics are listed in Table 6.3.6 and labeled as desirable. The importance of these characteristics may vary. When some of them cannot be fulfilled but the separation is good and the results are reproducible, the column is still preferred to a similar one that offers a better property in the desirable group but notr, for example, such a good separation. At this point, the choice of the column is more related to individual preferences and needs.

The Use of Guard Columns and Cartridges

Some of the column qualities, especially the capability to maintain identical separation characteristics for a large number of injections, can be improved with the use of guard columns or guard cartridges (e.g., SecurityGuard™ from Phenomenex). Even if the frits that are used in the path of the mobile phase eliminate solid contaminants that may plug the analytical column, when the column is used for a longer period of time, certain strongly retained materials can accumulate on the column, may never be eluted, and may dramatically reduce column lifetime. By modifying the surface of the stationary phase, these retained materials can cause shifts in peak retention, loss of resolution and efficiency, and degradation of peak shape.

Another factor contributing to column degradation can be the use of "harsh" mobile phases that can destroy in time the stationary phase (e.g., when a silica-based stationary phase is subjected to pH < 2 or pH > 8 solutions). Even when common solvents are used, some dissolution of silica (for silica-based columns) may occur, leading to the formation of a void volume at the head of the column and subsequent reduced column resolution. A common procedure to protect the analytical column from these problems is to use a guard column. Guard columns are short (or very short) columns or cartridges, packed with a stationary phase having an identical or a slightly less retentive bonded-phase as the analytical column. The guard column should have a very small dead volume such that it should not affect the column performance. At the same time, the guard column retains the noneluting contaminants and provides saturation of the mobile phase with silica by "bleeding" silica into the mobile phase instead of from the analytical column. The guard column must be replaced from time to time, before the analytical column starts to degrade. The replacement time interval is determined by the specific experimental conditions and can be done either at a fixed schedule (such as 100 injections) or when signs such as increase in the backpressure or some loss of resolution begin to be noticed.

Column Protection, Cleaning, Regeneration, and Storing

Commercially available chromatographic columns are typically received filled with a specific solvent that is specified by the vendor, and their storage is recommended to be done in a similar solvent as received. The columns for reversed phase are usually received filled with acetonitrile/water 65:/35 v/v; silica-based ion-exchange columns are received filled with methanol or methanol/water; normal-phase columns are received filled with isopropanol or hexane. Other columns such as those used for ion exclusion can be received filled with water. Typically, buffers are avoided as storage solvent, although some columns are received filled with a specific solvent with buffer, such as HILIC columns that can be received in acetonitrile/water (e.g., 90:10 v/v) that contains 100 mM of $HCOONH_4$. The columns must be properly installed, and pre-filtering and adequate guard columns are always recommended. The mobile phase that is selected with the goal of achieving separation must comply with specific restrictions regarding the nature of the solvent and the pH range. When using gradients, the solvents used as mobile phase must be miscible. Also, switching from one solvent to another should be done considering solvent miscibility. The solvent composition must be kept within the limits recommended by the vendor. For example, the water content used with an RP column should not exceed the recommended limit in order to avoid phase collapse by dewetting. The addition of salts for the modification of pH or ionic strength must be done at the lowest possible concentration, and salt precipitation in the column will destroy it. The pH range where the column is stable should be strictly obeyed. The temperature at which the column is used should not exceed the vendor specifications. For RP columns this temperature is typically not higher than 60 °C. The maximum backpressure of the column (indicated by the vendor) should never be exceeded, and the flow should go in the recommended direction (indicated on the column). Sudden pressure changes must be avoided. Also, the samples injected in the column should be (as much as possible) completely eluted.

After a period of usage, some column degradation may be noticed, with broader peaks, lower retention times, decreased separation capability, and increased column backpressure. When any of these problems appear, a column cleaning may be useful. The cleaning is typically done by rinsing the column (e.g., 10- to 20-column volume) with a specific solvent or sequence of solvents at 1/5 to 1/2 typical flow rate. For example, for RP columns, a set of solvents covering a wide range of polarity is recommended, starting with 95:5 water/acetonitrile v/v, followed by tetrahydrofuran (THF), followed by 95:5 acetonitrile/water v/v. Columns used for protein separations may be washed with 0.1% trifluoroacetic acid (TFA) in water followed by 0.1% TFA in acetonitrile/isopropanol ½ v/v, followed by rinsing with a common mobile phase (columns should not be stored in THF). Other specific cleaning procedures may be recommended by the column manufacturer. In general, one should not use a flow in the opposite direction than recommended, in order to diminish the backpressure increase. However, this restriction seems to be less important for columns with small particles, when back-flushing with a strong solvent are in some cases used to wash materials accumulated at the head of the column.

Columns that are not used for a certain period of time must be stored appropriately. For this purpose, the clean column must be filled with a specific solvent. For RP-type columns, a mixture of 65% acetonitrile and 35% water is recommended. Before storing, the column must be cleaned from the presence of buffers. For normal-phase columns, isopropanol is typically recommended as storing solvent. Ion-exchange columns are usually stored in methanol or methanol/water. For size-exclusion columns, water with 0.05% NaN_3 (to avoid bacteria growth) or with 10% methanol is typically recommended. However, column manufacturers may provide specific recommendations for column storage that must be followed.

Selection of Columns for Orthogonal Separations

In HPLC practice, it is sometimes difficult to decide whether a peak belongs to a unique compound or to a mixture of unseparated compounds. The use of detectors that allow the measurement of a compound even if it is present in a mixture improves the capability of analysis even if the HPLC separation is not good. This is, for example, the case of the use of UV detection for a compound with a unique absorption wavelength; the use of fluorescence detection when the compound has unique excitation/emission wavelengths; or the use of MS detection. However, in many cases the use of selective detection is either not sufficient or not possible. In such cases, the use of two columns with very different selectivity may solve the problem. If the second column does not lead to the separation of additional peaks, it is more likely that the first separation is complete. Columns with very different selectivity are also desirable for two-dimensional separations, when segments from the eluate from one column are further separated on the second column (2-D or multidimensional separations). The HPLC separations performed on different types of columns and using different solvents that lead to a different separation are labeled as *orthogonal*. The "orthogonality" of one separation versus another one can be evaluated using procedures specifically developed for this purpose. These procedures are based on the use of test mixtures on the two columns and the comparison of separation results [83]. A potential procedure for testing orthogonality is to make a plot of retention times for the compounds from the test mixture for the separation on the two columns. A good correlation between the retention times indicates a poor orthogonality. Because the separation selectivity in the HILIC mode is complementary to that in reversed phase and other modes, combinations of the HILIC, RP and other systems are attractive for two-dimensional applications. Other criteria are based on comparing column properties such as hydrophobic character and hydrogen-bonding characteristics. Column differences are discussed further in Section 6.4.

Physicochemical Characterization of Stationary Phases

Because the quality of stationary phases for HPLC is determined by the phase physical and chemical properties, the measurement of some of these properties provides important information regarding the quality of the phase [84]. The investigation of physicochemical properties can have the following purposes [85]:

1. Determination of bulk properties of stationary phase, such as specific surface area, volume, and size distribution of the pores and of the particles.
2. Characterization of the surface structure of the packing materials considering the concentration, conformation, and mobility of the organic attached groups, as well as characterizationof the structure, chemistry, and accessibility to the phase support.

The procedures used to achieve these purposes can be destructive or nondestructive. In the destructive procedures, the packing material is subjected to chemical or physical modifications that alter the phase. The nondestructive procedures involve instrumental analysis, such as fluorescence, infrared spectrometry, or NMR techniques. These methods leave the packing material intact so that it is suitable for direct chromatographic characterization [86].

Among the destructive techniques is the measurement of carbon load (C%) of the stationary phase. For this measurement, the combustion of the stationary phase and the measurement of total ash and CO_2 can be performed using elemental analysis instruments. Metallic impurities in silica (before and after acidic washing) or modified silica materials are also determined by a destructive technique, typically using inductively coupled plasma−atomic emission spectrometry (ICP-AES). For example, the Fe^{2+} content of the silica in many packing materials is between 5 and 400 ppm, Ni^{2+} content is about 1−2 ppm, and Al^{3+} is about 100 ppm. Other elements are at the ppb levels or lower. Acid washing was proved by ICP to reduce the metal content of silica by 70%, while washing the support with ethylenediaminetetraacetic acid (EDTA) resulted in silica that is completely free of metal impurities [87].

Various adsorption studies are utilized to evaluate the surface area of the HPLC stationary phase. Surface area is one the most important physical parameters because HPLC retention is proportional to this parameter. Several common experimental methods are available for measurement of the surface area of porous silica, and nitrogen or argon adsorption isotherms at the temperature of liquid nitrogen (around 77 K) are typically used in accordance with BET theory for calculating the total surface area per unit of adsorbent weight [88, 89]. Inverse SEC has been utilized for the same purpose [90−92]. For example, retention data of polystyrene samples of narrow molecular size distribution and known average molecular mass were used to estimate the external, internal, and total porosities of monolithic columns (Chromolith Performance from Merck) and compared to those of the conventional packed column (Luna C_{18} from Phenomenex) [93].

Other techniques for studying the stationary phase involve scanning electron microscopy (SEM), infra red spectroscopy (IR), diffuse reflectance infrared Fourier transform spectrometry (DRIFTS), nuclear magnetic resonance (NMR), fluorescence spectroscopy, and thermogravimetric analysis (TGA). The DRIFTS technique is used in particular for evaluation of silanol density and water adsorption on silica surface. Usually, in the IR spectra of silicas (silica gels) there are two intense IR absorption bands situated at 3400 and 1600 cm^{-1}, which are assigned, respectively, to the stretching and deformation vibrations of absorbed water molecules. After the absorbed water molecules are removed, the infrared spectrum of silica is characterized by a sharp band near 3750 cm^{-1}, assigned to the vibration of the isolated silanol groups on the surface and a broad band in the range of 3900−3200 cm^{-1} for the H-bonded silanols and the residual water molecule. Also, TGA has found wide application in estimation of silanol number. Silanol activity and, indirectly, the coverage density with the bonded phase can also be characterized using microcalorimetry, based on the heat of adsorption of a solvent on the stationary phase material (heat of immersion) [94]. When the silica surface is covered with a bonded phase (e.g., C18), the number of free silanol groups decreases, and this will produce a decrease in the heat of adsorption for polar solvents. This effect is shown in Figure 6.3.2 representing the decrease in the heat of immersion for several C18 stationary phases obtained from the same silica (7.1 μmol/m^2 silanol groups) and having different coverage densities with C18 groups, for three different solvents (methanol, acetonitrile, and hexane). As seen from

Figure 6.3.2, on the low-coverage density of the bonded phases, the heat of immersion for hexane is lower than for methanol or acetonitrile, but for high-covered bonded phases, hexane adsorbs higher heat than acetonitrile because the interactions with the C18 chains start to have more contributions.

Hydride silica stationary phases that have the typical groups Si-H were studied using IR spectroscopy based on the characteristic stretching band situated at 2250 cm^{-1}. The decrease of the intensity of this band can be used in evaluating the substitution with alkyl group during derivatization of silica surface, with the C−H bonds from bonded phases characterized by strong stretching bands situated between 3000 and 2800 cm^{-1}. The Si-O bonds have absorption bands situated between 900 and 1300 cm^{-1}. Alumina surfaces are characterized by the same procedures [25].

Considerable attention has been given to the study, by spectroscopic techniques, of the stationary phase structure. For example, the conformational state of the alkyl chains in silica gel modified with alkyl chains of various lengths in the dry state was investigated over a temperature range of 123−353 K using

FIGURE 6.3.2 The variation of heat of immersion for methanol, acetonitrile and hexane on several C18 stationary phases obtained from the same silica (7.1 μmol/m^2 silanol groups) and having different coverage densities with C18 groups [94].

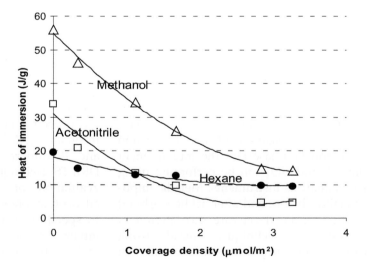

infrared spectroscopy [95]. The conformational state of the alkyl chains in their bound state via various conformation-sensitive vibrational bands of the symmetric and antisymmetric CH_2 stretching bands was monitored. CH_2 wagging type vibration bands were used to identify and determine the relative amounts of different conformers over the whole chains. The conformational state of the chemically attached alkyl chains was found to depend strongly on temperature, actual chain length, and chain position. Shorter alkyl chains, such as octyl and decyl chains, were reported to possess a high degree of conformational disorder, and increase in alkyl chain length was found to decrease the conformational disorder, most probably due to better chain packing.

High-resolution ^{29}Si-NMR has proved to be a powerful tool for structural elucidation of organic and inorganic silicon compounds. Because of the sensitivity of ^{29}Si chemical shifts, detailed information can be obtained by ^{29}Si-NMR on the structural surroundings of a given Si atom in a complex molecular framework. This information can be used to study microstructural details of Si-O-Si linkages connected in chains or branched and crosslinked structures. In solid-state samples, due to the limited motion, strong dipolar-dipolar and chemical shift anisotropy interactions occur. The line-broadening effects can be canceled using the magic angle spinning (MAS) technique. Other problems that arise in solid-state NMR spectroscopy related to the strong heteronuclear interactions and to the existence of very long spin-lattice relaxation times of the order of minutes to hours are solved through the cross polarization (CP) technique [96]. Solid-state ^{13}C-NMR using cross polarization and magic angle spinning (CP/MAS) can help characterize different carbon atoms from bonded phases. For example, this technique can be useful for characterization of different branched and unbranched alkyl groups resulting during silica derivatization and functionalization [97].

6.4. HYDROPHOBIC STATIONARY PHASES AND COLUMNS

Types of Hydrophobic Stationary Phases

The technique most often utilized in practice is reversed phase (RP), and this type of HPLC requires hydrophobic stationary phases. The related technique known as nonaqueous reversed-phase chromatography (NARP), as well as ion-pair reversed-phase chromatography, also use hydrophobic stationary phases. The most common hydrophobic phases are octadecyl (C18 or ODS) and octyl (C8), which consist of octadecyldimethylsilane groups or octyl-dimethylsilane groups, respectively, usually bonded to high-purity, spherical silica gel. Other alkyl phases and phenyl phases on silica are used occasionally, and long-chain phases such as C27 or C30 are sometimes used for special applications. Also considered as hydrophobic stationary phases but having some polarity are the fluorinated phases, and intermediate between hydrophobic and polar phases are the cyanopropyl type. Most bonded phases are made using so-called base-deactivated silicas (Type B purity). The deactivation consists of acid washing of the silica before bonding single or multiple reactant groups (ligands) to form the desired active phase.

Besides phases bonded on porous silica, core-shell particles are successfully utilized as support for hydrophobic phases. Also, monolithic silica-based materials with bonded hydrophobic phases are used in practice. (Various types of silica supports were discussed in Section 6.1, and basic procedures for silica surface derivatization were discussed in Section 6.2.)

Both octyl and octadecyl phases can have quite a large range of properties. These properties are a result of different reagents used for the silica derivatization (mono-, di-, or trifunctional), the reaction conditions (vertical or horizontal polymerization), the degree of derivatization (leading to different carbon loads), the type of silica, and the treatment of free silanols remaining on the silica surface (end-capping). Silanol coverage is related to phase resistance to the pH of the mobile phase, as well as the intensity of other interactions exhibited by the phase.

The common pH range for the stability of silica-based stationary phases is between 2 and 8. This range is not entirely satisfactory in some applications. Basic pH is able to deteriorate the silica structure, and pH below 2 produces hydrolysis of the fragments that make the active phase. Silica-based stationary phases resistant to a wider pH range are highly desirable, particularly when more acidic mobile phases are necessary. Lowering the pH of the aqueous component in the mobile phase to very acidic values, close to 1, may improve some of the retention parameters (retention time, peak shape, or selectivity), in particular for analytes having $pK_a < 2$. One reason for this effect is that at low pH values the interactions of residual silanols with amino containing solutes are suppressed, and thus peak tailing is no longer observable. However, such low pH values may not be acceptable for the stationary phase stability. The end-capping of silanol groups, with the purpose of reducing secondary interactions with the analytes, also has some effect on extending the acceptable pH range for the mobile phase. This effect is due to the blocking of the access of H^+ or OH^- ions toward the silica.

Besides the phases with the hydrophobic groups bound directly to a silica substrate, hydrophobic phases are also generated by bonding groups such as C8 or C18 on organic/inorganic silica particles (such as

ethylene-bridged hybrids or BEH phases). Also, polar-embedded phases with a polar group embedded between the silica surface and the hydrophobic moiety are successfully used. Among other advantages that may come from using the polar-embedded phases include a wider pH range of stability and less effect on the separation of the free silanol groups. The covering or encapsulating of the silica surface with polymeric materials is also used in some phases to reduce the activity of silanols. The results were phases showing minimal interaction and no or little tailing with strongly basic compounds (from where the name base deactivated).

Polymeric materials such as poly(styrene-divinylbenzene) (PS—DVB), with or with no phase attached to the polymer base, can also be used as reversed-phase media (these being more resistant to high pH mobile phases), but they typically have lower column efficiencies than those of silica gel-based packings. Special phases such as graphitized carbon, silicon hydride, and zirconia based are also known and have hydrophobic character.

Several hydrophobic types bonded on silica and polymeric materials used in RP-HPLC are presented in Table 6.4.1. As shown in the table, there are other types of interactions besides the hydrophobic interactions that are common to all columns.

End-capping of Free Silanols

A considerable difference can be seen in the characteristics of stationary phases that are end-capped and those that are not. This difference is illustrated, for example, in Table 6.3.5 where values for log K_{ow} are given for models for the two types of phase. Besides that, differences in peak shape and behavior toward polar compounds can be significant. Several strategies have been used to reduce the undesirable effect of residual silanols.

TABLE 6.4.1 Several types of hydrophobic stationary phases.

No.	Type of phase	Phase	Interactions
1	n-Alkyl on silica* end-capped	C2, C4, C8, C12, C14, C18, C20, C22, C27, C30	Medium or strong hydrophobic
2	n-Alkyl on silica not end-capped	C8, C12, C14, C18	Strong hydrophobic and weak polar
3	n-Alkyl on silica polar end-capped	C8, C18	Strong hydrophobic and medium polar
4	Polar embedded on silica	C8, C18, etc.	Strong hydrophobic and weak polar
5	Cyclic alkyl on silica	Cyclohexyl, n-C6 linked cyclohexyl	Medium hydrophobic
6	Mixed alkyl on silica	C18-short alkyl, C4-C18	Medium or strong hydrophobic
7	Aryl on silica	Phenyl, diphenyl, C2 linked phenyl, C6 linked phenyl	Medium hydrophobic and strong aromatic
8	Mixed alkyl-aryl on silica	C18-phenyl, C6-phenyl, cyclopropyl-phenyl, C14-phenyl	Strong hydrophobic and medium aromatic
9	Cyano on silica	Cyanopropyl, phenylcyanopropyl	Medium hydrophobic and strong polar
10	Fluorinated phases on silica	Pentafluorophenyl, perfluoroalkyl	Medium hydrophobic
11	Other on silica	Fulerene, cholesterol	Strong hydrophobic
12	Silicon hydride	C8, C18 bonded, silicon hydride partially graphitized	Special
13	Graphitized carbon	Graphytized carbon	Strong hydrophobic
14	Polymeric block	Polystyrene/divinylbenzene, ethylvinylbenzene/divinylbenzene	Strong hydrophobic
15	Phases bonded on organic polymers	C18 bonded on polymethacrylate, phenyl bonded on polymethacrylate, C18 bonded on divinylbenzene, pentafluorophenyl bonded on divinylbenzene	Medium or strong hydrophobic
16	Phases bonded on zirconia	C8, C18, etc.	Strong hydrophobic

Note: Phases indicated on silica include, besides silica, phases on organic/inorganic silica particles such as ethylene-bridged hybrids (BEH) or Gemini NX.

Among these strategies can be included the end-capping using trimethylsilyl groups by treating the phase with reagents such as trimethylchlorosilane or hexamethyldisilazane after the desired groups were already attached. This procedure was previously discussed. "Double end-capped" columns are reported as providing further coverage of silanol groups (e.g., Eclipse XDB). Another method used for increasing the stability of the silica-based stationary phase below $pH = 2$ relies on a special derivatization approach by which protecting groups are present close to the silica surface [98]. By steric hindrance, these groups provide a kind of shield for the area where the active phase is attached, protecting it from an acid attack from the mobile phase. This type of protection is achieved with voluminous groups, as shown schematically in the formulas in Figure 6.4.1. The figure shows the chemical structures as well as the models of molecules (with R = C4) picturing van der Waals surfaces obtained from simplified structures, with the geometry optimized using an Alchemy 2000 program (Tripos, Inc., MO). The larger volumes of isopropyl and isobutyl groups provide some protection to the bond connecting the silica surface, with the group acting as stationary phase. End-capping can also be used to further protect from hydrolysis stationary phases that are based on inorganic/organic base structures, such as those containing ethylene bridges. The reported results

show the excellent resistance of these columns to both higher and lower pH values compared to the stationary phase obtained on simple silica.

Another procedure to protect the silica surface is the use of very long hydrophobic alkyl chains (C27, C30) for derivatization. Because of the higher degree of surface shielding, long-chain stationary phases are stable over a wider pH range than C8 or C18 bonded phases.

Besides end-capping with small hydrophobic groups such as trimethylsilyl, end-capping with small fragments containing polar groups was found to protect the silica structure and extend the range of pH where the columns can be utilized. This procedure is also utilized for generating phases with some polar character in addition to the hydrophobic

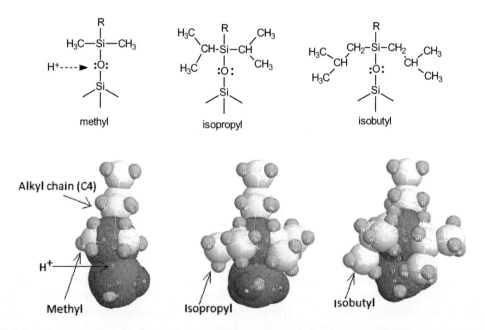

FIGURE 6.4.1 Schematic formulas for bonded phases having two methyl groups attached on the silicon bonded to the active fragment, and for phases sterically protected with two steric-hindrance diisopropyl groups or diisobutyl groups (the R chain was selected C4). Van der Waals surfaces with the geometry optimized using an Alchemy 2000 program are also shown.

character of long alkyl groups. One such column (Synergi 4u Hydro-RP 80A) can be used, for example, for the separation of amino acids by ion-pair chromatography (see Section 5.2). The polar-end-capping groups are usually amino or hydroxyl, bonded to a propyl chain. These polar-end-capped stationary phases possess hydrophobicity similar to non−end-capped ones, and they display enhanced hydrogen bond-type interactions, but the acidity of the polar groups is reduced, and it is better controlled than that of silanols. Polar end-capping makes the phases more compatible with highly aqueous mobile phases, and this type of column has useful specific applications. For example, use of a polar-end-capped C18 phase can retain more polar water-soluble compounds such as water-soluble vitamins, being a better choice for their separation. One additional procedure applied to have polar groups shielding the silanol groups from the silica surface is the use for derivatization of trifunctional reagents as previously shown in reactions 6.2.5 or 6.2.6 (see Section 6.2) and the generation of silanol groups that are not directly bonded to the silica gel surface. In reaction 6.2.5, these silanol groups are further reacted, but in special conditions (with a higher proportion of water) the final product may remain with intact OH groups. This type of reaction is schematically shown in reaction 6.4.1 for a tri-functional chlorinated reagent.

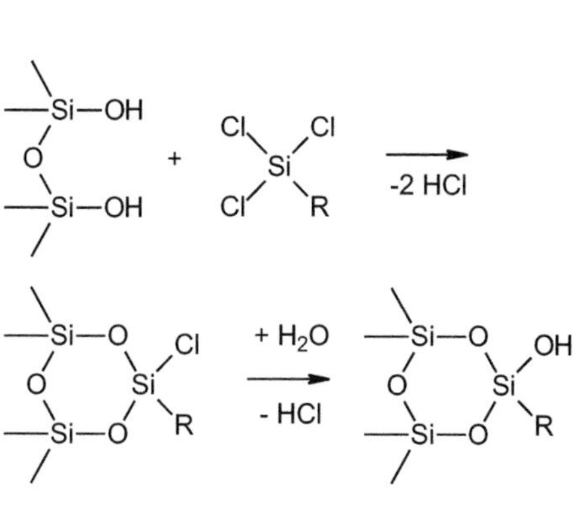

The silanol groups not connected to the silica gel surface may have lower acidity and a protection effect against the attack of strong bases or acids.

Other procedures utilized to shield the interaction of silanol groups with the analytes include the use of reagents having multi point reactive functionalities for attaching the desired group on silica surface [99]. Also, horizontal polymerization leads to better protection of silanol surface interactions [50]. The introduction of a small percentage of groups displaying positive charges on silica surface (charge surface hybrid or CSH) also has been used for blocking the interaction with the silanols and reducing peak tailing for basic compounds.

Preparation of Polar-Embedded Hydrophobic Phases

A special type of hydrophobic phase are those phases containing a polar-embedded moiety in the chain connecting the hydrophobic group with the silica. These phases are used in RP chromatography and have a number of advantages over common RP stationary phases (such as C8 or C18), including a wider range of selectivity versus the analytes, higher stability to phase collapse in eluents high in water content, reduced silanol activity with more symmetrical peaks, and better stability in low and high pH mobile phases (see Section 6.3).

The hydrophobic character of the columns with RP phases containing a polar-embedded group is close to that of a regular corresponding phase. The calculated log K_{ow} for the model of an amide-embedded phase shows, for example, close values to that of a not end-capped C18 phase (see Table 6.3.5.)

Polar-embedded stationary phases can be synthesized using a two-step process, as shown in reactions 6.2.11, 6.2.12, and 6.2.13. A single-step procedure can also be utilized for generating embedded phases. As an example, an embedded polar stationary phase

(6.4.1)

that contains a polar group within a long hydrophobic chain attached to the silica surface can be obtained by reaction 6.4.2:

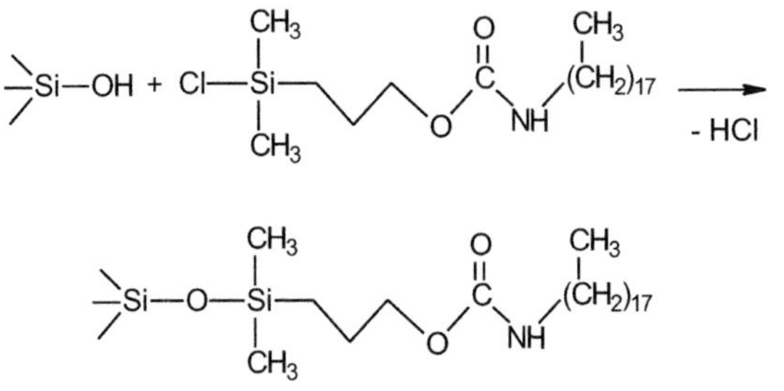

(6.4.2)

The advantages obtained with polar-embedded phases and the simplicity of a single-step procedure lead to the synthesis of a number of silanes with a polar-embedded functionality in a hydrophobic chain. They are currently used in various single-step reactions with the silica surface. This kind of reaction is schematically shown in reaction 6.4.3 for various embedded groups X:

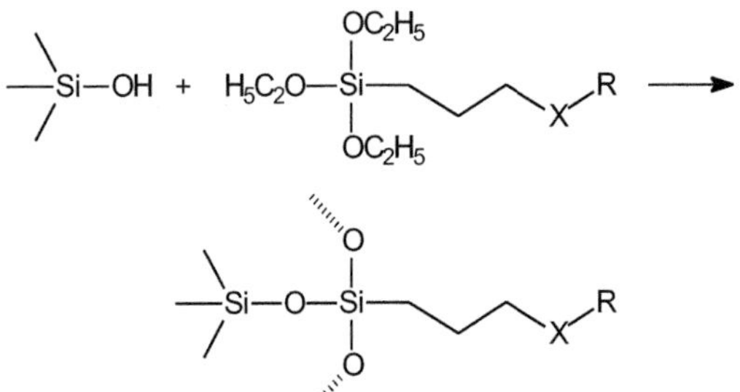

(6.4.3)

where −X− can be one of the following groups shown in Figure 6.4.2.

Polar embedded phases having two zones of polarity or an extended carbon linker (e.g., 5 or 6 atoms to the silica) are also known [100]. These phases reduce the possibility of interactions between the main polar group and silanol from the silica surface. Bidentate-embedded stationary phases have also been reported in the literature [101]. The additional end-capping of polar-embedded stationary phases can be done in the same manner as for other common hydrophobic phases.

Physical Characteristics of Hydrophobic Stationary Phases

In addition to the nature of the stationary phase, other physical and chemical characteristics of the phase are important for the column properties toward a separation. The main such characteristics and the range of their typical values are summarized in Table 6.4.2 (see also Table 6.3.1). Regarding the physical properties of stationary phases listed in Table 6.4.2 such as particle type, surface area, particle size, and pore size, some discussion can be found in Section 6.1. These properties, such as the pore size and the particle surface area, are somewhat interrelated, although no direct correlation exists between the two parameters since a wide variety of technologies are used in silica base manufacturing. The physical properties of stationary phase are important for the values of chromatographic parameters R, n (N), k, and α. For some applications, other parameters are

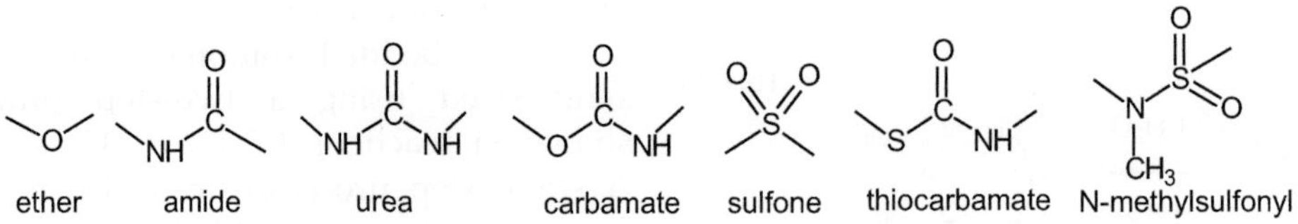

FIGURE 6.4.2 Bidentate groups that can be used in the synthesis of polar embedded stationary phases.

TABLE 6.4.2 Characteristics of the phase and of columns with hydrophobic stationary phases.

Property	Range
Phase type	Bonded on silica*, graphitized carbon, polymeric, bonded on polymer
Carbon load (for silica)	5% to 25 %
Coverage (for silica)	Depending on phase, 2.0 μmol/m^2 to 4.5 μmol/m^2
End-capping (for silica)	Yes or no, non-polar or polar
Hydrophobic character	Characterized by k for hydrophobic compounds or by α(CH$_2$)
Silanol activity (for silica)	Very low to high
Metal activity	Very low to high
Silica purity (for silica)	High to medium (Type B or Type A)
Wetting characteristics	Poor to good
Particle type (for silica)	Porous, core-shell, pellicular, monolithic
Particle shape (for silica)	Irregular or spherical
Surface area (for silica)	50 m^2/g to 500 m^2/g
Particle size	1 μm to 10 μm
Particle size distribution	Narrow (1 - 2% size variation) to larger distributions
Pore size	50 Å to 4000 Å
Column i.d.	2 mm to 10 mm (common i.d. in mm 2.1, 3.0, 4.6)
Column length	50 mm to 300 mm (or longer) (common length in mm 150, 250)

Note: Some of the phases indicated on silica include, besides silica, phases on organic/inorganic silica particles such as ethylene-bridged hybrids.

important such as the pore size of the stationary phase, which is very important when macromolecules are separated. For protein separation, for example, where the molecules are larger than about 2000 Da, the pore size of the column must be large (> 300 Å) in order to allow the analyte to penetrate the pores of the stationary phase and interact with the bonded moiety. Otherwise, the molecules are excluded from the interaction.

Retention and Separation Properties of Hydrophobic Stationary Phases

Retention in chromatography is a complex process whereby various interactions take place between the three participants in the separation:

(1) the analyte, (2) the stationary phase, and (3) the mobile phase. For this reason, in order to compare stationary phases regarding their capability of retention, the other two factors, which are the mobile phase and the analyte, must be specifically selected and kept unchanged within a comparison.

The *hydrophobic character* of the stationary phase is a very important property in RP-HPLC. A detailed discussion of hydrophobic interactions in general was given in Section 4.1, and the mechanism in RP-HPLC was described in Section 5.1. In Section 5.1 has been shown that under the name "hydrophobic interactions" are incorporated several types of contributions. They include the energy from the "ideal gas phase interactions", energy from van der Waals interactions, and the energy from the cavity creation in the mobile phase. All these interactions are indicated as "hydrophobic" since the cavity creation is the largest energy contributor, but they include polar and polarizability contributions (see rel. 5.1.26). As discussed in Section 4.1 (and schematically suggested in Figure 5.1.3), hydrophobicity is mainly a "response" to the polar character of the mobile phase the contribution from the van der Waals interactions being small. This aspect of hydrophobic interactions points to the major role of the solvent in RP-HPLC, which can be considered to "overrule" the role of the stationary phase. Most organic molecules have both hydrophobic moieties and polar moieties, and for a given solvent polarity, the solvophobic interactions depending mainly on the hydrophobic character of the molecule and of the stationary phase differentiate molecular retention in the HPLC process.

Several properties of the stationary phase are indicative of its hydrophobicity. For silica-based phases, these properties include (1) the nature of the bonded substituent (C8, C18, phenyl, etc.), (2) the carbon content (carbon load C%), (3) the density of surface coverage with the bonded phase, (4) the weakness of the silanol activity (increased, e.g., by the impurities in silica), and (5) the level of solvent molecules adsorbed on the stationary phase.

The nature of the bonded substituent influences the hydrophobicity by various characteristics. For saturated hydrocarbon substituents, the increase in the number of carbon atoms in the chain typically leads to an increased hydrophobicity, although this increase is not linear, as shown in Figure 5.1.6. In addition, a considerable overlapping in the hydrophobicity can be seen between columns with different numbers of carbons in the bonded moiety.

Another factor contributing to column hydrophobicity is the chemical nature of the substituent, in particular the polarity of the bonded ligand. This has been illustrated in Table 6.3.5, where $\log K_{ow}$ for models of different phases are listed. The values of $\log K_{ow}$ for models for phenyl and especially for cyanopropyl are significantly lower, for example, than those for C18 columns. However, physicochemical phase properties are difficult to be directly related to a value that gives the strength of the hydrophobic interactions, and only a higher or a lower resulting hydrophobicity can be suggested. Each of the physicochemical phase properties also depends on other stationary phase characteristics. For example, the carbon load (%C, as determined by elemental analysis), which is also important for phase hydrophobicity, is related to: (1) the nature and length of the hydrocarbon chain used in the bonded phase, (2) the degree of hydroxylation of silica surface subjected to the derivatization procedure, and (3) the derivatization yield giving the ratio between derivatized hydroxyl number to the initial number of OH groups on the silica surface. In general, a higher carbon load (which can reach up to 22—24% for RP stationary phases) is an indication of a more hydrophobic character. With new, more modern phases that have polar-embedded groups, bidentate attachments to the silica surface, or organic/inorganic solid support, it is even more difficult to predict hydrophobicity from the stationary phase construction.

To characterize the hydrophobicity of a stationary phase, one parameter that is typically used is the capacity factor k (or $\log k$) for hydrophobic-tested compounds [102]. The values of capacity factor k can be used to compare columns with different carbon contents, chain lengths, or other derivatization approaches. A higher k (for a hydrophobic analyte) indicates a more hydrophobic character for the phase. However, the value of k must be taken in a relative way, since it depends not just on the nature of the stationary phase, but also on the nature of the mobile phase and the particular hydrophobic analyte used in the test.

As is well known, RP-HPLC is commonly used for the separation of a variety of organic molecules that have a hydrophobic moiety, it may also contain other groups such as polar or even ionic ones. However, for the phase characterization regarding exclusively the hydrophobic interactions, the analyte must have only a hydrophobic character. Therefore, the retention must be measured for compounds that have very low or no polarity. Such compounds are characterized by little or no tendency to adsorb water, and water forms "beads" on their surface. Hydrophobic substances lack the groups that can participate in the formation of hydrogen bonds with water. Many of the hydrophobic substances are hydrocarbons, although several other moieties are not hydrophilic (halogen, ester, ether, and other less polar moieties). Specific hydrophobic compounds commonly used for the evaluation of hydrophobic character include toluene, ethylbenzene, naphthalene, and acenaphtene. When such compounds are used, a greater k (and $\log k$) will indicate a more hydrophobic stationary phase. For the selected compound, the calculation of k is performed based on retention time t_R and the value t_0 for the column, where $k = (t_R - t_0)/t_0$ and t_0 is typically measured using the retention time of uracil or of thiourea, which practically are not retained on hydrophobic columns but at the same time are not "excluded" from the pores of the packing material (as are, for example, certain ionic compounds).

The capacity factor for toluene (expressed as ln k) for a number of C18 columns, measured for a mobile phase consisting of 80% CH_3OH and 20% aqueous buffer 0.25 mM KH_2PO_4 at pH = 6 and at 24 °C (in isocratic conditions), are given in Table 6.4.3. The capacity factor k depends on several variables, the carbon load of the column being a contributor to the value of k [104]. For example, for the columns listed in Table 6.4.3, the carbon load is given in Figure 6.4.3, and the dependence of capacity factor on the carbon load is shown in Figure 6.4.4. As seen in Figure 6.4.4, a positive correlation exists between the capacity factor and the carbon load of the column, although only a $R^2 = 0.3942$ is obtained for this correlation. This poor correlation is, in part, the result of the fact that hydrophobicity is more a "response" to the polar character of the mobile phase than a force in itself.

As expected, other factors also contribute to the value of the capacity factor. One important such factor is the surface area of the stationary phase. As shown by rel. 2.1.13, $k = K\Psi$, and the value for Ψ depends on the pore volume (and/or the surface area) of the stationary phase; therefore, k should also depend on the surface area of the stationary phase. For the columns listed in Table 6.4.3, the surface areas are shown in Figure 6.4.5, and the dependence of capacity factor on surface area is shown in Figure 6.4.6. Similar to the carbon load, the correlation presented in Figure 6.4.6 is positive, but the R^2 value for it is not very good ($R^2 = 0.4779$). Some improvement in the R^2 value of a correlation is obtained if both carbon load and surface area are considered, but the correlation coefficient is still not very high, indicating that more factors contribute to the value of capacity factor. Even better results are obtained when a parameter is included to account for silanol activity of the column. The variation in silanol activity for the columns listed in Table 6.4.3 is shown in Figure 6.4.7.

The values of carbon load, phase surface area, and silanol activity (with arbitrary assigned values 1 for very low and 4 for high activity) allow an estimation of k, using a dependence equation of the form:

$$k_{calc} = a + b\,C\% + c\,\text{(Surface area)}$$
$$+ d\,\text{(Silanol activity)} \qquad (6.4.4)$$

A graph showing the variation of the calculated values k_{calc} for toluene as a function of measured k values is shown in Figure 6.4.8. As shown in this figure, there is a positive correlation between the measured and the calculated k values. However, the correlation is not very good. This indicates that the retention properties of a column are a function of a number of stationary-phase characteristics not accounted

TABLE 6.4.3   Capacity factor $\ln k$ for toluene on C18 stationary phases (5 μm particles) [103]. *Mobile phase consisted of 80% CH_3OH and 20% aqueous buffer 0.25 mM KH_2PO_4 at pH = 6.*

No.	Column	$\ln k$	No.	Column	$\ln k$
1	ACE C18	1.27	23*	LiChrosorb RP-18	1.095
2	ACE C18-300	0.59	24	LiChrospher RP-18	1.59
3	ACE C18-HL	1.85	25	Luna 5 C18(2)	1.58
4*	μBondapak C18	0.52	26**	Novopak C18	0.7
5	Capcell Pak AG C18	1.18	27	Nucleosil C18	1.42
6	Capcell Pak UG C18	1.26	28	Nucleosil C18 HD	1.405
7	Develosil ODS-HG	1.09	29	Nucleosil C18AB	1.02
8	Develosil ODS-MG	1.82	30*	Partisil ODS	0.3
9	Develosil ODS-UG	1.32	31*	Partisil ODS2	1.28
10	Exsil ODS	1.09	32*	Partisil ODS3	0.92
11	Exsil ODS1	0.65	33	Prodigy ODS2	1.39
12	Exsil ODSB	0.52	34	Prodigy ODS3	1.8
13	Gemini C18	1.41	35	Purospher RP18-e	2.05
14	Hichrom RPB	1.1	36	Resolve C18	1.15
15	Hypersil BDS C18	0.98	37	SunFire C18	1.8
16	Hypersil GOLD	0.69	38	Symmetry C18	1.81
17	Hypersil HyPurity C18	0.85	39	TSK ODS-120T	0.95
18	Hypersil ODS	0.98	40	TSK ODS-80TM	1.11
19	Inertsil ODS	1.29	41	Ultrasphere ODS	1.17
20	Inertsil ODS3	2.19	42	Vydac 218MS	0.4
21	Inertsil ODS2	1.47	43	Vydac 218TP	0.4
22	Kromasil C18	1.99	44	Vydac Selectapore 300M	0.33
45	Vydac Selectapore 300P	0.4	53	YMC ODS A	1.26
46	Vydac Selectapore 90M	0.77	54	YMC ODS AM	1.25
47	Waters Spherisorb ODS1	0.8	55	YMC Pro C18	1.26
48	Waters Spherisorb ODS2	1.29	56	Zorbax Extend C18	1.52
49	Waters Spherisorb ODSB	1.11	57	Zorbax ODS	1.7
50	XTerra MS C18	1.2	58	Zorbax Rx-C18	1.31
51**	YMC J'Sphere ODS H80	1.95	59	Zorbax SB-C18	1.15
52**	YMC J'Sphere ODS M80	0.98	60	Zorbax XDB-C18	1.4

* Note: Indicates 10 μm particle size
** Note: Indicates 4 μm particle size

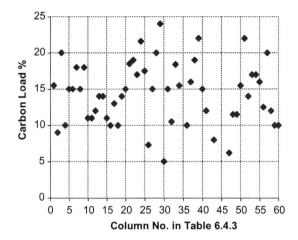

FIGURE 6.4.3 Values for carbon load % for the columns listed in Table 6.4.3.

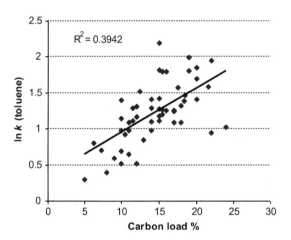

FIGURE 6.4.4 Variation of capacity factor ln k as a function of carbon load.

for in rel. 6.4.4, and/or the dependence on these parameters is not linear.

Several other studies showed that a much better correlation can be obtained for k for hydrophobic compounds on a large number of columns when an expression of the following type is used:

$$\log k = a + b\,\mathrm{n} + c\,\mathrm{n}^2 + d\,d_p + e\,d_p^2 + f\,c_L + g\,c_L^2$$
$$+ h\,E_c$$

(6.4.5)

where n is the number of carbons in the alkyl chain of the ligand (e.g., 8 for C8, 18 for C18), d_p is the pore diameter of the particles (related to the surface area of the phase), c_L is the density of the active phase (ligand density), and E_c is a parameter describing the end-capping ($E_c = 1$ for end-capped column and $E_c = 0$ otherwise) [104, 105]. The parameters a, b, c, and so forth, are coefficients obtained from the regression fit, and their values depend on the tested compound for the correlation. Even using a multiple and nonlinear correlation as in formula 6.4.5, when using a large number of columns, in order to obtain a R^2 value higher than 0.9 for the correlation between calculated and measured values for k, some outlier columns must be eliminated [105]. This indicates that either some characteristics of the outlier column are not reported correctly by the manufacturer or that these columns have unusual properties not captured by the parameters n, d_p, c_L, and E_c. Various other estimations for k are suggested in the literature [106] using similar or different parameters related to the physicochemical stationary-phase characteristics.

The use of capacity factor k for characterization of various chromatographic columns has considerable utility in particular because it has been shown that for two different hydrophobic compounds, an excellent linearity in the k values is obtained for a large number of RP columns [107]. This indicates that the particular choice of hydrophobic reference compound used for testing is not essential for comparison of column hydrophobicity, although the same compound must be used when comparing a set of columns. Depending on the selected compound, the value of k may cover a wider or a narrower scale. Because of the dependence of the value on the compound, the capacity factor k can be used only as a relative parameter for characterization of hydrophobicity. Ethylbenzene is commonly used as a standard compound for comparing column hydrophobicity.

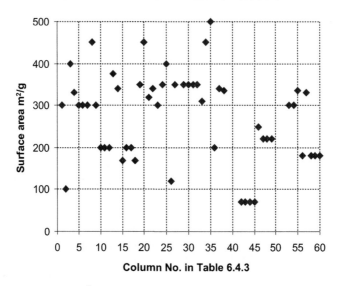

FIGURE 6.4.5   Values for surface area m^2/g for the columns listed in Table 6.4.3.

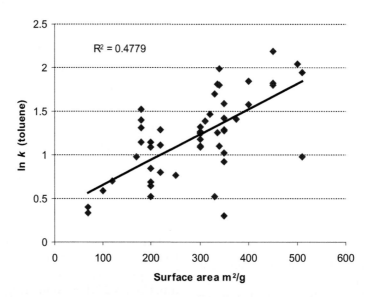

FIGURE 6.4.6   Variation of capacity factor ln k as a function of surface area of the particles.

RP-HPLC is a technique widely used not only for hydrophobic compounds separations. A range of molecules that have only a small hydrophobic moiety can also be successfully separated on this type of column. In a first approximation, it can be assumed that the separation of polar compounds containing hydrophobic moieties is achieved on RP columns only due to the interaction of their hydrophobic moieties with the hydrophobic

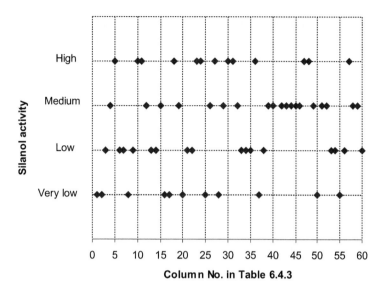

FIGURE 6.4.7 Silanol activity (qualitative) for the stationary phase for the columns listed in Table 6.4.3.

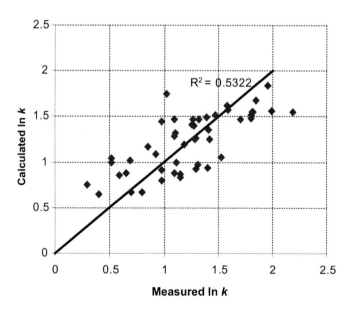

FIGURE 6.4.8 Dependence of calculated $\ln k_{calc}$ for toluene as a function of measured $\log k$ values for this compound, with $\ln k_{calc} = a + b\ C\% + c\ (\text{Surface area}) + d\ (\text{Silanol activity})$ ($a = 0.367045$, $b = 0.049043$, $c = 0.001141$, $d = -0.06382$).

stationary phase of the column (in equilibrium with the hydrophobic character of the mobile phase). For two different columns (indicated as c1 and c2) and the same mobile phase, an excellent correlation has been reported between $\log k_{c1}$ and $\log k_{c2}$ for a large number of compounds that have hydrophobic moieties [105]. This correlation between the retention on two columns can be written in the form:

$$\log k_{c1} = A_1 + A_2 \log k_{c2} \qquad (6.4.6a)$$

where A_1 and A_2 are constants depending on the two compared columns. Relation 6.4.6a can be easily generated if it is assumed that the capacity factor k_{c1} for a compound j is expressed by a formula of the type:

$$\log k_{c1}(j) = a_{c1} + \eta'(j)H_{c1} \qquad (6.4.7)$$

where $\eta'(j)$ is a parameter characterizing the hydrophobicity of compound j, H_{c1} is the parameter for the hydrophobicity of column c1, and a_{c1} is a constant. For a second column and the same evaluated compound, rel. 6.4.7 will be $\log k_{c2} = a_{c2} + \eta'(j) H_{c2}$. By replacing $\eta'(j)$ from one expression into the other, the following result is obtained:

$$\log k_{c1} = \left(a_{c1} - a_{c2} \frac{H_{c1}}{H_{c2}} \right) + \frac{H_{c1}}{H_{c2}} \log k_{c2}$$

$$(6.4.6b)$$

which is identical to rel. 6.4.6a when replacing the first term in rel. 6.4.6b with A_1 and the coefficient for $\log k_{c2}$ with A_2 (index j is omitted).

By selecting ethylbenzene (EB) as a standard reference ($\eta'(EB) = 0$), rel. 6.4.7 can be written for any given hydrophobic compound in the form:

$$\log k_{c1} = \log k_{c1}(EB) + \eta' H_{c1} \qquad (6.4.8)$$

For a compound j, rel. 6.4.8 (with some changes in notation) gives: $\log k_{c1}(j) - \log k_{c1}(EB) = \log (k/k_{EB}) = \log [\alpha_{c1}(j,EB)]$, which can be written in the form:

$$\log \alpha_{c1} = \log(k/k_{EB}) = \eta' H_{c1} \qquad (6.4.9)$$

Parameter H_c has the advantage of indicating the capability of a column to separate hydrophobic compounds (separation from EB in particular), while $\log k$ indicates only the retention capability. From rel. 6.4.8 it can be seen that H_c describes how much the hydrophobic character of the column influences the separation between ethylbenzene and another hydrophobic compound.

Similar to $\log k$, parameter H_c is dependent on column/stationary phase properties such as n the number of carbons in the alkyl chain of the active phase, the pore diameter d_p, the density of the active phase c_L, and the end-capping parameter E_c. For example, a formula successfully used for estimating H_c is the following:

$$H_c = a + b \log n + c \log d_p + d \log c_L + e E_c$$

$$(6.4.10)$$

As in the case of rel. 6.4.5, the coefficients a, b, c, d, and e are obtained from best regression fit for a large number of compounds. The value of H_c increases with the number of carbons in ligand that forms the bonded phase ($b > 0$), and it also increases for smaller pore diameters ($c < 0$), for higher ligand density ($d > 0$) and for end-capping ($e > 0$). Columns obtained by vertical polymerization having a high carbon content (large c_L) usually have a high H_c value.

Methylene Selectivity

Besides using the capacity factor and parameter H_c, the separation capability for hydrophobic compounds of an RP column is frequently described by another parameter that is also related to hydrophobicity, indicated as *methylene selectivity* and noted as $\alpha(CH_2)$. This parameter is defined as the selectivity α for two compounds that are different by a CH_2 group (one example being the pair toluene and ethylbenzene). The formula for $\alpha(CH_2)$ is as follows:

$$\alpha(CH_2) = \frac{k_{R-CH_2-R}}{k_{R-R}} \qquad (6.4.11)$$

The value of $\alpha(CH_2)$ depends as expected on the stationary phase and on the selected mobile phase. However, it has been found that for several classes of solutes with an adjacent number of carbons, $\alpha(CH_2)$ is a constant (for a given column and mobile phase). Even more, a linear dependence for log $\alpha(CH_2)$ on the number of CH_2 units was noticed for certain homologous series such as the one from benzene to amylbenzene. This linearity was verified for various columns [108] and also for various mobile phases [109]. Based on this observation, the values for log $\alpha(CH_2)$ can be obtained from the slope of the trendline of the graph of log k versus the number of methylene groups in the alkylbenzene:

$$\log \alpha(CH_2) = \log (k_i/k_j) = \log k_i - \log k_j$$
$$= \tan \beta$$
$$(6.4.12)$$

where i and j are two consecutive homologs $(i - j = 1)$.

The graph showing $\alpha(CH_2)$ calculation for the case of a Zorbax Bonus-RP column, with the mobile phase composed of acetonitrile and aqueous buffer (20 mM KH_2PO_4-K_2HPO_4 at pH = 7) in the ratio 65/35 (v/v) as reported in [108], is given in Figure 6.4.9. The methylene

selectivity can be used for comparing columns of the same type (e.g., C18 columns USP code L1) or even for comparing RP columns of different types such as C18 and C8, or C18 and C_{Phenyl}.

Although the capacity factor k and the methylene selectivity for hydrophobic compounds are related, their correlation is not very strong. As an example, the correlation between the values of k for butylbenzene and the values of $\alpha(CH_2)$ obtained for the homologous series from benzene to amylbenzene on different columns, when the mobile phase is composed of acetonitrile and aqueous buffer (20 mM KH_2PO_4-K_2HPO_4 at pH = 7) in the ratio 65/35 (v/v), is shown in Figure 6.4.10 (data from [108]). As seen from the graph, $R^2 = 0.6645$ indicates a modest positive correlation. The list of the columns on which the values for the graph in Figure 6.4.10 were measured is given in Table 6.4.4.

When a column is selected for an application requiring the separation of hydrophobic compounds, the preferred characteristic should be a high $\alpha(CH_2)$ for obtaining a better separation [110]. However, methylene selectivity depends on the characteristics of both the

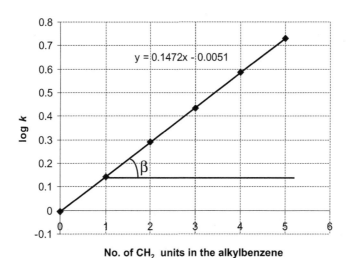

FIGURE 6.4.9 Calculation of $\alpha(CH_2)$ from the k values of a homologous series of alkylbenzenes [108].

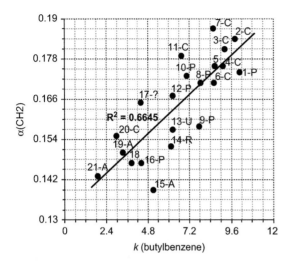

FIGURE 6.4.10 Correlation between the capacity factor k (for butylbenzene) and the methylene selectivity $\alpha(CH_2)$ for several columns described in Table 6.4.4.

TABLE 6.4.4 Various columns evaluated for the measurement of k (for butylbenzene) and $\alpha(CH_2)$ shown in Figure 6.4.10 [108].

No.*	Column name	Column type
1-P	Synergy Hydro-RP	Polar-endcapped ODS
2-C	Prodigy ODS(3)	Conventional ODS
3-C	Symmetry C18	Conventional ODS
4-C	Luna C18(2)	Conventional ODS
5-P	Aqua C18	Polar-endcapped ODS
6-C	Intersil ODS(3)	Conventional ODS
7-C	Zorbax XDB C18	Conventional ODS
8-P	YMC ODS-Aq	Polar-endcapped ODS
9-P	Metasil AQ	Polar-endcapped ODS
10-P	Prontosil C18 AQ	Polar-endcapped ODS
11-C	Zorbax SB C18	Conventional ODS
12-P	YMC Hydrosphere	Polar-endcapped ODS
13-U	Experimental Urea	Urea-linked C18
14-R	Symmetry Shield RP 18	Carbamate linked C18
15-A	Polaris Amide C18	Amide-linked C18
16-P	Keystone Aquasil	Polar-endcapped ODS
17-?	Polaris C18-A	Unknown linked C18
18-A	Zorbax Bonus-RP	Amide-linked C18
19-A	Discovery RP-Amide	Amide-linked C18
20-C	Hypurity Elite C18	Conventional ODS
21-A	Supelco ABZ+	Amide-linked C18

* *Note: the letters indicate P = polar-endcapped, C = Conventional, U = Urea linked, R = Carbamate linked, A = Amide linked.*

stationary and the mobile phase. Figure 6.4.11 shows the variation of $\alpha(CH_2)$ for several columns (LiChr. = LiChrospher 100 RP, Puro. = Purospher RP, Symm. = Symmetry-Shield RP), and two mobile phase systems, acetonitrile/water (ACN) and methanol/water (MeOH), at different concentrations [110]. All columns had 5 μm particle size, the column dimensions were 125 × 4 mm (LiChr. and Puro.), 150 × 3.9 mm (Symm), and except for LiChr. C8 the columns were end-capped.

Hydrophobic Subtraction Model

Most organic compounds contain in their molecules a variety of moieties with different polarities, besides the nonpolar ones. Besides hydrophobic interactions, all columns also interact with the solutes through other interaction types, not captured in the "hydrophobic" type. One of these interactions is caused by the presence of free silanol groups in the stationary phase. These groups interact either directly or through the adsorbed solvent on them, which can be water or other polar solvents [111, 112].

For RP-type columns, the nature of the interactions (besides the hydrophobic type covering solvent cavity formation, gas phase, and van der Waals interactions) include: hydrogen bonding, interactions of ion-dipol and ion-ion type, interactions with metal impurities from the silica, interactions with the adsorbed solvent, and steric interactions. Some authors

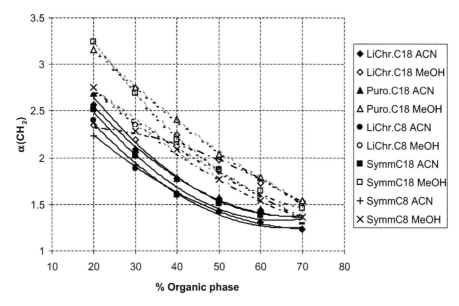

FIGURE 6.4.11 Variation of $\alpha(CH_2)$ for several columns and two mobile phase systems (acetonitrile/water and methanol/water) at different concentrations.

also include π-π interactions among those acting in RP, when the stationary phase contains phenyl or cyano groups [113]. In general, the π-π molecular interactions were not very important, as shown by ab-initio calculations, or by experimental results obtained for cyanopropyl columns [114]. All interactions may, or may not, play a role in the separation. The complexity of interaction process is schematically shown in Figure 6.4.12 [115].

The silanol groups on the silica have an opposite role in separation compared to the hydrophobic character of the bonded phase. It is generally accepted that the surface of modified silica still contains active sites given by Si—OH groups (silanols), regardless of the end-capping treatment of the material. The polar or apolar activities of the siloxane groups (Si-O-Si) are so low that they are not taken into consideration for the interactions between analytes and stationary phase. The silanols serve as attachment sites for the covalent silyl ether bonds that anchor-bonded phases to the silica support. For this reason, before the derivatization of the base silica gel (bare silica) it is desirable to have a large number of silanol groups. These groups are supposed to be derivatized with the bonded phase. However, because of steric hindrance during the derivatization of silica, the density of silanols on chromatographic-grade silica (about $8 \pm 1 \ \mu mol/m^2$) is much greater than the maximum possible concentration of alkyl groups in a bonded phase (about 4.5 $\mu mol/m^2$ for C18 ligands). Therefore, after the silica surface has been modified, numerous unreacted (residual) silanol groups are left within the bonded phase (about 50%). These residual silanols are weakly acidic, with pK_a values typically between 5 and 7, and they can interact with polar compounds through strong hydrogen bond and dipole-dipole interactions. At higher pH values of the mobile phase, the silanol groups may also have an ion-exchange type of contribution. As a result of the heterogeneous surface of the stationary phase, a mixed retention mechanism

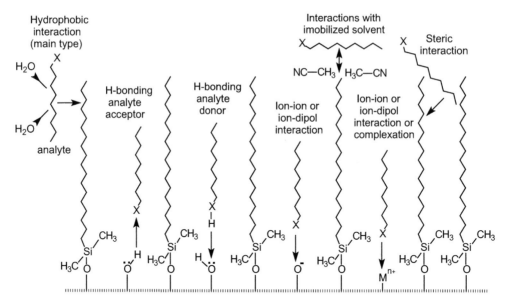

FIGURE 6.4.12 Schematic representation of various interactions in RP-HPLC.

is in place. This mechanism may influence the separation of all analytes. However, when only a small proportion of the analyte molecules participate in other interactions besides hydrophobic ones, peak tailing and loss of chromatographic resolution are seen. This happens especially when basic solutes are involved [116]. For this reason, silanol activity is sometimes characterized based on the values of As (j) or TF (j) (see rel. 2.2.16 and 2.2.17) for a specific polar compound j. The acidity of silanol groups can increase in the presence of transitional metal ions. The metal ions can also become adsorption sites for compounds that are able to form complexes, and the metals from silica can interfere in the retention process [117]. For this reason, high purity silicas are typically used for stationary phase manufacturing.

Besides some polarity resulting from the silanol groups (and adsorbed solvent), various columns are available having groups that confer some intentional polarity besides hydrophobic character. Examples are the columns that contain phenyl- or cyano-bonded phases. The columns with embedded polar groups may be included in this category. For these columns, besides the hydrophobic interactions, other forces contribute to the separation. When the separation is intended for highly hydrophobic compounds, these polar interactions are not important. However, the separation of organic compounds with hydrophobic and polar groups is more complex than that based on simple hydrophobic interactions, and in this case more parameters must be considered for describing the separation.

Due to various types of interactions that may take place between an analyte and the stationary phase, a deviation from rel. 6.4.9 is observed for particular analytes. In these cases the relation 6.4.9 can be written in the form:

$$\log \alpha_{c1} = \eta' H_{c1} + \Delta \qquad (6.4.13)$$

In rel. 6.4.13, $\Delta = 0$ for hydrophobic compounds, with no steric hindrance of interacting with the stationary phase. For other compounds, an extension of rel. 6.4.9 can be written for log α such that it will account in the value for a Δ, including all possible interactions. These interactions are: (1) steric interactions, (2) hydrogen bonding between a basic solute and an acidic column group (column acidity), (3) hydrogen bonding between an acidic solute and a basic column group (column basicity), and (4) cation exchange and/or ion-ion interactions. When solutes with a diversity of molecular structures were evaluated, it was possible to identify solutes for which the value of Δ is influenced (mainly) by only one type of the four listed interactions. Average values of Δ were then determined for each solute group (1 to 4), for a number of columns and using a specific mobile phase. For each column it was possible to generate a parameter S^*_{c1} accounting for steric interactions, A_{c1} accounting for hydrogen bonding between a basic solute and an acidic column group (column acidity), B_{c1} accounting for hydrogen bonding between an acidic solute and a basic column group (column basicity), and C_{c1} accounting for cation exchange or ion-ion interactions. Since these interactions can be assumed to be additive, for a column c1 and for the compound j the expression for $\alpha_{c1}(j)$ will be:

$$\log \alpha_{c1}(j) = \eta'(j)\,H_{c1} - \sigma'(j)S^*_{c1} + \beta'(j)A_{c1}$$
$$+ \alpha'(j)\,B_{c1} + \kappa'(j)C_{c1}$$

$$(6.4.14)$$

Since $\log \alpha_{c1}(j) = \log [k(j)/k_{EB}]$, rel. 6.4.14 is equivalent to the expression:

$$\log k(j) = \log k_{EB} + \eta'(j)H_{c1} - \sigma'(j)S^*_{c1}$$
$$+ \beta'(j)A_{c1} + \alpha'(j)B_{c1} + \kappa'(j)C_{c1}$$

$$(6.4.15)$$

where the hydrophobicity is described by log k_{EB} for the specific column, and the parameters $\eta'(j)$ for the solute and H_{c1} for the column (see rel. 6.4.9). Parameters $\eta'(j)$, $\sigma'(j)$, $\beta'(j)$, $\alpha'(j)$, and

$\kappa'(j)$ can be determined by multiple regression of values for log α_{c1} when H_{c1}, S^*_{c1}, A_{c1}, B_{c1}, and C_{c1} are known.

Parameters $\sigma'(j)$ and S^*_{c1} (this term is taken with a negative sign, although in some literature reports the positive sign is utilized [118]) are typically related to the steric resistance to insertion of a bulky solute characterized by $\sigma'(j)$ into a stationary phase characterized by S^*_{c1}. However, a different steric interaction process was observed for molecules that can penetrate the phase but have restricted orientation toward it, and the interaction with the phase is similar to a size-exclusion effect. Also, the shape of the molecule may affect the extent to which a hydrophobic moiety of the molecule can be "presented" to the stationary phase such that it will diminish the exposure to interactions with the mobile phase. The formula 6.4.14 for describing the selectivity is known as the *hydrophobic subtraction* model [119–121]. As further shown, the *hydrophobic subtraction* model is very useful for column characterization using a set of selected compounds, while the true calculation of log k for different analytes is not possible without knowing the specific parameters $\eta'(j)$, $\sigma'(j)$, $\beta'(j)$, $\alpha'(j)$, and $\kappa'(j)$. Expression 6.4.14 shows that by using rel. 6.4.14, it is possible to estimate how much a compound is separated from ethylbenzene on the specific column.

The hydrophobic subtraction model shows that when the analyte-depending parameter is zero or extremely small, that particular interaction does not contribute to the value of log $\alpha_{c1}(j)$. For example, formula 6.4.9 will result from rel. 6.4.14 for an analyte j with $\beta'(j)$, $\alpha'(j)$, $\kappa'(j)$, and $\sigma'(j)$ all zero. In this case, a hydrophobic compound with no polar groups and with very little steric interactions, such as toluene or ethylbenzene, will generate a log k value describing only hydrophobic interactions. This concept generated the idea that specific compounds in a specific mobile phase may lead to a unique contribution to log $\alpha_{c1}(j)$. For example, a mixture of uracil, toluene, ethylbenzene, quinizarin (or

1,4-dihydroxyanthraquinone), and amitriptyline or 3-(10,11-dihydro-5H-dibenzo[a,d]cycloheptene-5-ylidene)-N,N-dimethylpropan-1-amine, (available as Standard Reference Material SRM 870 [122]) was suggested for use in comparing RP columns, with a mobile phase 80% methanol and 20% buffer phosphate at pH = 7.0 [72], the retention of each compound being relevant for a specific type of interaction. The mixture will allow the characterization of the column regarding hydrophobicity, peak width (plate number N), peak asymmetry, and selectivity factor. This is achieved using several measurements. For example, uracil will be used for determining the void volume of the column, since all parameters $\eta'(j)$, $\beta'(j)$, $\alpha'(j)$, $\kappa''(j)$, and $\sigma'(j)$ are zero (or close to zero) for this compound. Ethylbenzene will generate the value for k_{EB}. (log k_{EB} can vary in the range of about 0.2 up to 3). Toluene and ethylbenzene will provide data for hydrophobicity by their k values and for $\alpha(CH_2)$ since these compounds do not have polar groups and $\beta'(j)$, $\alpha'(j)$, $\kappa'(j)$ and $\sigma'(j)$ in rel. 6.4.14 are negligible. For most C18 columns $\alpha(CH_2)$ has values between 1.1 and 1.6.

Both quinizarin and amitriptyline have hydrophobic and hydrophilic moieties, and they are retained by RP columns depending on the hydrophobicity of each column, and also on column polar character (for columns other than C8 or C18). Both of these compounds are used for evaluation of k indicative of retention of polar compounds with a hydrophobic moiety and also for estimation of peak asymmetry As. The capacity factors of quinizarin and of amitriptyline show a large variation for different types of columns. Also, the values for As on these two compounds show large variations from column to column. Table 6.4.5 indicates the values for As measured for amitriptyline and for quinizarin for several common columns, with the mobile phase 80% CH$_3$OH and 20% aqueous potassium phosphate buffer with conc. 5 mM and with pH = 7 [103].

Asymmetry is strongly influenced by other interactions besides hydrophobicity. Quinizarin is typically used for evaluation of metal activity in a column, and amitriptyline, which is a basic compound (pK$_a$ = 9.4), is used for characterization of silanol activity based on its peak shape. Since quinizarin is a metal-chelating reagent and its retention behavior is indicative of the presence or absence of metal ions in the stationary phase, low metal activity is characterized by symmetric peaks of quinizarin, and high metal activity by strong tailing. Amitriptyline has, besides the hydrophobic interaction with the stationary phase, an additional interaction with the silanol groups.

When all the molecules of an analyte interact equally with the stationary phase, even if other interactions besides the hydrophobic ones are at play, the typical Gaussian shape of the chromatographic peak should be expected. However, it is possible that a small percentage of the molecules of the analyte have a different type of interaction compared to the bulk of the molecules. If the amine is partly ionized, a small proportion of the molecules may have a stronger retention than the other, or only some molecules may have access to the interaction with the silanol groups. In this case, two different retentions will be applied to the same molecular species, and a peak with non-Gaussian shape will result. This is pictured in Figure 6.4.13, where the combination of two Gaussian peaks with slightly different retention times are combined, generating strong tailing in the resulting chromatogram.

Poor correlation is seen between the k for amitriptyline and the peak asymmetry (tailing) since the retention leading to the k value for most amitriptyline molecules is caused by hydrophobic interactions, while tailing is related to silanol activity. Also, poor correlation is seen between the k for quinizarin and its peak asymmetry (tailing). This indicates that the main retention process for this compound is still controlled by $\eta(j)$ and H$_c$ for the column and compound

TABLE 6.4.5 Values for asymmetry As for various columns as measured for amitripyline and for quinizarin [103].

	Column	*As* Amitriptyline	*As* Quinizarin
1	ACE C18	1.03	1.07
2	Xterra MS C18	1.12	1.12
3	Intersil ODS3	1.26	1.24
4	Luna C18(2)	1.52	1.46
5	Hypersil Elite C18	1.61	1.09
6	Discovery C18	1.78	1.37
7	Waters Spherisorb ODS1	1.88	2.27
8	Intersil ODS2	1.93	1.27
9	Symmetry C18	2.07	1.31
10	Hypersil BDS-C18	2.21	1.26
11	Partisil ODS2	2.74	3.07
12	Zorbax XDB-C18	3.05	1.07
13	Novapak C18	3.05	2.42
14	Zorbax SB-C18	3.07	1.43
15	Nucleosil C18	3.19	2.41
16	Vidac Selectapore 300M	6.50	1.34
17	Cosmosil C18 AR-II	8.30	1.16
18	μ Bondapack C18	8.99	2.56

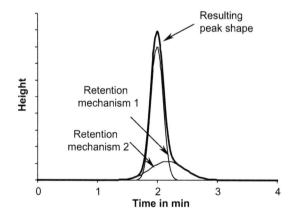

FIGURE 6.4.13 Combination of two Gaussian peaks for the same compound that undergoes two different retentions that generate strong tailing in the chromatogram.

hydrophobic moiety, and the secondary interaction affects only a small proportion of the analyte molecules. Peak asymmetry generated by quinizarin and amitriptyline is not correlated, since each compound has a different type of additional interaction with the stationary phase.

Different columns can be compared based on comparing parameters $\alpha(CH_2)$, H_c, S^*_c, A_c, B_c, and C_c or similar parameters. U.S. Pharmacopeia (USP) [123] offers a database and a comparison program that provide information for a large number of columns regarding H_c given as capacity factor k for ethylbenzene, chelating tailing factor CTF for quinizarin, capacity factor k for amitriptyline CFA, tailing factor for amitriptyline TFA, and bonding density BD in

$\mu mol/m^2$ of the stationary phase. The bonding density BD is assumed to be related to the steric parameter S*. The program also provides a "column distance F" (see rel. 6.4.17). An example of such information for several columns compared with an Ascentis C18 column is given in Table 6.4.6.

Other test solutions, more elaborate than those in SRM 870 mixture, have been suggested in the literature (PQRI approach). For example, the values for all the parameters from rel. 6.4.14 for a specific mobile phase were reported in the literature for 16 different analytes [118]. The parameters characterizing this set of compounds in a mobile phase consisting of acetonitrile/ aqueous 60 mM phosphate buffer 50/50 v/v are given in Table 6.4.7 (thiourea was used for measurement of t_0) [118]. The table also lists the

TABLE 6.4.6 Example of information from USP information regarding column description and comparison.

F	Column	H_c	CTF	CFA	TFA	BD	USP Code
0	Ascentis C18	2.7	1	6.2	1.3	3.6	L1
0.49	Discovery HS C18	2.4	1	5.4	1.3	3.8	L1
0.7	Zorbax Eclipse XDB C18	2.4	1	6.1	1.8	4	L1
0.71	PurospherSTAR RP-18e 5 μm	2.6	1.4	7.8	1.9	3.3	L1
0.75	Luna 5 μ C18(2)	2.2	1.2	5.3	1.1	3.4	L1
0.77	Acclaim 120 C18	2.3	1.2	5.6	1.6	3.2	L1
0.89	PurospherSTAR RP-18e 3 μm	2.1	1.2	5.7	1.5	3.3	L1
0.92	SepaxGP-C18	2.5	1.2	7.7	2.54	3.3	L1
0.98	BETASIL C18	3.2	1.6	7.6	2	3.6	L1
0.98	TSKgel ODS-100S	2.4	1.2	5.2	2	3	L1

TABLE 6.4.7 The parameters for solute hydrophobicity, steric effects, hydrogen bonding, ion exchange and log K_{ow} [114,118].

No.	Solute	η'	σ'	β'	α'	κ'	log K_{ow}
1	Acetophenone	−0.744	0.133	0.059	−0.152	−0.009	1.53
2	Benzonitrile	−0.703	0.317	0.003	0.080	−0.030	1.83
3	Anisole	−0.467	0.062	0.006	−0.156	−0.009	1.82
4	Toluene	−0.205	−0.095	0.011	−0.214	0.005	2.49
5	Ethylbenzene	0	0	0	0	0	2.93
6	4-Nitrophenol	−0.968	0.040	0.009	0.098	−0.021	1.61
7	5-Phenylpentanol	−0.495	0.136	0.030	0.610	0.013	2.83
8	5,5-Diphenylhydantoin	−0.940	0.026	0.003	0.568	0.007	2.15
9	cis-Chalcone	−0.048	0.821	−0.030	0.466	−0.045	3.89

(Continued)

TABLE 6.4.7 The parameters for solute hydrophobicity, steric effects, hydrogen bonding, ion exchange and log K_{ow} [114,118]. (*cont'd*)

No.	Solute	η'	σ'	β'	α'	κ'	log K_{ow}
10	*trans*-Chalcone	0.029	0.918	−0.021	−0.292	−0.017	3.89
11	*N,N*-Dimethylacetamide	−1.903	0.001	0.994	−0.012	0.001	−0.58
12	*N,N*-Diethylacetamide	−1.390	0.214	0.369	−0.215	0.047	0.13
13	4-*n*-Butylbenzoic acid	−0.266	−0.223	0.013	0.838	0.045	3.48
14	Mefenamic acid	0.049	0.333	−0.049	1.123	−0.008	5.4
15	Nortriptyline	−1.163	−0.018	−0.024	0.289	0.845	4.43
16	Amitriptyline	−1.094	0.163	−0.041	0.300	0.817	4.81

log K_{ow} values for each compound. The structures of the compounds from Table 6.4.7 are as follows:

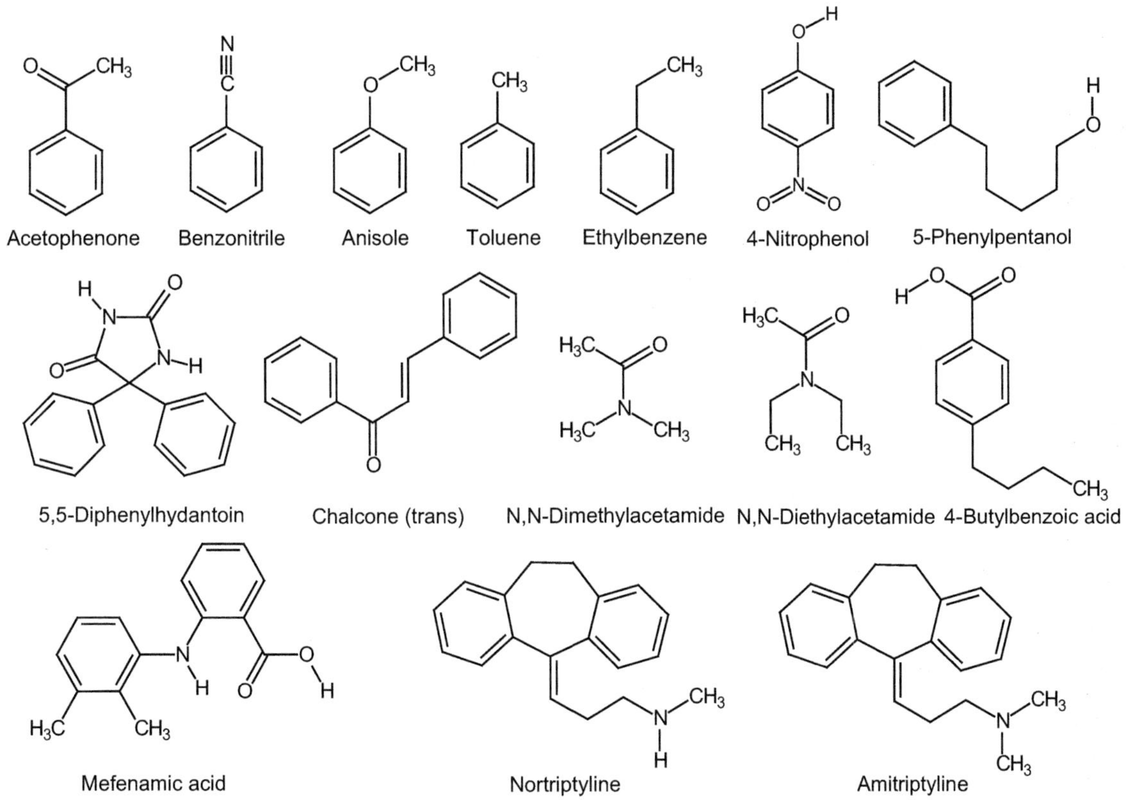

The parameters from Table 6.4.7 were obtained based on the values log α on a large number of columns for which the parameters

Acetophenone Benzonitrile Anisole Toluene Ethylbenzene 4-Nitrophenol 5-Phenylpentanol

5,5-Diphenylhydantoin Chalcone (trans) N,N-Dimethylacetamide N,N-Diethylacetamide 4-Butylbenzoic acid

Mefenamic acid Nortriptyline Amitriptyline

H_c, S^*_c, A_c, B_c, C_c for pH = 2.8 (C(2.8)), and C_c for pH = 7.0 (C(7.0)) were also calculated, using multiple regression. These values are given in Table 6.4.8, together with log k_{EB} for various columns 150 mm in length, 4.6 mm i.d., and with 5 μm particles and the same mobile phase acetonitrile/aqueous 60 mM phosphate buffer 50/50 v/v [118,124]. Assuming that all other parameters remain unaffected when changing the buffer solution from pH = 2.8 to pH = 7.0, it can be easily determined that the parameter C(7.0) can be obtained from C(2.8) from the following relation:

$$C(7.0) = C(2.8) + \log\left(\frac{k_{7.0}}{k_{2.8}}\right) \qquad (6.4.16)$$

However, rel. 6.4.16 is only an approximation [123,125].

The parameters from Table 6.4.8 are very useful in evaluating a specific column character. For example, a column such as Acclaim 120 C18 has a large H_c value and relatively low values for the other parameters, indicating a good separation of compounds based on their hydrophobic character, but lower efficiency for other interactions. Relation 6.4.14 allows the calculation of log α values for compounds with known parameters. For example for toluene, with the values from Table 6.4.7, the following log α value can be obtained (using C(2.8) value) for Acclaim 120 C18 column:

$$\log \alpha = -0.21156 + 0.00171 - 0.00156$$
$$+ 0.00578 + 0.00043$$
$$= -0.2052$$

TABLE 6.4.8 The values of H_c, S^*_c, A_c, B_c, $C_c(2.8)$, $C_c(7.0)$, and log k_{EB} for various columns.

No.	Column	H_c	S_c	A_c	B_c	$C_c(2.8)$	$C_c(7.0)$	log k_{EB}
1	Acclaim 120 C18	1.032	0.018	−0.142	−0.027	0.086	−0.002	1.002
2	Acclaim 120 C8	0.857	0.004	−0.274	0.012	0.086	0.016	0.780
3	Acclaim 300 C18	0.957	−0.018	−0.170	−0.170	0.261	0.222	0.462
4	Acclaim Organic Acid	0.833	0.063	−0.385	−0.001	−0.316	0.349	
5	Acclaim Polar Advantage C16	0.855	0.068	−0.116	0.023	−0.270	0.357	
6	Ace 5 CN	0.409	−0.107	−0.729	−0.008	−0.086	0.441	−0.019
7	Ace Phenyl	0.638	−0.145	−0.305	0.031	0.128	0.461	0.445
8	Ace 5 AQ	0.804	0.051	−0.129	0.034	0.009	0.167	
9	Ace 5 C18	1.000	0.026	−0.096	−0.006	0.143	0.096	0.895
10	Ace 5 C18-300	0.983	−0.025	0.046	0.012	0.262	0.237	
11	Ace 5 C8	0.834	0.007	−0.218	0.025	0.109	0.145	0.693
12	Allure C18	1.116	0.040	0.114	−0.044	−0.047	0.066	1.195
13	Allure PFP Propyl	0.732	−0.157	−0.179	−0.037	0.710	1.485	0.833
14	Atlantis dC18	0.917	−0.031	−0.192	0.001	0.036	0.087	0.908

(Continued)

TABLE 6.4.8 The values of H_c, S^*_c, A_c, B_c, $C_c(2.8)$, $C_c(7.0)$, and log k_{EB} for various columns. (*cont'd*)

No.	Column	H_c	S_c	A_c	B_c	$C_c(2.8)$	$C_c(7.0)$	log k_{EB}
15	BetaBasic Phenyl	0.571	−0.167	−0.422	0.054	0.099	0.753	0.234
16	Betasil Phenyl-Hexyl	0.693	−0.054	−0.323	0.021	0.038	0.341	0.637
17	Chromegabond WR C18	0.979	0.026	−0.159	−0.003	0.320	0.282	0.732
18	Chromegabond WR C8	0.855	0.025	−0.279	0.024	0.200	0.144	0.554
19	Chromolith RP18e	1.003	0.029	0.009	−0.014	0.103	0.187	0.493
20	COSMOSIL AR-II (C18)	1.017	0.010	0.127	−0.028	0.116	0.494	0.907
21	COSMOSIL MS-II (C18)	1.031	0.040	−0.131	−0.014	−0.118	−0.027	0.908
22	DeltaPak C18 100A	1.028	0.019	−0.017	−0.011	−0.051	0.024	0.956
23	DeltaPak C18 300A	0.955	−0.013	−0.104	0.016	0.235	0.286	0.481
24	Develosil C30-UG-5 (C30)	0.976	−0.036	−0.195	0.011	0.158	0.177	0.892
25	Develosil ODS-HG-5(C18)	0.980	0.015	−0.171	−0.008	0.187	0.221	0.911
26	Develosil ODS-MG-5 (C18)	0.963	−0.036	−0.164	−0.003	−0.012	0.051	1.051
27	Develosil ODS-UG-5 (C18)	0.997	0.025	−0.145	−0.004	0.150	0.154	0.926
28	Discovery BIO Wide pore C18	0.836	0.014	−0.253	0.028	0.121	0.119	0.528
29	Discovery BIO Wide pore C5	0.655	−0.019	−0.305	0.029	0.091	0.220	0.059
30	Discovery BIO Wide pore C8	0.840	0.018	−0.224	0.034	0.201	0.195	0.345
31	Discovery C18	0.984	0.027	−0.128	0.004	0.176	0.153	0.683
32	Discovery C8	0.832	0.011	−0.237	0.029	0.119	0.143	0.522
33	Discovery CN	0.397	−0.110	−0.615	−0.002	−0.035	0.513	−0.198
34	Discovery HS F5	0.631	−0.166	−0.325	0.023	0.709	0.940	0.603
35	Fluophase PFP	0.675	−0.129	−0.311	0.065	0.817	1.375	0.653
36	Fluophase RP	0.698	0.028	0.103	0.039	1.034	1.417	0.532
37	Genesis AQ 120A (C18)	0.960	−0.036	−0.157	0.007	0.060	0.233	0.981
38	Genesis C18 120A	1.005	0.004	−0.068	−0.007	0.139	0.125	0.993
39	Genesis C18 300A	0.975	0.005	−0.086	0.013	0.266	0.270	0.543
40	Genesis C4 300A	0.615	−0.057	−0.397	0.036	0.143	0.249	0.059
41	Genesis C4 EC 120A	0.646	−0.058	−0.330	0.027	0.063	0.400	0.526
42	Genesis C8 120A	0.829	−0.017	−0.081	0.018	0.055	0.300	0.795
43	Genesis CN 120A	0.424	−0.114	−0.681	−0.013	−0.001	0.573	0.134
44	Genesis CN 300A	0.397	−0.108	−0.645	−0.009	0.025	0.397	−0.340

(*Continued*)

TABLE 6.4.8 The values of H_c, S^*_c, A_c, B_c, $C_c(2.8)$, $C_c(7.0)$, and log k_{EB} for various columns. (*cont'd*)

No.	Column	H_c	S_c	A_c	B_c	$C_c(2.8)$	$C_c(7.0)$	log k_{EB}
45	Genesis EC C8 120A	0.864	0.005	−0.173	0.023	0.064	0.142	0.837
46	Genesis Phenyl	0.600	−0.147	−0.378	0.035	0.128	0.584	0.459
47	Hypersil Beta Basic-18	0.993	0.032	−0.099	0.002	0.163	0.126	0.808
48	Hypersil Beta Basic-8	0.834	0.016	−0.248	0.029	0.110	0.114	0.619
49	Hypersil BetamaxNeutral (C18)	1.099	0.035	0.068	−0.031	−0.038	0.012	1.231
50	Hypersil Bio Basic-18	0.975	0.025	−0.099	0.007	0.253	0.217	0.512
51	Hypersil Bio Basic-8	0.821	0.011	−0.232	0.029	0.231	0.211	0.253
52	Hypurity C18	0.981	0.020	−0.090	0.002	0.192	0.168	0.744
53	Hypurity C8	0.833	0.010	−0.200	0.034	0.157	0.161	0.546
54	Inertsil C8-3	0.830	−0.004	−0.267	−0.017	−0.334	−0.362	0.849
55	Inertsil CN	0.369	0.049	−0.808	0.083	−2.607	−1.297	0.050
56	Inertsil ODS-3	0.990	0.022	−0.145	−0.023	−0.474	−0.334	1.037
57	Inertsil ODS-P	0.978	−0.028	0.612	−0.038	0.234	–	1.048
58	Inertsil Ph-3	0.526	−0.179	−0.133	0.040	0.121	0.735	0.409
59	J'Sphere H80 (C18)	1.132	0.059	−0.023	−0.068	−0.242	−0.161	1.124
60	J'Sphere L80 (C18)	0.762	−0.036	−0.216	−0.001	−0.400	0.345	0.764
61	J'Sphere M80 (C18)	0.926	−0.026	−0.123	−0.004	−0.294	0.139	0.957
62	Jupiter 300 C18	0.945	0.031	−0.224	0.008	0.234	0.218	0.467
63	Jupiter 300 C4	0.698	0.008	−0.426	0.019	0.153	0.142	0.126
64	Jupiter 300 C5	0.729	0.021	−0.382	0.016	0.129	0.331	0.183
65	Kinetex XB-C18	0.975	−0.013	−0.083	0.023	−0.046	0.305	
66	Kinetex PFP 100A	0.688	0.089	−0.273	−0.038	0.943	1.538	
67	Kromasil 100-5C18	1.051	0.035	−0.070	−0.022	0.039	−0.057	1.098
68	Kromasil 100-5C4	0.734	0.002	−0.334	0.015	0.009	−0.003	0.700
69	Kromasil 100-5C8	0.864	0.013	−0.212	0.019	0.054	−0.001	0.881
70	Kromasil KR60-5CN	0.440	−0.135	−0.578	−0.014	0.216	1.036	0.306
71	Luna C18 (2)	1.002	0.024	−0.123	−0.007	−0.269	−0.174	0.983
72	Luna C5	0.800	0.030	−0.251	0.003	−0.277	0.115	0.770
73	Luna C8 (2)	0.889	0.041	−0.221	−0.001	−0.299	−0.169	0.859
74	Luna CN	0.452	−0.112	−0.323	−0.024	0.439	1.321	0.104

(*Continued*)

TABLE 6.4.8 The values of H_c, S^*_c, A_c, B_c, $C_c(2.8)$, $C_c(7.0)$, and log k_{EB} for various columns. (*cont'd*)

No.	Column	H_c	S_c	A_c	B_c	$C_c(2.8)$	$C_c(7.0)$	log k_{EB}
75	Luna Phenyl-Hexyl	0.775	−0.124	−0.284	−0.001	0.001	0.383	0.718
76	Nova-Pak CN HP 60A	0.362	−0.165	0.100	0.000	0.691	1.175	−0.413
77	Platinum EPS C18	0.616	−0.168	0.335	0.026	0.718	1.728	0.417
78	Platinum EPS C8	0.420	−0.152	0.151	0.026	0.509	1.369	0.022
79	PRECISION C18	1.003	0.003	−0.041	−0.009	0.079	0.340	0.976
80	PRECISION C8	0.821	−0.014	−0.179	0.022	0.095	0.241	0.692
81	Precision CN	0.431	−0.114	−0.485	0.019	−0.041	0.606	0.111
82	Precision Phenyl	0.587	−0.142	−0.304	0.030	0.094	0.504	0.420
83	Prevail C18	0.889	−0.070	0.316	0.022	0.107	1.205	0.975
84	Prevail C8	0.618	−0.089	0.040	0.041	0.081	1.072	0.530
85	Prodigy ODS (3)	1.023	0.025	−0.130	−0.012	−0.195	−0.134	1.003
86	Prodigy Phenyl-3	0.525	−0.198	0.051	0.024	0.228	1.465	0.358
87	ProntoSIL 120-5 C18 SH	1.032	0.018	−0.108	−0.024	0.114	0.403	0.938
88	ProntoSIL 120-5 C8 SH	0.739	−0.062	−0.081	0.013	0.076	0.526	0.687
89	ProntoSIL 120-5-C18 H	1.005	0.008	−0.105	−0.004	0.125	0.987	0.873
90	ProntoSIL 120-5-C18-AQ	0.974	−0.007	−0.083	0.003	0.137	0.224	0.910
91	ProntoSIL 120-5-Phenyl	0.557	−0.163	−0.217	0.022	0.167	0.706	0.387
92	ProntoSIL 200-5 C8 SH	0.761	−0.026	−0.194	0.024	0.125	1.443	0.439
93	ProntoSIL 200-5-C18 H	0.956	−0.002	−0.121	0.016	0.163	0.218	0.679
94	ProntoSIL 300-5 C8 SH	0.739	−0.041	−0.130	0.027	0.156	0.405	0.260
95	ProntoSIL 300-5-C18 H	0.956	−0.012	−0.089	0.015	0.238	0.249	0.511
96	ProntoSIL 60-5 C8 SH	0.929	−0.015	0.162	−0.017	−0.313	1.005	0.922
97	ProntoSIL 60-5-C18 H	1.158	0.041	0.067	−0.078	0.102	0.262	1.087
98	ProntoSIL 60-5-Phenyl	0.703	−0.196	−0.005	−0.009	0.410	1.509	0.649
99	ProntoSil CN	0.370	−0.114	−0.414	−0.028	0.168	0.668	−0.041
100	Purospher STAR RP18e	1.003	0.012	−0.070	−0.036	0.018	0.045	1.023
101	Restek Ultra C18	1.055	0.030	−0.068	−0.021	0.009	−0.066	1.101
102	RestekUltra C8	0.876	0.030	−0.229	0.018	0.043	0.011	0.883
103	Symmetry 300 C18	0.984	0.031	−0.051	0.003	0.228	0.202	0.549
104	Symmetry 300 C4	0.659	−0.017	−0.428	0.014	0.102	0.185	0.157
105	Symmetry C18	1.052	0.063	0.019	−0.021	−0.302	0.162	0.993

(*Continued*)

TABLE 6.4.8 The values of H_c, S^*_c, A_c, B_c, $C_c(2.8)$, $C_c(7.0)$, and log k_{EB} for various columns. (*cont'd*)

No.	Column	H_c	S_c	A_c	B_c	$C_c(2.8)$	$C_c(7.0)$	log k_{EB}
106	Symmetry C8	0.893	0.050	−0.205	0.021	−0.508	0.283	0.843
107	Synergy Fusion-RP	0.879	−0.030	−0.014	0.008	−0.238	0.362	
108	Synergy Hydro-RP	1.022	−0.006	0.169	−0.042	−0.077	0.260	
109	Synergy Polar-RP	0.654	−0.146	−0.257	−0.007	0.057	0.778	
110	Synergi Max-RP	0.989	0.028	−0.008	−0.013	−0.133	−0.034	0.976
111	Thermo CN	0.404	−0.111	−0.709	−0.009	−0.029	0.491	−0.088
112	Ultra PFP	0.501	−0.089	−0.228	−0.003	−0.033	0.588	0.289
113	Varian OmniSpher 5 C18	1.055	0.051	−0.033	−0.029	0.122	0.058	1.035
114	Xterra MS C18	0.985	0.012	−0.142	−0.015	0.134	0.051	0.803
115	XTerra Phenyl	0.690	−0.076	−0.374	−0.003	0.102	−0.033	0.409
116	XterraMS C8	0.803	0.004	−0.292	−0.005	0.058	−0.009	0.571
117	YMC Pro C18	1.015	0.013	−0.117	−0.006	−0.154	−0.005	0.939
118	YMC Pro C8	0.890	0.014	−0.214	0.007	−0.322	0.020	0.814
119	Zorbax Eclipse XDB-C18	1.077	0.024	−0.062	−0.033	0.055	0.089	0.958
120	Zorbax Eclipse XDB-C8	0.919	0.025	−0.219	−0.008	0.003	0.012	0.823
121	Zorbax Rx-18	1.077	0.040	0.310	−0.037	0.096	0.415	0.886
122	Zorbax RX-C8	0.792	−0.076	0.117	0.018	0.012	0.948	0.703
123	Zorbax StableBond 300A C18	0.906	−0.050	0.045	0.043	0.253	0.700	0.344
124	Zorbax StableBond 300A C8	0.701	−0.085	0.002	0.047	0.146	0.820	0.106
125	Zorbax StableBond 300A C3	0.526	−0.122	−0.194	0.047	0.057	0.711	−0.151
126	Zorbax StableBond 80A C18	1.008	−0.021	0.215	−0.002	0.077	0.822	0.884
127	Zorbax StableBond 80A C3	0.601	−0.124	−0.080	0.038	−0.084	0.810	0.450
128	Zorbax StableBond 80A C8	0.795	−0.079	0.138	0.018	0.014	1.020	0.710

A negative value for log α indicates that toluene elutes ahead of ethylbenzene (as expected). The contribution to the value for log α (and for log $k = 0.79680$) comes mainly from the hydrophobic effects.

Similar to the "rules" that lead to an increase in H_c, several observations can be made regarding the increase in the other column characteristics (some desirable and some not). For example, A_c decreases with the column end-capping, and columns with a higher acidity of the silica base have higher A_c values. Also, the presence of higher content of metals in the silica affects the increase in A_c. The value of B_c is in general affected by the nature of the bonded phase. The ion-exchange character is affected by the metal impurities in the silica and is very different for silica-based columns compared to zirconia-based columns. The steric interaction parameter S^*_c is typically increased for longer chain ligands that form the bonded phase and also increases with the density of the active phase. For phases indicated as monomeric (obtained with a monofunctional silanization reagent) or horizontal polymeric (see Section 6.2), the values for S^*_c are typically lower than for the phases obtained by vertical polymerization. These rules remain only directional, however, and no direct quantitation of a parameter can be obtained by a simple relation with the physicochemical characteristic of the stationary phase. Similar parameter values and results for various other types of columns are reported in the literature [114, 126, 127].

The values for parameters H_c, S^*_c, A_c, B_c, $C_c(2.8)$, $C_c(7.0)$, and $\log k_{EB}$ for various columns may vary significantly depending on column characteristics (e.g., the chemical nature of the bonded phase, carbon load, pore size, endcapping, etc.). Several typical values for these parameters for different types of columns are listed in Table 6.4.9. Comparison between

TABLE 6.4.9 Minimum, maximum or average* values for H_c, S^*_c, A_c, B_c, $C_c(2.8)$, $C_c(7.0)$, and $\log k_{EB}$ for various hydrophobic columns.

No.	Bonded phase	H_c	S_c	A_c	B_c	$C_c(2.8)$	$C_c(7.0)$	$\log k_{EB}$
1	C1 average	0.41	−0.08	−0.08	0.02	0.04	0.66	0.08
2	C3 min.	0.53	−0.12	−0.19	0.04	−0.08	0.71	−0.15
3	C3 max.	0.60	−0.12	−0.08	0.05	−0.06	0.81	0.45
4	C8 min.	0.42	−0.15	−0.29	−0.02	−0.51	−0.36	0.11
5	C8 max.	1.06	0.05	0.16	0.05	0.51	1.44	1.10
6	C18 min.	0.62	−0.17	−0.39	−0.17	−0.47	−0.33	0.34
7	C18 max.	1.16	0.06	0.61	0.04	0.72	1.73	1.23
8	C18 wide pore min.	0.98	0.00	−0.09	−0.01	−0.05	0.02	0.54
9	C18 wide pore max.	1.03	0.02	−0.02	0.01	0.27	0.27	0.99
10	C18 monolith average	1.01	0.02	0.12	−0.02	0.11	0.31	0.51
11	C18 embedded polar group min.	0.74	−0.01	−0.22	0.00	−0.27	0.17	0.77
12	C18 embedded polar group max.	0.97	0.07	−0.08	0.12	0.14	0.53	0.91
13	C30 average	0.96	−0.01	−0.10	0.02	0.24	0.29	0.48
14	Phenyl min.	0.53	−0.03	−0.19	−0.01	−0.05	0.02	0.23
15	Phenyl max.	1.03	0.02	0.05	0.05	0.41	1.51	0.99
16	Cyano min.	0.36	−0.09	−0.49	0.00	0.03	0.40	0.05
17	Cyano max.	0.45	0.05	0.10	0.08	0.69	1.32	0.31
18	Fluorinated min.	0.50	−0.07	−0.25	0.02	0.71	0.59	0.29
19	Fluorinated max.	0.73	0.03	0.10	0.07	1.03	1.49	0.83
20	Zirconia-based average	0.97	0.01	−0.62	0.00	2.01	2.01	−0.10

* Note: The minimum, maximum, and average values were obtained from a limited set of about 150 columns. Columns not included in the set may display other values.

columns can be obtained based on parameters described in Table 6.4.8 by different procedures. One visual procedure is the use of a "radar" graph. Figure 6.4.14 shows such a graph for four synergy-type columns. From Figure 6.4.14 it can be seen, for example, that Synergy Fusion-RP has strong hydrophobicity and moderate polar character, Synergy Hydro-RP has strong hydrophobic interactions and also strong polar interactions (strong hydrogen bonding and average cation-exchange character), Synergy Polar-RP has lower hydrophobic character but the strongest polar character, and Synergy Max-RP has strong hydrophobic interactions and low other interaction types.

The values from Table 6.4.8 can also be used to develop a formula for a "distance" between two different columns. This distance F can be calculated using an expression of the form:

$$F = \left\{ [f_H \, (H_{c1} - H_{c2})]^2 + [f_S \, (S_{c1}^* - S_{c2}^*)]^2 \right.$$
$$+ [f_A (A_{c1} - A_{c2})]^2 + [f_B (B_{c1} - B_{c2})]^2$$
$$\left. + [f_C (C_{c1} - C_{c2})]^2 \right\}^{1/2}$$

$$(6.4.17)$$

where f_H, f_S, f_A, f_B... are weighing factors selected based on the relative contribution of each parameter to the value of α. For the columns listed in Table 6.4.8, the weighing factors can be chosen with the following values: $f_H = 12.5$, $f_A = 30.3$, $f_B = 142.8$, $f_C = 83.3$, and $f_S = 100$ [118]. A selection of weighing factors all equal to 1 can also be made, and rel. 6.4.17 will in this case become:

$$F = [(H_{c1} - H_{c2})^2 + (S_{c1}^* - S_{c2}^*)^2 + (A_{c1} - A_{c2})^2$$
$$+ (B_{c1} - B_{c2})^2 + (C_{c1} - C_{c2})^2]^{1/2} \quad (6.4.18)$$

(Note: A similar distance between columns can be developed by using the parameters measured by USP test H_c, CTF, CFA, TFA, BD.)

Use of the distance parameter can allow the selection of columns to be similar in case of a need for a replacement with similar properties, or in case of search for an orthogonal separation, the distance may facilitate the choice of a very different column. The USP database [123] has a calculation that also allows an online comparison of numerous columns by displaying the F value and the characteristic

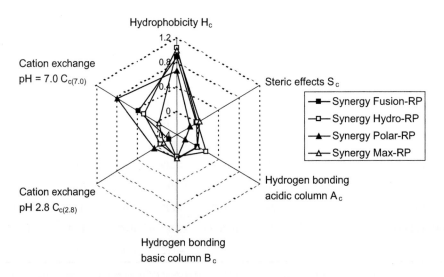

FIGURE 6.4.14 Radar plot displaying column properties for four different Synergy columns.

H_c, S^*_c, A_c, B_c, $C_c(2.8)$, and $C_c(7.0)$ values for the alternative columns. An example of such a comparison for an Acclaim 120 C18 column selected from more than 500 columns is shown in Table 6.4.10.

Other Tests for Comparing Hydrophobic Columns

Various other tests and analytes are reported in the literature for comparing different column characteristics [119, 127, 128]. For example, metal activity in a stationary phase can be evaluated using other reagents besides quinizarine. Measurement of the effect of metals is important since metallic ions can increase the acidity of adjacent silanols and can become adsorption sites for compounds that are able to form complexes. Overall, they can interfere in the retention process. For measuring metal activity on the surface of stationary phase, acetylacetone and also the couple of 2,2'-dipyridyl and 4,4'-dipyridyl were used. The retention behavior of 2,2'-dipyridyl and 4,4'-dipyridyl in the presence of metal ions on the silica surface is rather different: 2,2'-dipyridyl can form complexes with metallic species in particular with iron, whereas 4,4'-dipyridyl cannot form complexes. Therefore, the peak of 2,2'-dipyridyl will exhibit a post-tailing with an asymmetry depending on the concentration of metallic ions from the silica surface, while the peak of 4,4'-dipyridyl will remain symmetric.

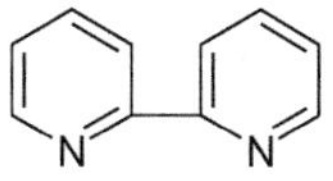

2,2'-dipyridyl 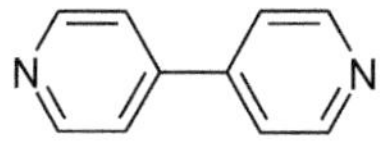   4,4'-dipyridyl

Although these dipyridyl derivatives are basic compounds that can be involved in strong interactions with dissociated silanol groups from silica surface, this inconvenience was solved by using buffered (pH = 7) mobile phase. [117].

TABLE 6.4.10 Example of column comparison as obtained from U.S. Pharmacopeia web program [123] based on PQRI test.

F	Column name	H	S*	A	B	C(2.8)	C(7.0)	USP Code
0	Acclaim 120 C18	1.032	0.018	−0.143	−0.027	0.086	−0.002	L1
0.24	TSKgel ODS-100Z	1.032	0.018	−0.135	−0.031	−0.064	−0.161	L1
0.67	Inertsil ODS-3	0.99	0.022	−0.146	−0.023	−0.474	−0.334	L1
0.74	LaChrom C18	0.993	0.013	−0.151	−0.006	−0.278	−0.12	L1
0.8	Prodigy ODS(3)	1.023	0.025	−0.131	−0.012	−0.195	−0.134	L1
0.83	YMC Pro C18	1.015	0.014	−0.12	−0.007	−0.155	−0.006	L1
0.84	Develosil ODS-UG-5	0.996	0.025	−0.146	−0.004	0.15	0.155	L1
0.85	XTerra MS C18	0.984	0.012	−0.143	−0.015	0.133	0.051	L1
0.91	Luna C18(2)	1.002	0.024	−0.124	−0.007	−0.269	−0.174	L1
0.96	Proto 300 C18	0.962	0.016	−0.132	0.005	0.224	0.147	L1
1.02	ProntoSIL 120 C18 SH	1.031	0.018	−0.109	−0.024	0.113	0.402	L1

Another type of test for comparing columns is related to the evaluation of stationary phase polarity. The ability of a reversed-phase column to retain polar compounds depends not only on the extent of hydrophobic interactions between the column and the hydrophobic moiety of the compound, but also on the two participants' polarity. For example, the selectivity difference between a conventional C18 and a polar-embedded phase results from their polarity differences. In general, higher polarity of the embedded group in a stationary phase results in a longer retention of polar compounds. One simple test to characterize the polarity of the stationary phase is based on using a mixture containing uracil, pyridine, phenol, N,N-dimethylaniline, p-butylbenzoic acid, and toluene in a mobile phase made of acetonitrile/aqueous buffer 60/40 (v/v) (the buffer consisted of 50 mM KH_2PO_4/KH_2PO_4 pH = 3.2). With this mixture, a column polarity index (P) was defined [128]:

$$P = k_{\text{p-butylbenzoic acid}} \times k_{\text{phenol}}/k_{\text{toluene}}^2 \quad (6.4.19)$$

This test shows that the polarity of polar-embedded stationary phases with the same hydrocarbon chain varies in the following order:

amide > carbamate > sulfonamide > alkyl

Another test for measuring polarity uses butyl paraben and dipropyl phthalate [129]. Retention studies on amide-embedded stationary phases showed that dipropyl phthlate elutes before butyl paraben, whereas the elution order on carbamate-embedded group phases is opposite. The order of polarity for the polar-embedded phases using this test is the following:

amine > amide > carbamate > ester > alkyl

The possibility of using ionic compounds in a 100% aqueous mobile phase for the characterization of RP columns has also been suggested [130]. In this test, various naphthalene disulfonic acids and naphthalene trisulfonic acids were employed as test compounds. These compounds are practically completely ionized in aqueous solution, and when using only water as a mobile phase, they are excluded from the pores of the hydrophobic packing material, having a retention time shorter than compounds such as uracil typically used for measuring the column dead time (or volume). This effect is attributed to ionic repulsion between the slightly negatively charged stationary phase surface (as a result of silanol group ionization) and the sulfonic acid anions. The test was performed using as mobile phase the solution 0.4 M of Na_2SO_4. With this mobile phase, the retention times of the test compounds increases compared to the pure water mobile phase, on all tested hydrophobic columns. However, the dead time t_0 for the columns as measured, for example, using uracil cannot be used for calculating a capacity factor, and the differences in retention must be compared versus the least retained test compound (1,5-naphtalenedisulfonic acid for disulfonic acids and 1,3,5,7-naphthalene-tetrasulfonic acid for trisulfonic acids). The test provides information on the hydrophobicity of the column and peak asymmetry.

Besides the nature of the column (and considering the nature of the test compounds), both the capacity factor, the methylene selectivity $\alpha(CH_2)$, and also other interactions are significantly influenced by the mobile phase used in the test [110, 131, 132]. However, the previous discussion did not explicitly include the nature of the mobile phase, except that the mobile phase must be kept the same for a pertinent comparison between columns. Since the selection of the mobile phase strongly affects the value of capacity factor, it can be concluded that rel. 6.4.14 and 6.4.9 are incomplete, and a coefficient dependent on solvent should be included in these expressions. Since the nature of each interaction is dependent on the nature of the mobile phase that may affect a particular interaction intensity (e.g., by the dissociation or suppressing dissociation of a given analyte), each interaction type should have a different solvent-dependent parameter. This would lead from rel. 6.4.15 to a formula for the capacity factor of the type:

$$\log k(j) = \log k_{EB} + \eta'(j)H_c - \eta''(j)H_{mo}$$
$$- \sigma'(j)S_c^* + \beta'(j)(A_c - A_{mo})$$
$$+ \alpha'(j)(B_c - B_{mo}) + \kappa'(j)C_c$$

$$(6.4.20)$$

In rel. 6.4.20, $\log k_{EB}$ depends the column and on the mobile phase, H_c, S^*_c, A_c, B_c, C_c depend only on the column, while H_{mo}, A_{mo}, and B_{mo} are parameters related only to a specific mobile phase and account for polar/polarizability, hydrogen bonding for a basic solute, and hydrogen bonding for an acidic solute, respectively. The sign of respective coefficients depends on the effect they have on separation (decreasing $\log k$ when the solvent interactions are stronger). Further discussion of the solvent contribution to separation can be found in Chapter 7.

The combination of the idea of choosing a specific analyte and a specific mobile phase for putting in evidence a specific character of the stationary phase led to the development of several tests allowing the characterization of a hydrophobic column from the point of view of hydrophobicity, polar interactions, hydrogen bonding, ion-exchange interactions, and steric interactions. Several such tests are further listed (see Test 1 to Test 5). Other tests were also suggested for characterization of RP chromatographic columns [137].

TEST 1. Mobile phase: MeOH/water = 55/45 (v/v), several buffers instead of water, column temperature: 40°C. [132,133]

Compound	Test
1 = thiourea	dead time $t_0(1)$
2 = toluene 3 = ethylbenzene	hydrophobicity from capacity factor $k(2)$, $k(3)$ methylene selectivity $\alpha(CH_2)$ = $k(3)/k(2)$
4 = phenol 5 = ethyl benzoate	capacity factor for polar solutes $k(4)$, $k(5)$
6 = aniline 7 = m-toluidine 8 = p-toluidine 9 = N,N-dimethylamine	capacity factor for basic solutes $k(6)$, $k(7)$, $k(8)$, $k(9)$ asymmetry As(6), As(7), As(8), As(9)

TEST 2. Mobile phase: A. 80% MeOH in water, B. 30% MeOH in phosphate buffer (0.02 M) pH = 7.6, C. 30% MeOH in phosphate buffer (0.02 M) pH 2.7, column temperature: 40°C [134]

Compound	Test
1 = alkylbenzenes	methylene selectivity $\alpha(CH_2)$
2 = amilbenzene 3 = butylbenzene	$\alpha(CH_2)$ = $k(2)/k(3)$
4 = triphenylene 5 = o-terphenyl	steric selectivity $\alpha = k(4)/k(5)$
6 = caffeine 7 = phenol	hydrogen bonding capacity $\alpha = k(6)/k(7)$
8 = benzylamine 9 = phenol	ion exchange capacity at pH > 7 $\alpha = k(8)/k(9)$ ion exchange capacity at pH < 3 $\alpha = k(8)/k(9)$

Other Properties of Reversed-Phase Columns Critical for Separation

Resolution of an HPLC column is a function of the number of theoretical plates N, in addition to the capacity factor k and the selectivity α. For this reason, an important parameter in column characterization is N. The number of theoretical plates is less dependent on the nature

TEST 3. Mobile phase: A. acetonitrile, B. n-heptane (dry), C: acetonitrile/water = 65/35 (v/v) [135]

Compound	Test
1 = uracil	dead time $t_0(1)$
2 = N,N-dimethyltoluamide 3 = anthracene	silanol index $\alpha = k(2)/k(3)$
4 = nitrobenzene	silanol index $k(4)$
5 = benzene 6 = anthracene	hydrophobicity $\alpha = k(6)/k(5)$
7 = toluene	theoretical plates N from $W_b(7)$

TEST 4. Mobile phase: A. 60% MeOH in water, column temperature, B. acetonitrile/water = 65/35 (v/v), column temperature: 30°C. [106,136]

Compound	Test
1 = toluene 2 = benzene	hydrophobicity from $(k(1) + k(2))/2$
3 = ethylbenzene	hydrophobicity $\alpha = k(3)/k(2)$
4 = aniline 5 = phenol	silanol index $1 + 3[k(4)/k(5) - 1]$

of the stationary phase (such as parameters n, d_p, c_L and E_c) and depends in most cases on the dimension of the particles and their homogeneous (spherical) shape.

The effective plate numbers (per meter) **n** for the columns in Table 6.4.3 are given in Figure 6.4.15. The measurement of **n** was performed for toluene as the analyte, with the mobile phase 90% CH_3OH 10% H_2O, isocratic at 24 °C [103]. The plate number depends on the particle dimensions as well as on other stationary phase characteristics. More modern columns with smaller particles than 5 μm have higher **n** values, and the columns with core-shell-type stationary phases have **n** values up to 300,000.

TEST 5. Mobile phase: A. 65% acetonitrile in 20 mM phosphate buffer pH = 7, B. 15% acetonitrile in 20 mM phosphate buffer pH = 7, C. 30% acetonitrile in 20 mM phosphate buffer pH = 2.5, D. 30% acetonitrile in 20 mM phosphate buffer pH = 7, E. 75% acetonitrile in 20 mM phosphate buffer pH = 2.5 [108]

Compound	Test
1 = butylbenzene 2 = amylbenzene	hydrophobicity $k(1)$, $k(2)$ (k_{BB} for capacity factor of butylbenzene: see Table 6.4.10)
3 = propylbenzene 4 = ethylbenzene 5 = toluene 6 = benzene	methylene selectivity $\alpha(CH_2)$ from the series benzene to amylbenzene
7 = caffeine 8 = phenol	hydrogen bonding capacity $\alpha = k(7)/k(8)$ ($\alpha_{caffeine/phenol}$)
9 = benzylamine	silanol activity index at pH = 2.5, $\alpha = k(9)/k(8)$ ($\alpha_{benzylamine/phenol\ pH = 2.5}$) silanol activity index at pH = 7, $\alpha = k(9)/k(8)$ ($\alpha_{benzylamine/phenol\ pH = 7.0}$)
10 = p-hydroxybenzoic acid 11 = sorbic acid 12 = benzoic acid 13 = salicylic acid 14 = p-toluic acid	retention and selectivity of polar acid $k(i)$, $i = 10 - 14$
15 = triprolidine 16 = chlorpheniramine 17 = diphenylhydramine	Peak asymmetry $As(15)$, $As(16)$, $As(17)$ at pH = 2.5
18 = tricyclic antidepressants such as desipramine, amitriptyline, etc.)	Peak asymmetry $As(18)$

Wetting characteristic of the RP stationary phases is another important parameter in the column utilization. Some analytes (samples) require for analysis a high content of water in the mobile phase to assure their solubility. As indicated in Section 6.3, phase collapse may

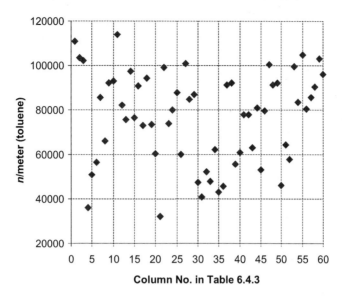

FIGURE 6.4.15 Effective plate numbers (per meter) **n** for the columns listed in Table 6.4.3.

occur if a column having a phase with low wettability is subject to a mobile phase too high in water. To avoid this problem, stationary phases that have higher wettability can be selected. Wettability can be estimated by measuring the maximum water concentration in common organic solvents (methanol, acetonitrile, and isopropanol) that allow the phase to remain wetted and not float to the surface of the solvent. All the brush phases were shown to tolerate higher concentrations of water in the mobile phase than the bulk phases before they become un-wettable. It was also proved that the more dispersive and less polar is the solvent, the more water the solvent can contain before any of the phases become unwettable. Stationary phases that are incompletely derivatized can be wetted even by pure water. This can be explained by the high content of residual silanol on their surfaces, which allows the polar interactions with water molecules. C18 phases are wetted by a lower content of water in organic modifier as compared to shorter chain phases such as C2, which becomes wetted

with solvents with a higher content of water (for a similar derivatization degree) [138].

Dewetting of the hydrophobic phases is possible when concentration of water in the mobile phase is higher than 80–90%. It is manifested by the loss of retention capability of the stationary phase, retention irreproducibility, increased tailing, and inability of the column to regenerate when returned to a high concentration of organic component. Study of this process [139] showed that two potential mechanisms are responsible for dewetting. One mechanism is related to the disturbance in the conformation of the alkyl chains in the presence of high water content in the mobile phase and is indicated as phase collapse. This explanation starts with the observation that short-chain stationary phases such as C4 and very long stationary phases such as C30 are less prone to dewetting. The short chains on the silica surface are not very well oriented, and the lack of effect from the water is expected since no specific conformation is to be disturbed. The very long-chain phases are in a strongly bundled form, and the water is not able to easily disturb the phase conformation, which may explain the resilience of this type of phase to higher water concentrations [140]. On the other hand, the interaction forces between the alkyl chains in C8 or C18 phases are easier to disturb (e.g., the melting point of $C_{18}H_{38}$ is 27−28 °C while $C_{30}H_{62}$ has a melting point of 64−67 °C), and the water perturbs the aligned conformation of the phase, reducing its capability of retaining the analyte. However, other studies [141] indicated that the alkyl-bonded phases are always in the most compact conformation, regardless of the concentration of the organic modifier in the mobile phase, and the change in conformation due to the water is not the main cause of the dewetting. This suggests that dewetting is very likely caused by the exclusion of the aqueous mobile phase from the pores covered with a hydrophobic material and the inability of the mobile phase to reenter the pores. Since most of the stationary phase surface is inside the pores, the exclusion of the mobile phase from the pores is the reason for a reduced active phase. The process is depicted in Figure 6.4.16.

A dry sorbent requires a specific pressure for the solvent to reenter the pores. This pressure can be estimated with the formula:

$$\Delta p = \frac{4\gamma'\cos\theta}{d} \qquad (6.4.21)$$

where Δp is the pressure required for the liquid to enter the pore, γ' is the surface tension of the liquid, d is the effective pore diameter, and θ is the contact angle between water and air on the adsorbent surface. The pressure necessary for the mobile phase to reenter the pores can be quite high (in an experiment using a C8 column and 0.1% CH_3COOH in water, a pressure of 270 bar was necessary to re-wet the column [142]). This proposed mechanism is in agreement with several experimental findings and can easily explain why short-chain alkyl phases are less prone to dewetting, their "exclusion" of water being much lower than that of C8 or C18 phases. The explanation for C30-type phases may be related to the fact that such phases are less densely packed, although the carbon content (C%) is high. However, it is more likely that the dewetting is a complex process whereby the exclusion of the highly aqueous mobile phase from the pores of the stationary phase plays an important role in the phase's loss of separation capability, but changes in the conformation of the alkyl chains from the silica surface under the influence of water is also likely to occur and to play a role in the modification of phase properties.

Dewetting/phase collapse can be reversible (not always 100%) by regenerating the column with a partially organic aqueous mobile phase. The process of regenerating the phase can require hours of flushing the column with partially organic phase (e.g., 60% CH_3CN and 40% H_2O). To avoid the dewetting problem, columns with special stationary phases [143] can be used in highly aqueous mediums and are preferred for specific applications such as the separation of very polar analytes. One such

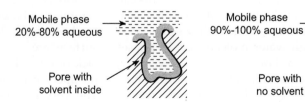

FIGURE 6.4.16 Schematic explanation of dewetting process when a mainly aqueous mobile phase does not penetrate the pores covered with a hydrophobic phase.

Mobile phase 20%-80% aqueous

Pore with solvent inside

Mobile phase 90%-100% aqueous

Pore with no solvent

column is, for example, the Synergy 4u-Hydro-RP (from Phenomenex), which has a polar end-capped surface.

Common Hydrophobic Stationary Phases

Reversed-phase chromatography is the HPLC type with the largest number of applications, and a wide variety of columns and stationary phases have been manufactured to be used for this technique. New columns are frequently introduced on the market, and their characteristics are described in numerous vendor publications (see, e.g.,

HPLC GC SPE Chromatography Product Guide 10/11, Phenomenex [144] and 2011–2012 Waters Chromatography Columns and Supplies Catalog, Waters [145]). Older columns are, however, still in use, and they offer characteristics that can be compared with those of newer columns that are less frequently evaluated. A list of common stationary phases for RP-HPLC that were produced between 1995 and 2010 is given in Table 6.4.11. The table contains mainly cyanopropyl (CN), phenyl (Ph), octyl (C8), and octadecyl (C18) phases. However, less common phases such as polar embedded or fluorinated are also included in the list.

TABLE 6.4.11 List of various columns used in RP-HPLC.

No	Description	No	Description
1	YMC Pack Pro C18 RS	32	Phenomenex Prodigy C18
2	Nomura Develosil ODS SR 5	33	Waters SymmetryShield RP18
3	YMC J'Sphere H80	34	Agilent Zorbax Eclipse XDB C18
4	GL Sciences Intersil ODS-EP	35	YMC ODS AQ
5	Shodex ODS Perfect	36	Agilent Zorbax Rx C18
6	Supelco Ascentis C18	37	Waters Spherisorb ODS-2
7	GL Sciences Intersil ODS 3V	38	Phenomenex Gemini C18
8	Azko Nobel Kromasil C18	39	ACT Ace C18
9	CICT L-Column ODS	40	Agilent Zorbax Stable-Bond C18
10	GL Sciences Intersil ODS-3	41	Merck Purosphere RP18
11	Alltech Alltima C18	42	Machery Nagel Nucleosil C18
12	ES Industries EPIC C18	43	Phenomenex Synergy Max RP
13	Imtakt Cadenza CD-C18	44	Waters Acquity UPLC HSS C18
14	Mackery Nagel Nucleodur Gravity C18	45	Thermo Hypersil Elite C18
15	Dionex Acclaim C18	46	Waters Atlantis dC18
16	Supelco Discovery HS C18	47	Waters Acquity UPLC HSS T3
17	Waters SunFire C18	48	Waters Nova-Pak C18
18	Shiseido Capcell Pak MGII	49	Keystone Fluophase PFP

(Continued)

TABLE 6.4.11 List of various columns used in RP-HPLC. (*cont'd*)

No	Description	No	Description
19	Waters Symmetry C18	50	TSK-Gel 80Ts
20	Agilent Zorbax Extend C18	51	Supelco Discovery HS F5
21	Agilent Zorbax Eclipse Plus	52	YMC J'Sphere M80
22	Merck Purosphere Rpe 18	53	Waters Spherisorb ODSB
23	Restek Allure PFP Propyl	54	Waters Xbridge C18 (Acquity UPLC BEH C18)
24	DionexAcclaim PA (Embedded)	55	Waters XTerra MS C18
25	GL Sciences Intersil ODS-2	56	YMC Hydrosphere C18
26	YMC Pack Pro C18	57	Phenomenex Aqua C18
27	Waters Acquity UPLC HSS C18	58	Machery Nagel Nucleodur Sphinx RP
28	Supelco Ascentis RP-Amide	59	Phenomenex Synergy Fusion RP
29	Nomura Devosil C30 UG 5	60	Waters XBridge Shield RP18
30	Phenomenex Luna C18	61	Waters Atlantis T3
31	Phenomenex Luna C18 (2)	62	Metachem Polaris C18-A
63	Supelco Supelcosil LC DB-C18	95	Merck Lichrosphere Select B
64	Phenomenex Luna Phenyl Hexyl	96	Alltech Alltima C8
65	Thermo Hypersil ODS	97	Agilent Zorbax Stable-Bond C8
66	Supelco Supelcosil LC-ABZ	98	Waters μBondapack C18
67	YMC Carotenoid C30	99	Alltech Platinum C18
68	Thermo Hypersil HyPurity C18	100	Waters Acquity UPLC HSS C18 SB
69	Supelco Discovery C18	101	Thermo Hypersil BDS C8
70	Varian Pursuit C18	102	YMC Basic
71	Supelco Supelcosil LC-ABZ+	103	ZirChrom PBD
72	Waters Symmetry C8	104	Supelco Supelcosil LC DB-C8
73	Keystone Prism	105	Agilent Zorbax Eclipse XBD Phenyl
74	Waters SunFire C8	106	Mac-Mod HydroBond AQ C8
75	Phenomenex Gemini C6-Phenyl	107	YMC J'Sphere L80
76	YMC Pro C8	108	Agilent Zorbax Rx C8
77	Agilent Zorbax Eclipse XDB C8	109	Waters Xterra RP8
78	Azko Nobel Kromasil C8	110	Waters Xterra MS C8
79	Phenomenex Synergi Polar-RP	111	Waters Xterra Phenyl
80	Merck Lichrososorb Select B	112	Phenomenex Prodigy C8

(Continued)

TABLE 6.4.11 List of various columns used in RP-HPLC. (*cont'd*)

No	Description	No	Description
81	Waters SymmetryShield RP8	113	Waters Xbridge C8 (Acquity UPLC BEH C8)
82	Thermo Hypersil BDS C18	114	Waters Nova-Pak Phenyl
83	Waters Xterra RP18	115	Agilent Zorbax SB- Phenyl
84	Agilent Zorbax Bonus RP	116	Restek Ultra Phenyl
85	Keystone spectrum	117	YMC Pack Ph
86	GL Science Intersil C8	118	GL Sciences Inertsil Ph3
87	GL Science Inertsil ODS-SP	119	Alltech Platinum EPS C18
88	Supelco Discovery RP Amide C16	120	Agilent Zorbax SB-Aq
89	Thermo Hypersil GOLD (C12)	121	GL Sciences Inertsil CN-3
90	ACT Ace C8	122	GL Science Inertsil 3 CN
91	Shiseido Capcell Pack C18	123	Keystone Fluophase RP
92	Waters Xbridge Phenyl (AcquityUPLC BEH Phenyl)	124	Thermo Hypersil Phenyl
93	Waters Nova-Pak C8	125	Supelco Discovery HS PEG
94	Restek Allure Ultra IBD	126	Varian Pursuit Diphenyl
127	Thermo Hypersil BDS Phenyl	133	Supelco Discovery Cyano
128	Agilent Zorbax SB-CN	134	Thermo Hypersil CPS CN
129	Restek Ultra PFP	135	Waters Spherisorb S5 Ph
130	YMC CN	136	Waters Spherisorb S5 CN RP
131	Phenomenex Luna CN	137	Waters Nova-Pak HP CN
132	Keystone Fluofix 120N		

A quick evaluation of these columns can be done using the two typical parameters that characterize a separation, namely, the capacity factor k and the selectivity factor α. As previously discussed, both k and α are compound dependent and mobile phase dependent. For a RP-type column, k is frequently indicated for specific non-polar compounds. The values of k for acenaphthene for the columns listed in Table 6.4.11 are given in Figure 6.4.17 (see Waters Reversed-Phase Column Selectivity Chart, Waters [145]). The mobile phase used for the evaluation was 65% CH_3OH and 35% aqueous buffer of 20 mM K_2HPO_4 - KH_2PO_4 at pH = 7 [78,107]. As shown in Figure 6.4.17, the range of k value starts around 0.25 (almost no retention) for columns with a CN-bonded phase and is as high as 33.5 for certain C18-bonded phases. The separation characterized by α values for the pair amitriptyline/acenaphthene, with the same mobile phase as the one used for measuring the k values, is shown in Figure 6.4.17 for a variety of columns (negative values for α indicate that amitriptyline elutes before acenaphthene). As seen from this figure, the best separation for the

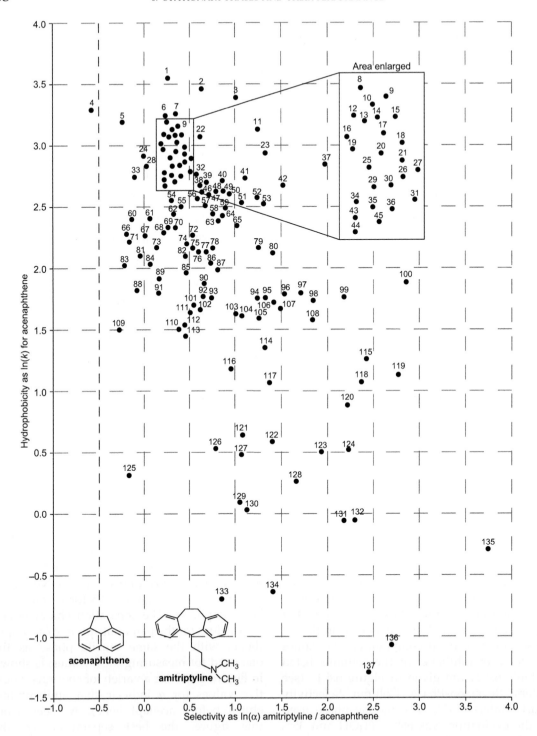

FIGURE 6.4.17 The values of ln k for acenaphthene and the values for ln α for the pair acenaphthene/amitriptyline for the columns listed in Table 6.4.11 [145].

pair ($\alpha \approx 44$) is obtained for a phenyl-type stationary phase.

The columns listed in Table 6.4.11 and displayed in Figure 6.4.17 include a variety of columns commonly used in RP chromatography. Among them are more standard columns with C8 or C18 stationary-phase and end-capped with methyl groups, along with olumns showing other bonded phases and some with polar-embedded functional groups and C8, C18, phenyl, alkyl chains of different lengths (e.g., C12). The nature of the hydrophobic bonded phase such as CN (with some polarity), phenyl, alkyl, and the length of the hydrocarbon fragment (for alkyl chains) influences the phase hydrophobic character, as characterized, for example, by the k values of acenaphthene. The separation of very "nonpolar" compounds is not achieved well on columns with low hydrophobicity (such as CN phases). However, very hydrophobic stationary phases leading to very high k values are also not appropriate for the separation, since they lead to unacceptably long retention times. Probably the best choices of columns for the separation of hydrophobic compounds are those with a C8 stationary phase.

For organic compounds that have a polar character, there are two choices. One is to use more hydrophobic columns that can point up small differences in the hydrophobicity of the molecules, and the other is to use columns with some polarity that involve additional interactions besides hydrophobic ones. For the choice of using highly hydrophobic columns, stationary phases with long alkyl chain bonded to silica surface (e.g., C27 to C30) are recommended. The separation of polar compounds on C8 or C18 stationary phases require a very low content of organic modifier in the mobile phase (less than 5%), which is sometimes a problem due to processes such as phase collapse and dewetting. The increase in the hydrophobicity of the stationary phase can increase the hydrophobic interactions with the hydrophobic moiety of the analyte molecules. In this way, the content of organic modifiers in the mobile phase could be used at levels above 5%, such that the columns are less prone to dewetting [146]. Some polar compounds are better separated on columns such as cyano or pentafluorophenylpropyl (e.g., Ascentis Express F5) that involve larger contribution from van der Waals interactions compared to solvent cavity formation energy. The separation of polar and nonpolar compounds together is best achieved on columns with an intermediate hydrophobic character as seen in Figure 6.4.17 for the separation of acenaphthene and amitriptyline, which is best achieved on a phenyl or CN-type column (with high values for α).

The list of columns from Table 6.4.11 is incomplete since more than 500 column types are commercially available (not considering different formats). These columns offer a wide range of characteristics allowing diverse separations. For example, very fast separations can be achieved by using shorter columns, with small particle size, higher flow rates, and possibly temperature of the column higher than ambient. One example of such a separation is given in Figure 6.4.18 for tenoxicam and piroxicam on a Zorbax Stable-Bond C18 column, 50 mm length, 4.6 mm i.d., and 1.8 μm particle size [147]. The separation was performed in gradient elution with mobile phase starting from 30% acetonitrile and 70% aqueous solution of 0.1% H_3PO_4 to 100% ACN in 1.5 min; flow rate: 2 mL/min and column temperature: 60 °C. The detection was performed by UV absorption at 368 nm

Hydrophobic Stationary Phases with Some Polarity

The common hydrophobic stationary phases are successfully used in a wide range of applications, and compounds with a large range of polarities and having only some hydrophobic moiety, are well separated using hydrophobic HPLC columns. However, some analytes are more problematic to separate. Examples of

FIGURE 6.4.18 Chromatogram for tenoxicam and piroxicam on a Zorbax StableBond C18 column, 50 mm length, 4.6 mm i.d. and 1.8 μm particle size [147].

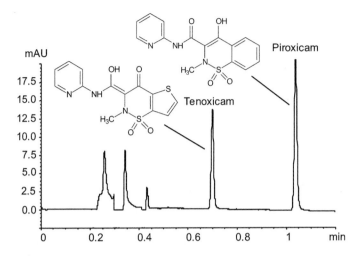

such analytes are compounds with a basic character such as amines that give peak tailing, large molecules that are not soluble in organic solvents such as proteins or saccharides, and very small polar compounds that are very little retained on hydrophobic stationary phases and require a low level of organic component in the mobile phase to be eluted from the column. Also, some isomers that differ minimally in hydrophobicity may be difficult to separate on conventional hydrophobic phases. For these "problem" analytes, either more polar stationary phases or special properties of common-type stationary phases were developed, such that the compounds can be successfully analyzed. Various such phases were obtained using procedures such as the use of a polar group in the bonded phase, sterically protected base silicas, bidentate bonded phases, polar-embedded stationary phases, the use of hybrid organic-inorganic particles, polar end-capped silica [49], or polymerization of specific organic monomers on the silica support [148]. Such phases may offer, besides hydrophobic interactions, other types of interactions such as hydrogen bonding, dipole-dipole interactions, aromatic π-π stacking, and steric selectivity, as pictured in Figure 6.4.12.

Some important types of phases used in RP-HPLC that have groups with some polarity are those containing cyano groups (e.g., cyanopropyl), phenyl groups connected on the silica surface with a hydrophobic linker, or those with an ether-linked phenyl (e.g., Synergy Polar-RP from Phenomenex). Several cyano and phenyl phases are listed in Table 6.4.11. The separation on such phases involves, besides hydrophobic interactions, a number of polar interactions that are stronger or weaker depending on the analyte. Besides better separation of compounds with some polarity, these phases can be very useful for the separation of isomers when hydrophobic interactions alone are not sufficient to assure separation.

A very good solution for obtaining hydrophobic stationary phases with the capability to have more contribution from polar interactions and withstand very high water content in the mobile phase is the use of polar-embedded hydrophobic stationary phases. As previously discussed, a variety of embedded phases can be made. Columns with various *embedded polar* groups such as ether, amide, urea, carbamate, sulfone, thiocarbamate, mixed amide and carbamate, were manufactured. Common commercially available columns contain embedded urea, amide, carbamate, or ether groups with C8 or C18 hydrophobic chains. Some of these phases are included in Table 6.4.8 (Phase No. 4, 15, 28, 33, 59, 60, 66, 73, 81, 83, 84, 85, 88, 109). The wide distribution in the k values for those phases indicates the versatility in the hydrophobicity that can be achieved with embedded materials. For some analytes, the phases provide the same type of retention as their C8 or C18 counterpart, as seen from Figure 6.4.17 when comparing the k values for the embedded phases with other conventional C18 or C8 phases from Table 6.4.11. However, a different selectivity compared to alkyl phases can be seen for other analytes, in particular for those with a stronger polar character. Also, with the incorporation in the alkyl chain of a polar functional group that is placed close to the surface of the silica gel, these phases retain the hydrophobic character, while the phase can remain solvated by water at low percentages of organic solvent, even with 100% water in the mobile phase. Under these conditions, the alkyl chains maintain

pyrimidine bases) can be achieved, even at intermediate pH values [143, 149, 150]. A comparison of conventional C18 columns with a number of polar-embedded C18 columns and a number of polar-end-capped columns reported in the literature [108] showed that, on average, considerable differences exist in the separation properties of these types of columns. The results of the study are summarized in Table 6.4.12 [108].

Other phases with hydrophobic character and some polar interaction capability are also available. For example, embedded phases containing two zones of polarity are known. One such type of phase contains one ether group, one sulfonamide moiety, and a C16 hydrophobic moiety. Such a phase can be used in a totally aqueous mobile phase and exhibits a high selectivity toward N-containing compounds. Another special phase has two anchor points to the silica support (it is a bidentate phase). The structures of a phase with two polar zones and that of a bidentate phase are further shown:

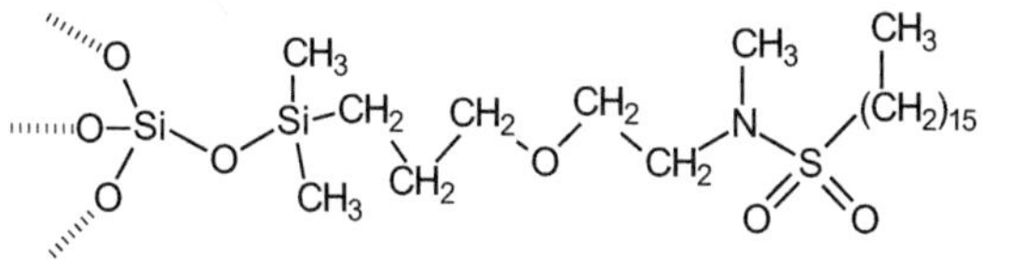

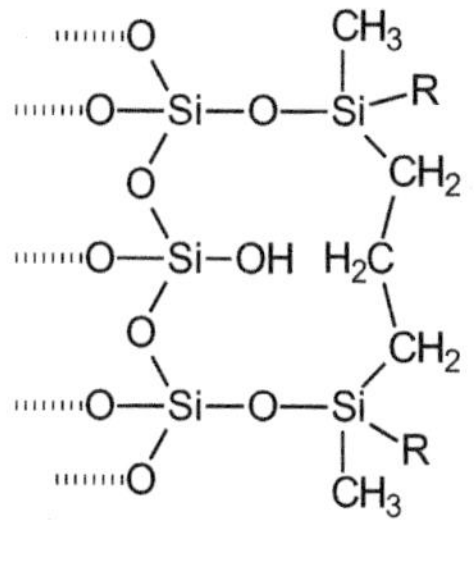

R = C$_8$H$_{17}$ or C$_{18}$H$_{37}$

their conformational freedom and can interact with polar analytes. The presence of the polar functionality close to the surface also shields the effects of unreacted silanol groups protecting the phase from the attack of stronger acids and stronger bases. Since the silanol activity is suppressed in polar-embedded phases, better peak shape and decreased tailing of basic compounds (such as amines, purines, and

The propylene bridges can sterically cover the silica matrix and the silanol groups, offering superior stability to low and high pH of mobile phase, and hindering the interaction of the analytes with basic character with the silanol groups that lead to better peak shape for these compounds [151].

The polar groups may also result from additional interactions besides those with the main

TABLE 6.4.12 Comparison on several test parameters for conventional, polar end-capped, and polar embedded columns [108]*.

Column type	N	k_{BB}	$\alpha(CH_2)$	$\alpha_{caffeine/}$ phenol	$\alpha_{benzylamine/}$ phenol pH = 2.5	$\alpha_{benzylamine/}$ phenol pH = 7
	plates m^{-1}	hydro-phobicity	methylene selectivity	hydrogen bonding	silanol activ. pH = 2.5	silanol activ. pH = 7
Conventional C18						
Luna 5mm C18(2)	116889	9.11	0.176	0.21	0.059	0.131
Inertsil ODS(3)	96669	8.63	0.171	0.261	0.045	0.147
Zorbax XDB C18	95997	8.57	0.187	0.213	0.088	0.134
Symmetry C18	91515	9.23	0.181	0.224	0.049	0.147
Hypurity Elite C18	87816	2.96	0.155	0.31	0.061	0.378
Zorbax SB C18	84568	6.74	0.179	0.283	0.079	0.412
Prodigy ODS(3)	108384	9.82	0.184	0.21	0.051	0.13
Mean		7.866	0.176	0.244	0.062	0.211
Polar end-capped						
Keystone Aquasil	109457	4.41	0.147	0.725	0.131	1.659
Aqua C18	87666	8.67	0.176	0.251	0.094	0.149
YMC Hydrosphere	109430	6.25	0.167	0.27	0.055	0.14
YMC ODS-Aq	90049	7.83	0.171	0.262	0.094	0.169
Prontosil C18AQ	92800	7.04	0.173	0.303	0.079	0.254
Metasil AQ	88846	7.77	0.158	0.217	0.073	0.162
Synergi Hydro-RP	113187	10.07	0.174	0.24	0.063	0.203
Mean		7.434	0.167	0.324	0.084	0.391
Polar embedded						
Zorbax Bonus-RP	110156	3.86	0.147	0.201	0	0.167
Polaris C18-A	105481	4.38	0.165	0.2	0.102	0.121
Discovery RP-Amide	115577	3.34	0.15	0.166	0.05	0.097
SymmetryShield RP18	109530	6.15	0.152	0.164	0.013	0.098
Supelco ABZ+	95551	1.88	0.143	0.182	0	0.23
Experimental urea	54458	6.25	0.157	0.133	0	0.8
Polaris Amide C18	87744	5.11	0.139	0.123	0	0.08
Mean		4.424	0.15	0.167	0.024	0.228

Note*: The columns were evaluated using *Test 5* at page 283 where parameters k_{BB}, $\alpha(CH_2)$, $\alpha_{caffeine/phenol}$, $\alpha_{benzylamine/phenol}$ pH = 2.5, $\alpha_{benzylamine/phenol}$ pH = 7 are defined.

bonded phase (e.g., C8 or C18). These polar groups may contribute to the separation and may also offer additional protection against uncontrolled acidic silanol groups in the stationary phase. These two effects are achieved with the use of polar end-capping. This technique protects the stationary phase from hydrolysis and offers a dual type of interaction with the analytes. The columns with polar end-capping can be considered to have a kind of mixed-mode character. This type of column may still show tailing when a mixed-mode type of interaction is not desirable. Columns with C8 or C18 hydrophobic phase and end-capped with polar substituents (e.g., NH_2 or OH on a propyl handle) are commercially available, and they offer an alternative-type column with its characteristics between a stationary phase non—end-capped and with short-alkyl chains and that of a more hydrophobic column (such as C8 or C18).

One special type of stationary phase offering additional polar interactions are those non—end-capped with short-alkyl chain bonded phase. As previously discussed, the unreacted silanol groups present on an alkyl-bonded silica impart a degree of polarity to the phase, and frequently this polarity is detrimental since it produces tailing on compounds with polar or ionic moieties. However, in some cases, the alkyl function alone provides insufficient separation selectivity, and the presence of silanols can cause polar interactions with the polar functionality on analytes, which can be useful. The resulting mixed mechanisms can yield improved separations, for example, for small polar molecules. Also, the phases with short alkyl chains (e.g., C4) have a low propensity for phase collapse when high concentrations of water are used in the mobile phase. The main problem with this type of phase is the difficulty in having a uniform and reproducible level of silanol groups. The activity of the free silanol groups on the phase surface is also dependent on the amount of water adsorbed from the

mobile phase. For this reason, this type of column may have reproducibility problems, depending on the water content in the mobile phase. Also, due to ion-exchange type interactions, the compounds with basic character may show severe tailing.

Other Types of Hydrophobic Phases

Other special hydrophilic stationary phases are those with very long alkyl bonded fragments. These phases are particularly useful for the separation of compounds with very little hydrophobic character that are easily eluted from typical C18 columns. For such compounds, the use of mobile phases with a very high content of water is necessary, imposing on the column the need to have resilience to dewetting. The long alkyl chain stationary phases, such as C27 or C30, were found useful for the separation of this type of compound. These phases are more retentive for polar and nonpolar analytes than are most polar-embedded and even high-coverage C18 phases. Because of a higher degree of surface shielding, long-chain phases also offer greater pH stability than do C8 and C18 phases. They are also more resistant to phase collapse under high aqueous conditions than are C18 phases. This behavior may be explained by the resistance of long chains to conformation changes in the presence of water at the column temperature, or it may be related to a lower density of C30 chains on the silica surface for similar C% content with a C18 phase [79].

Special stationary phases also include those that are based on core-shell (or fused core) technology. The main advantage of this type of phase is the higher number of theoretical plates compared to columns similar in dimensions but filled with porous particles. A variety of core-shell columns with different brand names are available from different commercial sources. One series of core-shell columns is the Kenetex (from Phenomenex), available in C8,

C18, XB-C18 (which has butyl side chains for protecting against silanol access), phenylhexyl, and pentafluoro-phenyl (PFP), with particle sizes of 2.6 μm and 1.7 μm [152, 153]. Another series is Ascentis Express (from Supelco), which is available in C18, phenyl-hexyl, C8, and pentafluorophenyl. Another series is Accucore from Thermo Scientific that offers C18 and PFP columns as well as a column end-capped with polar groups (Accucore aQ). Agilent offers Poroshell series.

One additional group of stationary phases consists of those with a fluorinated bonded fragment. Typically, the fluorinated fragment is attached to the silica surface with a short handle, and the phase has the following schematic structure:

$$\text{---}\underset{/}{\overset{\backslash}{\text{Si}}}\underset{\text{O}}{\searrow}\underset{\backslash}{\overset{|}{\text{Si}}}\text{---CH}_2\underset{\text{CH}_2}{\overset{\nearrow}{\searrow}}\text{R}$$

where R can be perfluorohexyl straight chain, perfluorohexyl branched chain, perfluorooctyl, perfluoropropyl, perfluorododecyl, pentafluorophenyl, pentafluorophenyl alkyl chain, pentafluorophenylpropyl, and other similar phases. In Table 6.4.8, phases No. 23, 43, 51, 123, and 132 are fluorinated. These phases cover a wide range for the k values, indicating that the fluorinated materials offer a wide range of hydrophobicity, some with very high log K_{ow} for the corresponding model as shown in Table 6.3.4. The fluorinated phases offer an alternative material for separations and were proven very useful in particular separations [154, 155].

A variety of other phases with hydrophobic character are also available, either commercially or at an experimental stage. As an example, EnviroSep™ type phases (from Phenomenex) contain a silica and polymer support and a hydrophobic bonded phase. Other columns are made using a variety of special technologies. For example, Capcell Pak type columns (Shiseido Co. Ltd.) involve a surface coating of the silica

base support with a silicone polymer that shields the silanol groups, the bonded phase being connected to the coated surface. Other columns take advantage of the ethylene bridged structure of the silica base but use it only on the surface of the silica particles (TWIN technology), as used, for example, in Gemini NX columns.

Silica monoliths can also be considered special columns. The C18-bonded phase on monoliths is, however, a relatively common type of column, such as Onyx Monolithic C18 or Chromolith Performance RP-18 columns, which are available in various formats. These columns have a typical dual porous structure with mesopores of about 130 Å and macropores of about 2 μm diameter. The nature of the bonded phase is similar to that for spherical particle C18 columns, and the silica surface is end-capped. This type of column allows a reduction in elution time up to nine times due to the capability to use higher flow rates without having problems with the column backpressure. Due to a rapid mass transfer of the solutes between the bonded phase and the mobile phase, the decrease in the number of theoretical plates at higher flow rates, as predicted by the van Deemter equation (see rel. 2.2.8), is not as intense. Shorter retention times also lead to better resolution [156]. For comparison, the variation of the theoretical plate height in various columns is shown in Figure 6.4.19. As shown in this figure, the flow rate for which the measurements were possible is higher for the monolith column, and at the same time the increase in the plate height is smaller.

An example of a separation of a monolith-type column is given in Figure 6.4.20 for the separation of norfloxacin, one of its metabolites and of ciprofloxacin. The column utilized for the separation is a Chromolith Performance RP-18e (Merck KGaA), 100 mm x 4.6 mm. Two overlaid chromatograms are shown in Figure 6.4.20, corresponding to a blank sample (A) and a sample collected from a same human volunteer using norfloxacin (B). Sample preparation was based on deproteinization of 200 μL human plasma

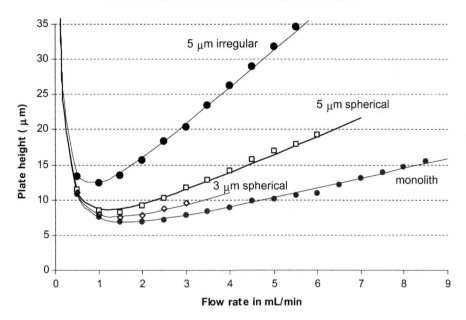

FIGURE 6.4.19 Variation in theoretical plate height for different types of stationary phases (irregular particles 5 μm, spherical particles 5 μm, spherical particles 3 μm, and monolithic).

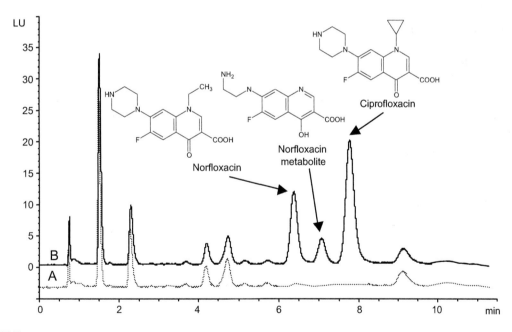

FIGURE 6.4.20 Separation on a Chromolith Performance RP-18e (Merck KGaA), 100 mm x 4.6 mm column of norfloxacin, one of its metabolites, and ciprofloxacin (added as internal standard); (A) blank plasma sample, (B) sample from human volunteer using norfloxacin [157].

with 50 μL acrylonitrile that also contained cipro-floxacin as internal standard, followed by mixing the supernatant resulting from deproteinization with 3400 μL mobile phase. A volume of 100 μL was then injected into the analytical column. The mobile phase for the separation was made of 1% triethylamine in aqueous solution brought to pH = 4.0 with phosphoric acid and methanol as organic modifier. The elution program consisted of: 85/15 (v/v) aqueous / methanol for 0−9 min, a linear gradient between 9 and 10 min to 40/60, followed by a linear gradient between 10 and 11 minutes to the initial composition (85/15), and kept constant up to 11.5 min. The flow rate was 2.5 mL/min, and temperature 25 °C. The detection was done with a FLD setup at 268 nm excitation and 445 nm emission [157].

Organic Polymer-based Hydrophobic Stationary Phases

Although most RP-HPLC columns are silica-based, the polymeric columns can be very useful for specific applications. This is particularly the case when a very low or a very high pH of the mobile phase is necessary. The main advantage of polymeric supports is that they are more resilient to very low and very high pH. Also, the polymeric substrate does not have a potential layer of unwanted polar groups as do the silica-based stationary phases, which even after derivatization still have unreacted silanol groups besides the desired ones. However, polymeric materials typically show lower theoretical plate numbers N for the same dimensions as the silica-based particles. Another disadvantage of polymeric supports is caused by the variation in the swelling of the polymeric particles when they are used in various solvents. This variation in swelling affects the volume occupied by the stationary phase in the column and can lead to the formation of void spaces and loss of efficiency. Organic polymeric materials are also less resilient to high pressures, and this makes polymeric

columns less adequate for the new developments toward UPLC. For these reasons, the use of polymeric columns, particularly for RP-HPLC, is limited. Some polymeric phases can operate at pH values as low as 1 and as high as 13. Some examples of RP polymeric phases are given in Table 6.4.13, where certain reported characteristics are also shown.

The stationary phase from these columns may consist of the polymer in itself or may have a specific bonded phase (such as C8, C18) on the polymeric support. As seen from Table 6.4.13, the polymeric columns display a wide range of acceptable pH for the mobile phase. Also, the reported values for N/m in the range of 80,000 are typical for many silica-based columns with 5 μm particles. The only parameter that may contribute to a lower rating of polymeric columns is the acceptable maximum pressure, which may bring some limitations to the flow rates that can be used. The pore system generated in the polymer is an important factor regarding the rate of mass transfer effects during the chromatographic process, and this is difficult to assess from data such as those given in Table 6.4.13. Besides stationary phases made with particles of rigid organic polymers, polymeric monoliths are also used as chromatographic media [37]. Other polymeric phases with hydrophobic bonded groups such as C18 or pentafluorophenyl are commercially available (e.g., Jordi phases from MicroSolv). Among different polymeric materials used as stationary phase in HPLC, several molecular imprinted polymers (MIPs) were also evaluated. The polymeric structure has been typically based on methacrylic acid and/or styrene crosslinked with ethyleneglycol dimethacrylate that were polymerized in the presence of a template molecule (e.g., nortriptyline [158]).

Special Hydrophobic Stationary Phases

Several special hydrophobic stationary phases have been used successfully in RP-HPLC.

TABLE 6.4.13 Examples of polymeric columns and their main characteristics.

Name	Polymer	Size μm	Pore μm	N/m	pH range	Funct. group	Max press(psi)	Equivalent
Shodex ODP2 HP	Polyhydroxymethacrylate	5	4	80,000	2–12	none	2250	C18
Shodex DE	Polymethacrylate	4	10	70,000	2–12	none	2250	C18
Shodex DS	Polymethacrylate	3.5	10	70,000	2–13	None	3000	Mixed mode
Shodex RP18-413	Styrene divinyl benzene	3.5	10	80,000	1–13	None	3300	Mixed mode
Shodex RP18-613	Styrene divinyl benzene	3.5	10	80,000	2–13	None	3300	Mixed mode
Shodex RP18-415	Styrene divinyl benzene	6	43	36,000	2–13	None	3300	Mixed mode
Asahipak ODP	Polyvinyl Alcohol	5	25	56,000	2–13	C18	2250	C18
Asahipak ODP40	Polyvinyl Alcohol	4	25	68,000	2–13	C18	1950	C18
Asahipak C8P	Polyvinyl Alcohol	5	25	45,000	2–13	C8	2250	C8
Asahipak C4P	Polyvinyl Alcohol	5	25	40,000	2–13	C4	2250	C4
Shodex NN	Polyhydroxymethylacrylate	10	10	40,000	2–12	none	2250	Mixed mode
Shodex JJ	Polyvinyl Alcohol	5	10	32,000	2–11	none	1400	Mixed mode

One of these phases is porous graphitic carbon (PGC). This material is obtained by decomposing organic matrices using silica as template. For example, the silica used as template is impregnated with a mixture of phenol and formaldehyde and then heated to 80–160 °C to initiate the polycondensation. The characteristics of the silica gel as template material determine the size and porosity of the particles that will be obtained. The polymer is then pyrolyzed under inert atmosphere (N_2) at 1000 °C. Thus, highly porous amorphous carbon is produced. This carbon corresponds to what is normally called carbon black. After this step, the silica template is dissolved with a hot aqueous NaOH solution. The graphitization is realized through a thermal treatment at high temperature (about 2300 °C) under inert atmosphere (Ar).

This operation eliminates certain surface attached functional groups, produces a structural rearrangement of C atoms, and removes the micropores. After cooling down to 1000 °C, the replacement of argon by hydrogen can induce reactions between hydrogen and free radicals or functional groups still present at the carbon surface. By deactivating the PGC surface becomes more uniform. The result is porous graphitic carbon, which is now commercialized (trade name of Hypercarb®) [159]. Unlike silica-based packing materials, carbon-based stationary phases have the advantages of being more resistant to hydrolysis, while the lack of swelling or shrinking makes them more useful than the polymeric materials. The efficiency of columns packed with these materials is comparable to modified silica-based columns. Some physical properties of PGC are given in Table 6.4.14 [160].

The PGC columns were successfully used in RP-HPLC practice. The retention of nonpolar compounds is similar to that of silica-based hydrophobic columns (C18 or C8), but the graphitic columns show a higher $\alpha(CH_2)$ value compared to any C18 or C8 column (using methanol as an eluent) [161]. However, the retention mechanism on PGC stationary phases is more complex than for silica bonded phases. These phases are able to interact with hydrophobic moiety from the analyte structure, but also with the polar groups from their structure. The strength of interaction depends on the molecular area of the analyte in contact with the graphite surface and on the nature and the type of functional groups at the point of interaction with the flat graphite surface. Under specific elution conditions, PGC behaves as a strong retentive stationary phase for both nonpolar analytes and polar compounds. Unlike silica-bonded phases when increased polarity of analyte determines a decreased retention behavior, the retention on PGC is less influenced by polarity, and in some circumstances it may increase with the increase of polarity of analyte. Consequently, the elution order on PGC is not necessarily the hydrophobicity order of analytes, as usually takes place in reversed-phase mechanism. Due to these characteristics, the PGC phases have the capacity to separate very polar compounds (carbohydrates and compounds with several hydroxyl, carboxyl, amino and other polar groups). Compounds with low or no affinity to C18 stationary phase can have strong retention and can be separated in accordance with their polarity in an order that is not specific to the RP mechanism [162, 163]. Also, the PGC phases have the ability to separate structurally related compounds (geometric and diastereoisomers) due to the flat and highly adsorptive surface of the graphite [164]. Several properties of graphitic columns such as resilience to a wide pH range, temperature resistance (up to 200 °C) [165], and the capability of their surface to be covered with surfactants behaving as an ion-exchanger make the PGC columns of considerable utility [166].

As an alternative to porous graphitic carbon, phases made of carbon-clad zirconia or titania have been suggested. This type of material can be produced either by high-temperature graphitization of organic polymers covering the zirconia or titania support or by chemical vapor deposition of carbon on zirconia. However, in practice, the surface of this type of stationary phase is not perfectly homogeneous, and a significant proportion of the surface of zirconia still remains uncovered by carbon, thus creating highly acidic residual zirconia groups [167].

Among other special stationary phases used in HPLC are those based on silica hydride. Bare silica hydride can be used for direct phase separations (in organic-normal phase HPLC) or for aqueous normal-phase separations. Commercial stationary phases with common dimensions such as 4 μm particle diameter and pore size of 10 nm are available. Bonded phases on silica with C18, C8, cholesterol groups, and the like are commercially available (e.g., Cogent Bidentate C18, Cogent Bidentate C8, Phenyl Hydride, Cogent UDC Cholesterol, etc.). These types of columns can be used in reversed-phase-type conditions [19].

TABLE 6.4.14 Some properties of graphitic carbon phases.

Property	Values
Particle shape	Spherical 3, 5, or 7 μm diameter
Porosity type	Porous 70–75 %
Specific surface area	Higher than 100 m^2/g
Pore volume	~0.7 m^3/g
Average pore diameter	~25 nm
Carbon load	100%
Mechanical strength	>400 bar

Selection of a Column for RP-HPLC Separations

The selection process of a column useful for an RP separation should be guided by the work previously performed and reported in the literature. An enormous body of information is available regarding numerous analytical procedures. Information may be available for compounds identical with those of interest (possibly in a different matrix) or regarding similar compounds, although not exactly the same. When such information is available, the problem of column selection remains open, but a much more limited search is necessary for finding a good fit. Starting with a recommended column, improvements can be made such that the chromatography will be done in a shorter time or with better resolution. This can be achieved, for example, by selecting the same stationary phase in a different format or with smaller particle size, in a narrower column, in a shorter or in a longer column depending on the improvement objectives. The criteria listed in Table 6.3.3 must be considered when format modifications are implemented. The type of available equipment influences the decision regarding the column selection. When UPLC equipment is available, columns with small particles and narrower diameter can be selected, since they provide higher N for the same length, in spite of their higher backpressure. Some of the advantages regarding column quality can be opposite to each other. For example, a column with small particle size and smaller diameter can lead to an increase in theoretical plate number N improving resolution and in reduction of the analysis time and use of solvents, which are very advantageous. On the other hand, the same parameters may lead to a reduction of the number of injections n that can be made on a column without affecting performance and to an unacceptable increase in column backpressure.

Athough a specific column is recommended, a change is sometimes needed or desired for a particular reason, for example, related to the availability or price of the column. In this case a column equivalent with the one recommended can be selected. For selecting a similar but not identical column, the calculated distance using rel. 6.4.17 or 6.4.18 [123] is very useful, and columns with a small distance F to the recommended one can be employed successfully.

The adequacy of a column depends mainly on achieving a good resolution R for the solutes. Capacity factor k plays an important role in achieving this goal. In general, for increasing k, for compounds with smaller hydrophobic moieties, columns with stronger hydrophobic character (large H_c values) are necessary. More hydrophobic compounds can be separated on less hydrophobic columns (see Table 6.4.8). This recommendation is based on the assumption that larger capacity factors k for a hydrophobic compound are obtained when H_c is larger. However, in order for a large H_c to contribute to a large k, the hydrophobicity parameter η' for the compound must be large. Since $\log k$ is well correlated with $\log K_{ow}$ (see, e.g., Figure 3.1.3 and [168]), a good correlation between η' values and $\log K_{ow}$ values would be expected. Figure 6.4.21 shows this correlation between η' and $\log K_{ow}$ for the compounds listed in Table 6.4.7. When nortriptyline and amitriptyline are not included in the correlation, the linear correlation gives $R^2 = 0.8277$, but when the whole set of compounds is used, $R^2 = 0.3657$, which indicates extremely poor correlation. Therefore, the assumption that large H_c and η' values alone lead to high $\log k$ values is not correct for certain compounds. In cases where besides hydrophobic moieties the compounds have polar groups, columns with additional interactions may be more appropriate for the separation. These columns include cyano of phenyl-bonded phases, or are C8, C18 columns with polar-embedded groups, have end-capping done using small polar fragments, or are not end-capped. As an example, Figure 6.4.17 displays

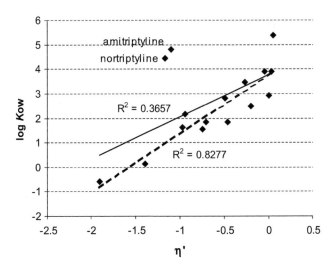

FIGURE 6.4.21 Correlation between η' values and log K_{ow} for the compounds given in Table 6.4.7 (including nortriptyline and amitriptyline $R^2 = 0.3657$, or excluding them $R^2 = 0.8277$).

the case of higher α for the pair amitriptyline versus acenaphthene, which is constantly higher for more polar columns.

Even for columns leading to high values for k for a certain type of compound, the separation is not necessarily good since the resolution R also depends on α. Most compounds analyzed by RP-HPLC have both hydrophobic and polar moieties and other column characteristics (S^*_c, A_c, B_c, $C_c(2.8)$, and $C_c(7.0)$) contribute to the separation, as previously discussed in this section. For this reason, the separation of compounds that have mainly a hydrophobic character can be evaluated based on column hydrophobic character. When polar groups are present (as is the case for most analytes), not only the hydrophobicity of the column is important, but other characteristics such as a partial polar character. This can be seen, for example, from the value for ln (α) for the pair acenaphthene versus amitriptylene shown in Figure 6.4.17. The largest α values are obtained for columns with some polar character (cyano and phenyl). Further discussion of column selection can be found in Section 9.4.

As previously discussed, the column nature together with the nature of the compounds to be separated also influences the peak tailing. The selection of columns with embedded polar groups can solve some of the tailing problems. The resilience to a wide pH range, also desirable in certain separations, is obtained using special end-capping or polar embedded groups. The use of pure base silica (Type B), which is less acidic, is another factor contributing to a wider pH range of use for silica-based columns. Also, dewetting problems are alleviated with special types of end-capping or with the use of hydrophobic columns with embedded polar groups.

6.5. POLAR STATIONARY PHASES AND COLUMNS

General Aspects

Polar stationary phases are used in normal-phase chromatography (NPC), in hydrophilic interaction chromatography (HILIC), and in aqueous normal-phase chromatography

(ANPC or ANP). More recently, HILIC has started to gain considerable popularity. Some delay in its application was caused by fewer available columns as compared to RP-HPLC. This lack of column availability is being rapidly eliminated, and various HILIC columns are now commercially offered [169]. The technique can be successfully applied for the analysis of highly polar organic molecules such as sugars, amino acids, and numerous pharmaceuticals.

Normal-phase chromatography is currently practiced mainly on silica, although other polar stationary phases were used in the past, such as alumina and magnesia. HILIC chromatography can also be performed on silica, but bonded phases containing terminal polar groups such as aminopropyl, diol bonded, amide bonded, and peptide bonded, are now common phases for HILIC. Cyano (e.g., cyanopropyl) phases can also be included in the polar phases category (see log K_{ow} in Table 6.3.4), although they are used in RP chromatography with a mobile phase that is more polar than the bonded phase of the column. Some cyano RP phases were indicated in Section 6.4. The polar phases are typically characterized by their polar groups (such as silanol in the case of silica)

and by their surface that is wettable with polar solvents such as water. Several types of polar-bonded moieties on silica and on polymeric substrates used in NPC and HILIC chromatography are presented in Table 6.5.1 [170]. The polarity of the phase is the highest for silica, followed by phases with weak ion-exchange character (anion, cation, zwitterions), such as amine phases, and followed by diol, amide, and cyano.

Specific Procedures for Polar-Phase Synthesis

The main attention in this chapter was given to synthesis of hydrophobic stationary phases for RP chromatography. However, as previously indicated in Section 6.2, some of the procedures used for synthesis of hydrophobic stationary phases can also be used for making polar phases by having a polar group in the radical R of chloro-R-silanes or alcoxy-R-silanes used for silica gel derivatization. Other procedures were specifically developed to generate polar phases. One such procedure starts with preparation of a bonded polysuccinimide on silica by reacting this compound with amino-propyl silica, as shown schematically in the reaction 6.5.1:

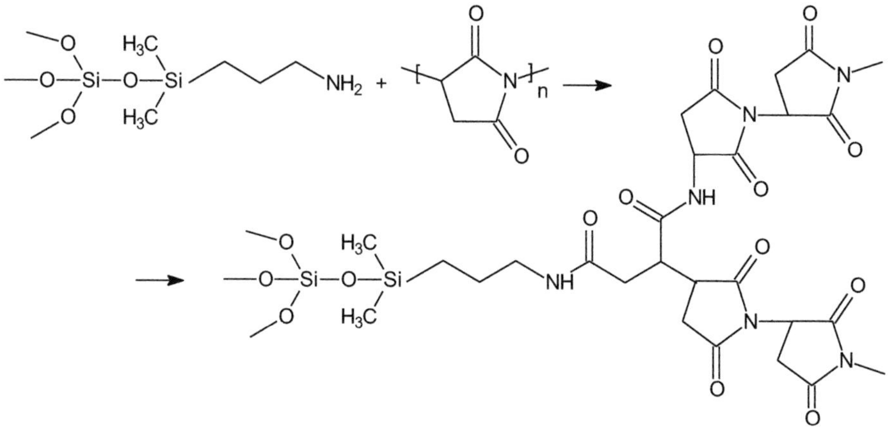

(6.5.1)

TABLE 6.5.1 Several types of polar stationary phases used in NPC and HILIC chromatography.

Type of phase	Phase
Silanol	Bare silica
Diol on silica	Diol, ether embedded and diol
Amide on silica	Amide terminal, polyamide
Weak polar on silica	Cyano, perfluorinated (also used in RP mode)
Weak anionic on silica	-C$_3$H$_6$-NH$_2$, diethylamine, triazole, etc.
Weak cationic on silica	Sulfonylethyl, etc.
Zwitterionic on silica	Amino-sulfonic, amino-carboxilic
Mixed mode on silica	Both polar and hydrophobic groups
Weak anionic polymeric	Different polar groups on porous polymers

Parts of the poly(succinimide) rings are opened to form amide bonds with the aminopropyl moieties. In this way a polymeric-type phase bonded on silica is generated. Only a fraction of the succinimide rings are engaged in this bonding; the rest of the rings remain intact and susceptible to reaction with nucleophiles, which can be used to produce a variety of functional silicas [170]. The poly(succinimide)-silica material can further undergo an alkaline hydrolysis in water to form a poly(aspartic acid) bonded phase, it can suffer hydrolysis in the presence of aminoethanol to form poly(2-hydroxyethyl)aspartamide bonded phase, and in the presence of aminoethansulfonic acid it can form poly(2-sulfonylethyl)aspartamide-bonded silica. The formation of different functional groups for poly(succinimide)-silica phases is schematically shown in reaction 6.5.2.

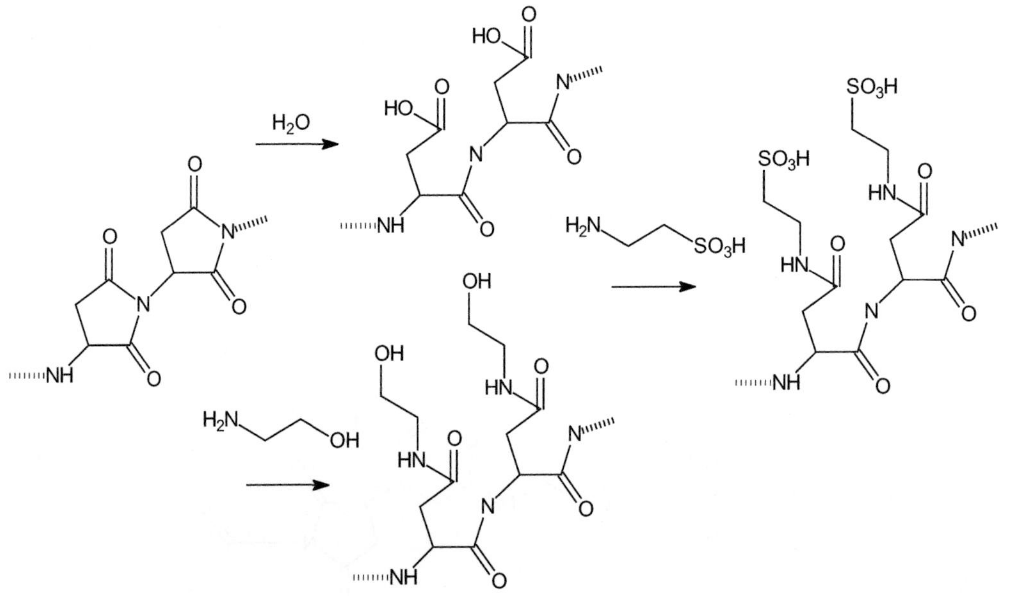

(6.5.2)

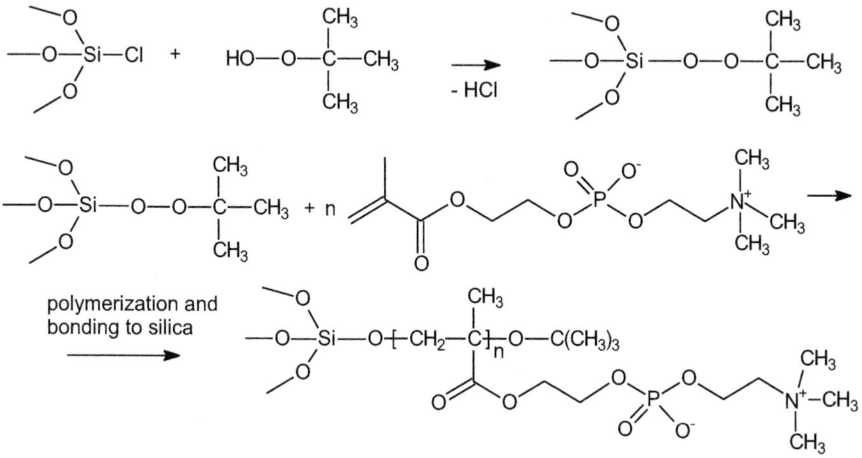

(6.5.3)

The poly(sulfonylethyl)aspartamide silica phase displays a zwitterionic character. Another phase with zwitterionic character can be obtained from a phosphorylcholine-type polymer connected to the silica surface that was initially derivatized to have a peroxy bridge bound on the silica surface [171]. The reactions leading to this polar stationary phase are given by 6.5.3.

groups, such as amide, carboxylic, sulfonic, amine, or quaternary ammonium in the presence of silica previously derivatized with methacryloyl or methacrylamido groups. The polymerization takes place using an initiator, such as 2,2'-azobisisobutyronitrile or ammonium peroxydisulfate [172]. This type of reaction is schematically indicated in reaction 6.5.4:

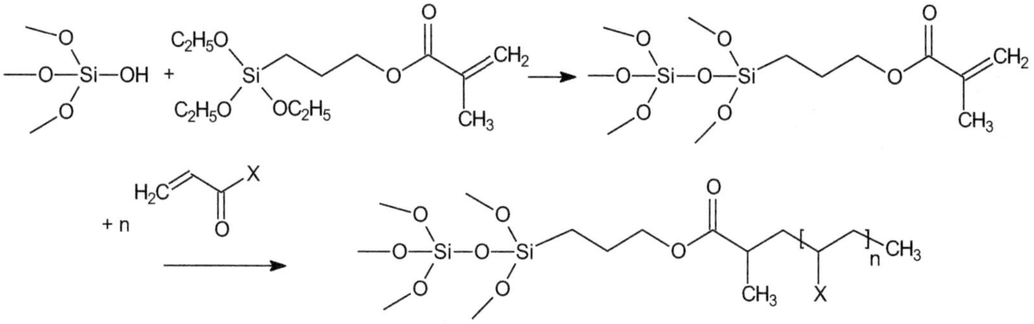

(6.5.4)

Another method for preparing polar phases leading to monolithic columns used for HILIC separations is based on the polymerization of vinyl monomers having various X functional

Synthesis of other active phases may include macrocycles such as perhydroxyl-cucurbit[6] uril. The macrocycle can be obtained following the reactions 6.5.5:

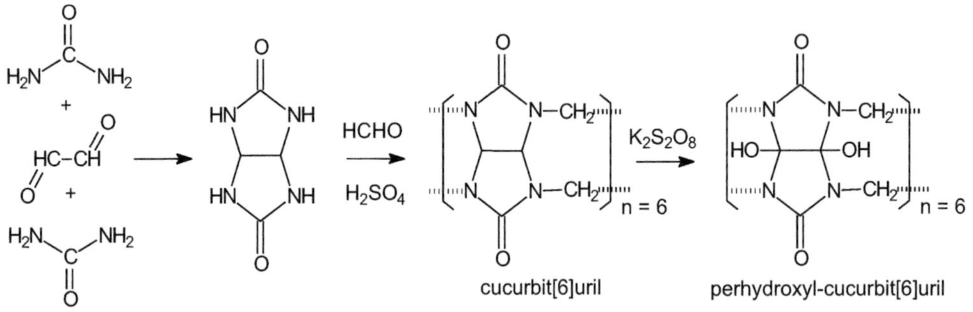

cucurbit[6]uril   perhydroxyl-cucurbit[6]uril

(6.5.5)

The bonding of perhydroxyl-cucurbit[6]uril can be done starting with silica gel derivatized with propylisocyanate groups [173]. The isocyanate further reacts with the OH groups of perhydroxyl-cucurbit[6]uril and generates a stable stationary phase that can be utilized for HILIC separations. Some of the stationary phases with polar groups previously described have found commercial applications and others have not.

Retention and Separation Properties of Polar Stationary Phases

As previously indicated, retention in chromatography depends on the nature of the three participants in the separation: (1) the analyte, (2) the stationary phase, and (3) the mobile phase. Polar phases have the capability to retain polar compounds, and the higher is the nonpolar character of the organic phase, the stronger the polar compounds are retained on the stationary phase (have higher capacity factors). On the other hand, the hydrophobic compounds are not well retained (or are not retained at all) on polar stationary phases.

In a similar manner as described for hydrophobic phases, several parameters can be used for the characterization and comparison of polar phases [174–181]. These include typical parameters for chromatographic column characterization such as height of theoretical plate H, peak asymmetry As, and dead time t_0

(obtained with a very nonpolar compound such as toluene $t_0 = t_{toluene}$). The interest for column characterization is geared in particular toward phases used in HILIC mode, due to their wider utilization. The value of capacity factor in HILIC was found to be significantly dependent on the mobile phase content in water [182]. For a number of columns, when tested versus polar compounds, the following formula describes this dependence of log k on the volume fraction of water ϕ_w in the mobile phase:

$$\log k = a + b \log \phi_w + c\, \phi_w \qquad (6.5.6)$$

where a depends on the analyte, and b and c are parameters that depend on both the analyte and the stationary phase. (The coefficients a and b in rel. 6.5.6 have negative values such that the increase in ϕ_w leads to a decrease in log k, which is in accordance with weaker retention of polar compounds when the water content increases in the mobile phase [180, 183]). However, a more detailed HILIC column characterization cannot be made using a simple parameter. HILIC type of separation is complex and involves hydrophylic, hydrophobic, and ion-exchange-type interactions (see Section 5.3). For this reason, HILIC column characterization requires a more elaborate series of tests that provide information on multiple characteristics of the stationary phase [179]. Among the tests suggested specifically for HILIC column characterization are the following:

1) Methylene selectivity $\alpha(CH_2)$ is a parameter that quantifies the hydrophobic character. As indicated in Section 5.3, HILIC separation involves hydrophobic effects, since the disruption of the interactions between the solvent molecules plays a noticeable role in the energy values involved in the separation. However, for HILIC columns, the evaluation of $\alpha(CH_2)$ cannot be done using hydrophobic compounds such as those recommended for RP columns (see rel. 6.4.11). Different compounds that are one CH_2 apart can be used for defining the $\alpha(CH_2)$ value. For example, a recommended comparison is the measurement of $\alpha(CH_2)$ as the ratio of capacity factors for uridine (Uri) and 5-methyluridine (5MeUri). The values for k are obtained in specific mobile phase (e.g., acetonitrile:aqueous buffer 90:10, the buffer being 20 mM ammonium acetate at pH = 4.7 [179]) and lead to rel. 6.5.7.

$$\alpha(CH_2) = \frac{k(Uri)}{k(5MeUri)} \qquad (6.5.7)$$

The choice of nucleosides for obtaining a value for $\alpha(CH_2)$ was made because these compounds are well retained on HILIC columns. The retention of uridine and 5-methyluridine is less affected by ion-exchange effects.

2) Hydroxy group selectivity $\alpha(OH)$ is a parameter that quantifies the hydrophilic character. A recommended comparison is the measurement of $\alpha(OH)$ as the ratio of capacity factors for uridine (Uri) and 2-deoxyuridine (2dUri), the values for k being obtained in the same mobile phase conditions as for $\alpha(CH_2)$ (acetonitrile: aqueous buffer 90:10, the buffer being 20 mM ammonium acetate at pH = 4.7 [179]) and given by rel. 6.5.8.

$$\alpha(OH) = \frac{k(Uri)}{k(2dUri)} \qquad (6.5.8)$$

The hydroxy group selectivity is not restricted to the comparison of the capacity factors for uridine and 2-deoxyuridine. Other compounds differing by an OH group can be used for calculating hydroxy group selectivity. The values for $\alpha(CH_2)$ and $\alpha(OH)$, obtained using formula 6.5.7 and 6.5.8 in the mobile phase acetonitrile:aqueous buffer 90:10 with the buffer being 20 mM ammonium acetate at pH = 4.7, were measured for several commercially available columns [179]. The list of these columns is given in Table 6.5.2, which also lists the values for the capacity factor k for uridine, values for $\alpha(CH_2)$, for $\alpha(OH)$, as well as other selectivities further defined. The values for selectivities $\alpha(CH_2)$ and $\alpha(OH)$ are also shown in Figure 6.5.1.

Besides the use of uridine and 2-deoxyuridine in formula 6.5.8, other compounds are also used for estimating an $\alpha(OH)$. As expected, the values for $\alpha(OH)$ depend on the nature of the analyte (as well as on the columns and the mobile phase). For this reason various $\alpha(OH)$ values can be assigned for the same HILIC column and the same mobile phase, but for different test compounds. For example, for the columns listed in Table 6.5.2, the values for $\alpha(OH)$ can be determined using other nucleoside/2′-deoxynucleoside pairs than those based on uracil, for example, where the base is thymine, adenine, cytosine, or guanine [179]. The values of $\alpha(OH)$ can also be obtained similarly for pairs 2′-deoxynucleosides/2′3′-dideoxynucleoside. The formulas of some of the nucleic bases and those of the compounds used for calculation of $\alpha(CH_2)$ and $\alpha(OH)$ with uracil-based nucleosides for HILIC columns are shown in Figure 6.5.2.

TABLE 6.5.2   List of several commercially available HILIC columns, the capacity factor k for uridine, and several α values.

No	Column	$k_{uridine}$	$\alpha(CH_2)$	$\alpha(OH)$	α_{dia}	α_{regio}	α_{shape}	α_{AX}	α_{CX}
1	ZIC-HILIC (5µm)	2.11	2.03	1.67	1.5	1.11	1.14	0.33	1.57
2	ZIC-HILIC (3.5 µm)	2.10	2.07	1.71	1.51	1.12	1.14	0.27	1.64
3	Nucleodure HILIC (3 µm)	2.20	1.55	1.28	1.46	1.08	1.14	0.34	0.95
4	Amide-80 (5 µm)	3.30	1.67	1.27	1.29	1.08	1.18	0.19	1
5	Amide-80 (3 µm)	4.58	1.64	1.27	1.28	1.08	1.18	0.41	2.75
6	XBridge Amide (3.5 µm)	2.55	1.7	1.29	1.3	1.07	1.16	0.47	1.2
7	PolySULFONYLETHYL (3µm)	1.58	2.13	1.48	1.21	1.06	1.24	0.06	0.35
8	PolyHYDROXYETHYL (3µm)	3.92	1.92	1.36	1.31	1.07	1.21	0.34	1.31
9	CYCLOBOND I (5 µm)	0.70	1.21	1.13	1.24	1.1	1.2	4.73	0.63
10	LiChrospher Diol (5 µm)	1.50	1.36	1.15	1.32	1.06	1.17	0.63	1.16
11	Chromolith Si	0.31	1	1.12	1.16	1.11	1.31	0.09	8.21
12	HALO HILIC (2.7 µm)	0.64	1.08	1.16	1.18	1.13	1.29	0.64	29.03
13	COSMOSIL HILIC (5 µm)	1.60	1.6	1.14	1.36	1.03	1.13	0.8	0.49
14	Sugar-D (5 µm)	1.58	1.74	1.44	1.45	1.1	1.22	1.9	0.25
15	NH2-MS (5 µm)	2.44	1.88	1.3	1.36	1.07	1.2	0.82	0.28

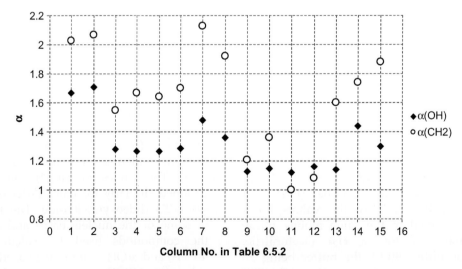

FIGURE 6.5.1   The values of $\alpha(CH_2)$ and $\alpha(OH)$ obtained using formula 6.5.7 and 6.5.8 for the columns indicated in Table 6.5.2 (mobile phase acetonitrile:aqueous buffer 90:10, the buffer being 20 mM ammonium acetate at pH = 4.7) [179].

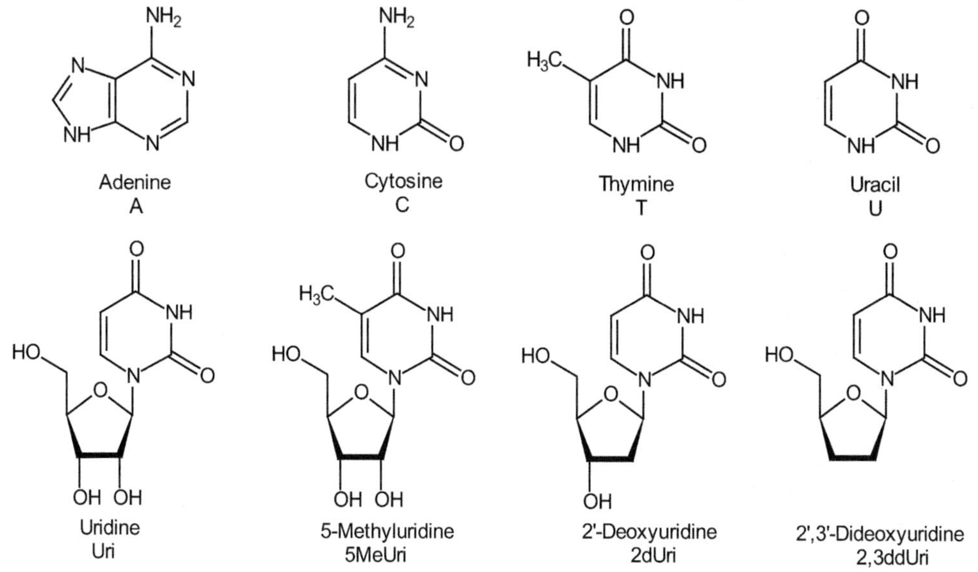

FIGURE 6.5.2 Formulas for nucleic bases and for some nucleosides based on uracil/thymine, used for the calculation of $\alpha(CH_2)$ and $\alpha(OH)$.

3) Isomer selectivity is another property that can be used for HILIC column characterization. Chiral isomers cannot be separated on phases without a chiral selector, and isomer selectivity of HILIC columns does not refer to chiral separations. However, the capability of separating structural isomers, cis-trans, and diastereoisomers is an important column characteristic. This selectivity has been suggested to be characterized by the values of α for two pairs of the following compounds [179]: (1) diastereoisomers vidarabine (Vid) and adenosine (Ade), and (2) regioisomers 2′-deoxyguanosine (2dGua) and 3′-deoxyguanosine (3dGua). The formulas for these compounds are given in Figure 6.5.3.

The expressions for selectivities $\alpha_{dia}(Vid/Ade)$ and $\alpha_{regio}(2dGua/3dGua)$ are given by rel. 6.5.9, and their values for the columns indicated in Table 6.5.2

(mobile phase acetonitrile: aqueous buffer 90:10, the buffer being 20 mM ammonium acetate at pH $= 4.7$) [179] are shown in Figure 6.5.4.

$$\alpha_{dia}(Vid/Ade) = \frac{k(Vid)}{k(Ade)}$$

$$\alpha_{regio}(2dGua/3dGua) = \frac{k(2dGua)}{k(3dGua)}$$

(6.5.9)

4) Molecular shape selectivity α_{shape} is another parameter that can be used for characterization of HILIC columns. The steric parameters α_{dia} and α_{regio} are related to shape of the molecule but in a more particular manner. A selectivity parameter recommended in the literature [179] for molecular shape characterization uses the two diastereoisomers 4−nitrophenyl-α-D-glucopyranoside (NPαGlu) and 4-nitrophenyl-β-D-glucopyranoside (NPβGlu). The formulas for these two compounds are

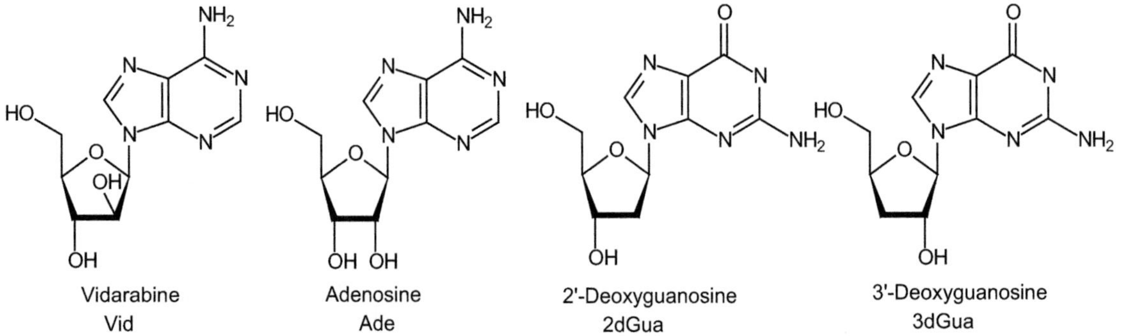

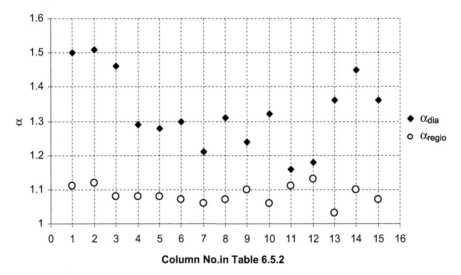

Vidarabine
Vid

Adenosine
Ade

2'-Deoxyguanosine
2dGua

3'-Deoxyguanosine
3dGua

FIGURE 6.5.3 The formulas of diasteroisomers vidarabine (Vid) and adenosine (Ade) and of regioisomers 2'-deoxy-guanosine (2dGua) and 3'-deoxyguanosine (3dGua).

FIGURE 6.5.4 The values of α_{dia}(Vid/Ade) and α_{regio}(2dGua/3dGua) obtained using formula 6.5.9 for the columns indicated in Table 6.5.2 (mobile phase acetonitrile: aqueous buffer 90:10, the buffer being 20 mM ammonium acetate at pH = 4.7) [179].

given in Figure 6.5.5, and the expression for α_{shape} is given by expression 6.5.10.

$$\alpha_{shape}(NP\alpha Glu/NP\beta Glu) = \frac{k(NP\alpha Glu)}{k(Np\beta Glu)}$$

(6.5.10)

For the columns listed in Table 6.5.2, and using the same mobile phase (acetonitrile:

aqueous buffer 90:10, the buffer being 20 mM ammonium acetate at pH = 4.7) [179], the values for α_{shape} are shown in Figure 6.5.6.

5) Two selectivity parameters can be used for evaluation of ion-exchange interactions. One selectivity is used for describing anion-exchange properties α_{AX}, and the other for

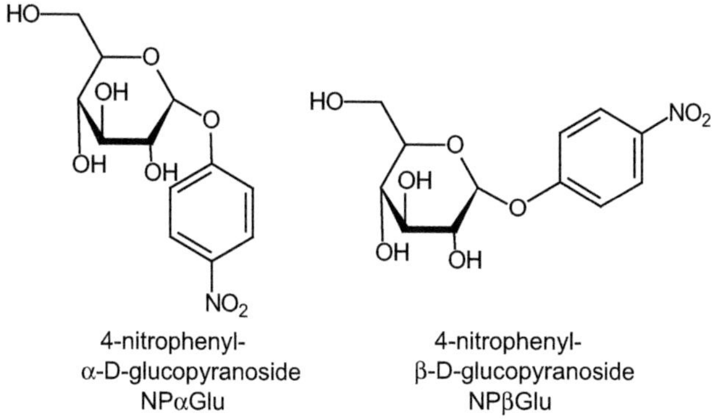

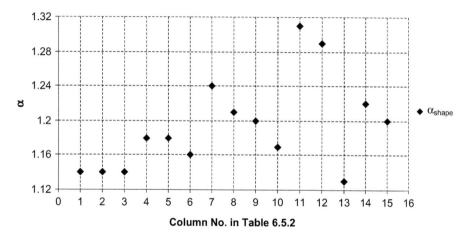

FIGURE 6.5.5 The formulas of diasteroisomers 4-nitrophenyl-α-D-glucopyranoside (NPαGlu) and 4-nitrophenyl-β-D-glucopyranoside (NPβGlu).

FIGURE 6.5.6 The values of for α_{shape} obtained using formula 6.5.10 for the columns indicated in Table 6.5.2 (mobile phase acetonitrile:aqueous buffer 90:10, the buffer being 20 mM ammonium acetate at pH = 4.7) [179].

cation-exchange properties α_{CX}. For these two selectivities the pairs of compounds recommended in the literature [179] are the following: (1) sodium p-toluenesulfonate (SPTS) and uridine (Uri) for characterization of anion exchange properties, and (2) N,N,N-trimethylphenylammonium chloride (TMPAC) and uridine (Uri) for characterization of cation-exchange properties. The formulas used to obtain these selectivity parameters are as follows:

$$\alpha_{AX}(SPTS/Uri) = \frac{k(SPTS)}{k(Uri)}$$

$$\alpha_{CX}(TMPAC/Uri) = \frac{k(TMPAC)}{k(Uri)} \qquad (6.5.11)$$

The values for α_{AX} and α_{CX} for the columns listed in Table 6.5.2 are shown in Figure 6.5.7 for a mobile phase acetonitrile:aqueous buffer 90:10, the buffer being 100 mM ammonium acetate at pH = 4.7 [179]. The chemical

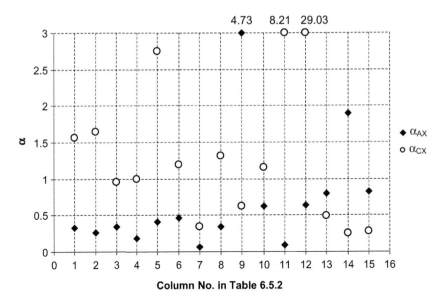

FIGURE 6.5.7 The values of for α_{AX} and α_{CX} obtained using formula 6.5.11 for the columns indicated in Table 6.5.2 (mobile phase acetonitrile:aqueous buffer 90:10, the buffer being 100 mM ammonium acetate at pH = 4.7) [179]. Note: the α_{AX} for column No. 9 is 4.73, and α_{CX} for column Nos. 11 and 12 are 8.21 and 29.03, respectively.

formulas for SPTS and for TMPAC are given in Figure 6.5.8.

6) Another parameter used to evaluate HILIC columns is α_t(theobromine/theophylline) [184]. This parameter differentiates columns regarding their basic, neutral, or acidic character with $\alpha_t < 1$ for basic, $\alpha_t \approx 1$ for neutral, and $\alpha_t > 1$ for acidic stationary phases. The formulas for theobromine and theophylline are given in Figure 6.5.8.

Unlike the case of hydrophobic columns, where the parameters for column characterization H_c, S^*_c, A_c, B_c, $C_c(2.8)$, and $C_c(7.0)$ were obtained using regression parameters for the calculation of α values, in the case of HILIC columns it was found more convenient to

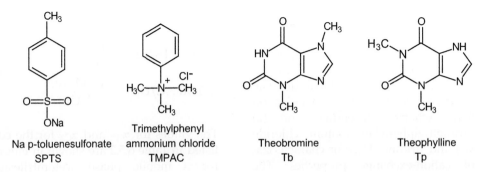

FIGURE 6.5.8 Formulas for sodium p-toluenesulfonate, N,N,N-trimethylphenylammonium chloride, theobromine and theophylline.

simply characterize the columns based on measured α parameters for specific pairs of compounds. Since the HILIC mechanism is more complex and less investigated than the mechanism for hydrophobic columns, further characterization of a multitude of columns is needed for a similar approach as that used in RP-HPLC.

HILIC column comparison can be done by different procedures based on the multitude of α parameters. One such procedure is the use of a radar graph similar to the graph from Figure 6.4.14 used for hydrophobic columns comparison. Such a graph that allows comparison of four selected columns is shown in Figure 6.5.9.

Bare Silica Stationary Phase

Silica stationary phase can be used in either NPC or HILIC mode. In NPC mode, the organic mobile phase that has no water does not imply the total absence of water from the silica surface. Anhydrous silica still contains a layer of water (possibly monomolecular) on its surface. In HILIC mode, silica is covered with a thicker layer of water, and a partition mechanism for the separation is more likely than on dry silica. For both NPC and for HILIC applications, silica of high purity (Type B purity) is commonly used. The extent of NPC utilization for analytical purposes is rather limited, and NPC is mainly applied in other chromatographic techniques, such as thin layer chromatography (TLC) and preparative chromatography (at lower pressure). There are, however, useful applications of HPLC on silica in NPC mode (nonpolar solvents), such as separation of very nonpolar compounds (for example, carotenoids or lipids) that require an organic mobile phase with no water or polar constituents, separation of compounds that are extremely polar and are not retained or are very weakly retained on RP columns, and also separation of certain isomers (achiral) since the silica in NPC conditions is more efficient than RP for such separations. As discussed in Section 5.3, the interactions between the stationary phase and the solute (analyte) in NPC and HILIC are stronger for more polar compounds. Since the polarity of the isomers (not chiral isomers) is frequently very different—for example,

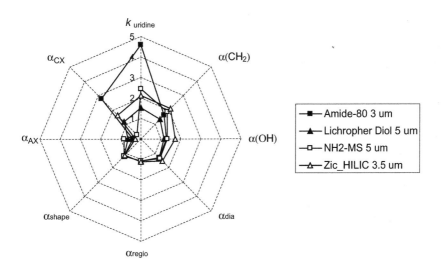

FIGURE 6.5.9 Radar graph for the comparison of columns using $k_{uridine}$ and several α parameters.

because the polar groups are positioned in different parts of the molecule—the interactions with the stationary phase are different from isomer to isomer. This explains the efficiency of polar phases in isomer separation. Besides the polar group position, other effects such as steric hindrance and additional effects such as propensity to form hydrogen bonds or to have electron donation or withdrawal interactions may affect the differences in retention.

Silica utilization in NPC mode has several disadvantages. The main problem is related to the reproducibility of such separations. Since the silica is always covered with a layer of water, the immobilized material acting as stationary phase can be considered to be water. The variation in the amount of water retained on the silica produces variations in the retention times. The different level of water can be caused by different levels of exposure of the stationary phase to water from solvents with different degrees of dryness or by the length of exposure of the column to a dry solvent that may dissolve some of the water forming the stationary phase. The sensitivity of column reproducibility to the water exposure is very high. Also, related to the water layer formation is the slow equilibration of the column when changing solvents and frequent irreproducibility in this equilibration process. In addition, some compounds are retained irreversibly on bare silica. Although the structure of Type B silica is highly homogeneous, silica columns frequently produce tailing of the peaks, indicating variability in the type of interactions on the stationary phase surface. Silica phases based on Type B (purity) specifically developed for work in HILIC mode were developed and are commercially available. Among these can be listed the columns indicated in Table 6.5.3. When used in HILIC mode, the layer of water on the silica surface is much thicker than in NPC mode, and the separations are less susceptible to mobile phase dryness. Also, other disadvantages of bare silica

as stationary phase are diminished, but some, such as peak tailing, may still be a problem.

HILIC Stationary Phases with a Bonded Surface

The development of silica-based stationary phases with a bonded material that has polar groups is the cause of a considerable expansion of applications using HILIC columns. The polarity of bonded-phase HILIC columns is lower than that of bare silica. However, their reproducibility and rapid regeneration when changing solvent corrected some of the main problems of bare silica columns. HILIC columns are widely used for the separation of important classes of polar organic compounds such as carbohydrates, amino acids, and peptides [185]. Also, due to the use of solvents containing both water and an organic solvent, HILIC columns are frequently recommended to be selected when MS or MS/MS detection is utilized, since these solvents provide good ionization media and lead to good sensitivity. A variety of types of stationary phases can be used for HILIC separations; some of the ones that are commercially available are discussed below. Most of these columns are silica-based. They include mainly columns made using porous particle materials, but several are made using core-shell (fused core) support.

1) Neutral HILIC stationary phases have the polar group amide, diol, or cyano. These derivatized silica phases are less polar than bare silica. The columns containing cyano groups have the lowest polarity (see Table 6.3.5 for the corresponding log K_{ow}), and this type of column can also be viewed (and used) as RP column. For this reason several cyano columns were discussed in Section 6.4. Some cyano columns are specifically recommended for HILIC work, and several examples are given in Table 6.5.4. Also

TABLE 6.5.3 Several columns indicated as usable in HILIC mode on underivatized silica.

Brand name	Manufacturer	Phase type/ pore size
Atlantis	Waters	Silica, 100 Å
Betasil	Thermo Hypersil	Silica, 100 Å
ChromoLith	Merck	Silica monolith
Hypersil	Thermo Hypersil	Silica, 120 Å
Kromasil	EKA Chemicals	Silica, 100 Å
Spheri-5 Silica	Brownlee (Alltech)	Silica, 80 Å
Promosil Silica	Bonna Agela	Silica
Venusil XBP Silica	Bonna Agela	Silica
Venusil XBP-L Silica	Bonna Agela	Silica
Inspire Silica	Dikma	Silica
Nucleodur unmodified	Macherey-Nagel	Silica
Nucleosil unmodified	Macherey-Nagel	Silica
COSMOSIL SL-II	Nacalai	Silica
YMC-Pack SIL	YMC	Silica
Zorbax Sil	Agilent	Silica
Ascentis Express HILIC	Supelco	Core-shell silica
Halo	Hichrom	Silica
Gold HILIC	Thermo	Silica

recommended for HILIC use is a fluorinated stationary phase (Epic HILIC FL with undisclosed structure), although this phase can be considered a RP type.

In HILIC stationary phases, the polar group is typically connected to the silica surface by a hydrocarbon chain (such as propyl or longer). One common polar group in HILIC stationary phases is amide. Amide polarity is not very high, but it is higher than cyano (see Table 6.3.4). An even more polar group is hydroxy, and common HILIC columns have diol groups in their bonded phase. Some diol phases may have a specific structure where ether groups are also present in the connecting chain (handle) to the silica surface (crosslinked diol). Polyethylene glycol bonded to silica is also used as a polar stationary phase. The retention mechanisms for neutral HILIC stationary phases consist of polar interactions and hydrogen bonding between the hydroxy or amide groups on the stationary media and the polar groups of the analytes. HILIC columns with amide groups have been proven to be useful for sugars, amino acids, and peptides analyses. Dihydroxypropyl (diol) groups are able to form stronger hydrogen bonds compared to the amide groups, and these columns are used when stronger polar interactions are necessary for the separation. Schematic structures of an amide and a diol stationary phase are shown in Figure 6.5.10.

More complicated structures that contain OH groups can be used as neutral HILIC stationary phases. Among these are phases with bonded cyclodextrin or bonded perhydroxyl-cucurbit-[6]uril groups that contain numerous OH functionalities and can act as HILIC stationary phases. An example of a schematic structure of a phase containing various polar groups, including a lactose moiety, triazole, and a phenyl spacer, is shown in Figure 6.5.11 [186].

Other materials such as those made from silica-bonded polysuccinimide [187] or polyhydroxyethyl-aspatamide can be used as neutral phases for HILIC. Phases containing a sulfur atom embedded in the chain bearing on OH group, such as mercaptoethanol silica and thioglycerol silica, have also been reported [169]. Several neutral HILIC columns that are commercially available are listed in Table 6.5.4. As indicated in this table, organic polymers are also used as a support for phases with neutral polar groups, such

TABLE 6.5.4 Several neutral HILIC columns that are commercially available.

Brand name	Manufacturer	Support/pore size	Nature of phase
Cogent Type C Silica	Microsolv	Silica, 100 Å	Silica hydride
Alltima Cyano	Grace Alltech	Silica	3-Cyanopropyl
Promosil CN	Bonna Agela	Silica	3-Cyanopropyl
Venusil XBP CN	Bonna Agela	Silica	3-Cyanopropyl
Nucleodur CN/CN-RP	Macherey-Nagel	Silica	3-Cyanopropyl
Nucleosil CN	Macherey-Nagel	Silica	3-Cyanopropyl
COSMOSIL CN-MS	Nacalai	Silica	3-Cyanopropyl
YMC-Pack CN	YMC	Silica, 120 Å	3-Cyanopropyl
Epic HILIC FL	ES Industries	Silica, 120 Å	Fluorinated
TSKgel Amide-80	Tosoh Bioscience	Silica, 100 Å	Amide
GlycoSep N	ProZyme	Silica	Amide
XBridge amide	Waters	Polyetoxysilane (BEH)	Amide
Unisol Amide	Bonna Agela	Silica	Amide
Venusil XBP Diol	Bonna Agela	Silica	Amide
Epic Diol HILIC	ES Industries	Silica	Diol
Luna HILIC diol	Phenomenex	Silica	Diol and ether embedded
Kintex HILIC Core-shell diol	Phenomenex	Silica core-shell	Diol
XBridge HILIC	Waters	Polyetoxysilane (BEH)	Diol
Inertsil Diol	GL Sciences	Silica	2,3-Dihydroxypropyl
Lichrospher Diol 100	Merck	Silica	2,3-Dihydroxypropyl
Silasorb Diol	Chemapol	Silica	2,3-Dihydroxypropyl
YMC-pack Diol 120 NP	YMC	Silica	2,3-Dihydroxypropyl
Inspire Diol	Dikma	Silica	Diol
Nucleosil OH (diol)	Macherey-Nagel	Silica	Diol
YMC-Pack PVA-Sil	YMC	Silica support	Polyvinyl alcohol polymerized on silica
Nucleodex β-OH α-PM, etc.	Macherey-Nagel	Silica	β-Cyclodextran
Cyclobond I 2000	ASTEC	Silica	β-Cyclodextran
Perhydroxyl-CB[6]uril silica	Custom synthesis	Silica	Perhydroxyl-cucurbit[6]uril
PolyHydroxyethyl A	PolyLC	Silica (polyaspartic acid)	Poly(2-hydroxyethyl-aspartamide)
PolyGlycoplex	PolyLC	Silica (polysuccinimide)	Poly(succinimide)
Hydrolyzed GMA-co-EDMA	Custom synthesis	Methacrylic copolymer	2,3-Dihydroxypropyl

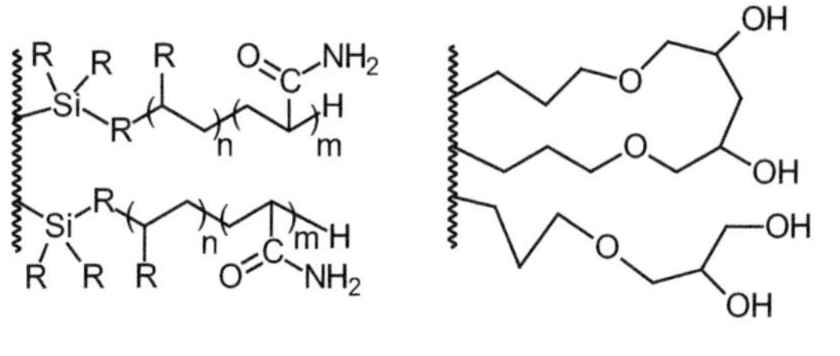

FIGURE 6.5.10 Schematic structures of an amide and a diol stationary phase.

as methacrylic polymers with 2,3-dihydroxypropyl groups [188], sorbitol bonded to a methacrylare polymer covering silica and other similar groups.

2) Anion-exchange-type stationary phases have the polar group amine or triazole. These phases are in fact weak anion-exchange stationary phases that can be used in HILIC mode being applied for the separation of neutral molecules. The propyl amino type stationary phase has been in use for a long time and applied in particular for the separation of carbohydrates. Besides amino groups attached to an aliphatic hydrocarbon chain, the amine group can also be attached to an aromatic ring. Several commercially available HILIC columns with weak anion-exchange properties are listed in Table 6.5.5.

3) Cation-exchange HILIC are cation-exchange columns that can be used in HILIC mode for the separation of neutral molecules. Among such columns some are weak cation exchangers and some strong cation exchangers (with sulfonic groups).

However, these columns can be successfully used in HILIC mode. Some columns with cation exchanger properties recommended for HILIC separations are listed in Table 6.5.6.

4) Zwitterionic HILIC stationary phases contain both an anionic and a cationic group in their structure. Two schematic structures of zwitterionic phases are shown in Figure 6.5.12.

Besides the two structures indicated in Figure 6.5.12, other zwitterionic structures have been made such as phases containing polypeptides bonded to silica. Several commercially available HILIC columns with a zwitterionic structure are listed in Table 6.5.7.

5) Mixed-mode HILIC columns have bonded phases with both nonpolar and anion exchange, cation exchange, or zwitterionic character. For example, Primesep N (Sielc) has embedded acidic groups (negatively charged, thus cation exchange) on a hydrocarbon chain bonded phase, Primesep AP (Sielc) has weak amino anion-exchange groups and nonpolar moieties, and Obelisc N (Sielc) has both negatively and positively charged groups (zwitterionic) on the same long chains of bonded phase. The ion-exchange groups give these columns enhanced selectivity in addition to RP character. An example of a phase containing an amide group, a tertiary amine and a long (C10) hydrocarbon spacer is shown in Figure 6.5.13 (schematic structure of Acclaim Mixed Mode from Dionex).

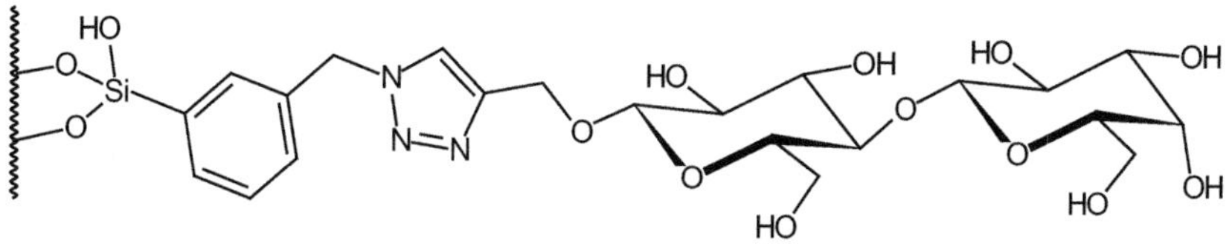

FIGURE 6.5.11 Example of a schematic structure of a phase containing various polar groups and a phenyl spacer.

TABLE 6.5.5 Several HILIC columns with weak anion exchange properties that are commercially available.

Brand name	Manufacturer	Support/pore size	Nature of phase
Luna Amino	Phenomenex	Silica, 100 Å	3-Aminopropyl
Hypersil APS2	Thermo Hypersil	Silica, 120 Å	3-Aminopropyl
Spherisorb NH$_2$	Waters	Silica, 80 Å	3-Aminopropyl
Zorbax NH$_2$	Agilent	Silica, 70 Å	3-Aminopropyl
Micropellicular AP Silica	Custom synthesis	Silica	3-Aminopropyl
EPIC-PI	ES Industries	Silica	Aromatic amine
TSK Gel NH$_2$-100	Tosoh Bioscience	Silica (endcapped)	Amino
Durashell NH$_2$	Bonna Agela	Silica	Amino
Promosil NH$_2$	Bonna Agela	Silica	Amino
Venusil XBP NH$_2$	Bonna Agela	Silica	Amino
Nucleodur NH$_2$/NH$_2$-RP	Macherey-Nagel	Silica	Amino
Nucleosil Carbohydrate	Macherey-Nagel	Silica	Amino
Nucleosil N(CH3)$_2$	Macherey-Nagel	Silica	Tertiary amine
Nucleosil NH$_2$	Macherey-Nagel	Silica	Amino
YMC-Pack PA	YMC	Silica	Amino
COSMOSIL HILIC	Nacalai USA	Silica	Triazole
Amino	Jordi	Silica	Amino
YMC-Pack Polyamine II	YMC	Silica	Polyamine (sec. and tert.)
Amino-bonded Zirconia	Custom synthesis	Zirconia	Mono-, di- and triamine
Asahipak NH$_2$ P	Shodex	Poly(vinyl alcohol) gel	Amine
COSMOSIL DEAE	Nacalai	Porous polymethacrylate	Diethylaminoethyl (DEAE)
GlycoSep C	ProZyme	Polymeric	DEAE
Astec apHera	ASTEC	PVA copolymer	Polyamine

Various other mixed-mode phases have been reported in the literature (see, e.g., [169]).

In addition to small particle phases, monolithic HILIC phases have been prepared. Besides bare silica Chromolith (see Table 6.5.3), polymer-coated monolithic silica was used as HILIC stationary phase, the coating being performed with 3-diethylamino-2-hydroxypropylmethacrylate, p-styrenesulfonic acid, etc. Also monolithic organic polymers with polar functionalities such as poly(hydroxymethacrylate) were reported in the literature [189].

Silica Hydride-based Phases

A number of silica hydride (silica Type C) phases are commercially available (e.g., from MicroSolv/Cogent) and can be used in HILIC

TABLE 6.5.6 Several HILC columns with cation exchange properties that are commercially available.

Brand name	Manufacturer	Support/pore size	Nature of phase	Type
SynChropak CM 300	Eichrom (Eprogen)	Silica, 300 Å	Carboxymethyl	WCX
Nucleosil NPE	Nacalai	Silica	Nitrophenyl	WCX
COSMOSIL CM	Nacalai	Silica	Carboxymethyl	WCX
PolyCAT A	PolyLC	Silica	Poly(aspartic acid)	WCX
PolySulfoethyl A	PolyLC	Silica	Poly(2-sulfonylethyl aspartamide)	SCX
Acrylamide CEC phase	Custom synthesis	4% Crosslinked polyacrylamide	Dodecyl chains and sulfonic acid	SCX
Excelpak CHA-P44	Yokogawa (Agilent)	Styrene/divinyl-benzene	Sulfonic acid	SCX
Sulfonated S-DVB	Custom synthesis	Styrene/divinyl-benzene	Sulfonic acid	SCX

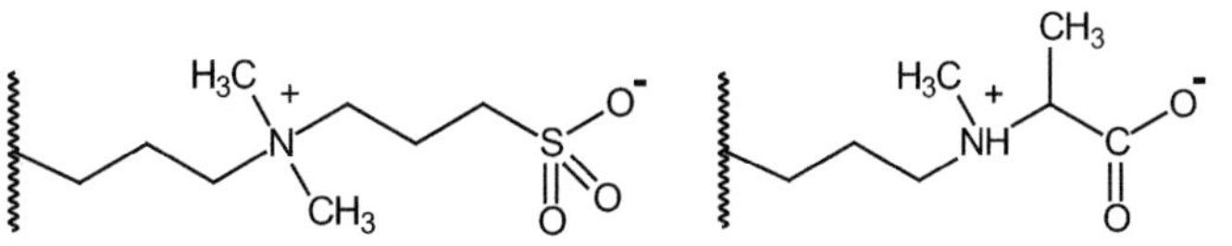

FIGURE 6.5.12 Schematic structures of two HILIC zwitterionic stationary phases.

TABLE 6.5.7 Several HILC columns with zwtterionic phases that are commercially available.

Brand name	Manufacturer	Support/pore size	Nature of phase
Nucleodur HILIC	Macherey-Nagel	Silica,110 Å	Dimethylamino and sulfonic
Syncronis HILIC	Thermo Fisher	Silica,100 Å, endcapped	Zwitterrionic
Obelisc R	Sielc	Silica	Zwitterrionic negative charged groups outside
Obelisc N	Sielc	Silica	Zwitterrionic positive charged groups outside
ZIC-HILIC	SeQuant	Silica	Polymeric sulfonyl-alkylbetaine
PolyCAT A	PolyLC	Silica	poly(aspartic acid)
ZIC-p-HILIC	SeQuant	Porous polymer	Polymeric sulfonyl-alkylbetaine

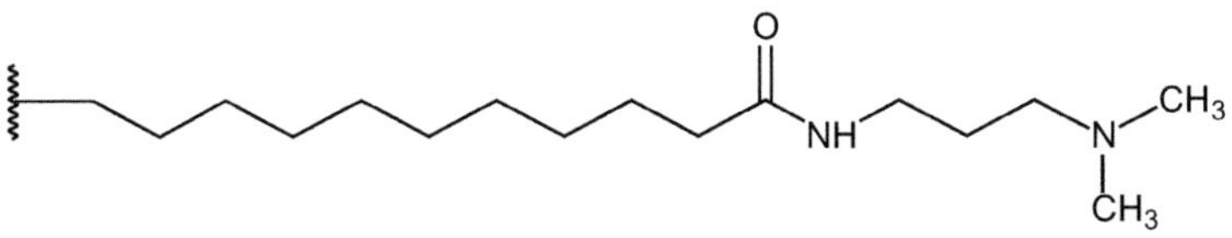

FIGURE 6.5.13 Schematic structure of a mixed mode stationary phase.

mode. These columns include a bare hydride column (Diamond Hydride) as well as columns with a bonded phase containing polar groups such as butylamino (Cogent HPS Amino). The types of chromatography that can be practiced on bare hydride columns include ANP and HILIC.

6.6. STATIONARY PHASES AND COLUMNS FOR ION-EXCHANGE, ION-MODERATED, AND LIGAND-EXCHANGE CHROMATOGRAPHY

General Aspects

Ion chromatography (IC) is a special type of HPLC used for the separation of ionic analytes. Depending on the charge of the ions, the phases can be either cationic or anionic. The dissociation capability of the ionic groups bound to the stationary phase leads to the classification of ion exchangers as strong (e.g., with groups $-SO_3^-$, $-NR_3^+$), medium (e.g., with groups $-PO(OH)O-$), or weak (e.g., with groups $-COOH$ or $-NH(CH_3)$). Zwitterionic phases (when used in totally aqueous buffers) that have both anions and cations bound to the stationary phase can act either as cationic or as anionic stationary phases, depending on the condition of the application. Such phases are also used in HILIC

chromatography with partially aqueous mobile phases. Application of zwitterionic ion chromatography (ZIC) as cation-exchange phases were reported for the separation of charged peptides. Amphoteric groups can also be present in an ion-exchange phase, and depending on the pH of the mobile phase, they may act as a weak cation or weak anion phase [190].

Based on their composition, the ion-exchange phases can be silica-based or based on an organic polymeric material (resin). Also, some inorganic oxides and silicates can be used as stationary phase in ion chromatography. A schematic diagram showing the types and the basic composition of the stationary phases used in IC is given in Figure 6.6.1.

Strong ion-exchange stationary phases are completely ionized over a wide range of pH. The degree of dissociation of weak ion exchangers depends on the mobile phase pH. For this reason, the ion-exchange behavior of strong ion-exchange phases, within typical range of pH operation, is not influenced by the pH of the mobile phase. The selection of a strong or weak ion-exchange stationary phase is determined by the nature of the analytes and by the purpose of analysis. The use of weak ion exchangers allows better modulation of the retention properties depending on the mobile phase. Also, besides ionic interactions, the

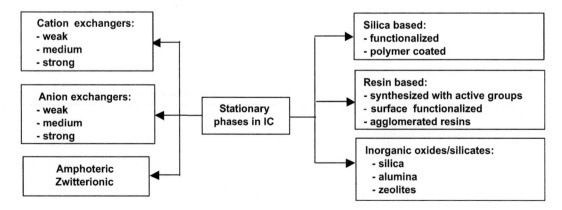

FIGURE 6.6.1 Diagram showing types of ion exchange (IC) stationary phases.

weak ion-exchange phases may develop other types of interactions (e.g., polar type).

One important characteristic of the IC columns is their ion capacity. The ion capacity of the ion exchanger is determined by the number of functional groups per unit weight of the stationary phase. The most commonly used unit is milliequivalents of charge per gram of dry packing or milliequivalents per mL of wet packing. Typical ion-exchange capacity in IC is 10-100 mEquiv/g. The counterion present in the stationary phase must be indicated, together with the ion capacity since it affects the degree of swelling of the packing and hence its volume. The ion-exchange capacity of a stationary phase plays a significant role in determining the concentrations of competing ions used in the mobile phase for elution. Higher capacity stationary phases generally require the use of more concentrated mobile phases. This is not a recommended feature in high-performance ion chromatography, since common conductometric detectors cannot function well with high salt concentrations. However, the lower ion-exchange capacity determines a low sample loading of the phase. For strong ion-exchange phases, with the ionization of the stationary phase not depending on the pH of the mobile phase, the loading capacity is the same regardless the pH. For weak ion exchangers, use of a pH of the mobile phase where the stationary phase is not ionized may considerably decrease the phase retention.

Another important characteristic of the IC columns is related to their hydrophobic character, which should be as low as possible. In many IC separations other molecules are present besides the inorganic ions, and the hydrophobic character of the analyte should not interfere with the separation. However, the ionic groups of the stationary phase are typically bonded to an organic fragment that is connected to either silica or an organic polymer. This part of the ionic stationary phase may influence the separation of the analytes by undesired hydrophobic interactions. Phases with low or ultra-low hydrophobicity have been manufactured, and this property is typically indicated for commercially available IC columns.

The cationic resins are commonly available in H^+ or in Na^+ form (exchangeable cation is H^+ or Na^+), and the anionic resins are available in Cl^- or OH^- form. The exchange of the counterion with another ion can be done by passing a (relatively) concentrated solution containing the desired ions through the stationary phase. The reconversion of an ion exchanger into its working form is known as regeneration. For this purpose, different possibilities are used depending on the nature of the exchanger and its acid−base properties. For instance, strong or weak acid cation exchangers are regenerated with HCl or H_2SO_4, and if the desirable form is Na^+, the regeneration of strong cation exchangers should be performed using NaCl, while the weak cation exchangers are regenerated with NaOH. Anion exchangers are regenerated using NaOH, NaCl, NH_4OH, HCl, H_2SO_4, Na_2SO_4, or Na_2CO_3. The operation of regeneration must be complete. Usually, the regeneration solution has a concentration of 1N and is used in a volume that is more than necessary for stoichiometric conversion. A special precaution must be taken to avoid the formation of precipitates in the resin (such as $CaSO_4$, etc.) during the regeneration process.

Besides a typical role as ion exchange, some phases can be used for the separation of neutral species between themselves and from ionic species. The ionic compounds from the solution are rejected by the selected resin, and the neutral molecules are separated through polar interactions. This type of separation is known as ion exclusion.

Silica-based Ion Exchangers

Various groups conferring an ion-exchange character can be bonded on silica by procedures similar to those used for RP, HILIC, or other stationary phases. The advantages of silica-based phases result from their large contact surface with the mobile phase, good mechanical properties, and lack of swelling and shrinking when the mobile phase is changed. Certain silica-derivatized phases can be used directly as a cation- or anion-exchange phase. One such example is aminopropyl silica. Another example is a strong cation exchange phase (SCX type) obtained by attaching to the silica surface ethylbenzene sulfonic acid ligands that lead to phase coverage with 0.4−0.7 meq/g capacity. However, it is more common to attach or form ionizable groups using a second reaction step performed on a previously derivatized silica. Reaction 6.2.10 (see Section 6.2) showed, for example, the formation of a SO_3H group from an SH group by oxidation. Similarly, strong anion exchangers with $N(C_2H_5)_3^+$ groups can be obtained from chloropropyl silica following the reaction 6.6.1:

(6.6.1)

Other ionic groups can be attached to the silica surface forming strong cation exchangers (SCX), strong anion exchangers (SAX), or weak ones (WCX or WAX). Some information on several common commercial silica-based stationary phases used for separation of anions (cation-exchange stationary phases) are given in Table 6.6.1 [191]. Several silica-based anion exchange IC columns are described in Table

6.6.2. Various other silica-based stationary phases for ion exchange chromatography were reported in the literature [192].

Synthesis of Organic Polymeric Ion Exchangers

Besides silica-based materials derivatized with organic moieties containing groups such as SO_3H, COOH, and NH_2, organic resins with specific active groups were frequently used as ion-exchange stationary phases. These materials consist of an organic polymeric (resin) backbone containing covalently bound groups that are able to exist in ionic form. Four types of polymeric packings are used for making stationary phases, including copolymers of styrene-divinylbenzene (PS-DVB), ethylvinylbenzene-divinylbenzene (EVB-DVB), polyvinyl copolymers such as those obtained from polyvinyl alcohol (PVA), and various polymethacrylates [193]. The presence of the ionic groups on these resins makes these resins act as polyelectrolytes. The main advantage of resin-based ion exchangers is their tolerance to eluents with extreme pH values, between 0−14, in contrast to the silica-based stationary phases, whose pH limits are 2−7. This wide range of pH values allows the use of selectivity effects on multicharged or weakly ionizable solutes. The use of polymeric resins has pressure limitations because they have lower mechanical resilience. Macroporous resins with a high degree of crosslinking are relatively more rigid and stable, and although they have lower ion-exchange capacity they are used in HPLC and can be included in longer columns, at higher flow rates. Monolith columns are also made from polymers such as polyacrylamide for IEC separations [194].

Several procedures are used for obtaining polymers with ion-exchange properties. One such procedure is the derivatization of an already polycondensed resin. This type of reaction was previously discussed in Section 6.2. One other procedure consists of the synthesis

TABLE 6.6.1 Some information for commercial silica-based cation exchange columns.

Column	Manufacturer	Type	Dimensions (length × i.d., mm)	Capacity (mEquiv g^{-1})	Particle size (μm)	Type of phase
Partisil 10 SCX	Whatman	strong	250 × 4.6	0.5	5, 10	-C$_6$H$_4$-SO$_3$H
Vydac SC	Separation Group	strong	250 × 4.6	0.1	30-44	-SO$_3$H
Luna SCX	Phenomenex	strong	150 × 4.6		5, 10	-C$_6$H$_4$-SO$_3$H
Phenosphere SCX	Phenomenex	strong	various	0.6	5, 10	-C$_6$H$_4$-SO$_3$H
IC YK-421	Shodex	weak	125 × 4.6			Coated silica polymer-COOH
SynChropak	Lab Unlimited	weak				

of the polymer already containing the desired ionic groups in the monomer, as shown in reaction 6.2.22. This procedure is commonly used, for example, for the synthesis of sulfonic resins, which can be obtained following the reactions 6.6.2 schematically shown here:

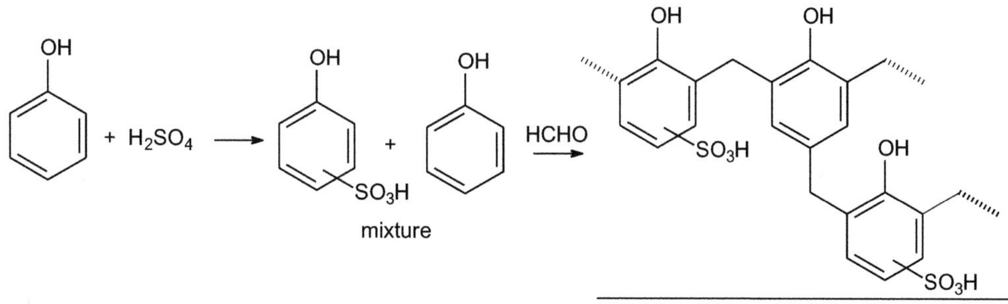

with methylenesulfonic acid groups. Also, phenols such as resorcinol, naphthol, phenoxyacetic acid, and various aldehydes are used in the condensation instead of simple phenol and formaldehyde. For example, the condensation of phenoxyacetic acid and formaldehyde leads

(6.6.2)

The sulfonation products of phenol can be further treated with formaldehyde without separation, and a highly crosslinked material can be obtained under appropriate conditions. Other resins can be prepared similarly to sulfonic resins. For example, starting with salicylic acid and formaldehyde, a resin with carboxylic groups is obtained.

The bonded ionic group also can be generated on the side chain of the resin. For example, condensation of sodium phenolate with Na$_2$SO$_3$ and HCHO leads to the formation of a resin

to a weak acid resin. Other variations of the condensation reaction are utilized, such as condensation of a phenol, a substituted benzaldehyde, and formaldehyde. Also, condensation of aromatic diamines such as *m*-phenylenediamine with formaldehyde is used to make a polymer with anion-exchange properties. Because the amino groups directly connected to the aromatic ring are weak bases, further methylation can be applied to form quaternary amines that are stronger bases. This condensation can be written as reaction 6.6.3:

TABLE 6.6.2   Some information for commercial silica based anion exchange columns.

Column	Manufacturer	Type	Dimensions (length × i.d., mm)	Capacity (mEquiv g^{-1})	Particle size (μm)	Type of phase
Vydac 302 IC 4.6	Separation Group	strong	50 × 4.6	0.1	10	Spherical particles with-NR$_3$]$^+$
Vydac 300 IC 405	Separation Group	strong	250 × 4.6	0.1	15	
Wescan 269-001	Wescan	strong	250 × 4.6	0.08	13	
Nucleosil 10 Anion	Macherey & Nagel	strong	250 × 4.0	0.06	10	
TSK Gel IC-SW	Toya Soda	strong	250 × 4.6	0.4	5	-N(C$_2$H$_5$)$_2$CH$_3$]$^+$
Partisil 10 SAX	Whatman	strong	250 × 4.6	0.5	5, 10	-NR$_3$]$^+$
Phenosphere SAX	Phenomenex	strong	various	0.4	5, 10	-NR$_3$]$^+$
Luna NH$_2$	Phenomenex	weak	various			-NH$_2$
SynChropak	Lab Unlimited	weak	various			

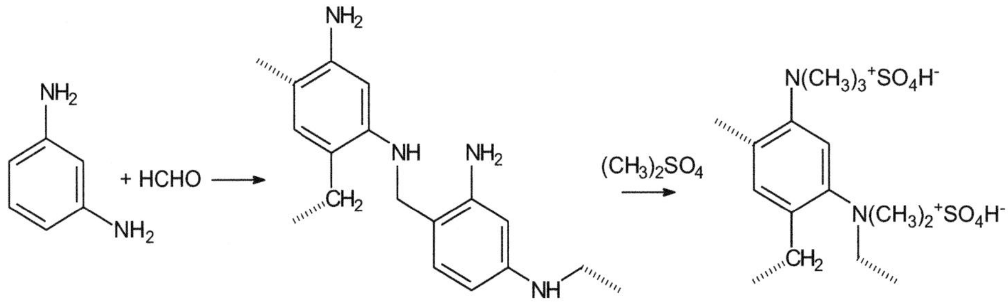

(6.6.3)

Condensation reactions using aliphatic polyamines, phenol, and formaldehyde also have been applied to generate anionic ion exchangers.

Polymerization of special vinyl monomers is another common procedure used to prepare ion-exchange resins. Most of these resins are made using styrene/divinylbenzene copolymers (PS-DVB or S-DVB). Among these, sulfonated polystyrene (which also may have sulfone bridges) and quaternary amines obtained by chloromethylation of polystyrene/divinylbenzene resins are the most common. Other anionic resins with slightly different properties have longer alkyl chain substituents at the quaternary nitrogen or groups such as -N(CH$_2$C$_6$H$_5$)(CH$_3$)$_2$$^+$, -N(C$_2H_4$OH)(CH$_3$)$_2$$^+$, -N(C$_2H_5$)$_2$, etc. . Weaker anion exchangers contain polyamine groups attached to the polystyrene chain. Other polymers used as backbone for ion-exchange materials include polymethacrylates (PMA or PM), polyhydroxymethacrylate (PHMA or PHM), and polyvinyl alcohol (PVA).

Anionic ion exchangers also can be prepared using specific monomers. For example, acrylonitrile can be polymerized, followed by reduction (with hydrogen and Ni catalyst) and the modification of the primary amine in quaternary amine, as shown in reaction 6.6.4:

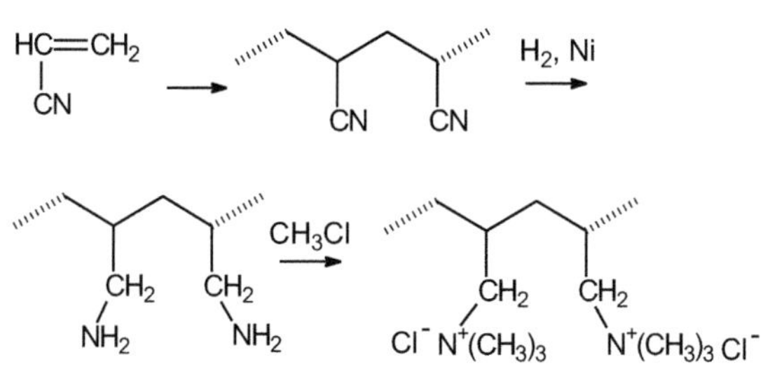

$$(6.6.4)$$

Many other synthetic paths have been either used or only explored for producing ion-exchange resins. Some resins are prepared to have more than one type of functional group, whereas others are made to contain unique structures. Amphoteric ion exchangers with both basic and acidic groups, as well as ion exchangers with specific chelating properties, are also available for various applications. An amphoteric ion exchanger, for example, can be obtained by copolymerization of styrene, divinylbenzene, and vinyl chloride, followed

$$(6.6.5)$$

by amination and sulfonation in a sequence described by reactions 6.6.5.

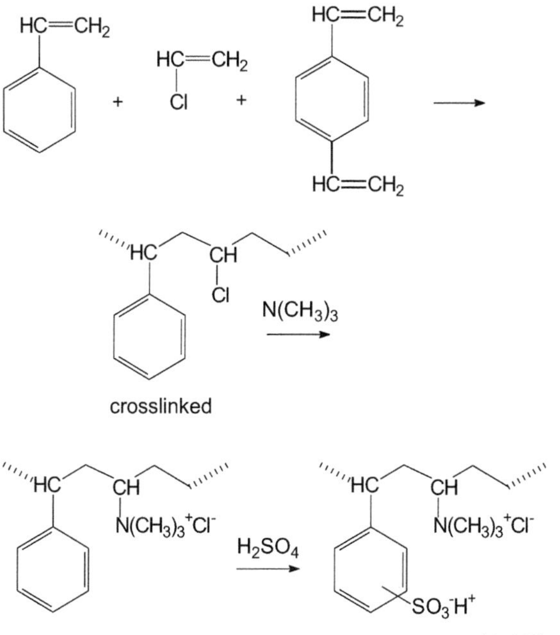

Using procedures similar to those previously described, other ionic groups can be attached to a polymeric base. A variety of phases with a synthetic organic polymer support and with various ion-exchange functionalities have been synthesized by these procedures. Some of these functionalities are listed in Table 6.6.3. From the groups indicated in the table, only the sulfonic groups confer strong acid properties, the other leading to weak acid properties. Quaternary amine functional groups form strong base exchangers. The anion exchangers containing trimethylamine groups are more

TABLE 6.6.3   Functional groups introduced in synthetic ion exchange resins.

Cation exchangers		
Functional group		*Type*
Sulfonic acid	$-SO_3^-H^+$	strong
Carboxylic acid	$-COO^-H^+$	weak
Phosphonic acid	$-HPO_3^-H^+$	medium
Phosphinic acid	$-HPO_2^-H^+$	weak
Arsonic acid	$-HAsO_3^-H^+$	weak
Selenonic acid	$-SeO_3^-H^+$	weak
Phenoxy group	$-C_6H_4-O^-H^+$	weak

Anion exchangers		
Functional group		*Type*
Quaternary amine	$-N(CH_3)_3]^+OH^-$	strong
Quaternary amine	$-N(CH_3)_2(CH_2CH_2OH)]^+OH^-$	medium
Tertiary amine	$-NH(CH_3)_2]^+OH^-$	weak
Secondary amine	$-NH_2(CH_3)]^+OH^-$	weak
Primary amine	$-NH_3]^+OH^-$	weak
Sulfides	$-SR_2]^+OH^-$	weak

basic than those containing dimethyl-β-hydrox-yethylamine. The most common groups in commercially available ion exchangers are sulfonic and carboxylic groups for cation exchangers and trimethylammonium group for anion exchangers.

Another procedure used to obtain materials with ion-exchange properties starts with cellulose. Strong alkali solutions acting on cellulose (at room temperatures) produce alkali cellulose. Studies on the structure of alkali cellulose obtained with 20–40% NaOH solutions indicated that the substance is not a true alcoholate but an addition complex, $R_{Cell}OH:NaOH$. The treatment of alkali cellulose, for example, with

pH of precipitation varies between 6 for low-substitution values to 1 for high substitution (D.S. of about 0.9). The material has ion-exchange properties.

Ion-exchange materials also can be obtained from dextrans. Dextrans are produced by certain bacteria growing on a sucrose substrate, and their structure is characterized by a $(1\rightarrow6)$-α-D-gluco-pyranosyl chain. Branches may occur at $(1\rightarrow3)$-linked points. The dextrans used for ion-exchange resins are usually crosslinked using epichlorhy-drin, followed by transformation in a cation exchange containing sulfonyl groups or an anion exchange as shown schematically in 6.6.7 for the anion-exchange preparation:

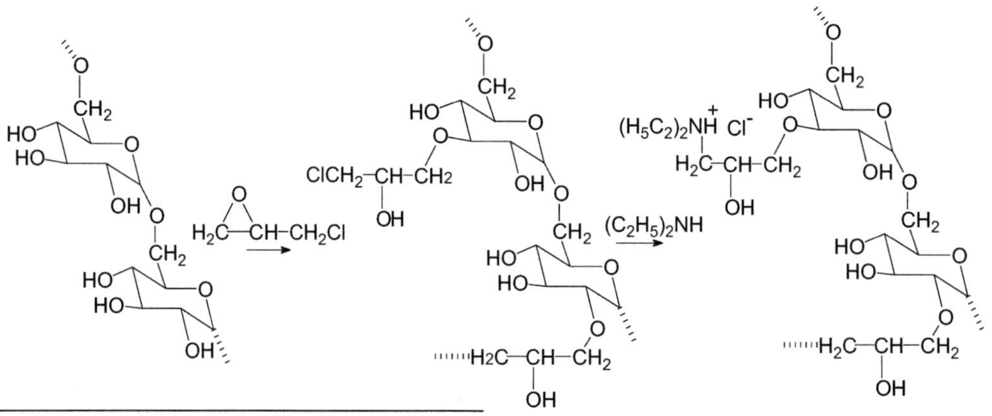

$$(6.6.7)$$

chloroacetic acid sodium salt, leads to the formation of carboxymethylcellulose (CMC), following the reaction:

$R_{Cell}OH:NaOH + ClCH_2COO^- Na^+$

$\rightarrow R_{Cell}O\text{-}CH_2COO^- Na^+ + NaCl + H_2O$

$$(6.6.6)$$

The degree of substitution (D.S.) that can be obtained for this product usually ranges between D.S. = 0.1 to D.S. = 1.2. Pure CMC is commercially available. Carboxymethylcellulose in itself is a weak acid that can be precipitated from solutions with a mineral acid. The

Besides the chemical structure of the ion-exchange material, its physical properties are also important. Many ion-exchange resins are synthesized for different purposes rather than to be used as a stationary phase in HPLC. The polymers are typically obtained in reactions involving water solutions and have the structure of a gel with various water contents. The elimination of this water generates a micro-porous structure of the polymer. In many applications, it is important to preserve part of this water, and complete drying of the polymer usually is not desired. However, the mechanical resistance of these polymers may preclude their utilization as stationary phases in HPLC. Besides the polymers with a microporous structure, highly porous polymers are useful in some applications. Resins to be used in solvents in which the polymer does not swell, for example, must have a higher porosity to allow better contact with the solvent. Porous polymers are, however, easily compressible and not always useful as a stationary phase for HPLC. The swelling properties of the polymer are also important since on one hand the variation in volume of the polymer is not desired, and on the other hand poor swelling properties do not allow a good contact of the resin with the mobile phase. Other types of physical structures for ion-exchange resins are also known, such as pellicular ion-exchange resins that have an inert core and may be useful in HPLC applications. Also, monolith polymeric phases can be used for HPLC purposes [195].

Latex-Agglomerated Ion Exchangers

The direct use of polymeric ion-exchange materials as stationary phase for HPLC encounters several problems. One problem is the relatively low mechanical stability even at moderate pressure, and the other is the swelling and shrinkage of the phase during the ion-exchange process. These problems are significantly alleviated using latex-agglomerated ion-exchange stationary phases. In addition to better mechanical stability and reduced swelling and shrinkage, these phases also offer high efficiencies and an appropriate ion-exchange capacity (between 0.03 and 0.1 mEquiv g^{-1}).

Latex-agglomerated exchangers contain an internal core particle (or support) that includes ionic functionalities on its surface. On this support is attached a monolayer of small diameter particles that carry functional groups consisting of bonded ions that are of the opposite charge with the functionality of the support. The groups of the outer particles have a double role: to attach the small particles to the core and to act as an ion exchanger for the ions in solution. The core particles are typically PS-DVB resin of moderate crosslinking, with a particle size in the 5–25 μm range. The outer microparticles consist of finely ground resin or monodisperse polymer (latex) with diameters up to 0.1 μm. These particles have very high porosity and are functionalized to contain an appropriate ion-exchange group. This group determines the ion-exchange properties of the composite particle. For example, an aminated latex produces agglomerated anion exchangers, while sulfonated latex produces agglomerate cation exchangers. A schematic view of this type of stationary phase is given in Figure 6.6.2 [191].

For a latex-agglomerated anion exchanger, the core particles are typically made of polystyrene-divinylbenzene with a sulfonated surface, which interacts electrostatically with quaternary groups from the fully aminated latex particles. The free quaternary amino groups that are not involved in interactions with sulfonil groups from the core particle are able to participate in the retention of anionic solutes or components from mobile phase. This kind of anion exchanger is chemically very stable, and even a high concentration of NaOH is unable to disrupt the ionic bonds between substrate particle and latex beads. Among other polymers used for latex formation are derivatized divynilbenzene—vinylbenzylchloride, polystyrene, and acrylate—but at a lower degree of crosslinking [196]. Columns with latex-agglomerated technology can be used at column backpressures as high as 3000 to 4000 psi, which is close to the use of typical silica-based columns.

Other types of latex-agglomerated stationary phases can be produced. For example, Nafion (copolymer of tetrafluoroethylene with perfluorovinylether with terminal sulfonic groups) can be coated directly to an octadecyl-modified silica being attached by simple hydrophobic interactions. The final material can be used as

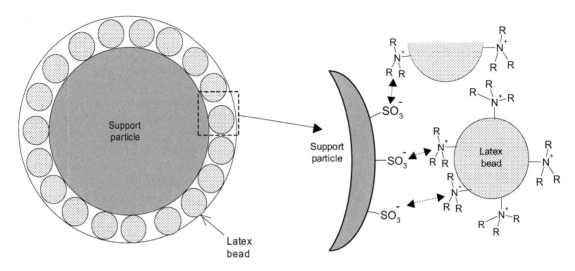

FIGURE 6.6.2 Schematic view of a latex-agglomerated type ion exchanger.

a cation exchanger, which has been proved to have a high selectivity. The schematic structure of Nafion is as follows:

$$\text{--}(CF_2\text{--}CF_2)_n\text{--}(CF_2\text{--}CF)_m\text{--}$$

Commercial Polymeric Stationary Phases Used in Ion-Exchange HPLC

A considerable number of stationary phases for ion-exchange HPLC have been synthesized and are commercially available. Also, various technologies have been implemented in the columns used for IC (e.g., latex-agglomerated materials extensively used by Dionex [197]). Several materials (columns) available for cation-exchange HPLC are listed in Table 6.6.4. Several anion-exchange IC columns are described in Table 6.6.5. Numerous other ion-exchange-type columns are commercially available (see, e.g., [199]). Also, columns with

multiple functionalities have been manufactured (e.g., IonPac CS5A from Dionex).

An important development in ion chromatography is the capillary IC. A capillary system (e.g., ICS-5000 from Dionex) can use IonPac capillary columns with 0.4 mm i.d. that are packed with the same materials as the equivalent analytical scale version (4 mm i.d. columns). Examples of such columns are Capillary IonPac CS12A for cation separation, IonPac AS18-Fast capillary or IonSwift MAX-100 Capillary monolithic for anion separation, and CarboPac PA20 for carbohydrate separations. These columns are used at significantly lower flow rates (10–20 μL/min). These capillary columns have also been developed for achieving shorter chromatographic runs.

Ion-Moderated Phases

Ion-moderated phases are typically made from a cation-exchange material in metal form (e.g., Na^+, K^+, Ca^{2+}, Pb^{2+}, and even H^+), which is used in specific conditions for the separation of neutral species based on a selective partition of

TABLE 6.6.4 Some cation exchange polymeric columns for HPLC use.

Column	Manufact.	Type	Dimensions (length × i.d., mm)*	Capacity (mEquiv g^{-1})	Technol.	Type of phase
PRP-X200, x400, etc.	Hamilton	strong	250 × 4.6			PS-DVB-SO$_3$H
IC YS-50	Shodex	weak	125 × 4.6			PVA-?
ES-502C	Shodex	strong				PVA-SO$_3$H
ICT-521	Shodex	strong	150 × 4.6			PS-DVB-SO$_3$H
LCA K02	Sykam	strong				PMA
IonPac CS-12, 12A, 14, 16, 17, 18, 19	Dionex	weak	250 × 4 or 250 × 2	various	Latex technol	PS-DVB-COOH
IonPac CS-15	Dionex	weak	250 × 4 or 250 × 2	various	Latex technol	PS-DVB-COOH, -PO$_3$H$_2$/crown ether
IonPac CS-12A	Dionex	medium	various	700-900	Latex	PS-DVB-COOH + PS-DVB-PO$_3$H$_2$
IonPac CS-11	Dionex	strong	250 × 2	35	Latex	PS-DVB- SO$_3$H
IonPac CS-10	Dionex	strong	250 × 2	80	Latex	PS-DVB- SO$_3$H
Diaion	Mitsubishi	strong	various			PS-DVB- SO$_3$H
Diaion	Mitsubishi Chemical	weak	various		highly porous	acrylic acid-methacrylate

Note: Columns with other dimensions are also available, and the corresponding capacity (in mEquiv/g) can be different. Vendor specifications must be reviewed (e.g. [197,198]).

the analyte between the liquid inside the resin and the mobile phase. The retention inside the resin is based on weak polar interactions between the analyte and the stationary phase. Some ion-moderated phases may be considered as functioning based on a ligand-exchange mechanism. Several columns that are commercially available for ion-moderated separations are listed in Table 6.6.6. Polymeric ion-moderated columns are typically used at pressures no higher than 300 psi for 4% crosslinked materials,

and no higher than 1000 psi for 8% crosslinked materials. Temperatures of 60−80 °C are commonly utilized during separation.

Besides ion-moderated polymers with ionic groups or salts of ionic groups, a different type of phase is based on crown ether moieties. The crown ether groups can be present in an organic-based solid support or immobilized on silica. For example, the immobilization of dibenzo-18-crown-6 ether leads to a structure like the following:

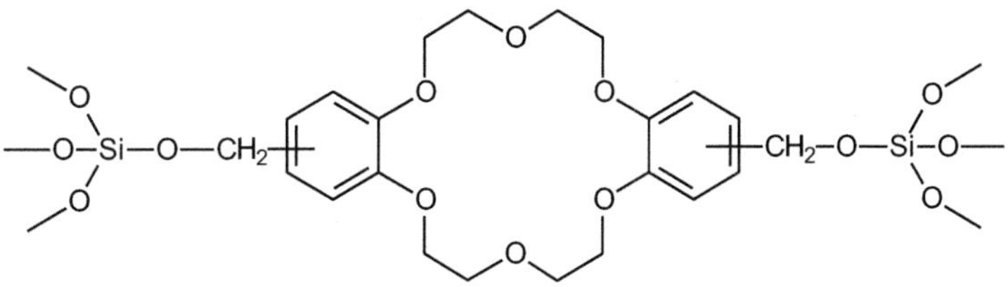

TABLE 6.6.5 Some anion exchange polymeric columns for HPLC use.

Column	Manufacturer	Type	Dim. (length × i.d., mm)	Type of phase
RCX-30	Hamilton		150 × 4.6	
PRP X100, X500	Hamilton		150 × 4.6	
Star-Ion A300	Phenomenex	strong	100 × 4.6	PS-DVB-NR_3^+
SP-502N	Shodex			PVA- NR_3^+
IonPac AS4A *	Dionex		various	
IonPac AS23, AS22,	Dionex	medium	"	PS-DVB-alcanol NR_3^+
IonPac AS14, 14A, 12A	Dionex	medium	"	PS-DVB-alcanol NR_3^+
IonPac AS9 HC, AS4A-SC	Dionex	medium	"	PS-DVB-NR_3^+ and alcanol
IonPac AS9 HC	Dionex	strong	"	PS-DVB-NR_3^+
IonPac AS10, 11, 16, 20, etc.	Dionex	various	"	PS-DVB-NR_3^+ and alcanol
IonPac AS18	Dionex	strong	"	Polyethylvinylbenzene-DVB-NR_3^+
CarboPac MA1	Dionex	strong	"	resin with tertiary amine
CarboPac SA10	Dionex	strong	"	latex nano beeds
CarboPac PA1, PA10, PA20, PA100, PA200	Dionex	strong	"	pellicular, nanoprous beeds, etc.
IC-SI-90 4E	Shodex	strong	250 × 4.6	PVA-NR_3^+
IC-SI-50 4E	Shodex	strong	250 × 4.6	PVA-NR_3^+
IC I-524A	Shodex	strong	100 × 4.6	PHMA gel-NR_3^+
IEC DEAE-825	NR_4^+	weak	75 × 8	PHMA-NR_2
Allsep	Grace – Alltech	strong	various	PMA gel- NR_3^+
BioSuite DEAE	Waters	strong	"	PMA gel- NR_3^+

BioSuite Q-PEEK	Waters	strong	"	PMA gel- NR_3^+
Cosmogel QA	Nacalai Tesque, Inc.	strong	"	PMA gel- NR_3^+
Discovery BIO PolyMA-WAX	Supelco	strong	"	PMA gel- NR_3^+
IC-Pak Anion	Waters	strong	"	PMA gel- NR_3^+
MetroSep Anion Dual 2	Metrohm-Peak, Inc.	strong	"	PMA gel- NR_3^+
Nucleogel SAX	Macherey-Nagel	strong	"	PMA gel- NR_3^+
Protein-Pak Q 8HR	Waters	strong	"	PMA gel- NR_3^+
PRP-X500	Hamilton Co.	strong	"	PMA gel- NR_3^+
Shodex IEC QA-825	Shodex	strong	"	PMA gel- NR_3^+
Super-Sep IC Anion	Metrohm-Peak, Inc.	strong	"	PMA gel- NR_3^+
TSKgel Q-STAT	Tosoh Bioscience	strong	"	PMA gel- NR_3^+
TSKgel DNA-STAT	Tosoh Bioscience	strong	"	PMA gel- NR_3^+
TSKgel SuperQ-5PW	Tosoh Bioscience	strong	"	PMA gel- NR_3^+
TSKgel BioAssist Q	Tosoh Bioscience	strong	"	PMA gel- NR_3^+
TSKgel IC-Anion-PW	Tosoh Bioscience	strong	"	PMA gel- NR_3^+
Zodiac IC Anion	Zodiac Life Sciences	strong	"	PMA gel- NR_3^+
IonPac AS7	Dionex	strong	250 × 4	PS-DVB-NR_3^+
IonPac AS5	Dionex	medium	250 × 4	PS-DVB-alkanol NR_3^+
Ion Swift Max 100 monolothic	Dionex	medium	various	PS-DVB-alkanol NR_3^+
Ion Swift Max 200 monolothic	Dionex	medium	"	PS-DVB-alkanol NR_3^+
Diaion	Mitsubishi Chemical	strong	"	styrenic/acrylic amine

* Note. IonPak AS columns have different degree of hydrophobicity in addition to anionic character [197].

TABLE 6.6.6 Some ion-moderated polymeric columns.

Column	Manufacturer	Type
HC-75 Ca^{2+}	Hamilton	PS-DVB sulfonic Ca^{2+}
HC-75 H^+	Hamilton	PS-DVB sulfonic H^+
HC-75 Pb^{2+}.	Hamilton	PS-DVB sulfonic Pb^{2+}.
RCM-Monosaccharide (L19 packing)	Phenomenex	8 % cross-linked PS-DVB sulfonic Ca^{2+}
RHM-Monosaccharide (L17 packing)	Phenomenex	8 % cross-linked PS-DVB sulfonic H^+
RAM-Carbohydrate	Phenomenex	8 % cross-linked PS-DVB sulfonic Ag^+
RSO-Oligosaccharide	Phenomenex	4 % cross-linked PS-DVB sulfonic Ag^+
RNO-Oligosaccharide	Phenomenex	4 % cross-linked PS-DVB sulfonic Na^+
RPM-Monosaccharide (L34 packing)	Phenomenex	8 % cross-linked PS-DVB sulfonic Pb^{2+}
RNM-Carbohydrate (L54 packing)	Phenomenex	8 % cross-linked PS-DVB sulfonic Na^+
ROA-Organic Acid (L22 packing)	Phenomenex	8 % cross-linked PS-DVB sulfonic H^+
RFQ-Fast Acid	Phenomenex	8 % cross-linked PS-DVB sulfonic H^+
RKP-Potassium	Phenomenex	8 % cross-linked PS-DVB sulfonic K^+
RCU-USP Sugar Alcohols (L19 packing)	Phenomenex	8 % cross-linked PS-DVB sulfonic Ca^{2+}
IonPac ICE-AS1	Dionex	PS-DVB sulfonic H^+
IonPac ICE-AS1	Dionex	PS-DVB sulfonic and carboxylic H^+
IonPac ICE-Borate	Dionex	PS-DVB sulfonic H^+
SUGARSH1011, SUGARSC1011, etc.	Shodex	
MCI CK/CA Series	Mitsubishi Chem.	PS-DVB sulfonic with various ions Ca^{2+} Ag^+ etc.

Crown ethers strongly bind certain cations, forming complexes. For example, 18-crown-6 ether has affinity for potassium ions, while 15-crown-5 has affinity for sodium. The crown ether can also display Donnan-type exclusion and act as an ion-exclusion phase. Some columns from this group can be used either as typical ion-moderated phase (CK08EH and CK08E Ser.) or for chiral separations (MCI Gel CRS10W, MCI Gel CRS15W).

Ligand-Exchange and Immobilized Metal Affinity Phases

Ligand-exchange phases are ion-exchange phases loaded with a transitional metal capable of forming bonds with the analyte. Special phases are used for immobilized metal affinity usually derived from iminodiacetic acid or tricarboxymethylethylenediamine. These phases are able to form very strong coordinative complexes with transitional metals. Similar-type phases with immobilized metal affinity have also been used for chiral separations where a chiral molecular unit is bonded on the stationary phase and is capable of forming complexes with transitional metal cations that are further complexing the analytes (see Section 6.7).

6.7. STATIONARY PHASES AND COLUMNS FOR CHIRAL CHROMATOGRAPHY

General Comments

Since most bioorganic molecules are chiral, the separation of chiral compounds is of considerable importance, for both analytical and preparative purposes. Chiral chromatography is therefore a field with numerous books (see, e.g., [200–204]) and journals, including the dedicated journal *Chirality* (Wiley). Chiral stationary phases and columns are also commercially available from a number of suppliers. This section presents only a summary of this rich volume of information.

Procedures for the Synthesis of Phases with Chiral Properties

A number of syntheses of chiral stationary phases are reported in the literature [200]. One important type of such a synthesis starts with a silica support on which specific *chiral selectors* are attached. Such phases can be obtained, for example, by a two-step silica derivatization similar to the procedure used for the synthesis of phases with embedded polar groups. A first step in these syntheses is the preparation of a silica-based material containing a reactive

functionality. The general reaction of this type is shown in 6.7.1:

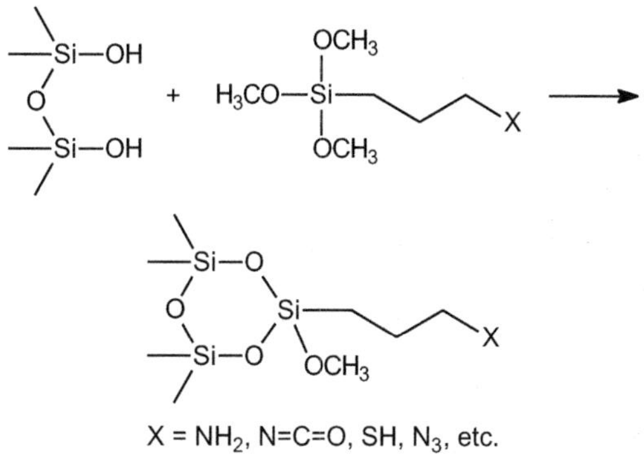

X = NH$_2$, N=C=O, SH, N$_3$, etc.

(6.7.1)

The reactive group can be amino, isocyanate, thio, or azide, and is further reacted with the desired compound that contains a chiral selector group for achieving the desired stationary phase. One such procedure starts with amino-propyl silica, which is further derivatized with N-(3,5-dinitrobenzoyl)-phenylglycine, with N-(3,5-dinitrobenzoyl)-isobutylglycine, or with N-(3,5-dinitrobenzoyl)-α-naphtyl-α-alanine. These reagents contain a chiral carbon and generate phases indicated as Pirkle type [205, 206]). The reaction with N-(3,5-dinitroben-zoyl)-phenylglycine is shown in 6.7.2:

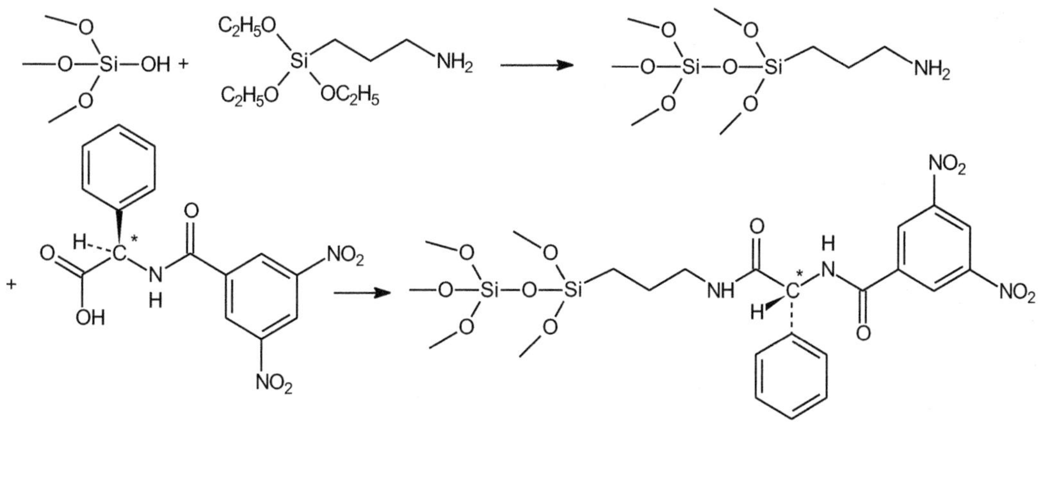

(6.7.2)

Another example of reactions used for attaching chiral groups on a silica surface starts with silica gel derivatized with triethoxy-isocyanato-propylsilane, followed by reaction with the desired amine, as shown in the set of reactions 6.7.3:

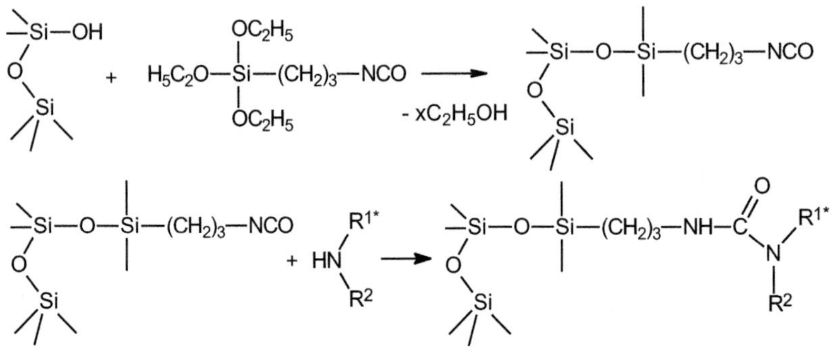

Another modified silica used as starting material contains azidopropyl groups. These groups can react with an alkyne group following a Huisgen cycloaddition ("click" chemistry). As an example, a quinine-type chiral stationary phase was

(6.7.3)

In reaction 6.7.3, the R^{1*} fragment contains a chiral carbon atom that imparts chiral properties to the stationary phase.

obtained by this type of reaction [207]. The schematics of this synthesis are shown in reaction 6.7.4.

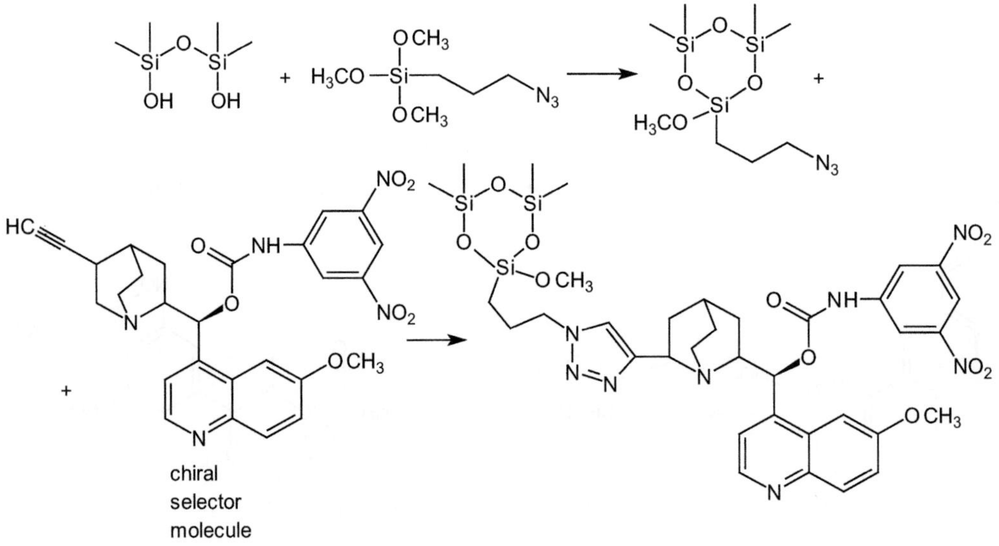

chiral
selector
molecule

(6.7.4)

The preparation of chiral selector molecule in reaction 6.7.4 starts with quinine that is transformed into 10,11-didehydroquinine. This molecule is reacted with 3,5-dinitrobenzeneisocyanate to introduce one additional functionality necessary for chiral recognition. The resulting molecule reacts with azidopropyl silica.

A multistep synthesis is used, for example, to obtain a chiral stationary phase containing a polymer bonded to silica with diaminocyclohexylacrylamide chiral selectors. This is done by attaching N-[2-(prop-2-enamido) cyclohexyl]prop-2-enamide (obtained from acryloyl chloride and cyclohexane-1,2-diamine) to a pre-derivatized silica, following the set of reactions shown in 6.7.5. Due to the active double bonds on the chiral monomer, the process leads to the formation of a polymeric active phase bonded to silica.

Polymers containing chiral selectors that are bonded to silica can also be obtained by synthesizing a monomer containing two ethoxysilane groups, as shown in reaction 6.7.6.

The resulting monomer can react with itself or with a silica gel surface, generating a silica-based material covered with a silica-based polymer with chiral properties.

A different type of chiral selector is offered by macrocycles. The macrocycles are rather voluminous, and the direct connection to the silanol groups is not possible. In these cases, a reactive "handle" or "spacer" is attached first to the silica surface, this handle further reacting with the desired bonded phase. For example, starting with amino silica a phase can be prepared having a crown ether moiety with attached tartaric acid functionalities that add asymmetry allowing the chiral separations. The reactions for this synthesis are schematically shown in reaction 6.7.7 [208].

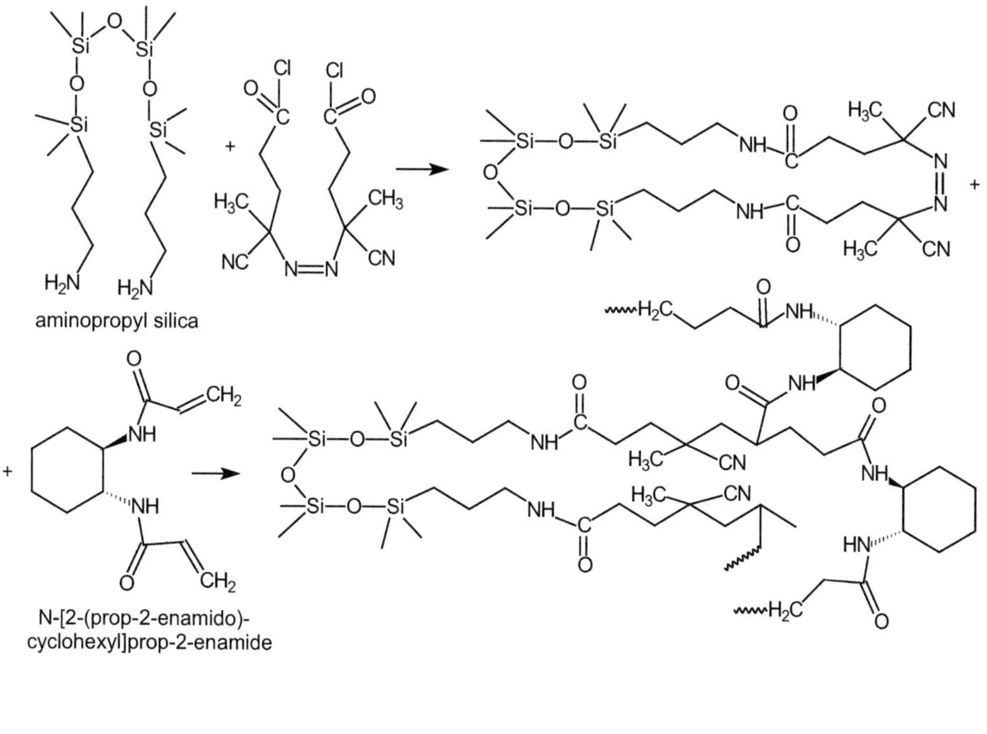

(6.7.5)

N-[2-(prop-2-enamido)-
cyclohexyl]prop-2-enamide

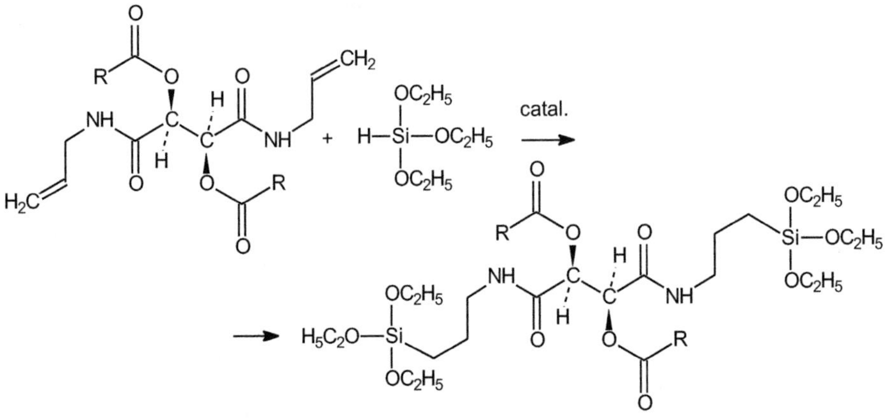

(6.7.6)

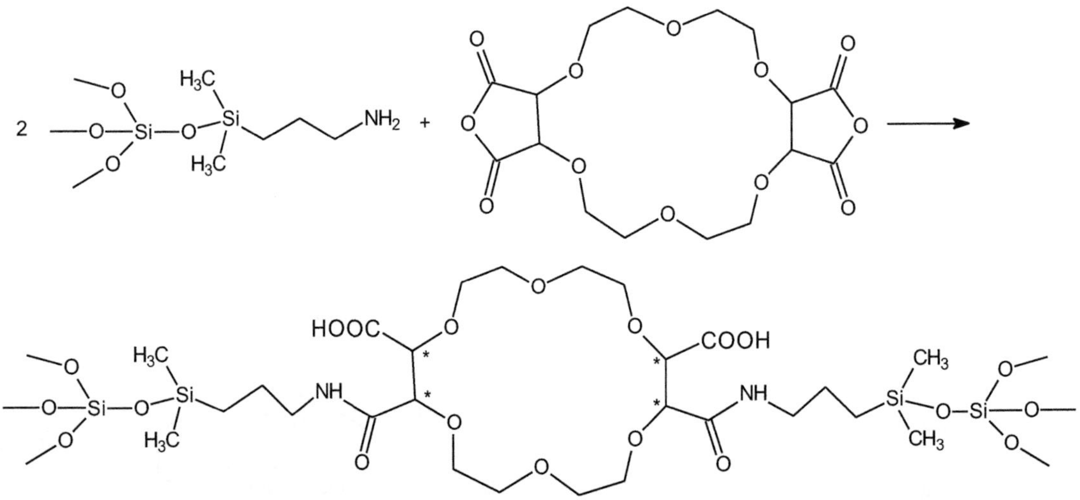

(6.7.7)

Besides tartaric acid, which has asymmetric carbons (indicated in reaction 6.7.7 as "*"), other functionalities such as binaphthyl, biphenanthryl, and carbohydrate, can be attached to the crown ether group for providing the necessary asymmetry. One synthesis with the goal of preparing a crown ether including a binaphthyl group and capable of being attached to the active silanols of silica surface follows the steps schematically shown in reactions 6.7.8 [209].

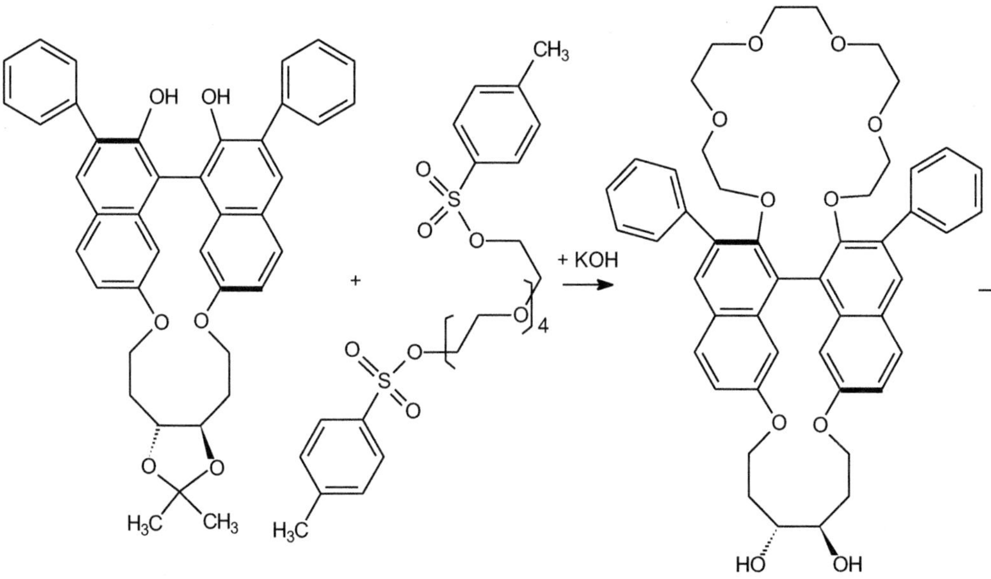

$$(6.7.8)$$

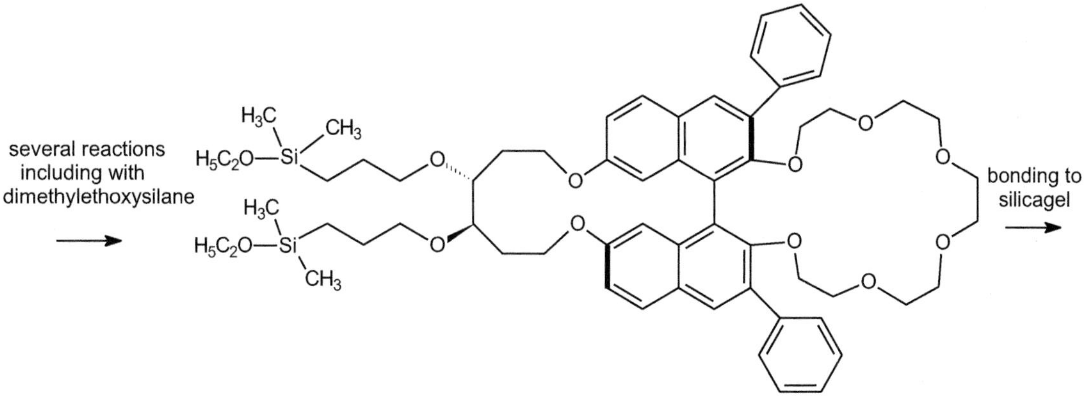

The bonded phase of the type shown in reaction 6.7.8 has chiral selectivity to numerous compounds, allowing the separation of their enantiomers.

Among other macrocycles that are bound to silica to be used for chiral separations are cyclodextrins and calixarenes. For example, a chiral phase containing cyclodextrin active phase can be obtained using the following set of reactions, where glycidyl groups are present in a pre-derivatized silica as shown in 6.7.9:

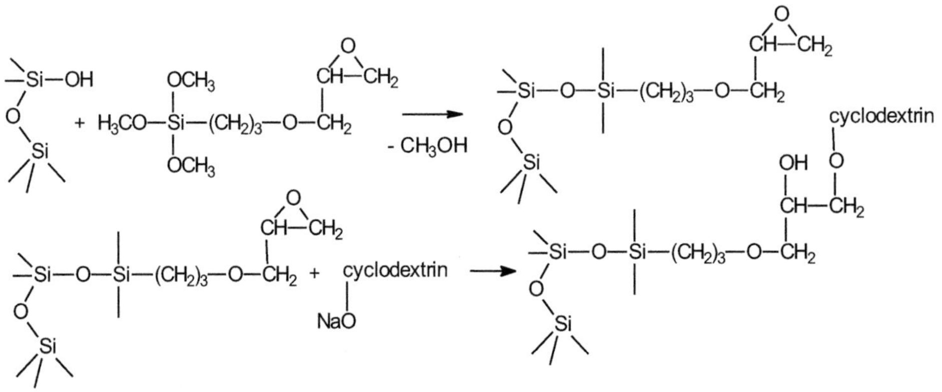

(6.7.9)

The same procedure of a reaction with glycidyl groups present in a pre-derivatized silica can be used to attach a preformed polymer such as derivatized cellulose. This procedure makes a bonded polymer coating on silica surface and is used, for example, to prepare chiral phases with cellulose derivatized with 3,5-dimethylphenylcarbamate. The derivatization of cellulose with 3,5-dimethylphenylcarbamate adds a moiety able to provide π-bond interactions.

The bonding of calixarenes to silica [210] starts with the attachment on the silica surface of a "handle" having glycidyl groups, for example, by treating active silica with γ-glicidoxypropyltrimethoxysilane. The presynthesized calyxarene can be attached to these groups, leading to

a stationary phase schematically shown in Figure 6.7.1. The groups attached to the calyx-[4]arene macrocycles may provide enough asymmetry to use such molecules for chiral separations. However, this type of phase can also be used for nonchiral separations.

Main Types of Stationary Phases Used in Chiral HPLC

Section 5.5 described several types of interactions that allow the separation of the enantiomers of chiral molecules. Based on these different interaction types, a considerable number of chiral stationary phases were developed. The main types of chiral stationary phases can be classified in specific groups, as follows [211].

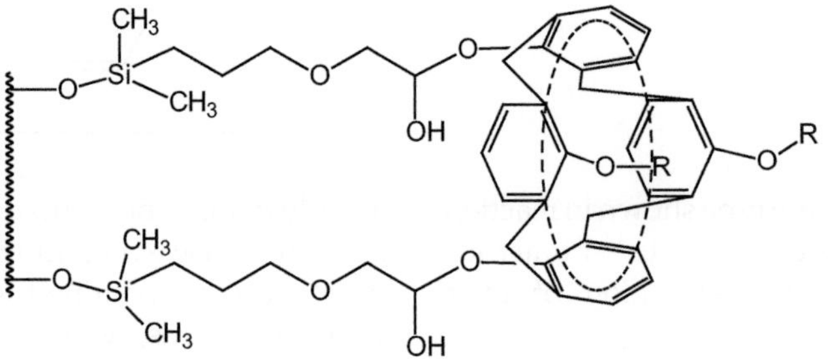

FIGURE 6.7.1 Schematic structure of a calyx[4]arene bonded phase on silica (R = tert-butyl).

1) A successful type of stationary phase is based solely on polar intermolecular interactions. These phases are silica-based and contain at least one substituted amide moiety that is able to establish both donor and acceptor hydrogen bonds, and a group able to establish either π-donor or π-acceptor, or both π-donor and π-acceptor intermolecular bonds. These phases are known as "brush" or Pirkle type [205, 206]. The bonded phase can be connected to the silica gel structure by single (monomeric functionalization) or multiple bonds (polymeric functionalization). The "handle" or "spacer" used for the connection to the silica surface is frequently an aminopropyl group, although other groups such as alkylglycidyl or alkylisocyanate can be used. The following structures are reported as phases containing groups capable of donor and acceptor hydrogen bonding and π-acceptor-type bonds: D- or L-Leucine, (R)- or (S)-Phenylglycine, (R,R)- or (S,S)- β-Gem 1, (R)- or (S) α-Burke 2, (3R,4S)- or (3S,4R)-Pirkle 1-J, (R,R)- or (S,S)-ULMO, (R,R)- or (S,S)-DACH-DNB. The schematic structure of these phases is given in Figure 6.7.2.

In order to obtain phases containing groups capable of donor and acceptor hydrogen bonding, and π-donor-type bonds, the active group of the stationary phase will typically include a naphthylamino fragment. Two commercially available phases of this type, the first with two chiral moieties of (S)-proline and (R)-1-(α-naphthyl)-ethylamine and urea linkage and another one having a L-naphthylleucine fragment, are shown in Figure 6.7.3. Phases containing both π-acceptor and π-donor fragments have also been developed, an example being (R,R)- or (S,S)-Whelk-O 1 (monomeric functionalization) and (R,R)- or (S,S)-Whelk-O 2 (polymeric functionalization) phases. The schematic structure of Whelk-O 1 phase is shown in Figure 6.7.4. Other Pirkle-type

stationary phases, such as BLAMO, quinine carbamates (see reaction 6.7.4) [212, 213], and others are available or reported to have been synthesized in laboratories [214, 215].

Pirkle (brush)-type phases are usable in NPC or HILIC-type mode, with the mobile phase less polar than the stationary phase. Common solvents are ethanol, isopropanol, hexane, and CH_2Cl_2. The phases can also be used in water–alcohol mixtures, but an excess of water will lead to the formation of strong hydrogen bonds with the mobile phase and interfere with separation. Also, the pH of the mobile phase must be in a narrower range (pH = 2.5 to 7.5) than for more robust columns typically used in RP-HPLC. Another limitation of these phases is that they are usable mainly with aromatic compounds (or compounds with a system of π electrons), since the π-donor or π-acceptor phase should interact with a π-acceptor or π-donor analyte, respectively. For analysis of compounds that do not have such groups, derivatization before separation with reagents introducing, for example, π-donor groups like naphthylamine can be performed. However, diastereoisomers can be separated without the help of a chiral phase, and when derivatization is necessary, it is preferable to perform it using chiral reagents and generate diasteroisomers which are easier to separate by RP-HPLC. Various suppliers offer Pirkle-type chiral chromatographic columns, including Advanced Separation Technologies (Astec) (3 types), IRIS Tecnol. (6 types), Machery-Nagel (2 types), Merck (2 types), Phenomenex (11 types), Regis Technologies (9 types), Sumika Chem. Anal. Service (11 types), and YMC (2 types).

2) Cellulose-based chiral stationary phases are another successful type (the principle of separation on these phases is briefly discussed in Section 5.5). These phases are made using cellulose derivatives such as microcrystalline triacetate-(MCTA), tribenzoate-, trisphenylcarbamate-,

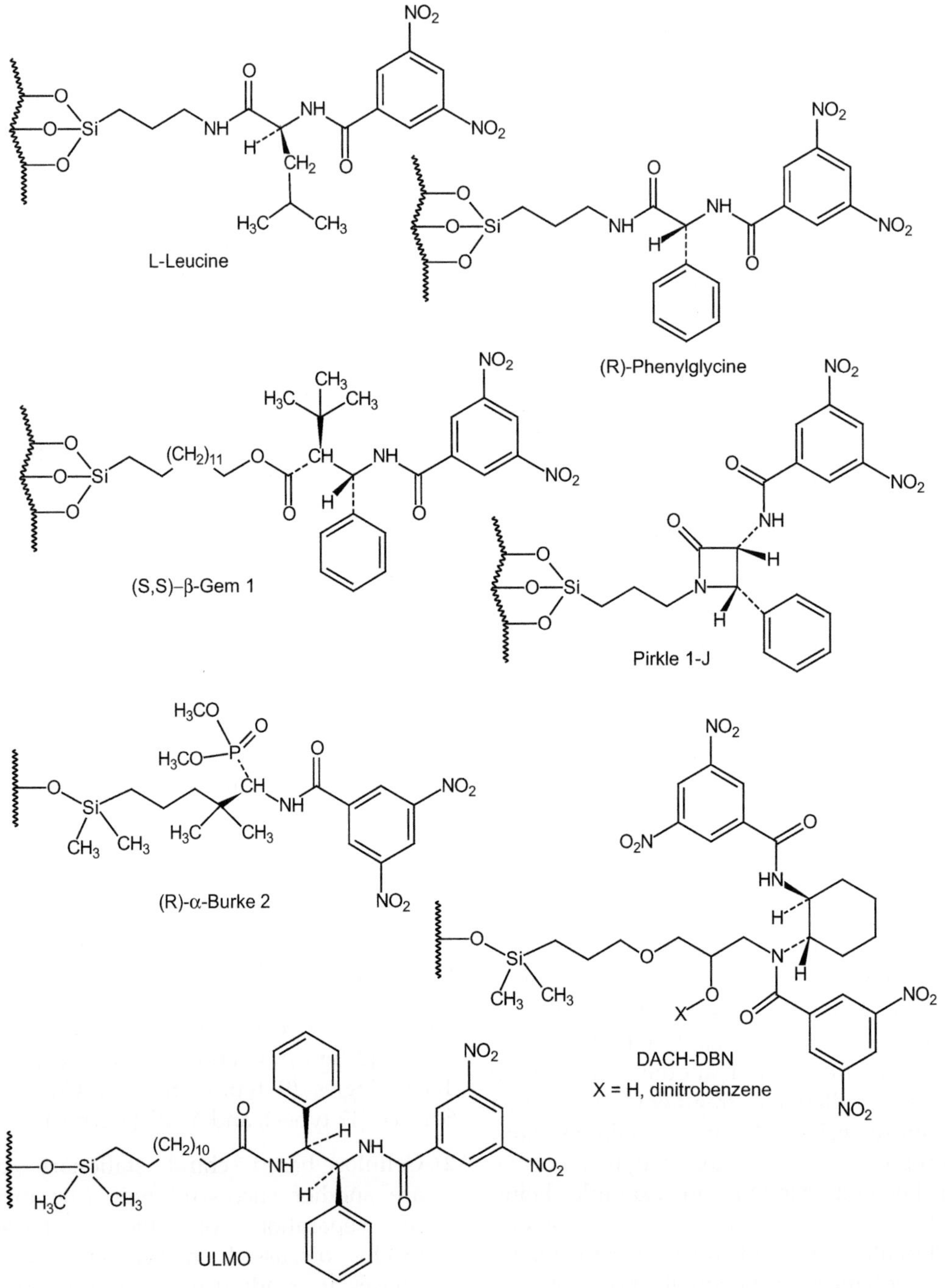

FIGURE 6.7.2 Commercially available chiral phases containing groups capable of donor and acceptor hydrogen bonding and π-acceptor type bonds.

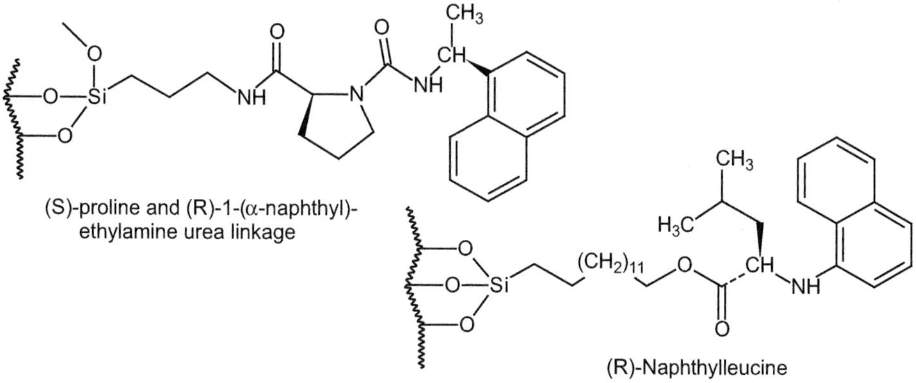

(S)-proline and (R)-1-(α-naphthyl)-
ethylamine urea linkage

(R)-Naphthylleucine

FIGURE 6.7.3 Commercially available chiral phases containing groups capable of donor and acceptor hydrogen bondin, and π-donor type bonds.

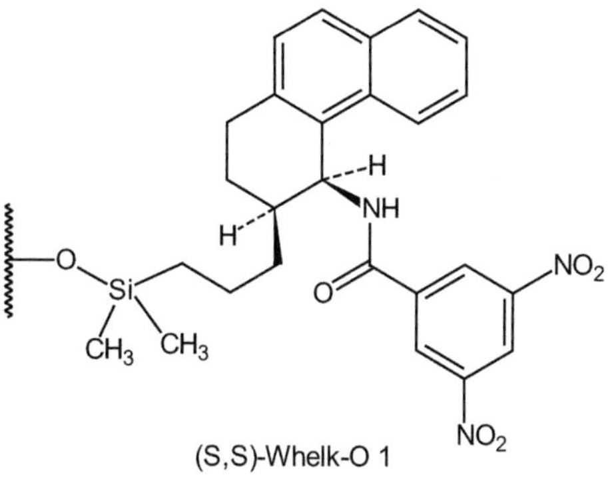

(S,S)-Whelk-O 1

FIGURE 6.7.4 Commercially available chiral phase containing groups capable of donor and acceptor hydrogen bonding and both π-donor and π-acceptor type bonds.

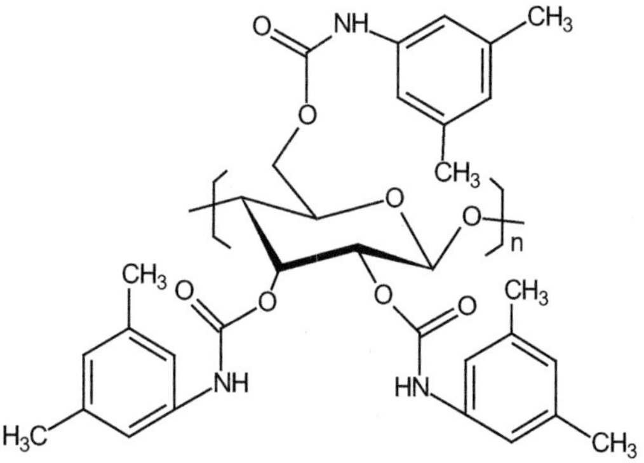

FIGURE 6.7.5 Schematic structure of tris(3,5-dimethyl-phenylcarbamate)-cellulose.

or tris(3,5-dimethylphenylcarbamate)-cellulose. The schematic structure of tris(3,5-dimethylphenylcarbamate)-cellulose is shown in Figure 6.7.5.

The derivatized cellulose material is typically coated on a silica support. The coating of silica with these materials can be achieved either by simple adsorption on silica surface using a proper solvent or by simultaneous forming of ether-bridges between cellulose and silica layer [216]. The process must preserve the porous structure of silica. Another

possibility to anchor cellulose is through glycidyl groups already attached to the silica surface (see Section 6.2). Coating of macroporous polymers such as poly(2-aminoethyl methacrylate-co-ethylenedimethacrylate) with derivatized cellulose has also been reported in the literature [217]. Several commercially available chiral columns based on cellulose derivatives are listed in Table 6.7.1. Utilization of cellulose-based columns requires a normal mobile phase, such as hexane-ethanol and hexane-isopropyl alcohol. The use of these

TABLE 6.7.1 Chiral columns based on derivatized cellulose.

Name	Structure	Applications
Chiralcel OA	-OCH$_3$	small aliphatic compounds
Chiralcel CA-1	-OCH$_3$	alcohols
Chiralcel OB	-OPh	small aliphatic and aromatic compounds
Chiralcel OC	-ONH-Ph	cyclopentenones
Chiralcel OK	-OC=CPh	aromatic compounds
Chiralcel OD	-ONH-3,5di-MePh	alkaloids, tropines, amines, beta blockers
Chiralcel OF	-ONH-*para*-ClPh	beta lactams, dihydroxypyridines, alkaloids
Chiralcel OG	-ONH-*para*-MePh	beta lactams, alkaloids
Chiralcel OJ	- O-*para*-MePh	aryl methyl esters, aryl methoxy esters
Cellulose DMP (Astec)	-ONH-3,5-di-MePh	beta blockers
Lux Cellulose-1	-ONH-3,5-di-MePh	
Lux Cellulose-2	-ONH-3-Cl-4-Me-Ph	

Note: Chiralcel (Daicel Chem. Ind.)

columns with a mobile phase containing some water can also be done, but addition of a salt in the mobile phase, such as a perchlorate, is necessary to prevent column degradation by the dissolution of the stationary active coating.

3) Cyclodextrins and chemically modified cyclodextrins represent another type of chiral stationary phase [218]. Cyclodextrins are produced by the action of enzymes cyclodextrin glycosyltransferase and amylase on starch. The main types of cyclodextrins produced are α, β, and γ (with, respectively, six, seven, and eight glucose residues), and they are generated by the enzymatic

cleavage of starch polymeric chain and formation of cyclic oligomers. The separation of α, β, and γ oligomeric products is based mainly on their different solubility in water, although other separation procedures can be utilized. Larger cyclodextrins are also known. Cyclodextrins offer a series of advantages to be used as chiral stationary phases. They can be anchored to silica (e.g., through reactions with glycidyl groups bonded to silica, as shown in Section 6.2), they offer a large number of chiral centers (30 for α, 35 for β, and 40 for γ-cyclodextrin), and they also provide the possibility to be derivatized at their free OH groups with less polar substituents. As discussed in Section 5.5, the separation on cyclodextrin type of phases is based on inclusion properties provided by their conical cavity of one part of the chiral molecule, and differences in the interaction with the chiral OH or OR groups of cyclodextrin with the enantiomer remaining substituents. The difference in the size of enantiomers is important for the choice of α, β, or γ cyclodextrin, and the molecular fit determines the range of analytes that can be separated. For the inclusion, α-cyclodextrins will host single phenyl groups or napthyl end-groups; β-cyclodextrins will accept naphthyl groups and heavily substituted phenyl groups; and γ-cyclodextrins are useful for bulky steroid-type molecules. Other interactions such as those involving the polar regions of an analyte and the surface hydroxyls of the stationary phase, and potential hydrophobic interactions in the cavity, provide the other two or more points of interaction required for chiral recognition.

Besides unmodified cyclodextrins, different modified cyclodextrins have been developed, and they expand the nature of compounds that can be separated. The derivatives are formed by bonding various groups onto the surface hydroxyls of the cyclodextrin cavity. The derivatives include acetyl (analogous to acetylated cellulose), (S)-hydroxypropyl ether, (S) or (R)-naphthylethylcarbamate (analogous to naphthyl Pirkle type columns), 3,5-dimethyl-phenylcarbamate (analogous to Chiralcell OD type column), and p-toluoylester. These modified cyclodextrin phases have several advantages. They offer different polarities from simple cyclodextrins, and they are more stable to different mobile phases, allowing the use of a wider range of solvents. Cyclodextrins are used mainly in normal-phase-type separations, with no or little water/buffers. It is possible to operate cyclodextrin columns in an alternative polar organic mode, with the mobile phase consisting of acetonitrile with up to 10% methanol plus up to 0.5% acetic acid and/or 0.5% triethylamine, and even in reversed-phase conditions using acetonitrile/ water mobile phase. Various suppliers offer chiral chromatographic columns based on derivatized cyclodextrins and resulting from α, β, or γ cyclodextrins. Among these suppliers are Advanced Separation Technologies (Astec) (Cyclobond 12 types), Diacel Chem. Ind. (Chiralpack AD derivatized with tris(3,5-dimethyl-phenylcarbamate) and Chiralpack AS), Machery-Nagel (4 types), Merck (2 types), Phenomenex (cyclodextrin with carboxymethyl functionalities bonded to methacrylate polymer), Showa Denko (4 types), Serva (3 types), Thermo Hypersil (2 types), and YMC (3 types). Besides cyclodextrins or modified cyclodextrins bonded to silica (typically with the help of a spacer), derivatized cyclodextrin (e.g., carboxymethyl cyclodextrin) can also be bonded to organic polymers.

To the class with chiral stationary phases similar with cyclodextrins can be added that based on amylose and that based on cyclofructans. Although amylose is a linear polymer, it can assume a helical form and can form inclusion compounds, such as with iodine, fatty acids, and aromatic compounds. Similar to both cellulose and cyclodextrins, it also offers the capability of hydrogen bonding and some hydrophobic interactions. Amylose can also be derivatized to generate chiral stationary phases with tris(3,5-dimethylphenyl-carbamate), with tris[(S)-α-methylbenzyl carbamate], and with tris(chloro-2-methylphanylcarbamate) (e.g., Lux Amylose-2 from Phenomenex).

4) Crown ether-based chiral stationary phases are also used for enantiomer separations [219]. In order to provide asymmetry to the crown macrocyclic group consisting of a polyether with the formula $[-(CH_2)_n-O-]_m$ (n = 2 or 3 and m 5 − 10), groups such as binaphthyl, biphenanthryl, tartaric acid, and carbohydrate are incorporated (see Section 6.2). Among the commercially available crown ethers used for chiral separations are Daicel Chem. Ind. (Crownpack CR 2 types), Regis Technol. (Chirosil RCA(+) and SCA(-)), and Phenomenex (Sumichiral O-8000).

5) Another type of chiral stationary phase is based on macrocyclic-type antibiotics immobilized on silica [220, 221]. The macrocyclic antibiotics contain numerous chiral centers, functionalities that allow bonding to silica (e.g., using prederivatized silica with reactive groups) and capability to offer π-π interactions, hydrogen bonding, inclusion/complexation, and ionic interactions. Several glycopeptide type antibiotics have been used for making stationary phases, including: Rifamycin(s), Vancomycin (18 chiral centers), Avoparcin, Ristocetin, glycopeptide A-40,926 (MDL 62,476), and Teicoplanin (23 chiral centers). Macrocyclic antibiotics used as a bonded stationary phase also include a family of thiopeptides, with the parent compound being thiostrepton (17 chiral centers). The antibiotics that are glycopeptides have dissociable groups (−OH of phenolic type; −NH_2; −COOH, and −OH from carbohydrate). Some of the groups are able to

exist in zwitterionic form and likely play a major role in the association with analytes during the chiral recognition. The chemical formula of vancomycin where the functional groups as well as three potential inclusion regions are seen (A, B, and C) and the formula of thiostrepton are given in Figure 6.7.6. The connection to silica can be made using the amino groups of the chiral antibiotic. The columns made with macrocyclic antibiotics, such as the one made with Vancomycin, can be used in normal phase as well as in reversed-phase mode. For normal-phase use, the typical solvents are hexane-ethanol mixtures, and for reversed-phase use, tetrahydrofuran-water mixtures. One company offering macrocyclic glycopeptide phases is Advanced Separation Technologies (Astec) (4 types of columns).

6) Proteins bonded to silica were also used as stationary phases for chiral separations. Proteins contain a large number of chiral centers and are known to interact with enantiomers to produce acceptable chiral

selectivity. Several types of proteins have been used so far for separation of enantiomers. One common protein used as chiral selector after immobilization on silica is bovine serum albumin (BSA) with Resolvosil BSA 7 and BSA-7PX (from Machery-Nagel), Ultron ES-BSA (from Shinwa Chem. Ind.), and Chiral BSA (from Shandon). Other proteins include human serum albumin (HSA) with Chiral-HSA (from Shandon and from Regis), a1-acid glycoprotein with Chiral AGP (from Regis), glycoproteins with Chirobiotic V, V2, T, T2, TAG and R (from Astec), ovomucoid with TSKgel Enantio-OVM (from Tosoh), ovoglycoprotein with Ultron ES-OVM (from Shinwa Chem. Ind.), avidin with Bioptic AV-1 (GL Science), cellobiohydrolase (Chiral CHB from Regis), and others (e.g. from ChromTech Ltd.). Some of these phases can be used in either normal-phase or reversed-phase mode with solvents such as isopropanol, ethanol, or acetonitrile in mixture with aqueous buffers.

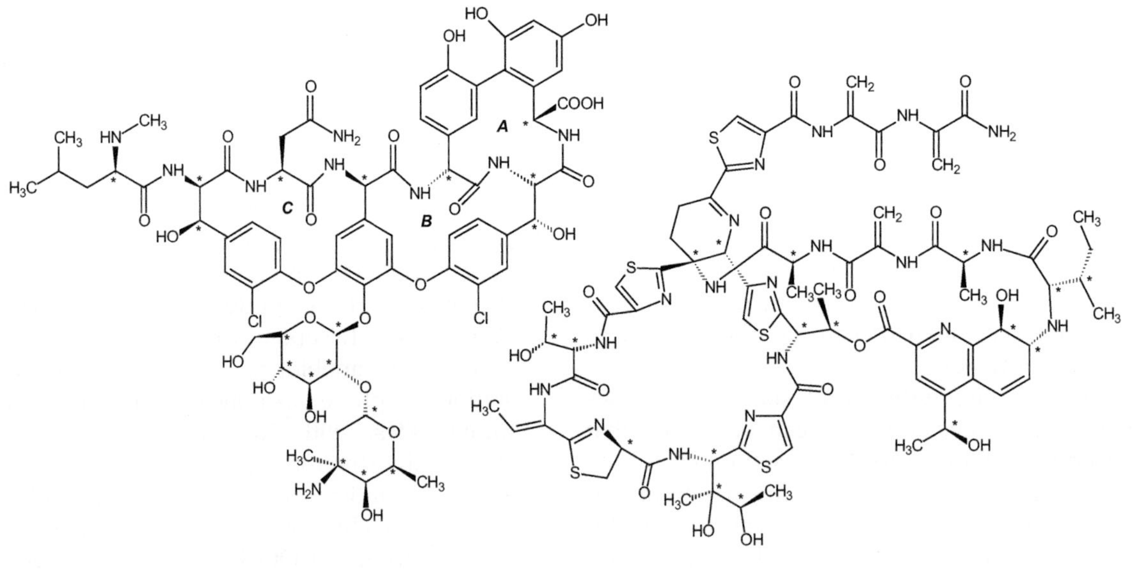

FIGURE 6.7.6 Formula of vancomycin with three potential inclusion areas (A, B, and C) and that of thiostrepton.

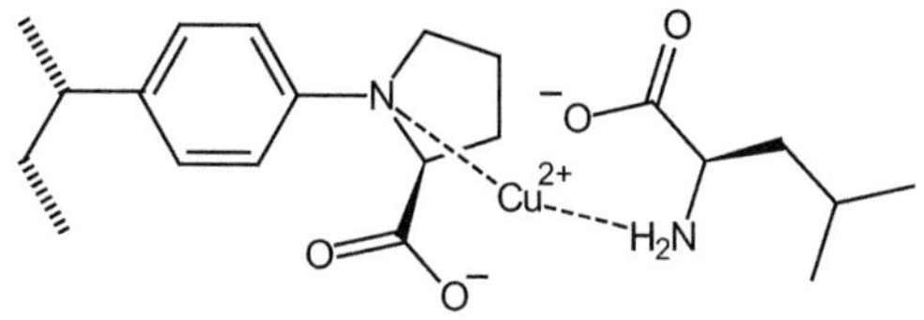

FIGURE 6.7.7 L-Proline bonded into a polystyrene type polymer and forming a ternary complex with Cu^{2+} and with L-leucine.

Besides bonding directly to silica, proteins were also bonded to organic polymers such as acrylates and further coated on silica or silica functionalized with amino groups that provide a stronger bonding [222]. Another possibility for bonding proteins to silica-based supports is to functionalize the silica surface to become an anion exchanger (e.g., quaternized polyvinylimidazole coated silica) and then to bond electrostatically the proteins to its surface. [223].

7) Ligand-exchange chromatographic columns contain chiral molecular units bonded on an organic polymer or on silica, the chiral units being capable of forming complexes with transitional metal cations such as Cu^{2+}, Ni^{2+}, or Zn^{2+}. When the metal ions can also participate in a ternary complex with an analyte from the mobile phase, the enantioseparation results from the difference between the stability constants of the ternary complexes formed by the two enantiomers present in solution. An example of such a stationary phase is given

in Figure 6.7.7, with (S)-proline attached to a polystyrene type polymeric structure.

The mobile phase for this type of separation must contain a constant concentration of metal ions (e.g., Cu^{2+}). This type of polymer was first obtained by bonding amino acids such as proline into polystyrene-divinylbenzene matrix, but other polymeric matrixes, such as cross-linked polyacrylamide and poly(glycidylmethacrylate), have been useful for immobilizing chiral amino acid selectors. The bonded amino acid can be (S)-phenylalanine or (S)-proline as typical chiral selectors, since they are able to form complexes with Cu^{2+}. Separation of α-amino acids as analytes has been performed successfully on such columns [224].

Another possibility of binding the chiral selector based on ligand exchange is to use silica as support. An example is the linking (2S,4R)-hydroxyproline to the silica matrix via a short spacer as shown in Figure 6.7.8 [225].

8) Chiral synthetic polymers are another type of material used as stationary phase for enantiomer separation. Some such phases can have the polymer bonded to silica in order to assure a high contact surface and mechanical rigidity. The polymer must contain chiral moieties to act as chiral selectors. One example is a polymer having diaminocyclohexyl-acrylamide chiral selectors with polymeric structure bonded to silica (P-CAP from Astec). The synthesis of this polymer was schematically shown in reaction 6.7.5. A similar phase, described as

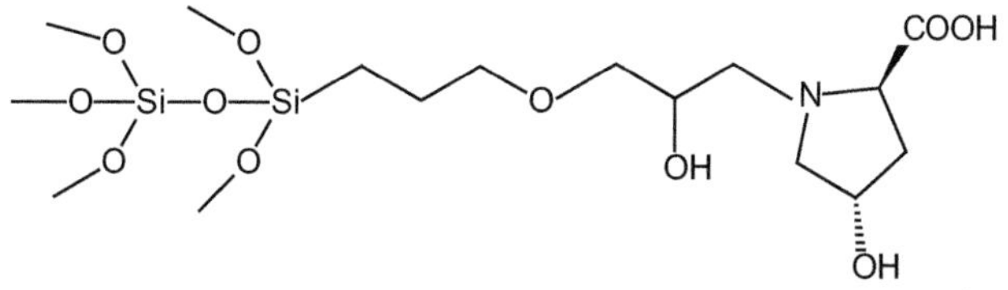

FIGURE 6.7.8 (2S,4R)-Hydroxyproline bonded to silica matrix following a reaction with silica derivatized with glycidyl groups.

poly(diphenyl-ethylenediamine-bis-acryloyl), is P-CAP-DP (from Astec). These stationary phases can be used in normal-phase (NP) mode (e.g., hexane + isopropanol, or methylene chloride + alcohol as a mobile phase), and also in NP with polar-organic mobile phase consisting of acetonitrile, methanol and a polar additive (trifluoroacetic acid, triethanolamine, etc.). Chiral synthetic polymers were also obtained using as chiral monomer O,O'-bis-(3,5-dimethylbenzoyl)-N-N'-diallyl-tartar diamide immobilized on silica (Kromasil CHI-DMB) and O,O'-bis-(4-tert-butylbenzoyl)-N-N'-diallyl-tartar diamide (Kromasil CHI-TBB), both commercialized by Eka Chemicals (see reaction 6.7.6). Synthesis and properties of other similar phases have been reported in the literature [226, 227]. Among other polymers evaluated for potential use in chiral separations are molecular imprinted polymers [228]. For example, mandelic acid enantiomers were successfully separated on a polymeric material that was obtained by copolymerization of methacrylic acid and ethylene glycol dimethacrylate at 4 °C with UV radiation in the presence of the template molecules R-mandelic acid, S-mandelic acid or R-phenylalanine, S-phenylalanine [229].

6.8. STATIONARY PHASES AND COLUMNS FOR SIZE-EXCLUSION CHROMATOGRAPHY

General Comments

Size-exclusion chromatography (SEC) is a technique applied mainly for the separation of polymers. The purpose of separation may be related to the purification of the polymer (e.g., of proteins) or analysis of polymers with the goal of assessing the molecular weight. Depending on the nature of the polymers, they may be soluble in an aqueous solvent with separation based on gel filtration (GFC) or soluble in an organic solvent with separation based on gel permeation (GPC). The technique may be performed at higher pressures and be considered part of the HPLC family, but it also can be performed at low pressure. Most SEC separations are performed for large molecules, with molecular weights that can be as high as 20×10^6 Daltons. However, smaller molecules can also be separated and quantitated using SEC.

Among the first materials used as stationary phases in size exclusion was dextran treated with 1-chloro-2,3-epoxypropane (epichlorhydrin), the resulting material being known as Sephadex. Numerous other materials are currently available, including some based on porous silica and others using semirigid polymeric materials such as polyhydroxymethacrylate, polyvinyl acetate, various dextrans, acrylamide-methylenebisacrylamide, agarose, poly(methyl-methacrylate), polystyrene, and styrene-divinylbenzene copolymer [230].

The separation in SEC is influenced by specific characteristics of the packing materials. Several such characteristics of SEC columns/packing materials are described in Table 6.8.1 [231]. Some of the characteristics listed in this table are relevant for any chromatographic column, such as resolution, column pressure, and chemical and thermal stability. Other characteristics are more relevant for SEC columns. One of these is stationary phase porosity. Porosity determines the range of molecular weights (MW) that can be covered by a specific column. The resolution of the columns regarding the separation of two analytes of different molecular weight depends on pore-size distribution. Columns with a wider range of pore diameters are able to separate a wider range of molecular weights for the analytes, but typically the resolution of these columns is lower, as compared to columns with a narrow pore-size distribution. However, columns with a narrow range of pore sizes (narrow pore-size

TABLE 6.8.1 Characteristics of SEC columns/packing materials.

Attribute	Variable	Importance	Common values	Range
Porosity	Pore size. Pore volume	Range of MW	$50-10^5$ Å	$50-10^5$ Å
Pore size distribution	Range of pore size	Resolution		
Resolution	Column efficiency (N)	Separation efficiency	18,000	7,000–25,000
Resolution	Particle size	Separation efficiency	5μm, 7μm	4–20 μm
Resolution	Particle shape	Separation efficiency	Spherical	Irregular-sphere
Column pressure	Particle size	Flow rate, analysis time	500 psi	600–2000 psi
Inertness	Column activity	Utility for MW evaluation	Inert	Some inertness
Recovery	Column activity	Recovery of the analyte	95%	Variable
Mechanical stability	Support type	Max working pressure	500, 2000 psi	Variable
Chemical stability	Support type. Phase type	Solvents, range of pH	Stable	Variable
Thermal stability	Support type.	Temperature of utilization	Up to 130 °C	20–160 °C
Load capacity	Amount of sample	Resolution	2–4 mg	2–4 mg

distribution) can be used only within a narrower range of molecular weights for the analytes.

For the separation of mixtures of molecules within a wide range of MW values (hydrodynamic volume), more than one SEC column in series can be used. In such cases, the column with the highest porosity is used first. For example, Phenogel columns that cover a wide range of porosities can be used up to four columns in series for achieving some separations. Since high backpressure can damage gel porosities, the maximum acceptable backpressure of the column should never be exceeded. A reduced flow rate of the mobile phase is sometimes necessary for maintaining the correct backpressure, in particular when more than one column in series is utilized for the separation.

Particle-size diameter of the columns is also important for resolution of the separation. For high column efficiency, packing materials in SEC should be fine, uniform, and spherical particles (3–20 μm), as in the common interactive HPLC. However, smaller particles (e.g., 3–5 μm), which lead to better resolution, also

increase the column backpressure. In SEC, the columns very frequently have a limited value for the backpressure in order to avoid collapse of the pores that damage the column. For this reason, when columns with small particles are used, a decrease in the flow rate of the mobile phase is typically recommended (below 1 mL/min, e.g., 0.2 mL/min). The increase in the column length, which also enhances resolution, is associated with an increase in column backpressure.

Two other characteristics are inertness and, a related one, recovery. The inertness of the SEC column is important in particular when the MW of polymers are estimated using SEC. When the column is not inert, other interactions besides size exclusion will influence the retention process (see rel. 3.4.9 where an enthalpic component affects separation), and the result for the MW evaluation may be incorrect. For preparative reasons, the recovery of the analytes is another important characteristic of the SEC columns. Proteins or other biologically active materials are often separated (e.g., for purification) on SEC columns. It is important that the material injected in the column be recovered and that its biological activity not be altered. When the biological activity of the analyte must be preserved, the nature of the mobile phase also plays an important role. For such GFC separations, the mobile phase is either water or a dilute buffer.

Silica-based SEC Phases and Glass Phases

Silica-based stationary phases are common for size-exclusion separations, their advantages being mainly related to the good mechanical properties and the lack of swelling, such that their pore size is independent of the nature of mobile phase. This typically is associated with good column efficiency. However, polymer-based phases are more frequently utilized than silica-based ones for SEC. This is due to the larger range of pore sizes offered by the polymeric materials (between 60 Å and 4000 Å) and the progress obtained regarding their good mechanical strength. Silica-based phases have been used in gel permeation-GPC (with organic-type mobile phase) but in particular in gel filtration-GFC (with aqueous-type mobile phase) modes.

Silica-based phases include bare silica as well as derivatized silicas, similar to the construction of other bonded phases on silica. Bonded-phase silica-based materials are made with groups such as diol, diol on Zr-clad silica, glycol ether, polyether, amide, and frequently with a propyl anchor group to the silica material. The purity of the silica material used for making the packing is very important in order to avoid other types of interactions with the analytes. The silica-based phases should be operated within the recommended range of pH, which is typically from 2.5 to 7.5. The silica-based phases are not stable outside this range. Several commercially available silica-based packings are listed in Table 6.8.2. Besides silica, porous glass has been used as

packing material for SEC columns [232]. The surface of the porous glass can be derivatized with diol groups, similar to that of porous silica [233].

Polymeric-based SEC Phases

Polymeric-based SEC phases are more common than silica-based ones. Some phases are designed for GPC, a very common material used for this purpose being polystyrene-divinylbenzene (PS-DVB). This polymer is typically prepared by suspension polymerization (see Section 6.1), and depending on the proportion of cross-linking reagent (DVB), either soft gels are obtained (2-12% DVB) or rigid polymers are generated (more than 20% DVB) [234]. Other phases are used for GFC (aqueous SEC). Examples include several soft gels based on dextran or agarose, more rigid polymers such as hydroxylated poly(methyl methacrylate) (HPMMA), and polyvinylalcohol (PVA) copolymers.

The porosity of the polymeric materials is controlled by the conditions in which the polymerization is carried out. When 20% or more of a nonpolymerizable compound is present, this compound dissolving the monomer but not the polymer, a product with a macroporous structure is obtained. This macroporous structure remains in the dry state of material. Porosity of the stationary phase is critical for the separation of compounds within a specific range of MW. However, the pore size of the polymer is difficult to measure when the polymer is swollen, and it is usually assessed with calibrants of known MW. Examples of calibrants are sets of polystyrenes with different known MW for GPC separations and sets of polyethylene oxide, dextrans, and proteins with different known MW. Polymeric packings with a narrow range of pore size have a separation capacity concentrated in a limited molecular weight range. Although the resolution of such columns is high, their use for SEC analyses is possible only for a narrow molecular weight

TABLE 6.8.2　Silica based SEC packings (for aqueous SEC or GFC)

Phase name (Supplier)	Pore size designation	Range of use	Particle size	Chemistry on silica
Shodex Protein KW (Showa Denko)	800 Series	1,000 to 10,000,000		
TSK-gel SW, SW$_{XL}$, SuperSW (Tosoh)	various G2000SW, G3000SW, G3000SW$_{XL}$, etc.	various 5,000 to 70,000	5 μm, 8μm, 10μm	diol
TSK-gel QC-PAK	TSK 200, TSK 300	5,000 to 50,000	5 μm	diol
TSK-gel BioAssist DS				
Zorbax GF Ser. (Agilent/ Crawford Scientific)	GF-250, GF-450	4,000-400,000, 10,000-900,000	4 μm 6 μm	Zr clad diol
Zorbax PSM (Agilent/ Thomas Scientific)	PMS 60, PMS 60S, PMS 300, PMS 1000, etc.	various	5 μm	silanized
LiChrosphere Si	Si60, Si100	also normal phase	5 μm, 10 μm	silica
UltraSpherogel (Grace/Beckman)	2000, 4000, etc.		5 μm	polyether
Bio-Sil , Bio-Select Bio-Rad)	SEC-125-5, 250-5, 400-5	1,000—1,000,000	5 μm, 10 μm	silica
Protein-Pak (Waters)	300SW		10 μm	diol
Bio-Sep (Phenomenex)	SEC-S2000, SEC-S3000, SEC-S4000	500—20,000,000		
SynChropak (Lab Unlimited)	GPC Peptide, GPC 100 to GPC 4000, CATSEC 100 to 4000	various ranges	5 μm, 10 μm	diol

distribution of the samples. In practice, SEC columns of different pore sizes are connected in series to provide a wider molecular weight separation range. This procedure of multiple columns is not very convenient, and unique columns in which the pore-size distribution of the stationary phase is broadened by blending together two or more phases are available. Mixed pore packings obtained by blending together several selected pore-size materials can be made such that the column exhibits a linear calibration for analytes in a wide range of MW.

The particle size of polymer-based packings is another parameter important for SEC column efficiency (height of theoretical plate). The packing materials must be as homogeneous as possible, and this is achieved by having particles of equal size and spherical shape, with uniform flow channels. Polymer particles with a highly mono-disperse particle-size distribution can be produced by a two-step microsuspension method. This process is based on using monosized polymer seed particles mixed with low-molecular-weight material during polymerization. This process allows the preparation of monosized compact or macroporous particles of predetermined size in the range of 1 to 100 μm. By coating the particle surface with a hydrophilic crosslinked polymer, supports for aqueous phase size-exclusion chromatography may be produced [235]. However, a separation by size

TABLE 6.8.3   Several common phases used in non-aqueous SEC (GPC).

Phase name (Supplier)	Pore size designation	Range of use*	Particle size
PLgel (Polymer Lab. Ltd./ Agilent)	various from 50 Å to 1,000,000 Å	various from 100–2,000 to 100,000–20,000,000	5 μm, 10 μm, 20 μm
PLgel multipore bed	various from MIXED-A to MIXED-E	various from 200–400,000 to 1,000–40,000,000	5 μm, 10 μm, 20 μm
PolySep (Phenomenex)	various from 1000 to 6000 and linear	various from 100–2,000 to 100,000–20,000,000	5 μm, 10 μm, 20 μm
Shodex K and KF,KD, KL (Showa Denko)	various indicated as 801, 802, … 807	various from 70,000 average to 200,000,000 average	Nominally 7 μm
Shodex K multipore bed	various indicated as 803L, to 807L	various from 1,500 average to 200,000,000 average	various depending on range (6 μm, 10 μm, 17 μm)
TSK-Gel H_{XL}, SuperHZ (Toyo Soda)	various indicated as G1000 to G7000	various from 1,000 average to 400,000,000 average	various depending on range (5 μm to 9 μm)
TSK-GEL H_{HR} and H_{XL} (Aldrich)	G1000H to G7000H, GMH-H, GMH-L, GMH-M	various from 1,500 to 1,000,000	
TSK-Gel H_{XL}, SuperHZ multipore bed	GMHXL, GMHXL-HT, GMHXL-L	400,000,000 average	9 μm, 13 μm, 6 μm
Styragel HR (Waters)	various indicated as HR 0.5, HR 1 to HR 4	various from small–1,000 to 5,000–600,000	5 μm
Styragel HR multipore bed	HR 4E HR 5E	50-10,000 2,000–4,000,000	5 μm
Styragel HT , Ultrastyragel (Waters)	various indicated as HT 3 to HT 6, Ultrastyragel	various from 500–30,000 to 200,000–10,000,000	10 μm
Styragel HMW (Waters)	HMW 7	500,000–100,000,000	20 μm
Styragel HT, Styragel HMW multipore bed	HT 6E HMW 6E	5,000–10,000,000	10 μm 20 μm
Bio-beades, S-X beads Bio-Rad		400–14,000	soft gel
Hydrocell (Biochrom Labs.)	GPC 3000, 3000HS	20,000–1,000,000	5 μm
Phenogel (Phenomenex)	50 Å, 100 Å, 500, 10^3 Å, 10^4 Å, 10^5 Å, 10^6 Å , and linear	various	5 μm, 10 μm, 20 μm

Note: the range of MW use based on polystyrene as analyte.

TABLE 6.8.4 Several common phases used in aqueous SEC (GFC).

Phase name (Supplier)	Pore size designation	Range of use, Da	Particle size	Chemistry
PL aquagel-OH (Polymer Lab. Ltd./ Agilent	AOH 30, 40, 50, 60	100,000, 1,000,000, 20,000,000	5 μm, 8 μm, 15 μm	polyhydroxyl surface
PL aquagel-OH mixed bed	AOH Mixed H, Mixed M	20,000,000, up to 600,000 respectively	8 μm,	polymer
Shodex OHpak (Showa Denko), Protein KW	various indicated as KB-802, to KB-806, KB-80M, SB-401, etc.	various from 4,000 to 20,000,000	various	HPMMA
TSK-GEL PW, PW$_{XL}$, PW$_{XL}$-CP (Toyo Soda)	various G1000 to G6000, GM, PW$_{XL}$ G5000PW, GMPW, etc.	various from 1,000 to 8,000,000	various	HPMMA
TSK-GEL H$_{HR}$ and H$_{XL}$ (Aldrich)	G1000H to G7000H, GMH-H, GMH-L, GMH-M	various from 1,500 to 1,000,000	7 μm	HPMMA
Toyopearl HW	40S, 40F, 40C, 50S, 50F, 55S, 55F, 65S, 65F, 75F (five pore sizes)	100 to 50,000,000	20—40, 30—60, 50—100	
TSK-GEL Alpha and Super AW,	Alpha-3000, Alpha-5000, Super AW2500 to Super AW6000	various up to 10,000,000	various	HPMMA
TSK-GEL Super-Multipore PW	PW-N, PW-M, PW-H	300-50,000 to 1000—10,000,000	4 μm, 5 μm, 8 μm	HPMMA
Ultrahydrogel (Waters)	various 120 Å to 2000 Å indicated as 250, 500, etc., linear	various from 5,000 to 7,000,000	10 μm	HPMMA
Asahipak (Asahi Chemical/ Phenomenex)	various GS-220, GS-320, GS520, GS-620, GS-710, GFA-30, GFA-7M, GF ser.	various from 3,000 to 10,000,000	various	PVA copolymer
Asahipak (Phenomenex)	GF-310 HQ to GF710 HQ, and multimode	various	5 μm, 6 μm, 9 μm	PVA copolymer
Suprema (PSS)	30 Å, 100 Å, 300 Å, 1000 Å, Linear S, M, XL	various from 20,000 to 10,000,000	various	
MCI Gel CQP (Mitsubishi Chemical)	various indicated as CQO06, CQP06G, CQP10, CQP30, etc.	various from 1000 to 1,000,000	10 μm	
Polyhydroxyethyl A (PolyLC Inc.)	200 Å, 1000 Å	100 to 1,000,000		
Bio-Prep SE (Bio-Rad)	SE100/17, SE1000/17	5,000-1,000,000	17 μm	agarose
PolySep-GFC-P (Phenomenex)	1000 to 6000 and Linear	various ranges		highly hydrophilic

of the synthesized particles is still necessary for obtaining particles with a narrow size distribution. This is achieved by using sedimentation or centrifugation [236]. Information regarding the particle shape and size can be obtained by microscopic methods. Particle diameters in the range 3–70 μm are commercially available. Smaller particles offer improved resolution but result in higher operating pressures and can prove more difficult to pack.

Mechanical stability of the polymer is important for establishing the limit of flow-rate operation. The maximum operating pressure of the packing should be below the compression point of the packing material. Since mobile phase viscosity is commonly increased in both GPC and GFC, particular attention must be given to this parameter.

The surface chemistry of the packing material is very important and must be selected so that any enthalpic contribution (interactions with the polymer surface) is minimal. For GFC, the packing materials are typically highly hydrophilic and should not possess charges. These effects are difficult to completely eliminate, and salts (buffer) as well as organic modifiers are typically added to the mobile phase to diminish such interactions.

The chemical and temperature stability of the packing is related to its inertness to specific solvents, and typically the crosslinked polymers are capable of being used with a wide range of solvents. Temperature stability of the packing is also necessary since some polymeric analytes are soluble in the mobile phase only at elevated temperature. Packing materials capable of being used at temperatures in the 50–120 °C range are common. Several phases used in GPC are listed in Table 6.8.3, and some used in GFC are listed in Table 6.8.4. A variety of column formats are available for specific applications [237]. Particular advantages and limitations as well as typical applications for each type of column are reported by the suppliers (see, e.g., [144, 238, 239]).

6.9. STATIONARY PHASES AND COLUMNS IN IMMUNOAFFINITY CHROMATOGRAPHY

General Aspects

Separation in immunoaffinity can be done as an HPLC technique [240, 241] and also at low pressure [242]. A large number of stationary phases, some of which are "custom made" for this technique, and only a few examples are presented in this section (for details regarding affinity chromatography phases, see, e.g., [243]).

Synthesis of Stationary Phases Used in Affinity and Immunoaffinity Chromatography

The two typical components of the stationary phases in immunoaffinity are the support and the active phase. Support for immunoaffinity chromatography should be chemically and physically stable, with a very good mechanical strength; should allow high flow rates of samples; and should have minimal nonspecific adsorption capability. The immobilization of ligands of interest should be such that high ligand accessibility is achieved, while the phase properties are not affected. For LC applications, particle-size and pore-size distribution of the support are very important parameters. A wide variety of materials, including organic and inorganic polymers, have been used for the design of these solid phase supports. One of the first materials used as support in immunoaffinity chromatography were natural polysaccharides such as agarose (the neutral gelling fraction of the complex natural polysaccharide agar), cellulose, and crosslinked dextran (sepharose, sephacryl, etc.). These materials are stable over a wide interval of pH (3–13) and are characterized by a high content of hydroxyl groups available for activation and derivatization. Their hydrophilic surface generally does not interact with proteins and exhibits only a low nonspecific adsorption. A major drawback of these materials is their poor mechanical strength related to their swelling ability. Other supports for the stationary phase may include silica as well as organic polymers such as methacrylates, polyacrylamide, and copolymers of polyacrylamide with other polyvinyl polymer. Although more resistant to pressure than polysaccharides, in comparison to inorganic materials, these organic supports exhibit a lower pressure tolerance. In addition, some of these materials show swelling differences in the presence of organic solvents, a broader pore-size distribution, decreased ligand-binding efficiency, and nonspecific interaction due to their hydrophobic character. Monolithic supports based on glycidyl methacrylate-co-ethylene dimethacrylate (GMA-EDMA) have been shown to be useful in modern affinity chromatography. Such columns are commercially available from BIA Separations (Slovenia) under the trade name of CIM® disk monolithic columns. The main advantage of the GMA-EDMA support is the significant concentration of chemically reactive epoxy groups, which can be used for immobilization of ligands, for example, containing amino groups. [244]. The stationary phase can also be porous, nonporous [245], or monolithic [241].

Silica is one of the most widely used inorganic supports for polymers in immunoaffinity chromatography. This support is very stable under pressure and can be easily derivatized in order to introduce different functional groups. Unfortunately, silica is unstable at mild alkaline pH values and dissolves significantly above pH of 8. In addition, nonspecific interactions can occur between silanol groups and the basic parts of the biomolecules. However, surface modification can be performed either by chemical modification or physical adsorption of ligands in order to minimize this nonspecific adsorption and to introduce a high density of functional groups [246].

The active phase in immunoaffinity can be very diverse, depending on the nature of the material immobilized on the stationary phase. The development of affinity chromatography depends on the construction of new stationary phases and affinity ligands, as well as new approaches to immobilization chemistry [247]. Stationary phases containing an immobilized enzyme or other compounds such as heparin [248], lectins, and nucleotides were successfully utilized in various chromatographic procedures, mainly low-pressure liquid chromatography, but also in HPLC. For silica-based stationary phases, a common material utilized as a starting material is silica pretreated with an aminosilane (e.g., γ-aminopropyltriethoxysilane, γ-aminopropylmethyldiethoxysilane, γ-aminopropyldimethylethoxysilane). The reaction of the propylamino silica surface with glutaraladehyde generates a reactive material on which an amino group of a protein (e.g., with lysine units in its structure) can be readily attached. This is schematically shown in reaction 6.9.1:

(further indicated as Sil—NH$_2$)

(6.9.1)

Other similar procedures use succinic anhydride instead of glutaraldehyde as an intermediate step of derivatization. In this case the reactions follow the path shown in 6.9.2:

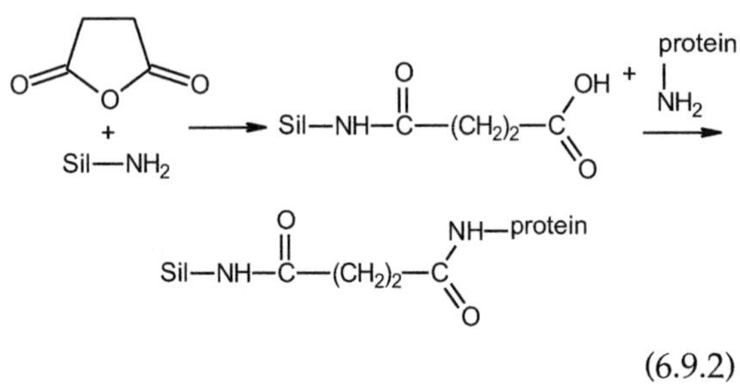

(6.9.2)

Some proteins that contain phenolic groups (e.g., with tyrosine units) can be bound to the aminosilica using azo-coupling, as shown schematically in reaction 6.9.3:

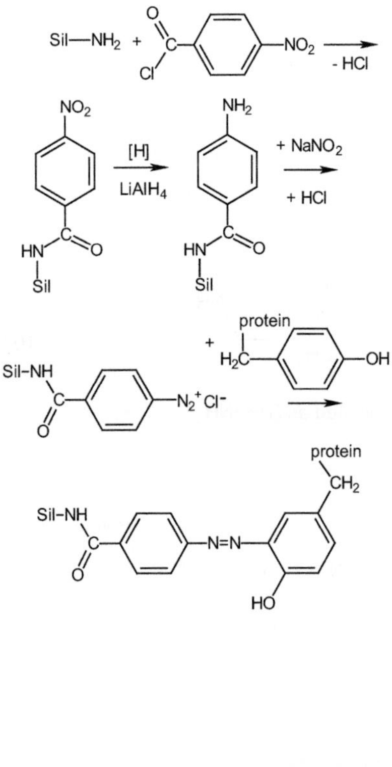

(6.9.3)

For an organic support, the linking process is usually done using an activating reagent such as cyanogen bromide when the molecule to bind contains a free primary amine, sulfhydryl, or hydroxyl groups for attachment. On the activated support, proteins (or other molecules) with specific binding capability are further immobilized. These can be selected by the user or can be general-purpose immobilized compounds. Immobilized heparin, for example, acts with a specific binding site to retain certain proteins, lectin resins can be used for the separation of glycoproteins from other glycoconjugate molecules, and nucleotide resins are used for the separation of specific proteins. Heparin resins, lectin resins, and others are commercially available. Specific immunoproteins also can be bound, for example, on agarose activated with cyanogen bromide or with other activation reagents such as 6-aminohexanoic acid, carbonyldiimidazole, and thiol [249]. These types of materials have a very high specificity for the specific antigen that generates the immunoprotein. Affinity resins containing immobilized sugars and sugar derivatives and resins with immobilized biotin or avidin are also available. Several activation reagents for agarose and crosslinked dextrans are shown in Table 6.9.1. Other resins are used having immobilized ligands such as p-amino benzamidine (ABA-5PW), m-aminophenyl boronic acid (Boronate-5PW), and iminodiacetic acid (CHelate 5PW), all from Sigma. Numerous other immobilization procedures are reported in the literature [249–251]. The affinity and immunoaffinity-type phases have excellent selectivity and work well in aqueous solutions, but each material must be developed for a specific analyte; the phases can be unstable toward organic solvents and may be stable only in a narrow pH range. Some bioaffinity phases are used for HPLC separations [252–254], while others must be used at lower pressures (1–2 bar) since some materials impose considerable limitations regarding the maximum pressure that can be applied to the phase.

TABLE 6.9.1 Reagents used to make activated resins able to bind proteins.

Activating reagent	Linkage to resin by	Available reactive group In the ligand	Specificity of the group to	Reaction conditions	Bond type to ligand	Stability
6-Aminohexanoic acid	isourea	carboxyl	amine, with carbodiimide coupler	pH 4.5−6.0	amide	good
6-Aminohexanoic acid N-hydroxy-succinimide ester	isourea	succinimidyl ester	amine	pH 6.0−8.0	amide	good
Carbonyldiimidazole	carbamate	imidazolyl carbamate	amine	pH 8.0−10.0	carbamate	good below pH 10
Cyanogen bromide	ester	cyanate	amine	pH 8.0−9.5	isourea	moderate
Epoxy	ether	epoxy	SH>NH	pH 7−8 SH pH 8−11 NH$_2$	SH: thioether, NH$_2$ amino ether	very good
N-hydroxy-succinimide ester	isourea	succinimidyl ester	amine	pH 6.0−8.0	amide	good
Periodate	oxidizes agarose, saccharides	aldehyde	amine	pH 4.0−10.0	reductive amination with NaBH$_3$(CN)	very good
Thiol	isourea	disulfide	sulfhydryl	pH 6.0−8.0	disulfide	good in nonreducing conditions
EDTA	metal		amino acids	wide pH range	chelate	good

References

[1] Neue UD. HPLC Columns, Theory, Technology, and Practice. New York: Wiley Blackwell; 1997.

[2] Vansant EF, van der Voort P, Vrancken KC. Characterization and Chemical Modification of the Silica Surface. Elsevier; 1995.

[3] Iler RK. The Chemistry of Silica. New York: John Wiley; 1979.

[4] Stöber W, Fink A. Controlled growth of monodisperse silica spheres in the micron size range. J. Colloid Interf. Sci. 1968;26:62−9.

[5] Chang SM, Lee M, Kim W-S. Preparation of large monodispersed spherical silica particles using seed particle growth. J. Colloid Interf. Sci. 2005;286: 536−42.

[6] Yoon SB, Kim J-Y, Kim JH, Park YJ, Yoon KR, Park S-K, Yu J-S. Synthesis of monodisperse spherical silica particles with solid core and mesoporous shell: mesopore channels perpendicular to the surface. J. Mater. Chem. 2007;17:1758−61.

[7] Iskandar F, Mikrajuddin, Okuyama K. In situ production of spherical silica particles containing self-organized mesopores. Nano Letters 2001;1: 231−4.

[8] Liteanu C, Gocan S, Bold A. Separatologie Analitica. Editura Dacia, Cluj-Napoca 1981.

[9] Cabooter D, Billen J, Terryn H, Lynen F, Sandra P, Desmet G. Detailed characterisation of the flow resistance of commercial sub-2 μm reversed-phase columns. J. Chromatogr. A 2008;1178:108−17.

[10] Gritti F, Guiochon G. Mass transfer mechanism in liquid chromatography columns packed with shell particles: Would there be an optimum shell structure? J. Chromatogr. A 2010;1217: 8167−80.

[11] Wyndham KD, O'Gara JE, Walter TH, Glose KH, Lawrence NL, Alden BA, Izzo GS, Hudalla CJ, Iraneta PC. Characterization and Evaluation of C18 HPLC stationary phases based on ethyl-bridged hybrid organic/inorganic particles. Anal. Chem. 2003;75:6781−8.

[12] Iraneta PC, Wyndham KD, McCabe DR, Walter TH. The Evolution in LC Column Performance. Waters Corp 2010.

[13] Yashima E, Sahavattanapong P, Okamoto Y. HPLC enantioseparation on cellulose tris(3,5-dimethylphe-nylcarbamate) as a chiral stationary phase: Influences of pore size of silica gel, coating amount, coating solvent, and column temperature on chiral discrimination. Chirality 1996;8:446−51.

[14] Tonhi E, Collins KE, Collins CH. High-performance liquid chromatographic stationary phases based on poly(dimethylsiloxane) immobilized on silica. J. Chromatogr. A 2005;1075:87−94.

[15] Melo LFC, Collins CH, Collins KE, Jardim ICSF. Stability of high-performance liquid chromatography columns packed with poly(methyloctylsiloxane) sorbed and radiation-immobilized onto porous silica and zirconized silica. J. Chromatogr. A 2000;869:129−35.

[16] Faria AM, Jardim ICSF, Collins KE, Collins CH. Immobilized polymeric stationary phases using metalized silica support. J. Sep. Sci. 2006;29:782−9.

[17] Silva RB, Collins CH. Chromatographic evaluation of radiation immobilized poly(methyloctylysiloxane) on titanium-grafted silica. J. Chromatogr. A 1999;845: 417−22.

[18] Gomez JE, Sandoval JE. New approaches to prepare hydride silica. Anal. Chem. 2010;82:7444−51.

[19] Pesek JJ, Matyska MT. Hydride-based silica stationary phases for HPLC: Fundamental properties and applications. J. Sep. Sci. 2005;28:1845−54.

[20] Pesek JJ, Matyska MT. Our favorite materials: Silica hydride stationary phases. J. Sep. Sci. 2009;32: 3999−4011.

[21] Sandoval JE, Pesek JJ. Synthesis and characterization of a hydride modified porous silica material as an intermediate in the preparation of chemically bonded chromatographic stationary phases. Anal. Chem. 1989;61:2067−75.

[22] Pesek JJ, Matyska MT. Hydride-based separation materials for high performance liquid chromatography and open tubular capillary electrochromatography. Chinese J. Chromatogr. 2005;23:595−608.

[23] Nawrocki J, Rigney MP, McCormick A, Carr PW. Chemistry of zirconia and its use in chromatography. J. Chromatogr. A 1993;657:229−82.

[24] Zizkovsky V, Kucera R, Klimes J, Dohnal J. Titania-based stationary phase in separation of ondansetron and its related compounds. J. Chromatogr. A 2007;1189:83−91.

[25] Pesek JJ, Matyska MT. Modified aluminas as chromatographic supports for high-performance liquid chromatography. J. Chromatogr. A 2002;952:1−11.

[26] Cabrera K. Applications of silica-based monolithic HPLC columns. J. Sep. Sci. 2004;27:843−52.

[27] Guiochon G. Monolithic columns in high-performance liquid chromatography. J. Chromatogr. A 2007;1168: 101−68.

[28] Chankvetadze B. Monolithic chiral stationary phases for liquid-phase enantioseparation techniques. J. Sep. Sci. 2010;33:305−14.

[29] Zou H, Huang X, Ye M, Luo Q. Monolithic stationary phases for liquid chromatography and capillary electrochromatography. J. Chromatogr. A 2002;954: 5−32.

[30] Gupta A, Biswas K, Basu Mallik A, Mukherjee S, Das GC. Preparation of silica monoliths via sol-gel route. Bull. Mater. Sci. 1995;18:497−501.

[31] Siouffi AM. Silica gel-based monoliths prepared by the sol-gel method: facts and figures. J. Chromatogr. A 2003;1000:801−18.

[32] Ellingsen T, Aune O, Ugelstad J, Hagen S. Mono-sized stationary phases for chromatography. J. Chromatogr. A 1990;535:147−61.

[33] Hjertén S, Liao J-L, Zhang R. High-performance liquid chromatography on continuous polymer beds. J. Chromatogr. 1989;473:273−5.

[34] Tennikova TB, Bleha M, Švec F, Almazova TV, Belenkii BG. High-performance membrane chromatography of proteins, a novel method of protein separation. J. Chromatogr. 1991;555:97−107.

[35] Wang QC, Švec F, Fréchet JMJ. Reversed-phase chromatography of small molecules and peptides on a continuous rod of macroporous poly (styrene-co-divinylbenzene). J. Chromatogr. A 1994;669:230−5.

[36] Švec F, Fréchet JMJ. Continuous rods of macroporous polymer as high-performance liquid chromatography separation media. Anal. Chem. 1992;64:820−2.

[37] Wang QC, Švec F, Fréchet JMJ. Macroporous polymeric stationary phase rod as continuous medium for reversed-phase chromatography. Anal. Chem. 1993;65:2243−8.

[38] Steinke JHG, Dunkin IR, Sherrington DC. Transparent macroporous polymer monoliths. Macromolecules 1996;29:5826−34.

[39] Vlakh EG, Tennikova TB. Preparation of methacrylate monoliths. J. Sep. Sci. 2007;30:2801−13.

[40] Nischang I, Teasdale I, Brüggemann O. Porous polymer monoliths for small molecules separations:

advancements and limitations. Anal. Bioanal. Chem. 2010;400:2289–304.

[41] Nawrocki J. Silica surface controversies, strong adsorption sites, their blockage and removal. Chromatographia 1991;31:193–205.

[42] Halasz I, Sebastian I. New stationary phase for chromatography. Angew. Chem. Internat. Ed. 1969; 8:453–4.

[43] Kirkland JJ. Method and apparatus for chromatographic separations with superficially porous glass beads having sorptively active crusts. United States Patent, 3,488,922 1970.

[44] Kirkland JJ. Superficially porous supports for chromatography. United States Patent, 3,505,785 1970.

[45] Kirkland JJ, Yates PC. Chromatographic packing with chemically bonded organic stationary phases. United States Patent, 3,795,313 1974.

[46] Verzele M, de Connink M, Dewaele C. On-column endcapping and derivatization in reverse-phase high performance liquid chromatography. Chromatographia 1984;19:443–7.

[47] Sudo Y. Optimization of end-capping of octadecyl-silylated silica gels by high-temperature silylation. J. Chromatogr. A 1997;757:21–8.

[48] Jiang Z, Fisk RP, O'Gara JE, Walter JE, Wyndham KD. U.S Patent Application No. 09/924,399 2001.

[49] Kirkland JJ. Development of some stationary phases for reversed-phase high-performance liquid chromatography. J. Chromatogr. A 2004;1060:9–21.

[50] Wirth MJ, Fatunmbi HO. Horizontal polymerization of mixed trifunctional silanes on silica. 2. Application to chromatographic silica gel. Anal. Chem. 1993;65:822–6.

[51] Courtois C, Pagès G, Caldarelli S, Delaurent C. Cholesteric bonded stationary phases for high-performance liquid chromatography: a comparative study of the chromatographic behavior. Anal. Bioanal. Chem. 2008;392:451–61.

[52] O'Sullivan GP, Scully NM, Glennon JD. Polar-embedded and polar-endcapped stationary phases for LC. Anal. Lett. 2010;43:1609–29.

[53] O'Gara JE, Walsh DP, Alden BA, Casellini P, Walter TH. Systematic study of chromatographic behavior vs. alkyl chain length for HPLC bonded phases containing an embedded carbamate group. Anal. Chem. 1999;71:2992–7.

[54] O'Gara JE, Alden BA, Walter TH, Petersen JS, Niederlaender CL, Neue U. Simple preparation of a C8 HPLC stationary phase with an internal polar functional group. Anal. Chem. 1995;67:3809–13.

[55] O'Sullivan GP, Scully NM, Glennon JD. , Polar-embedded and polar-endcapped stationary phases for LC. Anal. Lett. 2010;43:1609–29.

[56] Qui H, Liang X, Sun M, Jiang S. Development of silica-based stationary phases for high-performance liquid chromatography. Anal. Bioanal. Chem. 2011; 399:3307–22.

[57] Wang X, Cheng S, Chan JCC. Propylsulfonic acid functionalized mesoporous silica, synthesized in situ oxidation of thiol groups under template free conditions. J. Phys. Chem. C 2007;111:2156–64.

[58] Hasegawa I. Co-hydrolysis products of tetraethoxysilane (TEOS) and methyltriethoxysilane in the presence of tetramethylammonium ions. J. Sol-Gel Sci. Technol. 1993;1:57–63.

[59] Zhu G, Zhang L, Yuan H, Liang Z, Zhang W, Zhang Y. Recent development of monolithic materials as matrices in microcolumn separation systems. J. Sep. Sci. 2007;30:792–803.

[60] Pesek JJ, Patyska MT, Oliva M, Evanchic M. Synthesis and characterization of bonded phases made via hydrosilation of alkynes on silica hydride surfaces. J. Chromatogr. A 1998;818:145–54.

[61] Pesek JJ, Patyska MT, Muley S. Synthesis and characterization of a new type of chemically bonded liquid crystal stationary phase for HPLC. Chromatographia 2000;52:445–50.

[62] Pesek JJ, Matyska MT, Sharma A. Use of hydride-based separation materials for organic normal phase chromatography. J. Liq. Chromatogr. & Rel. Technologies 2008;31:134–47.

[63] Matyska MT, Pesek JJ, Shetty G. Type C amino columns for affinity and aqueous normal phase chromatography: synthesis and HPLC evaluation. J. Liq. Chromatogr. & Rel. Technol. 2009;33:1–26.

[64] Pesek JJ, Matyska MT, Hemphala H. HPLC evaluation of mono-ol, butylphenyl, and perfluorinated columns prepared via olefin hydrosilation on a silica hydride intermediate. Chromatographia 1996;43:10–6.

[65] Slater M, Snaulo M, Svec F, Fréchet JM. "Clic chemistry" in the preparation of porous polymer-based particulate stationary phases for HPLC separation of peptides and proteins. Anal. Chem. 2006;78:4969–75.

[66] Tanaka N, Hashizume K, Araki M. Comparison of polymer-based stationary phases with silica-based stationary phases in reversed-phase liquid chromatography: Selective binding of rigid, compact molecules by alkylated polymer gels. J. Chromatogr. A 1987;400:33–45.

[67] Platonova GA, Tennikova TB. Chromatographic investigation of macromolecular affinity interactions. J. Chromatogr. A 2005;1065:75–81.

[68] Kanazawa H. Thermally responsive chromatographic materials using functional polymers. J. Sep. Sci. 2007;30:1646–56.

[69] Gritti F, Guiochon G. The current revolution in column technology; How it began, where is it going? J. Chromatogr. A 2012;1228:2−19.

[70] Poole CF, Poole SK. Chromatography Today. 2nd ed. Amsterdam: Elsevier; 1993.

[71] Dong MW. Modern HPLC for Practicing Scientists. Hoboken, NJ: Wiley-Interscience; 2006.

[72] Bidlingmeyer B, Chan CC, Fastino P, Henry R, Koerner P, Maule AT, Marques MRC, Neue U, Ng L, Pappa H, Sander L, Santasania C, Snyder L, Woznyak T. HPLC column classification. Pharmacopeial Forum 2005;31:637−45.

[73] Marques M, editor. USP Chromatographic Columns. Rockville, MD: U. S. Pharmacopeia; 2009−2010.

[74] http://www.chemaxon.com

[75] Ye C, Terfloth G, Li Y, Kord A. A systematic stability evaluation of analytical RP-HPLC columns,. J. Pharm. Biomed. Anal. 2009;50:426−31.

[76] Engelhardt H, Blay C, Saar J. Reversed phase chromatography−the mystery of surface silanols. Chromatographia 2005;62:s19−29.

[77] Melander WR, Horvath Cs. Reversed-Phase Chromatography. In: Horvath Cs, editor. High-Performance Liquid Chromatography, Vol. 2. New York: Academic Press; 1980.

[78] Neue UD, Serowik E, Iraneta P, Alden BA, Walter TH. Universal procedure for the assessment of the reproducibility and the classification of silica-based reversed phase packings. I. Assessment of the reproducibility of reversed-phase packings. J. Chromatogr. A 1999;849:87−100.

[79] Rimmer CA, Sander LC, Wise SA. Selectivity of long chain stationary phases in reversed phase liquid chromatography. Anal. Bioanal. Chem. 2005; 382:698−707.

[80] McCalley DV. The challenges of the analysis of basic compounds by high performance liquid chromatography: Some possible approaches for improved separations. J. Chromatogr. A 2010;1217:858−80.

[81] Jerkovich AD, Mellors JS, Jorgenson JW. Recent developments in LC column technology. LCGC Europe 2003;16:20−3.

[82] Martin M, Guiochon G. Effect of high pressure in liquid chromatography. J. Chromatogr. A 2005;1090: 16−38.

[83] Pellett J, Lukulay P, Mao Y, Bowen W, Reed R, Ma M, Munger RC, Dolan JW, Wrisley L, Medwid K, Toltl NP, Chan CC, Skibic M, Biswas K, Wells KA, Snyder LR. Orthogonal separations for reversed-phase liquid chromatography. J. Chromatogr. A 2006;1101:122−35.

[84] Hetem MJJ, De Haan JW, Claessens HA, Van de Ven LJM, Cramers CA, Kinkel JN. Influence of alkyl

[85] Vasant EF, Van der Voort P, Vrancken KC. Characterization and Chemical Modification of the Silica Surface. Amsterdam: Elsevier; 1995.

[86] Rippel G, Alattyani E, Szepesy L. Characterization of stationary phases used in reversed-phase and hydrophobic interaction chromatography. J. Chromatogr. A 1994;668:301−12.

[87] Doyle CA, Dorsey JG. Reversed-phase HPLC preparation and characterization of reversed-phase stationary phases. In: Katz E, Eksteen R, Schoenmakers P, Miller N, editors. Handbook of HPLC. New York: Marcel Dekker; 1998. p. 293−323.

[88] Brunauer S, Emmett PH, Teller E. Adsorption of gases in multimolecular layers. J. Am. Chem. Soc. 1938;60:309−19.

[89] Lowell S, Shields JE, Thomas MA, Thomas M. Characterization of Porous Solids and Powders: Surface Area, Pore Size, and Density. The Netherlands: Kluwer Academic Publishers; 2004.

[90] Hagel L, Östberg M, Andersson T. Apparent pore size distributions of chromatography media. J. Chromatogr. A 1996;743:33−42.

[91] Guan-Sajonz H, Guiochon G. Study of physico-chemical properties of some packing materials. I Measurements of the external prosity of packed columns by inverse size exclusion chromatography. J. Chromatogr. A 1996;73:27−40.

[92] Guan-Sajonz H, Guiochon G, Davis E, Gulakowski K, Smith DW. Study of the physico-chemical properties of some packing materials: III Pore size and surface area distribution. J. Chromatogr. A 1997;773:33−51.

[93] Al-Bokari M, Cherrak D, Guiochon G. Determination of the porosities of monolithic columns by inverse size-exclusion chromatography. J. Chromatogr. A 2002;975:275−84.

[94] Buszewski B, Bocian S, Rychlicki G. Investigation of silanol activity on the modified silica surfaces using microcalorimetric measurements. J. Sep. Sci. 2011; 34:773−9.

[95] Jal PK, Patel S, Mishra BK. Chemical modification of silica surface by immobilization of functional groups for extractive concentration of metal ions. Talanta 2004;62:1005−28.

[96] Albert K, Bayer E. Characterization of bonded phases by solid-state NMR spectroscopy. J. Chromatogr. 1991;544:345−70.

[97] Dietrich B, Holtin K, Bayer M, Friebolin V, Kuhnle M, Albert K. Synthesis, characterization, and high-performance liquid chromatographic

evaluation of C14 stationary phases containing branched and unbranched alkyl groups. Anal. Bioanal. Chem. 2008;391:2627−33.

[98] Kirkland JJ, Glajch JL, Farlee RD. Synthesis and characterization of highly stable bonded phases for high-performance liquid chromatography column packages. Anal. Chem. 1989;61:2−11.

[99] Kirkland JJ, van Straten MA, Claessens HA. Reversephase high-performance liquid chromatography of basic compounds at pH 11 with silica-based column packings. J. Chromatogr. A 1998;797:111−20.

[100] Liu X, Bordunov A, Tracey M, Slingsby R, Avdolovic N, Pohl C. Development of polar embedded stationary phase with unique properties. J. Chromatogr. A 2006;1119:120−7.

[101] Liu X, Bordunov AV, Pohl CA. Preparation and evaluation of a hydrolytically stable amide-embedded stationary phase. J. Chromatogr. A 2006;1119:128−34.

[102] Ripple G, Alattyani E, Szepesy L. Characterization of stationary phases used in reversed-phase and hydrophobic interaction chromatography. J. Chromatogr. A 1994;668:301−12.

[103] Comparison Guide to C18 Reversed Phase HPLC Columns. MAC-MOD Analytical, Chadds Ford 2008.

[104] Snyder LR, Maule A, Heebsch A, Cuellar R, Paulson S, Carrano J, Wrisley L, Chan CC, Pearson N, Dolan JW, Gilroy J. A fast, convenient and rugged procedure for characterizing the selectivity of alkyl-silica columns. J. Chromatogr. A 2004;1057:49−57.

[105] Carr PW, Dolan JW, Neue UD, Snyder LR. Contribution to reverse phase column selectivity. I. Steric interaction. J. Chromatogr. A 2011;1218:1724−42.

[106] Galushko SV. The calculation of retention and selectivity in reversed-phase liquid chromatography II. Methanol-water eluents,. Chromatographia 1993; 36:39−42.

[107] Neue UD, Alden BA, Walter TH. Universal procedure for the assessment of the reproducibility and the classification of silica-based reverse phase packings. II. Classification of reverse phase packings. J. Chromatogr. A 1999;849:101−16.

[108] Layne J. Characterization and comparison of the chromatographic performance of conventional polar-embedded and polar-endcapped reverse phase liquid chromatography stationary phases. J. Chromatogr. A 2002;957:149−64.

[109] Kristulović AM, Colin H, Tchapla A, Guiochon G. Effects of the bonded alkyl chain length on methylene selectivity in reversed phase chromatography. Chromatographia 1983;17:228−30.

[110] Sándi Á, Szepesy L. Evaluation and modulation of selectivity in reversed-phase high-performance liquid chromatography. J. Chromatogr. A 1999;845:113−31.

[111] Sentell KB, Dorsey JG. Retention mechanisms in reversed-phase chromatography: Stationary phase bonding density and solute selectivity. J. Chromatogr 1989;461:193−207.

[112] Tchapla A, Héron S, Lesellier E, Colin H. General view of molecular interaction mechanisms in reversed-phase liquid chromatography. J. Chromatogr. A 1993;656:81−112.

[113] Snyder LR, Kirkland JJ, Dolan JW. Introduction to Modern Liquid Chromatography. 3rd ed. Hoboken, NJ: John Wiley; 2010.

[114] Marchand DH, Croes K, Dolan JW, Snyder LR. Columns selectivity in reversed-phase liquid chromatography VII. Cyanopropyl columns. J. Chromatogr. A 2005;1062:57−64.

[115] David V, Medvedovici A. Structure−retention correlation in liquid chromatography for pharmaceutical applications. J. Liq. Chromatogr. Rel. Technol. 2007;30:761−89.

[116] Dorsey JG, Cooper WT. Retention mechanisms of bonded-phase liquid chromatography. Anal. Chem. 1994;66. 857A-867A.

[117] Engelhardt H, Lobert T. Chromatographic determination of metallic impurities in reversed-phase HPLC columns. Anal. Chem. 1999;71:1885−92.

[118] Gilroy JL, Dolan JW, Snyder LR. Column selectivity in reversed-phase liquid chromatography IV. Type-B alkyl-silica columns. J. Chromatogr. A 2003;1000:757−78.

[119] Zhu PL, Dolan JW, Snyder LR, Djordjevic NM, Hill DW, . Lin J-T, Sander LC, Van Heukelem L. Combined use of temperature and solvent strength in reversed-phase gradient elution IV. Selectivity for neutral (non-ionized) samples as a function of sample type and other separation conditions. J. Chromatogr. A 1996;756:63−72.

[120] Dolan JW, Snyder LR, Djordjevic NM, Hill DW, Saunders DL, Van Heukelem L, Waeghe TJ. Simultaneous variation of temperature and gradient steepness for reversed-phase high-performance liquid chromatography method development: I. Application to 14 different samples using computer simulation. J. Chromatogr. A 1998;803:1−31.

[121] Dolan JW, Snyder LR, Djordjevic NM, Hill DW, Waeghe TJ. Reversed-phase liquid chromatographic separation of complex samples by optimizing temperature and gradient time: I. Peak capacity limitations. J. Chromatogr. A 1999;857: 1−20.

[122] Sander LC, Wise SA. A new standard reference material for column evaluation in reversed-phase liquid chromatography. J. Sep. Sci. 2003;26:283−94.

[123] http://www.usp.org/app/USPNF/columns.html

[124] Gonzalez A, Foster KL, Hanrahan G. Method development and validation for optimized separation of benzo[a]pyrene—quinone isomers using liquid chromatography—mass spectrometry and chemometric response surface methodology. J. Chromatogr. A 2007;1167:135—42.

[125] Wilson NS, Nelson MD, Dolan JW, Snyder LR, Carr PW. Column selectivity in reversed phase liquid chromatography: II. Effect of a change in conditions. J. Chromatogr. A 2002;961:195—215.

[126] Marchand DH, Snyder LR, Dolan JW. Characterization and applications of reversed-phase column selectivity based on the hydrophobic-subtraction model. J. Chromatogr. A 2008;1191:2—20.

[127] Marchand DH, Croes K, Dolan JW, Snyder LR, Henry RA, Kallury KMR, Waite S, Carr PW. Column selectivity in reversed-phase liquid chromatography. VIII. Phenylalkyl and fluoro-substituted columns. J. Chromatogr. A 2005;1062: 65—78.

[128] Valk K, Bevan C, Reynolds D. Chromatographic hydrophobicity index by fast-gradient RP-HPLC: A high-throughput alternative to log P log D. Anal. Chem. 1997;69:2011—9.

[128a] O'Sullivan GP, Scully NM, Glennon JD. Polar-embedded and polar-endcapped stationary phases for LC. Anal. Lett. 2010;43:1609—29.

[129] Engelhardt HR, Gruner R, Scherer M. The polarity selectivities of non-polar reversed phases. Chromatographia 2001;53:S154—61.

[130] Jandera P, Bunčeková S, Halama M, Novotná K, Nepraš M. Naphthalene sulphonic acids — new test compounds for characterization of the columns for reversed-phase chromatography. J. Chromatogr. A 2004;1059:61—72.

[131] Gilpin RK, Jaroniec M, Lin S. Dependence of the methylene selectivity on the composition of hydro-organic eluents for reversed-phase liquid chromatographic systems with alkyl bonded phases. Chromatographia 1990;30:393—9.

[132] Engelhardt H, Aranglo M, Lobert T. A chromatographic test procedure for reversed-phase HPLC column evaluation. LCGC 1997;17:856—65.

[133] Engelhardt H, Jungheim M. Comparison and characterization of reversed phases. Chromatographia 1990;29:59—68.

[134] Kimata K, Iwaguchi K, Onishi S, Jinno K, Eksteen R, Hosoya K, Araki M, Tanaka N. Chromatographic characterization of silica C18 packing materials. Correlation between a preparation pethod and retention behavior of stationary phase. J. Chromatogr. Sci. 1989;27:721—8.

[135] Walters MJ. Classification of octadecyl-bonded liquid chromatography columns. J. Assoc. Off. Anal. Chem. 1987;70:465—9.

[136] Moliková M, Jandera P. Characterization of stationary phases for reversed-phase chromatography. J. Sep. Sci. 2010;33:453—63.

[137] Kowalska S, Kupczyńska K, Buszewski B. Some remarks on characterization and application of stationary phases for RP-HPLC determination of biological important compounds. Biomed. Chromatogr. 2006;20:4—22.

[138] Scott RPW, Kucera P. Examination of five commercially available liquid chromatographic reversed phases (including the nature of the solute-solvent-stationary phase interactions associated with them). J. Chromatogr. A 1977;142:213—32.

[139] Rustamov I, Farcas T, Ahmed F, Chan F, LoBrutto R, McNair HM, Kazakevich YV. Geometry of chemically modified silica. J. Chromatogr. A 2001;913: 49—63.

[140] Pursch M, Strohschein S, Handel H, Albert K. Temperature dependent behavior of C30 interphases. A solid-state NMR and LC-NMR study. Anal. Chem. 1996;68:386—93.

[141] Kazakevich YV, LoBrutto R, Chan F, Patel T. Interpretation of the excess adsorption isotherms of organic eluent components on the surface of reversed-phase adsorbents: Effect on the analyte retention. J. Chromatogr. A 2001;913:75—87.

[142] Przybyciel M, Major RE. Phase collapse in reversed-phase LC. LCGC Europe 2002:2—5. October.

[143] Majors RE, Przybyciel M. Columns for reverse-phase LC separations in highly aqueous mobile phases. LCGC North America 2002;20:584—93.

[144] http://www.phenomenex.com

[145] http://www.waters.com

[146] Majors RE, Przybyciel M. Columns for reversed-phase LC separations in highly aqueous mobile phases. LCGC Europe December 2002:2—7.

[147] Sora I, Galaon T, Udrescu S, Negru J, David V, Medvedovici A. Fast RPLC-UV method on short sub-two microns particles packed column for the assay of tenoxicam in plasma samples. J. Pharm. Biomed. Anal. 2007;43:1437—43.

[148] Scomburg G, Deege A, Köhler J, Bien-Vogelsang U. Immobilization of stationary liquids in reversed- and normal-phase liquid chromatography: Production and testing of materials for bonded-phase chromatography. J. Chromatogr. 1983;282:27—39.

[149] Sándi A, Szepesy L. Characterization of various reverses-phase columns using the linear free energy relationship: I. Evaluation based on retention factors. J. Chromatogr. A 1998;818:1—17.

[150] O'Gara JE, Walsh DP, Phoebe Jr CH, Alden BA, Bouvier ESP, Iraneta PC, Capparella M, Walter TH. Embedded-polar group bonded phases for high performance liquid chromatography. LCGC 2001; 19:632–42.

[151] Kirkland JJ, Adams JB, van Straten MA, Claessens HA. Bidentate silane stationary phases for reversed-phase high-performance liquid chromatography. Anal. Chem. 1998;70:4344–52.

[152] Gritti F, Leonardis I, Shock D, Stevenson P, Shalliker A, Giochon G. Performance of columns packed with the new shell particles, Kinetex-C18. J. Chromatogr. A 2010;1217:1589–603.

[153] Gritti F, Guiochon G. Performance of columns packed with the new shell Kinetex-C18 particles in gradient elution chromatography. J. Chromatogr. A 2010;1217:1604–15.

[154] Przybyciel M. Novel phases for HPLC separations, LCGC. LC Column Technology Supplement 2006: 49–52. April.

[155] Reta M, Carr PW, Sadek PC, Rutan SC. Comparative study of hydrocarbon, fluorocarbon, and aromatic bonded RP-HPLC stationary phases by linear solvation energy relationship. Anal. Chem. 1999;71:3484–96.

[156] Lubda D, Cabrera K, Kraas W, Schaefer C, Cunningham D. New developments in the application of monolithic HPLC columns. LCGC Europe 2001:2–5. Dec.

[157] Medvedovici A, Sora DI, Ionescu S, Hillebrand M, David V. Characterization of a new norfloxacin metabolite monitored during a bioequivalence study by means of mass-spectrometry and quantum computation. Biomed. Chromatogr. 2008;22:1100–7.

[158] Vallano PT, Remcho VT. Affinity screening by packed capillary high-performance liquid chromatography using molecular imprinted sorbents I. Demonstration of feasibility. J. Chromatogr. A 2000; 888:23–34.

[159] Knox JH, Ross P. In: Grushka E, Brown PR, editors. Carbon-based packing materials for liquid chromatography, structure, performance and retention mechanism, Advances in Chromatography, Vol 37. Boca Raton: CRC Press; 1997. p. 74–119.

[160] Pereira L. Porous graphitic carbon as a stationary phase in HPLC: theory and applications. J. Liq. Chromatogr. Rel. Technol. 2008;31:1687–731.

[161] Mockel H, Braedikow A, Melzer H, Aced GA. A comparison of the retention of homologous series and other test solutes on an ODS column and a Hypercarb carbon column. J. Liq. Chromatogr. 1991;14:2477–98.

[162] Hanai T. Separation of polar compounds using carbon columns. J. Chromatogr. A 2003;989:183–96.

[163] Hanai T. Analysis of the mechanism of retention on graphitic carbon by a computational chemical method. J. Chromatogr. A 2004;1030:13–6.

[164] Xia YQ, Jemal M, Zheng N, Shen X. Utility of porous graphitic carbon stationary phase in quantitative liquid chromatography/tandem mass spectrometry bioanalysis: quantitation of diastereomers in plasma. Rapid Commun. Mass Spectrom. 2006;20:1831–7.

[165] Teutenberg T, Tuerk J, Holzhauser M, Giegold S. Temperature stability of reversed phase and normal phase stationary phases under aqueous conditions. J. Sep. Sci. 2007;30:1101–14.

[166] Nagashima H, Okamoto T. Determination of inorganic anions by ion chromatography using a graphitized carbon column dynamically coated with cetyltrimethylammonium ions. J. Chromatogr. A 1999;855:261–6.

[167] Poole SK, Poole CF. Retention of neutral organic compounds from solution on carbon adsorbents. Anal. Commun. 1997;34:247–51.

[168] Kaliszan R. Quantitative Structure-Chromatographic Retention Relationship. New York: JohnWiley; 1987.

[169] Jandera P. Stationary and mobile phases in hydrophilic interaction chromatography: A review. Anal. Chim. Acta 2011;692:1–25.

[170] Hemström P, Irgum K. Hydrophilic interaction chromatography. J. Sep. Sci. 2006;29:1784–821.

[171] Jiang W, Fischer G, Girmay Y, Irgun K. Zwitterionic stationary phase with covalently bonded phosphorylcholine type polymer grafts and its applicability to separation of peptides in the hydrophilic interaction liquid chromatography mode. J. Chromatogr. A 2006;1127:82–91.

[172] Ikegami T, Tomomatsu K, Takubo H, Horie K, Tanaka N. Separation efficiencies in hydrophilic interaction chromatography. J. Chromatogr. A 2008; 1184:474–503.

[173] Liu S-M, Xu L . Wu C-T, Feng Y-Q. , Preparation and characterization of perhydroxylcucurbit[6]uril bonded silica stationary phase for hydrophilic-interaction chromatography. Talanta 2004;64:929–34.

[174] Special issue: HILIC and mixed mode. J. Sep. Sci. 2010;33:679–997.

[175] Chirita R-I, West C, Finaru, C A-L. Elfakir, Approach to hydrophobic interaction chromatography column selection: Application to neurotransmitters analysis. J. Chromatogr. A 2010;1217:3091–104.

[176] Lämmerhofer M, Richter M, Wu J, Nogueira R, Bicker W, Linder W. Mixed-mode ion-exchange and their comparative chromatographic characterization in reversed-phase and hydrophilic interaction chromatography elution modes. J. Sep. Sci. 2008;31: 2572–88.

[177] Hao Z, Xiao B, Weng N. Impact of column temperature and mobile phase components on selectivity of hydrophilic interaction chromatography. J. Sep. Sci. 2008;31:1449–64.

[178] Dorpe SV, Vergote V, Pezeshki A, Burvenich C, Peremans K, de Spiegeleer B. Hydrophilic interaction LC of peptides: Column comparison and clustering. J. Sep. Sci. 2010;33:728–39.

[179] Kawachi Y, Ikegami T, Takubo H, Ikegami Y, Miyamoto M, Tanaka N. Chromatographic characterization of HILIC stationary phases: hydrophilicity, charge effects, structural selectivity, and separation efficiency. J. Chromatogr. A 2011;1219:5903–19.

[180] Guo Y, Gaiki S. Retention and selectivity of stationary phases of hydrophilic interaction chromatography (HILIC). J. Chromatogr. A 2011;1218: 5920–38.

[181] Marrubini G, Mendoza BEC, Massolini G. Separation of purine and pyrimidine bases and nucleosides by hydrophilic interaction chromatography. J. Sep. Sci. 2010;33:803–16.

[182] McCalley DV, Neue UD. Estimation of the extent of the water-rich layer associated with the silica surface in hydrophilic interaction chromatography. J. Chromatogr. A 2008;1192:225–9.

[183] Gin J, Guo Z, Zhang F, Xue X, Jin Y, Liang X. Study of the retention equation in hydrophilic interaction chromatography. Talanta 2008;76:522–7.

[184] Lämmerhofer M. HILIC and mixed-mode chromatography: the rising stars in separation science. J. Sep. Sci. 2010;33:679–80.

[185] Wang PG, He W, editors. Hydrophilic Interaction Chromatography (HILIC) and Advanced Applications. Boca Raton: CRC Press; 2011.

[186] Yu L, Li X, Guo Z, Zhang X, Liang X. Hydrophilic interaction chromatography based enrichment of glycopeptides by using click maltose: a matrix with high selectivity and glycosylation heterogeneity coverage. Chemistry 2009;15:12618–26.

[187] Alpert AJ. Cation-exchange high performance liquid chromatography of proteins on poly(aspartic acid) –silica. J. Chromatogr. A 1983;266:23–37.

[188] Xu M, Peterson DS, Rohr T, Svec F, Fréchet JM. Polar polymeric stationary phases for normal phase HPLC based on monodisperse macroporous poly(2,3-dihydroxypropyl methacrylate-co-ethylene dimethacrylate) beads. Anal. Chem. 2003;75:1011–21.

[189] Viklund C, Sjögren A, Irgum K. Chromatographic interactions between proteins and sulfoalkylbetaine-based zwitterionic copolymers in fully aqueous low salt buffers. Anal. Chem. 2001;73:444–52.

[190] Fritz JS, Gjerde DT. Ion Chromatography. Weinheim: Wiley-VCH; 2009.

[191] Weiss J. Ion Chromatography. 2nd ed. Weinheim: VCH; 1995.

[192] Auler LMLA, Silva CR, Collins KE, Collins CH. New stationary phase for anion exchange chromatography. J. Chromatogr. A 2005;1073:147–53.

[193] Buszewski B, Jaćkowska M, Bocian S, Kosobucki P, Gawdzik B. Functionalized polymeric stationary phases for ion chromatography. J. Sep. Sci. 2011;34:1–8.

[194] Savina IN, Galaev IY, Mattiasson B. Ion-exchange macroporous hydrophilic gel monolith with grafted polymer brushes. J. Molec. Recognit 2006;19: 313–21.

[195] Nordborg A, Hilder EF. Recent advances in polymer monoliths for ion exchange chromatography. Anal. Bioanal. Chem. 2009;394:71–84.

[196] Haddad PR, Jackson PE. Ion Chromatography. Principles and Applications. Amsterdam: Elsevier; 1990.

[197] http://www.dionex.com

[198] http://diaion.com

[199] Ion Exchange Chromatography. Montgomeryville: Tosoh Bioscience LLC; 2009.

[200] Beesley TE, W Scott RP. Chiral Chromatography. New York: John Wiley; 1998.

[201] Aboul-Enein HY, Ali I. Chiral Separations by Liquid Chromatography: Theory and Applications, Chromatographic Science, vol. 90. Boca Raton, FL: CRC Press; 2003.

[202] Ahuja S. Chiral Separations by Chromatography. Washington, DC: ACS Publication; 2000.

[203] Kazakevich YV, LoBrutto R. HPLC for Pharmaceutical Scientists. New York: Wiley-Interscience; 2007.

[204] Gübitz G, Schmid MG, editors. Chiral Separations: Methods and Protocols. Totowa, NJ: Humana Press; 2004.

[205] Pirkle WH, House DW. Chiral high-performance liquid chromatographic stationary phases. 1. Separation of the enantiomers of sulfoxides, amines, amino acids, alcohols, hydroxy acids, lactones, and mercaptans. J. Org. Chem. 1979;44:1957–60.

[206] Pirkle WH, Finn JM. Chiral HPLC stationary phases 3. General resolution of arylalkylcarbinols. J. Org. Chem. 1981;46:2935–8.

[207] Kacprzak KM, Lindner W. Novel Pirkle-type quinine 3,5-dinitrophenylcarbamate chiral stationary phase implementing click chemistry. J. Sep. Sci. 2011; 34:1–6.

[208] Hyan MH. Enantiomer separation by chiral crown ether stationary phases. In: Subramanian G, editor. Chiral Separation Techniques. 3rd ed. Weinheim: Wiley-VCH; 2007. p. 275.

[209] Hyun MH, Han SC, Lipshutz BH, Shin Y-S, Welch CJ. New chiral crown ether stationary phase

for the liquid chromatographic resolution of α-amino acid enantiomers. J. Chromatogr. A 2001;910:359—65.

[210] Mokhtari B, Pourabdollah K, Dalali N. Applications of nano-baskets of calixarenes in chromatography. Chromatographia 2011;73:829—47.

[211] Cavazzini A, Pasti L, Massi A, Marchetti N, Dondi F. Recent applications in chiral high performance liquid chromatography: A review. Anal. Chim. Acta 2011;706:205—22.

[212] Lämmerhofer M, Franco P, Lindner W. Quinine carbamate chiral stationary phases: systematic optimization of steric selector-select and binding increments and enantioselectivity by quantitative structure-enantioselectivity relationship studies. J. Sep. Sci. 2006;29:1486—96.

[213] Lämmerhofer M. Chiral recognition by enantioselective liquid chromatography: Mechanism and modern chiral stationary phases. J. Chromatogr. A 2010;1217:814—56.

[214] Armstrong DV, Zhang B. Chiral stationary phases for HPLC. Anal. Chem. 2001;73:557A—561A.

[215] Kato M, Fukushima T, Santa T, Nakashima S, Nishioka R, Imai K. Preparation and evaluation of new Pirkle type chiral stationary phases with long alkyl chains for the separation of amino acid enantiomers derivatized with NBD-F. Analyst 1998;123:2877—82.

[216] Huang W, Xing Y, Yu Y, Shang S, Dai J. Enhanced washing durability of hydrophobic coating on cellulose fabric using polycarboxylic acids. Appl. Surf. Sci. 2011;257:4443—8.

[217] Ling F, Brahmachary E, Xu M, Svec F, Fréchet JMJ. Polymer-bound cellulose phenylcarbamate derivatives as chiral stationary phases for enantioselectve HPLC. J. Sep. Sci. 2003;26:1337—46.

[218] Ward TJ, Armstrong DW. Improved cyclodextrin chiral phases: a comparison and review. J. Liq. Chromatogr. 1986;9:407—14.

[219] Hyun MH. Development and application of crown ether-based HPLC chiral stationary phases. Bull. Korean Chem. Soc. 2005;26:1153—63.

[220] Armstrong DW, Tang Y, Chen S, Zhou Y, Bagwill C, Chen JR. Macrocyclic antibiotics as a new class of chiral selectors for liquid chromatography. Anal. Chem. 1994;66:1473—84.

[221] D'Acquarica I, Gasparrini F, Misiti D, Pierini M, Villani C. HPLC Chiral Phases Containing Macrocyclic Antibiotics: Practical Aspects and Recognition Mechanism. In: Grushka E, Grinberg N, editors. Advances in Chromatography, vol. 46. Boca Raton: CRC press; 2007.

[222] Millot MC, Taleb NL, Sebille B. Binding of human serum albumin to silica particles by means of polymers: a liquid chromatographic study of the selectivity of resulting chiral stationary phases. J. Chromatogr. B 2002;768:157—66.

[223] Jacobson SC, Guiochon G. Enantiomeric separations using bovine serum albumin immobilized on ion-exchange stationary phases. Anal. Chem. 1992; 64:1496—8.

[224] Davankov VA. Ligand exchange chromatography. In: Wilson ID, Adlard ER, Cooke M, Poole CF, editors. Encyclopedia of Separation Science, vol. 5. Amsterdam: Academic Press (Elsevier); 2000. p. 2369—80.

[225] Davankov VA. Enantioselective ligand exchange in modern separation techniques. J. Chromatogr. A 2003;1000:891—915.

[226] Payagala T, Wanigasekara E, Armstrong DW. Synthesis and chromatographic evaluation of new polymeric chiral stationary phases based on three (1S,2S)-(-)-1,2-diphenylethylenediamine derivatives in HPLC and SFC. Anal. Bioanal. Chem. 2011;399:2445—61.

[227] Lee K-P, Choi S-H, Kim S-Y, Kim T-H, Ryoo JJ, Ohta K, Jin J-Y, Takeuchi T, Fujimoto C. , Comparison of monomeric and polymeric chiral stationary phases. J. Chromatogr. A 2003;987:111—8.

[228] Hosoya K, Tanaka N. Development of Uniform Sized, Molecular-Imprinted Stationary Phases for HPLC, in R. A. Bartsch, M. Maeda. Molecular and Ionic Recognition with Imprinted Polymers 1998;vol. 703. ACS Ser.

[229] Hung C-Y, Huang H-H, Hwang C-C. Chiral separation of mandelic acid by HPLC using molecular imprinted polymers. Eclet. Quim. 2005; 30:67—73.

[230] Wu C-S, editor. Handbook of Size Exclusion Chromatography. New York: Marcel Dekker; 1995.

[231] Barth HG, Boyes BE, Jackson C. Size exclusion chromatography and related techniques. Anal. Chem. 1998;70. 251R-278R.

[232] Dubin PL, Tacklenburg MM. Size exclusion chromatography of strong polyelectrolytes on porous glass columns. Anal. Chem. 1985;57:275—9.

[233] Mori S, Kato M. High performance aqueous size-exclusion chromatography with diol-bonded porous glass packing material. J. Chromatogr. A 1986;363:217—22.

[234] Huck CW, Bonn GK. Poly(styrene-divinylbenzene) based media for liquid chromatography. Chem. Eng. Technol. 2005;28:1457—72.

[235] Ellingsen T, Aune O, Ugelstad J, Hagen S. J. Chromatogr. 1990;535:147—61.

[236] Cheng CM, Micale FJ, Vanderhoff JW, El-Aasser MS. Pore structural studies of monodisperse porous polymer particles. J. Colloid and Interface Sci. 1992; 150:549—58.

[237] Caldwell JD, Cooke BS, Greer MK. High performance liquid chromatography-size exclusion chromatography for rapid analysis of total polar compounds in used frying oils. J. Am. Oil Chem. Soc. 2011;88:1669−74.

[238] MCI Gel, Technical Information; 2008-2009.

[239] http://technolab.no/pdf/shodex_catalogue_2008_2010_v2.pdf

[240] Clonis YD. Affinity chromatography matures as bioinformatic and combinatorial tools develop. J. Chromatogr. A 2006;1101:1−24.

[241] Sproß J, Sinz A. Monolithic media for applications in affinity chromatography. J. Sep. Sci. 2011;34:1−16.

[242] Hage DS, Thomas DH, Beck MS. Theory of a sequential addition competitive binding immunoassay based on high-performance immunoaffinity chromatography. Anal. Chem. 1993;65:1622−30.

[243] Affinity Chromatography, Principles and Methods. Uppsala: Amersham Pharmacia; 1981.

[244] Tennikova T, Strancar A. Short high-throughput monolithic layers for bioaffinity processing. LabPlus International February/March 2002:1−3.

[245] Bo C-M, Gong B-L, Hu W-Z. , Preparation of immobilized metal affinity chromatographic packings based on monodisperse hydrophilic non-porous beads and their application. Chinese J. Chem. 2008; 26:886−92.

[246] Ivanov AE, Kozlov LV, Shojbonov BB, Zubov VP, Antonov VK. Inorganic supports coated with N-substituted polyacrylamides: Application to biospecific chromatography of proteins. Biomed. Chromatogr. 1991;5:90−3.

[247] Rhemrev-Boom MM, Yates M, Rudolph M, Raedts M. Immunoaffinity chromatography: a versatile tool for fast and selective purification, concentration, isolation and analysis. J. Pharm. Biomed. Anal. 2001;24:825−33.

[248] Sasaki H, Hayashi A, Kitagaki-Ogawa H, Matsumoto I, Seno N. Improved method for the immobilization of heparin. J. Chromatogr. 1987; 400:123−32.

[249] Hermanson GT, Mallia AK, Smith PK. Immobilized Affinity Ligand Techniques. New York: Academic Press; 1992.

[250] O'Shannessy DJ, Wilchek M. Immobilization of glycoconjugates by their oligosaccharides: Use of hydrazido-derivatized matrices. Anal. Biochem. 1990;191:1−8.

[251] Domen PL, Nevens JR, Mallia AK, Hermanson GT, Klenk DC. Site-directed immobilization of proteins. J. Chromatogr. 1990;510:293−302.

[252] Jonker N, Kool J, Irth H, Niessen WMA. Recent developments in protein-ligand affinity mass spectrometry. Anal. Bioanal. Chem. 2011;399:2669−81.

[253] Winzor DJ. Determination of binding constants by affinity chromatography. J. Chromatogr. A 2004; 1037:351−67.

[254] Tetala KKR, van Beek TA. Bioaffinity chromatography on monolithic supports. J. Sep. Sci. 2010; 33:422−38.

7.1. CHARACTERIZATION OF LIQUIDS AS SOLVENTS

General Comments

Various liquids are used in HPLC as mobile phases. In addition to this important role, liquids are also necessary for dissolving the sample as it is injected in the HPLC system, as well as for other processes involved in sample preparation prior to HPLC analysis. For these reasons, the properties of liquids as solvents are of high interest in HPLC separations. Solvent characterization has been reported in the literature based on a number of parameters such as solubility parameter δ [1], polarity parameter P' [2], solvatochromic parameters [3], as well as other such solvent descriptors [4, 5]. Additional properties such as surface tension γ', dipole moment m, polarizability α, and other general molecular properties are also important for HPLC separations and characterize the solvent properties of liquids. These properties are further discussed in this chapter in connection with the selection of different solvents as HPLC mobile phases. Detailed description of various parameters characterizing the solvents is also available in the literature [2]. In HPLC, mixtures of solvents are frequently used. Particular properties of solvent mixtures and the way they result from the properties of pure solvents that are mixed are also presented. Besides organic solvents, aqueous solutions with the addition of buffers are frequently used in HPLC. This subject is also discussed in this chapter.

Besides the role as solvent, specific properties of the mobile phase are also important in HPLC. For example, solvent viscosity affects column backpressure and therefore separation conditions. As the mobile phase carries the separated analytes through the detector, other properties of the solvents used as a mobile phase become important. These properties include, for example, the wavelength of light absorption (e.g., when UV-Vis detection is used), volatility (e.g., when ELSD is used), ionization capability (e.g., when MS detection is used). The properties of solvents as they relate to the manner in which they affect detection will be further discussed in this chapter. The properties of liquids as solvents are presented in this section.

Solubility Based on Thermodynamic Concepts

One approach regarding the solubility of a nonpolar compound in a solvent can start with evaluation of the energy involved in the dissolution of a vaporized molecule i in a pure solvent S. Further comparison of solubility in different solvents A and B, can be done by comparing the energies involved in the transfer of vapor i to solvent A with that for the transfer to solvent B. In this way, this approach is useful for comparing solvents even if the "vapor phase" step remains only hypothetical.

The dissolution of a vaporized molecule can be viewed as the reverse of vaporization. The energy of vaporization can be separated in two terms, one accounting for the removal of a molecule from the liquid and the other for the collapse of the cavity originally occupied by the molecule. Assuming that a molecule in the liquid is surrounded by e other molecules and that the energy for each interaction is E_{ii}, the energy term of removal will be $e\,E_{ii}$. The collapse of the resulting cavity leads to the formation of new $i-i$ interactions of the same energy E_{ii}, but the number of such interactions will be $e/2$. The vaporization energy ΔE^{vap}_i is therefore given by the expression:

$$\Delta E^{vap}_i = \mathcal{N}\frac{e E_{ii}}{2} \qquad (7.1.1)$$

where $\mathcal{N}$ is the Avogadro number. When the process of dissolution of a molecule i in the solvent S takes place, $e/2$ interactions $S-S$ will be broken, and e' interactions $i-S$ will be created (with $e' \approx e$). Therefore, indicating by E_{iS} the interaction energy between a molecule of solute and one of solvent, the dissolution energy is given by the expression:

$$\Delta E^{sol}_{iS} = \mathcal{N}\left(\frac{e E_{SS}}{2} - e E_{iS}\right) \qquad (7.1.2)$$

In the solution, an equal number of intermolecular interactions per unit volume can be assumed, such that $\mathcal{N}e = C_t\,V_i$, where V_i is the molar volume of species i ($V_i = MW_i / \rho_i$, where ρ_i is the compound density) and C_t is a constant. With this assumption, rel. 7.1.1 takes the form:

$$\Delta E^{vap}_i = \frac{C_t}{2}V_i E_{ii} \qquad (7.1.3)$$

Relation 7.1 3 can be written in the following form:

$$E_{ii} = \frac{2}{C_t}\frac{\Delta E^{vap}_i}{V_i} \qquad (7.1.4)$$

Relation 7.1.4 indicates that the molecular interaction E_{ii} in the liquid is proportional to the ratio of vaporization energy per unit molar volume. This ratio can be used to give a specific solubility parameter with the notation δ_i that has the expression [1]:

$$\delta_i^2 = \frac{\Delta E_i^{vap}}{V_i} \qquad (7.1.5)$$

The units for δ are $(cal/cm^3)^{1/2}$. This solubility parameter can be used as a measure of the intermolecular interactions per unit volume of a pure liquid based on the following relation:

$$E_{ii} = \frac{2}{C_t}\delta_i^2 \qquad (7.1.6)$$

For solvent S, the parameter δ_S is similarly defined. The energy for the interaction $i-S$ can be approximated as the geometric mean of E_{ii} and E_{SS} such that:

$$E_{iS} = \sqrt{E_{ii}E_{SS}} = \frac{2}{C_t}\delta_i\delta_S \qquad (7.1.7)$$

Considering that the energy of mixing of a mole of i with a large quantity of pure S to form a dilute solution should be equal with the sum of ΔE^{vap}_i and ΔE^{sol}_{iS} given by rel. 7.1.1 and 7.1.2, respectively, the expression for the energy of mixing is given by the expression:

$$\Delta E_{iS}^{mix} = \mathcal{N}\left(\frac{e}{C_t}\right)(\delta_i^2 + \delta_S^2 - 2\delta_i\delta_S)$$
$$= V_i(\delta_i^2 + \delta_S^2 - 2\delta_i\delta_S) \qquad (7.1.8)$$

Assuming no volume variation during mixing at constant pressure, the energies E^{mix} can be taken as equal to the enthalpy (heat) of mixing ΔH^{mix}. In conclusion, rel. 7.1.8 can be written for one mole of solute i in the form:

$$\Delta H_{iS}^{mix} = V_i(\delta_i - \delta_S)^2 \qquad (7.1.9)$$

For a solution where the concentration of i and S are comparable, rel. 7.1.9 must be replaced by the similar relation [6]:

$$\Delta H_{iS}^{mix} = (x_iV_i + x_SV_S)(\delta_i - \delta_S)^2\phi_i\phi_S \qquad (7.1.10)$$

where x_i and x_S are the mole fractions and ϕ_i and ϕ_S are the volume fractions of i and S, respectively $(\phi_i = \frac{V_i}{V_i + V_S})$. For x_i very small, rel. 7.1.10 should be reduced to rel. 7.1.9, although the transformation is not straightforward due to the approximations involved.

Relations 7.1.9 and 7.1.10 indicate that the variation in the heat of mixing is always positive, which shows that the enthalpy of mixing is always unfavorable to the process. For a regular solution, the mixing enthalpy is different from zero, or $\Delta H \neq 0$ and $T\Delta S = -\Delta H - \sum n_iRT \ln x_i$. The expression for the chemical potential of a component i in a regular solution will be in this case given by the expression:

$$\mu_i = \mu_i^0 + \frac{1}{n_i}(H_i - H_i^0) + RT \ln x_i \qquad (7.1.11)$$

where $H_i - H_i^0$ is the change in enthalpy due to mixing $(H_i - H_i^0 = \Delta H^{mix}_i)$. From rel. 7.1.11, and using rel. 7.1.9 for ΔH^{mix}_i, the expression for the chemical potential of a component i in a regular (diluted) solution will be given by the expression $(R = 1.987$ cal deg^{-1} mol$^{-1})$.

$$\mu_i = \mu_i^0 + V_i(\delta_i - \delta_S)^2 + RT \ln x_i \qquad (7.1.12)$$

Relation 7.1.12 indicates that the mixing is favored by the entropy term $(T\Delta S = -RT \ln x_i)$ and that it is opposed by the enthalpy term. The enthalpy term is higher when the difference between the solubility parameters is larger, which is in agreement with the experimental observation that similar compounds dissolve one in the other, while different ones are less likely to mix.

Relation 7.1.9 or 7.1.10 also provides a useful expression for the activity coefficient γ_i of a compound i dissolved in a solvent S. The activity coefficient $a_i = \gamma_i x_i$ is related to the chemical potential by the well-known expression $\mu_i = \mu_i^0 + RT \ln a_i = \mu_i^0 + RT \ln x_i + RT \ln \gamma_i$. With this formula, and using 7.1.11, the following expression can be written for the activity coefficient:

$$\ln \gamma_i = \frac{1}{n_i RT} \Delta H_i \qquad (7.1.13)$$

From rel. 7.1.13, for the activity coefficient and rel. 7.1.9 for the enthalpy of mixing of a mole of compound i with solvent S (where $n_i = 1$), the following expression is obtained:

$$\ln \gamma_i = \frac{V_i (\delta_i - \delta_S)^2}{RT} \qquad (7.1.14)$$

(rel. 7.1.10 should be used instead of 7.1.9 when the concentrations of i and S are comparable). Tables of values for solubility parameter δ are available in the literature for a variety of compounds. Some of these values are indicated in Table 7.1.1, together with the molar volumes of the compound [2, 7].

Vaporization energy ΔE_i^{vap} of a solvent is determined by different types of internal molecular interactions (see Section 4.1 for various intermolecular forces). As a result, it can be concluded that δ is also composed of contributions from different types of interactions such as dispersion forces, polar interactions, proton-acceptor interactions, and proton donor interactions. Values for a "partial δ" describing this type of interaction were reported in the literature and are listed in Table 7.1.1. These values were obtained from extrathermodynamic properties, and for this reason their sum is not the same as δ [8]. These "partial δ" values are indicated as δ_d (for dispersion), δ_p (for polar), δ_a (for proton acceptor), δ_h (for proton donor).

Solubility parameter δ can be estimated using a number of procedures [1]. These include calculation from heats of vaporization ΔH^{vap} with the formula $\delta \approx \left(\dfrac{\Delta H^{vap} - RT}{V} \right)^{1/2}$ where

ΔH^{vap} is either measured or estimated, calculation from critical constants, calculation from superficial tension γ' with the formula $\delta = k \left(\dfrac{\gamma'}{V^{1/3}} \right)^{0.43}$ where $k \approx 4.1$, and by other procedures. Molar volume can be calculated using the simple formula $V = MW / \rho$ where ρ is the mass density.

Solubility parameter δ can be used for the evaluation of solubility (defined as the maximum amount of a solute that dissolves in a given volume of solvent; see, e.g., [9]), and of partition coefficient of a solute between two solvents. However, its main use in this book is related to the estimation of solvent properties related to their "elution" capabilities in RP-HPLC.

Miscibility of Solvents

Of particular interest for HPLC is the miscibility (solubility in any proportions) of one solvent in another. When using gradient HPLC, the pumps are delivering solvents in various proportions, and these solvents must be perfectly soluble one in another. This is the case for both mixtures of organic solvents and when buffers consisting of aqueous solutions of inorganic compounds are used. The use of buffers is discussed in Section 7.4. For various solvents, the miscibility is typically based on experimental results. A chart showing the miscibility of various solvents is given in Figure 7.1.1. Quantitative data for solubility are also available in the literature [3]. For the solubility in water of several solvents used in HPLC, such data are given in Table 7.1.2.

Estimation of Partition Constant and Selectivity in HPLC from the Solubility Parameter

The previous theory can be further applied for estimating the distribution coefficient for

TABLE 7.1.1 Molar volumes (cm^3/mol) and solubility parameters δ (cal/cm^3)$^{1/2}$ for some common compounds used as solvents (at 25 °C).

Compound	V	δ	δ_d	δ_p	δ_a	δ_h
Acetic acid	71.3	12.4	7	N.R.L.*	N.R.L.	N.R.L.
Acetone	73.8	9.4	6.8	5	2.5	0
Acetonitrile	52.7	11.8	6.5	8	2.5	0
Anisole	108.7	9.7	9.1	2.5	2	0
Benzene	89.2	9.2	9.2	0	0.5	0
Benzonitrile	103.3	10.7	9.2	3.5	1.5	0
Bromobenzene	105.3	9.9	9.6	1.5	0.5	0
1-Butanol	91.5	9.6	N.R.L.	N.R.L.	N.R.L.	N.R.L.
CCl_3-CF_3	119.3	7.1	6.8	1.5	0.5	0
CCl_4	96.9	8.6	8.6	0	0	0
$CFCl_2$-CF_3	80.02	6.2	5.9	1.5	0	0
CH_2Cl_2	64.4	9.6	6.4	5.5	0.5	0
$CHCl_3$	80.4	9.1	8.1	3	0.5	0
Chlorobenzene	102	9.6	9.2	2	0.5	0
CS_2	60.6	10	10	0	0.5	0
Cyclohexane	108.4	8.2	N.R.L.	0	0	0
Cyclohexanone	103.8	10.4	6.2	N.R.L.	N.R.L.	N.R.L.
Cyclopentane	93.21	8.1	N.R.L.	0	0	0
1,2-Dichloroethane	64.4	9.7	8.2	4	0	0
1,3-Dicyanopropane	94.6	13	8	8	3	0
Diethyl ether	104.4	7.4	6.7	2	2	0
Diethyl sulfide	109.98	8.6	8.2	2	0.5	0
Di-isopyl ether	142	7	6.9	0.5	0.5	0
Dimethylformamide	77.3	11.5	7.9	N.R.L.	N.R.L.	N.R.L.
Dimethylsulfoxide	70.9	12.8	8.4	7.5	5	0
Dioxane	85.3	9.8	7.8	4	3	0
Ethanol	58.6	11.2	6.8	4	5	5
Ethanolamine	60.4	13.5	8.3	Large	Large	Large
Ethyl acetate	98.1	8.6	7	3	2	0
Ethyl bromide	75	8.8	7.8	3	0	0

(Continued)

TABLE 7.1.1   Molar volumes (cm^3/mol) and solubility parameters δ (cal/cm^3)$^{1/2}$ for some common compounds used as solvents (at 25 °C). (*cont'd*)

Compound	V	δ	δ_d	δ_p	δ_a	δ_h
Ethylene glycol	55.8	14.7	8	Large	Large	Large
Formamide	39.7	17.9	8.3	Large	Large	Large
Isooctane	165.1	7	7	0	0	0
Methanol	40.6	12.9	6.2	5	7.5	7.5
Methyl acetate	79.5	9.2	6.8	4.5	2	0
Methyl benzoate	125.6	9.8	9.2	2.5	1	0
Methyl ethyl ketone	89.9	9.5	6.8	5	2.5	0
Methyl iodide	62.2	9.9	9.3	2	0.5	0
Methylene iodide	52.2	11.9	11.3	1	0.5	0
m-Xylene	123.4	8.8	8.8	0	0.5	0
n-Heptane	146.5	7.4	7.4	0	0	0
n-Hexane	131.1	7.3	7.3	0	0	0
Nitrobenzene	102.6	11.1	9.5	4	0.5	0
Nitromethane	53.7	11	7.3	8	1	0
n-Pentane	115.2	7.1	7.1	0	0	0
Octanol	158	10.3	5.2	N.R.L.	N.R.L.	N.R.L.
Perchloroethylene	102.2	9.3	9.3	0	0.5	0
Perfluoroalkanes	-	6	6	0	1	0
Phenol	87.9	11.4	9.5	N.R.L.	N.R.L.	N.R.L.
1-Propanol	75.1	10.2	7.2	2.5	4	4
2-Propanol	72.74	9.7	N.R.L.	N.R.L.	N.R.L.	N.R.L.
Propyl amine	81.7	8.7	7.3	4	6.5	0.5
Propyl chloride	88.2	8.3	7.3	3	0	0
Propylene carbonate	85.4	13.3	N.R.L.	N.R.L.	N.R.L.	N.R.L.
p-Xylene	123.4	8.8	8.8	0	0.5	0
Pyridine	80.6	10.4	9	4	5	0
Tetrahydrofuran	81.1	9.9	7.6	4	3	0
Toluene	106.6	8.9	8.9	0	0.5	0
Triethylamine	139.5	7.5	7.5	0	3.5	0
Water	18.02	21	6.3	Large	Large	Large

Note: N.R.L. = Not reported in the referenced literature

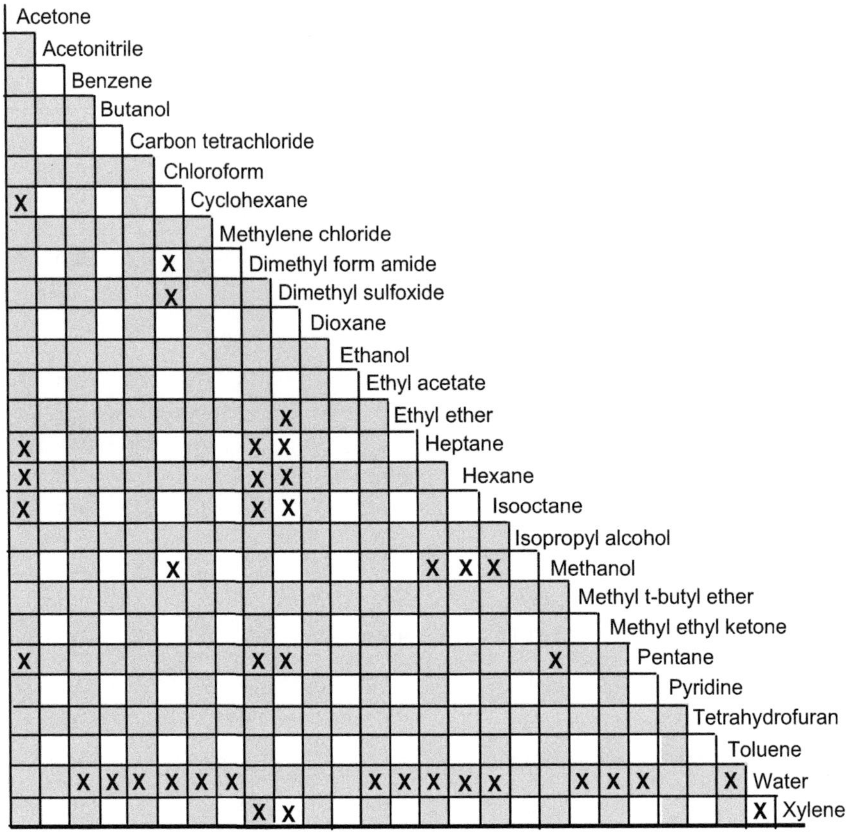

Miscibility chart (reading the triangular grid of Figure 7.1.1). Solvent labels lie on the diagonal; an "X" marks an immiscible pair between the row solvent and the column solvent. Columns are numbered as: 1 Acetone, 2 Acetonitrile, 3 Benzene, 4 Butanol, 5 Carbon tetrachloride, 6 Chloroform, 7 Cyclohexane, 8 Methylene chloride, 9 Dimethyl form amide, 10 Dimethyl sulfoxide, 11 Dioxane, 12 Ethanol, 13 Ethyl acetate, 14 Ethyl ether, 15 Heptane, 16 Hexane, 17 Isooctane, 18 Isopropyl alcohol, 19 Methanol, 20 Methyl t-butyl ether, 21 Methyl ethyl ketone, 22 Pentane, 23 Pyridine, 24 Tetrahydrofuran, 25 Toluene, 26 Water, 27 Xylene.

Row solvent	Immiscible with (column numbers)
Acetone	—
Acetonitrile	—
Benzene	—
Butanol	—
Carbon tetrachloride	—
Chloroform	—
Cyclohexane	1
Methylene chloride	—
Dimethyl form amide	4
Dimethyl sulfoxide	4
Dioxane	—
Ethanol	—
Ethyl acetate	—
Ethyl ether	8
Heptane	1, 8, 9
Hexane	1, 8, 9
Isooctane	1, 8, 9
Isopropyl alcohol	—
Methanol	4, 15, 16, 17
Methyl t-butyl ether	—
Methyl ethyl ketone	—
Pentane	1, 8, 9, 18
Pyridine	—
Tetrahydrofuran	—
Toluene	—
Water	3, 4, 5, 6, 7, 8, 13, 14, 15, 16, 17, 20, 21, 22, 25
Xylene	8, 9

FIGURE 7.1.1 Miscibility chart of various solvents used in HPLC.

a compound i between two solvents indicated as A and B. In case of equilibrium of the type:

$$i_B \rightleftarrows i_A$$

the constant $K_i = c_{i,A}/c_{i,B}$ that indicates the concentration ratio of i in the two phases is given by the expression 3.1.11, which can be written in the form:

$$K_i = [(\gamma_{i,B})/(\gamma_{i,A})]\exp[-\Delta\mu_i^0/(RT)]$$

For the choice of the standard state of a component in a solution, there are two common conventions. Since $\mu^0_{i,A}$ and $\mu^0_{i,B}$ are independent of solution concentration, one choice for the standard state is that of the pure liquid or solid i at the working temperature and pressure. In this case, $\mu^0_{i,A} = \mu^0_{i,B} = \mu_i^{Liquid}$, or $\mu^0_{i,A} = \mu^0_{i,B} = \mu_i^{Solid}$. For this choice, the result is $\Delta\mu_i^0 = 0$. Another convention is to choose the standard state to be that of the pure liquid only for a solvent, with all the other components having a "fictitious" state with the properties that pure i would have if its limiting low concentration properties in solution were to be retained in a pure substance. Selecting the choice of standard state $\mu^0_{i,A} = \mu^0_{i,B}$, the expression of K_i is reduced to the following form:

$$K_i = (\gamma_{i,B})/(\gamma_{i,A}) \qquad (7.1.15)$$

TABLE 7.1.2 Experimental solubility in water of various solvents used in HPLC.

Solvent	Sol. % in water	Solvent	Sol. % in water
water	100	iso-propanol	100
formamide	100	butyl acetate	7.81
dimethylsulfoxide	100	1,2-dichloroethane	0.81
dimethylformamide	100	methylene chloride	1.6
acetic acid	100	diethyl ether	6.89
acetonitrile	100	benzene	0.18
ethanol	100	methyl *tert*-butyl ether	4.8
methanol	100	xylene	0.018
acetone	100	toluene	0.051
dioxane	100	di-isopropyl ether	0.87
methyl ethyl ketone	24	carbon tetrachloride	0.08
ethyl acetate	8.7	trichloroethylene	0.11
chloroform	0.815	cyclohexane	0.01
tetrahydrofuran	100	pentane	0.004
n-butanol	0.43	hexane	0.001
n-propanol	100	heptane	0.0003

By including in rel. 7.1.15 the expression 7.1.14 for γ, the following expression results for K_i:

$$K_i = \exp\{V_i[(\delta_i - \delta_B)^2 - (\delta_i - \delta_A)^2]/(RT)\}$$
(7.1.16)

For an HPLC separation where a partition process is assumed between an immobilized liquid (stationary phase) A and a mobile phase B, it can be considered that the parameter δ_i corresponds to the analyte, δ_A corresponds to the immobilized liquid phase, and δ_B corresponds to the mobile phase. Relation 7.1.16 can be used in principle to predict the value for the capacity factor k when the phase ratio Ψ is known ($k = K\Psi$).

A simple rearrangement of rel. 7.1.16 leads to the formula:

$$K_i = \exp\{[(\delta_A - \delta_B)V_i(2\delta_i - \delta_A - \delta_B)]/(RT)\}$$
(7.1.17)

Relation 7.1.17 provides guidance regarding the choice of a solvent in an HPLC separation, in particular for RP-HPLC where nonionic solvents are prevalent. With the assumption that the stationary phase is very nonpolar and the mobile phase (eluent) is polar, it can be concluded from the values in Table 7.1.1 that $\delta_A < \delta_B$ and the term $(\delta_A - \delta_B)$ is always negative. Also, in order to be retained on the hydrophobic column, the solute must be rather hydrophobic and its δ_i value should be lower than δ_B and closer to δ_A. In conclusion,

$(2\delta_i - \delta_A - \delta_B)$ is also negative. Relation 7.1.17 shows that K_i for a solute is larger when the difference $(\delta_A - \delta_B)$ is larger, in other words, when the solvent used as eluent has a large δ_B value. This can be easily exemplified by considering a hydrophobic compound retained on a C18 column, which is not eluted by water. As δ_B of the eluent decreases, K_i decreases. When $2\delta_i > \delta_A + \delta_B$, the equilibrium is rapidly displaced toward having the analyte in the mobile phase (eluent). This process can be visualized in Figure 7.1.2 with a hypothetical example of two analytes with $\delta_i = 9.0$ (cal/cm³)$^{1/2}$ and $\delta_j = 9.3$ (cal/cm³)$^{1/2}$, respectively, both with $V_i = 100$ cm³/mol, retained on a stationary phase with $\delta_A = 6.0$ (cal/cm³)$^{1/2}$, as the value of δ_B of the solvent increases. It can be seen that for $\delta_B < 14$, neither compound is retained on the column (e.g., methanol will elute this compound). As δ_B increases (e.g., for a water/methanol mixture), the value of K_i increases rapidly, indicating retention on the stationary phase.

In gradient RP-HPLC, the initial polarity of the mobile phase is typically high, such that all analytes are retained on the column (large K_i).

As the content in the organic phase increases, δ_B decreases and the K_i values decrease. The compound slightly less hydrophobic ($\delta = 9.3$) is first eluted (K_i for $\delta = 9.3$ is low enough earlier than for $\delta = 9.0$ as shown in Figure 7.1.2). As δ_B continues to decrease, the K_i for the compound with $\delta = 9.0$ also decreases sufficiently that the compound is eluted.

The difference in δ between analytes is the base of their separation. For two analytes i and j the selectivity α is given by rel. 3.1.16 (equivalent to 2.1.49). The replacement in the formula for $\alpha = K_i/K_j$ of K_i and K_j with their expressions given by rel. 7.1.17, and assuming $n_i = n_j$, leads to the following formula:

$$\alpha = \exp\{(\delta_A - \delta_B)[V_i(2\delta_i - \delta_A - \delta_B) - V_j(2\delta_j - \delta_A - \delta_B)]/(RT)\}$$

$$(7.1.18)$$

Expression 7.1.18 allows estimation of the value of α for a given separation (when the molar volumes V and parameters δ are known). The formula shows that α increases as the difference between the δ parameters for the mobile phase and the stationary liquid phase increases.

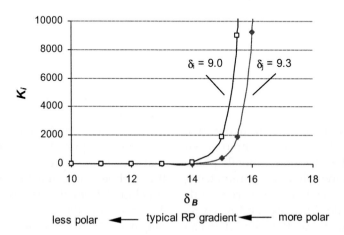

FIGURE 7.1.2 Exemplification of the variation of retention in a RP-HPLC separation (value of equilibrium constant K_i) as a function of the variation in δ_B of the eluent for two analytes with $\delta_i = 9.0$ (cal/cm³)$^{1/2}$ and $\delta_j = 9.3$ (cal/cm³)$^{1/2}$, respectively, both with $V_i = 100$ cm³/mol on a hypothetical stationary phase with $\delta_A = 6.0$ (cal/cm³)$^{1/2}$.

In other words, a stationary phase more different from the mobile phase will lead to a higher selectivity between the species to be separated. As expected, the more different are the analytes (different V and different δ), the better is the separation. The formula also shows that a (slightly) higher selectivity can be expected at lower temperatures than at a higher temperature.

Elution "Strength" of a Solvent

A mobile phase in HPLC capable of eluting the analytes at shorter retention times compared to another one is typically indicated as being "stronger." However, the "strength" of a solvent used as a mobile phase is dependent on the stationary phase. Use of rel. 7.1.17 to obtain a variation of the equilibrium constant K as indicated in Figure 7.1.2 requires the assumptions that the stationary phase is hydrophobic and that the mobile phase is more polar (e.g., in RP-HPLC) such that $\delta_A < \delta_B$. Also, in order to be retained on the hydrophobic column, the solute must be rather hydrophobic and its δ_i value should be lower than δ_B and closer to δ_A. In this case, the solvents with the lower δ are the "strongest," while those with high δ are "weak" solvents. For this reason, in RP-HPLC the stronger mobile phases made from an organic solvent and water have a high content of organic component and a low water content. However, the restrictions described for RP-HPLC are not applicable for HILIC, for example, where the situation is reversed and $\delta_A > \delta_B$. In HILIC (and NPC) the solvents with the lowest δ are the "weakest," and strong solvents have large δ, such as water or formamide. For this reason, parameter δ can be used for strength characterization, but this should be done depending on the context.

Since the theory of liquid–liquid partition is in many cases only an approximation of the real chromatographic process, use of parameter δ for characterization of the solvent and of the stationary phase (viewed as an immobilized liquid) must be done with caution. Even assuming that a pure liquid is adsorbed on a solid material and acts as a stationary phase, the direct use of the values δ for this liquid may not correspond to reality since the properties of the adsorbed liquid may be different from those of the pure one. Also, the interactions between the analyte and the mobile (and stationary) phase, not being completely described by the solubility parameter δ, it is important to consider in selecting a solvent the values of the other solubility parameters: δ_d (dispersion), δ_p (polar), δ_a (proton acceptor), δ_h (proton donor) for both analyte and eluent. Similar values of these parameters for solvent, analyte, and stationary phase indicate a "strong" eluent.

Solvent Characterization Using Octanol/Water Partition Constant K_{ow}

Characterization of a liquid based on its solubility parameter δ provides useful information on its solvent characteristics, but the true calculation of an equilibrium constant between a stationary phase and a mobile phase cannot be done accurately. Also, since the evaluation of δ is not straightforward, another parameter, which is readily available, can be useful for solvent characterization. This parameter is the K_{ow} value for the solvent. The values for K_{ow} are commonly found in the literature [10] and can be approximated using available computer programs (e.g., MarvinSketch 5.4.0.1 from ChemAxon Ltd. [11], EPI Suite [12], CLOGP [13], KOWWIN, SciLogP/Ultra [14], etc.). The values of log K_{ow} for a number of solvents are given in Table 7.1.3 (together with other parameters further discussed). The values for log K_{ow} can be obtained by a variety of procedures based on molecule π-substituents, additive fragment techniques, atomic contributions and solvent-accessible surface areas [15], physicochemical molecular properties, and solvatochromic parameters [16]. A higher K_{ow} indicates a less polar compound, and therefore the values for this parameter are growing in the opposite direction to δ. For a set of 36 common solvents, the correlation between log K_{ow} and δ is relatively weak, with R = −0.65575. For RP-type separations, log K_{ow} of the solvent can be used as guidance regarding its "strength." For example, the strength in RP-HPLC and log K_{ow} values (from Table 7.1.3) for several common solvents are in the order: n-hexane log K_{ow} = 3.13 > di-ethyl ether log K_{ow} = 0.84 > tert-butanol log K_{ow} = 0.54 > tetrahydrofuran log K_{ow} = 0.53 > isopropanol log K_{ow} = 0.25 > acetone log K_{ow} = 0.11 > dioxane log K_{ow} = −0.09 > ethanol log K_{ow} = −0.16 > acetonitrile log K_{ow} = −0.17 (experimental −0.34) > methanol log K_{ow} = −0.52 (experimental −0.77) > dimethylformamide log K_{ow} = −0.63 > water log K_{ow} = −0.65 (experimental −1.38). However, from the values of log K_{ow} of the solvent and based on their common use in the mobile phase in HPLC (e.g., in RP-HPLC), it can be concluded that solvents with much lower log K_{ow} are capable of eluting strongly retained compounds (with significantly higher log K_{ow} values). For example, acetonitrile can elute from a C18-type column compounds with log K_{ow} < 6 to 7 within relatively short retention times. This indicates that although the strength of different solvents can be compared based on their log K_{ow} it is not possible to use solvent log K_{ow} for quantitative predictions.

Solvent Characterization Based on Liquid-Gas Partition

The results obtained using thermodynamic concepts previously described were promising in interpreting separation in HPLC, but their use for the quantitative calculation of retention parameters was not very successful. For this reason, an experimental base closer to reality, equivalent to the one using thermodynamic concepts, was considered. For example, the energy of mixing a mole of solute i with a large quantity of pure solvent S to form a dilute solution as given by rel. 7.1.8 was shown to be equal with the sum of the energy of vaporization ΔE^{vap}_i and the dissolution energy ΔE^{sol}_{iS}. Therefore, it can be inferred that the characterization of a solvent can be based on measurement of how good the solvent is for dissolving a volatile solute. This can be done based on measurement, at equilibrium, of the solute concentration in the solvent (related to ΔE^{sol}_{iS}) and in the headspace above the solvent (related to ΔE^{vap}_i). This observation can be applied for establishing experimentally a scale for solvent characterization, when using a set of standard trial solutes i and measuring the distribution constant K_S between solvent and headspace for the solvent of interest. Experimentally, the distribution constants can be measured by adding a volume of test solution in a given volume of solvent S placed in a closed vial of a specific volume. A study of this type reported in the literature [17] used a 5 μL mixture of ethanol, dioxane, and nitromethane in a chamber of 13.4 mL with 2 mL solvent S at 25 °C. After equilibration, the composition of the gas phase and of liquid can be measured and K_S values calculated using the formula:

$$K_S = c_{i,S}/c_{i,g} \qquad (7.1.19)$$

The selection of the three test compounds ethanol, dioxane, and nitromethane was made on the assumption that they have different types of polar interactions with the molecules of tested solvent S. Dipole-dipole interactions are stronger with nitromethane, acidic-polar interactions are stronger with dioxane, and basic polar interactions are stronger with ethanol.

The K_S values are used in the calculation of a modified constant that is intended to eliminate the effect of the solvent molecular weight. The modification is done by use of the solvent molar volume V_S (mL/mole) in the expression:

$$K'_S = K_S V_S \qquad (7.1.20)$$

The values K'_S are then used to calculate the coefficients K'_S, which are obtained with further correction of K'_S values with the intention of correcting for nonpolar (dispersive) interactions. This is done with the relation:

$$K' = K'_S/K'_v \qquad (7.1.21)$$

where K'_v is the estimated K' value of an n-alkane whose molar volume is the same as that of the solute i. The values of K'_v are calculated using the expression:

$$\log K'_v = (V_i/163)\log K_{octane,S} \qquad (7.1.22)$$

In rel. 7.1.22 V_i is the molar volume of the solute (ethanol, dioxane, or nitromethane), and $K_{octane,S}$ is the experimental distribution coefficient of n-octane in the evaluated solvent. The constants K_S'' are further corrected to have zero value for n-hexane as a solvent. The resulting constants K_S'' are used to measure the excess retention of a solute relative to an n-alkane of equivalent molar volume.

For any evaluated solvent S, an experimental polarity parameter (or chromatographic strength) P' is then defined by the expression:

$$P' = \log K''_{ethanol,S} + \log K''_{dioxane,S}$$
$$+ \log K''_{nitromethane,S} \qquad (7.1.23)$$

Larger values for P' indicate a polar solvent (such as alcohol or water), and values close to zero show nonpolar solvents such as hexane and cyclohexane. Solvent polarity is used in selecting solvents in LC separations. However, parameter P' is not always sufficient for characterizing solvent properties. The types of interactions that dominate solvent behavior can be quite different between solvents with the same P'. For example, a polar solvent and a solvent forming hydrogen bonds, although they may have identical P', may not act in the same manner toward different solutes. An additional separation parameter

x_i was developed for solvent characterization, defined by the formula:

$$x_i = \log K''_{i,S}/P' \qquad (7.1.24)$$

where i can be ethanol (x_e), dioxane (x_d), or nitromethane (x_n). Other solvents were also used for obtaining an x_i value, such as toluene (x_t) or methyl ethyl ketone (x_m) [2]. It can be assumed that the larger the x_i value for a specific compound, the higher is the similarity with the comparing solvent. However, the value of x_i also depends on P', and relatively large K_S'' do not necessarily lead to large x_i values. For this reason, the values for x_i were used to group the solvents in nine main groups, the solvents in the same group having similar properties. These groups are (0) solvents with very low P' values (nonpolar), (1) aliphatic ethers, tetramethylguanidine, hexamethylphosphoric acid triamide, (2) aliphatic alcohols, (3) pyridine derivatives, tetrahydrofuran, amides, glycol ethers, sulfoxides, (4) glycols, benzyl alcohol, acetic acid, formamide, (5) methylene chloride, ethylene chloride, (6) tricresyl phosphate, aliphatic ketones and esters, dioxane, (7) aromatic hydrocarbons, halo-substituted aromatic hydrocarbons, nitro compounds, aromatic ethers, and (8) fluoroalkanols, m-cresol, water, chloroform. Values for P' for some solvents classified in these groups are given in Table 7.1.3. For comparison, the values of K_{ow} calculated using MarvinSketch [11] are also listed in the table.

The parameters x_e for ethanol, x_d for dioxane and x_n for nitromethane from Table 7.1.3 can be used to illustrate the separation of solvents in several groups, since in each group, the x_i values are close to each other. This can be illustrated in a triangular diagram. Since $x_e + x_d + x_n = 1$, their graphic representation can be done in a planar triangular diagram, and a representation in a tri-dimensional space is not necessary. The diagram is shown in Figure 7.1.3. The compounds in each group are indicated by the group number. The x_e, x_d and x_n values can be obtained from the graph. As shown in the

TABLE 7.1.3 Values of polarity (chromatographic strength) P' parameter, and of separation parameters x_e, x_d and x_n for several common solvents (e-ethanol, d-dioxane, n-nitromethane). The values for log K_{ow} are also listed.

	Compound	Group	P'	x_e	x_d	x_n	log K_{ow}
	Non-polar, hydrocarbons						
1	Carbon disulfide	0	0.3	-	-	-	1.95
2	Carbon tetrachloride	0	1.6	-	-	-	3
3	Cyclohexane	0	0.2	-	-	-	2.67
4	n-Decane	0	0.4	-	-	-	4.91
5	n-Hexane	0	0.1	-	-	-	3.13
6	Isooctane	0	0.1	-	-	-	3.71
7	Squalane	0	1.2	-	-	-	12.86
	Aliphatic ethers, other compounds						
8	Di-butyl ether	1	2.1	0.44	0.18	0.38	2.77
9	Di-ethyl ether	1	2.8	0.53	0.13	0.34	0.84
10	Di-isopropyl ether	1	2.4	0.48	0.14	0.38	1.67
11	Hexamethylphosphoramide	1	7.4	0.47	0.17	0.37	−1.4
12	Tetramethylguanidine	1	6.1	0.48	0.18	0.35	−0.16
13	Triethylamine	1	1.9	0.56	0.12	0.32	1.26
	Aliphatic alcohols						
14	n-Butanol	2	3.9	0.58	0.18	0.24	0.81
15	tert-Butanol	2	4.1	0.56	0.2	0.24	0.54
16	Ethanol	2	4.3	0.52	0.19	0.29	−0.16
17	Isopentanol	2	3.7	0.56	0.19	0.26	1.09
18	Isopropanol	2	3.9	0.55	0.19	0.27	0.25
19	Methanol	2	5.1	0.48	0.22	0.31	−0.52
20	n-Octanol	2	3.4	0.57	0.19	0.25	2.58
21	n-Propanol	2	4	0.54	0.19	0.27	0.36
	Various, amides, nitrogenous, etc.						
22	Diethylene glycol	3	5.2	0.44	0.23	0.33	−1.26
23	2,6-Dimethylpyridine	3	4.5	0.45	0.2	0.36	1.02
24	2-[2-(4-Nonylphenoxy)ethoxy]ethanol	3	-	0.38	0.22	0.4	5.15
25	N,N-Dimethylacetamide	3	6.5	0.41	0.2	0.39	−0.58

(Continued)

TABLE 7.1.3 Values of polarity (chromatographic strength) P' parameter, and of separation parameters x_e, x_d and x_n for several common solvents (e-ethanol, d-dioxane, n-nitromethane). The values for log K_{ow} are also listed. (*cont'd*)

	Compound	Group	P'	x_e	x_d	x_n	log K_{ow}
26	Dimethylformamide	3	6.4	0.39	0.21	0.4	−0.63
27	Dimethylsulfoxide	3	7.2	0.39	0.23	0.39	−1.41
28	Methoxyethanol	3	5.5	0.38	0.24	0.38	−0.57
29	Methyl formamide	3	6	0.41	0.23	0.36	−0.86
30	2-Methylpyridine	3	4.9	0.44	0.21	0.36	0.89
31	N-Methyl-2-pyrrolidone	3	6.7	0.4	0.21	0.39	−0.36
32	Pyridine	3	5.3	0.42	0.23	0.36	0.76
33	Quinoline	3	5	0.41	0.23	0.36	2.13
34	Tetrahydrofuran	3	4	0.38	0.2	0.42	0.53
35	Tetramethyl urea	3	6	0.42	0.19	0.39	−0.47
36	Triethyleneglycol	3	5.6	0.42	0.24	0.34	−1.3
	Various, acetic acid, formamide, etc.						
37	Acetic acid	4	6	0.39	0.31	0.3	−0.22
38	Benzyl alcohol	4	5.7	0.4	0.3	0.3	1.21
39	Ethylene glycol (ethan-1,2-diol)	4	6.9	0.43	0.29	0.28	−1.21
40	Formamide	4	9.6	0.37	0.34	0.3	−1.08
	Chlorinated aliphatic						
41	Ethylene chloride	5	3.5	0.3	0.21	0.49	1.5
42	Methylene chloride	5	3.1	0.29	0.18	0.53	0.84
	Nitriles, dioxanes, ketones, etc.						
43	Acetone	6	5.1	0.35	0.23	0.42	0.11
44	Acetonitrile	6	5.8	0.31	0.27	0.42	−0.17
45	Acetophenone	6	4.8	0.33	0.26	0.41	1.53
46	Aniline	6	6.3	0.32	0.32	0.36	1.14
47	Benzonitrile	6	4.8	0.31	0.27	0.42	1.83
48	Bis-(2-cyanoethyl) ether	6	6.8	0.31	0.29	0.4	−0.33
49	Bis-(2-ethoxyethyl) ether	6	4.6	0.37	0.21	0.43	0.03
50	γ-Butyrolactone	6	6.5	0.34	0.26	0.4	0.15
51	Cyano morpholine	6	5.5	0.35	0.25	0.4	−0.05

(*Continued*)

TABLE 7.1.3   Values of polarity (chromatographic strength) P' parameter, and of separation parameters x_e, x_d and x_n for several common solvents (e-ethanol, d-dioxane, n-nitromethane). The values for $\log K_{ow}$ are also listed. *(cont'd)*

	Compound	Group	P'	x_e	x_d	x_n	$\log K_{ow}$
52	Cyclohexanone	6	4.7	0.37	0.22	0.42	1.49
53	Dioxane	6	4.8	0.36	0.24	0.4	−0.09
54	Ethyl acetate	6	4.4	0.34	0.23	0.43	0.28
55	Formyl morpholine	6	6.4	0.37	0.25	0.39	−0.85
56	Methyl ethyl ketone	6	4.7	0.35	0.22	0.43	0.81
57	Propylene carbonate	6	6.1	0.31	0.27	0.42	0.79
58	Tetrahydrothiophene-1,1-dioxide	6	6.9	0.33	0.28	0.39	−0.59
59	Tricresyl phosphate	6	4.6	0.36	0.23	0.41	6.63
60	Tris-(2-cyanoethoxy)propane	6	6.6	0.32	0.27	0.41	−0.59
	Aromatic hydrocarbons, aromatic ethers, nitro compounds						
61	Benzene	7	2.7	0.23	0.32	0.45	1.97
62	Bromobenzene	7	2.7	0.24	0.33	0.43	2.25
63	Chlorobenzene	7	2.7	0.23	0.33	0.44	2.07
64	Di-benzyl ether	7	4.1	0.3	0.28	0.42	3.57
65	Di-pentyl ether	7	3.4	0.27	0.32	0.41	3.66
66	Ethoxybenzene	7	3.3	0.28	0.28	0.44	2.17
67	Fluorobenzene	7	3.2	0.24	0.32	0.45	2.12
68	Iodobenzene	7	2.8	0.24	0.35	0.41	2.9
69	Methoxybenzene	7	3.8	0.28	0.3	0.43	1.82
70	Nitrobenzene	7	4.4	0.26	0.3	0.44	1.91
71	Nitroethane	7	5.2	0.28	0.29	0.43	0.38
72	Nitromethane	7	6	0.29	0.32	0.4	0.02
73	Toluene	7	2.4	0.25	0.28	0.47	2.49
74	*p*-Xylene	7	2.5	0.27	0.28	0.45	3
	Various, water						
75	Chloroform	8	4.1	0.26	0.42	0.33	0.84
76	*m*-Cresol	8	7.4	0.38	0.37	0.25	2.18
77	1H,1H,7H-Dodecafluoroheptanol	8	8.8	0.33	0.4	0.27	3.45
78	Tetrafluoropropanol	8	8.6	0.34	0.36	0.3	0.65
79	Water	8	10.2	0.38	0.38	0.25	−0.65

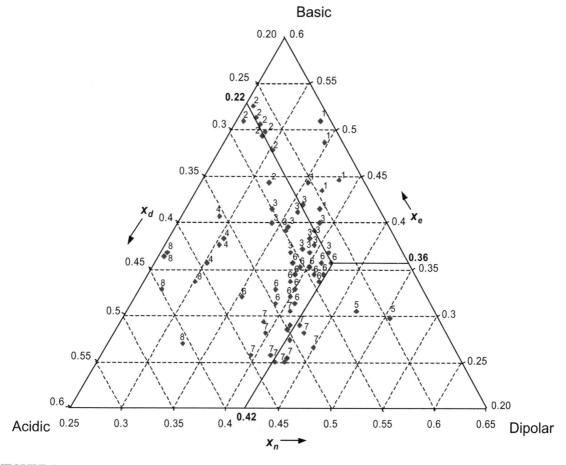

FIGURE 7.1.3 Triangular diagram showing groups of solvents clustered based on their x_e, x_d and x_n values.

diagram given in Figure 7.1.3, the solvents from the same group tend to cluster with x_e, x_d, and x_n values close to each other.

The polarity parameter P' is correlated, as expected, with the solubility parameter δ and with $\log K_{ow}$. The correlation with δ is positive, and for a set of 38 compounds gives $R^2 = 0.7486$. This correlation is shown in Figure 7.1.4. The correleation of P' and $\log K_{ow}$ leads to a negative slope, and for a set of 70 solvents gives the value $R^2 = 0.6612$. The graph showing the dependence of $\log K_{ow}$ on P' for 70 solvents is given in Figure 7.1.5. The graphs from Figures 7.1.4 and 7.1.5 show that parameters δ, $\log K_{ow}$

and P' provide basically similar information, but detailed solvent characterization is not possible using a unique parameter. Better indication regarding solvent properties is obtained when using x_e, x_d and x_n values for describing solvent similarities. The separation of solvent in classes based on these values indicate that specific types of interactions of analytes with solvent molecules, which are more prominent for a specific solvent than for another, are important in solvent characterization. These interactions include dispersion, dipole-dipole, hydrogen bonding, charge transfer, and ionic. Selection of solvents based on these properties

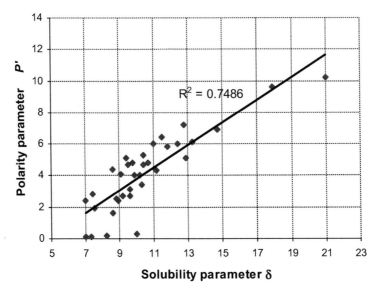

FIGURE 7.1.4 Correlation between the polarity parameter P' and the solubility parameter δ.

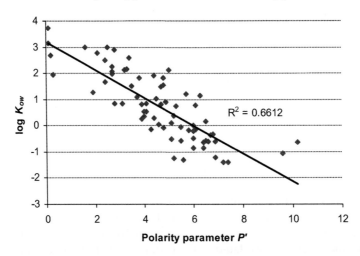

FIGURE 7.1.5 Graph showing the correlation between log K_{ow} and P' for 70 solvents.

and relation with the classification indicated in Table 7.1.3 will be further discussed in this chapter for specific HPLC types.

Solvatochromic Model

The polarity of a solvent can be assessed by different procedures, including spectroscopic measurements. Such measurements can be based, for example, on the chemical shift in an NMR experiment or a change in the absorption spectrum in IR or UV-vis for a compound used as a "molecular probe." For example, solvents with a polar character can produce a bathochromic effect on the UV spectrum generated from a $\pi \rightarrow \pi^*$ transition in a compound. Other such effects under the influence of a solvent are known. The change in the position, intensity, or bandwidth that occurs when a solute is transferred from the gas phase to a solvent is known as the *solvatochromic* effect. One common scale to evaluate solvent polarity known as $E_T(30)$ scale is based on the variations in the maximum wavelength of absorption in visible of 2,6-diphenyl-4-(2,4,6-triphenyl-N-pyridinium)phenolate (compound ET-30). This test compound has a large solvatochromic effect, changing the absorption from 453 nm for water to 810 nm for diphenyl ether as a solvent (solutions are red in methanol and blue in acetonitrile). The polarity $E_T(30)$ is further calculated from the expression [18]:

$$E_T(30) = \frac{28591}{\lambda_{max}(nm)} \qquad (7.1.25)$$

where λ_{max} is measured for the specific solvent. From $E_T(30)$ values (expressed in kcal/mol), a normalized parameter E_T^N can be obtained and is utilized for solvent characterization. This parameter is obtained from the formula:

$$E_T^N = \frac{E_T(30) - 30.7}{32.4} \qquad (7.1.26)$$

The normalized scale gives $E_T^N = 1$ for water and $E_T^N = 0$ for tetramethylsilane, which are the two extremes for $E_T(30)$ values.

The $E_T(30)$ polarity has been reported for a considerable number of solvents [19, 20]. The values for several common solvents are given in Table 7.1.4.

The $E_T(30)$ and E_T^N scales describe not only the polarity and polarizability but also the donor hydrogen-bond formation ability of a solvent. A π^* scale was developed accounting only for polarity and polarizability, not being affected by hydrogen bonding or ion dipole interactions. This scale was obtained using 4-ethyl-nitrobenzene as molecular probe. The $E_T(30)$ scale and the π^* scale are "single-parameter" polarity scales and are based on measurements of the property change of a single compound. Other such scales are known, some still based on the test of one compound (such as Nile Red), and others being "multiparameter" and based on measurements of more than one compound as a probe. For example, the solvent hydrogen-bond donor interactions can be described by an α scale developed on compounds such as several common dyes [21]. Hydrogen-bond acceptor interactions can be described by a β scale that was developed based on measurements of compounds such as 4-nitroaniline, N,N-diethyl-4-nitroaniline, 4-nitrophenol, and 4-nitroanisole [22].

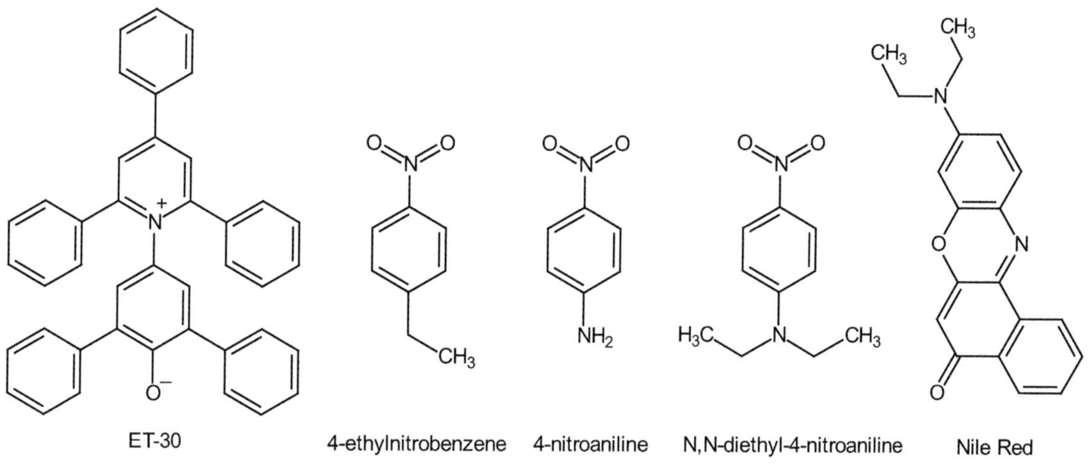

ET-30   4-ethylnitrobenzene   4-nitroaniline   N,N-diethyl-4-nitroaniline   Nile Red

TABLE 7.1.4 Values for $E_T(30)$ polarity for several common solvents.

Compound	$E_T(30)$	Compound	$E_T(30)$
Acetic acid	51.7	Formic acid	54.3
Acetonitrile	45.6	Glycerin	57
Benzene	34.3	n-Hexane	31
(tert-Butyl) methyl ether	34.7	n-Heptane	31.1
Carbon dioxide (40 °C/150 bar)	28.5	Methoxybenzene	37.1
Carbon tetrachloride	32.4	1-Methylpyrrolidin-2-one	42.2
Chloroform	39.1	Nitrobenzene	41.2
Cyclohexane	30.9	Nitromethane	46.3
Deuterium oxide	62.8	n-Octane	31.1
1,2-Dichlorethane	41.3	1,2-Propanediol	54.1
Dibenzylether	36.3	Propionic acid	50.5
Dichlormethane	40.7	Propionitrile	43.6
Diethylether	34.5	Pyridine	40.5
Diglyme	38.6	Styrene	34.8
N,N-Dimethylformamide	43.2	Tetrahydrofuran	37.4
Dimethylsulfoxide	45.1	Tetrahydropyran	36.2
1,4-Dioxane	36	Thiophene	35.4
1,3-Dioxolan	43.1	Toluene	33.9
Diphenylether (30 °C)	35.3	Trimethylphosphate	43.6
1,2-Ethanediol	56.3	p-Xylene	33.1
Formamide	55.8	Water	63.1

Tables with parameters α, β, and π^* are available in the literature, and some of these values are listed in Table 7.1.5 [23–25].

Solvents can be classified using parameters α, β, and π^* in a similar way to the one obtained using the x_e, x_d, and x_n values from Table 7.1.3. For this purpose, the solvents can be placed in a triangular diagram similar to the one shown in Figure 7.1.3 using the values for α, β, and π^* normalized to their sum Σ [26]. Also,

a tridimensional diagram with α, β, and π^* values can be generated, as shown in Figure 7.1.6. This figure allows visualization of the solvent "character" and potential closeness in the behavior of two different solvents. For example, it can be seen that solvents 7, 8, 31, 38, 40, 43, 46, 49, and 50 have rather similar properties, which is not surprising being all aliphatic alcohols.

Solvatochromic parameter π^* has some correlation with the polarity P' obtained from liquid

TABLE 7.1.5 Solvatochromic parameters α, β and π* for several common solvents.

No.	Compound	α	β	π*	No.	Compound	α	β	π*
1	Acetic acid	1.12	0.45	0.64	29	Ethanediol	0.9	0.52	0.92
2	Acetonitrile	0.19	0.4	0.75	30	1,3-Dioxolane	0	0.45	0.69
3	Aniline	0.26	0.5	0.73	31	Ethanol	0.86	0.75	0.54
4	Benzene	0	0.1	0.59	32	Ethyl acetate	0	0.45	0.55
5	Benzyl alcohol	0.6	0.52	0.98	33	Formamide	0.71	0.48	0.97
6	Butanoic acid	1.1	0.45	0.56	34	Formic acid	1.23	0.38	0.65
7	1-Butanol	0.84	0.88	0.47	35	Heptane	0	0	0
8	2-Butanol	0.69	0.8	0.4	36	Hexamethylphosphoramide	0	1.05	0.87
9	2-Butanone	0.06	0.48	0.67	37	Hexane	0	0	0
10	Carbon disulfide	0	0.07	0.61	38	Methanol	0.93	0.66	0.6
11	Carbon tetrachloride	0	0.1	0.28	39	Methyl acetate	0	0.42	0.6
12	Chlorobenzene	0	0.07	0.71	40	2-Methyl-2-propanol	0.68	1.01	0.41
13	1-Chlorobutane	0	0	0.39	41	Morpholine	0.29	0.7	0.39
14	Chloroform	0.44	0	0.58	42	Octane	0	0	0.01
15	Cyclohexane	0	0	0	43	Octanol	0.77	0.81	0.4
16	Cyclopentane	0	0	−0.087	44	Pentane	0	0	−0.08
17	1,2-Dichlorobenzene	0	0.07	0.67	45	Pentanoic acid	1.19	0.45	0.54
18	1,1-Dichloroethane	0.1	0.1	0.48	46	Pentanol	0.84	0.86	0.4
19	1,2-Dichloroethane	0	0.1	0.81	47	Piperidine	0	1.04	0.3
20	Dichloromethane	0.13	0.1	0.82	48	1,2,3-Propanetriol	1.21	0.51	0.62
21	Diethyl ether	0	0.47	0.27	49	1-Propanol	0.78	0.84	0.52
22	Diethyl sulfide	0	0.37	0.46	50	2-Propanol	0.76	0.95	0.48
23	Diethylamine	0.3	0.7	0.24	51	Propionitrile	0.10	0.37	0.71
24	Diisopropyl ether	0	0.49	0.27	52	2-Propanone	0.08	0.43	0.71
25	N,N-Dimethylacetamide	0	0.76	0.88	53	Propanoic acid	1.12	0.45	0.58
26	N,N-Dimethylformamide	0	0.69	0.88	54	Pyridine	0	0.64	0.87
27	Dimethyl sulfoxide	0	0.76	1	55	Pyrrolidine	0.16	0.7	0.39
28	Dioxane	0	0.37	0.55	56	Sulfolane	0	0.39	0.98

(Continued)

TABLE 7.1.5 Solvatochromic parameters α, β and π* for several common solvents. (*cont'd*)

No.	Compound	α	β	π*	No.	Compound	α	β	π*
57	Tetrahydrofuran	0	0.55	0.58	61	2,2,2-Trifluoroethanol	1.51	0	0.73
58	Tetramethylsilane	0	0.02	−0.09	62	m-Xylene	0	0.11	0.47
59	Toluene	0	0.11	0.54	63	p-Xylene	0	0.12	0.43
60	Triethylamine	0	0.71	0.14	64	Water	1.17	0.47	1.09

gas partition, but $R^2 = 0.6475$. Also, π* was found to be correlated with the Hildebrand solubility parameter δ^2 [24, 27]. However, the correlation remains acceptable only for compounds with low or medium polarity and is not applicable to polar compounds such as water or methanol. Figure 7.1.7 shows this correlation. As expected, solvatochromic parameter $E_T(30)$ is also related to π* and α, as well as to a polarizability correction parameter

δ* for a specific solvent. The following correlation equation has been verified:

$$E_T(30) = 2.8591\,[10.60 + 5.12(\pi^* - 0.23\,\delta^*) + 5.78\alpha]$$

$$(7.1.27)$$

Parameter δ* is 0.0 for nonchlorinated aliphatic solvents, 0.5 for chlorinated aliphatics,

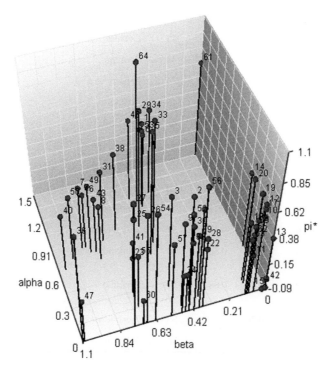

FIGURE 7.1.6 Tridimensional diagram showing the potential similarities between different solvents based on solvatochromic parameters α, β and π*. Solvent number is given in Table 7.1.5.

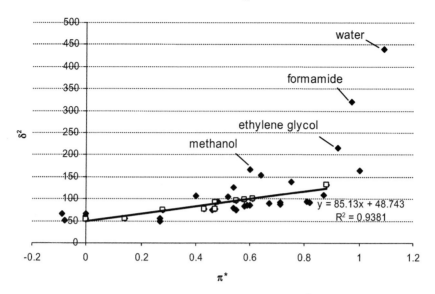

FIGURE 7.1.7 Correlation between π^* and Hildebrand solubility parameter δ^2. Trendline for the correlation of a limited number of compounds (less polar) is also shown.

0.75 for water, 1.0 for aromatic compounds. Several other correlations between solvatochromic parameters and the other solubility parameters have been described [25]. For example, various relations have been described between E_T^N and P', and also between π^*, P', and δ^* such as $P' = 0.83 + 6.31(\pi^* - 0.30\ \delta^*)$.

The values for α, β, and π^* were used in attempts to calculate capacity factors in RP-HPLC, although the precision of the results was not very satisfactory [28]. Nevertheless, a number of studies were reported regarding the use of these parameters for solvent characterization [3, 29] (see Section 9.3).

Eluotropic Strength

Besides solubility parameter δ and polarity P', some other parameters were developed for characterization of the behavior of certain solvents, in particular those related to HPLC applications. One such parameter is the eluotropic strength ε^0 [30, 31]. This parameter has been developed in connection with the adsorption-type equilibrium taking place in HPLC (see Section 3.2). For an adsorption equilibrium $i_{mo} \rightleftarrows i_{st}$ described by the equilibrium constant K'_i given by rel. 3.2.6, it was shown that the following formula can be written:

$$K'_i = \exp[-(n\ \mu^0_{S,st} - \mu^0_{i,st})/(RT)]C_S \quad (7.1.28)$$

where C_S is the quantity of adsorbed mobile phase per unit weight of adsorbent and n is the number of molecules of solvent displaced by one molecule of solute (analyte). The expressions for the standard chemical potential $\mu^0_{S,st}$ and $\mu^0_{i,st}$ can be obtained using the following formulas:

$$\mu^0_{i,st} = 2.302\ RT\ \alpha'\ S^0_i \quad (7.1.29)$$

and

$$\mu^0_{S,st} = 2.302\ RT\ \alpha'\ \varepsilon^0\ A_S \quad (7.1.30)$$

In these expressions, α' is a characteristic property of the solid phase that measures the ability of a unit of adsorbent surface to bind adsorbed molecules [30], and S^0_i is a (nondimensional) measure of adsorption energy of i onto a standard adsorbent surface (defined for

$\alpha' = 1$) from a standard mobile phase (for which eluotropic strength $\varepsilon^0 = 0$). Eluotropic strength ε^0 describes the adsorption energy of solvent S per unit area of standard adsorbent surface ($\alpha' = 1$) and characterizes the solvent "strength" (its units are cm^{-2}). The parameter A_S (in cm^2) is the surface area occupied on the adsorbent by a solvent molecule. Expressions 7.1.29 and 7.1.30 can be included in rel. 7.1.28. Following the replacement of the area occupied by the solvent A_S with the area occupied by the solute A_i, (where $A_i = n\, A_S$), the expression for log K'_i in the solvent S can be written as follows:

$$\log K_i = 2.303 \log C_S + \alpha' (S_i^0 - \varepsilon^0 A_i) \quad (7.1.31)$$

As shown in Sections 3.1 and 3.2, the partition coefficient K_i defined by rel. 3.1.7 and the equilibrium coefficient K'_i defined by rel. 3.2.6 and 7.1.31 are equivalent parameters describing the ratio of the concentrations of an analyte in the stationary phase versus the concentration in the mobile phase. For this reason, it is common in HPLC to indicate this coefficient as partition coefficient, regardless of the true type of equilibrium that it describes.

In expression 7.1.31, several parameters must be assessed. These parameters are specific for specific combinations of stationary phases and solvents. Some values of ε^0 for different solvents and different stationary phases are given in Table 7.1.6 (see, e.g., [32]).

For a different solid phase, the eluotropic strength of the same solvent is not the same as that seen in Table 7.1.6. On other polar phases the trend for the values of ε^0 remains similar

TABLE 7.1.6   Eluotropic strength on silica, alumina, florisil, MgO, and C18 stationary phases of various solvents used in HPLC.

Solvent	ε^0 $\equiv$SiOH	ε^0 Al(OH)$_3$	ε^0 Florisil	ε^0 MgO	ε^0 C18 *
Fluroroalkanes	-	−0.25	-	-	-
Pentane	0.00	0.00	0.00	0.00	
Hexane	0.00–0.01	0.00–0.01	-	-	-
Iso-octane	0.01	0.01	-	-	-
Petroleum ether	-	-	-	-	-
Cyclohexane	0.03	0.04	-	-	-
Cyclopentane	-	0.05	-	-	-
Carbon disulfide	-	0.15	-	-	-
Carbon tetrachloride	0.11	0.17–0.18	0.04	0.10	-
1-Chlorobutane	0.20	0.26–0.30	-	-	-
Xylene	-	-	-	-	-
Toluene	0.22	0.20–0.30	-	-	-
Chlorobenzene	0.23	0.30–0.31	-	-	-
Diisopropyl ether	-	0.28	-	-	-
Isopropyl chloride	-	0.29	-	-	-

(Continued)

TABLE 7.1.6   Eluotropic strength on silica, alumina, florisil, MgO, and C18 stationary phases of various solvents used in HPLC. (*cont'd*)

Solvent	ε^0 $\equiv$SiOH	ε^0 Al(OH)$_3$	ε^0 Florisil	ε^0 MgO	ε^0 C18 *
n-Propyl chloride	-	0.29	-	-	-
Benzene	0.25	0.32	0.17	0.22	
Diethyl ether	0.38–0.43	0.38			
Dichloromethane	0.30–0.32	0.36–0.42	0.23	0.26	
Chloroform	0.26	0.36–0.40	0.19	0.26	
Diethyl sulfide	-	0.38	-	-	-
1,2-Dichloroethane	-	0.44–0.49	-	-	-
Methyl ethyl ketone	-	0.51	-	-	-
Acetone	0.47–0.53	0.56–0.58			8.8
Dioxane	0.49–0.51	0.56–0.61			11.7
1-Pentanol	-	-	-	-	-
Tetrahydrofuran	0.53	0.45–0.62			3.7
Methyl *tert*-butyl ether	0.48	0.3–0.62	-	-	-
Ethyl acetate	0.38–0.48	0.58–0.62			-
Methyl acetate	-	0.60	-	-	-
Dimethyl sulfoxide	-	0.62–0.75	-	-	-
Diethylamine	-	0.63	-	-	-
Nitromethane	-	0.64	-	-	-
Acetonitrile	0.50–0.52	0.52–0.65			3.1
1-Butanol	-	0.70	-	-	-
Pyridine	-	0.71	-	-	-
2-Methoxyethanol	-	0.74	-	-	-
n-Propyl alcohol	-	0.78–0.82			10.1
Isopropyl alcohol	0.60	0.78–0.82			8.3
Ethanol	-	-	-		3.1
Methanol	0.70–0.73	0.95			1.0
Ethylene glycol	-	-	-	-	-
Dimethyl formamide	-	-	-	-	7.6

* Note: ε^0 for C18 are taken relative to $\varepsilon^0 = 1$ for methanol (Some values were not reported or measured).

from phase to phase. For example, proportionality can be seen for the values for silica and alumina where ε^0 ($\equiv$SiOH) $\approx 0.836\ \varepsilon^0$ (Al(OH)$_3$) ($R^2 = 0.9412$). On the other hand, on C18 phase the eluotropic strength is significantly different from that on polar phases, and the values decrease when those for alumina or silica increase.

The parameter S^0_i in rel. 7.1.31 can be estimated making the assumption that S^0_i is an additive property depending on the group of atoms in the molecule to be separated, but also on the stationary phase. This allows the calculation of S^0_i with the formula:

$$S^0_i = \sum_m E_m \qquad (7.1.32)$$

where E_m is a parameter (nondimensional) proportional to the free energy of adsorption for a group of atoms on a specific sorbent. The values are not totally independent of the neighbor or the structure of the rest of the molecule. Monovalent groups m have only one neighbor A (A-m) while divalent groups m have two neighbors, A and a (A-m-a), and the value for each E_m depends in the case of two neighbors if they are both aliphatic, one aliphatic and one aromatic, or both aromatic. Some values for E_m for different types of neighbors are given in Table 7.1.7.

The evaluation of the area occupied by the solute A_i, is the next problem in attempting to utilize rel. 7.1.31 for the evaluation of the equilibrium constant K'_i. This area can be considered as proportional to the van der Waals surface area of the molecule A_i (which can be calculated), with an estimated coefficient (e.g., 0.25). However, due to all the difficulties in a calculation of K'_i based on rel. 7.1.31, the estimation of its variation as a function of various parameters is more useful. This variation can be extended to the variation of the capacity factor k when considering rel. 3.2.9. For example, rel. 7.1.31 indicates that the increase in the eluotropic strength diminishes the value of capacity factor. In other words, use of a solvent with higher ε^0 decreases the retention time of a particular analyte. In general, for a polar stationary phase such as silica or columns used in HILIC, higher A_i values correspond to the more polar solutes. Therefore the effect of a solvent with higher eluotropic strength is more noticeable for solutes that are more polar. Eluotropic strength has an opposite trend for C18 phases compared to polar phases, as expected. This indicates that the solvents with higher ε^0 for polar phases are weaker solvents for reversed-phase ones. These types of observations have practical applications, and the choice of a solvent can be guided by its ε^0 values. A detailed discussion of these types of interactions can be found in several published studies (see, e.g., [30, 33, 34]).

Estimation of Selectivity Based on Eluotropic Strength

From rel. 2.1.49 for the selectivity α, and from rel. 2.1.13 between the equilibrium coefficient and the capacity factor, it can be seen that selectivity is given by the formula:

$$\alpha = \frac{K'_i}{K'_j} \qquad (7.1.33)$$

By introducing the expressions for K'_i and K'_j given by rel. 7.1.31, the result will be as follows:

$$\log \alpha = \alpha'[(S^0_i - S^0_j) - \varepsilon^0(A_i - A_j)] \qquad (7.1.34)$$

where both α' and ε^0 were considered the same for the two analytes i and j. Relation 7.1.34 shows that the selectivity increases as α' increases (increases the ability of a unit of adsorbent surface to bind molecules). Also, the selectivity increases when the difference between adsorption energies increases. On the other hand, the selectivity is decreased in solvents with high eluotropic strength (with the condition that $A_i > A_j$), this decrease being affected by the difference in the area occupied by each solute on the sorbent surface. The variation in α can be estimated using rel. 7.1.2 for

TABLE 7.1.7 Values for the additive parameter E_m for several common organic groups.

Group	A aliphatic, a aliphatic		A aliphatic, a aromatic		A aromatic, a aromatic	
	SiO_2	Al_2O_3	SiO_2	Al_2O_3	SiO_2	Al_2O_3
Methyl (-CH$_3$)	0.07	−0.03	-	-	0.11	0.06
Methylene (-CH$_2$-)	−0.05	0.02	0.01	0.07	0.07	0.12
Aromatic C (-CH=)	0.25	0.31	0.25	0.31	0.25	0.31
Fluoro (-F)	1.54	1.64	-	-	-0.15	0.11
Chloro (-Cl)	1.74	1.82	-	-	-0.20	0.20
Bromo (-Br)	1.94	2.00	-	-	-0.17	0.33
Iodo (-I)	1.94	2.00	-	-	-0.15	0.51
Disulfide (-S$_2$-)	1.90	2.70	0.94	1.10	-	-
Sulfide (-S-)	2.94	2.65	1.29	1.32	0.48	0.76
Ether O (-O-)	3.61	3.50	1.83	1.77	0.87	1.04
Tertiary amine (-N<)	5.80	4.40	2.52	2.48	-	-
Aldehyde (-CHO)	4.97	4.73	-	-	3.48	3.35
Nitro (-NO$_2$)	5.71	5.40	-	-	2.77	2.75
Nitrile (-CN)	5.27	5.00	-	-	3.33	3.25
Ester (-COO-)	5.27	5.00	3.45	3.40	4.18	4.02
Ketone (>CO)	5.27	5.00	4.69	3.74	4.56	4.36
Hydroxyl (-OH)	5.60	6.50	-	-	4.20	7.40
Imine (>C=N-)	-	6.00	-	4.46	-	4.14
Primary amine (-NH$_2$)	8.00	6.24	-	-	5.10	4.41
Sulfoxide (>SO)	7.20	6.70	4.20	4.00	-	-
Acid (-COOH)	7.60	21	-	-	6.10	19
Amide (-CO-NH$_2$)	9.60	8.90	-	-	6.60	6.20

(Note: e.g. for benzene $E_m = 6 (-CH=) = 1.86$; for nitrobenzne $E_m = 1.86 + 2.75 = 4.61$).

different solutes i and j by calculating $S^0{}_i$, $S^0{}_j$, A_i and A_j, and estimating α' and ε^0.

Solvent Properties of Liquid Mixtures

In HPLC, mixtures of liquids are commonly used as mobile phases. Even more, gradient separations are more common than isocratic ones. It is therefore of considerable interest to estimate a specific parameter that characterizes solvent properties for a mixture of liquids. However, this is not always possible for parameters such as Hildebrand solubility parameter δ, which is based on an individual molecular property (vaporization energy) or K_{ow} that is specific for one compound.

For other parameters that can be truly defined for solvent mixtures, such as polarity or solvatochromic parameters, approximations can be directly obtained. For the polarity P' of a solvent, a rough first approximation of its value for a mixture of solvents is given by the expression:

$$P'_{mix} = P'_1 \phi_1 + P'_2 \phi_2 + \dots P'_n \phi_n \qquad (7.1.35)$$

where P'_1, P'_2, and so on are the polarities of individual solvents, and ϕ_1, ϕ_2, and so on are the volume fractions of each component of the mixture. However, rel. 7.1.35 is only an approximation of polarity, and nonlinear variations of parameters with the composition are common for solvent mixtures.

For solvatochromic parameters α, β, π^*, E_T^N, the values for solvent mixtures can be obtained experimentally [35]. As an example, the variation of E_T^N for methanol/water and acetonitrile/water mixtures is shown in Figure 7.1.8 [36]. A quadratic trendline (best least squares fit) is shown in Figure 7.1.8 for the variation of E_T^N for the methanol/water mixture. The regression coefficient R^2 for the fit is very close to 1, indicating that an excellent approximation of

the variation can be expressed empirically. However, for the acetonitrile/water case, a more complicated dependence can be seen between E_T^N and solvent composition. The variation of solvatochromic parameters π^*, α and β is again nonlinear for methanol/water mixtures, as shown in Figure 7.1.9 and for acetonitrile/water, as shown in Figure 7.1.10 [37].

The eluotropic strength ε^0 can also be defined for a mixture of two pure solvents A and B. This value can be obtained from the eluotropic strength of the pure solvents ε^0_A and ε^0_B with the formula [38]:

$$\varepsilon^0 = \varepsilon^0_A + \frac{\log[x_B 10^{\mathscr{A}^{rel}_{B,st}(\varepsilon^0_B - \varepsilon^0_A)} + 1 - x_B]}{\mathscr{A}^{rel}_{B,st}} \qquad (7.1.36)$$

where x_B is the mole fraction of solvent B in the mobile phase and $\mathscr{A}^{rel}_{B,st}$ is the relative area of solvent B, reported to benzene for which $\mathscr{A}^{rel}_{B,st} = 6$ cm^2.

The main interest in evaluating specific parameters for characterizing a solvent mixture is related to predicting the influence of composition changes on the analytes retention (typically characterized by log k). A number of

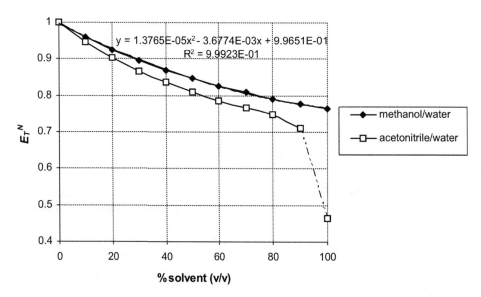

FIGURE 7.1.8 Variation of solvatochromic parameter E_T^N for methanol/water and acetonitrile/water mixtures.

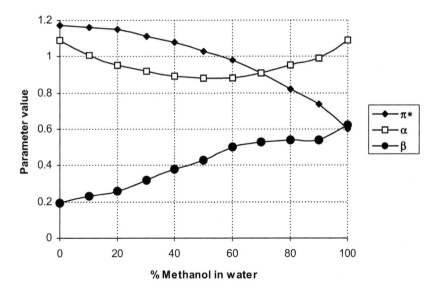

FIGURE 7.1.9 Variation of solvatochromic parameters π^*, α and β for methanol/water mixtures.

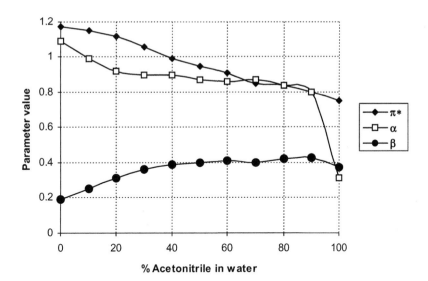

FIGURE 7.1.10 Variation of solvatochromic parameters π^*, α and β for acetonitrile/water mixtures.

experimental results showed that a linear dependence exists between the mobile phase composition and log k. This type of dependence is exemplified in Figure 7.1.11, where the values for log k are plotted versus the content in methanol in the mobile phase, for four compounds separated on a Purospher Star RP-18e column, 125 mm length, 4 mm inner diameter, 5 μm particle size [39]. The trendlines for the data given in Figure 7.1.11 show excellent linearity between log k and ϕ(%) values (Compound 1: log $k = 1.484 - 2.52 \ 10^{-2} \ \phi$, $R^2 = 0.9999$,

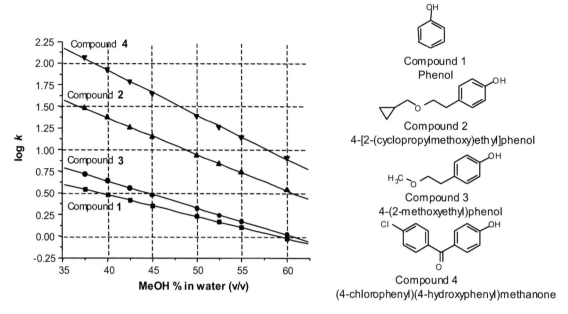

FIGURE 7.1.11   Variation of log k with the composition of mobile phase methanol/water for four analyte *(Purospher Star RP-18e, 125 mm length, 4 mm inner diameter, 5 μm particle size)* [39].

Compound 2: $\log k = 3.051 - 4.21\ 10^{-2}\ \phi$, $R^2 = 0.9994$, Compound 3: $\log k = 1.886 - 3.12\ 10^{-2}\ \phi$, $R^2 = 0.9993$, Compound 4: $\log k = 3.986 - 5.16\ 10^{-2}\ \phi$, $R^2 = 0.9991$). The linearity between $\log k$ and organic phase content (for RP-HPLC) is expressed by the formula 2.1.70, written once more below [33]:

$$\log k_{iX} = \log k_{iw} - S_i \phi \qquad (7.1.37)$$

where $\log k_{iw}$ is the capacity factor for the analyte i when separated with water as a mobile phase on a specific chromatographic column and $\log k_{iX}$ is the capacity factor for the same analyte in a mobile phase containing an organic modifier X with the volume fraction ϕ. The values for k_{iw} are not always measurable, and an extrapolation of the value can be obtained (still assuming relation 7.1.37 valid for $\phi \rightarrow 0$). Parameter S is assumed to be a constant specific for the given compound, given solvent system, and given chromatographic column. The approximation $S \approx 0.25\ (MW)^{1/2}$ where MW is the

molecular weight of the analyte was indicated in Section 2.1. Besides the dependence on the stationary phase, it is common that the calculated values of $\log k_{iA}$ also depend on the solvent system. Even if A is pure water and the values for k_{iw} should not depend on the other solvent, various values for k_{iw} are obtained when calculated by extrapolation, for example, from a methanol/water mobile phase or from a acetonitrile/water mobile phase. For two concentrations of the organic modifier ϕ_1 and ϕ_2, rel. 7.1.37 leads to the formula:

$$\log k_{iX2} = \log k_{iX1} - S(\phi_2 - \phi_1) \qquad (7.1.38)$$

Relation 3.1.38 allows the calculation of $\log k_{iX2}$ for the compound i in a solvent with the volume fraction ϕ_2 of organic component when the capacity factor k_{iX1} is known, and k_{iX1} corresponds to a mobile phase with the volume fraction ϕ_1 for the organic component. Variation of k in gradient conditions is discussed in Section 7.5.

Graphs such as those shown in Figure 7.1.11 can be used for calculating S, and by extrapolation for the calculating log k_{iw}. Several values for S for a water-methanol solvent mixture on different RP-type columns are reported in the literature [40] and are given in Table 7.1.8. The linear dependencies seen in Figure 7.1.11 are not always followed, and different compounds and mobile phase systems may show large deviations from linearity [41]. For example, for acetonitrile/water nonlinear variations of log k as a function of ϕ are frequently obtained, which are better approximated with a quadratic equation [42]. Such nonlinear dependence is illustrated in Figure 7.1.12 for the separation of chlorobenzene and 2,4-dinitrophenol on a Lichrospher 100 RP-10 column 250 × 4.0 mm with 5 μm particles with mobile phase acetonitrile/water at different compositions. As shown in Figure 7.1.12, the relation connecting the change in the log k with the change in the composition of mobile phase is better expressed by the formula:

$$\log k_{iX} = \log k_{iw} + c_{i1}\,\phi + c_{i2}\,\phi^2 \qquad (7.1.39)$$

The coefficients c_{i1} and c_{i2} in rel. 7.1.39 must be obtained experimentally similar to the case for the values for S. Although the nonlinear dependences of log k on solvent composition are not uncommon, linear relations are still frequently used to express the variation of the capacity factor log k as a function of organic content in the mobile phase ϕ.

An additional approach to estimate parameters for mixtures of solvents is to disregard the true meaning of parameters such as δ or K_{ow} as to be unique to a compound. In this case parameters δ or K_{ow} are assigned to a "hypothetical" liquid with identical property as the mixture. Using the same approach as that for the estimation of log k, estimation of an hypothetical δ for a solvent mixture can be easily obtained. From rel. 7.1.16 and 7.1.37 applied for the two solvents X and Y where a modifier at the volume fraction ϕ is added to X to generate Y, rel. 7.1.40 is obtained:

$$(\delta_Y - \delta_i)^2 = (\delta_X - \delta_i)^2 - S'\phi \qquad (7.1.40)$$

TABLE 7.1.8 Several values for S reported in the literature for water/methanol mixtures [40].

Analyte	Column				
	Monomeric C18 (end-capped)	Monomeric C18 (not end-capped)	Polymeric C18 (end-capped)	Polymeric C18 (not end-capped)	Monomeric C8 (not end-capped)
Phenol	2.21	2.97	2.52	2.35	3.13
Benzaldehyde	2.52	3.07	2.72	2.92	3.08
Acetophenone	2.82	3.63	3.04	3.08	3.39
Nitrobenzene	2.61	3.18	2.75	2.79	3.16
Methyl benzoate	3.17	3.82	3.46	344	3.78
Anisole	2.61	3.29	2.93	2.90	3.25
Fluorobenzene	2.70	3.27	3.07	2.90	3.28
Benzene	2.32	2.94	2.66	2.55	3.02
Toluene	2.9	3.52	3.23	3.13	3.56
Average	2.65	3.29	2.94	2.90	3.29

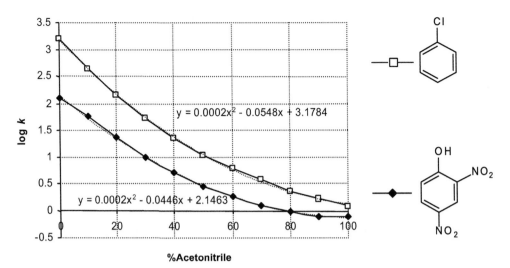

FIGURE 7.1.12 Variation of log k with the composition of mobile phase acetonitrile/water for two analytes, chlorobenzene and 2,4-dinitrophenol.

where $S' = S\dfrac{2.303n_iRT}{V_i}$. For relatively close values for δ_X and δ_Y, the following relation can be written:

$$\delta_Y^2 \approx \delta_X^2 - S'\phi \qquad (7.1.41)$$

For a liquid mixture X containing a pure solvent A plus an organic modifier B at the volume fraction ϕ, a hypothetical $K_{ow}^{hyp}(X)$ can also be defined by the expression:

$$\log K_{ow}^{hyp}(X) = \log K_{ow}(A) - S''\phi \qquad (7.1.42)$$

where S is specific for the solvent system and can be positive or negative. Assuming a linear variation of $K_{ow}^{hyp}(X)$ for different volume fractions of A and B, rel. 7.1.42 will give the expression:

$$\log K_{ow}^{hyp}(X) = (1 - \phi)\log K_{ow}(A)$$
$$+ \phi \log K_{ow}(B) \qquad (7.1.43)$$

The properties for the liquids and their mixtures discussed in this section are important criteria for selection of the mobile phase in HPLC, in particular in RP-HPLC and HILIC. A discussion regarding the use of these properties for the calculation of capacity factors can be found in Section 9.3

7.2. ADDITIONAL PROPERTIES OF LIQUIDS AFFECTING SEPARATION

General Comments

Depending on the HPLC type, specific physicochemical properties of the solvent used as mobile phase play a role in the separation. Among these properties are the following: (1) solvent viscosity η, (2) dielectric constant ε, (3) superficial tension γ', (4) solvent accessible surface area A^{SASA}, (5) dipole moment m, and (6) polarizability α. The role of these individual parameters in the separation was previously discussed in Chapters 4 and 5, and in part, their contribution is captured in some of the parameters discussed in Section 7.1 that characterize the properties of liquids as solvents. In this section, an overview regarding individual physicochemical parameters that may affect separation is given.

Solvent Viscosity

The effect of solvent viscosity on separation can be seen from the impact of this parameter on column plate number N. A more viscous solvent reduces the diffusion coefficient of sample components and slows down the mass transfer process. As shown in Section 2.2, various terms in the van Deemter equation depend inverse proportionally on the diffusion coefficient D of the solutes. As shown by rel. 2.2.15, D is inversely proportional to the viscosity η. Experimental studies [43] show that about a 2-fold loss in N takes place for a 2.5-fold increase in viscosity.

Besides affecting the column plate number N, viscosity also affects column backpressure (as shown by rel. 2.2.17). An increase in column backpressure limits the maximum flow rate in a chromatographic column and therefore the total runtime for a chromatogram. The choice of solvents with lower viscosity is therefore preferred in HPLC. The values of viscosity for several common solvents at 25 °C are given in Table 7.2.1.

Viscosity varies with temperature. Several models were developed for describing the variation of viscosity with the temperature, such as the Arrhenius model where the variation in the

TABLE 7.2.1 Several properties of common solvents important for the separation process.

Compound	Formula	Viscosity η (cP)	Dielectric constant	Dipole m (D) liq.	Dipole m (D) gas	Polariz. $4\pi \varepsilon_0$ (Å)^3
Acetic acid	$C_2H_4O_2$		6.2	1.92	1.75	5.33
Acetone	C_3H_6O	0.32	20.7	3.11	2.87	6.41
Acetonitrile	C_2H_3N	0.37	37.5	3.39	3.97	4.27
Benzene	C_6H_6	-	2.3	0	0	8.89
n-Butanol	$C_4H_{10}O$	2.95	17.8	2.96	1.6	9.21
Carbon disulfide	CS_2	-	2.6	0	0	6.79
Carbon tetrachloride	CCl_4	0.97	2.24	0	0	10.5
Chlorobenzene	C_6H_5Cl	-	5.5 - 6.3	1.39	1.72	11.06
Chloroform	$CHCl_3$	0.57	4.81	1.85	1.03	8.52
Cyclohexane	C_6H_{12}	1	18.5	0.2	0.61	11.07
Cyclohexanone	$C_6H_{10}O$	-	18.2	2.94	-	11.15
Cyclopentane	C_5H_{10}	0.47	2	0	0	9.2
Decane	$C_{10}H_{22}$	0.92	1.99	0	0	20.61
1,2-Dichloroethane	$C_2H_4Cl_2$	-	10.4	2.94	1.84	8.5
Di-ethyl ether	$C4H_{10}O$	0.23	4.34	1.27	1.13	9.33
Di-isopropyl ether	$C_6H_{14}O$	0.37	3.88	1.26	1.13	12.94
N,N-Dimethylformamide	C_3H_7NO	0.92	37.6	3.85	-	7.69

(Continued)

TABLE 7.2.1   Several properties of common solvents important for the separation process. (*cont'd*)

Compound	Formula	Viscosity η (cP)	Dielectric constant	Dipole m (D) liq.	Dipole m (D) gas	Polariz. $4\pi\,\varepsilon_0\,(\text{Å})^3$
Dimethylsulfoxide	C_2H_6SO	2.24	46.2	3.9	-	7.91
Dioxane	$C_4H_8O_2$	1.54	2.21	0.45	0.43	8.97
Ethanol	C_2H_6O	1.2	24.6	1.66	1.69	5.3
Ethyl acetate	$C_4H_8O_2$	0.45	6.02	2.05	1.78	9.28
Ethylene glycol	$C_2H_6O_2$	19.9	37.7	2.2	2.2	6.18
n-Heptane	C_7H_{16}	0.41	1.89	0	0	14.34
n-Hexane	C_6H_{14}	0.33	1.9	0	0	12.3
Isobutyl alcohol	$C_4H_{10}O$	4.7	16.68	2.96	1.64	9.07
Isooctane	C_8H_{18}	0.53	1.94	0	0	16.18
Isopropanol	C_3H_8O	2.37	19.9	3.09	1.59	7.14
Methanol	CH_4O	0.6	32.7	2.97	1.69	3.38
Methyl acetate	$C_3H_6O_2$	0.37	6.68	1.74	1.68	7.36
Methyl ethyl ketone	C_4H_8O	0.43	18.5	3.41	-	8.35
Methyl isobutyl ketone	$C_6H_{12}O$	-	13.11	2.68	-	12.04
Methyl *tert*-butyl ether	$C_5H_{12}O$	-	4	1.25	-	10.88
Methylene chloride	CH_2Cl_2	0.44	9.08	1.9	1.54	4.54
Morpholine	C_4H_9NO	7.42	-	1.75	-	9.47
Nitrobenzene	$C_6H_5NO_2$	-	34.8	3.93	4.28	11.22
Nitromethane	CH_3NO_2	-	35.9	4.39	3.44	4.66
n-Octanol	$C_8H_{18}O$	-	3.4	1.61	-	17.42
n-Pentane	C_5H_{12}	0.23	2.1	0	0	10.27
n-Propanol	C_3H_8O	2.27	20.1	3.09	1.65	7.23
Pyridine	C_5H_5N	-	12.4	2.31	2.15	8.25
Tetrahydrofuran	C_4H_8O	0.55	7.58	1.63	1.63	8.14
Tetrahydrothiopne-1, 1-dioxide	$C_4H_8O_2S$	-	43.3	4.69	-	11.46
Toluene	C_7H_8	0.59	2.37	0.38	0.37	10.97
Triethylamine	$C_6H_{15}N$	-	2.4	0.75	0.61	13.47
p-Xylene	C_8H_{10}	-	2.4	0.02	0	13.12
Water	H_2O	1	77.46	3.12	1.85	1.51

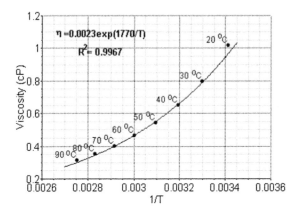

FIGURE 7.2.1   Variation of viscosity of water (in cP) with temperature.

viscosity follows an equation similar to that for the rate constant:

$$\eta(T) = \eta_0 \exp\left(E/RT\right) \qquad (7.2.1)$$

where η_0 and E are constants. As an example, the variation of viscosity of water (in cP) with inverse of the temperature (K) is given in Figure 7.2.1. The fit for the equation 7.2.1 is also shown in the figure. Viscosity also varies for solvent mixtures

depending on their composition. However this variation is difficult to predict and depends on the interactions at the molecular level between the solvents. As an example, the variation of viscosity as a function of composition for water/methanol, water/tetrahydrofuran, and water/acetonitrile at 25 °C is shown in Figure 7.2.2.

The increase in the viscosity of a solvent mixture must be taken into consideration when working with gradient elution. The increase in viscosity during the gradient creates an increase in the column backpressure, and this may be a problem when working close to the upper limit of backpressure acceptable for either the pumps or the column. As shown in Figure 7.2.2, for water/methanol mixtures in particular, the viscosity becomes more than 1.5 times higher than that of water during a gradient reaching 40 to 60% methanol.

Dielectric Constant

The role of dielectric constant ε of a solvent regarding its capability to dissolve ionized solutes was pointed out in Section 4.1 (the

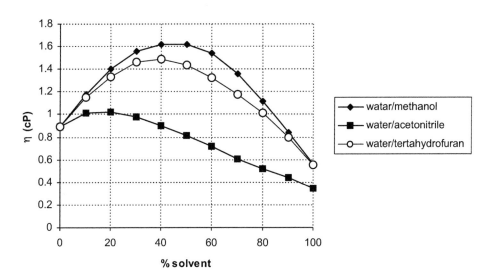

FIGURE 7.2.2   Variation of viscosity (in cP) for mixtures water/methanol, water/tetrahydrofuran and water/acetonitrile as a function of composition, at 25 °C.

dielectric constant is typically given as the relative dielectric constant to vacuum). A higher dielectric constant of the solvent correlates with a higher ability of the solvent to dissolve salts. This property is important when using buffers as eluant in HPLC. The dielectric constant of the solvent also affects interactions in solution that involve ions and polar molecules, decreasing the intermolecular energy when the dielectric constant increases, as shown by several formulas in Sections 4.1 and 4.2. The values of the dielectric constant for several common solvents at 25 °C are given in Table 7.2.1. The dielectric constant also depends on temperature. As an example, the variation of the dielectric constant of water as a function of temperature is shown in Figure 7.2.3.

Dipole Moment

The dipole moment of a molecule has been defined by rel. 4.1.12 (see Section 4.1). Since dipole moment describes the polarity of a molecule, characterization of a solvent polarity can be based on its dipole moment. The mechanism of a number of HPLC types of separation including HILIC and also, to a lower extent, RP involves polar interactions. For this reason, the dipole

moment m is a parameter useful in solvent characterization. Dipole moment varies as a function of temperature, and for some compounds it is different for the molecules in gas form and in liquid form. For molecular interactions used as a model in gas phase for understanding of the separation process, the values of m in the gas phase can be used. However, the values for the dipole moment in the liquid form seem more appropriate for characterization of solvent properties. Values for the dipole moment (in debye D) of several common liquids used in HPLC as a mobile phase (or mobile phase additives) are listed in Table 7.2.1 for both liquid and gas forms of the compound [44]. The reported values are for temperatures around 25 °C.

Polarizability

Similar to dipole moment, polarizability in an electric field of a solvent molecule provides some information regarding the intensity of interactions in solution. The dispersion forces between molecules make the largest contribution to the total interaction energy between nonionic molecules, being higher than dipole-dipole or dipole-induced dipole interactions. The intensity of dispersion interactions is dependent on molecular polarizability as shown by rel. 4.1.43. Polarizability also affects the interaction of polar molecules, as shown by rel. 4.1.46. As indicated in Section 4.1, polarizability can be defined as a tensor when the electrical field generates moments of dipole in different directions from that of the field $\vec{E}$. However, average molecular polarizability is the common value of interest [45]. The polarizabilities for several common solvents (expressed in $4\pi\varepsilon_0(\text{Å})^3$) are listed in Table 7.2.1. The results were obtained using MarvinSketch Plugin from ChemAxon Ltd. [11].

Superficial Tension

The importance of superficial tension in HPLC separations has been discussed in

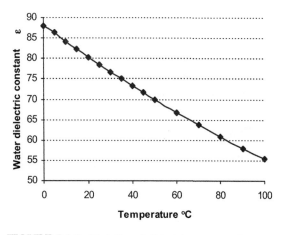

FIGURE 7.2.3 Variation of dielectric constant of water ε as a function of temperature.

relation to hydrophobic interactions in Section 4.1. Relation 5.1.26 showed that a higher value of γ' is associated with a "weaker" solvent in RP-HPLC. Solvents with higher γ' used as mobile phase lead to larger log k values in RP chromatography. When only the surface tension of the solvent used as mobile phase changes, rel. 5.1.26 can be reduced to an expression of the form:

$$\ln k = a + b\gamma' \qquad (7.2.2)$$

where a and b are approximated as constants.

Superficial tension decreases linearly with the temperature. This dependence can be obtained from Eötvös rule, which can be written as follows:

$$\gamma' = C_t \, V^{-2/3}(T_c - T) \qquad (7.2.3)$$

where C_t is a constant, V is the molar volume ($V = M \, / \, \rho$), T_c is critical temperature, and T is the temperature of interest. From rel. 7.2.3. the following expression of temperature dependence can be obtained:

$$\gamma'(T_2) = \gamma'(T_1) + C_t'(T_1 - T_2) \qquad (7.2.4)$$

The decrease of γ' with temperature is in agreement with the observation that the capacity factor k' decreases as temperature increases. Relation 7.2.4 can be further written in the form:

$$\gamma'(T_2) = \gamma\prime(T_1) + C_t''T_2 \qquad (7.2.5)$$

The values of γ' for several common solvents and the coefficient C_t'' for temperature dependence are given in Table 7.2.2.

Since mixtures of solvents are frequently used in HPLC as a mobile phase, the surface tension of these mixtures is typically of interest. Several studies are reported in the literature providing results on γ' for different solvent mixtures [46, 47]. As an example, the variation of surface tension with the organic phase concentration in water for several common solvents is shown in Figure 7.2.4.

TABLE 7.2.2   Values of surface tension γ' in mN/m (dyne/cm) and its temperature coefficient in mN/(m K) for several common solvents.

Compound	γ' 20 °C mN/m	Temp. coeff. mN/(m K)
Acetone (2-Propanone)	25.2	−0.112
Benzene	28.88	−0.1291
Carbon disulfide	32.3	−0.1484
Chlorobenzene	33.6	−0.1191
Chloroform	27.5	−0.1295
Cyclohexane	24.95	−0.1211
1,2-Dichloroethane	33.3	−0.1428
1,4-Dioxane	33	−0.1391
Ethanol	22.1	−0.0832
Isopropanol	23	−0.0789
Methanol	22.7	−0.0773
Methyl ethyl ketone (MEK)	24.6	−0.1199
N,N-dimethylformamide (DMF)	37.1	−0.14
n-Decane	23.83	−0.092
n-Heptane	20.14	−0.098
n-Hexane	18.43	−0.1022
Methylene chloride	26.5	−0.1284
Nitrobenzene	43.9	−0.1177
Nitromethane	36.8	−0.1678
1-Octanol	27.6	−0.0795
Propanol (25 °C)	23.7	−0.0777
Pyridine	38	−0.1372
Toluene	28.4	−0.1189
Water	72.8	−0.1514

Besides experimental values, the surface tension of liquid mixtures can also be estimated using various formulas based on thermodynamic principles [48, 49]. Several empirical

FIGURE 7.2.4 Variation of surface tension γ' (in mN/m = dyne/cm) with the organic solvent concentration in water for several common solvents.

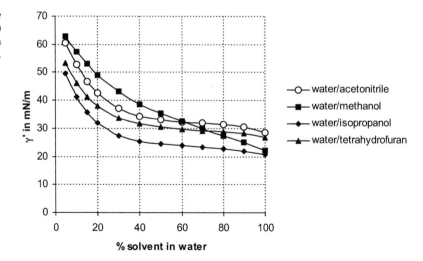

expressions were also suggested for estimation of γ'. For example, for water/acetonitrile mixtures γ' can be estimated using rel. 7.2.6, and for the mixtures for water/methanol the estimation can be done using rel. 7.2.7.

$$\gamma' = 40.898 \exp(-C_{org}/16.518) + 30.496$$
(7.2.6)

$$\gamma' = 49.042 \exp(-C_{org}/39.842) + 20.355$$
(7.2.7)

In these formulas, C_{org} is the % volume of organic component in the mobile phase (the results are in mN/m = dyne/cm).

7.3. PROPERTIES OF THE MOBILE PHASE OF IMPORTANCE IN HPLC, NOT RELATED TO SEPARATION

General Comments

The mobile phase is important in the separation process, and several other of its properties are important in the HPLC analysis. One such property is the mobile phase viscosity η, which was discussed in Section 7.2 and which affects the backpressure in the HPLC

system in accordance to rel. 2.2.17. Other properties are related in particular to the detection process. The mobile phase must differ from eluted molecules by certain physicochemical properties in order to be detected. These properties can be UV-absorption, fluorescence, molecular fragmentation in a mass spectrometer, refractive index, or others that make the analytes detectable. For this reason, some physicochemical properties of the mobile phase must be known, and the detection procedure must be selected such that the solvent (mobile phase) should not interfere with the detection. Other properties of the solvent may also be of interest, though not related to separation, such as boiling point. The values for the boiling point (at atmospheric pressure) of several common solvents are given in Table 7.3.1.

Refractive Index and UV Cutoff

Refractive index (RI) detection is commonly used for the analysis of compounds that do not have good absorption bands for UV light (see Section 1.3). For such compounds, other detection techniques, such as MS or ELSD, can also be used. However, RI is a convenient detection

TABLE 7.3.1 Physical properties of interest for detection for several common solvents.

Solvent	Refractive index	UV cut-off (nm)	Boiling point °C	Solvent	Refractive index	UV cut-off (nm)	Boiling point °C
Acetone	1.395	330	56.3	n-Hexane	1.375	195	68.7
Acetonitrile	1.344	190	81.6	Isobutyl alcohol	1.384	220	98
Benzene	1.501	278	80.1	i-Octane	1.404	210	126
n-Butanol	1.347	210	117.7	Methanol	1.329	210	64.7
sec-Butanol	1.397	260	107.7	Methyl acetate	1.362	260	57.5
t-Butyl methyl ether	1.369	210	55.5	Methyl ethyl ketone	1.381	330	80
Carbon tetrachloride	1.466	265	76.5	Methyl i-butyl ketone	1.394	330	117.5
Chloroform	1.443	245	61.2	Methylene chloride	1.424	245	39.8
Cyclohexane	1.427	200	80.7	Morpholine	-	285	129
Cyclopentane	1.406	200	49.3	i-Pentane	1.371	200	27.7
Declin	1.476	200	193	n-Pentane	1.358	210	36.1
n-Decane	1.412	210	174	i-Propanol	1.375	210	82.3
Dimethylformamide	1.427	270	153	n-Propanol	1.383	210	97.2
Dimethylsulfoxide	1.476	268	189	i-Propyl ether	1.368	220	69
Dioxane	1.422	220	101	Pyridine	1.510	330	115.3
Ethanol	1.361	210	78.3	Tetrahydrofuran	1.408	220	66
Ethyl acetate	1.370	260	77.1	Toluene	1.496	285	110.6
Ethyl ether	1.353	220	34.6	Triethylamine	1.400	-	88.8
Ethylene glycol	1.427	210	197.5	Trimethylpentane	1.389	215	115
n-Heptane	1.385	200	98.4	Water	1.323	<190	100

procedure since it offers good reproducibility and does not require expensive equipment. The sensitivity of the RI detection depends on the difference in the refractive index of the mobile phase and that of the analyte (besides other parameters related to the instrument construction). Table 7.3.1 presents a list of values for the refractive index for a number of common solvents. The refractive index depends on the wavelength of the incident beam, and the most accurate RI measurements are done with monochromatic light (usually 589 nm, the sodium D line), although white light is still commonly used for the measurements. The refractive index depends on temperature and typically decreases as temperature increases (for organic solvents this decrease is about 0.0005 for 1 °C and for water it is about 0.0001). When RI is used as a detector in HPLC, only isocratic separations can be applied. Premixed solvents are used as mobile phase as well. For RI detection, it is not recommended that the specific composition for the mobile phase be generated by using two pumps and a mixing device. Small fluctuations in the RI of the mobile phase typically generate large oscillations in the RI detector response. The refractive index of a binary solvent mixture is typically given by the expression:

$$\frac{n_{12}^2 - 1}{n_{12}^2 + 2} = \phi_1 \frac{n_1^2 - 1}{n_1^2 + 2} + \phi_2 \frac{n_2^2 - 1}{n_2^2 + 2} \qquad (7.3.1)$$

where n_1 and n_2 are the refractive indexes of the two components, n_{12} is the refractive index of the mixture, and ϕ_1 and ϕ_2 are the volume fractions of the two solvents. The value of ϕ_i can be obtained from the relation $\phi_i = x_i\, V_i / \Sigma\, x_i\, V_i$ where x_i is the mole fraction and V_i is the molar volume.

Measurement of the UV absorption of the analytes is probably the most frequent detection technique used in HPLC. The influence of the solvent on this type of measurement has two aspects. One is the requirement that the solvent be "transparent" (to absorb very little light) in the region where the absorption of the analyte is measured. This "transparency" is characterized by the UV cutoff value, defined as the wavelength at which the absorbance A of the solvent versus air, in a 1 cm cell, is equal to unity (see rel. 1.4.9). The UV absorption increases significantly when the wavelength decreases, and this cutoff value indicates that at lower wavelength the absorption of light is too strong to allow utilization. The UV cutoff values for several common solvents are given in Table 7.3.1. A more accurate description of solvent absorption is obtained from the UV spectrum of the solvent. A contribution to the cutoff value for a solvent may come not only from the solvent itself, but also from certain impurities or additives that may be present in the solvent (such as BHT or phthalates). The potential UV absorption of the additives used as buffers should also be considered when choosing the mobile phase.

The second aspect regarding the solvent influence on the UV detection in HPLC is related to modification of the absorption bands of the analyte. As previously discussed in Section 7.1, solvents with a more polar character produce a bathochromic (higher wavelength) effect on the UV spectrum of the analytes, with the absorption bands corresponding to a $\pi \rightarrow \pi^*$ transition. For many compounds, such absorption bands can increase with $1 - 20$ nm when changing, for example, from hexane to ethanol as solvent. Compounds that form H-bonds with the solvent molecule will also exhibit a bathochromic effect. The increase of the accepting capacity in H-bond of the solvent will increase the maximum wavelength in the absorption spectrum of the analyte. On the other hand, the bands corresponding to n $\rightarrow \pi^*$ transitions will suffer a hypsochromic effect when the polarity of the solvent increases. The n $\rightarrow \pi^*$ transitions can also be influenced by pH changes due to the structural modification of the analyte. For example, a bathochromic shift can be seen for compounds with phenolic hydroxyls when the pH is modified from acid to basic values due to the modification in the compound dissociation status. For aromatic amines the effect is hypsocromic when pH is modified from basic to acidic values (due to protonation).

Fluorescence

Fluorescence is also influenced by the solvent composition. Both emission wavelength and fluorescence intensity for many compounds containing dissociable functional groups are dependent on pH. The fluorescence intensity can be influenced with as much as one order of magnitude and may shift the emission maximum when strong interactions with solvent occur. A shift of the emission band toward higher wavelengths is observed with the increase of the dielectric constant of the solvent. When the solvent absorbs radiation at the emission or absorption wavelengths of the analyte(s), a significant decrease of detection sensitivity can be noticed. The presence of impurities in the mobile phase, noticeable mainly the O_2, can produce a quenching of the fluorescence signal, noticeable mainly when the analyte concentration is close to the detection limit.

Solvent Selection for MS Detection

Solvent (mobile phase) selection for MS detection (electrospray ionization [ESI] and atmospheric pressure chemical ionization [APCI]) is very important since the ionization efficiency is strongly influenced by the solvent [50]. However, selection of the mobile phase with the purpose of obtaining good sensitivity in MS detection is a complex problem, and the subject is beyond the purpose of the present book. For this reason, only general aspects of the problem are further discussed.

Several characteristics of the mobile phase influence ionization. Among these characteristics are volatility, surface tension, viscosity, conductivity, ionic strength, dielectric constant, electrolyte concentration, pH, and potential of gas-phase ion-molecule reactions. These properties must be considered in relation to the chemical and physical properties of the analyte, including pK_a, hydrophobicity, surface activity, ion solvation energy, and proton affinity. Also, the operational parameters of the instrument such as the flow rate and temperature of the mobile phase are important. An arbitrary selection of the mobile phase composition in LC-ESI/MS is not possible since only polar solvents and volatile additives can be used for obtaining good sensitivity. The selection of the mobile phase often must be balanced between ESI response and LC separation efficiency.

The following solvents are compatible with both interfaces commonly used in MS detection (APCI and ESI): water, alcohols, acetonitrile, tetrahydrofuran, acetone, dimethylformamide, dicholoromethane, chloroform, and others. For the positive ionization mode, the addition of a volatile acid to the mobile phase, usually formic acid at 0.1−0.2% levels, is common, while for the negative ionization mode the addition of a volatile salt such as ammonium formate or ammonium acetate at 40 − 50 mM level is typically practiced. An organic solvent in mixture with water is common as a mobile phase. Generally, the use of aprotic solvents is not suitable with the ESI interface. Pure water is also a poorer solvent for ESI than water mixed with organic solvents such as methanol or acetonitrile. Due to pure water's higher viscosity, the electrophoretic mobility of ions is lower, leading to inefficient charge separation and difficulties in producing a stable spray. Also, the evaporation of water from the charged droplet is slower than the evaporation of an organic solvent.

Unlike ESI, which requires mobile phases based on polar or medium polar solvents, in APCI both polar and nonpolar solvents can be used. For this reason, APCI can be chosen as the interface in NP-LC with MS detection, while ESI is not recommended for this type of HPLC. The solvents commonly used in NP-LC/MS-APCI include n-hexane, 2-propanol, methanol, ethanol, iso-hexane, iso-octane, tetrahydrofuran, chloroform, ethoxynonafluorobutane, with additives such as formic acid, acetic acid, trifluoroacetic acid, or ammonia, diethylamine, triethylamine, and dimethylethylamine (depending on positive or negative ionization mode). The possible suppression effect of strongly acidic or basic additives depends on the analytes, but in general it must be avoided. In APCI it is recommended, when possible, to replace acetonitrile with methanol in order to enhance detection sensitivity. This can be explained by the stronger basicity character of acetonitrile compared to that of methanol, which competes with target analytes for protonation. Additionally, acetonitrile tends to polymerize in APCI plasma, coating the corona needle with an insulating layer after several hours in operation. It is recommended that dimethylformamide content in the mobile phase be lower than 10% when using the API electrospray, and a high signal background can be noticed when using this solvent with the APCI interface. Chlorinated hydrocarbons can enhance the ionization yield only for the APCI interface.

The formation of multiple molecular ions, especially due to the formation of sodium ion adducts, is commonly observed in electrospray mass spectrometry and may make it difficult to obtain good reproducibility and sensitive quantitation. In negative ionization mode, alkylamine additives could improve detection by suppression of multiple molecular ions through preferential formation of a predominant alkylamine adduct ion [51].

Besides ESI and APCI-MS, other detection techniques depend on an evaporative process (evaporative light scattering detector (ELSD), corona charged aerosol detector (cCAD), etc.). In general, the easier it is to evaporate the mobile phases, the higher is the sensitivity for those detectors. However, the detection process is more complex than simple evaporation, and the sensitivity depends on a number of additional factors [52]

Solvent Purity in HPLC

Solvents used in HPLC must be very pure unless a known impurity is present in the solvent and it does not affect the HPLC analysis in any way. Solvent impurities may affect the HPLC analysis in various ways. The impurity may generate: (1) interaction with the analytes, (2) problems with the separation, (3) problems with the detection, and (4) deterioration of the HPLC equipment.

Solvent impurities may come from additives or solvent decomposition (e.g., in case of peroxide formation in ethers, or hydrolysis of

esters), or they can be present from the solvent synthesis. As an example, chlorinated compounds may hydrolyze to form HCl, which can produce some decomposition of the analytes. The presence of impurities may also affect the separation by producing changes in the retention time compared to a pure solvent. Detection is frequently the main part affected by solvent impurities, either by increasing the signal background or by quenching the signal (such as in case of fluorescence). Equipment deterioration is also possible, for example, from the HCl present in some old chlorinated solvents.

Besides the impurities dissolved in solvents, small insoluble particles may be present in some solvents. These impurities must be eliminated by filtration, which is typically performed using 0.45 μm pore filters. The filters must be made from materials perfectly inert to the solvent (e.g., PTFE). However, many prefiltered solvents for HPLC are commercially available.

7.4. BUFFERS AND OTHER ADDITIVES IN HPLC

General Comments

Retention in HPLC of organic compounds that have acidic, basic, or amphoteric character is highly dependent on pH. This is mainly caused by the change in the compound structure due to the pH changes (see Section 3.5). It is common that within 2 pH units, the structure of a compound is changed, for example, from a completely neutral state into an ionized form. The change in the retention of a compound at different pH values of the mobile phase can be an undesirable effect, and then an effort has to be initiated to maintain a constant pH during the separation. In many cases, however, the pH change of the mobile phase can be used to the advantage of achieving a specific separation.

The value of pH of the mobile phase is frequently controlled with buffers. The variation of the pH of a mobile phase is done with a gradient separation and can be achieved by different procedures. For RP-HPLC, for example, one procedure is the modification of the ratio of two solutions: (1) an aqueous (or partially aqueous) buffer solution A, and (2) an organic (or partially organic) solution B that does not contain buffers or additives. Another procedure consists of modifying the ratio of two partially aqueous solutions A and B, each one buffered at a different pH. These solutions with different pH may have the same or different contents of the organic phase. Gradients with different pH and different organic contents can also be achieved by mixing more than two solutions. However, three or four solution gradients, though sometimes utilized, are not common. The ionization status of a compound can be even more important regarding its retention in a polar (HILIC) or in an ion-exchange separation than it is in RP-HPLC. When the mobile phase does not contain an organic component (e.g., in ion chromatography), the gradient is achieved by mixing two solutions with different pH (and possibly different inorganic additive contents).

Although the buffers are added in order to maintain a specific pH of the mobile phase, the stability in time of the pH of the buffer solutions is not always good. This is, for example, the case of buffers that contain NH_4OH as a base. Due to the volatility of ammonia and its elimination from the solution, the pH of such buffers may decrease (significantly) in time or during the solvent sparging (if this is performed). In such situations, the pH of the buffer is not the same at different times during its utilization, which can produce significant variations in the

separation. The pH of buffers containing NH_4OH, for example, can vary as much as one pH unit in several days. Fresh buffers must be prepared at set time intervals in order to avoid this problem. Besides the variation in the pH, the stability of buffers in time may also be related to the growth of microorganisms. Some buffers may act as media for microorganism growth, and the same solution of a buffer has only a limited utilization time. This growth can be delayed by the addition of small amounts of NaN_3 and/or by refrigerating the buffer solutions during storage. The growth of microorganisms may slightly modify the buffer pH, but the main problem is related to the clogging of the filters (frits) present along the path of the mobile phase, even when the mobile phase was initially free of particles.

In a number of HPLC separations, added to the mobile phase are compounds that do not directly affect the pH but are necessary for other purposes related either to separation or to other aspects of the HPLC analysis. These compounds are indicated as additives. In case of ion pair, hydrophobic interaction, or displacement chromatography, the additives are part of the separation process, and such additives are discussed in association with each specific technique. Other aspects related to additives are further discussed in this section.

Buffer pH

A solution of a weak acid HA and its conjugate base A^-, or a solution of a weak base B and its conjugated acid BH^+, has the capability to show resistance to pH changes upon the addition of a strong acid or base. These types of solutions are known as buffers. The weak acids and bases involved in buffer preparation can be monoprotic or polyprotic and can be inorganic or organic. In water solutions, acids, bases, and salts are, at least in part dissociated, and their analytical concentrations are not equal with the true concentration in solution. For a mixture of an acid HA and its salt (e.g. as Na^+ salt), the mass balance for the solution requires that:

$$c_{HA} + c_{NaA} = [HA] + [A^-] \qquad (7.4.1)$$

where c_{HA} and c_{NaA} are the analytical molar concentrations and [HA] and [A$^-$] are the molar concentrations in the solution upon dissociation and after equilibrium is established. On the other hand, the electrical neutrality (for a monoprotic acid) of the solution requires that:

$$[Na^+] + [H^+] = [A^-] + [OH^-] \qquad (7.4.2)$$

Since the salt is assumed to be completely dissociated, $[Na^+] = c_{NaA}$ and rel. 7.4.2 gives:

$$[A^-] = c_{NaA} + [H^+] - [OH^-] \qquad (7.4.3)$$

Relation 7.4.1 and 7.4.3 give the expression for [HA]:

$$[HA] = c_{HA} - [H^+] + [OH^-] \qquad (7.4.4)$$

(It is assumed that the notations $[H^+]$ and $[H_3O^+]$ are equivalent and for the simplicity of writing the notation $[H^+]$ is adopted). From water dissociation constant $K_w = [H^+][OH^-]$, rel. 7.4.3 and 7.4.4 can be easily expressed as a function of $[H^+]$ only. Since the molar concentration of an acid and its conjugate base in practical applications is typically much larger than the difference $[H^+] - [OH^-]$, it is common to use for rel. 7.4.3 and 7.4.4 the simplifications:

$$[A^-] = c_{NaA} \quad \text{and} \quad [HA] = c_{HA} \qquad (7.4.5)$$

With these simplifications, the expression for the dissociation constant of an acid can be written in the form:

$$K_a = \frac{[H^+][A^-]}{[HA]} = [H^+]\frac{c_{NaA}}{c_{HA}} \qquad (7.4.6)$$

Relation 7.4.6 gives the pH value for a buffer made from a weak acid and its salt and can be written in the form:

$$pH = pK_a + \log\frac{c_{NaA}}{c_{HA}} \qquad (7.4.7)$$

Relation 7.4.7 (known as Henderson-Hasselbach equation) also explains the properties of buffers to resist to pH changes upon addition of strong bases or acids, as long as both HA and NaA are still present. The pH of such solution is given by rel. 7.4.7, while strong acids and bases are completely dissociated and $pH = -\log c_{HA}$ for the acids and $pH = 14 + \log c_B$ for the bases.

In analogy to rel. 7.4.6, the following expression is valid for bases:

$$pH = 14 - pK_b - \log\frac{[BH^+]}{[B]} \qquad (7.4.8)$$

where the following replacements can be made: $[BH^+] = c_{BHX}$ and $[B] = c_B$. Formulas similar to 7.4.7 can be developed for polyprotic acids and bases.

The pH of a selected buffer in HPLC is basically determined by the separation need but also by the stability of the stationary phase, which is frequently limited to the range 2 to 8. This is also the range of most common buffers, although buffers in the range 1 to 12.9 are reported in the literature (e.g., HCl + glycine for pH as low as 1.04 and NaOH + glycine for pH as high as 12.97) [53].

Buffer Capacity

The resistance of buffers to pH changes is characterized by *buffer capacity*. Several definitions of this property are known. Buffer capacity β can be defined as the number of moles of a strong acid or base that causes 1.00 L of buffer to change the pH by one unit. Another definition that describes buffer capacity even for small changes is given by the formula:

$$\beta = \frac{dn}{dpH} \qquad (7.4.9)$$

where n is the number of equivalents (moles for monoprotic bases) of a strong base added to the buffer in the infinitesimal amount dn to change the pH by d pH. It can be assumed that the addition of a base (e.g., NaOH) leads to the increase of c_{NaA} and $dn = dc_{NaA}$. With the notation $c_{buf} = [HA] + [A^-]$ and the value for [HA] obtained from rel. 7.4.6, the value for c_{buf} can be written as follows:

$$c_{buf} = \frac{[H^+][A^-]}{K_a} + [A^-] \qquad (7.4.10)$$

Relation 7.4.10 can be rearranged to give the concentration of $[A^-]$ as a function of c_{buf}, $[H^+]$ and K_a. From $dn = dc_{NaA}$ where c_{NaA} is given by rel. 7.4.3, and by replacing in c_{NaA} the value for $[A^-]$ from 7.4.10, it can be concluded that:

$$\beta = \frac{dc_{NaA}}{d\text{pH}} = \frac{d\left(\dfrac{K_w}{[H^+]} - [H^+] + \dfrac{c_{buf}K_a}{K_a + [H^+]}\right)}{d[H^+]} \frac{d[H^+]}{d\text{pH}}$$

$$(7.4.11)$$

The calculation of the derivatives leads to the formula:

$$\beta = 2.303\left(\frac{K_w}{[H^+]} + [H^+] + \frac{c_{buf}K_a[H^+]}{(K_a + [H^+])^2}\right)$$

$$(7.4.12)$$

Relation 7.4.12 indicates that the buffer capacity has a relatively complicated dependence on pH and acid dissociation constant K_a, and shows that a higher buffer concentration is associated with a higher buffer capacity. A smaller increase in pH is produced by the addition of a base (and a smaller decrease in pH is produced by the addition of an acid) when β is larger. Figure 7.4.1 shows the variation of buffer capacity β with the pK_a and with the buffer concentration c_{buf}. From Figure 7.4.1 it can be seen that buffer capacity β has a maximum for the value of pK_a and, as expected, is proportional with c_{buf}. The high values of β at very low or very high pH seen in Figure 7.4.1 are caused by the invariability of pH of a concentrated solution of acid or base when a small addition of acid or base is made and is irrelevant for buffer capacity.

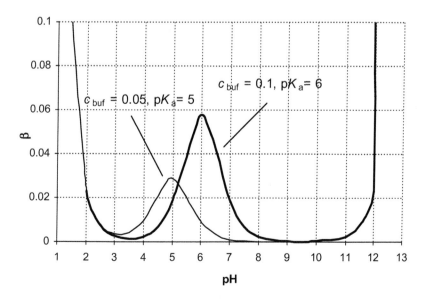

FIGURE 7.4.1 Variation of β with the solution pH and concentration.

Buffers in Partially Aqueous Solvent Mixtures

In practice, for different HPLC separations in which one of the mobile phase components has a high water content, the desired pH is obtained by using the aqueous phase as a buffer with a known pH value. However, the buffer pH calculated for a totally aqueous solution is not the true value of the pH in the solutions that are partially composed of water and an organic miscible solvent (such as methanol, ethanol, or acetonitrile). For water the dissociation constant K_w is given by $K_w = [H^+][OH^-] = 10^{-14}$ at 25 °C. Neutral is defined as the state at which $[H^+]$ equals $[OH^-]$, which occurs when $[H^+] = 10^{-7}$ equivalent with a pH of 7. For methanol, for example, the auto-protolysis constant is $K_{CH3OH} = [H^+][CH_3O^-] = 10^{-16.6}$ [54]. In methanol, neutral pH should be taken when $[H^+]$ equals $[CH_3O^-]$, which occurs when $[H^+] = 10^{-8.3}$ or a pH of 8.3. In conclusion, methanol–water mixtures have auto-protolysis constants $K_{H2O/CH3OH}$ between 10^{-14} (water) and $10^{-16.6}$ (methanol), and the neutral in these mixtures ranges from pH = 7 to pH = 8.3. In aqueous basic solutions, the anion is OH^-, and in basic solutions that contain high concentrations of methanol, the anion will be a mixture of OH^- and CH_3O^- [55].

A common procedure for determining the pH of aqueous/organic solutions is to directly measure it after mixing the aqueous buffer and the organic modifier. For a pH measured in a 100% aqueous solution with the electrode calibrated with aqueous standard buffers, the pH is accurate (and can be indicated as $_w^w pH$). For the pH measured in an organic/aqueous mixture with the electrode calibrated with aqueous buffers, a value indicated as $_w^s pH$ is obtained. However, this is not a correct value. A correct value would be obtained only if the pH is measured with an electrode calibrated with buffers prepared in the same solvent as the one used for the mobile phase. In such a case,

a $_s^s pH$ value would be obtained. This measurement requires knowledge of the pH value of the reference buffers prepared at different partial aqueous compositions, which is not usually available [56]. A correction can be obtained for $_w^s pH$ since between the values $_s^s pH$ and $_w^s pH$ there is a difference given by a term δ that is constant for each mobile phase composition. The following relation can be used for $_s^s pH$ calculation:

$$_s^s pH = _w^s pH - \delta \qquad (7.4.13)$$

The parameter δ for methanol–water based mobile phases can be estimated from the solvent composition (volumetric organic phase content ϕ expressed as $\phi = \dfrac{v_{organic}}{v_{organic} + v_{water}}$) with the following empirical equation [57]:

$$\delta = \frac{0.09\phi_{MeOH} - 0.11\phi_{MeOH}^2}{1 - 3.15\phi_{MeOH} + 3.51\phi_{MeOH}^2 - 1.35\phi_{MeOH}^3} \qquad (7.4.14)$$

The parameter δ for an acetonitrile-water-based mobile phase is the following:

$$\delta = \frac{-0.446\phi_{ACN}^2}{1 - 1.316\phi_{ACN} + 0.433\phi_{ACN}^2} \qquad (7.4.15)$$

The organic content ϕ in a mobile phase also influences the dissociation of acids and bases. For example, methoxide ion is a more potent nucleophile than hydroxide, such that basic methanol-water mixtures can show different chemical behavior than water alone, even when the hydrogen ion activity is the same. This fact is relevant to column stability and sample stability in basic methanol/water mixtures. When the content of the organic solvent in solution increases, the dielectric constant and the activity coefficients decrease. In the presence of the organic component from the mobile phase, the acidic/basic properties of solutes are modified in different proportions,

TABLE 7.4.1 The fitting parameters for predicting the slope (α_s) and the intercept (β_s) of the correlation between $pK_a(w)$ and $pK_a(s)$, as given in formula 7.4.16 in water/acetonitrile [58].

Compounds	α_1	α_2	α_3	α_4	β_1	β_2	β_3	β_4
Aliphatic carboxylic acids	9.97	−8.59	8.83	−8.72	−0.68	9.94	8.45	−8.59
Aromatic carboxylic acids	−2.42	3.14	−1.98	2.12	9.97	−9.12	5.96	−6.90
Phenols	10.05	−10.04	7.97	−8.37	−5.33	9.95	0.19	−0.70
Amines	−0.73	−0.27	−0.87	−0.12	−1.82	2.25	−1.75	0.90
Pyridines	−1.67	0.67	−1.66	0.67	−1.78	1.89	−0.58	−0.40

depending on solute structure and the nature of the organic modifier. The linear relationship between $pK_a(w)$ of a solute in pure water and $pK_a(s)$ of the same solute in a solvent s (s may be pure or a mixture), can be written as the following empirical relation:

$$pK_a(s) = \alpha_s \cdot pK_a(w) + \beta_s \qquad (7.4.16)$$

In rel. 7.4.16 the intercept β_s is related to the differences in basic character, dielectric constants, and specific solvation interactions (e.g., hydrogen bonding) of the solute and the solvent s and water, respectively. The slope α_s is related to the differences between specific solvation interactions, which depend on the solvent and solute. Thus, for an organic modifier the values of α_s and β_s are calculated for sets of compounds according to the following equations dependent on the volume fraction of the solvent (ϕ_s) in the mobile phase composition:

$$\alpha_s = \frac{1 + \alpha_1 \phi_s + \alpha_2 \phi_s^2}{1 + \alpha_3 \phi_s + \alpha_4 \phi_s^2} \qquad (7.4.17)$$

and

$$\beta_s = \frac{1 + \beta_1 \phi_s + \beta_2 \phi_s^2}{1 + \beta_3 \phi_s + \beta_4 \phi_s^2} \qquad (7.4.18)$$

where α_1, α_2, α_3, α_4, β_1, β_2, β_3, and β_4 are parameters obtained by numerical best fit techniques for all acids or bases from the same family. For example, the values of the best fit parameters for several classes of compounds and a mobile phase consisting of water/acetonitrile are given in Table 7.4.1 [58].

Besides dependences as given by formula 7.4.16, other approximations were proposed to estimate the $pK_a(s)$ values in water/organic solvent mixtures. Simpler formulas such as linear or quadratic dependencies were suggested for this purpose, with expressions dependent on $pK_a(w)$ and volume fraction of the solvent (ϕ_s) as given in the following equations:

$$pK_a(s) = A' + B'\phi_s \qquad (7.4.19)$$

$$pK_a(s) = A' + B'\phi_s + C'\phi_s^2 \qquad (7.4.20)$$

where $A' \approx pK_a(w)$ and where B' or B' and C' are empirical parameters dependent on a specific compound for which $pK_a(s)$ is needed [59]. Examples of the values of these parameters are given in Table 7.4.2 for methanol/water solutions, as a function of methanol content. Other such estimations $pK_a(s)$ based on $pK_a(w)$ and ϕ are reported in the literature [58].

As previously indicated (see Section 3.5), another parameter influencing the pH is the temperature. As some chromatographic separations are performed with mobile phase at higher temperatures, it is of interest to see how the pH varies with this parameter. For example, the pH

TABLE 7.4.2 Polynomial equations describing the dependences of $pK_a(s)$ of the some buffers currently used in RP-LC as a function of the MeOH content in mobile phase ϕ_{MeOH}.

pK_a	Equation in solvent/water	R^2
pK_1 H$_3$PO$_4$	$pK(s) = 2.127 + 2.16 \cdot 10^{-2}\phi_{MeOH} + 6.81 \cdot 10^{-5}\phi^2_{MeOH}$	0.9982
pK_2 H$_2$PO$_4^-$	$pK(s) = 7.202 + 1.27 \cdot 10^{-2}\phi_{MeOH} + 2.14 \cdot 10^{-4}\phi^2_{MeOH}$	0.9997
pK_1 citric acid	$pK(s) = 3.121 + 1.56 \cdot 10^{-2}\phi_{MeOH} + 6.47 \cdot 10^{-5}\phi^2_{MeOH}$	0.9992
pK_2 citric acid	$pK(s) = 4.756 + 1.62 \cdot 10^{-2}\phi_{MeOH} + 9.33 \cdot 10^{-5}\phi^2_{MeOH}$	0.9986
pK_3 citric acid	$pK(s) = 6.391 + 2.15 \cdot 10^{-2}\phi_{MeOH} + 7.65 \cdot 10^{-4}\phi^2_{MeOH}$	0.9996
pK acetic acid	$pK(s) = 4.757 + 1.26 \cdot 10^{-2}\phi_{MeOH} + 1.09 \cdot 10^{-4}\phi^2_{MeOH}$	0.9998
pK NH$_4^+$	$pK(s) = 9.238 - 5.78 \cdot 10^{-3}\phi_{MeOH} + 2.22 \cdot 10^{-5}\phi^2_{MeOH}$	0.9991

of a standard solution of 0.050 M of pure potassium hydrogen phthalate in water is 4.0 at 15 °C [60]. At any other temperature T °C between 0 °C and 55 °C, its pH is defined by the following dependence [61]:

$$pH_T = 4.0 + \frac{1}{2}\left(\frac{T-5}{100}\right)^2 \qquad (7.4.21)$$

Between 55 °C and 60 °C the dependence of pH on T is the following:

$$pH_T = 4.0 + \frac{1}{2}\left(\frac{T-5}{100}\right)^2 - \frac{T-55}{500} \qquad (7.4.22)$$

The combined effects of temperature and organic content in solvating medium are more complex, and the values of pH must be determined experimentally. Some examples of variations of pH with temperature and methanol, ethanol, or acetonitrile content for a potassium hydrogen phthalate buffer are illustrated in Table 7.4.3 [62]. From Table 7.4.3, it can be seen that the solvent and the temperature can influence the pH significantly. When used in HPLC separations, the pH for the buffer solutions is typically indicated for room temperature (25 °C). Since the pH varies with the

temperature, this factor must be taken into consideration when performing a separation at a different temperature than the one where the buffer pH was measured [63, 64].

Besides pH, the addition of an organic modifier also affects the buffer capacity β. Organic solvents have the effect of decreasing buffer capacity and also of shifting the pH where β has a maximum. For methanol, for example, this shift is toward higher pH values [55].

Solubility of Buffers in Partially Organic Mobile Phases

The choice of a mobile phase composition based on an aqueous component containing buffers must take into consideration the solubility of the buffer in the presence of the organic modifier. Since buffers are made using acids, bases and salts, and these can sometimes be inorganic compounds, their solubility in the organic or partially organic phase can be low. This solubility depends on the nature of the buffer, its concentration, and the nature and percentage of the organic modifier in mobile-phase composition. For inorganic salts, the solubility depends mainly on the nature of the cation, and their solubility trend in partial

TABLE 7.4.3 Values of pH for a reference standards solution of 0.05 M potassium hydrogen phthalate in various aqueous organic solvents mixtures at different temperatures (pH = 4 in 100% water and 25 °C) [62].

Temperature (°C)	Methanol			
	10%	20%	50%	64%
	pH	pH	pH	pH
10	4.254	4.490	5.151	5.488
25	4.243	4.468	5.125	5.472
40	4.257	4.472	5.127	5.482

Temperature (°C)	Acetonitrile			
	5%	15%	30%	50%
	pH	pH	pH	pH
15	4.163	4.533	5.001	5.456
25	4.166	4.533	5.000	5.461
35	4.178	4.542	5.008	5.475

Temperature (°C)	Ethanol			
	10%	20%	40%	70%
	pH	pH	pH	pH
10	4.235	4.513	5.026	5.469
25	4.236	4.508	4.976	5.472
40	4.260	4.534	4.978	5.493

TABLE 7.4.4 Solubility (S, mM/L) of potassium phosphate based buffers* in different mobile phase compositions [65].

Organic content (%, vol)	S acetonitrile	S MeOH	S tetrahydrofuran
0–40	>50	>50	>50
50	>50	>50	25
60	45	>50	15
70	20	35	10
80	5	15	<5
90	0	5	0

* Note: Solutions containing KH_2PO_4 and K_2HPO_4 are used as buffers in the range of pH = 4.8 to 8.0 (measured in water).

in buffer capacity). Particular attention must be given to buffer solubility when used in gradient separations, and the organic content of the mobile phase is increased during the chromatographic run (e.g., in RP-HPLC). A higher content of organic component is usually associated with a decrease in buffer solubility, and the buffer concentration must be adjusted appropriately to avoid any precipitation of salts in the chromatographic system.

Influence of the Buffer on Column Stability and Properties

The properties of the stationary phases are critical regarding their resilience to the extreme pH values of the mobile phase (see Chapter 6). However, the chemical stability of the silica-based stationary phase is also affected by the type and concentration of the used buffers. The effect of different buffers leading to the deterioration of silica backbone may be caused by the combined effects of pH and the capacity of complexation of the buffer. In an experiment performed to evaluate column stability, 6 L of eluent containing 50% CH_3OH and 50% buffer with pH = 10 (v/v) were passed through

organic solvents follows approximately the solubility in water: $NH_4^+ > K^+ > Na^+$. The alkylammonium cation has a higher solubility in organic solvents, such as methanol or acetonitrile, due to the affinity of the alkyl chain toward these solvents. As an example, several solubility values for potassium phosphate-type buffers in solvents that are commonly used in HPLC are given in Table 7.4.4 [65].

As a result of the lower solubility of many buffers in organic solvents, the buffer concentration should be selected at the lowest acceptable concentration (although this leads to a decrease

a column packed with C18 stationary phase [66]. The effect on column chemical stability was measured as the amount of dissolved silica. A mobile phase based on phosphate buffer dissolved about 110 mg/column, a carbonate-based buffer dissolved about 40 mg/column, and for borate- and glycine-based buffers the dissolution was almost unobservable. This showed that the nature of the anions in an eluent buffer can influence substantially the longevity of RP columns. Other studies [67] have shown that column deterioration is best prevented by sodium as the buffering cation. In contrast, potassium is a more aggressive buffering cation than sodium and ammonium. These effects seem to be caused by an increase in the pH for the aqueous/organic solution upon the addition of the organic phase, which is different for buffers with different chemical nature, although in water their pH is the same [68].

The longevity of silica-based RP columns also depends on the concentration (ion strength) of the buffer. Even at pH = 7 it was proved that by passing 10 L of mobile phase containing buffer/ACN = 50/50 (v/v), the dissolution of silica support is about 175 mg/column for a buffer concentration of 10 mM/L, 220 mg/column for a concentration of 50 mmoles/L, and 325 mg/column for 250 mM/L. Therefore, in order to prevent early column failure, low buffer concentrations are recommended [67].

In conclusion, in order to maintain the longevity of silica-based columns apart from the choice of the resistant columns (see Chapter 6), the selection of the nature and concentration of the buffer in the eluent is very important. Typically, borate and organic-based buffers such as glycine, in low concentration, combined with the properly selected counterion substantially prevent early column degradation of silica-based RP columns. On the other hand, the use of phosphate and carbonate as buffering anions leads to a faster dissolution of the silica support.

Specific columns may display a different type of interaction depending on the mobile phase pH. This is, for example, the case of weak cation exchange columns and weak anion exchange columns. The ionic character of the stationary phase containing, for example, carboxyl groups can be basically eliminated by adjusting the pH of the mobile phase within two pH units of pK_a of the acidic groups. The column will act as a polar-type column and can be used in HILIC type separations.

Suitability of the Buffers for Detection in HPLC

Besides their role in the separation, buffers must be suitable for the detection selected in an HPLC separation. Even if a separation is done preferably with the mobile phase containing a specific buffer, the buffer is rendered useless if the detection cannot be performed with it in the mobile phase. Most types of detection can be affected by buffers, including the main detection techniques such as UV, fluorescence, and all techniques based on evaporative processes (MS, ELSD, and cCAD). Even in RI detection, a high concentration of a buffer may reduce the sensitivity of the technique.

For UV detection, there are two separate aspects regarding the use of buffers. One is related to the change in the wavelength of absorption for specific compounds when they are present in solutions of different pH values. During the development of a specific analytical method, it should be verified that the detection is performed at the correct wavelength for the specific pH of the mobile phase. The second aspect is related to the transparency of the mobile phase containing the buffer. Specific buffers may shift the UV cutoff wavelength to higher values compared to the pure solvent. The UV cutoff values for a number of buffers typically used in HPLC are given in Table 7.4.5.

For fluorescence detection, solution pH is also very important. Both emission wavelength

TABLE 7.4.5 Common buffers and their UV cut-off values (for a specific concentration).

Buffer	pK_a	pH working range	UV cut-off
Potassium formate/formic acid	3.8	2.5–5.0	210 nm (10 mM)
Potassium acetate/acetic acid	4.8	3.8–5.8	210 nm (10 mM)
Ammonium formate	3.8; 9.2	2.8–4.8; 8.2–10.2	210 nm (50 mM)
Ammonium acetate	4.8; 9.2	3.8–5.8; 8.2–10.2	210 nm (50 mM)
Trifluoroacetic acid	0.5	1.5–2.5	210 nm (10 mM)
KH_2PO_4/H_3PO_4	2.12	1.1–3.1	<200 nm (0.1%)
KH_2PO_4/K_2HPO_4	7.2	6.2–8.2	<200 nm (0.1%)
Ammonium hydroxide./ammonia	9.2	8.2–10.2	200 nm (10 mM)
Diethylamine·HCl/diethylamine	10.5	9.5–11.5	200 nm (10 nm)
Trietylamine HCl/triethylamine	11	10–12	200 nm (10 mM)
Borate ($H_3BO_3/Na_2B_4O_7 \cdot 10H_2O$)	9.24	8.2–10.2	210 nm
Glycine·HCl/glycine	9.8	8.8–10.8	230 nm
Tri-K-Citrate/HCl	3.06; 4.7; 5.4	2.1–4.1; 3.7–5.7; 4.4–6.4	230 nm (10 mM)
1,3- Bis-tris propane HCl / Bis-tris propane [bis(tris(hydroxymethyl)methylamino) propane]	6.8; 8.3	6.5–9	225 nm (10 mM)
1-methylpiperidine·HCl/1-methylpiperidine	10.1	9.1–11.1	215 nm (10 mM)
Pyrollidine·HCl/pyrollidine	11.3	10.2–12.5	210 nm (10 mM)

and fluorescence intensity for many compounds are dependent on pH. When different structures are possible for a compound, the fluorescence of one species is usually different from that of the other. For this reason, the mobile phase pH must be carefully controlled in most methods using fluorescence detection.

For all detection techniques that use an evaporative step, the use of nonvolatile buffers in the mobile phase is not acceptable. MS detection, in particular, being a common detection technique and having excellent qualities regarding sensitivity and selectivity, requires volatile buffers. Chemical properties and concentration of the buffers, as well as pH, have a significant effect on analyte response in ESI. Many of the buffers commonly used in LC such as phosphate and borate are not compatible with ESI-MS. When used, even at low concentration, they produce baseline instability, increased background, signal suppression, and rapid contamination of the ion source, resulting in a rapid decrease in sensitivity and stability. For these reasons, methods developed, for example, for UV detection cannot be always directly transferred for MS detection. Modification of the buffer is sometimes necessary.

Ammonium acetate or formate is typically used to make buffers for MS detection. Acidity can be modified with formic or acetic acid in positive API electrospray, while ammonia, triethylamine (TEA), or N-methylmorpholine can be used to adjust basic buffers in negative API electrospray. Also the strong volatile acids, such as trifluoroacetic acid (TFA) commonly used in LC-UV analysis of peptides and proteins, may cause significant signal suppression in ESI. In practice, some of these volatile buffers are still used, but their concentrations should not exceed 10 mM in order to avoid suppression of ionization and reduced sensitivity. The suppression effect on sensitivity of TFA and of other fluorinated acids may be diminished by the use of an increased level of volatile organic acids (e.g., HCOOH). These acids may displace the TFA anions from the ion pairs formed with analyte cations during the ionization process [69].

Other Additives in the Mobile Phase

Besides buffers, other additives are used in the mobile phase for HPLC analyses. These may include acids, bases that help the separation by modifying the pH (without buffering), or other compounds that modify the ion strength of the mobile phase, compounds necessary for maintaining a specific structure of the stationary phase (e.g., in ion-exclusion chromatography where the stationary phase is a cation-exchange material in metal form, the mobile phase must contain a low level of that metal ion), chaotropic compounds (some used in IP and some used for facilitating the analytes solubility), as well as preservatives that either stop the modification of the mobile phase (e.g., the formation of epoxides in tetrahydrofuran) or stop the growth of microorganisms in buffers (such as the addition of small amounts of NaN_3). The addition of strong acids such as CF_3COOH, CCl_3COOH, or even H_2SO_4, H_3PO_4, may be necessary in some separations for adjusting the pH (a buffer may be formed in situ by the addition of such acids). The pH increase can be achieved with strong bases such as $(C_2H_5)_3N$, ethanolamine, or quaternary ammonium hydroxides. Certain neutral additives may also influence the pH

by changing the ion strength of the mobile phase. Additives are also necessary in the mobile phase for specific detection techniques. Typical is the addition of HCOOH in the mobile phase for MS detection in positive mode or the addition of $HCOONH_4$ or CH_3COONH_4 for MS detection in negative mode. Details regarding the presence of additives in the mobile phase are described in each specific method that uses them. Among other additives in the mobile phase, organic modifiers that affect the "strength" of the mobile phase have been reported in the literature [70]. For example, small amounts of alcohols with a long aliphatic chain can be added in a mobile phase containing methanol and water, the result being a "stronger" mobile phase for RP separations [71].

7.5. GRADIENT ELUTION

General Comments

Gradient separations in HPLC are those in which the composition of the mobile phase is changed during the chromatographic run (see Section 1.1). The composition change may be in the % of one organic component in water, in pH, in pH and % of the organic component, in ionic strength, and the like. More complicated gradients with changes in the nature and composition of the mobile phase are also applied for some separations. The major applications of gradients in HPLC are for RP, HIC, and HILIC [42]. In other separation techniques such as IC and NPC, the gradients are also used, but in size-exclusion and chiral separations, the gradients are not usually applied. The change in the mobile phase composition in gradient HPLC can be done in a linear mode (see rel. 1.4.1), nonlinear mode (see rel. 1.4.3 and 1.4.4), or even as step gradient (sudden change in mobile phase composition). Multiple gradient ramps are also possible during one

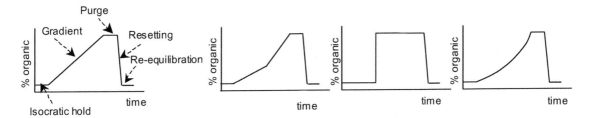

FIGURE 7.5.1 Examples of different common gradient shapes with the increase in organic component in the mobile phase.

chromatographic run. Examples of different common gradient shapes with the increase in organic component in the mobile phase as practiced in RP-HPLC are illustrated in Figure 7.5.1.

In Section 1.4, it was indicated that in practice the change in the mobile phase composition takes place with a certain delay (dwell time t_D) caused by small dead volumes existent in the HPLC instruments before the column itself, and that actual changes are less sharp than is shown in Figure 7.5.1 due to various mixing effects.

As shown in Figure 7.5.1, a gradient commonly includes at the beginning of the run a short time section where the mobile phase composition is kept constant (isocratic). This isocratic hold also includes the dwell time. This section of isocratic hold is chosen as having the "weakest" solvent in the chromatogram, and it is designed for the retention of the least retained compounds in the sample (for RP-HPLC the lowest organic content in mobile phase). The gradient slopes are designed to achieve the desired separation in the optimal (or close to optimal) retention times. Following the gradient slope, the gradient program typically includes a section with the "strongest" mobile phase, with the role of eluting all sample components from the stationary phase. After this section, the mobile-phase composition is restored to the initial conditions to make the system ready for the next injection. A re-equilibrating time is necessary for restoring the stationary phase to the initial conditions, which are not attained instantaneously. This re-equilibration time depends on the column dimensions as well as on the type of column and mobile phase. The volume necessary for column re-equilibration is generally divided into two portions: the system washout and the "true" column re-equilibration volumes. One empirical rule to establish these necessary volumes uses the following formula:

$$V_{reeq} = 3V_{total} + 5V_{column} \quad \text{with}$$

$$V_{column} = \frac{\pi \varepsilon*}{4} d_{col}^2 L \quad \text{and} \quad V_{total} \approx 650$$
$$- 3000 \ \mu L \tag{7.5.1}$$

where $\varepsilon*$ is the column porosity and $\varepsilon* \approx 0.7$ (for typical 5 μm columns) as shown in Table 2.1.1. (When the column i.d. and the column length are expressed in mm and the volume is expressed in mL (cm³), a factor of 10^{-3} must be included in the calculation.) The dwell volume for common HPLC systems is between 500 μL and 3 mL. The re-equilibration time is obtained from V_{reeq} based on the flow rate from the formula $t_{reeq} = V_{reeq}/U$. For example, for a column 2.1 x 150 mm and a small system volume, an equilibration time of minimum 4 min is necessary at 1 mL/min flow rate. However, rel. 7.5.1 seems to give an overestimation of the necessary volume of solvent to re-equilibrate a column. In many cases, only two column volumes of the initial eluent are sufficient for the column re-equilibration.

TABLE 7.5.1　Time table for the gradient separation of hydroxybenzenes from cigarette smoke.

Time min	Solvent B %	Solvent C %	Solvent D %	Flow rate mL/min	Max pressure Barr
0.0	0	0	0	1.4	400
0.5	0	0	0	1.4	400
10.5	31	0	0	1.4	400
15.5	100	0	0	1.4	400
20.0	100	0	0	1.4	400
20.2	0	0	0	1.4	400
22.0	0	0	0	1.4	400

In modern HPLC instruments the gradients are obtained from the solvent supply system by computer-controlled pumping based on a gradient time table. One example of a gradient timetable that was used for the separation by RP-HPLC of several hydroxybenzenes from cigarette smoke is given in Table 7.5.1. The separation was performed on a Beckman Ultrasphere ODS column 15 cm x 4.6 mm i.d., 5 μm particle size. Mobile phase was composed of solution A: water with 4% acetonitrile and with 1% acetic acid, and solution B: acetonitrile with 1% acetic acid. The diagram of this gradient change is shown in Figure 7.5.2. The detection for the analytes was done using a FLD, and the resulting chromatogram is shown in Figure 7.5.3 [72].

Parameters Characterizing the Gradient Separation

A considerable number of studies have been reported regarding the parameters characterizing the gradient separation [73–77]. One problem

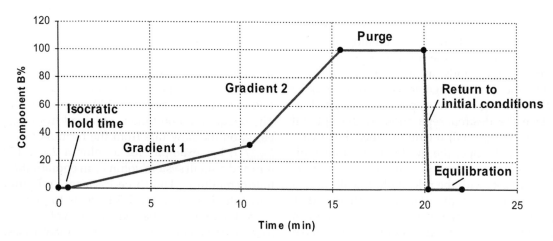

FIGURE 7.5.2　Gradient profile for a separation of hydroxybenzenes on a Ultrasphere ODS column 15 cm × 4.6 mm i.d., 5 μm particle size.

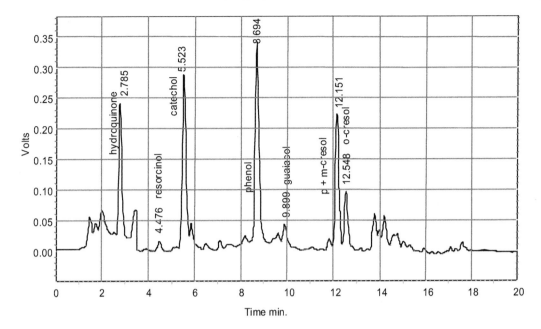

FIGURE 7.5.3 Chromatogram showing the separation of several hydroxybenzenes from cigarette smoke using gradient elution.

related to the characterization of gradient separation originates in the variability of the capacity factor k_i during the cycle, as previously discussed in Section 2.1 (see Figure 2.1.6). The need for a constant parameter specific for the separation, which is the *gradient retention factor k**, was indicated. A common procedure to obtain an expression for k^* starts with the assumption that the analyte j is present in the chromatographic column as a very narrow zone, moving with the linear velocity u_j. This velocity during the gradient is given by the expression:

$$u_j = \frac{dx}{dt} = \frac{u_0}{1 + k(\phi)} \qquad (7.5.2)$$

where x is the distance in the column, $k(\phi)$ is the instantaneous capacity factor dependent on mobile phase composition ϕ (and the nature of the compound and stationary phase), and u_0 is the linear flow rate of the mobile phase. The mobile phase composition $\phi(t)$ changes with time, and a simplifying assumption is that the solute is injected at $x = 0$ and $t = 0$. In rel. 7.5.2, the following substitution can be made:

$$z = t - \frac{x}{u_0} \qquad (7.5.3)$$

With this substitution, $\phi(t)$ becomes a function of z ($\phi(z)$), and equation 7.5.2 can be written in the form:

$$dt = \frac{1 + k(\phi(z))}{k(\phi(z))} dz = \left(\frac{1}{k(\phi(z))} + 1\right) dz \quad (7.5.4)$$

The integration of rel. 7.5.4 from $z = 0$ to $z = t_R - t_0$ (where t_0 is the dead time for the column and t_R is the peak retention time) leads to the expression:

$$t_0 = \int_0^{t_R - t_0} \frac{1}{k(\phi(z))} dz \qquad (7.5.5)$$

In order to integrate rel. 7.5.5, the expression of k as a function of ϕ must be known. Assuming that the change in the mobile phase composition is linear with time, the expression for ϕ is given by $\phi = \phi_0 + \dfrac{b}{t_0}t$ as shown from rel. 2.1.71, and b is the *gradient steepness* $b = \dfrac{\Delta\phi S}{t_{grad}}t_0$ (given by rel. 2.1.74). The dependence of capacity factor k on ϕ is given by the expression 2.1.73 (further discussed in Section 7.1 where it was shown that this relation is only an approximation). For a compound j (index j omitted) rel. 2.1.73 can be written in the form:

$$\log k = \log k_0 - S\phi' \qquad (7.5.6)$$

where k_0 (for compound j) is the capacity factor for an isocratic separation obtained with the mobile phase having the organic content given by ϕ_0. The expression for $\log k_0$ is given by the formula:

$$\log k_0 = \log k_w - S\phi_0 \qquad (7.5.7)$$

and the expression for ϕ' is given by formula:

$$\phi' = \frac{b}{St_0}t \qquad (7.5.8)$$

In natural logarithms, rel. 7.5.6 can be written as follows:

$$\ln k = \ln k_0 - S'\phi' \qquad (7.5.9)$$

where $S' = 2.303\,S$. The dependence of mobile-phase composition ϕ' on variable z can be expressed by the formula:

$$\phi' = \frac{b}{St_0}\left(z + \frac{x}{u_0}\right) \qquad (7.5.10)$$

With these substitutions, the integration performed in rel. 7.5.5 gives the expression:

$$t_R - t_0 = \frac{t_0}{b}\log(2.303k_0b + 1) \qquad (7.5.11)$$

In the process of obtaining expression 7.5.11 for t_R, it was assumed that the gradient starts changing at the head of the chromatographic column at $t = 0$, or in other words that the dwell volume V_D and as a result the dwell time $t_D = V_D/U$ are zero. However, the dwell volume in common HPLC should not be disregarded. Considering the dwell time of the system, the correct expression for t_R will become:

$$t_R = \frac{t_0}{b}\log\left[2.303k_0b\left(1 - \frac{t_D}{t_0k_0}\right) + 1\right] + t_0 + t_D \qquad (7.5.12)$$

By analogy with rel. 2.1.7 used for isocratic separations, rel. 7.5.11 allows the calculation of the capacity factor k_e (specific for a given compound i) for the peak eluting at t_R, based on the formula:

$$t_R = t_0 k_e + t_0 \qquad (7.5.13)$$

where

$$k_e = \frac{1}{b}\log(2.303k_0b + 1) \qquad (7.5.14)$$

The gradient retention factor k^* usually applied for characterization of gradient separations (see Section 2.1), being an "average" value, corresponds to a peak that would migrate halfway through the chromatographic column in isocratic conditions. For this reason, the value for k^* must be double that of k_e, and one approximation for it is given by the formula:

$$k^* = 2\,k_e \qquad (7.5.15)$$

The theory of parameters describing the gradient elution has been the subject of numerous developments that attempted either to improve the approximations used for obtaining previous relations or to extend the results for different gradient profiles (e.g., gradient starting at $t \neq 0$ or multiple gradient slopes). Expressions other than 7.5.14 that approximate a k^* for gradient separations are reported in the literature [78, 79].

The parameter k^*_j can be used for peak characterization similar to k_j for isocratic separations, as shown in Section 2.1 where the selectivity α^* and resolution R^* were defined by the formulas:

$$\alpha^* = \frac{k_i^*}{k_j^*} \qquad (7.5.16)$$

$$R^* = \frac{1}{4}\left(\frac{\alpha^* - 1}{\alpha^*}\right)\left(\frac{k^*}{1 + k^*}\right)N^{1/2} \qquad (7.5.17)$$

The number of theoretical plates N is also compound dependent, and rel. 7.5.17 would be more correctly written as dependent on N^*. However, $N^* \approx N$ with good approximation.

Other developments related to gradient retention factor k^* are common in the dedicated literature [76, 80, 81]. For example, the peak width in gradient HPLC has been studied for explaining a "narrowing" effect due to the acceleration of elution as the concentration of stronger eluent increases. The peak width in gradient HPLC is given by the expression:

$$W_b = 4Gt_0(1 + k_e)N^{-1/2} \qquad (7.5.18)$$

where G is a "compressing factor" approximated by the expression:

$$G \approx \frac{1 + k^*}{1 + 2k^*} \qquad (7.5.19)$$

Except for the compression factor G, rel. 7.5.18 is equivalent to rel. 2.1.44.

During a gradient separation, one additional parameter that changes is the viscosity of the mobile phase. This viscosity change may lead to undesirable increases in the column backpressure. The column backpressure can be calculated for a column filled with porous particle using rel. 2.2.17. Replacing in this formula the linear flow rate u with the volumetric flow rate U given by rel. 2.2.3, the expression becomes:

$$\Delta p = \frac{4\eta U \phi_r L}{\pi \varepsilon^* d^2 d_p^2} \approx \frac{2500\,\eta UL}{d^2 d_p^2}(psi) \qquad (7.5.20)$$

where common notations were used (η is the solvent viscosity, d is column diameter d_p is the particle diameter, and ϕ_r is the column flow resistance factor). The maximum pressure during the separation is obtained using maximum η during the gradient (rel. 7.5.20 tends to predict lower pressure values than experimentally obtained). As shown in Section 7.2, the viscosity of solvent mixtures can be significantly higher than that of pure solvents, and an increase in the backpressure during the gradient is very common (see Figure 7.2.2).

Usefulness of Gradient versus Isocratic Elution

Gradient elution is basically used for three main purposes: (1) Reduction of the total run-time of separations (see Section 1.4), (2) modification of retention times in a separation that does not provide a good separation between specific compounds, and (3) cleaning and/or regeneration of the chromatographic column. Other more uncommon utilizations of gradient can be mentioned, such as loading of the column with a specific reagent.

(1) In common HPLC separation, the values for the capacity factors k_i for different solutes should be in the range $2 \le k_i \le 10$ (values between 1 and 2 are sometimes acceptable). These values are necessary for performing the separation at acceptable retention times ($t_R = t_0 + k\,t_0$), with typical t_0 values in the range given in Table 2.1.1 (for $U = 1$ mL/min). It is not uncommon that a sample contains some compounds with "low affinity" for the stationary phase and others with "very high affinity." For RP-HPLC, for example, in mobile phases with a higher organic content, some compounds with small hydrophobic moieties are poorly retained and may have $k < 1$. Their retention is very sensitive to small (unintentional) changes in the mobile phase composition, leading to changes in the retention times. Also, interferences from un-retained materials

from the sample matrix are possible when peaks of interest elute too early. For these cases, a "weak" mobile phase is initially necessary. However, such a mobile phase may not be acceptable for the elution of more hydrophobic compounds with larger k_i. With a polar mobile phase these compounds are eluted late in the chromatogram and therefore have broad peaks. Although these peaks are well separated, the peaks are wide and their integration is less accurate, and the runtime is too long. In these cases, a gradient elution is necessary to shorten the separation time. A reduction in the capacity factor, and therefore of the retention time, can be achieved using gradients, when the content in the stronger solvent is increased in the mobile phase, as indicated be rel. 7.5.6 (for larger ϕ). In this way, the early eluting peaks in a chromatogram are subject to the "weak" solvent and they

have $k_i > 2$. The peaks that would elute at long retention times in a "weak" solvent are subject to a "stronger" solvent such that they have $k_j < 10$. The shortening of the retention time for late eluting peaks is illustrated in Figure 7.5.4 for RP-HPLC.

One additional advantage of shortening the retention time by using gradient elution is related to the peak shape in the chromatogram. At longer retention times, the peaks are broadened, and shorter times help in obtaining sharper peaks. A detailed discussion of peak shape in chromatography is found in Chapter 2. Gradient elution also provides good repeatability of the retention times and also of the peak area, such that no adverse effects appear regarding quantitation with gradient separation, as compared to isocratic elution [42].

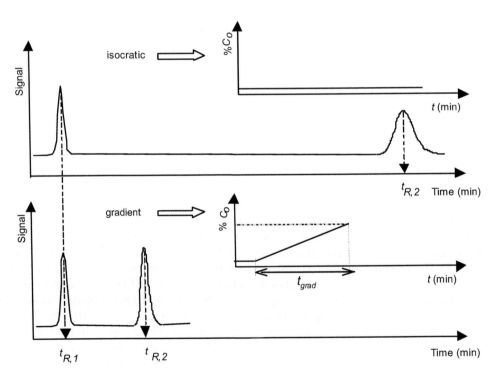

FIGURE 7.5.4 Difference in the retention of two hypothetic compounds in isocratic conditions and in gradient conditions, in an RP separation when the concentration of the organic phase C_o is increased.

(2) The reduction of k_i values for different analytes must be done such that the values for resolution remain $\alpha_{i,j} > 1.2$ for all the components of interest in the sample, and the value for resolution $R > 1$ (or better $R > 1.5$). However, R depends on k_i (k^*) in addition to α and N, and smaller k_i values may lead to poor separations. For this reason, when decreasing k_i values by using a "stronger" mobile phase, this cannot be done without restrictions, and the gradient must maintain a good separation. The choice of a convenient gradient is relatively simple when an initial isocratic separation (or a gradient with small composition changes) provides a good separation. Starting with this separation, the slope of gradient is increased such that the change from a "weak eluting" mobile phase to a "strong eluting" mobile phase is done more rapidly. The new separation must remain acceptable regarding selectivity and resolution. The process can also start in reversed order, with a relatively poor separation in a rapid gradient and short retention times, which are changed to slower composition changes that would allow a better separation.

The expressions connecting k_i to various parameters describing the mobile phase developed in Section 7.1 (e.g., 7.1.37 or 7.1.38) also show that the variation in k_i depends not only on the mobile phase composition, but also on the nature of the stationary phase and of the analytes. When the composition of the mobile phase is changed (within a range) from "weak" to "strong" or when the pH and the ionic strength are changed, different components from the sample may be affected differently, and their retention times $t_{e,i}$ and $t_{e,j}$ (and corresponding k^* for each analyte) may vary in different ways [82]. For this reason, it is possible that specific values for $\alpha_{i,j}$ increase, although both k^*_i and k^*_j decrease. Different effects of increasing the mobile phase strength (% of organic component for RP-HPLC) are illustrated in Figure 7.5.5a and 7.5.5b for two couples

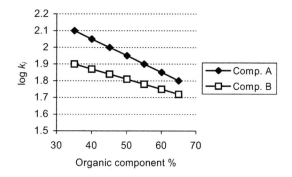

FIGURE 7.5.5a Variation of log k_i as the content in organic phase increases and the separation worsens.

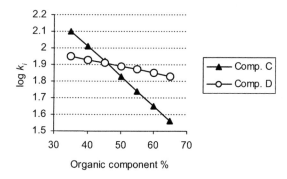

FIGURE 7.5.5b Variation of log k_i as the content in organic phase increases and the separation first decreases and then improves.

of hypothetic compounds (assuming linear variation of log k with the % of organic solvent). For the first couple of compounds (A and B) shown in Figure 7.5.5a, as the % of organic component in the mobile phase increases, log k_A − log k_B decreases, and the separation is worsened. For the second couple of compounds (C and D) shown in Figure 7.5.5b, as the % of organic component in the mobile phase increases, log k_A − log k_B has initially a decrease, but further on increases and the separation is improved.

The effect shown in Figure 7.5.5b is not uncommon, and improvements in the separation can sometimes be obtained by increasing the "strength" of the mobile phase, which lead at the same time to shorter retention times. Predictions of such changes are in general difficult to make, and descriptions of analytical methods from the literature provide details on specific mobile phase compositions and gradients that assure a good separation. This is the main cause for the need of a "trial-and-error" strategy for optimizing some separations. The selection of a specific stationary phase in obtaining a desired separation is very important in such cases.

(3) In most gradients, a purge region is included for cleaning the column from any remaining solutes, not necessarily analytes. This region may have the highest concentration of the "strong" eluent. For example, in RP-HPLC the purge region may consist of a high concentration of organic phase or even of pure organic component. When pure organic component is not necessary for cleaning the column, a lower concentration should be used such that the re-equilibration of the column is done faster.

Cleaning the stationary phase from all injected sample components is part of the good care of the chromatographic column. However, this is not always convenient, and in some applications it is recommended that cleaning be done after a number of injections (analyzed samples) and not after each sample. In such cases, a separate gradient run (or isocratic run) with strong (possibly pure) solvents is applied, followed by a longer re-equilibration of the stationary phase to the initial condition.

For practical purposes, the reduction of the runtime of a separation can be very useful. Contribution to achieving this goal, in RP, for example, can be obtained not only from using a mobile phase with a higher content of organic phase, but also from using sharper gradient changes. The slope of gradient, or its speed, can be optimized in order to assure best results such that the separation is not compromised and the analysis runtime is shortened [83, 84].

In practice, mainly when the gradients are very sharp, a gradient distortion can be observed. This is manifested by large deviations from the sharpness of a required gradient and may be different from instrument to instrument. Low-pressure mixing instruments tend to have more gradient distortions than high-pressure mixing systems.

Besides the advantages, some problems may appear in gradient elution. One problem is related to the inability to use RI detection with gradients. Another problem may be the appearance of a drift in the baseline of the chromatogram. This drift depends on the selected solvents and on the detector, and it is not a common problem. Also, the gradient is not necessary when the analytes elute close to each other in isocratic separation.

7.6. MOBILE PHASE IN REVERSED-PHASE CHROMATOGRAPHY

General Aspects

The characterization of solvents regarding their properties, as previously described in this chapter, should be used as guidance in choosing the mobile phase for a specific HPLC separation. However, calculation of the capacity factor or of selectivity based on expressions like 7.1.38 is rather complicated, and the precision of the result is not always very good. The use of expression 7.1.38 with known S (see Table 7.1.8) and $\log k_A$ is applicable for calculating $\log k_{AB}$ for a mixture of solvents A and B. However, most separations (in RP-HPLC and HILIC in particular) are performed in gradient conditions. For this reason, the formulas developed in Section 7.1 are useful for obtaining a general view on solvent selection, and less so in the direct calculation of capacity factors. Within the principle of reversed-phase chromatography, the mobile phase for this technique should be more polar than the stationary phase. Given the very large number of applications of RP-HPLC, only general guidance can be provided regarding the selection of solvents and solvent mixtures in RP-HPLC. For each specific method reported in the literature, detailed descriptions of the mobile phase and of gradient conditions (when used) are provided. The general types of solvents used in RP-HPLC can be discussed based on classification of solvents and their characteristics given in Section 7.1 (see, e.g., Table 7.1.3 and Figures 7.1.3, 7.1.6).

Water and Mobile Phases with High Water Content

Water is the most common solvent in RP-HPLC and is typically used in the mobile phase as a mixture with organic solvents. Figure 7.1.1 provides a guide regarding the miscibility of water with other solvents. The only exception of water not being used in an RP type separation is in NARP. Pure water is seldom utilized, and the dewetting of hydrophobic columns in the presence of pure water or of solvents with a too high proportion of water is discussed in Chapter 6. Water provides a higher polarity of the mobile phase; it is also the typical vehicle for buffers, acids, bases, or salts used in RP-HPLC. As a solvent, besides its high polarity (log $K_{ow} = -0.65$ or experimental -1.38 , $P' = 10.2$, $\pi^* = 1.09$), water has a high hydrogen-bond donor capability and an average hydrogen-bond acceptor capability (see Table 7.1.5). With these qualities, water is an excellent solvent for polar samples, including amino acids, carbohydrates, proteins, and many other compounds. As expected, for RP-HPLC, water is a "weak" solvent. In gradient elution HPLC, mobile phases with a high water content are used for poorly retained analytes or at the beginning of gradient programs for the same purpose. The purity of water for HPLC is very important, in particular when high water content is necessary in the mobile phase and

when sensitive detection is utilized (e.g., MS). Besides acting as a polar solvent with "weak" elution character in RP-HPLC, water is the ideal solvent for buffers and ionic additives in the mobile phase (see Section 7.5).

Alcohols

One group of solvents that is very common in RP-HPLC are the alcohols. They are frequently utilized in mixture with water. The polarity of alcohols is lower than that of water, but alcohols have a relatively high capability to form hydrogen bonds, both as donor and as acceptors, usually with higher hydrogen bond acceptor capability compared to water. Among the alcohols, the most frequently used is methanol, followed by isopropanol and ethanol. For the decrease of mobile phase polarity (making the mobile phase "stronger" for RP), it is sufficient in many separations only to increase the content of the organic phase (ϕ), not changing the chemical nature of the alcohol. Methanol is typically used as organic modifier. However, if methanol does not provide enough "strength" for the mobile phase, alcohols with a higher number of carbon atoms in the molecule can be used (C2, C3, C4). Alcohols, as the length of their hydrocarbon chain increases, also have higher solubility for less polar solvents such as hexane and cyclohexane. This allows the use of mixtures of alcohol/hydrocarbon as a solvent phase in NARP when very hydrophobic analytes have to be separated. Alcohols in general have a low UV cutoff value.

Acetonitrile

Acetonitrile is another solvent that is very commonly used in RP-HPLC. This solvent has medium polarity, a weak hydrogen-bond acceptor, and very weak donor capability. Compared to methanol, acetonitrile has only slightly higher values for P' and π^*, but a higher K_{ow}. It typically acts as a stronger eluent

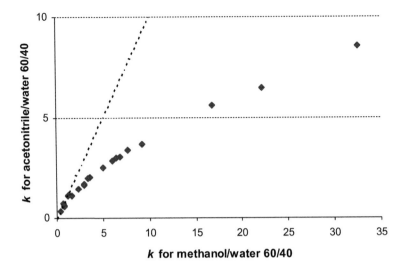

FIGURE 7.6.1 Plot of capacity factor k in methanol/water 60/40 and in acetonitrile/water 60/40 on a Spherisorb ODS-2 column, for a number of monosubstituted aromatic compounds.

compared to methanol. In Figure 7.6.1 are plotted the values for the capacity factor k in methanol/water 60/40 and in acetonitrile/water 60/40 on a Spherisorb ODS-2 column, for a number of compounds including a range of monosubstituted aromatics [85]. As shown in Figure 7.6.1, the values for k in methanol/water are significantly higher than those in acetonitrile/water for most compounds, especially as the capacity factors increase.

As indicated in Section 7.1, $\log K_{ow}$ for acetonitrile is -0.17 (calculated) and -0.34 experimental, but pure acetonitrile can elute from a hydrophobic chromatographic column even molecules with $\log K_{ow}$ as high as 7. For this reason, water/acetonitrile mixtures are common solvents in RP-HPLC, and the eluting strength of this mobile phase increases upon the addition of acetonitile. From the $\log K_{ow}$ of the analyte an empirical formula can be given, which provides guidance regarding the volume fraction of water in acetonitrile for the elution within a 25-min window from a common 150 mm RP column:

$$\phi \approx \frac{7 - \log K_{ow}}{7.5} \qquad (7.6.1)$$

Relation 7.6.1 does not provide any definite value for the retention time (compounds with low $\log K_{ow}$ will elute close to the dead volume of the column, while those with large $\log K_{ow}$ toward the end of the time window), but indicates the range of necessary composition for acetonitrile/water mixtures. The UV cutoff of acetonitrile is very low (see Table 7.3.1), which is an important characteristic when UV detection is utilized.

Other Solvents Used in RP-HPLC and NARP

Ethers are compounds typically used in RP-HPLC for further decreasing the polarity of the mobile phase compared to short-chain aliphatic alcohols or acetonitrile. Tetrahydrofuran, dioxane, *tert*-butyl methyl ether, and diethyl ether are among the common ethers used in RP-HPLC. Ethers also have very low donor hydrogen-bonding capability and medium acceptor hydrogen-bonding capability. Except for NARP where ethers may be more frequently used, in conventional RP-HPLC, the ethers, when used, are typically at relatively low proportions. These compounds have a UV-cutoff value around 210–215 nm.

Ketones are not very different in their solvent capabilities compared to ethers. However, ketones typically have relatively high UV cutoff values (around 330 nm) and therefore are less amenable for UV detection, which is one of the most common detection techniques. For this reason, the use of ketones as modifiers of the mobile phase in RP-HPLC is less common. However, since some ketones may have strong eluting properties in RP-HPLC, they can be used, for example, for the elution of hydrophobic compounds from graphitic columns.

Esters, such as ethyl acetate are sometimes used as mobile phase modifiers in RP-HPLC. However, their low water solubility, their potential to hydrolyze in acidic or basic media, and their relatively high UV cutoff value makes these solvents less useful in RP-HPLC.

Hydrocarbons are an important group of solvents with applications in particular in NARP. Their immiscibility with water makes them useless in conventional RP-HPLC, which frequently uses water as a polar component in the mobile phase. However, in NARP, mixtures of alcohols and hydrocarbons are common components of the mobile phase added with the purpose of lowering the polarity of the mobile phase. Hydrocarbons such as hexane, heptane, or cyclohexane are good solvents for highly hydrophobic analytes such as triglycerides or carotenoids.

Halogenated hydrocarbons have medium polarity (except for CCl_4) but low hydrogen-bonding capability. They are very good solvents for many compounds, but are not water soluble, which makes them of little use in RP-HPLC when water is present in the mobile phase. However, they may be used in NARP. Among the solvents used more frequently in this technique are methylene chloride, chloroform, and carbon tetrachloride. Other chlorinated solvents are seldom utilized. However, for some solubility reasons, special chlorinated solvents such as chlorobenzene can be used, for example, when the separation must be performed at temperatures higher than 70 °C and a less volatile solvent must be used as an eluent. Some concerns related to the use of chlorinated compounds as mobile phase may come from their potential hydrolysis with the formation of HCl.

Mobile Phase Composition and Detection in RP-HPLC

The properties of the mobile phase are important not only for the separation in HPLC but also for the type of utilized detection. The UV cutoff of different solvents is given in Table 7.3.1, and water and acetonitrile have the lowest UV cutoff values. Short-chain alcohols as well as hexane and heptane also have low UV cutoff values. On the other hand, ketones, and esters have a higher UV cutoff value.

The use of various solvents in connection with MS and MS/MS detection has been discussed in Section 7.3. The presence of water is typically required in the mobile phase for this detection technique, and acetonitrile or short-chain alcohols are frequently used in the mixture.

7.7. MOBILE PHASE IN ION-PAIR LIQUID CHROMATOGRAPHY

General Aspects

In ion pairing (IP), the mobile phase contains two components with different roles. One is the solvent, and the other is the ion-pairing agent (IPA) or the hetaeron (see Section 5.2). In practice, it is common to dissolve IPA in water, possibly with other additives such as a buffer, to which is added the necessary organic modifier for the separation. Restrictions to the range of pH values for the buffer depend on the stationary phase and are identical to those for reversed phases, usually within the interval

2−8. IP can be performed either in isocratic or in gradient conditions. For the gradient condition, all the components in the mobile phase can be modified, including organic modifier, IPA concentration, as well as the pH of the mobile phase [86].

Ion-Pairing Agents (Hetaerons)

As the applicability of ion-pair chromatography is extended from polar acidic and basic organic molecules to ionic species, even including inorganic ions, a large variety of ionic compounds were used as ion-pair additives (IPAs) able to form molecular association with such analytes. The IPAs are always selected to have the opposite charge to the analyte. The most common IPAs for acidic analytes are quaternary amines. The counterion of the quaternary amine can be sulfate, chloride, bromide, iodide, hydroxide, or dihydrogen-phosphate. A number of alkyl substituents (hydrophobic chains) at the nitrogen can be chosen for these IPA molecules, and the most common ones are trimethylalkyl, with alkyl ranging from C4 to C10. For basic compounds, typically used IPAs are sulfonates $R-SO_2-O^-$ and sulfates $R-O-SO_2-O^-$, with the R group alkyl, aryl, aryl with alkyl side chain such as alkylbenzenesulfonates, and in some cases with other substituents [87]. Also, fluorinated carboxylic acids $CF_3(CF_2)_nCOOH$ with $n = 0$ (trifluoroacetic acid), $n = 1, 2, 3, 4$ are common IPA for basic compounds. The counterions for these reagents are usually either H^+ or Na^+ ions, but other counterions are possible.

The mechanism of IPA retention as discussed in Section 5.2 shows that the hydrophobic chain plays a similar role regarding retention as the hydrophobic moiety of the solutes in RP-HPLC. As the hydrophobic character of the IPA increases, the retention of the complex with the analyte (solute) is stronger. However, the analyte structure, in particular on columns where different interactions besides the hydrophobic ones are possible, also plays a significant role in the separation (see Figure 5.2.4).

Besides the typical IPAs, numerous other IPAs have been reported in the literature [88]. For the separation of inorganic ions, for example, crystal violet can be used as IPA, the eluting order obtained on a hydrophobic column being in this case: $Cl^- < NO_2^- < Br^- < NO_3^- < I^- < SO_4^{2-} < S_2O_3^{2-}$, which is the same as that observed in anion-exchange LC [89].

Several studies reported in the literature indicate the use of ionic liquids in the IP mechanism (playing either the role of IPA or that of additive) [90]. Both the cationic and the anionic participants of the ionic liquid (IL) contribute to the improvement of peak shape and of resolution. The ion-pairing process can be affected such that changes in the retention time are seen upon the addition of ionic liquids in an IPA separation. This type of effect is illustrated in Figure 7.7.1, which shows the separation of four pharmaceutical compounds (metamizole as Na salt, 4-methylaminoantipyrine, fenpiverine bromide, and pitofenone hydrochloride) on a C18 column with an aqueous mobile phase (pH adjusted to 3) and sodium hexane sulfonate at 10 mM/L as IPA (trace B in Figure 7.7.1). The addition of 1-butyl-1-methyl-pirrolidinium tetrafluoroborate (an ionic liquid) at 10 mM/L in the separation produced the change shown in trace A in Figure 7.7.1. In the separation shown in trace B, sodium hexane sulfonate does not affect the retention of metamizole as Na salt, while the addition of the ionic liquid increases its retention. The effect of the addition of the ionic liquid on the other analytes is a shortening of the retention time.

Organic Modifiers of the Mobile Phase in IP

The organic modifiers used in IP are very similar to those used in RP-HPLC. Methanol

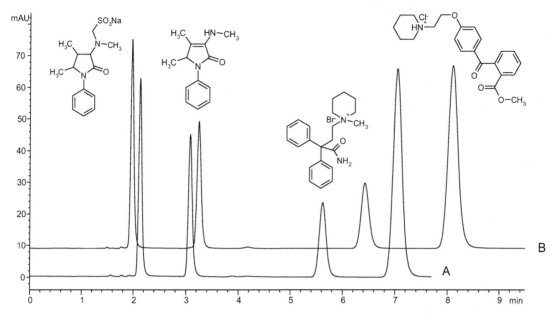

FIGURE 7.7.1 Comparison between retention of target compounds when the aqueous mobile phase (pH = 3) contains: (A) both sodium hexane sulfonate and 1-butyl-1-methyl-pyrrolidinium tetrafluoroborate (ionic liquid) at 10 mM/L and (B) only 10 mM/L sodium hexane sulphonate (Elution order: metamizole as Na salt; 4-methylaminoantipyrine; fenpiverine bromide; pitofenone hydrochloride).

and acetonitrile are the most utilized organic component of the mobile phase, although other solvents reported in the literature are useful for IP separations [91]. The effect of the organic modifier is similar but more complex compared to that of the organic component in RP separations. The change of organic content in the mobile phase also causes changes in the mobile phase dielectric constant, which can influence the strength of the ionic interactions between charged analyte and IPA. According to the electrostatic model presented in Section 5.2, the effect of the nature and concentration of the organic modifier in the mobile phase is found implicitly in the values of $k_i(0)$ and K_{IPA} in rel. 5.2.30, which gives the value for the capacity factor in IP. For most situations this dependence of log k on % organic modifier is linear (see rel. 3.2.12), but for some compounds the dependence is polynomial. Two examples are illustrated in Figure 7.7.2 where three biguandines

separated by IP mechanism follow a linear dependence between log k and the content of methanol in the mobile phase, while the fourth analyte (ranitidine) follows a quadratic dependence. The column used for the separation shown in Figure 7.7.2 was a Zorbax Eclipse XDB 150 mm length, 4.6 mm i.d. and 5 μm particle size, the IPA agent was 10 mM $C_8H_{17}SO_3Na$ in water adjusted to pH = 2 with H_3PO_4, the flow-rate was 1 mL/min, the injection was 20 μL of a solution containing 200 μg/mL of each analyte.

The dependence of capacity factor log k on the content of the organic modifier in the mobile phase can be significantly deviating from linear. As an example, the bile acids sodium taurocholate, sodium taurodeoxycholate, and sodium taurochenodeoxycholate can be used as IPA for analysis of pralidoxime, obidoxime, and pyridostigmine [92]. The dependence of log k for pralidoxime on the content of methanol in the

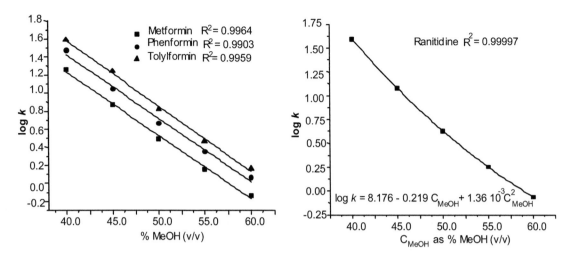

FIGURE 7.7.2 Linear and polynomial dependences of the capacity factor log k on the methanol percentage (v/v) in mobile phase composition for IP mechanism applied to the retention of four polar compounds.

mobile phase for these three different bile acids used as IPA is shown in Figure 7.7.3. Dependencies with a U shape of $\log k$ as a function of organic phase composition $C_o\%$ have been encountered in other ion-pair separations [93]. This nonlinear dependence of $\log k$ with increase in organic mobile phase content is caused by the different effects of the solvent on the values of various terms in rel. 5.2.30, such as K_{IPA}, $n_{IPA,max}^{s.p.}$, and κ (Debye length given by rel. 5.2.20 and modified through ε_{mo}).

The pH of the Mobile Phase, Chaotropes and Other Additives in IP

The pH of the mobile phase plays an important role in the formation of ion pairs and in their stability. As a result, the capacity factor k in IP is strongly dependent on pH, as shown in Section 5.2. For this reason, the mobile phase pH must be properly adjusted for achieving the optimum separation. This adjustment can be done using buffers or by simple addition of acids or bases. The addition of chaotropes in the mobile phase for IP is also practiced in certain separations [94]. Other additives have

been used in the mobile phase for improving IP separation. Among these can be mentioned the addition of EDTA or of potassium tetrakis (1H-pyrazolyl) borate as chelating agents [86].

Accordance of Mobile Phase in IPC with the Detection Mode

Detection based on UV absorption is widely used in IPC. Therefore ion-pair agents must have a low UV cutoff wavelength. The UV absorption of sodium alkanesulfonates and quaternary ammonium salts is low for typical UV range of detection in HPLC. A higher absorption takes place when ion liquids containing pyridinium or imidazolium moieties are used. Fluorescence is also frequently used, and most ion-pairing reagents do not affect it negatively.

A number of ion-pair reagents used in IP-LC separations of highly polar analytes are not suited for use in LC-MS. These reagents are nonvolatile and reduce the detection sensitivity. When MS detection is employed, volatile IP reagents such as aliphatic amines or perfluorinated organic acids at the lowest possible concentration in

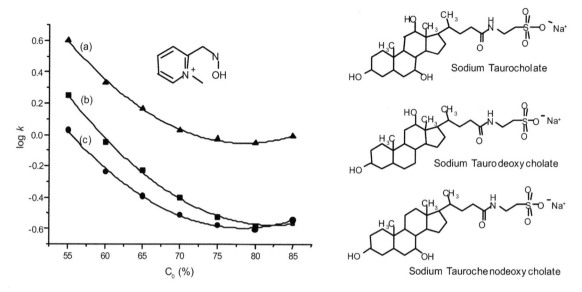

FIGURE 7.7.3 Dependences of log k on methanol concentration (C_o) in mobile phase for pralidoxime cation (2-[(hydroxyimino)methyl]-1-methylpyridin-1-ium), and different ion pairing agents: (a) sodium taurochenodeoxycholate; b) sodium taurodeoxycholate; c) sodium taurocholate.

mobile phase are recommended. Trifluoroacetic acid (TFA), pentafluoropropanoic acid (PFPA), and heptafluorobutanoic acid (HFBA) have commonly been used in the analysis of polar basic compounds. These ion-pairing agents form relatively stable ion pairs with basic compounds, decreasing the secondary interactions between free silanol groups of the stationary phase and resulting in decreased peak tailing, improved resolution, and better retention. However, acidic ion-pairing agents may suppress ionization in the MS source. Alkylamines (triethylamine, N,N-dimethyl-n-butylamine, N,N-dimethylhexylamine, or tri-n-butylamine) are commonly used as ion-pairing reagents in RP-LC with negative ion ESI/MS detection of acidic compounds, such as nucleoside mono-, di- and triphosphates and sulfonates, sulfates, sulfonated dyes, and halogenated acids [50].

The evaporative light scattering detector (ELSD) is another example of a detector that requires volatile mobile phases, and the choice of IPA and buffers with this detection

must avoid nonvolatile additives such as alkylsulfonates.

7.8. MOBILE PHASE IN HILIC AND NPC

General Comments

The mobile phase in HILIC and NPC is characterized by its lower polarity compared to the stationary phase. Typical mobile phases in HILIC are made of an aqueous buffer (10−40% of the mobile phase) with a controlled pH and an organic solvent such as acetonitrile or methanol, although other solvents such as ethanol or 2-propanol are sometimes used as organic modifiers. The solvent strength in HILIC has the following order: water > methanol > ethanol > 2-propanol > acetonitrile > acetone > tetrahydrofuran. For this reason, in HILIC separations using gradient conditions, the separation starts with a solvent low in water, and the water content is increased for the elution of the analytes that are more strongly retained [95].

One particular aspect of the mobile phase in HILIC is that compared to the mobile phase in RP, it typically contains a lower level of water (maintaining a lower polarity of the mobile phase compared to the stationary phase). This characteristic may pose problems with buffer solubility. For this reason, buffers with good solubility in mixtures containing a higher content of organic solvents are necessary for HILIC separations. For example, ammonium formate or acetate are preferred to phosphate buffers. Some other studies suggest that tri-fluoroacetic acid can also be used for the control of pH of aqueous buffer. The common concentration range for the buffers is 5–20 mmoles/L. The range of values of pH for the buffers that can be used in HILIC is limited by the nature of the stationary phases and is similar to that for RP-HPLC (pH = 2–8) [96, 97].

Acetonitrile is the preferred organic solvent in HILIC applications, while the other solvents may lead to insufficient sample retention and broad or nonsymmetrical peak shapes. This is, for example, the case of methanol, which is less used as the organic component in HILIC separations. The relatively poor performance of methanol in HILIC may be due to its similarity to water, both methanol and water being protic solvents. Methanol can compete to the solvation of the surface of silica or of other polar stationary phases used in HILIC and provide strong hydrogen-bonding interactions [98]

The solvents or additives used in NPC can be alkanes, cycloalkanes (n-pentane, n-hexane, n-heptane, i-octane, cyclopentane, cyclo-hexane), fluoroalkanes, chlorinated alkanes (dichloromethane, chloroform, carbon tetrachloride, propylchloride), ethers (diethyl ether; di-i-propyl ether), esters (methyl acetate, ethyl acetate), alcohols (methanol, ethanol, 1-propanol, 2-propanol), amines (pyridine, propylamine, triethylamine), and carboxylic acids or their derivatives (e.g., dimethylformamide). Some

values for the eluotropic strength of these solvents are listed in Table 7.1.6.

Double Role of the Mobile Phase in HILIC and NPC

The mobile phase in HILIC and NPC seems to play a more important role than simply elution medium (see rel. 6.5.6 that shows the variation of k with the water content in the mobile phase). The polar solvent molecules from the mobile phase (e.g. water) can be adsorbed onto the polar sites of the surface of the stationary phase, changing its interacting properties with analytes. In HILIC this effect can be diminished when enough water is present in the mobile phase to be adsorbed on the stationary phase surface, assuring that no significant changes in surface nature takes place during the separation. The same role as the water can be played in HILIC by other components of the mobile phase. As an example, an acidic buffer with an amino-propyl-based stationary phase produces the protonation of the amino groups, which then participate in the HILIC mechanism.

Since the mobile phase in HILIC (and NPC) plays a role in the nature of the stationary phase surface, its content in water (for HILIC) or other additives must be carefully considered. For example, during gradient separations in HILIC, the content in a buffer or an added salt may be changed if the additive is present only in the aqueous phase and absent in the organic modifier. The change in the mobile phase pH or ionic strength may be avoided in cases of gradient separations with two solvents A and B, where A is mainly organic and B mainly aqueous, when the same buffer/additive content should be used in both solutions. For the gradient separations with the initial mobile phase with no buffer and with the buffer content increasing during the run, changes in the retention mechanism may occur. The change in the retention mechanism can be exemplified by the variation in capacity factor k shown in Figure 7.8.1 for the

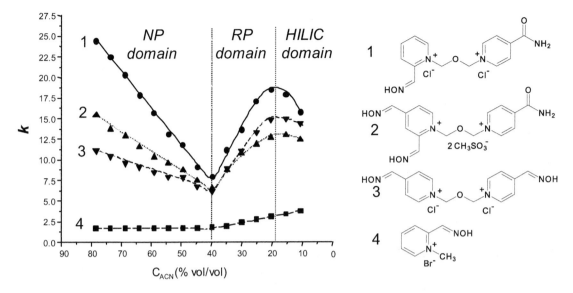

FIGURE 7.8.1 Change in the value of capacity factor k for the separation of four cationic-type oximes on a zwitterionic stationary phase upon the change in the mobile phase content in organic component (acetonitrile, ACN) and in ionic strength (KBr). 1. HI-6, 2. HLö-7, 3. Obidoxime, 4. Pralidoxime.

separation of several cation-type oximes (HI-6, HLö-7, obidoxime, and pralidoxime) on a stationary phase containing a sulfobetaine ligand bound on silica matrix (ZIC-HILIC, Merck, $150 \times 4.6 \times 5$; 200 Å pore size), with the change in the acetonitrile content and keeping only the aqueous phase with a content of 10 mM/L KBr. The variation in the values of k suggests the switch of the retention mechanism by changing the mobile phase composition and ionic strength [99].

An expected decrease in the values of capacity factor k is shown in Figure 7.8.1 for the organic phase content decreasing from 80 to 40% (and an increase in KBr concentration). However, below 40% can, an unexpected increase in k can be noticed. Between 40% and about 20% ACN, the value of k increases as in a RP-type HPLC. For even lower organic phase content, a typical decrease in k for a HILIC separation is seen. The same separation as shown in Figure 7.8.1 performed under constant ionic strength shows a unique NP type of retention mechanism.

In case of separations with an NP mechanism practiced on a silica column, the control of water content in the mobile phase is essential to maintain a constant silica activity. Dry solvents may dissolve some of the water present on the silica surface and modify its structure such that column reproducibility is not very good, exposing to the analytes either immobilized water or silanol groups. To achieve column stability, the silica columns are usually equilibrated with a standardized solvent (ethyl acetate containing 0.06% water) [100].

The Gradient Elution in HILIC Separations

Gradient elution is commonly used in HILIC. Although hydrophobic interactions are involved in HILIC separations, the main process is based on polar interactions. Therefore, better separations are obtained for compounds with large differences in polarity and not in the values for log K_{ow}. Gradient elution in HILIC starts from a composition of a mobile phase rich in organic constituents (e.g., acetonitrile or sometimes methanol), and the concentration of the polar (aqueous) component is increased in time. Buffers are frequently used in the mobile phase, and a good practice is to have the same buffer in the two phases that make the gradient such that the pH of the mobile phase is kept constant during the run.

The high initial acetonitrile content in the mobile phase will assure sufficient retention for the analytes with low affinity for the stationary phase. Running the gradient toward a high water concentration (e.g., 90%) favors the desorption of strongly retained analytes on the stationary phase. The gradient type depends on the nature of the stationary phase, organic component chosen for mobile phase composition, and on the nature of the analytes in the injected sample [101].

The pH of the buffers for HILIC may be selected to enhance the dissociation of sample analytes (pH > 7 for acids and pH< 7 for bases). Selectivity in HILIC can be well controlled with the pH of the mobile phase, the difference in retention of different analytes in HILIC being more affected by pH as compared to RP-HPLC [98]. The hydrophobic interactions are less sensitive to pH modifications because hydrophobic surfaces are usually not changed with pH changes. On the other hand, the pH of the mobile phase can strongly affect molecular polarity.

Influence of Mobile Phase on Detection in HILIC and NPC

The conclusions regarding the influence on UV detection of the mobile phase composition used in HILIC or NPC are not different from those for RP-LC (solvents must have a cutoff UV value below the wavelength selected for detection). A particular discussion can be made in the case of MS detection. The polar analytes are typically eluted with higher organic modifier content in HILIC than in RP-LC, which may improve the ESI/MS response. Often the best sensitivity in ESI is achieved when the analyte is already ionized in a liquid phase by using an acidic mobile phase for basic analytes, such as amines (pH two units below pK_a of the analyte), and basic conditions for acidic analytes, such as carboxylic acids and phenols (pH two units above pK_a of the analyte) [50]. The buffer concentrations should not exceed a level of 10 mM/L in order to avoid suppression of ionization and reduced sensitivity. Both polar and nonpolar solvents can be used in APCI.

Highly volatile NP-LC solvents are also well suited for atmospheric pressure photonionization (APPI). Lower vaporization temperatures can be used with easily vaporizable solvents, and this may be useful when analyzing thermolabile compounds. Many NP-LC solvents possess ionization energies below the 10.6 eV photons (e.g., 2-propanol 10.17 eV, n-hexane 10.13 eV, i-octane 9.89 eV, tetrahydrofuran 9.40 eV) and can be directly ionized by a krypton discharge lamp without any dopant addition. The use of low-proton affinity NP-LC solvents (hexane, chloroform) with toluene as a dopant can enhance the ionization through charge exchange, and thereby they improve the ionization efficiency for nonpolar compounds. NP-LC solvents successfully applied to APPI analysis include ethanol, 2-propanol, hexane, heptane, cyclohexane, i-octane, tetrahydrofuran, ethylacetate and chloroform [50].

The use of a mobile phase rich in organic volatile solvents, as practiced in HILIC and NPC, may also be favorable for enhancing the sensitivity of detection in techniques such as evaporative light scattering detection (ELSD) and corona-charged aerosol detection (cCAD), which involve an evaporative process (see Section 7.3). This has been proved for specific analytes as reported in the literature [52].

7.9. MOBILE PHASE IN ION-EXCHANGE AND ION-MODERATED CHROMATOGRAPHY

General Comments

The main component of the mobile phase in IC is typically water in which a specific buffer, acid, or base is dissolved. The dissolved additives in the aqueous mobile phase are selected depending on whether anionic or cationic separation is practiced, and must provide competing (driving) ions X that replace reversibly the analyte retained on the stationary phase. For a wide range of pH values for the mobile phase, the retention capacity for strong ion-exchange stationary phases remains unchanged. However, the analytes are significantly influenced by pH values 2 pH units larger than the analyte pK_a. However, for weak ion exchangers, the mobile phase also influences the stationary phase ionization and therefore its retention capacity. The capacity factor for an analyte A is given in IC by rel. 5.4.12, written once more below:

$$\log k_A = \alpha - \beta \log C_X \qquad (7.9.1)$$

As shown in Section 5.4, C_X is the concentration of the driving ion in the eluent. The values for the parameters α and β in rel. 7.9.1 (or 5.4.12) depend explicitly on the equilibrium constant $K_{A,X}$ for the analyte versus driving ions (see rel. 5.4.11), the charges of the ions involved in the separation, the ion-exchange capacity of the stationary phase, the weight of the stationary phase, and the volume of mobile phase in the column. The nature of the stationary phase, mobile phase, and the analytes determines these parameters. The mobile phase properties can be modified by using additives, by selecting the pH, and even by adding an organic modifier such as acetonitrile or methanol. The column in IC is commonly preconditioned with the same mobile phase used initially for elution. In case of a gradient, the composition of the mobile phase is modified, increasing the concentration of the driving ions C_X. Also, addition of ligands that can interact with the analytes may be used for facilitating the analyte elution.

A special type of solvent delivery system can be used in ion chromatography. This system is known as eluent generator (Dionex). When an eluent generator is used, the pump(s) deliver water while the reagents necessary for the elution are generated electrochemically from special cartridges. For this purpose, DC current is applied to a special cartridge to produce either KOH for anion-exchange eluents or methanesulfonic acid for cation-exchange eluent. This type of eluent generator is typically offered as a whole assembly, including other parts such as a degasser, and it is installed before the injector in the LC system (see Section 1.4).

The mobile phase in IC must be carefully selected in relation to the detection system. The most common detection technique in IC is based on conductivity, and the presence of acids, bases, or salts in the mobile phase may produce a large background for this type of detection. For this reason, two types of detection based on conductivity are practiced in IC: (1) with chemical suppression of the background signal, and (2) without chemical suppression of the background signal. Chemical suppression can be achieved using a resin or a semipermeable membrane that eliminates mobile phase ions that cause high-conductivity background (see Section 1.4). The mobile phase must be selected in accordance with the type of conductivity detection [102].

Mobile Phase in Cation-Exchange Chromatography

Various mobile phases are utilized in cation-exchange chromatography, depending on the analytes, the selected column, and the detection system. Most mobile phases are aqueous, frequently containing diluted HCl, HNO_3, H_2SO_4, or CH_3SO_3H (concentration range between 2 and 50 mM/L). As discussed in Section 5.4, separation of different ions can be achieved based on their different retention constants $K_{C,M}$. Since the retention follows the order $Li^+ < H^+ < Na^+ < NH_4^+ < K^+ < Rb^+ < Ag^+$, the elution can be done for monovalent ions using acids (H^+ driving ion) that generate a high enough concentration of H^+ to assure elution (see rel. 7.9.1). However, the retention of divalent ions (or trivalent ions) is much stronger than that of monovalent ions, and the elution with acids requires a higher concentration of H^+ ions. It is difficult to apply this requirement because in detections without or even with chemical suppression, the background signal (conductance) caused by the mobile phase is very high. The use of a solution of $AgNO_3$ as mobile phase may elute divalent ions since Ag^+ does have a higher affinity than H^+ for the stationary phase. This procedure requires a suppressor column in Cl^- form for eliminating Ag^+ ions [103]. In case of strongly retained ions (e.g., transitional metallic ions), the elution is frequently achieved using a complexing agent, such as a solution of a weak organic acid (e.g., tartaric, or α-hydroxyisobutiric acid) that does not generate a large signal background (low H^+ concentration) and forms complexes with the analyte. This procedure has even been applied for the elution of ions such as Ca^{2+}, Sr^{2+}, Pb^{2+}, and Ba^{2+} with 2,3-diaminopropionic acid and HCl in the mobile phase.

Besides metallic cations, IC has been applied successfully for the analysis of organic cations, such as those from biogenic amines (e.g., putrescine, cadaverine, and histamine). In this case a diluted solution of H_2SO_4 (5.0 mM/L) can be used as a mobile phase [104]. The extension of ion chromatography to the analysis of various ionic molecules frequently requires a specific

TABLE 7.9.1 Some buffers recommended for cation exchange LC.

Buffer/counterion	pK_a (at 25°C)	pH interval
Formic acid/Na^+, or Li^+	3.75	3.5–4.5
Acetic acid/Na^+, or Li^+	4.75	4.8–5.2
Maleic acid/Na^+	2.00	1.5–2.5
Malonic acid/Na^+, or Li^+	2.88	2.35–2.40
Citric acid/Na^+	3.13	2.60–3.60
Lactic acid/Na^+	3.80	3.60–4.30
Succinic acid/Na^+	4.20	4.30–4.80

pH of the mobile phase. This pH is specifically chosen to perform the separation as well as to provide a mobile phase with electric conductivity that can be easily suppressed. Specific buffers are recommended for use in IC; some of these are listed in Table 7.9.1.

Mobile Phase in Anion-Exchange Chromatography

A variety of mobile phases are used in anion-exchange chromatography. Electrochemical generation of OH^- ions (e.g., produced with a Dionex EG40 eluent generator) allows the use of water as a mobile phase. Buffers based on CO_3^{2-}/HCO_3^- are also widely used for separation of inorganic and organic anions. The major advantage of this buffer is related to the suppressor reaction, which leads to H_2CO_3, which is weakly dissociated, and consequently its contribution to the background signal is very low. There are alternatives to carbonate/bicarbonate buffer, such as solutions of amino acids. For pH > 7 only the carboxy group is dissociated, and thus it plays its role in anion separation. Besides that, the suppressor reaction carried out at a pH corresponding to the isoelectric point of the amino acid converts it into a zwitterionic form with low contribution to the background signal. Another additive in anion-exchange mobile phases is the tetraborate ion ($B_4O_7^{2-}$). The suppression of tetraborates is based on its change to H_3BO_3, which is weakly dissociated; thus it does not contribute to the background conductivity. However, $B_4O_7^{2-}$ has a low affinity for the stationary phases, and for this reason it is only used for the elution of F^- and short-chain R-COO⁻ anions.

The mobile phases with low background conductivity can be used for non-suppressed anion chromatography and are usually based on diluted aqueous solutions of organic salts, such as benzoates, phthalates, or sulfobenzoates. These anions are also characterized by a significant affinity to the stationary phase, and meanwhile they produce a relatively low conductivity of the mobile phase. The pH of mobile phase must be adjusted to 4–7 in order to favor the dissociation of the weak acid groups of the additive, which at their turn influence the retention process of the inorganic anionic species (e.g., F^-, Cl^-, Br^-, I^-, NO_2^-, NO_3^-, PO_4^{3-}, SO_4^{2-}, $S_2O_3^{2-}$, SCN^-). However, the background conductivities of these mobile phases are higher than the conductivity of carbonate/bicarbonate buffer after passing the suppressor column, which is of the 15–20 µS/cm level. Thus, at concentration of 0.5–1 mM/L of the mentioned organic salts, the background conductivity is situated between 60 and 160 µS/cm, which is high and affects the detection performances.

Besides inorganic ions, IC is also used for the separation of numerous other analytes that can be present in ionic form. For these compounds, the use of buffers that assure the formation of the ionic form of the analytes is necessary. Similar to the case of cations, the easy elimination of the conductivity created by the mobile phase components using suppressors is one criterion for selecting such buffers. Table 7.9.2 lists several buffers used in anion-exchange LC. Other mobile phases in anion-exchange chromatography have been reported. Some

TABLE 7.9.2 Some buffers recommended for anion exchange LC.

Buffer/counterion	pK_a (at 25°C)	pH interval
Piperidine/Cl⁻	11.12	10.6–11.6
Piperazine/Cl⁻, HCOO⁻	5.67	5.0–6.0
N-methylpiperazine/Cl⁻	4.75	4.5–5.0
Ethanolamine/Cl⁻	9.50	9.0–9.5
Diethanolamine/Cl⁻	8.88	8.4–8.8
Triethanolamine/Cl⁻	7.75	7.3–7.7
Histidine/Cl⁻	5.95	5.5–6.0
1,3-Diaminopropane/Cl⁻	10.46	9.8–10.3

such phases may have a multipart composition that can favor a complex separation process with ion-exchange, ion-exclusion, and ion-pairing principles for the separation [105]

When UV detection is used in ion chromatography, solutions of salts of phosphoric, sulfuric, or perchloric acid are suitable as mobile phase because these anions do not strongly absorb radiation in this spectral domain. When amperometric detection is chosen, the mobile phase acts as a support electrolyte and the electrolyte concentration must be about 50–100 higher than concentration of the anion analytes. In this case, hydroxide, chloride, chlorate, or perchlorate of alkali metals is used as the supporting electrolyte for anion elution. Ethylenediaminotetraacetic acid (EDTA) can be used for the elution of very strongly retained polyvalent anions such as polyphosphates. Besides that, EDTA can form anionic complexes with many metallic ions by the control of pH of mobile phase, a property that can be used in separating metallic cations by anion exchange LC [106].

Gradient Elution in Ion Chromatography

The use of gradient elution in IC is less common. The conductivity detection, which is usually employed in IC, is sensitive to changes in mobile phase composition, and isocratic separations are more convenient since they produce a constant background. In non-suppressed IC the use of a so-called isoconductive gradient, in which the conductances of the starting and finishing eluents used to obtain the gradient are equal, partially overcomes this difficulty, but the variation in eluotropic strength for isoconductive gradients is quite limited. In suppressed IC, the use of gradient elution is based on the availability of suppressors with sufficient capacity to ensure that the background conductance of the suppressed eluent remains essentially constant over the course of gradient [107]. On the other hand, the eluotropic strength of an eluent with constant composition can be quite limited. For achieving separation, gradients are necessary in many cases, and they can be conveniently used in suppressed IC when good suppressors are available.

In gradient IC, similarly to other HPLC types of separation, an effective gradient capacity factor k_e can be used for describing retention. For a linear gradient, the expression for the effective capacity factor can be approximated by an expression of the form:

$$\log k_{e,A} = \alpha' - \beta' \log\left(\frac{\Delta C_X}{\Delta t}\right) \qquad (7.9.5)$$

where the intercept α' and slope β' are specific for the given separation system, the analyte A^-, and the competing (driving) ion X^- of the IC separation. The gradient ramp $\Delta C_X/\Delta t$ is expressed in mM/min, and it is assumed to start as the analyte reaches the head of the column. This dependence is illustrated in Figure 7.9.1 for three anions (formate, acetate, and methansulfonate) eluted using OH$^-$ driving ions under linear gradient mode for a separation on a Dionex AS11 IonPak column with a mobile phase flow rate of 1.00 mL/min water, Dionex EG40 eluent generator that creates HO$^-$ ions (KOH), an ASRS-II suppressor, and the detection based on conductivity [108].

Chromatofocusing

Cromatofocusing is a procedure typically used for separation of amphoteric compounds,

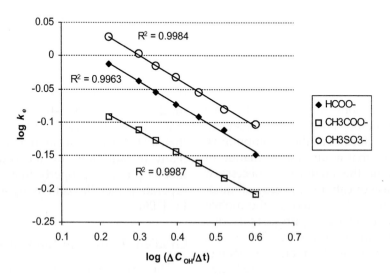

FIGURE 7.9.1 Linear dependence of log k_e for three anions on linear gradient ramp.

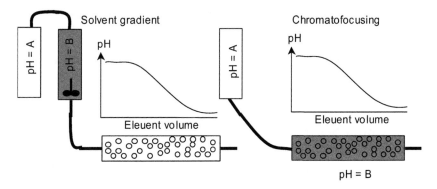

FIGURE 7.9.2 Schematic diagram of creating a pH gradient within a ion exchange chromatographic column for chromatofocusing.

most commonly proteins, on an ion-exchange column that has a pH gradient within the column (not an externally applied solution gradient) [109, 110]. A pH gradient can be produced within an ion-exchanger column by combining the buffering capacity of the ion exchanger with that of a buffer in the mobile phase, in a similar manner as a pH gradient is formed if two buffers at different pH are gradually mixed. For this purpose, the resin, which is, for example, an anion exchanger, is initially adjusted to a high pH value. The mobile phase has a different pH, for example, slightly acidic, and as it flows it changes the pH across the column. A schematic diagram of the process is shown in Figure 7.9.2 [111]. The separation is based on the initial retention of the analyte (protein) at the high pH of the column. An eluent at a lower pH typically made from several buffer species (with a range of pK_a values) is passed through the column. This creates a physical gradient that migrates down the column and the protein moves with it. The mobile phase elutes the protein when the buffer exiting the column has a pH equal with pI of the protein. The range of pK_a values in the eluent must be selected to provide buffer capacity across the entire pH range of intended gradient [112].

Mobile Phase in Ion-Moderated Chromatography

The choice for mobile phase in ion-moderated chromatography is rather limited. The simplest mobile phase is pure de-ionized water, which has been proved useful, for example, in the analysis of carbonate ion. For some columns, a specific ion must be present in the mobile phase for maintaining column integrity. Examples are ions such as H^+, Ca^{2+}, and Pb^{2+} (e.g., diluted H_2SO_4 is used to generate H^+ ions). For the separation of organic acids, solutions of inorganic acids are usually used as the mobile phase. This also depends on the type of suppressor column employed in case of conductivity detection. In case of UV-detection, H_2SO_4 solution is also frequently used. The high retention of some aliphatic and aromatic carboxylic acids can be prevented by the addition in mobile phase of a small content of miscible solvents, such as methanol, ethanol, i-propanol, or acetonitrile.

7.10. MOBILE PHASE IN CHIRAL CHROMATOGRAPHY

General Comments

Enantiomer separation (not assuming derivatization and transformation of enantiomers in diastereoisomers) can be done by means of two different procedures, each one offering a chiral environment: (1) use of chiral stationary phases and a nonchiral mobile phase, and (2) use of an achiral stationary phase with a mobile phase containing a chiral additive [113].

Mobile Phase for Chiral Separations on Chiral Stationary Phases

Many of the stationary phases used in chiral LC can be employed either in RP or in NP mode, with the mobile phase more polar than the stationary phase, or less polar than the stationary phase, respectively. Among the phases that can be used in both modes are polysaccharides (cellulose and modified celluloses), cyclodextrins, modified cyclodextrins, macrocyclic glycopeptides, crown ethers, and Pirkle-type phases. Even when the phase is used in RP mode, the content of the polar solvent is typically low and water is frequently absent in the mobile phase. Commonly used solvents are acetonitrile, methanol, tetrahydrofuran, 2-propanol, and ethanol. For cyclodextrin-based stationary phases, dimethylformamide and dimethylsulfoxide are also mentioned as possible polar components of the mobile phase. The solvent selection may determine which interaction type is more important for the separation, for example, π-π stacking, H-bonding, or hydrophobic [114]. Enantioseparations can be strongly influenced by the pH of the mobile phase [115, 116]. The basic character of the mobile phase can be obtained with basic additives such as diethylamine or triethylamine, while the acidic character can be obtained using trifluoroacetic acid, for example. The pH influences molecular structure as well as the formation/elimination of hydrogen bonds and plays an important role in the separation. Column stability is also affected by the mobile phase pH, many chiral phases being less resilient than RP columns to extreme pH values of the mobile phase. The typical pH range of utilization of many chiral phases is between 3 and 7.

For normal-phase utilization of chiral stationary phases, the mobile phase is made from apolar solvents such as hexane or heptane with polar components such as methanol, ethanol, 2-propanol, chloroform, methylene chloride, lower ethers, or esters. A certain level of other additives can be used for the separation, such as diethylamine (DEA), triethylamine, butylamine, or trifluoroacetic (TFA) acid, to control the pH. The pH can also be controlled with buffers such as ammonium acetate, citrate buffers, and triethylammonium phosphate [117]. In general, better separations are obtained at longer retention times, and therefore, lower levels of polar components in the mobile phase are typically used. The same effect of longer retention times is obtained using as polar solvents compounds such as chloroform, methylene chloride, tetrahydrofuran, or ethyl acetate, which are less polar than methanol, ethanol, or acetonitrile [118]. Similar to the case of RP utilization, the pH of the mobile phase is very important for determining selectivity [119–122].

Besides the control of pH, the buffers may have additional roles in chiral separations. For example, the use of triethylamine/acetate buffer is preferred for separations on most cyclodextrin-based stationary phases, leading to better separations and better peak shapes. This can be explained by the special effect of the buffer on various types of interactions determining enantiomer separation. Some buffers such as triethylamine/acetate, formate, or citrate are considered capable of forming inclusion complexes with the cyclodextrins and affect separation not only due to pH control. Buffer concentration is also important for achieving specific separations, and at high enough concentrations (e.g., higher than 1.5%) some buffers affect enantioselectivity negatively [115].

Mobile Phase for Chiral Separations on Achiral Stationary Phase

The separation of two enantiomers on an achiral stationary phase requires a chiral additive in the mobile phase to create a chiral environment. The mobile phase can be similar to that used in a nonchiral separation of RP or NP type. The solvents for the mobile phase may include water, acetonitrile, methanol, and ethanol, or nonpolar solvents such as hexane and heptane. Cyclodextrin has been frequently used as a chiral additive. One example of using this chiral additive is the separation of (±)-camphor and (±)-α-pinene using a RP column and α-cyclodextrin in the mobile phase [123]. In case of cyclodextrin used as a chiral modifier, solvents with higher hydrophobicity such as ethanol and iso-propanol are typically used for achieving faster elution. Among other examples of chiral additives are cationic β-cyclodextrins [124], bovine serum albumin (BSA), and alpha-1-acid glycoprotein (ORM). In many cases, the influence of the additive on the separation is considered to be caused by the formation of complexes in the solution [125]. However, in the case of the additive N-alkyl-L-hydroxyproline plus copper acetate, used for the separation of certain enantiomers of amino acids on a C18 stationary phase, the chiral selector is strongly adsorbed onto the C18 surface, effectively forming a chiral stationary phase. For an aqueous mobile phase used for elution, there is virtually no column bleed when the additive is left out from the mobile phase after column conditioning. In this case, the equilibrium takes place between the chiral analyte and the chiral surface of the stationary phase, with the separation caused by differences in the interactions on the solid phase for each enantiomer.

The previous examples show that two limiting models for the separation of enantiomers with a chiral additive in the mobile phase are possible: (A) the chiral additive interacts with the enantiomers in solution, and the adduct is selectively retained on the achiral column, (B) the chiral additive is retained on the column, and the stationary phase acts as a chiral medium. A simple theory can be developed to evaluate the factors influencing the separation in each case. For case A − solution interactions, the two enantiomers will be noted E and E^*, and the chiral additive will be called Q. The equilibria for the enantiomer E are the following:

(1) between E in the mobile phase (m) and in the stationary phase (s) described by the constant K_1 given by rel. 7.10.1:

$$E_m \leftrightarrows E_s \qquad K_1 = \frac{[E]_s}{[E]_m} \qquad (7.10.1)$$

(2) between E and Q in the mobile phase to form the adduct EQ described by the constant K_2 given by rel. 7.10.2:

$$E_m + Q_m \leftrightarrows EQ_m \qquad K_2 = \frac{[EQ]_m}{[E]_m[Q]_m} \quad (7.10.2)$$

(3) between the complex EQ in the mobile phase and in the stationary phase described by the constant K_3 given by rel. 7.10.3:

$$EQ_m \leftrightarrows EQ_s \qquad K_3 = \frac{[EQ]_s}{[EQ]_m} \qquad (7.10.3)$$

Since the enantiomer E is present both free and as a complex with Q in both the mobile and stationary phases, the distribution coefficient D for the enantiomer E between the mobile phase and the stationary phase is given by the expression:

$$D = \frac{[E]_s + [EQ]_s}{[E]_m + [EQ]_m} \qquad (7.10.4)$$

The value of $[E]_s$ can be obtained as a function of $[E]_m$ from rel. 7.10.1, and the value of $[EQ]_s$ can be obtained as a function of $[E]_m$ and $[Q]_m$ from rel. 7.10.3 and 7.10.2. Also, $[EQ]_m$ can be expressed as a function of $[E]_m$ and $[Q]_m$ from

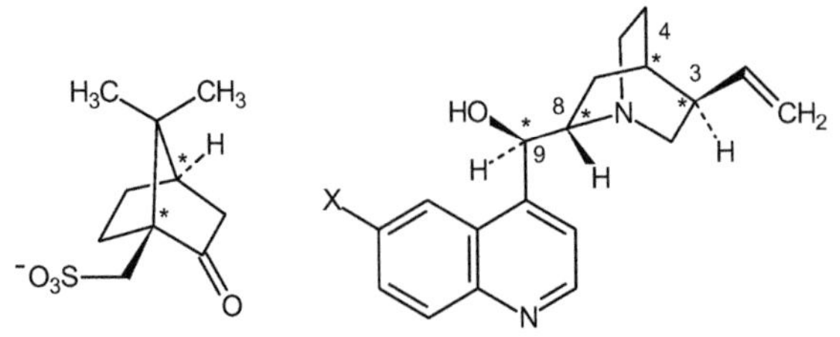

X = -OCH₃ 3(R), 4(S), 8(S), 9(R) quinine

X = -OCH₃ 3(R), 4(S), 8(R), 9(S) quinidine

X = -H   3(R), 4(S), 8(S), 9(R) cinchonidine

X = -H   3(S), 4(R), 8(S), 9(R) cinconine

FIGURE 7.10.1   Formulas for (+)-10-camphor-sulphonic acid, and for quinine, quinidine, cinchonidine and cinconine.

rel. 7.10.2. Including all these values in rel. 7.10.4, the expression of D becomes:

$$D = \frac{K_1 + K_2 K_3 [Q]_m}{1 + K_2 [Q]_m} \qquad (7.10.5)$$

Relation 7.10.5 indicates that the distribution constant D for the enantiomer E is equal with K_1 if the chiral additive is absent ($[Q] = 0$) or if K_2 and K_3 are very small or zero. The increase in K_2, K_3, and $[Q]$ lead to the increase in D.

The corresponding distribution constant for the enantiomer E^* involves constants K^*_1 (and $K_1 = K^*_1$), K^*_2 and K^*_3 (similarly defined as K_1, K_2, and K_3). Constants K^*_2 and K^*_3 can be assumed different from the corresponding ones for the enantiomer E because Q is assumed to interact differently with each enantiomer, and because EQ and E^*Q can be viewed as two diastereoisomers, their retention on the achiral stationary phase being different. With these constants being different, the selectivity α given by $\alpha = D/D^*$ will be different from 1, and the separation is possible. In general, an increase in $[Q]$ leads to an increase in α, indicating that more chiral additive favors the separation.

For case B—solid phase interaction, the additive Q is retained on the solid phase based on the equilibrium:

$$Q_m \leftrightarrows Q_s \qquad K_4 = \frac{[Q]_s}{[Q]_m} \qquad (7.10.6)$$

Since the enantiomer E is assumed to interact only with the adsorbed additive, the next equilibrium can be written as follows:

$$E_m + Q_s \leftrightarrows EQ_s \qquad K_5 = \frac{[EQ]_s}{[E]_m [Q]_s} \qquad (7.10.7)$$

The distribution coefficient D is given by the expression 7.10.4 where $[EQ]_m = 0$. In this case, by substituting in D the values for $[E]_s$ (from rel. 7.10.1), for $[EQ]_s$ (from rel 7.10.7), and for $[Q]_s$ (from rel. 7.10.6), the following expression is obtained:

$$D = K_1 + K_4 K_5 [Q]_m \qquad (7.10.8)$$

This expression shows that the capacity factor of the separation $k = D\Psi$ depends on the additive Q concentration and constants K_1, K_4, and K_5. Among these constants, only K_5 is different from K^*_5. The selectivity α depends only on the ratio K_5/K^*_5, and an increase in $[Q]$ leads to an increase in α. In every separation, one model or the other better explains the process, but in most cases each type of interaction likely plays a role.

Ion-Pairing Mechanism for Enantioseparation

The use of achiral stationary phases for separating the enantiomers of compounds that contain ionizable or strongly polar groups can be done by ion-pair HPLC using chiral API. In principle, the chiral API will form with the analyte associates that act as diastereoisomers [126]. Among the more common chiral API agents are (+)-10-camphor-sulphonic acid or (-)-10-camphor-sulphonic acid for the analysis of cationic compounds, and quinine, quinidine, cinchonidine, and cinchonine as counterion for the separation of acids [127,128] (see Figure 7.10.1). The mobile phases typically used in such separations should be high in the organic-phase component. The ion-pair separation in NP mode can be performed on a silica stationary phase [129].

7.11. MOBILE PHASE FOR SIZE-EXCLUSION SEPARATIONS

General Comments

In most size-exclusion cases (in both known modes GFC and GPC), the mobile phase serves mainly to dissolve the sample and carry it through the column. Compared to RP-HPLC or HILIC, the mobile phase in SEC has a smaller role in modulating the separation (see Section 5.6). However, polymer dissolution can be a challenging task, which is worsened by the fact that in SEC the samples are usually more concentrated than in other HPLC types (e.g., several mg/mL). In solution, macromolecules can assume various shapes (conformations) such as globules, rods, and coils. Dissolution of polymers is mainly determined by enthalpic interactions, since the gain in mixing entropy for polymers during dissolution is typically small. The intermolecular interactions between the polymer chain segments and solvent molecules have an associated energy that can be positive or negative. For a "thermodynamic good solvent,", interactions between polymer segments and solvent molecules are energetically favorable and will cause polymer coils to expand. For a "poor solvent," polymer-polymer self-interactions are preferred, and the polymer coils will contract. The thermodynamic "quality" of the solvent depends on the chemical composition of both the polymer and solvent molecules, and also on the solution temperature. In one special type of solvent (and at

a specific temperature), the polymer behaves as in the bulk polymer without changing its conformation. Such a solvent is called a theta solvent. The enthalpy of interactions with theta solvent must only compensate the effect of excluded volume expansion. Detailed discussion regarding polymer solutions and the influence of solvent on polymer conformation are reported in the literature [130, 131].

As indicated in Section 5.6, the equilibrium constant K_{SEC} for the separation, and as a result the capacity factor k_{SEC}, depend on the effective mean radius $\bar{r}$ of the polymer. Since the shape of the macromolecule in solution and macromolecular hydrodynamic volume depend on the solvent choice, when SEC is used for the determination of molecular weight of the polymers, the resulting value for MW may be influenced by the solvent choice [132]. Also, in the case of protein separation, the solvent may affect biological activity by changing the conformation of the protein. This can have a detrimental effect in cases where biological activity must be preserved. For other SEC utilizations, such as polymer identification, changes in molecular shape are less important.

In order to obtain reliable information regarding the macromolecular size (molecular weight) by SEC, besides the choice of a solvent that does not change conformation, it is essential that the separation is controlled exclusively by the entropy of the adsorption/desorption process (when the goal of the separation is, for example, only polymer purification, this is not an essential factor). Enthalpic interactions with the stationary phase lead to retention that is not a function of the molecular size of the analyte. For such cases where enthalpic interactions must be diminished (or even eliminated), the solvent can be selected specifically for this purpose. This solvent selection is typically done based on experimental trials and depends on the analyte and the stationary phase. Additives in the solution may sometimes affect the interaction between the analyte and the stationary phase, in particular for charged macromolecules. Due to the potential change of the polymer shape caused by the dissolution solvent—because the stationary phase can sometimes be affected when the solvent is changed, and also because the solvent composition has in general little effect on SEC separation process—most SEC separations are performed in isocratic conditions.

Typical Solvents for Gel Filtration (GFC)

A variety of solvents are used for water-soluble (hydrophilic) polymers. For nonionic polymers, pure water can be used successfully as eluent. For less hydrophilic polymers that may have some solubility problems in pure water, a certain percentage of organic solvent in water can be used, and for ionic polymers buffers and additives can be beneficial for the separation. Maintaining a constant ionic strength of the mobile phase is also sometimes necessary for reproducible results of molecular weight measurements. Some typical solvents used in water-soluble polymers are given in Table 7.11.1.

The salts used for constant ionic strength may include Na_2SO_4, $NaNO_3$, and CH_3COONa. The buffers that can be used include acetic acid/sodium acetate and other common buffers. The organic solvents may be methanol, acetonitrile, isopropanol, and other similar solvents. The manufacture of the SEC columns usually indicates restrictions regarding the solvents that can be used for a separation.

Specific hydrophilic polymers may not be soluble in water, although they also do not dissolve in organic solvents. This is the case, for example, for cellulose and other polysaccharides, as well as for specific proteins. In such cases, special solvents are necessary. For example, cellulose can be solubilized in dimethylacetamide + LiCl, certain cationic polymers can be solubilized in formamide + LiCl or in dimethylformamide/triethylamine/pyridine.

TABLE 7.11.1   Common solvent used in gel filtration.

Type of polymer	Typical eluents
Nonionic	Water, Water with 0.1 to 0.2 M salt for constant ionic strength Water with buffer at pH = 7
Nonionic with some hydrophobicity	Water with up to 20% organic modifier Water, organic modifier, 0.1 to 0.2 M salt and/or buffer pH = 7
Anionic	Water with 0.1 to 0.2 M salt for constant ionic strength and with buffer at pH = 7–9
Anionic with some hydrophobicity	Water, organic modifier up to 20%, 0.1 to 0.2 M salt and/or buffer pH = 7–9
Cationic	Water with 0.3 to 1.0 M salt for constant ionic strength and with buffer at pH = 2–7
Cationic with some hydrophobicity	Water with organic modifier, 0.3 to 1.0 M salt for constant ionic strength and with buffer at pH = 2–7

In case of proteins, specific restrictions are sometimes imposed on the mobile phase in order to promote solubilization, avoid protein adsorption on the stationary phase, and prevent protein denaturation. For example, addition of salts such as NaCl, or Na_2HPO_4 , of amino acids (glycine, alanine, or arginine), or of organic solvents (methanol, acetonitrile) may preclude protein adsorption, but higher content than 5–10% of organic solvents or more than 0.1–0.4 mM/mL salts must be avoided in order to maintain protein biological activity. Sometimes, for protein solubilization a detergent such as sodium dodecylsulfate (SDS) is added. This compound leads to the dissociation of protein aggregates, but it also increases protein hydrodynamic volume by formation of protein–SDS complexes [133–135].

Typical Solvents for Gel Permeation (GPC)

For GPC, the solubilization of the polymer can also be a considerable challenge. Tetrahydrofuran is a common solvent for many polymers such as polystyrene, poly-(methylmetacrylate), epoxy resins, polycarbonates, polyvinylchloride, and polystyrene/acrylonitrile. Other solvents used in polymer analysis include toluene, chloroform, and benzene [136]. Some common synthetic polymers such as polyethylene or polypropylene are not easily solubilized, and they are soluble only in solvents such as trichlorobenzene or methylcyclohexane at temperature above 90—100 °C [137]. The solvents used in SEC separations should not affect the structure of the stationary phase. Column manufacturers typically indicate the range of solvents allowed on a specific column.

References

[1] Hildebrand JH, Scott RI. The Solubility of Non-Electrolytes. New York: Dover Pub.; 1964.

[2] Snyder LR. Classification of the solvent properties of common liquids. J. Chromatogr. Sci. 1978;16:223—34.

[3] Sadek PC, Carr PW, Doherty RM, Kamlet MJ, Taft RW, Abrams MH. Study of retention process in reversed-phase high-performance liquid chromatography by the use of the solvatochromic comparison model. Anal. Chem. 1985;57:2971—8.

[4] Vitha M, Carr PW. The chemical interpretation and practice of linear solvation energy relationships in chromatography. J. Chromatogr. A 2006;1126: 143—94.

[5] Abraham MH, Ibrahim A, Zissimos AM. Determination of sets of solute descriptors from chromatographic measurements. J. Chromatogr. A 2004;1007: 29—47.

[6] Scatchard G. Equilibria in non-electrolyte solutions in relation to the vapor pressures and densities of the components. Chem. Rev. 1931;8:321—33.

[7] Shinoda K. Principles of Solution and Solubility. New York: Marcel Dekker; 1978.

[8] Snyder LR. The role of the mobile phase in liquid chromatography. In: Kirkland JJ, editor. Modern Practice of Liquid Chromatography. New York: Wiley-Interscience; 1971.

[9] Moldoveanu SC, David V. Sample Preparation in Chromatography. Amsterdam: Elsevier; 2002.

[10] Hansch C, Leo A, Hoekman D. Exploring QSAR, Hydrophobic. Electronic and Steric Constants. Washington, DC: ACS; 1995.

[11] http://www.chemaxon.com

[12] http://www.epa.gov/oppt/exposure/pubs/episuite.htm

[13] http://www.daylight.com

[14] http://www.scivision.com/gProdPage/tReviews/AciLogPUltra/TestYourOwn.html

[15] Iwase K, Komatsu K, Hirono S, Nakagawa S, Moriguchi I. Estimation of hydrophobicity based on the solvent-accessible surface area of molecules. Chem. Pharm. Bull. 1985;33:2114—21.

[16] Erős D, Kövesdi I, őrfi L, Takács-Novák K, Acsády G, Kéri G. Reliability of log P predictions based on calculated molecular descriptors. A critical review. Curr. Med. Chem. 2002;9:1819—29.

[17] Rohrschneider L. Solvent characterization by gas-liquid partition coefficients K_S of selected solutes "i". Anal. Chem. 1973;45:1241—7.

[18] Johnson BP, Khaledi MG, Dorsey JG. Solvatochromic solvent polarity measurements and retention in reversed-phase liquid chromatography. Anal. Chem. 1986;58:2354—65.

[19] Reichardt C, Asharin-Fard S, Blum A, Eschner M, Mehranpour A-M, Milart P, Niem T, Schäfer G, Wilk M. Solute/solvent interactions and their empirical determination by means of solvatochromic dyes. Pure & Appl. Chem. 1993;65: 2593—601.

[20] http://www.uni-marburg.de/fb15/ag-reichardt/et-30-values?language_sync=1

[21] Taft RW, Kamlet MJ. The solvatochromic comparison method 2. The alpha scale of solvent hydrogen-bond donor (HBD) acidities. J. Amer. Chem. Soc. 1976;98:2886—94.

[22] Kamlet MJ, Taft RW. The solvatochromic comparison method 1. The beta scale of solvent hydrogen-bond acceptor (HBa) basicities. J. Amer. Chem. Soc. 1976; 98:377—83.

[23] http://www.stenutz.eu/chem/solv26.php

[24] Kamlet MJ, Abboud JLM, Abraham MH, Taft RW. Linear solvation energy relationships 23. A comprehensive collection of the solvatochromic parameters, π^*, α and β, and some methods for simplifying the generalized solvatochromic equation. J. Org. Chem. 1983;48:2877—87.

[25] Luehrs DC, Chesney DS, Godbole KA. Correlation of the chromatographic eluting strength of solvents with solvent polarity and polarizability parameters. J. Chromatogr. Sci. 1991;29:463—6.

[26] Snyder LR, Carr PW, Rutan SC. Solvatochromatically based solvent selectivity triangle. J. Chromatogr. A 1993;656:537—47.

[27] Chong WJ, Carr PW. Limitations of all empirical single parameter solvent strength scales in reversed-phase liquid chromatography. Anal. Chem. 1989;61:1524—9.

[28] Snyder LR, Kirkland JJ, Dolan JW. Introduction to Modern Liquid Chromatography. 3rd ed. Hoboken: Wiley; 2010.

[29] Lochmüller CH, Marshall DB, Wilder DR. An examination of chemically modified silica surface using fluorescence spectroscopy. Anal. Chim. Acta 1981;130:31—43.

[30] Karger BL, Snyder LR, Horvath C. An Introduction to Separation Science. New York: Wiley; 1973.

[31] Patel HB, Jefferies TM. Eluotropic strength of solvents, Prediction and use in reversed-phase high-performance liquid chromatography. J. Chromatogr. 1987;389:21—32.

[32] Snyder LR. Principles of Adsorption Chromatography. New York: Marcel Dekker; 1968.

[33] R., Scott PW, Kucera P. Solute interactions with the mobile and stationary phases in liquid-solid chromatography. J. Chromatogr. A 1975;112:425—42.

[34] Liteanu C, Gocan S, Bold A. Separatologie Analitica". Editura Dacia, Cluj-Napoca; 1981.

[35] Habibi-Yangjeh A. A model for correlation of various solvatochromic parameters with composition of aqueous and organic binary solvent systems. Bull. Korean Chem. Soc. 2004;25:1165—70.

[36] Rosés M, Bosch E. Linear solvation energy relationship in reversed-phase liquid chromatography. Prediction of retention from a single solvent and a single solute parameter. Anal. Chim. Acta 1993;274: 147—62.

[37] Rosés M, Bosch E. Linear solvation energy relationship in reversed-phase liquid chromatography. Prediction of retention from a single solvent and a single solute parameter. Anal. Chim. Acta 1993;274: 147—62.

[38] Sadek PC. The HPLC Solvent Guide. 2nd ed. New York: John Wiley; 2002.

[39] Galaon T, Medvedovici A, David V. Hydrophobicity parameter (log K_{ow}) estimation for some phenolic compounds of pharmaceutical interest from retention studies with mobile phase composition in RP-LC. Sep. Sci. Technol. 2008;43:147—63.

[40] Dolan JW, Gant JR, Snyder LR. Gradient elution in high-performance liquid chromatography. II. Practical application to reversed-phase systems. J. Chromatogr. 1979;165:31—58.

[41] Sándi Á, Szepesy L. Evaluation and modulation of selectivity in reversed-phase high-performance liquid chromatography. J. Chromatogr. A 1999; 845:113—31.

[42] Schellinger AP, Carr PW. Isocratic and gradient elution chromatography: A comparison in terms of speed, retention reproducibility and quantitation. J. Chromatogr. A 2006;1109:253—66.

[43] Snyder LR. Column efficiencies in liquid adsorption chromatography: past, present and future. J. Chromatogr. Sci. 1969;7:352—60.

[44] McClellan AL. Tables of Experimental Dipole Moments. San Francisco: W. H. Freeman and Co.; 1963.

[45] Miller KJ, Savchik J. A new empirical method to calculate average molecular polarizabilities. J. Am. Chem. Soc. 1979;101:7141—440.

[46] Cheong WJ, Carr PW. The surface tension of mixtures of methanol, acetonitrile, tetrahydrofuran, isopropanol, tertiary butanol and dimethyl-sulfoxide with water at 25°C. J. Liq. Chromatogr. 1987;10:561—81.

[47] Vazquez G, Alvarez E, Navaza JM. Surface tension of alcohol water + water from 20 to 50°C. J. Chem. Eng. Data 1995;40:611—4.

[48] Escobedo J, Mansoori GA. Surface tension prediction for liquid mixtures. AIChE J 1998;44:2324—32.

[49] Yongqi H, Zhibao L, Jiufang L, Yigui L, Young J. A new model for the surface tension of binary and ternary liquid mixtures based on Peng-Robinson equation of state. Chinese J. Chem. Eng. 1997;5: 193—7.

[50] Kostiainen R, Kauppila TJ. Effect of eluent on the ionization process in liquid chromatography - mass spectrometry. J. Chromatogr. A 2009;1216:685—99.

[51] Gao S, Zhang ZP, Edinboro LE, Ngoka LC, Karnes HT. The effect of alkylamine additives on the sensitivity of detection for paclitaxel and docetaxel and analysis in plasma of paclitaxel by liquid chromatography-tandem mass spectrometry. Biomed. Chromatogr. 2006;20:683—95.

[52] Mitchell CR, Bao Y, Benz NJ, Zhang S. Comparison of the sensitivity of evaporative universal detectors and LC/MS in the HILIC and the reversed-phase HPLC modes. J. Chromatogr. B 2009;877:4133—9.

[53] Lourié Y. Aide-Mémoire de Chimie Analitique. Editions Mir, Moscou; 1975.

[54] Marcus Y, Glikberg S. Recommended methods for the purification of solvents and test for impurities in methanol and ethanol. Pure Appl. Chem. 1985;57:855—64.

[55] Subirats X, Bosch E, Roses M. Retention of ionisable compounds on HPLC. XVII. Estimation of the pH variation of aqueous buffers with the change of the methanol fraction of the mobile phase. J. Chromatogr. A 2007;1138:203—15.

[56] Canals I, Portal JA, Bosch E, Roses M. Retention of ionizable compounds on HPLC. 4. Mobile-phase measurement in methanol/water. Anal. Chem. 2000;72:1802–9.

[57] Canals I, Oumada FZ, Roses M, Bosch E. Retention of ionizable compounds on HPLC. 6. pH measurements with the glass electrode in methanol–water mixtures. J. Chromatogr. A 2001;911:191–202.

[58] Espinoza S, Bosch E, Roses M. Retention of ionizable compounds in HPLC. 14. Acid-base pK values in acetonitrile-water mobile phases. J. Chromatogr. A 2002;964:55–66.

[59] David V, Albu F, Medvedovici A. Structure–retention correlation of some oxicam drugs in reversed-phase liquid chromatography. J. Liq. Chromatogr. Rel. Technol. 2004;27:965–84.

[60] Hetzer HB, Dust RA, Robinson RA, Bates RG. Standard pH values for potassium hydrogen phthalate reference buffer solution from 0 to 60 °C. J. Res. Natl. Bur. Std., A, Phys. Chem. 1977;81A:21–4.

[61] Bates RG, Guggenheim EA. Report on the standardization of pH and related terminology. Pure Appl. Chem. 1960;1:163–8.

[62] Rondinini S, Mussini PR, Mussini T. Reference values standards and primary standards for pH measurements in organic solvents and water + organic solvent mixtures of moderate to high permittivities. Pure Appl. Chem. 1987;59:1549–60.

[63] Wiczling P, Markuszewski MJ, Kaliszan R. Determination of pK_a by pH gradient reversed-phase HPLC. Anal. Chem. 2004;76:3069–77.

[64] Neue UD, Méndez A, Tran KV, Diehl DM. pH and selectivity in RP-chromatography. In: Kromidas S, editor. HPLC Made to Measure: A Practical Handbook for Optimization, Wiley-VCH Verlag GmbH & Co. Weinheim: KGaA; 2006. p. 71.

[65] Schellinger AP, Carr PW. Solubility of buffers in aqueous–organic eluents for reversed-phase liquid chromatography. LCGC North America 2004;22: 544–8.

[66] Claessens HA, van Straten MA, Kirkland JJ. Effect of buffers on silica-based column stability in reversed-phase high-performance liquid chromatography. J. Chromatogr. A 1996;728:259–70.

[67] Claessens HA, van Straten MA. Review on the chemical and thermal stability of stationary phases for reversed-phase liquid chromatography. J. Chromatogr. A 2004;1060:23–41.

[68] Tindall GW, Perry RL. Explanation for the enhanced dissolution of silica column packing in high pH phosphate and carbonate buffers. J. Chromatogr. A 2003;988:309–12.

[69] Kromidas S. More Practical Problem Solving in HPLC. Weinheim: Wiley-YCH; 2005. p. 187–207.

[70] David V, Galaon T, Caiali E, Medvedovici A. Competitional hydrophobicity driven separations under RP-LC mechanism: application to sulphonylurea congeners. J. Sep. Sci. 2009;32:3099–106.

[71] Nikitas P, Pappa-Louisi A, Agrafiotou P, Fasoula S. Simple models for the effect of aliphatic alcohol additives on the retention in reversed-phase liquid chromatography. J. Chromatogr. A 2011;1218:3616–23.

[72] Moldoveanu SC, Kiser M. Gas chromatography/ mass spectrometry vs. liquid chromatography/fluorescence detection in the analysis of phenols in mainstream cigarette smoke. J. Chromatogr. A 2007; 1141:90–7.

[73] Hao W, Zhang X, Hou K. Analytical solutions of the ideal model for gradient liquid chromatography. Anal. Chem. 2006;78:7828–40.

[74] Nikitas P, Pappa-Louisi A. Expressions of the fundamental equation of gradient elution and a numerical solution of these equations under any gradient profile. Anal. Chem. 2005;77:5670–7.

[75] Quarry MA, Grob RL, Snyder LR. Prediction of precise isocratic retention data from two or more gradient elution runs. Analysis of some associated errors. Anal. Chem. 1986;58:907–17.

[76] Liteanu C, Gocan S. Gradient Liquid Chromatography. New York: Wiley; 1974.

[77] Snyder LR, Dolan JW. High-Performance Gradient Elution. The Practical Application of the Linear-Solvent-Strength Model. Hoboken, NJ: Wiley-Interscience; 2007.

[78] Snyder LR, Kirkland JJ, Dolan JW. Introduction to Modern Liquid Chromatography. 3rd ed. Hoboken, NJ: John Wiley; 2010.

[79] Dolan JW, Snyder LR. Maintaining fixed band spacing when changing column dimensions in gradient elution. J. Chromatogr. A 1998;799:21–34.

[80] Dolan JW. How fast can a gradient be run. LC/GC Europe 2011;24:406–10.

[81] Meyer VR. Practical High-performance Liquid Chromatography. Chichester: Wiley; 2010.

[82] Kaliszan R, Wiczling P. Theoretical opportunities and actual limitations of pH gradient HPLC. Anal. Bioanal. Chem. 2005;382:718–27.

[83] Dolan JW. How fast can a gradient be run. LC/GC Europe 2011;24:406–10.

[84] Majors RE. Method translation in liquid chromatography. LC/GC Europe 2011;24:412–7.

[85] Smith RM, Burr CM. Retention prediction of analytes in reversed-phase high performance liquid chromatography based on molecular structure. J. Chromatogr. 1989;475:57–74.

[86] Cecchi T. Ion pairing chromatography. Crit. Rev. Anal. Chem. 2008;38:161–213.

[87] Szasz G, Tokacs-Novak K, Kökösi J. HPLC study on ion-pairing ability of deoxycholic acids epimers. J. Liq. Chromatogr. Rel. Technol. 2001;24:173–85.

[88] Woollard DC, Indyk-Harvey E. Rapid determination of thiamine, riboflavin, pyridoxine, and niacinamide in infant formulas by liquid chromatography. J. A.O.A.C. Int. 2002;85:945–51.

[89] Tonelli D, Zappoli S, Ballarin B. Dye-coated stationary phases: a retention model for anions in ion-interaction chromatography. Chromatographia 1998;48:190–6.

[90] Fernandez-Navarro JJ, García-Álvarez-Coque MC, Ruiz-Ángel MJ. The role of the dual nature of ionic liquids in the reversed-phase liquid chromatographic separation of basic drugs. J. Chromatogr. A 2011;1218:398–407.

[91] Pang XY, Sun HW, Wang YH. Determination of sulfides in synthesis and isomerization systems by reversed-phase ion-pair chromatography with a mobile phase containing tetramethylene oxide as organic modifier. Chromatographia 2003;57:543–7.

[92] Radulescu M, Voicu V, Medvedovici A, David V. Retention study of some cation-type compounds using bile acid sodium salts as ion pairing agents in liquid chromatography. Biomed. Chromatogr. 2011;25:873–8.

[93] Radulescu M, Iorgulescu EE, Mihailciuc C, David V. Comparative study of the retention of pyridinium and imidazolium based ionic liquids on octadecylsilica stationary phase under ion pairing mechanism with alkylsulphonate anions. Rev. Roum. Chim. 2012;57:61–7.

[94] Flieger J. The effect of chaotropic mobile phase additives on the separation of selected alkaloids in reversed-phase high performance liquid chromatography. J. Chromatogr. A 2006;1113:37–44.

[95] Hemström P, Irgum K. Hydrophilic interaction chromatography. J. Sep. Sci. 2006;29:1784–821.

[96] Karatapanis AE, Fiamegos YC, Stalikas CD. A revisit to the retention mechanism of hydrophilic interaction liquid chromatography using model organic compounds. J. Chromatogr. A 2011;1218: 2871–9.

[97] McCalley DV. Is hydrophilic interaction chromatography with silica columns a viable alternative to reversed-phase liquid chromatography for the analysis of ionisable compounds? J. Chromatogr. A 2007;1171:46–55.

[98] Jandera P. Stationary and mobile phases in hydrophilic interaction chromatography: a review. Anal. Chim. Acta 2011;692:1–25.

[99] Medvedovici A, Sora ID, Radulescu M, David V. Discontinuous double mechanism for the retention of some cation-type oximes on hydrophilic stationary phase in liquid chromatography. Analytical Methods 2011;3:241–4.

[100] Caude M, Jardy A. Normal phase liquid-chromatography. In: Katz E, Eksteen R, Schoenmakers P, Miller N, editors. Handbook of HPLC. New York: Marcel Dekker; 1998. p. 325.

[101] Dejaegher B, Mangelings D, Vander Heyden Y. Method development for HILIC assays. J. Sep. Sci. 2008;31:1438–48.

[102] Weiss J. Ion Chromatography. Weinheim: VCH; 1995. p. 66.

[103] Small H, Stevens TS, Bauman WC. Novel ion exchange chromatographic method using conductometric detection. Anal. Chem. 1975;47:1801–9.

[104] Liao BS, Sram J, Cain TT, Halcrow KR. Aqueous sulfuric acid as the mobile phase in cation ion chromatography for determination of histamine, putrescine, and cadaverine in fish samples. J. A.O.A.C. Int. 2011;94:565–71.

[105] Doyle JM, Miller ML, McCord BR, McCollam DA, Mushrush GW. A multicomponent mobile phase for ion chromatography applied to the separation of anions from the residue of low explosives. Anal. Chem. 2000;72:2302–7.

[106] Haddad PR, Jackson PE. Ion Chromatography. Principles and Applications. Amsterdam: Elsevier; 1990.

[107] Chen Y, Srinivasan K, Dasgupta PK. Electrodialytic membrane suppressors for ion chromatography make programmable buffer generators. Anal. Chem. 2012;84:67–75.

[108] Madden JE, Avdalovic N, Haddad PR, Havel J. Prediction of retention times for anions in linear gradient elution ion chromatography with hydroxide eluents using artificial neural network. J. Chromatogr. A 2001;910:173–9.

[109] Sluyterman LAAE, Elgersma O. Chromatofocusing: isoelectric focusing on ion exchange columns: 1 General principles. J. Chromatogr. 1978;150:17–30.

[110] Sluyterman LAAE, Wijdenes J. Chromatofocusing: Isoelectric focusing on ion exchange columns: 2. Experimental verification. J. Chromatogr. 1978;150: 31–44.

[111] Chromatofocusing with Polybuffer and PBE. Uppsala: Amarsham Pharmacia Biotech; 2001.

[112] http://www.validated.com/revalbio/pdffiles/chrfocus.pdf

[113] Gübitz G, Schmid MG. Chiral separation by chromatographic and electromigration techniques. A review, Biopharm. Drug Dispos. 2001;22:291–336.

[114] Ilisz I, Berkecz R, Peter A. Retention mechanism of high-performance liquid chromatographic enantio-separation on macrocyclic glycopeptide-based chiral stationary phases. J. Chromatogr. A 2009;1216: 1845−60.

[115] Gubitz G, Schmid MS, editors. Chiral Separations: Methods and Protocols. Totowa, NJ: Humana Press; 2004.

[116] Matthijs N, Maftouh M, Vander Heyden Y. Screening approach for chiral separation of pharmaceuticals IV. Polar organic solvent chromatography. J. Chromatogr. A 2006;1111:48−61.

[117] O'Brien T, Crocker L, Thompson R, Thompson K, Toma PH, Conlon DA, Feibush B, Moeder C, Bicker G, Grinberg N. Mechanistic aspects of chiral discrimination on modified cellulose. Anal. Chem. 1997;69:1999−2007.

[118] Lynam KG, Stringham RW. Chiral separations on polysaccharide stationary phases using polar organic mobile phases. Chirality 2006;18:1−9.

[119] Ye YK, Stringham RW. The effect of acidic and basic additives on the enantioseparation of basic drugs using polysaccharide-based chiral stationary phases. Chirality 2006;18:519−30.

[120] Change SC, Reid GL, Chen S, Chang CC, Armstrong DW. Evaluation of a new polar-organic high performance liquid chromatographic mobile phase for cyclodextrin bonded chiral stationary phases. Trends Anal. Chem. 1993;12:144−53.

[121] Tang Y, Zielinski WL, Bigott HM. Separation of nicotine and nornicotine enantiomers via normal phase HPLC on derivatized cellulose chiral stationary phases. Chirality 1998;10:364−9.

[122] Younes AA, Mangelings D, Vander Heyden Y. Chiral separations in normal phase liquid chromatography: Enantioselectivity of recently commercialized poly-saccharide-based selectors. Part I: enantioselectivity under generic screening conditions. J. Pharm. Biomed. Anal. 2011;55:414−23.

[123] Bielejewska A, Duszczyk K, Sybilska D. Influence of organic solvent on the behaviour of camphor and α-pinene enantiomers in reversed-phase liquid chromatography systems with α-cyclodextrin as chiral additive. J. Chromatogr. A 2001;931:81−93.

[124] Xiao Y, Tan TT, Ng SC. Enantioseparation of dansyl amino acids by ultra-high pressure liquid chromatography using cationic β-cyclodextrins as chiral additives. Analyst 2011;136:1433−9.

[125] Rodríguez-Bonilla P, López-Nicolás JM, Méndez-Cazorla L, García-Carmona F. Development of a reversed phase high performance liquid chromatography method based on the use of cyclodextrins as mobile phase additives to determine pterostilbene in blueberries. J. Chromatogr. B 2011;879:1091−7.

[126] Pettersson C, Schill G. Separation of enantiomers in ion-pair chromatographic systems. J. Liq. Chromatogr. 1986;9:269−90.

[127] Subramanian G, editor. Practical Approach to Chiral Separations. Weinheim: VCH; 1994.

[128] Karlsson A, Pettersson C. Separation of enantiomeric amines and acids using chiral ion-pair chromatography on porous graphitic carbon. Chirality 1992;4:323−32.

[129] Pettersson C. Chromatographic separation of enantiomers of acids with quinine as chiral counter ion. J. Chromatogr. A 1984;316:553−67.

[130] Brandrup J, Immergut EH, Gruelke EA, Abe A, Bloch DR, editors. Polymer Handbook. New York: Wiley; 1999.

[131] Billmeyer Jr FW. Textbook of Polymer Science. New York: Wiley-Interscience; 1962.

[132] Berek D. Size exclusion chromatography—A blessing and a curse of science and technology of synthetic polymers. J. Sep. Sci. 2010;33:315−35.

[133] Arakawa T, Ejima D, Li T, Philo JS. The critical role of mobile phase composition in size exclusion chromatography of protein pharmaceuticals. J. Pharm. Sci. 2010;99:1674−92.

[134] Potschka M. Size-exclusion chromatography of polyelectrolytes: Experimental evidence for a general mechanism. J. Chromatogr. 1988;441:239−60.

[135] Mizutani T, Mizutani A. Prevention of adsorption of protein on controlled-pore glass with amino acid buffer. J. Chromatogr. 1975;111:214−6.

[136] Polymer and Hydrocarbon Processing Solutions with HPLC. Agilent Solutions Guide, Publ. No. 5968−7020E; 1999.

[137] Rao B, Balke ST, Mourey TH, Schunk TC. Methyl-cyclohexane as a new eluting solvent for the size-exclusion chromatography of polyethylene and polypropylene at 90°C. J. Chromatogr. A 1996;755: 27−35.

8

Solutes in HPLC

8.1. NATURE OF THE SOLUTE

General Aspects

The chemical nature of solutes in a sample is a critical factor in selecting and performing an HPLC analysis. The classification of analytes based on their chemical nature can involve several criteria. One such classification can be based on the functional groups of the analytes (e.g., similar to that of an organic textbook). Another classification can be based on the role of the analyte in everyday life. This section presents a short description of these two classifications.

Essential in Modern HPLC Separations
http://dx.doi.org/10.1016/B978-0-12-385013-3.00008-2

Solutes Classification Based on Their Chemical Structure

Depending on the molecular weight, the solutes/analytes are initially divided into small molecules and macromolecules. Small molecules are further classified based on their chemical structure. First, the organic and inorganic compounds are separately classified. For organic compounds, the simplest class is that of saturated hydrocarbons (linear or branched). Following this class are saturated cyclic hydrocarbons, unsaturated hydrocarbons with one or more double bonds, unsaturated hydrocarbons with triple bonds, aromatic hydrocarbons, and so on. Combinations of all these structures are possible.

Various functional groups can be attached on the hydrocarbon backbone. They can be classified based on the nature of atoms in the functional group or other criteria (such as monofunctional and bifunctional). This procedure will differentiate halogenated compounds, alcohols, enols, phenols, ethers, peroxy compounds, thiols, sulfides, amines, imines, a variety of other nitrogenous compounds (nitro, oximes, etc.), aldehydes, ketones, carboxylic acids, various derivatives of organic acids (such as esters, lactones, and acyl chlorides), derivatives of carbonic acid (such as ureas and cyanates), sulfonic acids, and other less common types. More than one type of functional group can be attached on the hydrocarbon backbone. Specific classes of compounds are generated from this large variety of combinations, such as carbohydrates, amino acids, and lipids.

A special group of compounds is that of heterocycles (both aromatic and nonaromatic). Heterocycles can be classified based on the heteroatoms in the cycle such as oxygen (furans, pyrans), nitrogen (pyrole, pyrazole, imidazole, triazole, pyridine, pyrazines, etc.), sulfur (thyophene), or different heteroatoms (oxazoles, thiazole, oxadiazoles, etc.).

In addition to all the classes of compounds containing C, H, O, N, or S, other types of organic compounds must be included, such as those containing boron, silicon, arsenic, stibium, and metallic elements (organometallic compounds).

For macromolecules, the chemical structure also can be an important criterion of classification. Specific groups such as polymeric carbohydrates, lignins, tannins, Maillard browning polymers, proteins, and nucleic acids, as well as various types of synthetic polymers, can be differentiated based on their structure.

The information regarding the chemical nature of the analytes (solutes) is very important since the selection of the type of separation, the best type of stationary phase to be used, the nature and pH of the mobile phase, and so on can be determined by it.

Classification Based on the Role of the Solute in Everyday Life

Depending on the role/function in everyday life, analytes are frequently classified in various categories. No specific criteria are set for such classification, and only some of the most important categories are indicated here. It should also be mentioned that the classification based on the role in everyday life is frequently combined with that based on the analyte structure.

One common classification of samples is that of biological and nonbiological. The two groups are frequently subdivided into multiple subgroups. Another classification is based on criteria related to environmental issues, and groups such as environmental pollutants, pesticides, herbicides, and fungicides are considered. Several large groups of compounds are classified on the basis of their role in health issues. These are the groups of pharmaceuticals, metabolites, and biodegradation products [1]. Numerous classes of pharmaceuticals are further differentiated such as antibiotics, anticancer drugs, antiepileptics, steroids, analgesics, and vitamins. Metabolites and biodegradation products are also further classified based on their source, mechanism of production, and so on. A classification based on the compound's role in toxicology or in forensic science is also utilized, with groups such as carcinogens, illicit drugs, poisons, and other toxins.

Analysis of food and agricultural products plays an important role in everyday life, and as a result, classifications in groups related to these fields are also common. Analyte groups such as nutrients, flavors, toxins in food, and main food components are frequently differentiated. Numerous other groups of compounds not listed previously are common. Among these can be mentioned flavors, polymer additives, solvents, surfactants, and dyes.

Although such classifications as those based on the role of the analyte in everyday life may appear to be unrelated to the HPLC separation technique, this classification is in fact extremely important. Most real-life samples are complex mixtures that must be separated and their components detected. Depending on the group to which they belong, the analytes may be associated in specific ways. The matrix that must be separated from the analytes depends very frequently on the group of the sample classified based on its everyday life criteria. For example, biological samples such as blood/plasma will contain proteins, while food products may contain large amounts of lipids or of carbohydrates. The selection of a separation frequently depends on the matrix that is described by the information regarding the class of sample. Also, the specificity of the separation and sensitivity of the analysis is frequently determined by the type of sample as classified based on its role in everyday life.

8.2. PARAMETERS FOR SOLUTE CHARACTERIZATION IN THE SEPARATION PROCESS

General Aspects

The chemical nature of the solute determines a number of physicochemical parameters that are important for the HPLC separation. These parameters must be known in order to help in the choice of type of chromatography, column, mobile phase, and separation conditions. It should be noticed that solute parameters are predetermined (depending on the sample), while those of the column and mobile phase can be selected to fit the analysis needs. Solute properties are important not only for the choice of column and mobile phase, but also for decisions regarding the whole process of analysis: sampling, sample preparation, selection of the type of detection, and the quantitation procedure (see Section 1.3). It should be underlined that the solutes in a sample include all the species that must be separated, and not only the analytes. For this reason, besides the properties of the analytes, the properties of the matrix components must be considered in an HPLC analysis.

Molecular Weight

The molecular mass of one molecule is its mass expressed in unified atomic mass units (1/12 the mass of one atom of the isotope carbon-12, sometimes named dalton or Da). The relative molecular mass is frequently (and not correctly) indicated as molecular weight (MW) of a molecule. MW is the ratio of the mass of the molecule to 1/12 of the mass of isotope carbon-12 (MW is dimensionless). MW and molecular mass are numerically equal, but they are not the same parameter, although the terms are frequently used interchangeably.

The molecular weight (relative molecular mass) represents an important parameter in molecular characterization. It is related to many other molecular properties, as shown in Chapter 4. Also, a common purpose of MW in HPLC is its use in the differentiation between small molecules and macromolecules. In fact, as a common definition, macromolecules are chemical compounds formed from at least 1000 atoms linked by covalent bonds. However, instead of number of atoms, a MW higher than about 5000 is commonly used to indicate a macromolecule, and a MW lower than about 2000 to indicate a small molecule. Between these two limits is a gray area where, among others, the molecules known as oligomers are placed.

Acidic or Basic Character of Solutes

A considerable number of molecules contain some specific functional groups likely to lose or gain protons (e.g. in aqueous solutions). As shown in Section 3.5, basic compounds are protonated as positive ions, while acidic compounds are protonated as neutral molecules, the process depending on the pH of the solvation medium. The nonionizable, acidic, basic, or amphoteric character of molecules is important for their separation. Tables with acidity constants are readily available in the literature for common compounds. For more complex molecules, various techniques are available for pK_a calculation. A common procedure for this purpose is based on the calculation of partial charges of atoms in the molecule [2,3]. Computer programs are also available for the calculation of pK_a (e.g., MarvinSketch [4]). For macromolecules with multiple ionizable groups, the individual pK_a values for each functionality are not anymore a relevant parameter.

Isoelectric Point

The isoelectric point (pI) is the pH value at which the molecule carries no electrical charge (in a solution). The concept is particularly important for zwitterionic molecules such as amino acids, peptides, and proteins. For an amino acid, the isoelectric point is the average of pK_a values for the amine and the carboxilic group. In the case of amino acids with multiple ionizable groups (e.g., lysine with two amino groups or aspartic acid with two acid groups), the isoelectric point is given by the average of the two pK_a of the acid and base that lose/gain

a proton from the neutral form of the amino acid. This can be extended to the definition of pI of peptides and proteins. The pI value can be used to indicate the global basic or acidic character of a zwitterionic molecule; compounds with pI > 7 can be considered basic, and those with pI < 7 can be considered acidic. For complex molecules such as proteins, isoelectric point is useful in the description of acidic or basic character, where individual pK_a values are not relevant

Molecular Polarity and Octanol/Water Partition Constant

The asymmetrical charge distribution in a molecule (separation of the center of positive charges from that of negative charges), which causes the molecule to act as an electric dipole, is defined as polarity (see Sections 1.1 and 4.1). However, the term *polarity* is frequently used with a wider meaning, including polarity caused by the presence of a dipole moment m_i in the molecule, as well as the molecule polarizability α_i. These two parameters are involved in the calculation of various interactions between solutes, solutes and stationary phase, and solutes and mobile phase molecules. Besides a quantitative characterization of molecules based on m_i and α_i values, it is common to qualitatively assess a polar character by the presence in the molecules of polar functional groups such as -OH, -COOH, -NH$_2$, >NH, and -SO$_3$H since these groups bring both dipole moments and polarizability to molecules.

Molecules can also be characterized by the opposite of polar character, which is the hydrophobic character. Octanol/water partition constant K_{ow} is commonly used for characterizing the hydrophobicity of a compound (see rel. 1.1.1). For this reason, K_{ow} values for solutes are important parameters in HPLC, in particular for RP-HPLC, but also for other chromatographic types, such as HILIC. Octanol/water parameters have, besides chromatography, a widespread utilization in other important fields of science, such as drug design and environmental studies. For this reason, values for K_{ow} for a large body of compounds are experimentally available [5,6], can be calculated using computer programs, for example, MarvinSketch 5.4.0.1, ChemAxon Ltd. [4], EPI Suite [7], and can be calculated using an additive fragment methodology [8]. Various calculation procedures for log K_{ow} are described in the literature [9–13]. The polar and/or hydrophobic character of a molecule can be related to its K_{ow} value. Positive values for log K_{ow} indicate some hydrophobic character, and larger values show more hydrophobicity. Molecules with low or negative values for log K_{ow} are frequently indicated as polar, although there is not a direct relation between K_{ow} and the charge distribution in the molecule.

Experimental K_{ow} (or log K_{ow}) as reported in the literature [6] may be listed for the same compound as having several values (that are close or relatively close). Also, different estimation methods may not always provide identical values or values identical to the experimental ones for K_{ow}. This should not create any significant confusion regarding K_{ow} since large variations in its value for a unique compound are uncommon. The dependence between calculated K_{ow} values using two different computer programs (MarvinSketch [4] and EPI Suite [7]) and K_{ow} experimental values is shown in Figure 8.2.1. For allowing a uniform comparison between compounds, the K_{ow} values used in this book were the calculated ones based on the MarvinSketch 5.4.0.1 program.

Another aspect regarding octanol/water partition is related to the characterization of ionic compounds using a distribution coefficient D_{ow} instead of K_{ow}, as defined in Section 3.5. The distribution coefficient D_{ow} for ionizable species depends on pH, and only for nonionizable compounds $D_{ow} = K_{ow}$. Variation of D_{ow} with the pH can also be obtained using computing programs (e.g., MarvinSketch 5.4.0.1). For the partition processes, the value of D_{ow} better

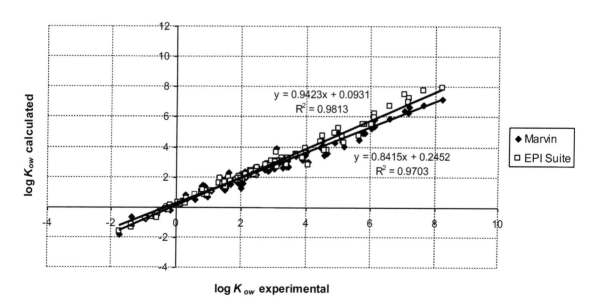

FIGURE 8.2.1 The dependence between the calculated K_{ow} (MarvinSketch [4] and EPI Suite [7]) and experimental K_{ow} values.

describes the affinity of a compound to one of the phases involved in partition.

The values of $\log K_{ow}$ are very useful for characterization of polarity in small molecules. However, the same technique cannot be applied for characterization of the polarity of polymers, where the $\log K_{ow}$ values lose their meaning. Proteins, for example, have the capability of folding, and the polar groups have the tendency to congregate in such a manner as to maximize electrostatic interactions. In a polar solvent like water or aqueous solutions of acids, the protein may change its tertiary and quaternary structure and expose polar side chains toward the solvent, the hydrophobic moieties being congregated toward a more hydrophobic core. The opposite effect may take place in the presence of organic solvents. This behavior would lead to a variable octanol/water partition.

Solubility in Water from Octanol/Water Partition Constant

Water solubility is an important parameter since mobile phases usually contain water. The evaluation of solubility of different organic compounds in water based on thermodynamic considerations is typically much less precise compared to the estimation based on the values of parameter K_{ow}. Values for the octanol/water partition constant K_{ow} are readily available. The estimation of solubility of organics in water from K_{ow} values is typically done using regression equations of the form:

$$\log C_i = a \log K_{ow} + b \qquad (8.2.1)$$

or of the form:

$$\log (1/C_i) = a' \log K_{ow} - b' \qquad (8.2.2)$$

where C_i is the solubility in water and a, a', b, b' are empirical parameters. Some values for these parameters are given in Table 8.2.1 [14].

Van der Waals Molecular Volume and Surface

Van der Waals molecular volume $\mathcal{V}$ and van der Waals surface $\mathcal{A}$ are important parameters for understanding molecular interactions

TABLE 8.2.1 Regression Equation for the Estimation of Water Solubility Based on $\log K_{ow}$ [14]

Equation	Units for C_i	Used for chemical class
$\log C_i = -1.37 \log K_{ow} + 7.26$	$\mu M/L$	Aromatics, chlorinated hydrocarbons
$\log (1/C_i) = 1.113 \log K_{ow} - 0.926$	mol/L	Alcohols
$\log (1/C_i) = 1.229 \log K_{ow} - 0.720$	mol/L	Ketones
$\log (1/C_i) = 1.013 \log K_{ow} - 0.520$	mol/L	Esters
$\log (1/C_i) = 1.182 \log K_{ow} - 0.935$	mol/L	Ethers
$\log (1/C_i) = 1.221 \log K_{ow} - 0.832$	mol/L	Alkyl halides
$\log (1/C_i) = 1.294 \log K_{ow} - 1.043$	mol/L	Alkynes
$\log (1/C_i) = 1.294 \log K_{ow} - 0.248$	mol/L	Alkenes
$\log (1/C_i) = 0.966 \log K_{ow} - 0.339$	mol/L	Aromatics
$\log (1/C_i) = 1.214 \log K_{ow} - 0.850$	mol/L	Various
$\log (1/C_i) = 1.237 \log K_{ow} + 0.248$	mol/L	Alkanes

TABLE 8.2.2 Van der Waals Radii r for Several Elements in Å [15]

H 1.20						
Li 1.82	Be 1.53	B 1.92	C 1.70	N 1.55	O 1.52	F 1.47
Na 2.27	Mg 1.73	Al 1.84	Si 2.10	P 1.80	S 1.80	Cl 1.75
K 2.75	Ca 2.31	Ga 1.87	Ge 2.11	As 1.85	Se 1.90	Br 1.85
Rb 3.03	Sr 2.49	In 1.93	Sn 2.17	Sb 2.06	Te 2.06	I 1.98

during a separation. In RP-HPLC in particular, van der Waals molecular area A plays an important role in the expression of the free energy for the partition equilibrium and directly to the value of capacity factor for an analyte, as shown in Section 5.1 (see rel. 5.1.26). The calculation of both van der Waals molecular volume and area starts with the concept of van der Waals radius of an atom. This is the radius of an imaginary sphere used to model the atoms describing its finite size. The radius is obtained based on results from gas kinetic collision cross sections, gas critical volumes, crystal densities (extrapolated at 0 K), liquid state properties, X-ray diffraction data, and the like [15,16]. Several "mean" van der Waals radii (in Å) are indicated in Table 8.2.2.

For most organic molecules, the atoms are placed at covalent-bond distance, which is shorter than the sum of the van der Waals radii of the connected atoms. For this reason, the van der Waals molecular volume is smaller than the sum of the volume of each component atom. This is also true for the molecular surface. For a diatomic molecule with two atoms with r_1 and r_2 van der Waals radii, and the covalent-bond distance l, the calculation of the molecular volume is obtained starting with the two interlocked spheres as shown in Figure 8.2.2.

The volume of one sphere V_2 can be calculated as $V_2 = (4/3)\pi (r_2)^3$. The volume of a spherical segment $\Delta V_1 = \pi h_1^2(r_1 - h_1/3)$ must be added to V_2, and the volume of another spherical segment with $\Delta V_{2-1} = \pi h_2^2(r_2 - h_2/3)$ must be subtracted from V_2 to obtain the total volume. The values for h_1 and h_2 are obtained

as $h_1 = r_1 + l - m$, $h_2 = r_2 - m$ and $m = (r_2^2 - r_1^2 + l^2) / (2l)$. The total volume will be $V = V_2 + \Delta V_1 - \Delta V_{2-1}$. *Note:* The volume of a molecule $v_j = V_j/N$ (where V is the molar volume $V = MW/\rho$) is not the same as van der Waals volume V since the volume of voids and the changes in the volume due to interactions are included in the value for v.

The area of the molecule can be obtained as the sum of two spherical segments with $A_1 = 2\pi r_1 h_1$, and $A_2 = 2\pi r_2 (2r_2 - h_2)$. Since direct calculations of van der Waals molecular volumes and areas are difficult for larger molecules, various approximations were developed for this purpose [17]. Computer programs are available for calculating van der Waals volume and areas (e.g., MarvinSketch [4]). (Molar volume V can also be obtained using computer programs [18])

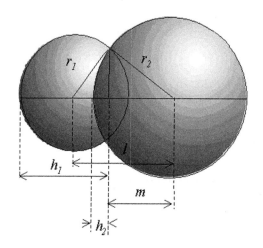

FIGURE 8.2.2 Description of the calculation of the volume of a diatomic molecule.

Correlation between van der Waals Molecular Surface and Octanol/Water Partition Constant

An expression for the octanol/water partition constant K_{ow} can be obtained based on the solvophobic theory developed for RP-HPLC in Section 5.1. For liquid-liquid partition the compound j is distributed between two nonmiscible liquid phases A and B in an equilibrium of the type:

$$j_B \rightleftharpoons j_A \qquad (8.2.5)$$

When equilibrium is attained for "distributing" the compound j between phases A and B, the difference between the chemical potentials $\mu_{i,A}$ and $\mu_{i,B}$ of the component j in each of the two phases A and B must be zero. As shown in Section 3.1, this condition (expressed by rel. 3.1.2) is equivalent to the following expression for the equilibrium constant (see rel. 3.1.12):

$$K_j = \exp[-\Delta G_j^0/(RT)] \qquad (8.2.6)$$

Assuming that no volume changes occur during the process, the free enthalpy ΔG^0 is taken as equal to the free energy of the process ΔA^0, and rel. 8.2.6 can be written in the form:

$$K_j = \exp\left[-\Delta A_j^0/(RT)\right]$$
$$= \exp\left[-(A_{j,A} - A_{j,B})/(RT)\right] \qquad (8.2.7)$$

In rel. 8.2.7, the value for the free energy $A_{j,S}$ (where $S = A$ or B) is given by the change in standard free energy $A^{sol}_{j,S}$ necessary for placing a molecular species j into a solution formed by molecules S (symbol Δ and the index "0" for standard expressions are omitted). This energy is given by rel. 4.1.58. In Section 4.1 it was shown that the expression for $A^{sol}_{j,S}$ is the following:

$$A^{sol}_{j,S} = A^{cav}_{j,S} + A^{es}_{j,S} + A^{disp}_{j,S}$$
$$+ RT \ln(RT/p_0 V_S) \qquad (8.2.8)$$

where the expressions for the terms in rel. 8.2.8 are given by rel. 4.1.74, 4.1.75, and 4.1.76, respectively. With rel. 8.2.8 used in the expression for $\ln K$ given by rel. 8.2.7, the following formula is obtained:

$$RT \ln K_j = (A^{cav}_{j,B} - A^{cav}_{j,A}) + (A^{es}_{j,B} - A^{es}_{j,A})$$
$$+ (A^{disp}_{j,B} - A^{disp}_{j,A})$$
$$+ RT \ln (V_A/V_B)$$
$$\qquad (8.2.9)$$

Each set of terms in rel. 8.2.9 can be further estimated based on the solvophobic theory developed in Section 4.1. For the difference in the free energy for the cavity formation, making the assumption that $W_S \approx 1$, the following formula is obtained:

$$A^{cav}_{j,B} - A^{cav}_{j,A} = \mathcal{N} (\gamma'_B - \gamma'_A)\mathcal{A}_j + \mathcal{N}\left[(\kappa^e_B\right.$$
$$- 1) \gamma'_B(V_B)^{2/3} - (\kappa^e_A$$
$$\left. - 1)\gamma'_A(V_A)^{2/3}\right]\mathcal{A}_j/(V_j)^{2/3}$$
$$\qquad (8.2.10)$$

The estimation in the difference in electrostatic forces based on rel. 4.1.64 gives the expression:

$$A^{es}_{j,B} - A^{es}_{j,A} = -\frac{\mathcal{N}m_j^2}{2v_j}(\mathcal{D}_B P_{j,B} - \mathcal{D}_A P_{j,A})$$
$$\qquad (8.2.11)$$

The estimation of the differences in the dispersion forces obtained from rel. 4.1.66 using the assumption that $Q''_{j,S} \approx 0.1 Q'_{j,S}$ gives:

$$A^{disp}_{j,B} - A^{disp}_{j,A} = -\frac{16.75 D_j}{8\pi}(Q'_{j,B}\Upsilon_{j,B}D_B$$
$$- Q'_{j,A}\Upsilon_{j,A}D_A)$$
$$\qquad (8.2.12)$$

With the terms given by rel. 8.2.10, 8.2.11, and 8.2.12, it can be seen that the expression of the equilibrium constant should have the following form:

$$\log K_j = a \mathcal{A}_j + b(V_j)^{-2/3}\mathcal{A}_j + c \qquad (8.2.13)$$

where the values for a and b depend only on the solvents A and B, while the value for c depends on the solvents and the solute as well. For the two solvents A = octanol and B = water, it can be concluded that a and b are the same for all analytes j. Since in this case $\gamma'_B - \gamma'_A > 0$, and also $(\kappa^e_B - 1)\,\gamma'_B\,(V_B)^{2/3} - (\kappa^e_A - 1)\,\gamma'_A\,(V_A)^{2/3} > 0$, from rel. 8.2.10 it can be seen that a larger $\mathcal{A}_j$ will generate a larger K_{ow}.

The term c in rel. 8.2.13 includes the contribution of electrostatic and the dispersion forces to the free energy. It is therefore expected that more polar analytes will have a negative c, larger in absolute value. Predictions of this theory were very clearly verified when plotting $\log K_{ow}$ values as a function of $\mathcal{A}_j$ for a number of compounds. This type of plot is exemplified in Figure 8.2.3 for several classes of mono-functional compounds. The results from Figure 8.2.3 show that rel. 8.2.13 can be well approximated by a relation of the form:

$$\log K_{j,ow} = a'\mathcal{A}_j + c \qquad (8.2.14)$$

where a' is a constant (for $\mathcal{A}_j$ in Å^2, $a' \approx 1.46$ 10^{-2} with K_{ow} calculated based on MarinSketch; other values are $a' \approx 1.57\ 10^{-2}$ with K_{ow} calculated based on EPI Suite, and $a' \approx 1.63\ 10^{-2}$ with experimental K_{ow}). The value for c depends only on the nature of the substituent (polar or nonpolar) on the hydrophobic moiety of the compound j. Further investigation of rel. 8.2.14 indicates that it can be extended with excellent agreement to the following formula:

$$\log K_{i,ow} = a'\mathcal{A}_j + \sum_n c_n \qquad (8.2.15)$$

where c_n are constants for each substituent. Values for c_n for different organic groups attached to an aliphatic hydrocarbon chain are given in Table 8.2.3 [19].

The calculation of $\log K_{ow}$ based on rel. 8.2.15 with the values for c indicated in Table 8.2.3, for a set of 147 compounds, including mono, bi, tri,

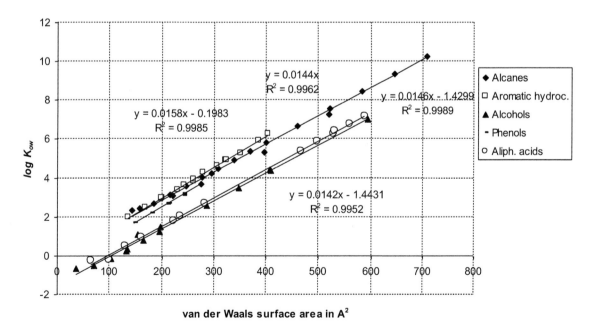

FIGURE 8.2.3 Dependence of $\log K_{ow}$ on van der Waals molecular surface $\mathcal{A}_j$ for different classes of mono-functional compounds.

TABLE 8.2.3   Values for Constant c in the Calculation of log K_{ow} (MarvinSketch Values) from van der Waals Surface Area of the Molecule* ($a' \approx 1.46\ 10^{-2}$).

Group	c_n	Group	c_n
Aromatic ring**	0.055	Aliphatic secondary amine	−1.897
Alcohol	−1.444	Aliphatic tertiary amine	−2.200
Phenol	−0.470	Aromatic primary amine	−0.998
Aliphatic ether	−1.581	Aromatic secondary amine	−1.320
Aromatic/aliphatic ether	−0.825	Ketone	−1.512
Aliphatic acid	−1.375	Nitro aromatic	−0.612
Aromatic acid	−0.852	Chloro aromatic	0.357
Aliphatic primary amine	−1.650	Bromo aromatic	0.481
Aromatic ester	−1.100	Nitrile	−1.072

* Note: each group is counted as many times as present.
** Note: Separated rings are counted individually, condensed rings are counted as one.

tetra, and pentafunctional ones, with identical or different functional groups, produced an excellent agreement with log K_{ow} target values obtained using the MarvinSketch program (for which parameter c was optimized). This is shown in Figure 8.2.4 where the excellent correlation with $R^2 = 0.9961$ is displayed.

Optimization of c_n paramaters based on log K_{ow} values calculated with EPI Suite or with experimental values leads to values close (but

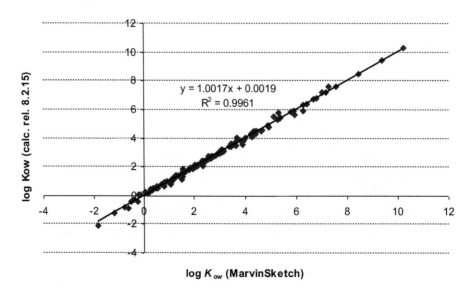

FIGURE 8.2.4   Correlation between log K_{ow} calculated and target values obtained using MarvinSketch. Calculation based on rel. 8.2.15 with $a' = 1.46\ 10^{-2}$ and the values for c_n indicated in Table 8.2.3.

not identical) to those from Table 8.2.3, and also to very good correlations between calculated values and the target values.

Based on the theory developed for the evaluation of log K_{ow} the expression for c_n should be given by:

$$c_n = [(A^{es}_{j,B} - A^{es}_{j,A}) + (A^{disp}_{j,B}$$
$$- A^{disp}_{j,A})]/(2.303\ RT) + \log(V_A/V_B)$$

$$(8.2.16)$$

where A = octanol and B = water. Relation 8.2.16 indicates that the value for c can provide useful information regarding the electrostatic and dispersion forces between the molecule of the analyte and the two solvents.

Partial Charge Distribution

Analysis of the electronic population is a subject of considerable interest for understanding molecular properties (see, e.g., [20,21]). The partial charge distribution determines physicochemical properties such as dipole moment m, ionization constants, and reactivity. Various approaches and procedures are reported in the literature for calculating partial charges [21–23]. Computer programs are also available for partial charges calculation (e.g,. MarvinSketch [4,24,25]). The values for partial charges (point charges) are useful for understanding molecular polarity, although they are not used frequently in specific calculations related to separations. The potential application of this parameter could be in those LC separations based on electrostatic interactions, such as HILIC, NP, ANP, and the chiral mechanism.

Parameters for Solute Hydrophobicity, Steric Effects, Hydrogen Bonding, and Ion-Exchange Character

In the development of certain models for the characterization of stationary phases, it was necessary to include parameters for characterization of solutes (analytes), with which the stationary phase interacts. For example, in Section 6.4 it was shown that the capacity factor for a compound j is given by the expression (see rel. 6.4.15):

$$\log k(j) = \log k_{EB} + \eta'(j)H_{c1} - \sigma'(j)S^*_{c1}$$
$$+ \beta'(j)A_{c1} + \alpha'(j)B_{c1} + \kappa'(j)C_{c1}$$

$$(8.2.17)$$

where log k_{EB} is the capacity factor for ethylbenzene, parameters H_{c1}, S^*_{c1}, A_{c1}, B_{c1}, and C_{c1} describe the column characteristics. The corresponding parameters $\eta'(j)$, $\sigma'(j)$, $\beta'(j)$, $\alpha'(j)$, and $\kappa'(j)$ describe solute properties: $\eta'(j)$-hydrophobicity, $\sigma'(j)$-steric interactions, $\beta'(j)$-hydrogen-bonding acceptance, $\alpha'(j)$-hydrogen-bonding donation, and $\kappa'(j)$-cation exchange or ion-ion interactions. The values of these parameters would be very useful for solute characterization in HPLC, and some were reported in the literature for selected analytes [26–29]. However, these parameters were obtained for a specific mobile phase and were established with the specific purpose of column characterization. They were obtained from best-fit regression lines and are not readily available for solutes that were not in the "test" group.

Molar Volume, Hildebrand Solubility Parameter

In several expressions for the calculation or estimation of capacity factor k and therefore of α, the molar volume V and the solubility parameter δ for the solvent, the stationary phase, and the analyte were involved. The discussion regarding these parameters as given for the solvent molecule is applicable for the solute molecule. The calculation of the molar volume V and of δ was discussed in Section 7.1.

Capacity factor k can be calculated in principle by knowing Ψ for the stationary phase and K_i for the solute i ($k_i = K_i \Psi$). Relation 7.1.17 shows, for example, how the equilibrium constant K_i depends on δ_i and V_i, and on solvent and stationary-phase properties. However, the direct calculation of a specific retention factor k based on that formula or of a selectivity α based on rel. 7.1.18 does not generate accurate values and can be useful only as a directional information.

Evaluation of Solubility of Nonpolar Compounds from Hildebrand Solubility Parameter

The dissolution process of nonelectrolytes can be viewed hypothetically as being formed from two steps. The first step is melting of the compound i to form a supercooled liquid (a liquid cooled below its freezing point without solidifying). The second step is the mixing of this liquid with the solvent S. For a transformation at constant pressure and temperature at equilibrium, the free enthalpy $\Delta G = 0$, and this holds true for melting at fusion temperature T_f. Therefore, the free enthalpy of melting (fusion) is $\Delta G_f = 0$ and $\Delta H_f = T_f \Delta S_f$, where ΔH_f is the heat of fusion and ΔS_f is the entropy of fusion. At a given temperature T, the free energy of melting will be given by the expression:

$$\Delta G_f = \Delta H_f - T\Delta S_f = \Delta H_f - (T/T_f)\Delta H_f \tag{8.2.18}$$

The free energy of mixing for ideal solutions does not take place with any heat change, and therefore:

$$\Delta G^{mix} = -T \, \Delta S^{mix} \tag{8.2.19}$$

The ideal behavior of a solution assumes that each component has its molecules behaving as if it is surrounded by molecules of the same kind (to follow Raoult law). The variation in

the entropy of mixing ΔS^{mix} of ideal solutions is given by the expression [30]:

$$\Delta S^{mix} = -R \sum_j n_j \ln x_j$$

$$\left(\text{with } x_i = n_i \bigg/ \sum_{j=1}^{r} n_j \right) \tag{8.2.20}$$

For the dissolution process of i in S, it can be assumed that $x_S = 1$ for the solvent, and therefore $\Delta G^{mix} = - RT \, n_i \ln x_i$, (the index S from $x_{i,S}$ and $n_{i,S}$ is omitted), or for $n_i = 1$ the expression for ΔG^{mix} becomes:

$$\Delta G^{mix} = -RT \ln x_i \tag{8.2.21}$$

The total free enthalpy of dissolution $\Delta G = - \Delta G_f + \Delta G^{mix}$, and at the equilibrium $\Delta G = 0$. Including expressions 8.2.18 and 8.2.19 in the formula for ΔG leads to the expression:

$$(1 - T/T_f)\Delta H_f + RT \ln x_i = 0, \tag{8.2.22}$$

which can be rearranged in the form:

$$\ln x_i = [\Delta H_f(1/T_f - 1/T)]/R \tag{8.2.23}$$

Relation 8.2.23 gives the formula for solubility (expressed as the maximum mole fraction of a solute in the solvent) for the formation of an ideal solution. As shown by rel. 8.2.23, the solubility increases with the temperature T. This is true for many compounds, including both nonelectrolytes and electrolytes, although there are exceptions.

In rel. 8.2.23 the mole fraction x_i can be changed into molar concentration c_i using the approximation:

$$x_i = \frac{n_i}{n_i + n_S} \approx \frac{n_i}{n_S} = \gamma_i c_i \frac{M_S}{1000\rho_S} \tag{8.2.24}$$

where γ_i is the activity coefficient included to correct for the deviation from ideal solutions, M_S is the molecular mass (weight) MW of the solvent, and ρ_S is its density. In this case, rel. 8.2.23 can be written in the form [31]:

$$\ln \frac{\gamma_i c_i M_S}{1000 \rho_S} = \frac{\Delta H_f}{RT} \frac{T - T_f}{T_f} \qquad (8.2.25)$$

From rel. 8.2.25, the solubility for nonelectrolytes can be estimated based on solubility parameters δ, using for γ_i its expression given by rel. 7.1.12. Rearranging rel. 8.2.25 and using rel 7.1.12 for $\ln \gamma_i$, the expression for $\ln c_i$ is given by the formula:

$$\ln c_i = \frac{\Delta H_f}{RT} \frac{T - T_f}{T_f} - \frac{V_i}{RT}(\delta_i - \delta_S)^2 + \ln \frac{1000 \rho_S}{M_S}$$

$$(8.2.26)$$

Considering $\Delta H_f \approx T_f \Delta S_f$, and taking $\Delta S_f \approx$ 13 cal mole^{-1} deg^{-1} (the fusion entropy being relatively constant for many compounds), rel. 8.2.26 can be estimated by the formula:

$$\ln c_i = 6.54 \frac{T - T_f}{T_f} - \frac{V_i}{RT}(\delta_i - \delta_S)^2 + \ln \frac{1000 \rho_S}{M_S}$$

$$(8.2.27)$$

Relation 8.2.27 is useful for evaluation of a solid nonelectrolyte solubility into a solvent.

Solvatochromic Parameters for the Solute

Solvatochromic parameters for the solute can be developed in the same manner as for a solvent (see Section 7.1). Parameters α, β, and π^* for a number of compounds were listed in Table 7.4.1, and more such parameters are available in the literature [32,33]. However, except for a limited number of analytes, estimation of these parameters is difficult or impossible. This reduces the utility of formulas using such parameters to only directional information and comparison for analogous cases where the solvent is changed while the analytes and the stationary phase remain the same. Accurate

calculations, for example, of the capacity factor for a specific analyte, are not typically successful.

Other Solute Properties Affecting Separation

Depending on the type of HPLC analysis, specific physical properties of the solute play various roles in the separation process. A short enumeration of some of these properties used for calculating the capacity factor is as follows for the main HPLC types.

1) For RP-HPLC, the solvophobic theory shows that the capacity factor for an analyte is expressed by rel. 5.1.26. In that expression, all terms depend on analyte characteristics except for log Ψ and the term that accounts for the change in the free volume of the system during the retention process. Rel 5.1.26 includes the term representing the interactions in an ideal gas system consisting of solute molecule j and the ligand L of the stationary phase $E_T(j, L)$, the terms describing the dispersion and electrostatic forces between the molecule j and the molecules of the surrounding solvent S, a term involving polar/polarizability interactions, and two terms related to the energy necessary for cavity formation by the analyte in the mobile phase. The interaction between the solute j and ligand L in the ideal gas phase is related to the energy of intermolecular interactions (as given by rel. 4.1.46) where the dipole moment m_j, polarizability $\alpha_{0,j}$, and ionization potential I_j are implicated. Indirectly, the molecular dimensions are also a factor in the expression of $E_T(j, L)$. The same parameters of the solute m_j, $\alpha_{0,j}$, and I_j, plus molar volume V_j and molecular diameter d_j are important for the term describing the dispersion and electrostatic forces with the surrounding solvent. The terms related to the energy necessary for cavity formation in

the mobile phase are related in particular to the van der Waals solute area $\mathcal{A}_j$.

The detailed contribution of each of these parameters is rather difficult to assess, since most interactions take place competitively between the solute and the stationary phase on one hand, and between the solute and the mobile phase on the other. Only general comments can be made indicating that higher dipole moment and polarizability are conducing to stronger polar interactions, and that a larger van der Waals area $\mathcal{A}_j$ of the solute molecule is an indication of stronger hydrophobic interactions.

2) In ion-pair chromatography (IP) the theory of separation is basically expressed by rel. 5.2.30 (see Section 5.2). According to this formula, the capacity factor (and therefore the separation) in IP is determined by a capacity factor $k_j(0)$ in the absence of IPA (hetaeron) and by K_{IPA}, and by the charge and concentration of IPA. The values for $k_j(0)$ depend on the same parameters as in RP-HPLC.

3) In polarity-based separations, basically the same factors as in RP-HPLC determine the capacity factor of analyte. As discussed in Section 5.3, the weight of each term in the expression of the change in free energy of the separation process, and therefore in the expression of the capacity factor, is different from that in RP-HPLC, with the polar interactions playing a more important role. For this reason, the more polar compounds are more strongly retained than the less polar ones. Similar to the case of RP-HPLC, the direct calculation of capacity factors is not a practical way for assessing solute (analyte) behavior in the separation process.

4) In ion-exchange chromatography (IC), the separation process is described by equation 5.4.7, which indicates that the exchange equilibrium depends on an osmotic term, the charges of the ions, and the activity coefficients of each species. Again, since evaluation of the terms involved in equation 5.4.7 is

difficult (e.g., the activity coefficient of an ion in the resin), practical observations are used for predicting the separation, such as the knowledge that cations typically separate in the order single charge (Na^+, K^+, NH_4^+, etc.) followed by double charge (Mg^{2+}, Ca^{2+}, etc.), while small anions typically follow the sequence F^-, Cl^-, Br^-, NO_3^-, SO_4^{2-}, $Cr_2O_7^{2-}$. Organic anions such as small organic acids can also be separated using IC, specific methods for separation being reported in the literature (see, e.g., [34]).

5) In size exclusion, the physical dimensions of the molecule are important for characterization of the separation (see rel. 5.6.1 and 5.6.2). In this separation technique, the hydrodynamic volume of the solute (or the effective mean radius) is involved in calculating separation parameters.

8.3. OTHER PARAMETERS FOR SOLUTE CHARACTERIZATION

General Aspects

Various properties of the solute molecule are involved in the chromatographic process, some of them, in particular being those that are directly related to the separation process (discussed in Section 8.1). In addition to those, other properties are of importance. One such property is the diffusion coefficient D_j of the analyte. As shown in rel. 2.2.2, the longitudinal diffusion in an HPLC column is proportional with D_j. Different theories and also empirical relations were developed for estimation of diffusion coefficients. For example, Stokes theory empirically modified for better prediction gives the following formula for the diffusion coefficient for nonelectrolytes j in liquids B:

$$D_{j,B} = 7.4 \cdot 10^{-8} \frac{(\psi_B M_B)^{0.5} T}{\eta V_j^{0.6}} \qquad (8.3.1)$$

where V_j is the molar volume of solute j (in cm^3 $mole^{-1}$), M_B is the molecular weight of solvent B, T is temperature in Kelvin degrees, η the viscosity of the solution (in 10^{-4} g cm^{-1} s^{-1}, or centipoise), and ψ_B an "association" factor for the solvent (ψ_B is 1 for nonpolar solvents, 1.5 for ethanol, 1.9 for methanol, 2.6 for water).

Other formulas describing properties of the solutes involve parameters such as the refractive index (see rel. 4.1.68), diameter of the molecule, solvent-accessible area SASA, Kihara parameter [35], and rate of hydrolysis. These parameters can be either found in the literature or estimated [14].

Physical Properties of the Analyte Determining Detection

The choice of detection in HPLC is determined by a particular physicochemical property of the analyte that allows its measurement at very low levels. The selected property must have a value significantly different from that of the mobile phase. Examples of such properties include UV absorption, refractive index, fluorescence, molecular mass, and fragmentation in a mass spectrometer. The subject of properties determining the choice of detection in HPLC is vast and beyond the purpose of this book. Such properties are typically discussed in detail in each individual method of analysis.

References

[1] Kowalska S, Krupczyńska K, Buszewski B. Some remarks on characterization of application of stationary phases for RP-HPLC determination of biologically important compounds. Biomed. Chromatogr. 2006;20:4–22.

[2] Clarke FH, Cahoon NM. Ionization constants by curve fitting: Determination of partition and distribution coefficients of acids and bases and their ions. J. Pharm. Sci. 1987;76:611–20.

[3] Dixon SL, Jurs PC. Estimation of pK$_a$ for organic oxyacids using calculated atomic charges. J. Comput. Chem. 1993;14:1460–7.

[4] http://www.chemaxon.com

[5] Hansch C, Leo A. Exploring QSAR, Fundamentals and Applications in Chemistry and Biology. Washington, DC: ACS; 1995.

[6] Hansch C, Leo A, Hoekman D. Exploring QSAR, Hydrophobic, Electronic and Steric Constants. Washington, DC: ACS; 1995.

[7] http://www.epa.gov/oppt/exposure/pubs/episuite.htm

[8] Moldoveanu SC, David V. Sample Preparation in Chromatography. Amsterdam: Elsevier; 2002.

[9] Klopman G, Li J-Y, Wang S, Dimayuga M. Computer automated log P calculations based on an extended group of contribution approach. J. Chem. Inf. Comput. Sci. 1994;34:752–81.

[10] Viswanadhan VN, Ghose AK, Revankar GR, Robins RK. Atomic physicochemical parameters for three dimensional structure directed quantitative structure-activity relationships IV. Additional parameters for hydrophobic and dispersive interactions and their application for an automated superposition of certain naturally occurring nucleoside antibiotics. J. Chem. Inf. Comput. Sci. 1989;29:163–72.

[11] Csizmadia F, Tsantili-Kakoulidou A, Panderi I, Darvas F. Prediction of distribution coefficient from structure. 1. Estimation method. J. Pharm. Sci. 1997;86:865–71.

[12] Erős D, Kövesdi I, Őrfi L, Takács-Novák K, Acsády G, Kéri G. Reliability of log P predictions based on calculated molecular descriptors. A critical review,. Curr. Med. Chem. 2002;9:1819–29.

[13] Iwase K, Komatsu K, Hirono S, Nakagawa S, Moriguchi I. Estimation of hydrophobicity based on the solvent-accessible surface area of molecules. Chem. Pharm. Bull. 1985;33:2114–21.

[14] Lyman WJ, Reehl WF, Rosenblatt DH. Handbook of Chemical Property Estimation Methods. Washington, DC: ACS; 1990.

[15] Bondi A. Van der Waals volumes and radii. J. Phys. Chem. 1964;68:441–51.

[16] http://www.webelements.com/periodicity/van_der_waals_radius/

[17] Zhao YH, Abraham MH, Zissimos AM. Fast calculation of van der Waals volume as a sum of atomic and bond contributions and its application to drug compounds. J. Org. Chem. 2003;68:7368–73.

[18] http://www.aim.env.uea.ac.uk/aim/density/density.php

[19] Moldoveanu SC, David V. unpublished results

[20] Mulliken RS. Electronic population analysis on LCAO MO molecular wave function. I, J. Chem. Phys. 1955;23:1833–41.

[21] Moldoveanu SC, Savin A. Aplicatii in Chimie ale Metodelor Semiempirice de Orbitali Moleculari, Edit. Academiei RSR, Bucuresti 1980.

[22] Heinz H, Suter UW. Atomic charges for classical simulations of polar systems. J. Phys. Chem. B 2004;108:18341–52.

[23] Rappe AK, Goddard III WA. Charge equilibration for molecular dynamics simulations. J. Phys. Chem. 1991;95:3358–63.

[24] Stewart JJP. MOPAC-7, QCPE 113. Bloomington: Indiana University; 1994.

[25] Frisch MJ, Frisch A, Foresman JB. Gaussian 94. Pittsburgh: Gaussian Inc.; 1995.

[26] Gilroy JL, Dolan JW, Snyder LR. Column selectivity in reversed-phase liquid chromatography IV. Type-B alkyl-silica columns. J. Chromatogr. A 2003;1000:757–78.

[27] Marchand DH, Snyder LR, Dolan JW. Characterization and applications of reversed-phase column selectivity based on the hydrophobic-subtraction model. J. Chromatogr. A 2008;1191:2–20.

[28] Marchand DH, Croes K, Dolan JW, Snyder LR, Henry RA, Kallury KMR, Waite S, Carr PW. Column selectivity in reversed-phase liquid chromatography. VIII. Phenylalkyl and fluoro-substituted columns. J. Chromatogr. A 2005;1062:65–78.

[29] Bidlingmeyer B, Chan CC, Fastino P, Henry R, Koerner P, Maule AT, Marques MRC, Neue U, Ng L, Pappa H, Sander L, Santasania C, Snyder L, Woznyak T. HPLC column classification. Pharmacopeial Forum 2005;31:637–45.

[30] Hildebrand JH, Scott RI. The Solubility of Non-Electrolytes. York: Dover, New; 1964.

[31] Gmehling JG, Anderson TF, Prausnitz JM. Solid-liquid equilibria using UNIFAC. Ind. Eng. Chem. Fundam. 1978;17:269–73.

[32] Kamlet MJ, Abboud JLM, Abraham MH, Taft RW. Linear solvation energy relationships. 23. A comprehensive collection of the solvatochromic parameters, π^*, α and β, and some methods for simplifying the generalized solvatochromic equation. J. Org. Chem. 1983;48:2877–87.

[33] http://www.stenutz.eu/chem/solv26.php

[34] Roger WJ, Michaux S, Bastin M, Bucheli P. Changes to the content of sugars, sugar alcohols, myo-inositol, carboxylic acids and inorganic anions in developing grains from different varieties of Robusta (*Coffea canephora*) and Arabica (*C. arabica*) coffees. Plant Sci. 1999;149:115–23.

[35] Tee LS, Gotoh S, Stewart WE. Molecular parameters for normal fluids. Ind. Eng. Chem. Fundamen. 1966;5:363–7.

CHAPTER

9

HPLC Analysis

OUTLINE

Essentials in Modern HPLC Seperations
http://dx.doi.org/10.1016/B978-0-12-385013-3.00009-4

9.1. CHEMICAL NATURE OF THE ANALYTES AND THE CHOICE OF HPLC TYPE

General Aspects

The chemical nature of the sample and physicochemical parameters of the analytes and other solutes from the matrix are crucial factors in selecting not only the type of HPLC technique as discussed early in Section 1.3, but also the stationary phase and the mobile phase. In other words, the whole HPLC method is determined mainly by the nature of the sample, its analytes (within the instrumentation availability), as well as by the nature and composition of the matrix. A preliminary scheme related to HPLC choice is shown in Figure 9.1.1.

The first step in selecting the HPLC type is the evaluation of the information regarding the sample. Small molecules (e.g., with MW < 5000 Dalton) and large molecules (polymers with MW > 5000) are treated differently. Another important parameter regarding the sample is its solubility. Samples soluble in water or water/organic-polar solvents are treated differently from samples soluble only in organic-nonpolar solvents. Another sample characteristic is the ionic or nonionic character. This type of evaluation of sample properties is visualized in Figure 9.1.1. Using as guidance the description of the sample given in Figure 9.1.1, a preliminary selection of an HPLC type appropriate for the analysis is also suggested. The choice of the HPLC type is a very important step in a successful analysis. The selection is made considering various sample properties, analysis requirements, instrumentation availability, and the like. The scheme is very simplistic, since specific techniques can be used for other types of molecules than those suggested in Figure 9.1.1. For example, RP-HPLC can be used successfully for protein analysis, and GPC can be used, if necessary, for the separation of small molecules from large molecules. Several comments are made regarding the choice of HPLC type.

1) Organic nonpolar small molecules that are not soluble in water and possibly not soluble in organic polar solvents can be analyzed using NPC or NARP. Normal-phase chromatography (NPC) and nonaqueous reversed-phase chromatography (NARP) use solvent that can dissolve molecules such as

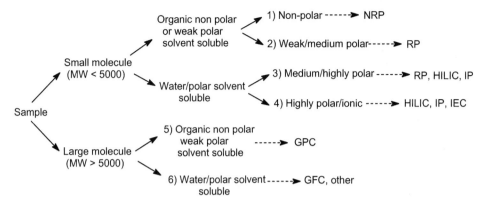

FIGURE 9.1.1 Diagram showing the sample properties useful for the preliminary selection of HPLC type of analysis.

carotenoids, triglycerides, and sterols. Molecular solubility as well as types of interactions with the stationary phase are proper criteria for the separation of this type of molecules.

2) Small molecules having some polarity and solubility in organic solvents (polar or nonpolar) represent a very large class of compounds. This class includes pharmaceuticals, biological small molecules, and compounds to be analyzed in food and beverages, in the environment, and in agricultural products. The main type of chromatography applied for analysis of these compounds is RP-HPLC. This technique is extremely versatile, and a large number of columns and combinations of mobile phases were developed for RP-HPLC applications. In case the separation of enantiomers is necessary for this group of analytes, chiral chromatography must be applied.

3) Small molecules soluble in water or polar solvents that have polar groups but are not ionic also represent a very large class of compounds. This class also includes many pharmaceuticals, biological small molecules, and compounds to be analyzed in food and beverages, in the environment, in agricultural products. RP-HPLC is frequently used for the separation of these types of molecules, when some part of their structure contains hydrophobic moieties. For molecules with numerous polar groups such as amino acids or carbohydrates, HILIC chromatography can be used for the analysis. Also, in some cases these types of molecules may be analyzed using ion-pair (IP) chromatography by adding an ion-pairing agent (IPA) in the mobile phase and using RP-HPLC type columns.

4) Small ionic molecules that are water soluble can be analyzed by different types of HPLC. One type is ion chromatography, either cation-exchange or anion-exchange chromatography, depending on the particular type of ion. Highly polar or ionic molecules can also be analyzed using IP chromatography. Even RP-HPLC can be used on special columns that allow very high content of water in the mobile phase.

5) Organic soluble large molecules are frequently separated for differentiation based on their molecular weight. This can be achieved using gel permeation chromatography. GPC is also used when the molecular weight of a polymeric material must be evaluated.

6) The type labeled as large water-soluble molecules can include a wide variety of

compounds such as polymeric carbohydrates, proteins of different types, and certain synthetic polymers. Depending on the purpose of analysis and only if a separation based on molecular weight is necessary, the HPLC of choice will be gel filtration SEC. Other techniques can be used depending on sample properties, such as RP-HPLC, HIC (hydrophobic interaction chromatography), displacement, or bioaffinity chromatography.

Based on the selected HPLC type, further choices are made including sample preparation, column selection, mobile phase selection, detection type, and quantitation method. Some of these selections are discussed further in this chapter.

Use of Polarity of the Analyte in the Choice of HPLC Type

The polar or nonpolar (hydrophobic) character of the analyte is a simple criterion that provides guidance regarding the choice of the chromatography type that could be the most appropriate for a specific separation. This character can be estimated using octanol/water partition constant $\log K_{ow}$ or coefficient $\log D_{ow}$ for the analyte. As described in Section 3.5, for neutral molecules, $\log D_{ow} = \log K_{ow}$, but for molecules that can be present in ionic form, the molecules will have different structures depending on pH and different $\log D_{ow}$. In such cases, $\log D_{ow}$ at isoelectric point must be used instead of $\log K_{ow}$. Figure 9.1.2 suggests different chromatographic types depending on $\log K_{ow}$ of the analyte.

Details regarding the choice of the HPLC type are further discussed in this chapter. The simple scheme shown in Figure 9.1.2 is unable to capture many aspects of HPLC analysis. For example, a specific HPLC type can cover a much wider range of $\log D_{ow}$ values, and more than one technique can be applied for a specific class of molecules with a specific polarity [1]. Also, samples may contain complex mixtures of molecules, which may have a wide range of $\log K_{ow}$ values. In such cases, one alternative is to analyze one group of solutes by one technique and another group by a different technique. The use of a more versatile column with a wide range of polarity of solvents is another alternative. Multimode separations also can be applied in cases of very different analytes.

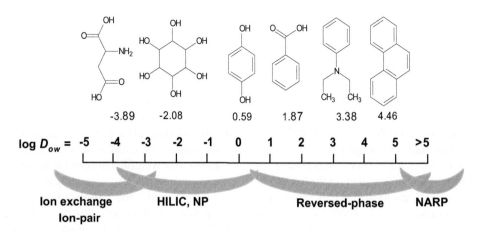

FIGURE 9.1.2 Different HPLC types preferentially utilized depending on $\log K_{ow}$ of the analyte molecule.

9.2. THE QUANTITY OF SAMPLE INJECTED FOR HPLC ANALYSIS

General Aspects

HPLC as a separation technique has a variety of uses, the main ones being analytical, but some are semipreparative (e.g. flash chromatography) and preparative. Semipreparative and preparative HPLC are of considerable practical and theoretical interest, and these techniques use large samples. However, the subject of large samples in HPLC is beyond the purpose of the present book, and the literature should be consulted for information in this field (see, e.g., [2,3]). In analytical HPLC, the sample is usually a solution of analytes together with their matrix, which is injected as a small volume in the mobile phase for performing the separation. Several aspects of the injection important for the separation include: (1) the volume of the injected sample, (2) the sample concentration (amount of solutes in the dissolved sample), and (3) the nature of the solvent used for sample dissolution. These subjects related to sample injection are discussed further in this section.

Sample Volume and Amount

The injected sample is characterized by its volume, usually in the range of 5 μL to 25 μL, for standard analytical HPLC. However, other injection volumes are used. For certain UPLC applications or for micro HPLC, volumes in the range of 20–500 nL are common. Large volume injections, up to 1 mL, can also be used in special applications [4,5]. The injection volume must be precise, and the injection must be reproducible. The injection volume V_{inj} is directly related to the amount of sample $m = c_i V_{inj}$ delivered to the HPLC system. Since the area A_i of the chromatographic peak (detector response) is proportional to the amount of sample m (see rel. 2.1.77), a larger sample volume will generate a larger signal

(peak area). This indicates that for samples of lower concentration, a larger injection volume may be desirable. However, a large injection volume may affect the chromatographic peak shape. In "ideal" conditions, when an extremely small (narrow) injection is made, the "ideal" peak broadening (for a compound j) is given by rel. 2.1.44, which can be written in the form:

$$W_b = \frac{4t_{R,j}}{N_{0,j}^{1/2}} \qquad (9.2.1)$$

where the notation N_0 is used to indicate the theoretical plate number in "ideal" conditions. Relation 9.2.1 allows the calculation of "ideal" *peak volume*, which is given by the formula:

$$V_{0,peak} = W_b U \qquad (9.2.2)$$

For a sample volume V_{inj} the peak volume increases and is typically obtained using the expression [6]:

$$V_{peak} = (1.333 V_{inj}^2 + V_{0,peak}^2)^{1/2} \qquad (9.2.3)$$

From rel. 9.2.3 regarding the peak volume, the calculation of N_j as it results from an expression of the form 9.2.1 gives the following formula:

$$N_j = \frac{16(t_{R,j}U)^2}{1.333 V_{inj}^2 + V_{0,peak}^2} \qquad (9.2.4)$$

From rel. 9.2.1 and 9.2.2, the expression of $N_{0,j}$ can be easily obtained and the ratio of $N_{0,j}$ and N_j gives the formula:

$$\frac{N_j}{N_{0,j}} = \frac{V_{0,peak}^2}{1.333\, V_{inj}^2 + V_{0,peak}^2} \qquad (9.2.5)$$

This expression indicates that a loss in efficiency of less than 10% is caused by an injection volume $V_{inj} < 0.3\, V_{0,peak}$, and a loss of efficiency of less than 1% is caused by an injection volume $V_{inj} < 0.1\, V_{0,peak}$. A value for V_{inj} can be further obtained when the value of $V_{0,peak}$ is known. An evaluation of $V_{0,peak}$ can be obtained starting with rel. 9.2.2 and the calculation of W_b. Assuming that the height of the theoretical plate H can be approximated by the formula $H \approx 2\,d_p$ (see Section 2.2 following rel. 2.2.13) where d_p is the diameter of particles in the column, the value for $N_{0,j}$ can be estimated as $N_{0,j} \approx L\ /\ 2d_p$. The value for t_R is given by rel. 2.1.7 as $t_R = t_0\,(1 + k)$, and t_0 is estimated based on rel. 2.2.3 as $t_0 = (\varepsilon^*\pi\,d^2\,L)\ /\ 4U$. (where the diameter of the column d and the length L are in mm and $\varepsilon^*\pi/4 \approx 5\ 10^{-4}$). These expressions lead to the following formula:

$$V_{0,peak} = \sqrt{2}\varepsilon^*\pi\,d^2 d_p^{1/2} L^{1/2}(1 + k) \qquad (9.2.6)$$

With estimations $k > 1$, $\varepsilon^* \approx 0.64\ 10^{-3}$, and $V_{inj} < 0.3\ V_{0,peak}$, the allowable sample volume is given by the following approximation:

$$V_{inj} < 0.14\,L^{0.5} d^2 d_p^{0.5} \qquad (9.2.7)$$

with (V_{inj} in µL, L and d in mm, d_p in µm). For example, for a column with $L = 150$ mm, $d = 4.6$ mm, $d_p = 5$ µm, rel 9.2.7 indicates $V_{inj} < 80$ µL for a loss in efficiency of less than 10%, and for a column with $L = 100$ mm, $d = 2$ mm, $d_p = 3$ µm, rel 9.2.7 indicates $V_{inj} < 17$ µL. Although the increased injection volume may produce a reduction of column efficiency, larger volumes than recommended are sometimes used when the resolution is still satisfactory and an increase in analysis sensitivity is necessary for the measurement of compounds present in traces.

The amount of material injected in the chromatographic column is an additional parameter that affects the separation. Excessive amount of sample in the chromatographic column leads to stationary phase overload. In such cases, the stationary phase becomes saturated with a specific analyte, and the retention cannot take place within the narrow region occupied by the zone in the column containing the sample. The mobile phase carries the unretained solute on a "fresh" portion of the column where the analyte is retained (depending on its specific

equilibrium constant K), but this is associated with an apparent lowering of k values (shorter retention times) and tailing. This effect is suggested in Figure 9.2.1 where the shapes of overloaded peaks are indicated. The amount (in µg) that can be loaded in an analytical chromatographic column can be estimated using the expression:

$$m_i \approx AV_0(1 + k)\left(\frac{NL}{1000}\right)^{-1/2} \qquad (9.2.8)$$

where A is a constant depending on the nature of the stationary phase, V_0 is the dead volume (volume of the mobile phase in the column), k is the capacity factor, N is the number of theoretical plates per meter, and L (in mm) is the column length. For conventional analytical columns $A \approx 0.05$ to 0.2 (for m in µg) and largely depends on the nature of the stationary phase. Larger samples than are accommodated by the column result in column overload with increase in peak width and decrease in resolution. Relation 9.2.8, as well as the values given in Table 6.3.2 indicate approximate values for sample loading in the conventional analytical columns.

Considering an injection volume of 20 µL in a column with $L = 150$ mm and $d = 4.6$ mm and an acceptable sample loading of 50 µg, the resulting concentration of the sample should not exceed 2.5 mg/mL. This type of concentration generates very large peaks for most detectors, and lower (or significantly lower) concentrations of the analyte are used in practice. For detectors such as fluorescence or MS, even low µg/mL

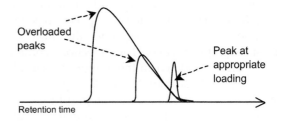

FIGURE 9.2.1 Shortening of the retention time and tailing typical for column overloading.

concentrations may produce signal overloading since these detectors typically work in ng/mL range. When the sample amount is in the range of 10–20 mg for a conventional column, peaks as shown in Figure 9.2.1 can be obtained with an RI detector, while other detectors generate overloaded peaks (with upper part of the peak cutoff) that are not suitable for quantitation.

The estimations of sample volume and concentration discussed in this section give rather conservative estimates of the maximum volume and amount of sample to be injected in the chromatographic column. For very diluted samples, larger injection volumes are sometimes necessary in order to have good detection. Also, the injected samples often contain a low level of the analyte of interest, but a large level of other solutes that form the sample matrix. In such cases, if the matrix reduction is not possible by sample preparation, the amount of sample injected in the chromatographic column may need to exceed up to 10 to 20 times the column-loading capacity regarding the matrix. Even in these conditions, the sample concentration may remain low for the analytes but at acceptable levels for detection. The early overloading compounds may or may not have any effect on the peak shape of low-level analytes eluting after the matrix peak. Large peaks, however, may have long tails that can interfere with the separation, and the conditions must be selected such that the peaks of interest elute beyond this tail. Late overloading peaks may contribute to the shortening of the retention time of the peaks that elute earlier and have a low (in acceptable range) concentration. These types of interference from one compound to another are difficult to predict, and specific details are given in individual method descriptions (see, e.g., [7,8]).

Nature of Sample Solvent and Its Importance in Separation

The solvent used to dissolve the sample must be selected so that it completely dissolves the sample and no denaturation occurs upon dissolution. The solution of the sample must be stable and free of any particles (if necessary filtered through 0.45 μm filters). Also, the solvent used for sample dissolution must be soluble in the mobile phase. In many applications, the solvent is selected as close as possible to the mobile phase. Even when all these requirements are fulfilled, the separation can be affected by the sample solvent [9]. When the sample solvent is selected identical with the mobile phase (in gradient condition, the same as the initial solvent), its nature does not negatively affect the separation. However, when it is different and the injection volume is larger, the solvent that dissolves the sample may lead to a different retention value for an analyte as compared to the retention in the mobile phase. In this way, both the peak shape and even the separation can be affected.

In practice, rel. 9.2.5 (and therefore 9.2.7) may not always be valid. For larger injection volumes, N_j may decrease when V_{inj} increases as rel. 9.2.5 predicts, but this may depend on the retention of the analyte on the column. When a compound is not eluted from the head of the column at the initial composition of the mobile phase, it will concentrate as a very narrow plug that will be eluted only when the mobile phase composition changes. In such cases, the volume of the injection may not influence the column efficiency. This effect is known as *adsorption compression* [10]. In conditions where adsorption compression takes place, larger injection volumes can be used, with the advantage of increasing the sensitivity of the measurement at no or little loss of efficiency. In some cases, when adsorption compression takes place and the injection volume increases, even a slight increase in the column efficiency can be detected.

The retention at the head of the column may be influenced by the nature of sample solvent when the injected solvent is different from the mobile phase and it is not immediately diluted with the mobile phase [9]. For solvents that are "weaker" than the mobile phase, the solvent sample plug may allow focusing of the analytes at the head of the column, and so an adsorption compression may take place. For solvents that elute the analyte more strongly than the mobile phase, the sample may migrate differently at the head of the column because of the sample solvent, and the peak shape may be adversely affected, in particular at larger injected volumes. This is illustrated in Figure 9.2.2 for a set of six analytes (metil-2-hydroxybenzoate, salicylic acid, piroxicam, propyl 2-hydroxybenzoate, meloxicam, and 4-chlor-4'-hydroxy benzophenone, each at 0.25 μg) when the injection is

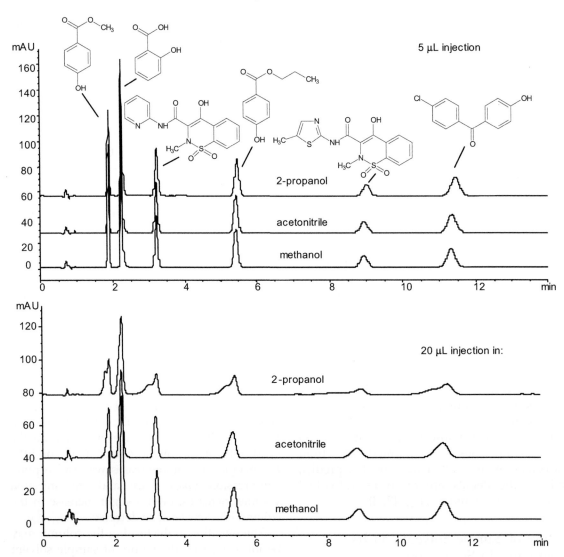

FIGURE 9.2.2 The influence of the injection volume and nature of the sample diluent (2-propanol, acetonitrile and methanol) on the shape of the peaks belonging to six model compounds (equal amounts of 0.25 μg compound were injected, at two different dilutions).

done in three different sample solvents and two sample dilutions. The separation was done on a Zorbax Eclipse XDB-C18, 150 mm × 4.6 mm i.d. × 5 μm p.d.; 25°C, with a mobile phase 35% acetonitrile and 65% aqueous solution of 0.2% H_3PO_4; flow-rate: 2 mL/min; 237 nm detection. In Figure 9.2.2, the peak shape is good at 5 μL injection for all three sample solvents (2-propanol, acetonitrile, and methanol). The effect of the solvent of the sample is not very important at low injection volumes (in conventional analytical chromatography of less than 5 μL). For 20 μL injection, the peak shape is affected for 2-propanol and acetonitrile, which are "stronger" solvents than the mobile phase.

Peak shape seems also to be affected by the difference in viscosity between the sample solvent and the mobile phase. This effect is dominant in certain separations [11]. While strong eluting solvents of equal viscosity with the mobile phase cause only peak broadening, the difference in viscosity between the sample solvent and the mobile phase causes distortion of the shape of early eluting peaks. Other studies were also performed regarding the influence of solvent samples on the peak shape [12].

Selection of the mobile phase as the sample solvent is not always feasible in RP-LC. One reason can be the low solubility of target analytes in the mobile phase. Moreover, the injection step is almost always correlated with the sample processing. Often the analyzed samples are subject to liquid-liquid extraction (LLE) or solid-phase extraction (SPE), and after the cleanup step the analytes are present in a diluted organic solution. Evaporation of the diluted sample in order to concentrate it can be applied at this point of sample preparation. However, this operation can be time-consuming and can affect the recovery of the analytes (e.g., by analyte evaporation simultaneously with the solvent or by decomposition when the temperature used is too high). In these cases, it is preferable to inject directly in the HPLC system a part of the organic layer containing the analytes. However, in cases when the solution of the analytes is too diluted, higher volumes of solutions (in other solvent than the mobile phase) may be required for the injection.

Injections of larger volumes than is recommended by rel. 9.2.7 are possible when focusing of the analytes takes place at the head of chromatographic column. However, another procedure can also be used to allow much larger volume injections. This procedure is based on focusing the sample solvent at the column head. For successful use of large injection volumes by this procedure, the following conditions must be fulfilled: (1) the sample solvent must be hydrophobic (such as hexane, heptane, octane, and upper alcohols) in order to be focused together with the analytes at the head of the column after injection; (2) the mobile phase must have a high content of water in order to avoid the dissolution of the solvents; and (3) the analytes must have a lower hydrophobicity than the sample solvent such that after solvent focusing, they can participate in the separation process without interference from the sample solvent [13–15]. An example of several separations of two analytes by injecting up to 500 μL of samples in hexane as sample solvent are shown in Figure 9.2.3. The separation conditions are as follows: the column was a Zorbax Eclipse XDB-C18, 150 x 4.6 mm, 5 μm particle size, mobile phase contained 20% ACN and 80% water, at 30°C; flow rate of 1 mL/min. Column preparation for a consecutive injection in a sequence consisted of the following operations: a fast step gradient to 100% ACN (in 0.05 min), 10 min needed for elimination of the hydrophobic sample solvent loaded to column during the previous run, a fast step gradient back to initial elution conditions in 0.05 min, 10 min for adequate column re-equilibration. These operations are needed for column conditioning and for the removal of all interferences transferred from previous separation stages. The injected samples with

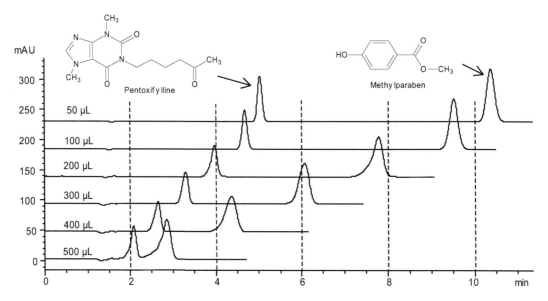

FIGURE 9.2.3 Overlaid chromatograms for different sample volumes containing 0.8 μg pentoxifylline and 0.6 μg methylparaben in different volumes of n-hexane on a C18 stationary phase [14].

different volumes contained the same amount of analytes (0.8 μg for pentoxifylline and 0.6 μg for methylparaben). The results from Figure 9.2.3 indicate that the injection of a large volume of hydrophobic solvent is also associated with a reduction in the retention time of the analytes. A linear dependence between the injected volume V_{inj} of hydrophobic solvent and the retention time t_R for each analyte can be noticed in the separation for a V_{inj} larger than a specific small volume V_{min}. The capacity factor k_i for each analyte follows the expression:

$$k_i(V_{inj}) = k_i - b \, V_{inj} \quad \text{for} \quad V_{inj} > V_{min} \quad (9.2.9)$$

where k_i is the capacity factor for an injection volume $V_{inj} < V_{min}$, and b is a constant for each separation system (analyte, sample solvent, column, mobile phase). Relation 9.2.9 suggests that the hydrophobic solvent of the sample covers a portion of the stationary phase, making it unavailable for the retention of the analyte.

For HILIC separations, it is common to use pure organic solvents as the sample solvent.

Organic solvents are weak eluents under HILIC conditions, so that polar analytes are accumulated in a narrow zone on the head of the column (sample on-column focusing). This is an advantage for the analysis of polar drugs in biological samples, for instance, when the proteins from the sample matrix are precipitated with acetonitrile or with methanol. The organic supernatant can then be directly injected into the HILIC column, avoiding the step of solvent evaporation and the residue reconstitution. Injection of a large volume of aqueous sample solvent of high elution strength should be avoided in HILIC. The volume of polar solvents for the sample should be limited by the same rules as nonpolar solvents in the case of RP-HPLC. The use of large volumes of polar solvents for the sample injection may lead to broad or split peaks. An organic content of more than 50% for the sample solvent is typically recommended in HILIC [16,17].

For specific types of chromatography, such as size exclusion, sample viscosity and volume play an even more important role in the separation than for RP or HILIC. In SEC, sample concentration can be high and polymers can significantly influence the viscosity of the sample. Also, larger injection volumes than in other types of chromatography are typically utilized in SEC. For this reason, some specific restrictions are necessary regarding injection in SEC. For example, the relative viscosity of the sample in SEC should not exceed double that of the mobile phase (for a dilute aqueous buffer, this corresponds to a concentration of protein of about 70 mg/mL). The volume load in SEC should not exceed 1–5% of the total column volume, although larger volumes are sometimes injected. Only injections that do not exceed 2% of the column volume were usually proven to maintain good resolution [18].

Another aspect regarding the nature and volume of sample solvents is related to the way they influence the response of specific detectors. For example, the signal of RI or of electrochemical detectors can be significantly influenced by the sample solvent. In some cases, the "elution" of sample solvent can generate a large signal (sometimes negative) that can adversely affect the signal measurement. The problem of detector response to the sample solvent is addressed in various studies (see, e.g., [19]).

9.3. ESTIMATION OF PARAMETERS DESCRIBING THE SEPARATION

General Comments

The typical characterization of a separation is done using the capacity factor k and selectivity α. For this reason, considerable effort has been made and reported in the literature, with the goal of predicting k and α from the physico-chemical properties of the analyte, of the column, and of the mobile phase involved in the separation [20,21]. The contribution of these three participants to the separation coming

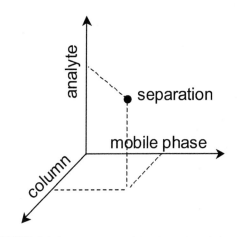

FIGURE 9.3.1 Illustration of the dependence of separation on column, mobile phase, and analyte.

together is suggested in the diagram given in Figure 9.3.1. In the diagram from this figure, each of the axes (column, mobile phase, analyte) can be further characterized by a number of parameters previously discussed in this book. For example, for RP-HPLC parameters such as $\alpha(CH_2)$, H_c, S^*_c, A_c, B_c, and C_c characterize the column (see Section 6.4). For HILIC, the column is characterized by $\alpha(CH_2)$, $\alpha(OH)$, α_{dia}, α_{regio}, α_{shape}, (see Section 6.5). The mobile phases are characterized by parameters such as Hildebrand solubility δ, polarity P', solvatochromic parameters $E_T(30)$, π^*, α, and β, (see Section 7.1) and other physicochemical parameters such as viscosity η, dielectric constant ε, surface tension γ', A^{SASA}, dipole moment m. The analytes have their own physicochemical characterizing properties, including MW, charge, K_{ow}, surface area $\mathcal{A}$, solvatochromic parameters π^*, α, and β, dipole moment m_j, polarizability α_j, ionization potential I_j, (see Section 8.1). Additional parameters such as temperature and mobile phase flow rate were also considered among the factors that influence separation. In spite of numerous attempts to calculate k and α based on some of the previously listed parameters, most of the results remained only estimative. In cases of good agreement with the experiment, the results were limited to a small group of analytes. Nevertheless, important information can be obtained from the predicted values for capacity factor and selectivity, either for comparison purposes between the analytes of a sample or for directing the separation in the desired direction.

The procedures for calculating k (or log k) can be grouped in the following types: (1) Estimation of capacity factor k for similar systems at different mobile phase compositions, (2) evaluations from correlations with the octanol/water partition constant K_{ow}, (3) calculation of the capacity factor from the van der Waals molecular surface of the analyte, (4) empirical prediction based on solute, mobile phase, and stationary phase characteristics, and (5) attempts to calculate log k based on the interactions during separation. Each of these types is discussed in this section.

Estimation of Capacity Factor k for Similar Systems at Different Mobile Phase Compositions

The simplest expression for the value of k for a separation, where only the composition of the mobile phase is changed, is given by rel. 7.1 37, which can be written as follows:

$$\log k_{iAB} = \log k_{iA} - S_i \phi \qquad (9.3.1)$$

where S_i is a constant specific for a given compound i, a solvent system, and a given chromatographic column (see Table 7.1.8), ϕ is the volume fraction of the organic solvent with $\phi = \dfrac{V_B}{V_A + V_B}$ and k_{iA} is the (extrapolated) value of k_i for pure solvent A (usually water when the notation for k_{iA} is k_{iw}). Relation 9.3.1 provides many systems with a good estimation of log k_i. The values for log k_{iA} and S_i can be calculated (if unknown) when the values for log k_{iAB} are known for at least two solvent concentrations (a least-squares deviation technique can be used when more than two values k_{iAB} are known).

The applicability of rel. 9.3.1 is exemplified in Figure 9.3.2, which shows the correlation between the experimental values of log k_i and calculated values of log k_i using rel. 9.3.1 for methanol/water 20/80 v/v mobile phase ($\phi = 0.2$), for a number of compounds listed in Table 9.3.1. The compounds were separated on

FIGURE 9.3.2 Correlation between the experimental values of log k_i and calculated values of log k_i using rel. 9.3.1, for a number of compounds listed in Table 9.3.1, separated on a LiChrosphere 100 RP-8 column with methanol/water 20/80 v/v mobile phase [22].

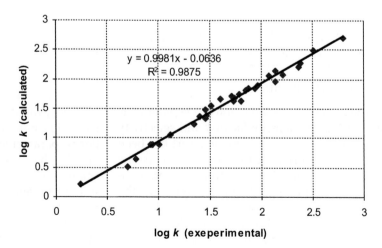

y = 0.9981x - 0.0636
R² = 0.9875

TABLE 9.3.1 Example of compounds showing good linearity for log k vs. ϕ, as described by rel. 9.3.1.

Compound	Compound	Compound	Compound
Aniline	o-Toluidine	N,N-Dimethylaniline	Benzyl alcohol
Methylbenzoate	Benzyl cyanide	p-Cresol	p-Ethylphenol
Toluene	α-Naphtylamine	2,6-Dimethylphenol	α-Naphtol
Ethylbenzene	o-Nitrotoluene	Hydroquinone	Methylparaben
Ethylbenzoate	Acetophenone	Phenol	Ethylparaben
Chlorobenzene	Dimethyl phthalate	o-Cresol	Propylparaben
Bromobenzene	Pyridine	3,5-Dimethylphenol	Butylparaben
Caffeine	Anisole	β-Naphtol	

a LiChrosphere 100 RP-8 column 125 × 4.0 mm 5.0 μm particle size. The values for k_{iA} and S_i were obtained using the least-squares fitting technique based on experimental k_i values for five solvent concentrations in the range 30 to 70% methanol [22].

For two concentrations of the organic modifier ϕ_1 and ϕ_2, rel. 9.3.1 leads to the formula 9.3.2 (identical with rel. 7.1.38):

$$\log k_{iX2} = \log k_{iX1} - S(\phi_2 - \phi_1) \qquad (9.3.2)$$

Relation 9.3.2 allows the calculation of log k_i in a solvent with the organic component at the concentration ϕ_2 when the capacity factor is known for the volume fraction ϕ_1 for the solvent.

Even when rel. 9.3.1 or 9.3.2 is obtained for systems where good linearity is verified for several organic component concentrations, the calculated k_{iA} (in the pure solvent A) may depend on both solvents A and B. As indicated in Section 7.1, when solvent A is, for example, water, k_{iw} should not depend on the other solvent. However, different values for k_{iw} may be obtained when calculated by extrapolation, for example, from a methanol/water mobile phase, or from a acetonitrile/water mobile phase, for the same stationary phase.

For some separation systems, rel. 9.3.1 (and 9.3.2) does not provide accurate predictions. For example, when acetonitrile/water is used as a mobile phase, the linear dependence 9.3.1 is less frequently obeyed as compared to the case of methanol/water mobile phase. A better fit with experimental data has been reported for some systems by an equation of the form 7.1.39 [23,24]. This formula is written below with the specification that it is valid only for compound i (in the specific solvent system and chromatographic column):

$$\log k_{iAB} = \log k_{iA} + c_1\phi + c_2\phi^2 \qquad (9.3.3)$$

The values for the parameters log k_{iA}, c_1 and c_2 in rel. 9.3.3 can be obtained similar to log k_{iA} and S_i for rel. 9.3.1, by using a least-squares fitting technique for known log k_i values at other concentrations ϕ. An example of a dependence of log k_i on ϕ given by an equation of the form 9.3.3 is shown in Figure 9.3.3 for 2,5-dimethylphenol separated on a Zorbax ODS column with acetonitrile/water as a mobile phase [25].

The procedure based on rel. 9.3.1 or 9.3.3 can give good results in predicting log k, but it is limited to one compound, one column, and one solvent system, and it requires k_i values for other concentrations of the solvent. In these

FIGURE 9.3.3 Equation for the calculation of log k for 5-dimethylphenol in
a separation on a C18 column with
acetonitrile/water as mobile phase [25].

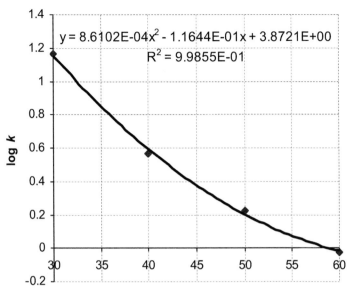

$$y = 8.6102E\text{-}04x^2 - 1.1644E\text{-}01x + 3.8721E\text{+}00$$
$$R^2 = 9.9855E\text{-}01$$

% Acetonitrile in water

cases, good predictions for log k_{iAB} can be obtained when parameters c_1, c_2, and log k_{iA} are obtained from data on the same stationary-phase/mobile phase system (when a set of log k_{iAB} are known for other solvent concentrations). Similar to the case of applicability of rel. 9.3.1, the extrapolated values for log k_{iA} depend on the other solvent (B). This is exemplified in Table 9.3.2 where the values log k_{iw} (extrapolated to water) are given for toluene separated on five different columns using methanol/water or acetonitrile/water as a mobile phase [22]. The calculation of log k_{iw} was performed based on six organic phase concentrations (20 to 70%) using a least-squares deviation technique to generate the quadratic dependence.

Since a considerable number of HPLC separations are performed in gradient conditions, direct calculation of log k_i by rel. 9.3.1 or 9.3.3 is useful in this case only for understanding the expected separation results. Predictions for the values of log k_{iAB} during gradient separations can be obtained using rel. 7.5.14 (for linear gradients) when the value k_0 is known.

Evaluation of Capacity Factor k from Octanol/Water Partition Constant K_{ow}

Various studies have been reported that attempted to evaluate capacity factors k_i or selectivities α_i using $K_{i,ow}$ values [26–29]. Use

TABLE 9.3.2 Calculated values for log k_{wi} (quadratic dependence) for toluene, using data for two different mobile phase systems.

Column	log k_{iw} from AcCN/H$_2$O	log k_{iw} from MeOH/H$_2$O
LiChrospher 100 RP-18e (125 × 4 mm, 5 μm)	3.62	3.94
Purospher RP-18e (125 × 4 mm, 5 μm)	3.76	4.76
LiChrospher 100 RP-8 (125 × 4 mm, 5 μm)	3.38	3.35
SymmetryShield RP-C18 (150 × 3.9 mm, 5 μm)	3.77	4.48
Symmetry-Shield RP-C8 (150 × 3.9 mm, 5 μm)	3.32	4.28

of K_{ow} for estimating k has its origin in several studies on liquid-liquid extraction, which show that partition coefficients K_{iX} and K_{iY} for a compound i in two systems—solvent X/water and solvent Y/water—is given by the expression 3.1.24 (see, e.g., [30]). The extension of rel. 3.1.24 to two systems: (1) stationary phase/mobile phase and (2) octanol/water, generated expression 3.1.26, once more written as follows:

$$\log k_i = a \log K_{i,ow} + b \qquad (9.3.4)$$

For a given column and mobile phase, knowledge of the capacity factor $\log k_i$ for at least two compounds (and of $\log K_{i,ow}$ from literature) allows the estimation of coefficients a and b and in principle the calculation of $\log k_i$ for any other compound. Use of a larger set of compounds with known $\log k_i$ and $\log K_{i,ow}$ for the calculation of a and b (by a least-squares deviation technique) leads to more reliable results. The experimental verification of rel. 9.3.4 has been done in several studies [26,27,29]. However, the complexity of the chromatographic processes makes it impossible to use a single parameter like K_{ow} to describe a wide range of separations. For a specific column and mobile phase system a good correlation can be obtained between calculated $\log k_i$ by rel. 9.3.4 and experimental K_{ow} values, as it was shown in Figure 3.1.3 ($R^2 = 0.9355$) for 72 mono and disubstituted aromatic compounds. Those results were obtained for a C18 stationary phase with water/methanol 50/50 (v/v) as a mobile phase [26]. Other correlations between $\log k_i$ and $\log K_{iow}$ were not that good. For example, the correlation reported in [29] for 76 very different compounds gave only an $R^2 = 0.5539$. Another example of the dependence between $\log k_i$ and $\log K_{ow}$ (experimental) is further illustrated in Figure 9.3.4 for a set of compounds listed in Table 9.3.1. These compounds were separated on a Symmetry-Shield RP-C8 column, 150×3.9 mm with 5.0 μm particles endcapped, using methanol/water 60/40 v/v mobile phase. As shown in Figure 9.3.4, the

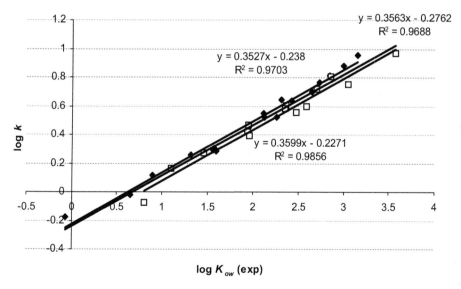

FIGURE 9.3.4 Correlation between $\log k$ and $\log K_{ow}$ for compounds listed in Table 9.3.1 separated on a Symmetry-Shield RP-C8 column. Trendline equations listed on figure for phenolic compounds (--□-- $R^2 = 0.9856$), other aromatic (--◆-- $R^2 = 0.9703$), and all compounds ($R^2 = 0.9688$).

correlation between log k and log K_{ow} is in this case very good even if a variety of compounds was considered. Even better results are obtained if the correlation is done separately for phenolic compounds and for compounds with no phenolic groups (as shown in Figure 9.3.4, $R^2 = 0.9856$ for phenolic compounds, $R^2 = 0.9703$ for other aromatic compounds, and $R^2 = 0.9688$ for all compounds). For a different stationary phase, a graph similar to that shown in Figure 9.3.4 is given in Figure 9.3.5. In this case, the chromatographic column was a LiChrospher 100 RP-8, 125 × 4.0 mm with 5.0 μm particles not end-capped, with the same mobile phase as used for Figure 9.3.4, for comparison.

Comparison of Figures 9.3.4 and 9.3.5 shows that the stationary phase is important for the values of the parameters a and b in rel. 9.3.4 and also indicates that different classes of compounds behave differently depending on the stationary phase. As shown in Figure 9.3.5, a better correlation between log k and log K_{ow} is obtained for compounds without phenolic

groups than for those with phenolic groups, which is an indication that chemical structure of the solute plays an important role in the applicability of rel. 9.3.4. The accuracy of rel. 9.3.4 increases when it is applied on compounds with similar chemical structure.

The nature of the mobile phase also affects parameters a and b in rel. 9.3.4. An exemplification of this effect is further discussed for a number of aromatic compounds listed in Table 9.3.3, with the separation on a Lichrospher 100 RP-18 column 250 × 4.0 mm with 5 μm particles. The two mobile phases used for the separation were methanol/water and acetonitrile/water. Table 9.3.3 gives the values for log K_{ow} as well as the values for the capacity factor log k_w when the mobile phase is water. The values for K_{ow} were obtained using the computer package MarvinSketch 5.4.0.1 (ChemAxon Ltd. [31], see Section 8.2 for a discussion of various sources for K_{ow} values).

For pure water as mobile phase, the correlation between log k_w and log K_{ow} is given in

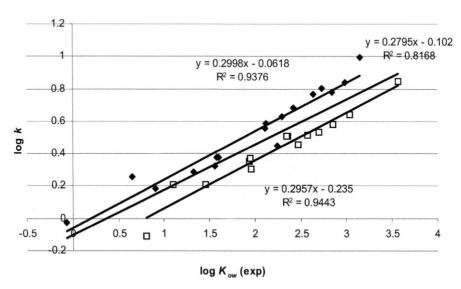

FIGURE 9.3.5 Correlation between log k and log K_{ow} for compounds listed in Table 9.3.1 separated on a LiChrospher 100 RP-8 column. Trendline equations listed on figure for phenolic compounds (--□-- $R^2 = 0.9443$), other aromatic (--◆-- $R^2 = 0.9376$), and all compounds ($R^2 = 0.8168$).

TABLE 9.3.3 Compounds investigated for the correlation of their $\log k_w$ with $\log K_{ow}$ (MarvinSketch).

Compound	$\log k_w$	$\log K_{ow}$	Compound	$\log k$	$\log K_{ow}$	Compound	$\log k_w$	$\log K_{ow}$
Phenol	1.60*	1.67	2,6-Dichlorophenol	2.80	2.88	Chlorobenzene	3.20	2.58
4-Nitrophenol	2.04*	1.61	4-Chloro-3-methylphenol	3.19	2.79	Naphthalene	3.66	2.96
3-Nitrophenol	2.09*	1.61	2,4-Dichlorophenol	3.14	2.88	p-Xylene	3.58	3
2-Methylphenol	2.17*	2.18	3,5-Dichlorophenol	3.44	2.88	Propylbenzene	4.10	3.38
2-Chlorophenol	2.21*	2.27	2,4,6-Trichlorophenol	3.54	3.48	Biphenyl	4.31	3.62
2,4-Dinitrophenol	2.10	1.55	Pentachlorophenol	4.85	4.69	Butylbenzene	4.49	3.82
2-Nitrophenol	2.13	1.61	Benzene	2.20	1.97	Anthracene	4.79	3.95
3-Chlorophenol	2.65	2.27	Nitrobenzene	2.16	1.91	Pyrene	5.13	4.28
4-Chlorophenol	2.59	2.27	Toluene	3.16	2.49	Chrysene	5.79	4.94
2,4-Dimethylphenol	2.44	2.7	Ethylbenzene	3.48	2.93			

Note: indicates measured values in water, the other $\log k_w$ being extrapolated to water.

Figure 9.3.6. The figure shows one trendline for all compounds, and also the trendlines for the data corresponding only to compounds with phenolic groups ($R^2 = 0.8985$) and without phenolic groups ($R^2 = 0.9906$). The results from Figure 9.3.6 are in good agreement with the predictions of rel. 9.3.4 (for the particular system with water mobile phase $a = 1.247$, $b = 0.0597$). Further verification of rel. 9.3.4 can be made for other mobile phases. The same type of correlation shown in Figure 9.3.6 is given in Figure 9.3.7A for a mobile phase 50% methanol in water, and in Figure 9.3.7B for a mobile phase 50% acetonitrile in water. As seen in Figure 9.3.7, a relation of the form 9.3.4 is still valid for various mobile phases, but parameters a and b are changed, and they are different as the mobile phase changes. The variation of parameters $a(\phi)$ and $b(\phi)$ with the modification in mobile phase composition is illustrated in Figure 9.3.8 for the separation using the Lichrospher 100 RP-18 column with mobile phase methanol/water or acetonitrile/water. Based on the shape of the graphs shown in Figure 9.3.8, the estimation of $\log k$ from $\log K_{ow}$ when the mobile phase composition is changing should be obtained from an equation of the form:

$$\log k_i = a(\phi) \log K_{i,ow} + b(\phi) \qquad (9.3.5)$$

with:

$$a(\phi) = a_0 + a_1\phi + a_2\phi^2 \qquad (9.3.6)$$
$$b(\phi) = b_0 + b_1\phi + b_2\phi^2 \qquad (9.3.7)$$

In rel. 9.3.6 and 9.3.7, the parameters a_0, a_1, a_2, and b_0, b_1, b_2, are independent of the compounds separated. This indicates that rel. 9.3.5 can be used for estimation of $\log k_i$ when $K_{i,ow}$ is known for a range of compounds and the coefficients involved in the calculation of $a(\phi)$ and $b(\phi)$ are also known. These coefficients can be obtained from best-fit curves, but this requires values for k_i at several mobile phase concentrations. Formula 9.3.5 extends the applicability of calculation of $\log k_i$ to a range of solvent compositions (besides a variety of compounds). However, the use of rel. 9.3.4 and/or 9.3.5 for any compound must be done with caution. Larger deviations from the correct value of calculated $\log k_i$ are

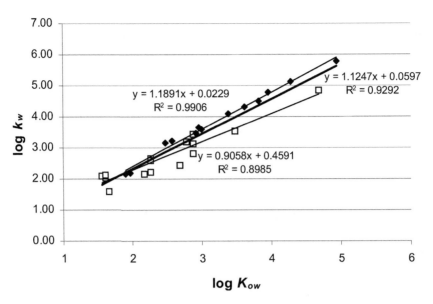

FIGURE 9.3.6 Correlation between log k_w and log K_{ow} on a Lichrospher 100 RP-18 column with mobile phase water. Trendline equations listed on figure for phenolic compounds (-- □ -- $R^2 = 0.0.8985$), other aromatic (-- ◆ -- $R^2 = 0.9906$), and all compounds ($R^2 = 0.9292$).

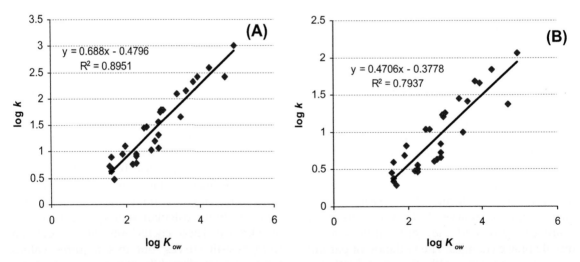

FIGURE 9.3.7 Correlation between log k and log K_{ow} on a Lichrospher 100 RP-18 column with mobile phase 50% methanol in water (A) and 50% acetonitrile in water (B).

seen when the compounds used for the calculation of parameters *a* and *b* are obtained from compounds very different from the one with unknown capacity factor. Since K_{ow} values are readily available, it is very convenient to use K_{ow} for the description of chromatographic processes, in spite of various shortcomings discussed in this section.

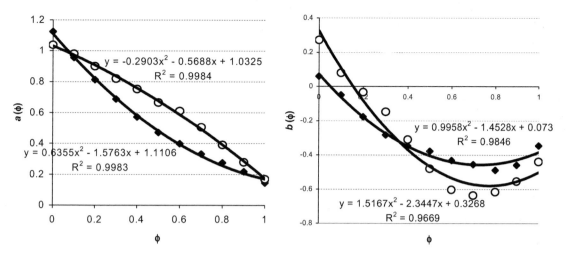

FIGURE 9.3.8 Variation in parameters $a(\phi)$ and $b(\phi)$ in expression 9.3.5 when the composition of the mobile phase changes for methanol/water (--O--) and for acetonitrile/water (--◆--).

Calculation of Capacity Factor from van der Waals Molecular Surface of the Analyte

The van der Waals molecular area of the analyte A_j is a parameter shown to influence the value of the free energy for the equilibrium, in particular for RP-HPLC, but also for IP separations, and to a certain extent for HILIC and NPC. Unfortunately, as shown in Section 5.1, the detailed calculations of capacity factors k_i using formulas of the type 5.1.26, in which A_j is present, are relatively difficult. This difficulty comes from the lack of knowledge regarding the values of other parameters necessary in the formulas and from the fact that the formulas are too complicated. Since rel. 5.1.26 predicts a dependence between $\log k$ and A, it should be possible to establish empirical correlations between the two values. Such correlations were already verified between $\log K_{ow}$ and A (see Section 8.2). Also, good experimental correlations were proven between $\log K_{ow}$ and $\log k$. It is therefore expected to find good correlations between $\log k$ and A within the same class of compounds. From rel. 9.3.4 and rel. 8.2.15, the following expression for $\log k_i$ can be immediately generated

$$\log k_i = a(a' A_i + \sum_n c_n) + b \qquad (9.3.8)$$

where a and b depend on the stationary phase and the mobile phase, $a' = 1.46 \cdot 10^{-2}$, and c_n are given in Table 8.2.3 (for K_{ow} calculated using MarvinSketch). The results for the calculated $\log k_i$ using rel. 9.3.8 for the compounds listed in Table 9.3.1 (not including caffeine) compared to experimental values reported in the literature are shown in Figures 9.3.9 and 9.3.10. Figure 9.3.9 shows the results for the separation on a Symmetry-Shield RP-C8 column, 150 × 3.9 mm with 5.0 μm particles end-capped, with methanol/water 50/50 v/v as a mobile phase ($a = 0.5731, b = -0.3715$). The same results are shown in Figure 9.3.10 for methanol/water 70/30 v/v as a mobile phase ($a = 0.3203, b = -0.3499$).

The results shown in Figures 9.3.9 and 9.3.10 indicate very good agreement between the experimental and calculated values, considering the variety of compounds evaluated and the limited number of parameters used

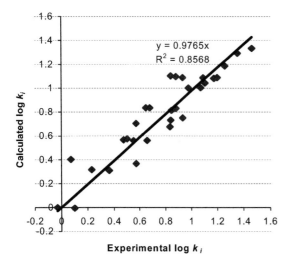

FIGURE 9.3.9 Calculated vs. experimental log k_i values (rel. 9.3.8) for methanol/water 50/50 v/v on a Symmetry-Shield RP-C8 column.

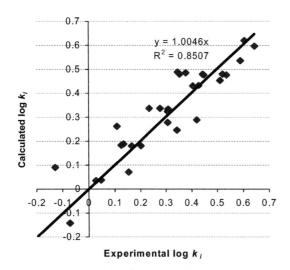

FIGURE 9.3.10 Calculated vs. experimental log k_i values (rel. 9.3.8) for methanol/water 70/30 v/v on a Symmetry-Shield RP-C8 column.

in the calculation. For the use of rel. 9.3.8, the values for a and b must be known for each column and each mobile phase, but the values for a' and c_n are given, and $\mathcal{A}_i$ can be easily

calculated for any compound (e.g., using the MarvinSketch package). This type of calculation can be of considerable utility when the values for log k_i are known for two or more other compounds for a given mobile phase and a given column. These known log k_i values together with the corresponding $\mathcal{A}_i$ can be used for the calculation of a and b, and further for the estimation of log k_i for any other compound separated in the same conditions (the same column and the same mobile phase). For the compounds listed in Table 9.3.1, the average relative standard deviation of calculated log k_i versus experimental value was found to be about 17% for methanol/water 50/50 v/v mobile phase, and about 13% for methanol/water 70/30 v/v mobile phase case. Similar to the case of the use of log K_{ow} for the calculation of log k_i, the use of a larger set of compounds with known log k_i for the calculation of a and b (by a least-squares deviation technique) leads to more reliable results. The calculation of log k_i by rel. 9.3.8 can be further used in relations of the type 9.3.1 for the same solvent but at a different concentration (when S_i is known).

Prediction of log k Based on Solute, Mobile Phase, and Stationary Phase Characteristics

A considerable number of attempts were made to develop a general model for the characterization of any free energy A^0 related property of a separation. For HPLC, this energy-related property is typically log k. As discussed in Section 6.4, the interactions in RP separations must include (1) hydrophobic interactions, (2) steric interactions, (3) hydrogen bonding between a basic solute and an acidic column groups (column acidity), (4) hydrogen bonding between an acidic solute and a basic column groups (column basicity), and (5) cation exchange and/or ion-ion interactions. Since the contribution of these interactions is additive, a general formula was developed [21] with the following general expression:

$$\log k = h_0 + hH + sS + aA + bB + cC + vV$$

(9.3.9)

Parameters h, s, a, b, c, and v are solute dependent and come from scales related to (h) solute's hydrophobic character, but also including polarizability and dipolarity, (s) molecular shape, (a) hydrogen-bond donating ability, (b) hydrogen-bond accepting ability, (c) cation-exchange capability, and (v) molecular size [20,23,32]. The parameters h_0, H, S, A, B, C, and V are dependent on both the stationary phase and mobile phase, but for a specific mobile phase they depend only on the stationary phase. This concept was used for column characterization as described in Section 6.4. Solutes capable of interacting by a unique type of interaction (in addition to hydrophobic) were identified, and their contribution to the capacity factor was measured and averaged for a number of columns obtaining solute characterizing parameters [33]. For each column c1, it was than possible to generate parameters for column characterization. This result was expressed by the formula 6.4.15, written below:

$$\log k(j) = \log k_{EB} + \eta'(j)H_{c1} - \sigma'(j)S^*_{c1}$$
$$+ \beta'(j)A_{c1} + \alpha'(j)B_{c1} + \kappa'(j)C_{c1}$$

(9.3.10)

Comparing formula 9.3.9 with 9.3.10, $h_0 = \log k_{EB}$, solute parameters h, s, a, b, and c are represented by $\eta'(j)$, $\sigma'(j)$, $\beta'(j)$, $\alpha'(j)$, and $\kappa'(j)$ and column parameters H, S, A, B, and C are represented by H_{c1}, S^*_{c1}, A_{c1}, B_{c1}, and C_{c1}. The success of rel. 9.3.10 for the characterization of the chromatographic column with a set of test compounds and a well-specified mobile phase has been remarkable [34, 35]. However, the main role of expression of the type 6.4.15 is not for solute characterization and cannot be used to predict $\log k$ without having the values

$\eta'(j)$, $\sigma'(j)$, $\beta'(j)$, $\alpha'(j)$, and $\kappa'(j)$ for a specific solute j. These values are available for the set of "test compounds" but not for other analytes that were not in the test set.

The success of column characterization using an expression of the form 9.3.9 (an approach known as linear solvation energy relationships, LSER) has been further investigated for the general use of predicting capacity factor for an analyte j [20,36,37]. For this purpose, each term in rel. 9.3.9 can be estimated not from purely empirical correlations, but based on solvent and solute solvatochromic characteristics [20,23,32]. One such approach, leading to a formula analogous to 9.3.9 but also including solvent contributions, suggests the following expression for the calculation of $\log k_j$:

$$\log k_j = \log k_0 + r\frac{(\delta^2_{mo} - \delta^2_{st})}{100}V_j$$
$$+ e(\pi^*_{mo} - \pi^*_{st})\pi^*_j + a(\beta_{mo} - \beta_{st})\alpha_j$$
$$+ b(\alpha_{mo} - \alpha_{st})\beta_j$$

(9.3.11)

In rel. 9.3.11, the index j indicates the solute, and mo and st stand for mobile and stationary phases. Several parameters such as Hildebrandt solubility parameter δ, molar volume V, and solvatochromic (normalized) parameters π^*, α, and β were previously described in Section 7.1. The coefficients r, e, a, and b are numbers generated from correlations, and $\log k_0$ is a reference value [38]. Relation 9.3.11 is similar to rel. 6.4.20.

For a specific stationary phase (and assuming that its properties do not change with the mobile phase), rel. 9.3.11 can be reduced to the following:

$$\log k_j = c_{st,j} + r_j\delta^2_{mo} + e_j\pi^*_{mo} + a_j\beta_{mo} + b_j\alpha_{mo}$$

(9.3.12)

where $c_{st,j}$ depends on the stationary phase through k_0 (reference), δ_{st}, π^*_{st}, α_{st}, and β_{st} and also depends on the solute j, while r_j, e_j, a_j, and b_j depend on the solute only. The other parameters, δ_{mo}, π^*_{mo}, α_{mo}, and β_{mo} depend only on the mobile phase. A linear correlation has been demonstrated to exist between δ^2 and π^* for some solvents (not including polar solvents as shown in Figure 7.1.7) [39]. Therefore, rel 9.3.12 can be further simplified to the following:

$$\log k_j = c_{st,j} + e^1_j \pi^*_{mo} + a_j \beta_{mo} + b_j \alpha_{mo} \quad (9.3.13)$$

(where e^1_j is an adjusted coefficient). Another simplification is possible, since the retention in RP-HPLC does not depend strongly on the interaction of the solute hydrogen bond acidity a_j with the solvent hydrogen bond basicity β_{mo}, such that rel. 9.3.13 can be written in the form:

$$\log k_j = c_{st,j} + e^1_j \pi^*_{mo} + b_j \alpha_{mo} \quad (9.3.14)$$

The terms depending on π^*_{mo} and α_{mo} can be described together with a certain approximation by E^N_T since this parameter incorporates both types of interactions described by parameters π^* and α (see rel. 7.1.27). For this reason, the following expression can be expected to be valid for the capacity factor:

$$\log k_j = c_{st,j} + p_j E^N_{T\,mo} \quad (9.3.15)$$

The application of rel. 9.3.15 is of limited use for a true calculation of the capacity factor k, since the value for p_j and for $c_{st,j}$ must be known. However, such values are available in the literature for several series of compounds [32,36,38].

The results expressed by rel. 9.3.15 show the usefulness of parameter E^N_T of the mobile phase for solvent characterization, showing that compounds with larger E^N_T lead to larger k values and therefore to larger retention times in RP-HPLC. This can be a criterion for selecting a specific solvent (or solvent mixtures) as a mobile phase in RP-HPLC.

Relation 9.3.15 can be used to estimate the variation in k_j when the mobile phase is changed. Using the notation $k_j(1)$ and $k_j(2)$ for the capacity factors in two different mobile phases (1) and (2), from rel. 9.3.15 can be immediately obtained:

$$\log k_j(1) - \log k_j(2) = p_j[E^N_{T\,mo}(1) - E^N_{T\,mo}(2)] \quad (9.3.16)$$

Expression 9.3.16 indicates that when the solvent is changed (e.g., from methanol to acetonitrile), the value for log k changes proportional to the change in E^N_T of the two solvents.

Relation 9.3.15 also points out a certain disagreement with rel. 9.3.1 regarding the linear variation of log k_i with the mobile phase composition. In Section 7.1 (see Figure 7.1.8) it was shown that the variation of E^N_T for a solvent mixture is not linear. Substituting such a variation for $E^N_{T\,mo}$ in rel. 9.3.15, a disagreement with rel. 9.3.1 is obtained. However, both expressions 9.3.1 and 9.3.15 being approximations, it can be concluded that for specific systems they may provide useful information.

Besides expression of the form 9.3.9 used to predict log k, other procedures based on molecular properties were suggested [40]. One of these procedures is based on the additive fragment concept where the calculation uses a relation of the form:

$$\log k = \sum f_n + \sum F_m \quad (9.3.17)$$

where f_n is a constant for the particular fragment and F_m is a correction factor for a specific structural feature in the molecule. The procedure can also be extended to the calculation of a log k (new) for a compound similar in structure to another compound that has a log k(known) value using the expression:

$$\log k(\text{new}) = \log k(\text{known}) \pm \sum f_n \pm \sum F_m \quad (9.3.18)$$

where the fragments f_n and the corrections F_m are subtracted and/or added appropriately to change the known compound into the new one. Although this type of procedure provides good results, it requires information on f_n and F_m values, which are available for a limited range of compounds [41,42].

Evaluation of the Energies of Interaction in the Separation System

Procedures for the direct calculation of log k from the evaluation of interaction energies in a separation system were previously presented in Chapter 5. For PR-HPLC, the solvophobic theory [43–46] offers a method for such calculation. The role of the solvent (mobile phase) in the separation results from its physical parameters, including surface tension γ', dielectric constant ε, molar volume V, polarizability α, and ionization potential I, that must be used for the calculation of k (see rel. 5.1.26). The problem with such calculations is that particular parameters for solute, solvent, and stationary phase are not always available. This is in particular the case for the stationary phase and for mixtures of solvents used as mobile phases. For this reason, approximations and results for systems with restricted values for some parameters were reported in the literature [44,47]. The formula developed in Chapter 5 for explaining the retention is useful mainly for the understanding of the role of various solvent parameters in influencing the separation.

9.4. STEPS IN DEVELOPMENT AND IMPLEMENTATION OF AN HPLC SEPARATION

General Comments

The typical path in HPLC practice was described in Section 1.3. It is also appropriate to follow this path for developing and implementing a new HPLC method. The development starts with the collection of information about the analysis and the analytes. Based on this information, a decision can range from adapting a method already described in the literature with no or with minor changes, to the development of a completely novel analytical procedure. This choice is based on the requirements of the analysis (analytes or class of analytes to be separated, sample matrix, scope of analysis, etc.), the access to specific instrumentation, and the availability in the literature of an adequate method. Following this start, a decision must be made on the type of HPLC. Once the HPLC type is known, the instrumental setup is put in place. The next step is usually the selection of the chromatographic column. Since a change in the chromatographic column during a method development is a "noncontinuous step" and implies other changes in the method development, it is important to select a column as correctly as possible from the beginning. After the column is selected, the mobile phase composition that will be first tried is identified. Changes in the mobile phase can be made easily, and further modifications in the mobile phase are typically done for improving the initial separation. The analysis should begin with analysis of a set of standards representing the analytes in the sample. Generally, the method improvements are initially done also using standards. Following these steps, the "real" samples are analyzed and the results are evaluated. Validation of the analytical procedure is usually the last step in method development, and it is frequently required for the official acceptance of an analytical method. Each of these steps is discussed further in this section. However, it should be emphasized that method development in HPLC is a complex process with numerous variants and so can only be sketched here.

Information for Starting the Development of an HPLC Method

The development of an analytical method in general and of an HPLC method in particular should start with the collection of information regarding: (1) analytes to be measured and purpose of the analysis, (2) type of sample, and sample characteristics, (3) analysis requirements, (4) methods of analysis reported in the literature, and (5) available instrumentation, suppliers, and funding.

1) Information regarding the purpose of the analytical method is very important, since a wide range of requests can be made for an analysis. This information should contain the list of analytes or class of analytes that must be analyzed (if known), the nature of the sample (type of material, origin, etc.), whether or not qualitative information for the sample is required, the further use of the results, and the like. When too little information about the sample is available, preliminary analyses of the sample should be performed, not necessarily using HPLC. For example, a GC/MS analysis (with or without derivatization) may provide some valuable qualitative information. Other techniques can be used for the same purpose, such as infusion of a solution of the sample in a LC/MS system that may provide information about the molecular weight of the sample components.

2) The information regarding the type of sample should include as many details as possible about the nature of the sample matrix (matrix composition), the amount of sample available, the value of the sample, and the need/availability of sample preparation. For example, it is necessary to know if the analytes are small molecules or polymeric ones. In the case of small analyte molecules, data regarding volatility, solubility, and reactivity are necessary. For macromolecules, a general characterization is always useful. Other data regarding the sample are helpful. Such information may include an estimated level of analytes in the sample (trace, medium levels, major constituent), sample perishability, and safety concerns about the sample

3) Important data to be considered before starting to develop a new analytical method also include the necessary precision and accuracy for the results, the number of samples that will be analyzed, and the time requested for delivering results.

4) Methods of analysis are available for almost any analyte and are described in the literature. The already reported methods may not be intended for the specific type of sample on which the new method should be applied or may not satisfy certain requirements for the new method. It is also possible that the method described in the literature cannot be implemented, for example, due to the unavailability of instrumentation. However, the literature always offers a valuable starting point for developing a new method. When a method is available and can be implemented, it is frequently beneficial to implement the method and directly evaluate its adequacy for the new purpose. In certain cases, where no method is available in the literature, procedures for the analysis of similar compounds as those planned for analysis are helpful.

5) The information on instrument availability is another critical starting point in developing a new analytical method. Depending on the analysis requirements, a selection should be made for a specific pumping system or a detector type. Several criteria for selecting a detector were described in Section 1.4. Information regarding most suppliers is available on the web.

After collecting the above information, the analyst may be able to decide what analysis should be developed, which characteristics of this method are necessary, and what HPLC instrument is available (for the method development and/or for running the samples).

Choice of HPLC Type

The HPLC type as described in Section 1.2 is chosen in connection with the steps required in an analysis before the separation (see Section 1.3). These steps include sample collection and sample preparation, which must be performed in close coordination with the choice of the HPLC type. The amount of collected sample, the collection procedure, as well as sample dissolution, cleanup, concentration, and possible derivatization, are performed based on an established goal regarding the form in which the sample will be subject to HPLC analysis. A considerable volume of information is available in the literature regarding sample collection and preparation (see, e.g., [30]). A "processed sample" is therefore entering the HPLC analysis, and depending on its nature, a choice of the HPLC type is made.

The most common type of HPLC used in practice is RP-HPLC because of its versatility, reliability, and ease of use [48]. When RP-HPLC is appropriate as the method of analysis, it should be preferred to other methods, even if these other methods are also applicable. For molecules that do not separate well on RP phases, for example, for highly polar molecules, HILIC may be a better choice than RP. For ionic analytes, ion chromatography is typically utilized. Depending on the purpose of the analysis, other techniques may be selected, such as size exclusion in case of separation of macromolecules, or a chiral phase in case enantiomers must be separated. Various discussions of HPLC type with regard to the type of processed sample/analytes can be found throughout the present book and are summarized in Section 9.1.

Choice of the Chromatographic Column

Based on the type of separation selected for the analysis, the next step in developing a new method is to select the analytical column. When RP-HPLC is selected, a large range of RP chromatographic columns are commercially available. RP stationary phases range from those with strong hydrophobic character (e.g., C8 and C18 columns end-capped with hydrophobic groups) to those that manifest some polar interactions (see Section 6.4). Valuable information regarding column characterization and tools for column comparison and evaluation are available (e.g., PQRI evaluation approach [49]). Also, RP columns are available in numerous formats and with different physical characteristics of the stationary phase (pore size, particle size, number of theoretical plates, core-shell type, monolithic, etc.). For the development of a new method, it is typically recommended a start with columns offering a large number of theoretical plates and good stability in a wide pH range. It is also better to avoid starting with "extreme columns" such as columns with a special stationary phase, very small particle size, with narrow i.d., or columns that are very short. In the opposite case when enough information regarding the analytical method is available and only improvements to the method are sought, more "extreme values" for various parameters should be explored in order to improve resolution, shorten analysis time, reduce volume of mobile phase, and the like. A detailed discussion of RP columns is given in Section 6.4.

A variety of HILIC columns are also commercially available. The polarity of HILIC columns ranges from the very polar such as bare silica to the less polar (e.g., amide columns). Also, columns in different formats and with specific properties (e.g., core-shell columns) are available for this technique. A detailed discussion on HILIC columns characteristics is presented in Section 6.5. Columns should be selected based on the information that is potentially available in the literature for the analytes of interest, and should be compared with the more general information presented in Chapter 6 of this book. Physical properties of the column (particle dimensions, size of the column, etc.) should be selected based on sample composition (including the matrix), but also on the available instrumentation (e.g.

HPLC or UPLC). Columns used in ion chromatography, in chiral separations, or in size exclusion are also presented in Chapter 6.

Together with the selection of the analytical column, the selection and use of a guard column and/or guard cartridge is highly recommended. The guard column or cartridge typically matches the nature of the stationary phase in the analytical column and is recommended by the column manufacturer.

Choice of Mobile Phase and Achieving Separation of Standards

Because mobile phase composition can be easily changed, and also because it is common in HPLC to use mobile phase gradients, changes in the mobile phase are commonly used for modifying and improving a separation. The basic composition of the mobile phase is, as expected, related to the type of HPLC and of the column used for the separation. Descriptions regarding solvent properties and utilization in different HPLC types are given in Chapter 7. For RP-HPLC, common solvents are water with an organic modifier such as acetonitrile or methanol. However, various other organic solvents can be used as organic modifier, and in NARP no water is used in the mobile phase. The two objectives of the first trials for a separation are: (1) identify that all the analytes are retained on the stationary phase at the initial mobile phase composition, and (2) verify that all the analytes are eluted when the mobile phase strength increases sufficiently. In order to retain the analytes, a weak solvent is initially used as the mobile phase, which in RP means a low organic modifier in the aqueous mobile phase. As discussed in Section 6.4, for many chromatographic columns it is recommended that, in order to avoid column dewetting, more than 80 to 90% water not be used in the mobile phase. A larger content of organic solvent in the mobile phase is associated in RP with a lower value for the capacity factor, as indicated by rel. 2.1.70 or 7.1.37. The solvent can be chosen

based on the analyte log K_{ow} value; for compounds with log K_{ow} in the range 0.0 to 4 or 5, a simple choice is to use as the mobile phase acetonitrile/water with a gradient between 80% water to 0% water in a 15–25 min run. Ideally, in these conditions all the analytes are initially retained (do not elute at the dead retention time), and all analytes are eluted during the chromatographic run.

Once the mobile phase has been selected, evaluation of the method should begin. When the analytes are known, a common practice is to attempt the separation of a synthetic mixture containing the analytes and not the real sample. The concentration of the components in the synthetic mixture should be in the range of that for the analytes in the real sample (if known), or higher than that in the case of trace component analysis. The results of the first separations will constitute the base for further improvements (optimization). In the event that not all the analytes are retained and the column can be used at lower content of organic modifier, this content can be lowered to 3–5%. Also, for stronger retention acetonitrile can be replaced with methanol. If this attempt fails, the column usually should be replaced with one with higher retention capability (or the type of HPLC must be changed). The problem of too strongly retained compounds typically can be solved by using stronger organic modifiers such as tetrahydrofuran or isopropanol. If this change is not successful, the column should be replaced with a less retaining one. Besides the changes in the organic modifier, the changes in the mobile phase offer a wide range of other possibilities, including pH change, addition of additives, modification in the flow rate, and modification of the temperature of the solvent. When practiced in types of HPLC other than RP, the elution power of the mobile phase is different from that in RP. For example, in HILIC less polar solvents have a lower elution power and polar solvents have a "stronger" one, leading to faster elution.

Besides modifying the retention times to bring them in a desired window, changes in the mobile phase composition and nature are very important for improving (modifying) the separation, since some of the solutes may not be well separated. Different compounds in the sample may react differently to the changes in the mobile phase, such that two peaks that are poorly separated may become farther apart when the mobile phase is changed (undesired merging of peaks is also possible). This type of mobile phase modification is known as *separation optimization*. For separation optimization, several schemes for solvent composition changes have been proposed (e.g., use of scout gradients, triangulation algorithms) [50–56]. The success of a good separation can be followed using, for example, a resolution map or a surface response where the values of *R* or α are plotted in function of a parameter (or two) that are modified [48]. Achieving separation is highly dependent on the differences in the structure of the compounds to be separated. Typical problems occur when the analytes have similar structures—for example, when they are isomers (enantiomers are separated only with chiral phases). In such cases, even when RP-HPLC is used for separation, the use of columns with some polarity may improve the result, and therefore this aspect must be figured into the choice of the column.

The choice of mobile phase is not solely related to the separation process. The mobile-phase composition must also be selected consonant with the detector utilized for the analysis since some detectors cannot be used with certain mobile phases or with a mobile phase that changes the composition during gradient elution (e.g., in RI). A discussion of choosing the detector in an HPLC analysis can be found in Section 1.4. The separation and the choice of the detection in HPLC are highly interrelated. For example, when the detection is selective enough that two compounds can be easily measured even when they coelute in a chromatogram, less effort is needed to obtain a good separation for the two compounds. On the other hand, when two compounds are not differentiated by the detector, they should be well separated.

Application of the Method to Real Samples

The acceptable separation of the components of a synthetic mixture must be further applied to real samples. Real samples often contain matrix compounds that may coelute with the analytes or that are not sufficiently separated from the analytes. The peaks with shapes different from Gaussian are in particular suspected of containing an interference, especially when the same peak was Gaussian for the standards. For this reason, the separation must be further improved or optimized in order to obtain a good separation and eliminate potential interferences. Verifying that no interference from the sample matrix affects the analyte peaks involves a challenging process. When selective detectors such as MS or MS/MS are used, interferences are more easily eliminated by the detector. However, in other cases, verification is not so simple. One alternative to verifying that there is no interference is to analyze a "blank" sample that does have the same matrix as regular samples but contains no analyte. Such samples, however, are not always available. Other procedures involve modifications of the mobile phase composition, or even of the column, with the goal of shifting retention times to verify that the peak areas for the analytes in the same sample but with a different separation are not changed. This can be done, for example, using a weak mobile phase with very long retention times, which would hopefully offer a better separation, although it is not practical for routine sample analysis. When standard materials with a matrix close to that of the real samples are available, their analysis and generation of correct quantitation represent another proof of lack of interference.

Besides the improvement of the separation, during the application of the method on real samples, other parameters of the method must be verified, including: determining the optimum amount of sample to be injected in the HPLC system such that the separation is not affected and the detector provides sufficient signal; verifying that after the separation all sample components are eluted from the column; and evaluating the modifications in resolution R after a number of samples were injected to prove column stability.

Method Validation

Once the separation is considered acceptable for use, a thorough verification process is usually practiced, indicated as *validation*. Validation covers much more than separation itself and may involve qualification of the sample and sampling (sample collection, chain of custody, etc.), certification of chemical standards, qualification of instrumentation (operational qualification, performance qualification, calibration specifications, necessary documentation, etc.), validation of the analytical method, data validation, scheduling of audits, training, and operator qualifications. Formal validation also includes

TABLE 9.4.1 Parameters used for the validation of an analytical method.

Parameter	Short description
Specificity/ selectivity	Specificity refers to a method that produces response for a single analyte in the presence of other components in the matrix. Selectivity refers to a method responding to a limited number of chemical compounds.
Precision	Precision refers to the reproducibility of measurement within a set, indicating the scatter or dispersion of the set about its central value (mean). The scatter is characterized by the standard deviation.
Reproducibility, intermediate precision, and repeatability	Reproducibility is typically considered the precision between different laboratories. Intermediate precision refers to long-term variability within a single laboratory. Repeatability refers to precision obtained over a short period of time with the same equipment (in the same lab.) when using different matrices (at least 3) and different concentrations of the analyte.
Accuracy	Accuracy can be considered as an experimental value that approximates the bias. Bias is the difference between an accepted (or true) value for an amount or a concentration analyzed and the result of the analysis.
Linearity	Linearity indicates the linear dependence between the signal and the concentration or amount and is characterized by the standard deviation for the slope and the standard deviation for the intercept.
Range, linear range	The range is the interval between the upper and lower levels that have been demonstrated to be determined with precision and accuracy. The range with linear response is the linear range.
Limit of detection	The concentration (or amount) corresponding to the average signal that is with 4.66 standard deviations higher than the average of the blank signal.
Limit of quantitation (or of determination)	The minimum concentration (or amount) that produces quantitative measurements with acceptable precision and accuracy (signal usually about 10 times higher than the blank).
Recovery	Recovery is the ratio (in percent) between a known added amount of an analyte and the measured amount.
Robustness	Robustness refers to the quality of an analysis to not be influenced by small experimental modifications during the performance of the process.
Ruggedness	Degree of reproducibility under a variety of conditions such as different laboratories, analyses, or instruments.
Stability	Stability indicates that the same results are obtained in time and under different conditions.

documented evidence indicating that the analytical process generates consistent results. The validation involves internal confirmation or external confirmation by other laboratories, use of other methods, and use of reference materials in order to evaluate the suitability of the chosen methodology [57]. There are various levels of formalized validation. Detailed information on validation is available in many publications (see, e.g., [58,59]), and only a short summary of validation steps for the analytical method is given in Table 9.4.1.

Method Transfer

Once a method has been developed and verified, it is sometimes necessary to transfer it to another laboratory or to implement it on more than one instrument. The method transfer may involve some problems, although in principle it should be a simple operation. The problems may be caused by differences in the instrumentation such as in the pumping system (e.g., change from high-pressure mixing to low-pressure mixing) or in the detection (different sensitivities of the detector), or in the tubing/connections when larger dwell volumes are present in the system. Other problems may come from column-to-column variability. Also, the transfer of the method can be performed with changes from the initial method, such as choice of a column of the same type but in a different format (shorter and/or narrower) and an increase in the flow rate of the mobile phase. These changes are not supposed to affect significantly the separation, but this claim must be verified. After the transfer of a method, at least part of the validation should be repeated.

9.5. SEPARATIONS BY RP-HPLC

General Comments

Various aspects related to RP-HPLC have been presented in this volume. The mechanism in RP-HPLC was discussed in Section 5.1, details about the columns were given in Section 6.4, and mobile phases used in RP-HPLC were described in Section 7.6. Because RP-HPLC is widely used, a number of other aspects general for HPLC were exemplified and discussed in reference to this method. The applicability of this technique covers a wide range of compounds, and an enormous volume of dedicated literature describing particular methods of analysis is available. A general overview regarding the selection of columns and of mobile phases in RP-HPLC based on the log K_{ow} value of the analyte is schematically shown in Figure 9.5.1. Regarding the mobile phase, the use of gradients is widespread in RP-HPLC, and a range of solvent polarities can be used for elution.

Only very general comments regarding the analysis of various classes of compounds by RP-HPLC are further provided. Depending on the molecular properties of the analyte, certain particularities of RP-HPLC must be considered; these are discussed later in the chapter for several classes of compounds.

Application of RP-HPLC to Analysis of Small Polar Molecules

RP-HPLC can be successfully used in the analysis of small polar (nonionic) molecules. Uracyl, which is not retained on common C18 columns has log $K_{ow} = -1.07$ (experimental). Other compounds even with small hydrophobic moieties but with log $K_{ow} > -0.86$ can be retained on hydrophobic columns. For example, acrylamide with log $K_{ow} = -0.35$ can be separated on a C18 column using a low organic content in the mobile phase. This type of procedure can be applied, for example, for the analysis of acrylamide in cigarette smoke [60]. The chromatogram obtained using MS/MS detection (MRM mode with transition $m/z = 72$ to 55) after separation on two Gemini-NX 5μ C18 150 × 2 mm columns

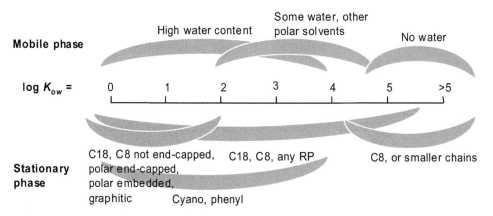

FIGURE 9.5.1 Overview of columns and mobile phases used in RP-HPLC based on log K_{ow} of the analyte.

with 95% water (with 0.1 % formic acid) and 5% methanol and at 0.3 mL/min flow rate is shown in Figure 9.5.2 (retention time for acrylamide is 3.68 min).

For small molecules with low hydrophobic character and with polar groups in their structure, columns that exhibit additional interactions

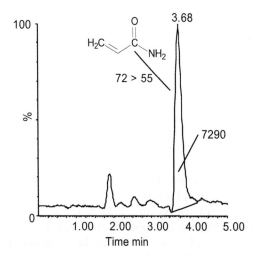

FIGURE 9.5.2 Chromatogram for the analysis of acrylamide in cigarette smoke on two Gemini-NX 5u C18 150 × 2 mm columns in series with mobile phase 95% water and 5% methanol at 0.3 mL/min flow. Peak area indicated on the graph corresponds to about 25 ng/mL analyte.

to hydrophobic ones may be more appropriate. For example, on a Synergy-4μ Hydro-RP 80 Å, 250 × 4.6 mm column, which is made with polar end-capping groups, acrylamide elutes at 9.4 min when using the same mobile phase and flow rate as on Gemini-NX columns. On this column, 100% water can be used as mobile phase, and compounds with log D_{ow} as low as -3.5 can be separated. Small polar molecules also can be analyzed successfully using other HPLC techniques such as HILIC.

Analysis of Small Molecules with Average or Strong Hydrophobic Character

This type of molecule includes a very large number of compounds, which are typically analyzed very successfully using RP-HPLC. Compounds that have positive log K_{ow} values can be separated easily using RP columns. Numerous particularities are related to specific physical and chemical characteristics of the compound. Columns with other interaction capabilities besides hydrophobicity may be necessary for better separations in some cases. Neutral compounds as well as basic or slightly acidic compounds may be best separated on hydrophobic columns with the polar activity of silanol groups reduced as much as possible, silanol activity being reduced by various procedures such as the use of ultra pure silica, double end-capping, the use of silica support with ethane cross-linked bridges (BEH technology), and use of CSH technology (see Section 6.4).

Compounds that have very high log K_{ow} may show low solubility if any water is present in the mobile phase, and their separation must be done using only organic mobile phase (NARP technique). One example of an application for NARP is the analysis of solanesol ($C_{45}H_{74}O$) in cigarette smoke using a Spherisorb 5μ ODS (2) 250 × 4.6 mm column [61]. Solanesol has a log $K_{ow} = 14.12$ and has virtually no solubility in water or polar solvents. For this reason, the separation must use a nonaqueous mobile phase and can be done using a gradient of three solvents, A- methanol, B- ethanol, C- heptane (ethanol is needed between heptane and methanol since these two solvents are not miscible). The gradient (linear) starts with 100% methanol reaching 100% heptane at 10 min and returning to initial conditions starting at 11 min. The detection is done in UV using the absorption at 205 nm for the analyte in the range 9 min to 12 min and at 365 nm outside this time range.

The resulting chromatogram is shown in Figure 9.5.3.

Analysis of Small Sugar Molecules, Oligo, and Polysaccharides

Small sugar molecules are very polar compounds. The values for experimental log K_{ow} of a few simple sugars are as follows: arabinose log $K_{ow} = -3.02$, ribose log $K_{ow} = -2.32$, glucose log $K_{ow} = -3.24$, glucosamine log $K_{ow} = -3.14$, fructose log $K_{ow} = -3.04$, and sucrose log $K_{ow} = -3.70$. From these low log K_{ow} values it is obvious that regular RP-HPLC is not the technique of choice for sugars analysis. However, similar to amino acids, sugars can be derivatized with various reagents and made amenable for RP-HPLC [30, 62]. Without derivatization, free sugars can be analyzed using HILIC or ion-moderated chromatography.

Oligosaccharides are polar molecules with low or very low log K_{ow} values. These molecules are not at all separated on hydrophobic columns. However, excellent separations are obtained on HILIC columns and on ion-moderated columns. Larger polysaccharides are typically separated based on their MW, using the gel filtration technique (GFC).

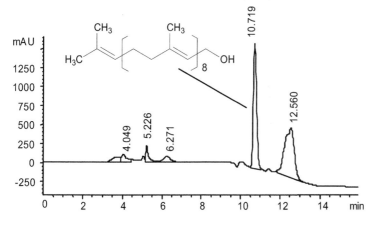

FIGURE 9.5.3 Solanesol ($C_{45}H_{74}O$) eluting at 10.719 min on a Spherisorb 5μ ODS column.

Analysis of Amino Acids, Peptides, and Proteins

Amino acids represent an important class of compounds that are present in nature either as free molecules or as building blocks for peptides and proteins. Table 9.5.1 lists several common amino acids present in proteins. The table also lists the experimental values for log D_{ow} at the pI point of the amino acid. Since the values of log D_{ow} for all amino acids are lower than log K_{ow} for uracil, this indicates that amino acids cannot be separated using typical C18 columns.

Various alternatives for amino acid analysis have been developed in practice. One possibility is the derivatization of amino acids, such

TABLE 9.5.1   Common amino acids that are building blocks for proteins, the pK$_a$'s and log D_{ow} at pI values

Name	Abbrev. (three-letter)	Abbrev. (one-letter)	Radical R* 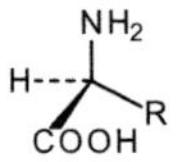 H---$\overset{NH_2}{\underset{COOH}{\mid}}$R	pK$_a$α− COOH	pK$_a$ α−NH$_3^+$	pK$_a$ side	Experim. log D_{ow} at pI
L-Alanine	Ala	A	CH$_3$-	2.34	9.69	-	−2.85
L-Arginine	Arg	R	$\underset{H_2N}{\overset{HN}{>}}$C−NHCH$_2CH_2CH_2$	2.17	9.04	12.48	−4.20
L-Asparagine	Asn	N	H$_2$N-CO-CH$_2$-	2.02	8.84		−3.82
L-Aspartic acid	Asp	D	HOOC-CH$_2$-	2.09	9.82	3.86	−3.89
L-Cysteine	Cys	C	HS-CH$_2$-	1.71	10.78	8.33	1.71
L-Cystine	Cys-Cys		-CH$_2$-S-S-CH$_2$-				−5.08
L-Glutamic acid	Glu	E	HOOC-CH$_2$CH$_2$-	2.19	9.67	4.25	−3.69
L-Glutamine	Gln	Q	H$_2$N-CO-CH$_2$CH$_2$-	2.17	9.13	-	−3.64
Glycine	Gly	G	H-	2.34	9.60	-	−3.21
L-Histidine	His	H	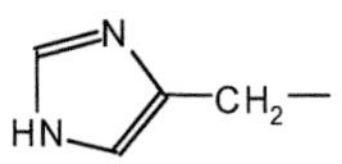 (imidazole)−CH$_2$−	1.82	9.17	6.04	−3.32
L-Hydroxylysine	Hyl	-	H$_2$N-CH$_2$-CH(OH)-(CH$_2$)$_2$-	2.13	8.62	9.67	−4.88
L-Hydroxyproline	Hyp	-	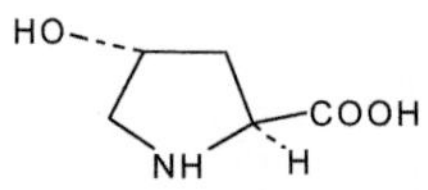 HO-(pyrrolidine-COOH)	1.92	9.73	-	−3.17
L-Isoleucine	Ile	I	$\underset{CH_3CH_2}{\overset{CH_3}{>}}$CH−	2.36	9.68	-	−1.72
L-Leucine	Leu	L	(CH$_3$)$_2$CHCH$_2$-	2.36	9.6		−1.52

TABLE 9.5.1 Common amino acids that are building blocks for proteins, the pK_a's and log D_{ow} at pI values (*Cont'd*)

Name	Abbrev. (three-letter)	Abbrev. (one-letter)	Radical R* $H-\underset{\underset{COOH}{\mid}}{\overset{\overset{NH_2}{\mid}}{C}}-R$ 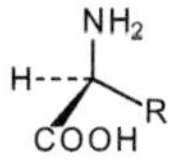	$pK_a\alpha-$ COOH	pK_a $\alpha-NH_3^+$	pK_a side	Experim. log D_{ow} at pI
L-Lysine	Lys	K	$H_2N-(CH_2)_4-$	2.18	8.95	10.79	−3.05
L-Methionine	Met	M	$CH_3SCH_2CH_2-$	2.28	9.21	-	−1.87
L-Phenylalanine	Phe	F	$(C_6H_5)-CH_2-$	1.83	9.13	-	−1.38
L-Proline	Pro	P	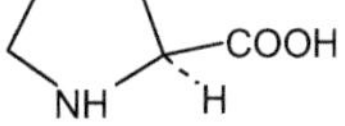	1.99	10.6	-	−2.54
L-Serine	Ser	S	$HO-CH_2-$	2.21	9.15	-	−3.07
L-Threonine	Thr	T	$CH_3-CH(OH)-$	2.63	9.10	-	−2.94
L-Tryptophan	Trp	W	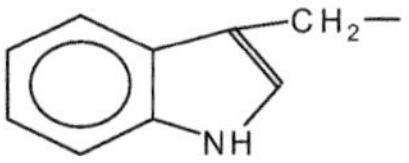	2.38	9.39	-	−1.05
L-Tyrosine	Tyr	Y	$HO-(C_6H_4)-CH_2-$	2.2	9.11	10.07	−2.26
L-Valine	Val	V	$(CH_3)_2CH-$	2.32	9.62	-	−2.26

* Note: Whole formula shown for proline and hydroxyproline.

that larger hydrophobic moieties are attached to their molecule. This is achieved, for example, using a reaction with o-phthalaldehyde (OPA) for amino acids containing a primary amine group and with fluorenylmethyl chloroformate (FMOC) for amino acids containing a secondary amine group. The reactions with these reagents are as follows:

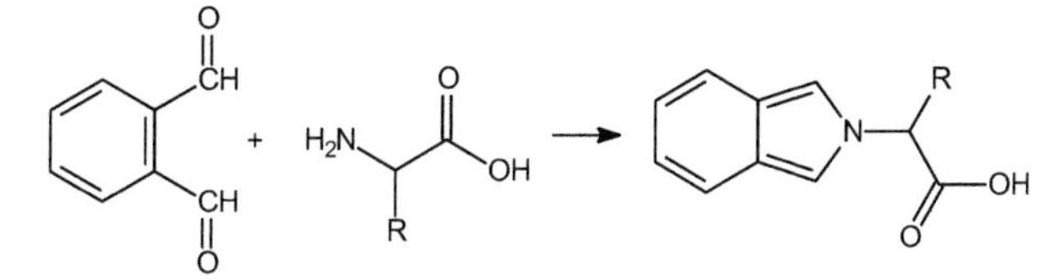

o-phthaldialdehyde (OPA)

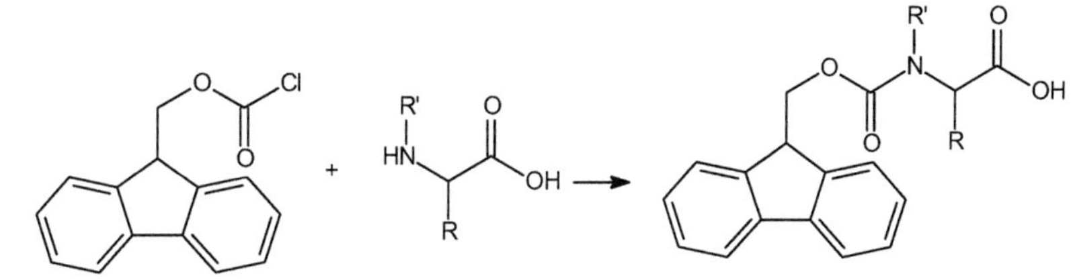

9-fluorenylmethyl chloroformate (FMOC)

The derivatization can be automated online in certain HPLC instruments (e.g., Agilent 1100 or 1200 series). The separation can be done on a Zorbax Eclipse Plus C-18 column using gradient with two solutions: (sol A) 10 mM Na_2HPO_4, 10 mM $Na_2B_4O_7$, 1% tetrahydrofurane in water at pH = 8.17; and (sol. B) methanol/acetonitrile/water 50/30/20 v/v/v. A chromatogram obtained by this procedure is illustrated in Figure 9.5.4 for FLD detection (Ex = 340 nm, Em = 450 nm, and Ex = 266 nm, Em = 305 nm) [63,64]. The use of narrower columns with smaller particles (3 μm) allows for the separation shown in Figure 9.5.4 to be obtained in less than 15 min. The separation can be further accelerated to less than 7 min on a Zorbax Rapid Resolution HT Eclipse Plus C18 column 4.6 × 50 mm with 1.8 μm particles [65]. Other derivatization techniques are also reported in the literature [30,66]. Other

alternatives for amino acid analysis include the use of ion pairing followed by separation on hydrophobic columns (e.g., C8 or C18), use of HILIC, or use of ion-exchange chromatography (IEC).

The connection of amino acids through peptide bonds leads to peptides (typically containing less than 50 amino acids) and proteins. Certain proteins may also contain more than one polypeptide in its structure. RP-HPLC is used successfully for protein and peptide analysis [67]. Several aspects regarding the column and mobile phase selection are, however, specific for this type of analysis. Different proteins may have a wide range of polarities, and their retention on an RP column may vary considerably. Proteins and peptides are usually polar molecules, but they also may exhibit various degrees of hydrophobicity. The most common hydrophobic phases used

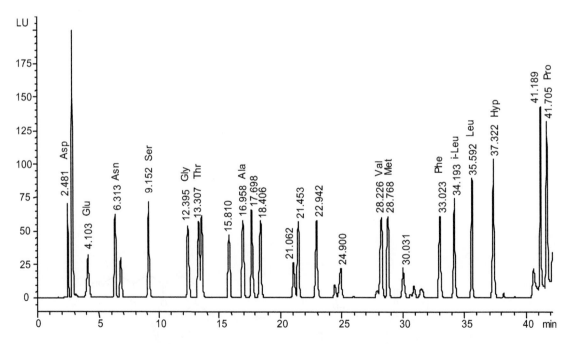

FIGURE 9.5.4 Chromatogram of 21 amino acids after derivatization with OPA and FMOC. The separation was performed on a Zorbax Eclipse Plus C-18, 4.6 × 250 mm, 5 μm column and detection was done using FLD.

for protein separations are C8 and C18. Ranges for column hydrophobicity were described in Table 6.4.9 (based on the column H_c value), and guidance for the selection of a column can be obtained from Tables 6.4.8 and 6.4.9. Columns that may have lower hydrophobicity but additional interactions with polar groups from the protein molecule may be better suited for some separations. Phenyl and cyano columns may show some special selectivity and are used in specific applications where the differences in hydrophobicity between different proteins are not sufficient for their separation. The same is true for columns end-capped with polar groups or not end-capped (see Table 6.4.12). However, some proteins with higher hydrophobicity can be separated more easily using short-chain bonded phases such as C1 to C3 since they may be retained too strongly on C8 or C18 columns.

One particularly important aspect in protein analysis using RP-HPLC is the selection of the pore size of the stationary phase. This parameter is related to the molecular weight of the analyte. The analyte's free access to the bonded phase is very important since retention on the stationary phase depends on the accessible surface area of the packing. For small molecules, such as small peptides, the stationary phase with 8−12 nm pore size is well suited, but for larger molecules with the MW higher than 1,000−2,000 Da, as it is the case for proteins, stationary phases with pore size around 30 nm (or larger) are necessary. As shown in Table 6.4.9, this type of column does not differ significantly regarding hydrophobic and other separation properties from columns with 8−12 nm pore size, but in the case of proteins they offer a better separation.

Since the interaction of large molecules with the hydrophobic bonded phase is probably based mainly on an adsorption process (not on distribution), in protein separation, monomeric-type bonded phases or phases obtained by horizontal polymerization display a similar capacity factor as vertical polymeric bonded phases (see Section 6.2). Monomeric bonded phases and those with horizontal polymerization typically show better reproducibility in protein/peptide separations. For this reason, monomeric phases are preferred to phases obtained with vertical polymerization, although the ones with vertical polymerization may have a larger carbon load.

The diffusion of proteins in the chromatographic column is in general slower than that of small molecules (see rel. 2.2.5 to 2.2.7). When the diffusion coefficients of the analyte in the mobile phase D and in the stationary phase D_s are small, several contributor terms to the plate height (H_C, H_T, H_S) are large, and for this reason the resulting peaks of the proteins may be significantly wider than for small molecules. As shown by rel. 2.2.5 to 2.2.7, a lower flow rate u in the column has the effect of diminishing plate height components H_C, H_T, and H_S, but longer separation times result. For this reason, columns with a high theoretical plate number N (as measured for a test small molecule compound) are preferred for protein separation (although the N value for the protein is much smaller than that obtained with the test compound). Columns with small particles, with core-shell, or even pellicular (non porous) particles, that have high N values can be useful for protein separation. The flow rate is typically maintained at a constant value in the range 0.5 to 2 mL/min.

Another aspect related to column selection involves the working pH of the mobile phase in protein/peptide separation. Mobile phases with low pH (2 to 3.5) may be necessary in protein separation, and stationary phases with good resilience to low pH are very useful. For this reason, and also because they may be easier to obtain with large pores, organic polymer-based columns are successfully used in proteins/peptide separations. The mobile phase in protein separation is typically made from an aqueous buffer of a specific pH (solution A) and an organic modifier (solution B). Separations frequently start with a high percentage of aqueous mobile phases, and typically the level of organic modifier is not necessary to increase to more than 60–70%. However, a higher level of the organic phase may be used for column cleaning.

The buffer or the pH modifiers are usually based on a dilute acid solution (0.05 to 0.1 %). The acids used in the buffer (pH modifier) may include phosphoric, trifluoroacetic (TFA), formic, or acetic. Since the proteins are amphiprotic molecules, the pH of the mobile phase determines the ionization state of the molecule and therefore is very important parameter for the retention (log D_{ow} for the peptide PAFKTLVKAW varies as a function of pH between −4.30 and −12.20). For cases when denaturation of the protein is not a concern, buffers with low pH in the range 2 to 3.5 are frequently used. At low pH, the hydrophobic character of the protein may be enhanced due to suppression of ionization of carboxylic groups. Also, the silanol activity of the column is reduced, and the typical tailing of basic compounds is diminished. The acids used in the mobile phase for proteins and peptides, in particular TFA, may act as an ion pair to the proteins with basic character (pI > 7). The ion pair formed with TFA has a stronger retention (larger log k) than expected for the free compound.

For the case of proteins/peptides that must be separated without denaturing them, pH values of the buffer must be around neutral. For this purpose, compounds such as ammonium acetate, formate, bicarbonate, or trimethylammonium phosphate are used as buffers. Related to protein denaturing, care must be given to the column temperature. Since a decrease in mobile phase viscosity can be achieved by increase in the column temperature, it is common to use temperatures around 50°C for increasing column efficiency. However, higher temperatures of the column may lead to protein denaturing, and room temperatures must be used to avoid this effect.

The organic modifier in protein/peptide separation can be methanol, isopropanol, acetonitrile, or other similar solvents. Acetonitrile is probably the most common organic modifier, since it has lower viscosity and leads to less peak tailing. The variation in log k for proteins and peptides is typically very sensitive to the content of organic modifier in the mobile phase. Based on rel. 9.3.1, $\log k = \log k_w - S \phi$, with S approximated by $S \approx 0.25 \, (M)^{1/2}$, where M is the molecular weight of the analyte. It can be seen from rel. 9.3.1 that the dependence of log k on ϕ (the proportion volume of the organic component) has a larger slope when M is larger. The strength of the solvents follows the rules described in Section 7.1, and in general the value for log K_{ow} is a good indicator for the solvent strength. For example, the strength follows the order: isopropanol > dioxane > ethanol > acetonitrile > methanol > water.

The elution can be performed in isocratic or in gradient modes, with the typical increase in the organic phase content. Specific parameters of the separation must be carefully monitored, in particular during the gradient, including protein solubility and stability in an increased organic phase content. Since ion-pair formation with proteins is also possible when using acids in the mobile phase, the separation may be unexpectedly affected during the increase in the organic phase content and the decrease, for example, in the content of TFA. In such cases, variations in the selectivity may depend on gradient profile. In order to avoid variation in the pH and acid content during the gradient, the organic phase is recommended to have the same content of acid as the aqueous one. For example, a common gradient is obtained using a content of 0.1% TFA in both water and acetonitrile solvents.

In addition to the buffer and the organic modifier, for some protein separations surfactants are added in the mobile phase. These are necessary in cases where the protein solubility in the mobile phase is poor. A variety of surfactants used in protein separations are reported in the literature [68], and they include ionic, nonionic, or zwitterionic surfactants. However, the addition of surfactants may modify the retention properties of the stationary phase, and the addition must be done considering this effect.

Another aspect of protein/peptide separation is related to the recovery and biological activity of the separated compounds. Due to the use in RP-HPLC of eluting solvents that may produce denaturing of the proteins (acids, organic compounds), this technique is not the best for protein separation when the biological activity must be preserved. Also, due to potential irreversible adsorption of proteins on the RP stationary phase, the recovery of the initial sample can be incomplete. Deactivation of sites in the stationary phase producing irreversible adsorption can be achieved, for example, by pretreating the column with a "conditioning" biopolymer such as bovine serum albumin. Even when the recovery is complete and no strong acids or large concentrations of organic modifier are used in the mobile phase, changes in the protein tertiary and quaternary structure may take place during RP separation (protein unfolding). Other HPLC techniques (typically based on SEC) must be used when the separation must preserve bioactivity.

The solvents used for protein/peptide separation must be selected taking into consideration the detection technique. For UV detection, the solvents/buffers must have a cutoff value lower than the wavelength where protein can be detected. For mass spectrometry detection, some solvents or additives such as TFA may decrease the ionization yield. Addition of an excess of HCOOH in the solvent may improve sensitivity for basic proteins (analyzed in negative ionization mode).

Reversed-phase HPLC is sometimes used for protein separation as part of a multidimensional separation in conjunction, for example, with a cation exchange column, with a HILIC column, or with a SEC column, with which RP columns are orthogonal [69]. Typically, the eluent from the cation-exchange column which is aqueous (or mostly aqueous) is further passed into a RP column where initially it is strongly retained (since no organic phase is present and eluted by the addition of the organic component). Such a separation can be achieved discontinuously, using a fraction collector, where fractions from the first column are passed to the second column. Direct-coupled multidimensional liquid chromatography (MDLC) also can be used for separations, and columns consisting of successive segments of two orthogonal stationary phases (cation exchange followed by hydrophobic phase) are reported in the literature [70,71]. Multiple columns and column switching during separation is another technique used in the analysis of complex protein samples. Various arrangements of the orthogonal columns are possible. For example, the proteins are separated first on a cation-exchange column and distributed to one or more RP columns [72]. A different instrumental setup uses several SEC columns for the initial separation followed by RP columns [73].

Analysis of Nucleobases, Nucleosides, Deoxynucleosides, Nucleotides, and Nucleic Acids

Nucleobases adenine, guanine, thymine, uracyl, and cytosine (see Figure 6.5.3) are polar molecules, and their separation by regular RP-HPLC is not typically used. The values for log K_{ow} for these molecules are given in Table 9.5.2. Uracyl, for example, is even used in many RP-HPLC separations to determine the dead time t_0 of the column. Nucleosides that

TABLE 9.5.2 Values for log K_{ow} for nucleobases, nucleosides, and deoxynucleosides.

Nucleobase	Nucleoside	Deoxynucleoside
Adenine log $K_{ow} = -0.09$	Adenosine log $K_{ow} = -1.05$	Deoxyadenosine log $K_{ow} = -0.55$
Guanine log $K_{ow} = -0.91$	Guanosine log $K_{ow} = -1.90$	Deoxyguanosine log $K_{ow} = -1.30$
Thymine log $K_{ow} = -0.62$	5-Methyluridine log $K_{ow} = -1.60$	Thymidine log $K_{ow} = -0.93$
Uracyl log $K_{ow} = -1.07$	Uridine log $K_{ow} = -1.98$	Deoxyuridine log $K_{ow} = -1.51$
Cytosine log $K_{ow} = -1.73$	Cytidine log $K_{ow} = -2.51$	Deoxycytidine log $K_{ow} = -1.77$

have a deoxyribose sugar linked via a beta-glucosidic linkage to one nitrogen in the nucleobase have even lower log K_{ow} values. The same is true for 2-deoxynucleosides, and although they have slightly higher log K_{ow} values as compared to the nucleosides, they still have negative values for log K_{ow}.

Nucleotides have a phosphate group bonded to the 2, 3, or 5 carbon of the sugar in the molecule of nucleoside or deoxynucleoside (to form deoxynucleotides). The phosphate can be a monophosphate, diphosphate, or triphosphate. These phosphate groups bring ionic character to the molecule, and the separation by RP-HPLC of these molecules is not possible. Nucleic acids are highly polar polymers, and their separation is not possible by RP-HPLC. The small compounds from this class are typically analyzed using different HPLC techniques such as HILIC or IEC. Nucleic acids can also be separated using SEC-type techniques.

9.6. SEPARATIONS BY ION-PAIR CHROMATOGRAPHY

General Comments

Various aspects related to ion-pair chromatography were presented previously. The mechanism was discussed in Section 5.2, and details about the mobile phases used for ion pairing were given in Section 7.7. Ion pairing is a common separation technique for certain important classes of molecules, including numerous organic acids, bases, and zwitterionic compounds that can generate ions. The technique takes advantage of the exceptional versatility of RP-HPLC and adjusts the hydrophobic character of polar or ionic molecules such that they are amenable for separation on hydrophobic columns. Numerous applications of this technique are present in the literature. However, ion-pair chromatography may pose some problems since it uses a more complicated mobile phase and typically shows slower column equilibration.

One example of a separation on a C18 column using as IP agent sodium octanesulfonate is given in Figure 9.6.1. The figure shows the chromatogram for several guanidines and biguanidines. The analyzed compounds are cyanoguandine (log $K_{ow} = -0.70$), melamine (log $K_{ow} = -0.59$), 4,6-diamino-1,3,5-triazine-2-yl guanidine (log $K_{ow} = -0.79$), N,N-dimethyl melamine (log $K_{ow} = 0.34$), 1,1-dimethylbiguandine (metformin) (log $K_{ow} = -0.64$), and 1-methyl-biguandine (log $K_{ow} = -0.90$). In spite of the strong hydrophilic character of the separated compounds (some with negative log K_{ow}), the separation on a hydrophobic column in the presence of sodium octanesulfonate in the mobile phase is very good (biguanidines are likely to form double ion pairs with the ion-pairing agent).

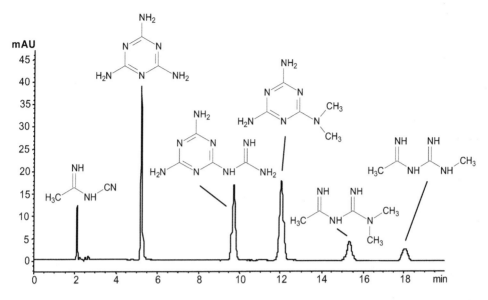

FIGURE 9.6.1 Chromatogram of several guanidines and biguanidines using IP separation. (Column: Inertsil ODS 250 × 4.6 × 5 μm; mobile phase: 20% acetonitrile; 80% phosphate aqueous buffer with pH = 2.5 containing 10 mM $C_8H_{17}SO_3Na$; 30°C; flow-rate = 1.2 mL/min; detection at 218 nm) [74].

Ion pairing has been utilized successfully for the separation of amino acids without derivatization. As an example, a good separation can be achieved for several amino acids in the presence of C_2F_5COOH as IPA on a polar end-capped Synergi 4u Hydro-RP 80A column (Phenomenex, Torrance CA 90501, USA) (see Chapter 6). This column offers additional retention for polar compounds. The separation is shown in Figure 9.6.2. The mobile phase used gradient with solution A: 98% water + 2 % methanol with 0.2% C_2F_5COOH (pentafluoropropanoic acid) and 0.1% HCOOH, and solution B: 50% water + 50% methanol with 0.1% HCOOH. Initial condition was 60% A, hold 0.5 min, to 0% A at 15 min hold for 3 min, followed by resetting and equilibration for 3 min. The flow rate was 1 mL/min, and detection was done using an API-5000 MS/MS system (AB Sciex). The amino acids were standards at concentrations of around 6 nmol/mL and the injection volume was 3 μL. As shown in Figure 9.6.2, some amino acids were not well separated, and only 15 peaks

can be visually distinguished in the chromatogram, but the difference in the ions used for MS/MS detection still allows individual identification and measurement avoiding any interference. The large difference in k_j (implicitly in retention times) between different amino acids is caused by the difference in their structure. Both the increase in the content of organic phase used during the gradient and the decrease in IPA concentration act toward decreasing k_i, and much longer (unusable) retention times of the late eluting peaks would be noticed in isocratic conditions. Instead of pentafluoropropanoic acid as IPA, other fluorinated organic acids were reported in the literature, such as nonafluoropentanoic acid with separation on a graphitic column [75], nonafluoropentanoic acid and trifluoroacetic acid with separation on a Dikama Diamonsil C18 column [76], and tridecafluoroheptanoic acid with separation on an Acquity UPLC BEH C18 column [77].

Ion-pair chromatography can be used for the separation of analytes following a derivatization

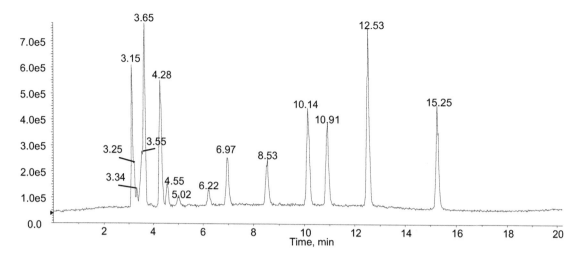

FIGURE 9.6.2 Chromatogram of 24 amino acids (only 15 peaks separated) using pentafluoropropanoic acid as IPA. The amino acids are: hydroxyproline 3.15 min, asparagine 3.19 min, serine 3.23 min, glycine 3.25 min, aspartic acid 3.34 min, β-alanine 3.45 min, threonine 3.47 min, sarcosine 3.52 min, glutamic acid 3.55 min, alanine 3.6 min, proline 3.65 min, glutamine 3.82 min, cysteine 3.88 min, histidine 4.28 min, ornithine 4.32 min, lysine 4.55 min, arginine 5.02 min, valine 6.22 min, methionine 6.97 min, tyrosine 8.53 min, isoleucine 10.14 min, leucine 10.91 min, phenylalanine 12.53 min, tryptophan 15.25 min.

step. Such derivatization can be useful, for example, for quantitation purposes. The use of isotopically labeled standards for each amino acid necessary for quantitation using MS detection is difficult since it requires a considerable number of labeled amino acids. Alternative procedures (isobaric tagging

isotope for the standards. For the amino acids, a common such reagent is 2-(4-methylpiperazine)acetic acid N-hydroxysuccinimide ester (NHS) or 2,5-dioxopyrrolidin-1-yl-2-(4-methylpiperazin-1-yl)acetate. The reaction with an amino acid of this reagent is as follows:

$$\text{(9.6.2)}$$

reagent iTRAQ and aTRAQ) use a derivatization with a labeled reagent for the sample and with the same reagent containing no

The NHS reagent is available unlabeled (MW = 255, Δ = 0) and also as $^{13}C_6$, $^{15}N_2$ labeled on the methylpiperazine group

(MW $= 263$, $\Delta = 8$). The derivatization adds 112 amu for the internal standards amino acids and 121 amu for the sample amino acids. This difference in the molecular weight can be easily detected with MS instrumentation. The derivatized amino acids can be separated on an AB Sciex C18 column using heptafluorobutyric acid as API, and methanol/water with 0.1 formic acid as a mobile phase [78,79].

9.7. SEPARATIONS BY HILIC AND NPC

General Comments

Molecules with high polarity, having negative log K_{ow}, are well separated on polar stationary phases. The technique that is very well suited for such separations is HILIC. Considerable progress was made in HILIC separations when the number of available stationary phases started to grow (see Section 6.5). The main advantage of HILIC in comparison with NPC is the use of water in the mobile phase, which allows the mobile phase to reach a wide range of polarities (even if they must remain at lower polarity than that of stationary phase) and also offers excellent solubility for a large number of polar compounds (see Section 7.8). A general overview regarding the use of column and of mobile phase in HILIC and NPC based on log K_{ow} of the analyte is schematically shown in Figure 9.7.1.

Analysis of Small Polar Molecules Using HILIC

A considerable number of small molecules are polar and have negative log K_{ow} values. Such molecules include many classes of organic compounds such as organic acids, carbohydrates, and amino acids, to list only three important groups of simple molecules. Since solubility in water is associated with polar groups, and a large number of pharmaceuticals are water-soluble molecules, HILIC is an important technique used for analysis of pharmaceuticals. In the case of MS detection, the presence of water and of an organic solvent in the mobile phase facilitates the ionization process. In case of MS with electrospray ionization (ESI), this represents an advantage compared to NPC where no water is present in the mobile phase and ESI is difficult or not possible to use. One example of the use of a HILIC column with MS detection is shown in Figure 9.7.2 for the separation of sucralose, quinic acid (1,3,4,5-tetrahydroxycyclohexan-carboxylic acid), and inositol (1,2,3,4,5,6-cyclohexanehexol) from a tobacco extract. A TSK gel amide-80 4.6 × 150 mm, 3 mm column was used for the separation, the mobile phase being

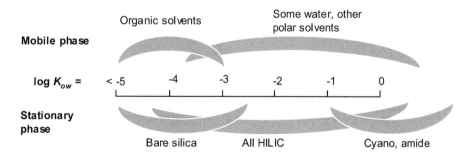

FIGURE 9.7.1 Overview of columns and mobile phases used in RP-HPLC based on log K_{ow} of the analyte.

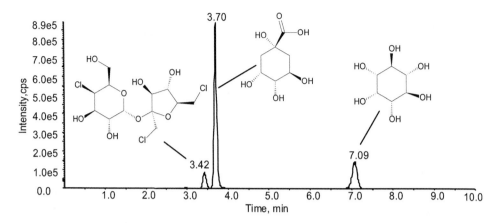

FIGURE 9.7.2 Chromatogram showing the peaks for sucralose (added), quinic acid, and myo-inositol (cis-1,2,3,5-trans-4,6-cyclohexanehexol) from a tobacco sample on a HILIC column with MS/MS detection in negative ionization mode.

55/45 acetonitrile/aqueous solution 50 mM CH_3COONH_4. The detection was done by ESI MS/MS in negative ionization mode [80].

Analysis of Small Sugar Molecules, Oligo, and Polysaccharides

HILIC offers an excellent way of separating small sugar molecules. An example of an analysis of small carbohydrates done on a HILIC column is shown in Figure 9.7.3 for fructose, glucose, and sucrose from a tobacco extract. The separation was done on a YMS-Pack Polyamine II 250 x 4.6 mm column with 5 mm particles, 120 Å pore size. The mobile phase consisted of a solution 70% acetonitrile in water (v/v) with a flow rate of 1 ml/min. The detection was done using RI.

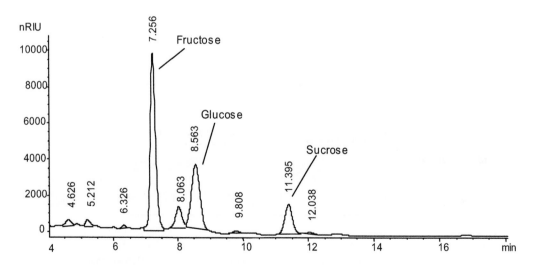

FIGURE 9.7.3 Chromatogram for a tobacco extract showing the separation of fructose, glucose and sucrose on an YMS-Pack Polyamine II column.

Oligosaccharides and larger polysaccharides are also very well separated on HILIC-type columns [81,82]. As an example, acidic, basic, neutral, and higher molecular weight oligosaccharides with degrees of polymerization DP10 up to DP 55 can be separated on a Nucleodex β-OH column.

Analysis of Amino Acids, Peptides, and Proteins

Amino acids were successfully analyzed using separation on a HILIC column. For example, excellent separation can be achieved on a ZIC-HILIC 2.1 × 250 mm, 5 μm particles column, with the gradient mobile phase consisting of water/acetonitrile and 10 mM acetic acid [83]. Besides amino acids, peptides and proteins can be separated very successfully on HILIC columns. Various types of HILIC columns such as bare silica, amide, amine, diol and polyhydroxyethyl aspartamide have been used for this purpose. Examples of separations include those on TSKgel Amide-80, with the mobile phase a mixture of acetonitrile (MeCN)–water containing 0.1% trifluoroacetic acid (TFA) [84], on Atlantis silica HILIC, BEH silica HILIC, BEH amide glycan [85–87]. More complex proteins such as glycoproteins are also separated using HILIC columns [88,89].

Separations Using the eHILIC (ERLIC) Technique

A number of successful applications of HILIC separations on ion-exchange columns are reported in the literature [90]. This technique can be practiced on cation-exchange columns or anion-exchange columns. The technique has been used, in particular, for the separation of basic and strongly acidic peptides in a mixture. For example, phosphopeptides are acidic compounds, while other peptides are basic, and their separation is difficult using typical ion-exchange or HILIC separations. An ion-exchange column with a mobile phase containing acetonitrile allows very good separations.

The Use of NPC Technique

The use of normal-phase chromatography should be in principle as valuable as that of HILIC. The capability of silica to separate isomers that are difficult to separate by other HPLC techniques makes NPC irreplaceable in some cases. However, problems with NPC reproducibility due to the different coverage of bare silica with water, slow column equilibration also related to water adsorption on the silica surface, solvent demixing with use of gradients, and peak tailing caused by various types of OH groups on the silica surface, some from silanol and some from adsorbed water, rendered this technique less attractive. Some of these problems were diminished, for example by keeping a constant percentage of water in the mobile phase, or by conditioning the stationary phase with methanol instead of water and maintaining a small amount of methanol in the mobile phase.

9.8. SEPARATIONS BY ION-EXCHANGE CHROMATOGRAPHY

General Comments

Ion-exchange chromatography is utilized for the separation and analysis of small ionic molecules such as inorganic anions and cations, for analysis of a variety of ionic organic molecules such as organic acids and amino acids, for analysis of some neutral molecules such as carbohydrates and alcohols, as well as for analysis of polymeric ionic molecules such as proteins and nucleic acids. A detailed discussion on stationary phases used in ion chromatography was given in Section 6.6, and a discussion on mobile phases was given in Section 7.9. Only general comments about the applicability of this HPLC technique are given in this section.

Separation of Small Ions by IC

Small inorganic ions are successfully separated using ion chromatography [91]. For example, Na^+, NH_4^+, K^+, Mg^{2+}, and Ca^{2+} can be easily separated and determined using a Ion-Pac CS16 column (carboxylic acid type) with the analytes eluted isocratically with methanesulfonic acid (MSA) that is electrochemically generated at a concentration of 30 mM (see Section 7.9). Similarly, anions such as F^-, Cl^-, NO_2^-, SO_4^{2-}, Br^-, NO_3^-, PO_4^{3-} can be separated on strong anion-exchange columns, such as IonPac AS18 using as eluent a 33 mM KOH solution in isocratic conditions at 1 mL/min [91]. A similar analysis of small anions on an IC column is shown in Figure 9.8.1. The separation was performed on a SI-90-4E column from Shodex, 250×4.0 mm that is a PVA gel-NR_4^+. The eluent used was a solution of 3.2 mM Na_2CO_3/$NaHCO_3$. The flow rate of the mobile phase was 0.7 mL/min. The separation was achieved in 25 minutes, and the detection was done using a conductivity detector (with suppression). A considerable number of other separations of small inorganic ions are reported in the literature, with similar types of columns [19].

Separation of Ionic Organic Molecules

Many organic ionic molecules can be separated successfully using ion-exchange chromatography. These molecules include among others organic acids, amines, and amino acids. One example of a separation for a standard mixture of 17 amino acids is shown in Figure 9.8.2. The separation was achieved using a Shim-pack Amino Na 100 x 6.0 mm analytical column and a Shim-pack ISC 38 x 4.6 precolumn using gradient elution (Solution A sodium citrate buffer at pH = 3.15, Solution B sodium acetate at pH 7.40, and Solution C, NaOH solution at pH = 12.4) [92,93]. The detection was done using post-column derivatization with NaOCl, *o*phthalaldehyde (OPA), and N-acetyl-L-cysteine in carbonate buffer. The sample consisted of a mixture of amino acids, each at 0.1 µM/mL with injection volume of 10 µL, and the detection was done using an FLD, with Ex = 350 nm and Em = 450 nm. Similar analyses were done using post-column derivatization with ninhidrin and UV detection (at 440 nm and 570 nm) [94]. Numerous methods using ion chromatography for a variety of ionic molecules are reported in the literature (see, e.g., [95]).

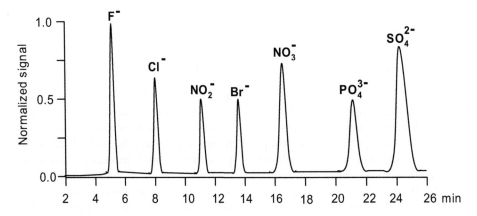

FIGURE 9.8.1 Chromatogram of a standard mixture of F^-, Cl^-, NO_2^- all 5 µg/mL and Br^-, NO_3^-, PO_4^{3-}, and SO_4^{2-} all 10 µg/mL, on a SI-90-4E from Shodex, 250×4.0 mm, elution with 3.2 mM Na_2CO_3/$NaHCO_3$ solution in water (conductivity detector).

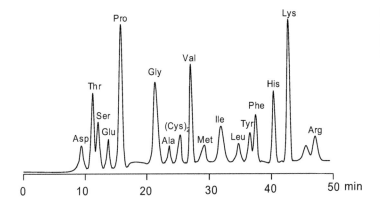

FIGURE 9.8.2 Chromatogram of a standard mixture of 17 amino acids using a Shimpack Amino Na column and gradient elution, with fluorescence detection.

Separation of Neutral Organic Molecules

Organic molecules that are considered neutral, such as carbohydrates and alcohols, can be successfully separated by IC. Carbohydrates have a very weak acidic character (with $pK_a >$ 12), and using a basic mobile phase, it is possible to separate carbohydrates on an anion-exchange column. For quantitation, pulsed amperometric detection is usually employed. An example of such separation on a CarboPac PA20 capillary column 150×0.4 mm is given in Figure 9.8.3. The separation was done in isocratic conditions with eluent 10 mM KOH solution at 10 μL/min and PAD detection for a set of standards with 10 mM concentration and an injection of 0.4 μL.

A common procedure for the separation of carbohydrates and organic acids is the use of ion-moderated stationary phases (see Section 6.6). Specific columns are recommended for different types of analytes from this class. For example, the resins in Ca^{2+} form are recommended for monosaccharides and sugar alcohols, in H^+ form for organic acids, in Ag^+ form for oligosaccharides separation. Temperatures at 60 to 80 °C are typically used in these separations.

Separation of Proteins and Nucleic Acids

Ion chromatography has been used successfully for protein and nucleic acids separations.

For proteins, since they can be either positively or negatively charged depending on the pH of the mobile phase, both cation- and anion-exchange columns have been used for separation [96]. Polymer-based columns are in particular useful for such separations, and they have been used for both analytical and preparative purposes. In principle, a protein will be positively charged at pH lower than isoelectric point (pI) and negatively charged at pH higher than pI. However, in practice the values of the pH of the mobile phase bracketing the pI value at 1.5–2 pH units are avoided. Some proteins show anomalous behavior around pI due to independent dissociation of different groups at different parts of the molecule. For this reason cation exchangers are often used in mobile phases with 1.5–2 pH units lower than pI, and anion exchangers in mobile phases more basic with 1.5 −2 units than pI. Both strong and weak cation- and anion-exchange columns have been used for the separation. Similar to the case of RP columns, large pore materials are typically used in protein separations. Separation of proteins on ion-exchange columns is also done using the chromatofocusing technique (see Section 7.9).

Nucleic acid, and in particular oligonucleotides, were successfully separated using ion-exchange columns. Due to the phosphate groups in their molecules, these compounds

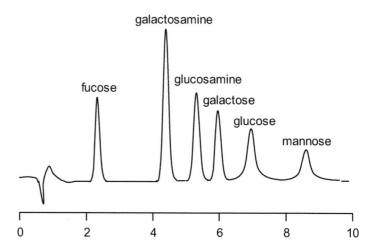

have an ionic acidic character. Similar to the case of proteins, for the separation of larger nucleic acid molecules, stationary phases with large pores must be utilized.

9.9. CHIRAL SEPARATIONS

General Comments

Various mechanisms involved in chiral separations were discussed in Section 5.5, chiral columns were presented in Section 6.7, and mobile phases for chiral separations were described in Section 7.10. Chiral separations are especially necessary in the pharmaceutical field, since biological activity is frequently different between enantiomers. Various procedures are used for the separation, a common one being derivatization of the enantiomers with a chiral reagent that will change the enantiomers into diastereoisomers (see, e.g., [30]), which can be separated by nonchiral chromatographic procedures. The derivatization approach avoids the use of expensive columns with chiral stationary phases, and the use of common stationary phases is often simpler.

However, derivatization represents an additional step in HPLC analysis, requires the presence of derivatizable groups in the analyte, and may involve undesirable side reactions, formation of decomposition products, and even potential racemization of the enetiomers. Also the procedure must be performed with derivatization reagents of high enantiomeric purity.

The enantiomer separation approach using columns with chiral stationary phases is very common and is also applicable for separations on preparative scale. However, it requires special expensive columns that may be utilized only to a limited types of compounds. Other difficulties in the use of chiral stationary phases result from the fact that most chiral columns are designed for addressing specifically the separation of the enantiomers and may not separate well other matrix components. For this reason, it is frequently necessary to use the chiral columns on samples with a simple matrix, containing besides the enantiomers only a few other components. This may require different analysis steps before enantiomer separation, which would produce the enantiomer mixture in a simpler matrix. A large volume of information is available in the literature regarding many particular chromatographic chiral separations (see, e.g., [97–101]).

The chiral mobile phase approach for enantiomer separation represents a simple and flexible alternative. However, this approach is not always applicable. Since the mobile phase containing the chiral selector cannot be reused, this technique cannot be applied using expensive chiral additives in the mobile phase, and it is not used for preparative purposes [102].

One example of a simple separation is shown in Figure 9.9.1 for (+) and (−) anatabine. The separation was performed on a Chirex (S)-VAL and (R)-NEA column, 250 x 4.6 mm, with the mobile phase (isocratic) 80% hexane, 10% methanol, 10% i-propanol, and a flow rate of 1.5 mL/min. The solution contained 140 μg/mL enantiomers mixture, and the injection was 5 μL. The measurement was done using absorption in UV at 260 nm. This example also illustrates the use of a weak-polar mobile phase: it is known that water in the mobile phase for chiral separations is not a good additive since it interacts strongly in different ways with the analytes and the stationary phase (through polar interactions and hydrogen bonds) and degrades the differentiation between enantiomers.

Separation of enantiomers in biological matrices is particularly challenging. Even after a matrix is simplified, for example, by protein precipitation, liquid-liquid extraction, or solid-phase extraction, one or more matrix components may remain in the processed sample and may co-elute with the enantiomers [103]. Some of these interferences were resolved in various studies using selective detection such as MS or MS/MS. Also, bidimensional LC separations with two online columns (with or without heart-cut) can be applied in more complicated analyses (see, e.g., Figure 1.4.7). For such analysis, the first column is used to separate the racemic from the other matrix components, and the chiral phase is used for separation of the enantiomers. Even in this type of separation, restrictions are applied to the mobile phase that must be acceptable for both separations (some chiral columns are not usable in phases with high water content).

One example of a bidimensional separation with heart-cut is that of (+) and (−) carvedilol in plasma samples after protein precipitation with acetonitrile. For this separation, the first column was a monolithic Chromolith Performance RP-18e column (Merck), and the second column was a cellulose tris-(3,5-dimethylphenylcarbamate). The mobile phase in the

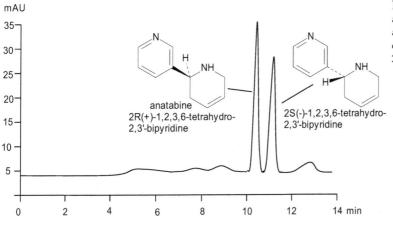

FIGURE 9.9.1 Separation of (+) anatabine and its (−) enantiomer on a Chirex (S)-VAL and (R)-NEA column with UV measurement at 260 nm.

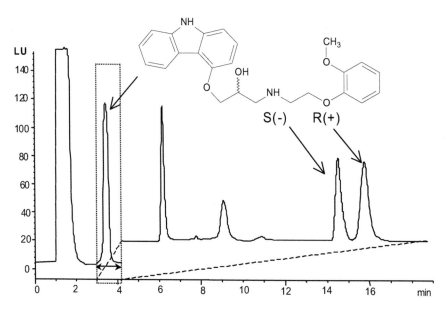

FIGURE 9.9.2 Bidimensional FLD-HPLC chromatogram of a spiked plasma samples containing 150 ng/mL racemic mixture of carvedilol. Enantiomer separation was done on a cellulose tris-(3,5-dimethyl-phenylcarbamate) stationary phase. First part of the figure shows the chromatogram resulting from the first column, and the second part shows the separation for carvedilol enantiomers.

first separation (between 0 and 4.2 min) on monolithic column was 35% acetonitrile and 65% (v/v) aqueous buffer containing 50 mM CH_3COONa at pH = 6. The second separation (after 4.2 min) was performed with a mobile phase containing 90% acetonitrile and 10% buffer with 50 mM CH_3COONa at pH = 6. The flow-rate of the first separation was set to 2 mL/min, and for the second separation it was set to 1 mL/min. Transfer of the heart-cut from the achiral column to the chiral one and the simultaneous fluorescence detection over both HPLC dimensions during a single chromatographic run were possible by using two high-pressure six-port switching valves and a 2 mL loop for storing the sample after passing from the first column and the detector. The resulting chromatogram for a spiked plasma sample with a racemic mixture of carvedilol is shown in Figure 9.9.2 [104]. The first part of the figure shows the detector response for the

efluent coming from the first column, a heart-cut being stored in the 2 mL loop. The portion of chromatogram after 4 min indicated the detector response for the eluent coming from the second column with the carvedilol enantiomers separated.

9.10. SEPARATIONS BY SIZE-EXCLUSION CHROMATOGRAPHY

General Comments

The applications of size-exclusion chromatography include: (1) separation of small molecules from polymeric molecules, (2) separation and purification of polymeric molecules, and (3) evaluation of molecular weight of polymers. As previously indicated, depending on their solubility in aqueous/polar solvents or in organic nonpolar solvents, size-exclusion chromatography is differentiated in aqueous or GFC and nonaqueous or GPC, respectively. A considerable number of applications, in particular for polymeric molecules such as synthetic macromolecules, natural polymers (e.g., cellulose, starch, lignins, proteins, nucleic acids, etc.) and even viruses, are reported in the literature [105−107].

Specific problems associated with the analytes such as solubility, denaturation, and recovery from the HPLC column, are discussed in many applications. For example, avoidance of molecular degradation in SEC is an important issue especially for ultra-high molar mass polymers (MW > 5000 kDa) and those with biological activity [108]. Molecular weight determination of polymers is also a complex issue, the SEC measurements being sometimes corroborated with other MW determination methods [109]. A particular problem in the use of SEC for the measurement of MW is related to the dimensions of the pores and pore-size distribution of the phase. As previously indicated (see Section 6.8), columns with a narrow pore-size distribution allow a more precise measurement of MW of analytes, but they can be applied only in a narrow range of MW values. On the other hand, columns with a wide pore-size distribution can be used in a wider MW range but do not provide such accurate MW measurements. In order to achieve both characterics, SEC can be conducted using two to four columns of large dimensions (7.8 mm I.D. × 300 mm) connected in series. The range of utilization of SEC columns is also related to the nature of the analytes that are separated, and the same column can be used within different MW ranges depending on the nature of the analyte molecule. For example, the recommended use of Toyopearl columns [110] (Tosoh Bioscience) for different MW ranges and for different types of analytes is illustrated in Figure 9.10.1.

The use of large column banks (two to four columns in the order of decreasing pore size) achieves high resolution and accurate MW measurement. However, it requires long analysis time and significant solvent consumption. In recent years, increasing interest has been shown in developing high-speed SEC using a single column of small dimensions. High-speed SEC improves sample throughput

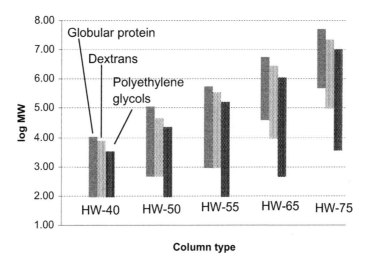

FIGURE 9.10.1 Utilization ranges (log MW) for different Toyopearl columns (Tosoh Bioscience) for different types of analytes.

and also reduces solvent usage. Moreover, small SEC columns allow the use of low flow rates, which is very important for online SEC mass spectrometry applications [111].

Utilization of ultra-high-pressure liquid chromatography (UPLC) in SEC is another attractive alternative to solve the problem of large solvent volumes and long analysis time. However, implementation of UPLC for analysis of macromolecules poses several challenges, including: (1) development of packing materials with large pore diameters and pore volumes that are mechanically stable at ultra high pressure, (2) avoidance of high shear stress generated by the flow of polymer solutions, which may affect the conformation of the polymer chains, and (3) assurance of proper diffusion of the polymers in the pore of the stationary phase during the separation. Effort is continually made in solving such problems [112].

Examples of Applications of SEC

From the multitude of applications of SEC, only two examples are given here. One such example is for an aqueous SEC (GFC) with the separation of glucose, maltose (DP2), maltotriose (DP3), maltotetrose (DP4), and maltopentose (DP5), which can be applied for the quantitation in hydrolyzed starch. The separation is shown for standards and was performed on two columns in series SB401-4E (from Shodex) 250 × 4.6 mm, 10 μm particles. The mobile phase for the separation was water at a flow rate of 0.1 mL/min with a refractive index detector. Figure 9.10.2 shows the chromatogram and the plot of MW versus the retention time. (The dependence of retention time of different carbohydrates versus MW is not shown in the typical logarithmic form that is expected to give a linear dependence).

Another example is for a nonaqueous SEC (GPC) application that allows the separation of a set of polystyrenes of different MW. The separation is shown in Figure 9.10.3 and was obtained on three Phenogel 10 μm GPC columns, 10^5, 10^4, and 10^3 Å, 300 × 7.8 mm, with tetrahydrofuran (THF) mobile phase, at a flow rate of 1 mL/min. The detection was done using RI. Injection volume was 100 μL of a solution 1% of seven standards of polystyrene. The plot of log MW as a function of retention time is also shown

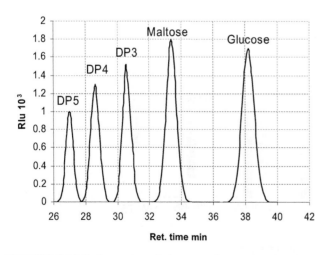
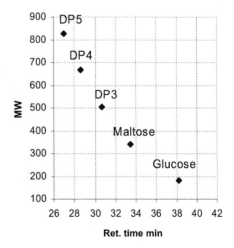

FIGURE 9.10.2 Separation of glucose and several maltopolyoses (DP2 to DP5) on two SB401-4E columns in series. Column dimensions were 250 × 4.6 mm, 10 μm particles, and the mobile phase was water at 0.1 mL/min. The plot of MW as a function of retention time is also shown.

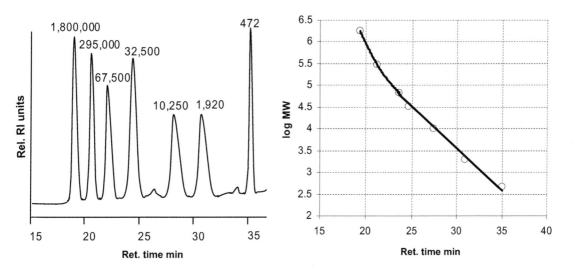

FIGURE 9.10.3 GPC separation of seven polystyrene standards (MW indicated above each peak) on three Phenogel 10^5, 10^4, and 10^3 Å, 300 × 7.8 mm columns. The plot of log MW as a function of retention time is also shown.

in the figure. As shown in Figure 9.10.3, the plot of log MW as a function of retention time is not perfectly linear, but after a calibration with standards, the measurement of MW can be done with good accuracy. However, for correct results it is necessary that the standards used for calibration and the analyte macromolecules for which the MW is measured share a similar chemical nature. When the standards and the analytes have different chemical natures, differences in the molecular shape or different enthalpic interactions with the stationary phase may affect the elution and therefore the accuracy of the MW measurement.

References

[1] McCalley DV. The challenges of the analysis of basic compounds by high performance liquid chromatography: Some possible approaches for improved separations. J. Chromatogr. A 2010;1217:858–80.

[2] Guiochon G, Felinger A, Katti AM, Shirazi DG. Fundamentals of Preparative and Nonlinear Chromatography. 2nd ed. Amsterdam: Elsevier; 2006.

[3] Wellings D. A Practical Handbook of Preparative HPLC. Amsterdam: Elsevier; 2006.

[4] Holm A, Molander P, Lundanes E, Greibrokk T. Determination of the pesticide rotenone in river water utilizing large-volume injection packed capillary column switching liquid chromatography with UV and time-of-flight mass spectrometric detection. J. Chromatogr. A 2003;983:43–50.

[5] Hogenboom AC, Hofman MP, Kok SJ, Niessen WM, Brinkman UT. Determination of pesticides in vegetables using large-volume injection column liquid chromatography-electrospray tandem mass spectrometry. J. Chromatogr. A 2000;892:379–90.

[6] Snyder LR, Kirkland JJ, Dolan JW. Introduction to Modern Liquid Chromatography. 3rd ed. Hoboken, NJ: John Wiley; 2010.

[7] Lunn G, Hellwig LC. Handbook of Derivatization Reactions for HPLC. New York: John Wiley; 1998.

[8] Chromatography Product Guide, 11/12 Torrance: Phenomenex; 2011.

[9] Layne J, Farcas T, Rustamov I, Ahmed F. Volume-load capacity in fast-gradient liquid chromatography. Effect of sample solvent composition and injection volume on chromatographic performance. J. Chromatogr. A 2001;913:233–42.

[10] Kazakevich Y, LoBrutto R. Stationary phases, in HPLC for Pharmaceutical Scientists. In: Kazakevich Y, LoBrutto R, editors. Hoboken, NJ: John Wiley; 2007. p. 124.

[11] Keunchkarian S, Reta M, Romero L, Castells C. Effect of sample solvent on the chromatographic peak shape of analytes eluted under reversed-phase liquid

chromatographic conditions. J. Chromatogr. A 2006; 1119:20–8.

[12] Ruta J, Rudaz S, McCalley DV, Veuthey J-L, Guillarme D. A systematic investigation of the effect of sample diluent on peak shape in hydrophilic interaction liquid chromatography. J. Chromatogr. A 2010;1217:8230–40.

[13] Udrescu S, Sora ID, Albu F, David V, Medvedovici A. Large volume injection of 1-octanol as sample diluent in reversed phase liquid chromatography: application in bioanalysis for assaying of indapamide in whole blood. J. Pharm. Biomed. Anal. 2011; 54:1163–72.

[14] Udrescu S, Medvedovici A, David V. Effect of large volume injection of hydrophobic solvents on the retention of less hydrophobic pharmaceutical solutes in RP-LC. J. Sep. Sci. 2008;31:2939–45.

[15] Loesser E, Babiak S, Drumm P. Water-immiscible solvents as diluents in reversed-phase liquid chromatography. J. Chromatogr. A 2009;1216:3409–12.

[16] Jandera P. Stationary and mobile phases in hydrophilic interaction chromatography: A review. Anal. Chim. Acta 2011;692:1–25.

[17] Johnson JR, Karlsson D, Dalene M, Skarping G. Determination of aromatic amines in aqueous extracts of polyurethane foam using hydrophilic interaction liquid chromatography and mass spectrometry. Anal. Chim. Acta 2010;678:117–23.

[18] Kromidas S. HPLC Made to Measure. A Practical Handbook for Optimization. Weinheim: Wiley-VCH; 2006. 395.

[19] Fritz JS, Gjerde DT. Ion Chromatography. 4th ed. Weinheim: Wiley-VCH; 2009.

[20] Torres-Lapasio JR, Garcia-Alvarez-Coque MC, Rosés M, Bosch E. Prediction of the retention in reversed-phase liquid chromatography using solute-mobile phase-stationary phase polarity parameters. J. Chromatogr. A 2002;955:19–34.

[21] Vitha M, Carr PW. The chemical interpretation and practice of linear solvation energy relationship in chromatography. J. Chromatogr. A 2006;1126: 143–94.

[22] Sándi Á, Szepesy L. Evaluation and modulation of selectivity in reversed-phase high-performance liquid chromatography. J. Chromatogr. A 1999;845:113–31.

[23] Torres-Lapasio JR, Rosés M, Bosch E, Garcia-Alvarez-Coque MC. Interpretative optimization strategy applied to the isocratic separation of phenols by reversed-phase liquid chromatography with acetonitrile-water and methanol-water mobile phases. J. Chromatogr. A 2000;996:31–46.

[24] Schoenmakers PJ, Billiet HAH, De Galan L. Influence of organic modifiers on the retention behaviour in reversed-phase liquid chromatography and its consequences for gradient elution. J. Chromatogr. 1979;185:179–95.

[25] Dolan JW, Gant JR, Snyder LR. Gradient elution in high-performance liquid chromatography. II. Practical application to reversed-phase systems. J. Chromatogr. 1979;165:31–58.

[26] El Tayar N, van de Waterbeemd H, Testa B. , The prediction of substituent interactions in the lipophilicity of disubstituted benzenes using RP-HPLC. Quant. Struct.-Act. Relat. 1985;4:69–77.

[27] Kaliszan R. Quantitative Structure-Chromatographic Retention Relationship. New York: John Wiley; 1987.

[28] David V, Galaon T, Caiali E, Medvedovici A. Competitional hydrophobicity driven separations under RP-LC mechanism: application to sulphonylurea congeners. J. Sep. Sci. 2009;32:3099–106.

[29] Valk K, Bevan C, Reynolds D. Chromatographic hydrophobicity index by fast gradient RP-HPLC: A high throughput alternative to logP/logD. Anal. Chem. 1997;69:2022–9.

[30] Moldoveanu SC, David V. Sample Preparation in Chromatography. Amsterdam: Elsevier; 2002.

[31] http://www.chemaxon.com

[32] Abraham MH, Ibrahim A, Zissimos AM. Determination of sets of solute descriptors from chromatographic measurements. J. Chromatogr. A 2004;1007: 29–47.

[33] Gilroy JL, Dolan JW, Snyder LR. Column selectivity in reversed-phase liquid chromatography IV. Type-B alkyl-silica columns. J. Chromatogr. A 2003;1000: 757–78.

[34] Snyder LR, Maule A, Heebsch A, Cuellar R, Paulson S, Carrano J, Wrisley L, Chan CC, Pearson N, Dolan JW, Gilroy J. A fast, convenient and rugged procedure for characterizing the selectivity of alkyl-silica columns. J. Chromatogr. A 2004;1057:49–57.

[35] Marchand DH, Snyder LR, Dolan JW. Characterization and applications of reversed-phase column selectivity based on the hydrophobic-subtraction model. J. Chromatogr. A 2008;1191:2–20.

[36] Rosés M, Bosch E. Linear solvation energy relationships in reversed-phase liquid chromatography. Prediction of retention from a single solvent and a single solute parameter. Anal. Chim. Acta 1993;274: 147–62.

[37] Sadek PC, Carr PW, Doherty RM, Kamlet MJ, Taft RW, Abraham MH. Study of retention process in reversed-phase high-performance chromatography by the use of the solvatochromic comparison method. Anal. Chem. 1985;57:2971–8.

[38] Bosch E, Bou P, Rosés M. Linear description of solute retention in reversed-phase liquid chromatography

by a new mobile phase polarity parameter. Anal. Chim. Acta 1994;299:219−29.

[39] Cheong WJ, Carr PW. Limitations of all empirical single parameter solvent strength scales in reversed-phase liquid chromatography. Anal. Chem. 1989;61:1524−9.

[40] Smith RM, Burr CM. Retention prediction of analytes in reversed-phase high-performance liquid chromatography based on molecular structure. J. Chromatogr. 1989;475:57−74.

[41] Smith RM, Burr CM. Retention prediction of analytes in reversed-phase high-performance liquid chromatography based on molecular structure, III. Monosubstituted aliphatic compounds. J. Chromatogr. 1989;481:71−84.

[42] Smith RM, Burr CM. Retention prediction of analytes in reversed-phase high-performance liquid chromatography based on molecular structure, V. CRIPES (Chromatographic retention index prediction expert system). J. Chromatogr. 1989;485:325−40.

[43] Sinanoğlu O. The C-potential surface for predicting conformations of molecules in solution. Theor. Chim. Acta 1974;33:279−84.

[44] Horvath C, Melander W, Molnar I. Solvophobic interactions in liquid chromatography with nonpolar stationary phases. J. Chromatogr. 1976;125:129−56.

[45] Horvath Cs, Melander W. Liquid chromatography with hydrocarbonaceous bonded phases; theory and practice of reversed phase chromatography. J. Chromatogr. Sci. 1977;15:393−404.

[46] Galushko SV. Calculation of retention and selectivity in reversed-phase liquid chromatography. J. Chromatogr. 1991;552:91−102.

[47] Galushko SV. The calculation of retention and selectivity in reversed-phase liquid chromatography II. Methanol-water eluents. Chromatographia 1993; 36:39−42.

[48] Dolan JW. "The Perfect Method". Santa Monica, CA: Advanstar; 2008.

[49] http://www.USP.org/app/USPNF/columns.html

[50] Sadek PC. The HPLC Solvent Guide. New York: John Wiley; 2002.

[51] Schoenmakers PJ, Blaffert T. Effect of model inaccuracy on selectivity optimization procedures in reversed-phase liquid chromatography. J. Chromatogr. 1987;384:117−33.

[52] Coenegracht PMJ, Smilde AK, Benak H, Bruins CHP, Metting HJ, DeVries H, Doornbos DA. Multivariate characterization of solvent strength and solvent selectivity in reversed-phase high-performance liquid chromatography. J. Chromatogr. 1991;550:397−410.

[53] Heinisch S, Puy G, Barrioulet M-P, Rocca J-L. Effect of temperature on the retention of ionizable

compounds in reversed-phase liquid chromatography: Application to method development. J. Chromatogr. A 2006;1118:234−43.

[54] Billiet HAH, Drouen ACJH, de Galan L. Rapid optimization of the concentration of the ion-pairing reagent in ion-pairing reversed-phase liquid chromatography. J. Chromatogr. A 1984;316:231−40.

[55] Nikitas P, Pappa-Louisi A, Agrafiotou P. Effect of the organic modifier concentration on the retention in reversed-phase liquid chromatography: II. Tests using various simplified models. J. Chromatogr. A 2002;946:33−45.

[56] Nikitas P, Pappa-Louisi A, Agrafiotou P, Fasoula S. Simple models for the effect of aliphatic alcohol additives on the retention in reversed-phase liquid chromatography. J. Chromatogr. A 2011;1218:3616−23.

[57] Huber L. Validation and Qualification in Analytical Laboratories. Buffalo Grove, IL: Interpharm Press; 1999.

[58] U.S. FDA, Technical Review Guide: Validation of Chromatographic Methods, Center for Drug Evaluation and Research (CDER), Rockville, MD, 1993.

[59] AOAC Peer-Verified Methods Program, Manual on Policies and Procedures, Arlington, VA, 1993.

[60] Moldoveanu SC, Gerardi AR. Acrylamide analysis in tobacco, alternative tobacco products, and cigarette smoke. J. Chromatogr. Sci. 2011;49:234−42.

[61] Moldoveanu SC, Coleman III W. A pilot study to assess solanesol levels in exhaled cigarette smoke. Beitr. Tabakforsch. Int. 2008;23:144−52.

[62] Ciucanu I, Kerek F. A simple and rapid method for the permethylation of carbohydrates,. Carbohydr. Res. 1984;131:209−17.

[63] Henderson JW, Ricker RD. Bidlingmeyer BA. Woodward C. Rapid, accurate, sensitive, and reproducible analysis of amino acids; Agilent, Part No. 5980−1193E

[64] Henderson JW, Brooks A. Improved amino acid method using Agilent Zorbax Eclipse Plus C18 columns for a variety of Agilent LC instruments and separation goals, Agilent Part No. 5990−4547EN.

[65] Woodward C, Henderson JW, Wielgos T. High speed amino acid analysis (AAA) on 1.8 reversed-phase (RP) columns, Agilent, Part No. 5989−6297EN.

[66] Molnár-Perl I. Quantitation of Amino Acids and Amines by Chromatography, Methods and Protocols. Amsterdam: Elsevier; 2005.

[67] Neverova I, Van Eyk JE. Role of chromatographic techniques in proteomic analysis. J. Chromatogr., B 2005;815:51−63.

[68] Deutscher MP. Guide to Protein Purification, Methods in Enzymology, vol. 182. London/San Diego: Academic Press; 1990.

[69] Issaq HJ, Chan KC, Janini GM, Conrads TP, Veenstra TD. Multidimensional separation of peptides for effective proteomic analysis. J. Chromatogr. B 2005;817:35–47.

[70] Link AJ, Eng J, Schieltz DM, Carmack E, Mize GJ, Morris DR, Garvik B, Yates III JR. Direct analysis of protein complexes using mass spectrometry. Nat. Biotechnology, 1999;17:676–82.

[71] Wolters DA, Washburn MP, Yates III JR. An automated multidimensional protein identification technology for shotgun proteomics. Anal. Chem. 2001;73: 5683–90.

[72] Davis MT, Beierle J, Bures ET, McGinley MD, Mort J, Robinson JH, Spahr CS, Yu W, Luethy R, Patterson SD. Automated LC–LC–MS–MS platform using binary ion-exchange and gradient reversed-phase chromatography for improved proteomic analyses. J. Chromatogr. B 2001;752:281–91.

[73] Opiteck GJ, Jorgenson J, Anderegg R. Two dimensional SEC/RPLC coupled with mass spectrometry for the analysis of peptides. Anal. Chem. 1997;69:2283–91.

[74] David V, Albu F, Medvedovici A. Retention behavior of metformin and related impurities in ion-pairing liquid chromatography. J. Liq. Chromatogr. Rel. Technol. 2005;28:81–95.

[75] Petritis K, Chaimbault P, Elfakir C, Dreux M. Parameter optimization for the analysis of underivatized protein amino acids by liquid chromatography ionspray tandem mass spectrometry. J. Chromatogr. A 2000;896:253–63.

[76] Qu J, Wang Y, Luo G, Wu Z, Yang G. Validated quantittaion of underivatized amino acids in human blood samples by volatile ion-pair reversed-phase liquid chromatography coupled to isotope dilution tandem mass spectrometry. Anal. Chem. 2002;74:2034–40.

[77] Waterval WAH, Scheijen JLJM, Ortmans-Ploemen MMJC, Habets-van der Poel CD, Bierau J. Quantitative UPLC-MS/MS analysis of underivatized amino acids in body fluids is a reliable tool for diagnosis and follow-up of patients with inborn errors of metabolism. Clin. Chim. Acta, 2009;407:36–42.

[78] Held PK, White L, Pasquali M. Quantitative urine amino acid analysis using liquid chromatography tandem mass spectrometry and aTRAQ reagents. J. Chromatogr. B 2011;879:2695–703.

[79] Kaspar H, Dettmer K, Chan Q, Daniels S, Nimkar S, Daviglus ML, Stamler J, Elliott P, Oefner PJ. Urinary amino acids analysis: A comparison of iTRAQ-LC-MS/MS, GC-MS, and amino acid analyzer. J. Chromatogr. B 2009;877:1838–46.

[80] Moldoveanu SC, Davis MF. Analysis of quinic acid and of *myo*-inositol in tobacco, Beitr. Tabak. Intern., submitted for publication.

[81] Alpert AJ, Shukla M, Shukla AK, Zieske LR, Yuen SW, Ferguson MAJ, Mehlert A, Pauly M, Orlando R. Hydrophilic-interaction chromatography of complex carbohydrates. J. Chromatogr. A 1994;676:191–202.

[82] Fu Q, Liang T, Zhang X, Du Y, Guo Z, Liang X. Carbohydrate separation by hydrophilic interaction liquid chromatography on a 'click' maltose column. Carbohydr. Res. 2010;345:2690–7.

[83] Kato M, Kato H, Eyama S, Takatsu A. Application of amino acid analysis using hydrophilic interaction liquid chromatography coupled with isotopic dilution mass spectrometry for peptide and protein quantification. J. Chromatogr. B 2009;877:3059–64.

[84] Yoshida T. Peptide separation by hydrophilic-interaction chromatography: a review. J. Biochem. Biophys. Meth. 2004;60:265–80.

[85] Gilar M, Jaworski A. Retention behavior of peptides in hydrophilic-interaction chromatography. J. Chromatogr. A 2011;1218:8890–6.

[86] Alpert AJ. Hydrophilic-interaction chromatography for the separation of peptides, nucleic acids and other polar compounds. J. Chromatogr. 1990;499:177–96.

[87] Jandera P. Stationary and mobile phases in hydrophilic interaction chromatography: A review. Anal. Chim. Acta 2011;692:1–25.

[88] Gilar M, Yu Y-Q, Ahn J, Xie H, Han H, Ying W, Qian X. Characterization of glycoprotein digests with hydrophilic interaction chromatography and mass spectrometry. Anal. Biochem. 2011;417:80–8.

[89] Zhu B-Y, Mant CT, Hodges RS. Hydrophilic-interaction chromatography of peptides on hydrophilic and strong cation-exchange columns. J. Chromatogr. A 1991;548:13–24.

[90] Alpert AJ. Electrostatic repulsion hydrophilic interaction chromatography for isocratic separation of charged solutes and selective isolation of phosphopeptides. Anal. Chem. 2008;80:62–76.

[91] http://www.dionex.com

[92] Joergensen L, Thestrup HN. Determination of amino acids in biomass and protein samples by microwave hydrolysis and ion-exchange chromatography. J. Chromatogr. A 1995;706:421–8.

[93] Principles and Applications of the Prominence Amino Acid Analysis System, Shimadzu HPLC All. Report No 26.

[94] Macchi FD, Shen FJ, Keck RG, Harris RJ. Amino acid analysis using postcolumn ninhydrin detection in a biotechnology laboratory, in C. Cooper, N. Packer, K. Williams. In: Methods in Molecular Biology, vol. 159. Totowa, NJ: Humana Press; 2000.

[95] http://www.dionex.com/en-us/documents/acclaim-library/lp-71591.html

[96] Mant CT, Hodges RS. In: High-Performance Liquid Chromatography of Peptides and Proteins, Separation, Analysis, and Conformation. Boca Raton, FL: CRC Press; 1991.

[97] Aboul-Enein HY, Ali I. Chiral Separations by Liquid Chromatography and Related Technologies (Chromatographic Science, Vol. 90. New York: Marcel Dekker; 2003.

[98] Welch CJ. Microscale chiral HPLC in support of pharmaceutical process research. Chirality 2009;21:114–8.

[99] Anderson ME, Aslan D, Clarke A, Roeraade J, Hagman G. Evaluation of generic chiral liquid chromatography screens for pharmaceutical analysis. J. Chromatogr. A 2003;1005:83–101.

[100] Ilisz I, Berkecz R, Peter A. Retention mechanism of high-performance liquid chromatographic enantioseparation on macrocyclic glycopeptide-based chiral stationary phases. J. Chromatogr. A 2009;1216:1845–60.

[101] Ilisz I, Berkecz R, Péter A. HPLC separation of amino acid enantiomers and small peptides on macrocyclic antibiotic-based chiral stationary phases: A review. J. Sep. Sci. 2006;29:1305–21.

[102] Gübitz G, Schmid MG. Chiral separation by chromatographic and electromigration techniques. A review, Biopharm. Drug Dispos. 2001;22:291–336.

[103] Mislanova C, Hutta M. Role of biological matrices during the analysis of chiral drugs by liquid chromatography. J. Chromatogr. B 2003;797:91–109.

[104] Medvedovici A, Albu F, Georgita C, Sora DI, Galaon T, Udrescu S, David V. Achiral-chiral LC/ LC-FLD coupling for determination of carvedilol in plasma samples for bioequivalences purposes. J. Chromatogr. B 2007;850:327–35.

[105] Mori S, Barth HG. Size Exclusion Chromatography. Berlin: Springer Verlag; 1999.

[106] Wu C-S. In: Handbook of Size Exclusion Chromatography and Related Techniques. New York: Marcel Dekker; 2004.

[107] Churms SC. Modern size-exclusion chromatography of carbohydrates and glycoconjugates (Chapter 8). In: El Rassi Z, editor. Carbohydrate Analysis by Modern Chromatography and Electrophoresis, 66. Amsterdam: Elsevier; 2002. p. 267–303.

[108] Aust N. Application of size-exclusion chromatography to polymers of ultra-high molar mass. J. Biochem. Biophys. Meth. 2003;56:323–34.

[109] Kostanski LK, Keller DM, Hamielec AE. Size-exclusion chromatography—a review of calibration methodologies. J. Biochem. Biophys. Meth. 2004;58: 159–86.

[110] http://wolfson.huji.ac.il/purification/PDF/Gel_Filtration/TOSOH_GelFiltration.pdf

[111] Han Y, Lee SS, Ying JY. Spherical siliceous mesocellular foam particles for high-speed size exclusion chromatography, Studies in Surface. Sci.. and Catal. 2007;165:829–32.

[112] Uliyanchenko E, Schoenmakers PJ, van der Wal S. Fast and efficient size-based separations of polymers using ultra-high-pressure liquid chromatography. J. Chromatogr. A 2011;1218:1509–18.

Symbols*

A	area, free energy, proportionality constant, absorbance, molecule or fragment A, etc.		h	molar heat of mixing
			H_c	column hydrophobic parameter
A	area		i	index for component, etc.
a	activity, coefficient in an equation, radius, etc.		I	intensity, ionic strength
$\mathcal{A}$	area (various)		j	index for component
As	asymmetry		k	capacity factor, retention factor
B	molecule or fragment B, etc.		K	equilibrium constant, distribution constant
b	coefficient in an equation, gradient steepness, etc.		*K*	thermodynamic distribution constant
C	constants, molecule or fragment, Coulomb, concentration, thermal expansion coefficient, etc.		k^*	average retention factor
			k_B	Boltzmann constant $k_B = R/\mathcal{N} = 1.3806504$ 10^{-23} J K^{-1}
c	concentration (molar when indicated)		K_{ow}	octanol/water partition coefficient (constant)
d	diameter, depth, distance		L	length
D	diffusion coefficient, distribution coefficient, Clausius-Mosotti function etc.		M	molecule or fragment M, molecular mass
$\mathcal{D}$	dilution		MW	molecular weight
e	electron charge		m	mass, amount, dipole moment
E	electrode potential		n	number, refractive index
E	energy		N	theoretical plate number
f	function		*n*	effective plate number
F, f	force, Faraday constant, factor, etc.		$\mathcal{N}$	Avogadro number ($\mathcal{N} = 6.02214179 \times 10^{23}$ mol^{-1})
G	free enthalpy (Gibbs)		p	pressure
h	height		*P*	peak capacity
H	height equivalent to a theoretical plate (HETP)		Q	heat
H	enthalpy			

(Continued)

---

* Note: This list includes symbols utilized more frequently in the book. For all other symbols, explanations are given in the text.

(*Cont'd*)

q	electric charge, quantity, etc.		z	electrical charge
r	radius, distance		α	selectivity, polarizability, coefficient
R	gas constant ($R = 8.31451\,\mathrm{J\,deg^{-1}\,mol^{-1}} = 1.987\,\mathrm{cal}$ $\mathrm{deg^{-1}\,mol^{-1}}$)		$\alpha(CH_2)$	methylene selectivity
R	resolution		α'	parameter measuring the ability of a unit of adsorbent surface to bind molecules
Rm	molar refraction		β	coefficient
s	parameter		δ	solubility parameter, Dirac delta function
S	parameter, entropy, etc.		ε	molar absorbance, elutropic strength, dielectric constant, etc.
S^0	measure of adsorption energy		ε^*	column porosity
t	time, other meanings		ϕ	volume fraction, flow resistance
T	temperature, transmittance, etc.		Φ	proportionality constant, fluorescence efficiency (yield), etc.
t'_R	reduced retention time		γ	obstruction factor, activity coefficient
t_0	void time, holdup time, dead time		γ'	surface tension
TF	tailing factor		η	viscosity, etc.
t_R	retention time		η'	hydrophobicity parameter
u	linear flow rate		κ	Debye length
U	volumetric flow rate		λ	wavelength
V	volume		μ	chemical potential
V'_R	reduced retention volume		$\bar{\mu}$	electrochemical potential
V_0	void volume, dead volume		ν	frequency, number of molecules in a phase, fraction of molecules, volume of a molecule
V_D	dwell volume		Π	osmotic pressure
V_R	retention volume		σ	standard deviation
W	Width, work		σ^2	variance
W_b	width at the base of Gaussian curve/ chromatographic peak		υ	reduced linear velocity
W_h	width at half height of Gaussian curve/ chromatographic peak		ξ	distance
W_i	width at the inflexion of Gaussian curve/ chromatographic peak		Ψ	phase ratio
X	molecule or fragment X		Ω_0	momentum zero for a statistical population
x	length, distance, molar fraction		Ω_n	momentum "n" for a statistical population
Y	molecule or fragment Y		Ω	ion-exchange capacity

Index

Note: Page numbers with "f" denote figures; "t" tables.

Printed and bound by CPI Group (UK) Ltd, Croydon, CR0 4YY

08/05/2025

01864827-0007